WATER SUPPLY AND
POLLUTION CONTROL

WATER SUPPLY AND POLLUTION CONTROL

THIRD EDITION

JOHN W. CLARK
New Mexico State University

WARREN VIESSMAN, Jr.
*Environment and Natural Resources
Policy Division,
Library of Congress*

MARK J. HAMMER
*Director of Environmental Engineering
Kirkham, Michael and Associates*

Thomas Y. Crowell
HARPER & ROW, PUBLISHERS
New York Hagerstown Philadelphia San Francisco London

WATER SUPPLY AND POLLUTION CONTROL, Third Edition

Library of Congress Cataloging in Publication Data
Clark, John William.
  Water supply and pollution control.
  (IEP series in civil engineering)
  Includes index.
  1. Sanitary engineering. I. Viessman, Warren, joint
author. II. Hammer, Mark J., 1931- joint author.
III. Title.
TD145.C55 1977   628   76-57126
ISBN 0-7002-2495-5

To   Jacqueline, Jeanne, and Audrey

CONTENTS

CHAPTER FIVE
Transportation and Distribution of Water 138

CHAPTER SIX
Wastewater Systems 198

CHAPTER ELEVEN
Biological-Treatment Processes 497

PREFACE

The third edition of WATER SUPPLY AND POLLUTION CONTROL has been modernized to reflect state-of-the-art practices and recent trends in federal, environmental, and pollution control legislation. Emphasis is on application of scientific methods to solve problems related to the development, transportation, distribution, processing, and treatment of water and liquid wastes.

The chapters on "Water Quality" and "Systems for Treating Water and Wastes" have been updated to reflect changes in attitude toward environmental issues as reflected in the National Environmental Policy Act of 1969 and the Federal Water Pollution Control Act Amendments of 1972 (PL92–500). These chapters act as integrating mechanisms to acquaint students with the nature of water and waste treatment systems before presenting the details of systems components.

A new chapter on "Water-Quality Models" is designed to relate the treatment of waste flows to the overall environment of the receiving body of water. The material is organized to acquaint students with techniques for combining individual pollution control facilities into integrated systems for regional water quality control and management. Both model structuring and conceptualization are presented.

The treatment of principles of physical, chemical, and biological treatment processes necessary for designing and managing modern water and waste-water facilities has been extensively revised. "Processing of Sludges" now reflects the increasing importance of solid waste handling in the pollution control field.

"Advanced Waste-Water Treatment Processes" has been rewritten to focus on the removal of pollutants not affected by conventional processes. Changes in subject matter reflect the goals of PL92–500 and the trend toward a higher degree of waste-water reuse. Chapters on water reuse and water law have been revised to incorporate advances in technology and changes in federal laws.

The authors wish to acknowledge the advice and assistance of many students and teachers who have reviewed and commented on previous editions. Specific recognition is given to Helen Hudson and Mary Lou Wegener for their help in preparing the manuscript and to Russell C. Brinker and Norma Jean Viessman for editorial assistance.

1

Introduction

Air, water, food, heat, and light constitute the five essentials for human existence. Environmental engineering concerns itself, to some degree, with all of these. This book is primarily concerned with the development, transportation, processing, and disposal of water and waste.

Water and liquid wastes must be considered simultaneously, as there is but a fine line of distinction between them. One community's waste may constitute part of another's water supply. The ultimate goal in water management is the maximum economic use of the total water resource.

1-1

HISTORY

Man's search for pure water began in prehistoric times. Much of his earliest activity is subject to speculation. Some individuals might have led water where they wanted it through trenches dug in the earth. Later, a hollow log was perhaps used as the first water pipe.

Thousands of years must have passed before our more recent ancestors learned to build cities and enjoy the convenience of water piped to the home and drains for water-carried wastes. Our earliest archeological records of central water supply and wastewater disposal date back about 5000 years, to Nippur of Sumeria. In the ruins of Nippur there is an arched drain with the stones set in full "voussoir" position, each stone being a wedge tapering downward into place.[1]* Water was drawn from wells and cisterns. An extensive system of drainage conveyed the wastes from the palaces and residential districts of the city.

The earliest recorded knowledge of water treatment is in the Sanskrit medical lore and Egyptian Wall inscriptions.[2] Sanskrit writings dating about 2000 B.C. tell how to purify foul water by boiling in copper vessels, exposing to sunlight, filtering through charcoal, and cooling in an earthen vessel.

There is nothing on water treatment in the sanitary and hygienic code of the early Hebrews in the Old Testament, although three incidents may be cited as examples of the importance of fresh water. At Morah, Moses is said to have sweetened bitter waters by casting into them a tree shown him by God.[3] During the wandering in the wilderness, the Lord commanded Moses to bring forth water by smiting a rock.[4] At a much later date, Elisha is said to have "healed unto this day" the spring water of Jericho by casting "salt" into it.[5]

The earliest known apparatus for clarifying liquids was pictured on Egyptian walls in the fifteenth and thirteenth centuries B.C. The first picture, in a tomb of the reign of Amenhotep II (1447–1420 B.C.), represents the siphoning of either water or settled wine. A second picture, in the tomb of Rameses II (1300–1223 B.C.), shows the use of wick siphons in an Egyptian kitchen.

The first engineering report on water supply and treatment was made in A.D. 98 by Sextus Julius Frontinus, water commissioner of Rome. He produced two books on the water supply of Rome. In these he described a settling reservoir at the head of one of the aqueducts and pebble catchers built into most of the aqueducts. His writings were first translated into English by the noted hydraulic engineer Clemens Herschel in 1899.[2]

In the eighth century A.D. an Arabian alchemist, Geber, wrote a rather specialized treatise on distillation that included various stills for water and other liquids.

The English philosopher Sir Francis Bacon wrote of his experiments on the purification of water by filtration, boiling, distillation, and clarification by coagulation. This was published in 1627, one year after his death. Bacon also noted that clarifying water tends to improve health and increase the "pleasure of the eye."

The first known illustrated description of sand filters was published in 1685 by Luc Antonio Porzio, an Italian physician. He wrote a book on conserving the health of soldiers in camps, based on his experience in the Austro-Turkish War. This was probably the earliest published work on mass sanitation. He described and illustrated the use of sand filters and sedimentation. Porzio also stated that his filtration was the same as "by those who built the Wells in the Palace of the Doges in Venice and in the Palace of Cardinal Sachett, at Rome."[2]

* Superscript numbers refer to references at the end of the chapter.

The oldest known archeological examples of water filtration are in Venice and the colonies she occupied. The ornate heads on the cisterns bear dates, but it is not known when the filters were placed. Venice, built on a series of islands, depended on catching and storing rainwater for its principal freshwater supply for over 1300 years. Cisterns were built and many were connected with sand filters. The rainwater ran off the house tops to the streets, where it was collected in stone-grated catch basins and then filtered through sand into cisterns (see Fig. 1-1).

FIGURE 1-1 Venetian cistern head located at Dubrovnik, Yugoslavia, showing a stone grating.

A comprehensive article on the water supply of Venice appeared in the *Practical Mechanics Journal* in 1863.[6] The land area of Venice was 12.85 acres and the average yearly rainfall was 32 inches (in.). Nearly all of this rainfall was collected in 177 public and 1900 private cisterns. These cisterns provided a daily average supply of about 4.2 gallons per capita per day (gpcd). This low consumption was due in part to the absence of sewers, the practice of washing clothes in the lagoon, and the universal drinking of wine. The article explained in detail the construction of the cisterns. The cisterns were usually 10 to 12 feet (ft) deep. The earth was first excavated to the shape of a truncated inverted pyramid. Well-puddled clay was placed against the sides of the pit. A flat stone was placed in the bottom and a cylindrical wall was built from brick laid with open joints. The space between the clay walls and the central brick cylinder was filled with sand. The stone surfaces of the court yards were sloped toward the cistern, where perforated stone blocks collected the water at the lowest point and discharged it to the filter sand. This water was always fresh and cool, with a temperature of about 52 degrees Fahrenheit (°F). These cisterns continued to be the principal water supply of Venice until about the sixteenth century.

Many experiments were conducted in the eighteenth and nineteenth centuries in England, France, Germany, and Russia. Henry Darcy patented filters in France and England in 1856 and anticipated all aspects of the American rapid sand filter except coagulation. He appears to be the first to apply the laws of hydraulics to filter design.[7] The first filter to supply water to a whole town was completed at Paisley, Scotland, in 1804 but this water was carted to consumers.[2] In Glasgow, Scotland, in 1807 filtered water was piped to consumers.[8]

In the United States little attention was given to water treatment until after the Civil War. Turbidity was not as urgent a problem as in Europe. The first filters were of the slow sand type, similar to British design. About 1890 rapid sand filters were developed in the United States and coagulants were introduced to increase their efficiency. These filters soon evolved to our present rapid sand filters with slight modification.

The drains and sewers of Nippur and Rome are among the great structures of antiquity. These drains were intended primarily to carry away runoff from storms and the flushing of streets. There are specific instances where direct connections were made to private homes and palaces, but these were the exceptions, for most of the houses did not have such connections. The need for regular cleansing of the city and flushing of the sewers was well recognized by commissioner Frontinus of Rome, as indicated in his statement, " I desire that nobody shall conduct away any excess water without having received my permission or that of my representatives, for it is necessary that a part of the supply flowing from the water-castles shall be utilized not only for cleaning our city but also for flushing the sewers."

It is astonishing to note, from the days of Frontinus to the middle of the nineteenth century there was no marked progress in sewerage. In 1842, after a fire destroyed the old section of the city of Hamburg, Germany, it was decided to rebuild this section of the city according to modern ideas of convenience. The work was entrusted to an English engineer, W. Lindley, who was far ahead of his time. He designed an excellent collection system that included many of the ideas presently used. Unfortunately, the ideas of Lindley and their influence on public health were not recognized.

The history of the progress of sanitation in London probably affords a more typical picture of what took place in the middle of the nineteenth century. In 1847 a royal commission was appointed to look into the sanitary conditions of London following an outbreak of cholera in India which had begun to work westward. This royal commission found that one of the major obstacles was the political structure, especially the lack of central authority. The city of London was only a small part of the metropolitan area, comprising approximately 9.5% of the land area and less than 6% of the total population of approximately 2.5 million. This lack of central authority made the execution of sewerage works all but impossible. The existing sewers were at different elevations, and in some instances the wastes would have had to flow uphill. Parliament, in 1848, followed the advice of this commission and created the Metropolitan Commission of

Sewers. That body and its successors produced reports that clearly showed the need for extensive sewerage works and other sanitary conditions.[9] Cholera appeared in London during the summer of 1848 and 14,600 deaths were recorded during 1849. In 1854 cholera claimed a mortality of 10,675 people in London. The connection was established between a contaminated water supply and spread of the disease, and it was determined that the absence of effective sewerage was a major hindrance in combatting the problem.

In 1855 Parliament passed an act "for the better local management of the metropolis," thereby providing the basis for the Metropolitan Commission of Sewers, which soon after undertook an adequate sewerage system. It will be noted that the sewerage system of London came as a result of the cholera epidemic, as was true of Paris.

The natural remedy for these foul conditions led to the suggestion that human excrement be discharged into the existing storm sewers and that additional collection systems be added. This created the combined sewers of many older metropolitan areas. These storm drains had been constructed to discharge into the nearest watercourse. The addition of wastes to the small streams overtaxed the receiving capacities of the waters and many of them were covered and converted into sewers. Much of the material was carried away from the point of entry into the drains, which in turn overtaxed the receiving waters. First the smaller and then the larger bodies of water began to ferment and create a general health problem, especially during dry, hot weather. The solution has been the varying degrees of treatment as presently practiced, dependent upon the capabilities of the receiving stream or lake to take the load.

The work on sewerage in the United States closely paralleled that of Europe, especially England. Some difficulty was experienced because of the variation in rainfall patterns in America as compared with those of England. The English rains are more frequent but less intense. Our storm drains must be larger for like topographical conditions. The more intense rains tend to have a better cleansing action, and in general the receiving streams carry a larger volume of water. This, together with the lower population densities, tends to produce less nuisance than is being experienced in Europe. The density of population in England and the small amount of land suited for liquid waste farming led to interest in methods of treating wastewater before it is discharged into fresh water.

More recent developments in water supply and wastewater disposal are discussed in later chapters under the appropriate headings.

1-2

CURRENT STATUS

Increased demands currently being placed on water supply and waste disposal have necessitated far broader concepts in the application of environmental engineering principles than those originally envisioned.

The average rate of water use for the urban population of the United States is approximately 165 gpcd; peak demands have developed considerably beyond past design practices. The standards for water quality have significantly increased with a marked decrease in raw-water quality available. Considerable research is being directed toward the ultimate use of brackish water or seawater for domestic supplies.

Waste-treatment-plant effluents normally discharge into a stream, lake, ocean, or other body of water. The degree of treatment required is determined by the ability of the receiving waters to assimilate the wastes, and the uses to which the receiving waters are put.

In general practice, large bodies of water or rivers in good condition receive wastes with limited treatment. Expensive treatment is necessary where receiving waters are unable to assimilate additional pollution, or where the body of water is immediately used as a raw-water source for domestic purposes or satisfies extensive recreational demands. Effluent chlorination is usually required where receiving waters are to be used for water supply or bathing.

Land disposal of waste effluents is practiced in the United States, especially in the semiarid Southwest. Removal of settleable and floating solids is usually required prior to the effluent being distributed over the land. Many state health departments regulate the use of effluents on crops, especially vegetables that might be eaten raw. Some sewage effluents are being used to recharge ground-water reservoirs and to check saltwater intrusion. Waste-plant effluents are being utilized by industry with varying degrees of treatment.

Surface waters used as a raw-water supply are normally treated by coagulation, filtration, and disinfection. The degree of treatment is determined by the health hazards involved and by the quality of the raw water. Well waters are normally not treated, except for disinfection. Groundwaters are becoming polluted with increasing frequency, however, and require additional surveillance.

Water supply and wastewater disposal are interrelated activities of the community. Although they are closely associated, the primary accent has been on providing a safe water supply. The reasons for this are threefold: (1) the effects of an unsatisfactory water supply are usually detectable immediately; (2) the unsafe water supply affects the community served; and (3) water systems are income-producing, while waste systems normally derive most of their revenue from taxation. Upon the informal recommendations of state and federal health services, water systems have been constructed and willingly improved by the community served. The construction and improvement of waste-treatment facilities have, in many instances, come about because of and after formal complaints and court action.

Public water must be palatable and wholesome. It must be attractive to the senses of sight, taste, and smell and must be hygienically safe.

Liquid waste-disposal systems must collect the wastes from homes and industry and convey those wastes without nuisance to hygienic disposal.

1-3
___

PROJECTED PROBLEMS

Today, as populations throughout the world multiply at an alarming rate, it is evident that environmental control is a critical factor. Land and water become increasingly important as the population increases. Most European and Asian nations have reached the maximum population that their land areas can bear comfortably. They are faced with the problem of providing for more people than the land will conveniently support.

There are important lessons to be learned from the countries of Europe and Asia—populations increase, but water resources do not. The use and control of our water resources must be nearly perfect to maintain our way of life.

Environmental engineering needs are far greater than the available supply of trained personnel, and future needs are certain to be even greater. The future potential of any profession is usually determined by the basic factors of demand and/or need. Demand is the less reliable guide because it is subject to change, due both to technological advances and to the instability of social trends. A need is a more dependable guide. It is born of a requirement and thrives when the requirement determines the welfare of a nation.

Environmental engineering will continue to grow in importance because it fills a definite need. The services provided by water-resources engineers are growing in importance in a world staggering under the weight of the greatest population it has ever known.

___

REFERENCES

1. W. Durant, *Our Oriental Heritage* (New York: Simon and Schuster, Inc., 1954), p. 132.

2. M. N. Baker, *The Quest for Pure Water* (New York: American Water Works Association, 1949), pp. 1–3, 6, 11.

3. *Old Testament* (King James version), Exodus 15:22–27.

4. *Old Testament* (King James version), Exodus 17:1–7.

5. *Old Testament* (King James version), II Kings 2:19–22.

6. Anonymous, "The Water Cisterns in Venice," *J. Franklin Inst.*, 3rd ser. 70 (1860): 372–373.

7. H. Darcy, *Les Fontaines publiques de la ville de Dijon; distribution d'eau et filtrage des eaux* (Paris: Victor Dalmont, 1856).

8. D. Mackain, "On the Supply of Water to the City of Glasgow," *Proc. Inst. Civil Engrs.* (*London*) 2 (1842–1843): 134–136.

9. First Report of the Metropolitan Sanitary Commission (London: 1848).

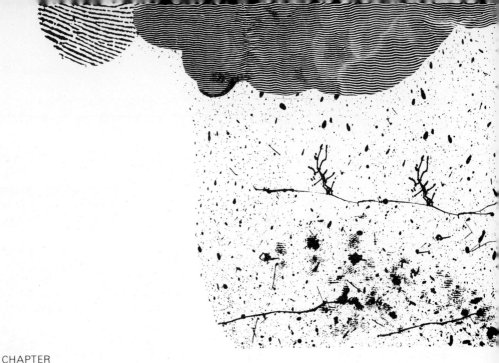

2
Environmental Quality

Environmental pollution is of such scope that it is impossible to write a text-book that comprehensively covers all its extensions. The purpose of this chapter is to call some of the more obvious problems to the reader's attention and hopefully indicate that there are strong interrelations among all environmental quality problems. Although the remainder of this book emphasizes those aspects of environmental engineering concerned with the development, transportation, processing, and treatment of water and liquid wastes, the reader should not lose sight of the close relationship of water to all other aspects of the environment.

Environment

Some eight thousand years ago man made a discovery which eventually transformed the whole of human life. For perhaps 1 million years prior to this time, man had wandered individually or in small bands in search of whatever food nature might provide. The discovery that food could be produced, either by cultivating plants or by taming animals, changed man's entire way of life. With an assured control of his food supply, he built the great civilizations of antiquity. Development in a purely agrarian society was slow and its benefits were thinly distributed. The lot of a large segment of society in ancient civilizations was hard, a condition due chiefly to the fact that all work was performed by muscle power.

In the second or first century B.C., man made his second revolutionary discovery by finding that much of the muscle power could be replaced by natural forces. The first machine to use these natural forces was water power.

Around the tenth century, people began to use water power for other than agricultural purposes. They invented water-driven machines which could hammer metals, saw wood, and run oil presses. As the supply of usable energy increased, there was rapid development of an industrial base.

Near the end of the eighteenth century, other great forward steps took place. Man invented a number of intricate water-power machines which could carry on such delicate processes as spinning and weaving. At about the same time, men harnessed fossil fuel (coal) to heat water to steam and then forced the steam under pressure to drive pistons of work engines.

In the nineteenth century, oil and natural gas were added to the list of energy-producing fuels, and internal combustion engines were designed to use gasoline from the oil. By the end of the nineteenth century, man developed electrical energy from water power, transmitted it great distances, and then converted the electrical power to mechanical energy to do work.

Finally, in the twentieth century, man found a new and tremendous power source, nuclear energy. Fissionable minerals were added to man's supply of energy-producing fuels and gave him a vast new reservoir of potential energy. As a result of fresh scientific knowledge with its technological consequences and their utilization by industry, man strode from discovery to discovery, from invention to invention, at a consistent and ever-quickening pace. That development has brought us to the situation as it exists today. There is no turning back. Progress under the influence of industrialization has produced a demand for material goods and services that continues to grow. Total world energy consumption in 1972 represented 1% of world energy reserves (Table 2-1).

In this rapidly accelerating process, nature has not received her due. She has been shamelessly exploited but, as long as nature yielded what man clamored for, there has scarcely been serious question raised about what was happening. We have reached the point in environmental pollution where nature can no longer yield all that is desired and the immediate future discloses many shortcomings.

There are symptoms of impending shortages in many areas of our natural resources. These shortages are generally not caused by nature but by man's pollution and misuse. They are a consequence of human population explosion and the massive resultant effect our species is having on everything which occurs on this planet. The central problem is that there are more people, all requiring more of the material things of life. There has been a relaxation in man's struggle for simple survival along with a rapid drop in death rates all over the world. There have not been corresponding drops in birth rates.

TABLE 2-1
MEASURED WORLD NONRENEWABLE ENERGY RESERVES
(quadrillion Btu)ᵃ

Area	Fossil Fuels				Uranium (nonbreeders)	Total
	Solid Fuels	Crude Oil	Natural Gas	Oil Shale and Tar Sands		
Africa	361.7	526.6	201.7	81.4	198.1	1,369.5
Asia (less USSR)	2,608.7	2,211.4	432.6	870.2	3.1	6,126.0
Europe (less USSR)	2,581.5	57.1	153.6	117.0	46.4	2,955.6
USSR	3,325.5	333.6	577.9	139.0	Unknown	>4,376.0
North Americaᵇ	5,070.9	301.0	380.6	9,111.0ᵇ	422.7	15,286.2
South America	49.8	311.5	60.6	23.7	11.9	457.5
Oceania	459.8	9.4	24.9	9.2	99.1	602.4
TOTAL	14,457.9	3,741.2	1,831.9	10,351.5	>781.3	>31,173.2

ᵃ 1 quadrillion Btu = 500,000 barrels petroleum per day for a year
= 40 million tons of bituminous coal
= 1 trillion cubic feet of natural gas
= 100 billion kilowatt-hours (based on a 10,000-Btu/kWh heat rate).

ᵇ According to the Bureau of Mines, North American tar sands and oil shale reserves may be severely overstated. Development of most of these reserves is not economically feasible at the present time.

Source: *Energy Perspectives* (Washington, D.C.: U.S. Department of the Interior, February 1975), p. 2.

Environmental engineers have been directly involved in the reduction of death rates, especially in the underdeveloped countries. One of the first programs available through the United Nations has been in the area of preventive medicine. This includes a safe water supply and some sanitary means of waste disposal. The objectives of these programs have usually been achieved in a short period of time through outside effort with little participation by the country involved. The net result has been a sudden increase in population because of the reduced death rate, unaccompanied by the time-dependent problems of economic efficiency and social adjustment.

The population of the United States was about 212 million in 1974 and increased 0.7% during the preceding year (see Fig. 2-1). While the total population of the United States has increased throughout American history, the rate of growth has undergone a long-term decline. The projection in Fig. 2-1 indicates that the population would grow to 262 million in 2000, or by 24%. This is based on a growth rate of 1.0% during the early 1980s and then would decline to 0.6% by the end of this century as it converged toward zero.

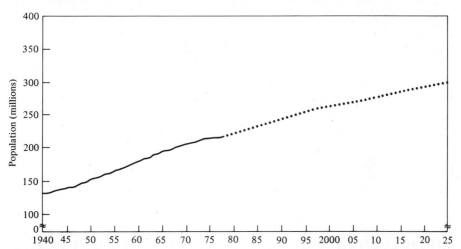

FIGURE 2-1 Population of the United States, 1940–2025. [Source: *Population Estimates and Projections* (Washington, D.C.: U.S. Department of Commerce, Bureau of the Census, October 1975).]

Environmental pollution is related to the numbers of people, but population control is not the entire answer. Population concentration is also a leading factor. More important, the degree of pollution depends upon the standard of living and the state of technological development. To sustain this standard, we are organized to collect widely scattered resources, process them, change them in form, and distribute useful goods to the consumer. In so doing, contaminants are introduced by nearly all human activities.

Contaminants are also introduced by many natural phenomena—forest fires, collapsing river banks, and volcanic eruptions. A contaminant is considered a " pollutant " only if it can adversely affect something that man values and is present in high-enough concentration to do so. These pollutants are either unwanted by-products of our activities or spent substances that have served intended purposes. They tend to impair our economy and the quality of our life. Many pollutants can be carried long distances by air, or water, or on articles of commerce, threatening the health, longevity, livelihood, recreation, cleanliness, and happiness of citizens who have no direct stake in their production but cannot escape their influence. The continued strength and welfare of our nation depend on the quantity and quality of our resources and on the quality of the environment in which our people live.

The industrial potential for pollution is almost staggering to the mind. The nation experienced a remarkable growth rate of gross national product (GNP) of 4.7% per year for the period 1960–1965. Projecting this growth forward at the rate of 4.0% indicates a GNP of about \$4500 billion (1954 dollars) in 2020 compared with \$550 billion in 1965. Therefore, the GNP has been increasing at about 2.5 times the rate of population increase in recent years.

When future generations write of our era, they will note how much of our technical capability was devoted to producing commodities inevitably destined to contaminate the environment. The most spectacular example must be the automobile. Its exhaust and fumes pollute our atmosphere; it has brought about the need for ever-spreading highways, and it is ultimately discarded in junkyards which detract from the beauty of the landscape.

There is nothing simple about controlling environmental quality, but the authors believe that the needs of modern society can be met without seriously affecting the balance of nature and that the interests of technology and ecology need not conflict. We must look realistically at our social and economic responsibilities and do everything feasible to eliminate pollution where it can be eliminated and to control it where its prevention is not possible. The crucial requirement is that sufficient energy and support be devoted to the task.

2-1

AIR POLLUTION

Early in December 1952, the city of London, England, went through a 4-day period of still air during which pollution accumulated in a dense fog. Some months later a review of mortality statistics revealed that 4000 excess deaths took place in the city during the 7-day period beginning with the first day of the fog. The cardiorespiratory illness rate increased to more than twice the normal rate for that time of the year and did not return to normal until 3 weeks later. London went through a similar episode in 1962 and New York went through much the same type of disaster in 1953 and 1962.

There are many other cases on record and probably some of less populated areas that are unrecorded. All the recorded episodes have one thing in common, an abnormal atmospheric condition known as a *temperature inversion* (Fig. 2-2).

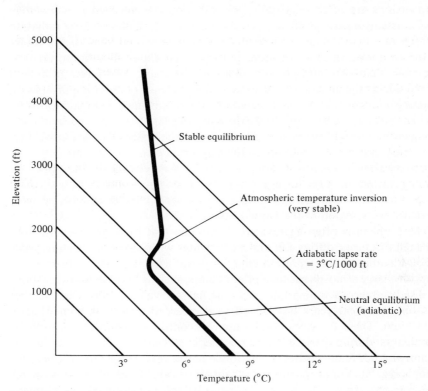

FIGURE 2-2 Atmospheric temperature distribution.

In the latitudes of the continental United States during autumn and parts of the winter, high-pressure cells often became stationary over large areas. These high-pressure cells lead to widespread stagnation of the air where pollutants can accumulate and rapidly reach dangerous levels. Normally the time to reach such high levels of pollution is several days; however, there are circumstances when even a few hours may prove sufficient to cause severe adverse effects.

Where little wind exists for extended periods, vertical transport is the major vehicle in the dilution of pollutants. The change in air temperature with altitude (adiabatic lapse rate) is an important factor in vertical transport. In dry air, the adiabatic lapse rate is 5.4°F per 1000 ft [0.98 degree Celsius (centigrade; °C) per 100 meters (m)]; in other words, the air temperature decreases 5.4°F for each 1000-ft increase in altitude. When the lapse rate is less than adiabatic, the atmosphere is more stable and vertical transport of pollutants is hindered. A negative lapse rate or increase in air temperature with increase in altitude (inversion) results in extremely stable air. An inversion can take place within a few hundred feet of the ground surface and hold most of the air pollutants in a thin film in contact with the earth's surface.

Inversions are of two general types, radiation and subsidence. A *radiation inversion* usually occurs at night when the earth's surface is cooled by radiation at a faster rate than the air. This results in the air near the ground being cooler than the air above. In some locations in the United States, radiation inversions occur more than two-thirds of the nights of the year.[3] Fog layers in valleys observed during the morning are usually visible evidence of radiation inversions. *Subsidence inversions* are normally caused by the sinking motion in high-pressure areas. The subsidence inversion over the Los Angeles area is caused by air circulating around a high-pressure area located over the eastern Pacific Ocean. This air descends as it moves south along the California coast. As this air descends, it compresses and increases in temperature. This may result in the upper air becoming warmer than the lower air, thus producing an inversion near the land surface. Temperature inversions occur a considerable number of days per year in virtually every region of the United States.

The problem is often aggravated by the fact that these atmospheric conditions tend to occur during fall and early winter when domestic heating needs add considerably to the pollution emission load. This is a special problem in areas with a large atmospheric load of pollutants, such as London, or in locations with disadvantageous topography, such as narrow valleys. Improved weather forecasting in recent years has helped to decrease the problem somewhat in certain areas. These forecasts can be used to control emission of atmospheric contaminants during periods of atmospheric stagnation. This procedure is common practice in connection with atomic power reactor operation, where the effluent is regulated in proportion to the dispersal capabilities of the atmosphere at a given time. This provides the distinct advantage of a large nuclear power generator as compared with smaller fossil-fuel plants. Another aspect of meteorologic forecasting is individual self-protection. Persons who are potential sufferers from air pollutants can take special precautions prescribed by their physicians in times of increased risk associated with predictable meteorologic conditions.

Health and Air Pollution

Hazards of air pollution have been demonstrated most vividly in episodes of acute exposure. There are many reliable reports of individuals and small groups of persons made ill or even killed by exposure to toxic fumes, vapors, pesticide aerosols, smoke, or high concentrations of dust. In some instances, major community catastrophes have resulted from air pollution during periods of air stagnation, temperature inversion, and large amounts of ordinary combustion-produced air pollutants. In these tragic episodes, fatalities occurred primarily among the elderly and those with preexisting cardiac or pulmonary disease.

Many scientists believe that long-continued exposure to low levels of air pollution has unfavorable effects on human health. Attempts to identify possible effects of ordinary urban air pollution on longevity or on the incidence of serious disease have generally been inconclusive. The health of the country has traditionally been measured by the rate at which people die; a decreasing death rate is taken as an indication of increasing health. The U.S. death rate declined rapidly in the first half of this century as infectious diseases were controlled, but it has remained almost constant since 1950 (Fig. 2-3). It is difficult to be specific about general indicators such as deaths per unit of population, but we have apparently reached a rather abrupt place on the curve where the advantages of medical progress are being offset by the disadvantages of the environment.

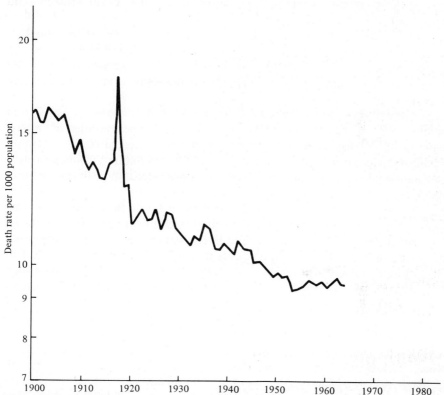

FIGURE 2-3 United States death rate, 1900–1974. [Source: *U.S. Vital Statistics* (Washington, D.C. U.S. Public Health Service).]

Disease and Air Pollution

"The development of freeway systems can increase the amount and alter the pattern of pollution in any particular area. This is true also of the location of power plants and the choices of fuel for thermo-electric power. It is likely that prolonged low level exposures from such sources are deleterious to health, but it may be exceedingly difficult to establish such relationships, particlarly as these effects may become manifest only over a prolonged period of time, possibly many years. At the present time it is difficult to predict the effects of prolonged low level exposures so that it might be better to avoid the likeihood of such effects by locating plants in such a way or utilizing such fuels as will minimize exposures."[5]

Air pollution is mostly associated with diseases of the bronchial tree—from the common cold to lung cancer. Air pollution also irritates the eyes; and pollutants such as lead can build up in the body, reaching harmful levels. Others, such as carbon monoxide, are not cumulative but can cause health problems and even death in high concentrations.

Emphysema is the fastest growing cause of death in the United States. It is a progressive breakdown of air sacs in the lungs usually brought about by chronic infection of the bronchial tubes. The disease progressively diminishes the ability of the lungs to transfer oxygen to the bloodstream and carbon dioxide from it.

In the 10-year period between 1950 and 1959, deaths among males from emphysema rose over 500%. In 1962 more than 12,000 persons died of emphysema in the United States and 1000 workers were forced to retire early onto Social Security rolls because of the disease.[6] Air pollution is indicated as a contributing factor as deaths from emphysema are twice as high in the city as in the country.

Bronchitis has been given much attention in Great Britain and has been found to vary directly with such air pollution measures as population density, amount of fuel burned, sulfur dioxide levels, settled dust, airborne dust, and decreased visibility. Their studies indicate that 10% of all deaths and more than 10% of illnesses were attributed to this disease. Studies indicate that more than 13% of adult males in the United States have the disease when the British criteria are used. Known sufferers of the disease have been systematically observed and showed a worsening of their symptoms on days of high air pollution.

Bronchial asthma attacks have been associated with a long list of stimuli, and it is difficult to define precisely the role of air pollutants. It is a disease in which trachea and bronchi show an increased responsiveness to various stimuli, as manifested by a widespread narrowing of the airways. It has long been known that occupational exposure to certain dusts and vapors, including many that are sometimes found in substantial quantities in the air over our cities, can bring about attacks.

A special form of asthma was studied among American troops and their dependents living in Yokohama, Japan, during 1946. The disease later appeared around Tokyo and was given the name *Tokyo-Yokohama asthma*. The disease did not yield to the usual treatment, but patients normally recovered if they were promptly removed from the area upon becoming afflicted. Studies indicated that the incidence of the disease and the severity of attacks correlated with increased air pollution, which had its origins in the adjacent heavily industrialized area.

In large-scale studies of absenteeism at widely scattered industrial plants in the United States, it was found that acute infections of the upper respiratory track increased as levels of air pollution increased. Investigations in Great Britain, Japan, and the Soviet Union have confirmed these American findings. Deaths from lung cancer have been increasing rapidly in recent years and air pollution appears to be involved. The rate of lung cancer is double in large cities as compared to rural areas. The death rate for lung cancer is proportional to city size, and the level of air pollution also generally increases with population.

It has generally been accepted that air pollution can affect health, and it has been demonstrated conclusively that under severe conditions, people were made ill and human deaths occurred. There is also a growing body of circumstantial evidence which indicates that long-term low-level air pollution exposure can contribute to and aggravate chronic diseases that affect large numbers of our population.

The Pollutants

Air pollutants may be solids such as dust or soot particles, liquid droplets such as sulfuric acid mist, or gases such as sulfur dioxide, carbon monoxide, hydrogen sulfide, and oxides of nitrogen. They include fumes and particulates from lead, vanadium, arsenic, beryllium, and their compounds, and fluorine and phosphorus compounds. In addition, new pollutants can be created in the air by the interaction of these and other substances in the presence of sunlight. These substances may travel through the air, disperse, and react among themselves and with other substances, both chemically and physically. Eventually, whether or not in their original form, they can leave the atmosphere. Some substances, such as carbon dioxide, can enter the atmosphere faster than they leave it and thus gradually accumulate in the air.

Dry air at sea level contains 78.09% nitrogen by volume, and 20.94% oxygen. The remaining 0.97% of the gases that make up dry air includes small amounts of carbon dioxide, helium, argon, krypton, and zenon, as well as very small amounts of other gases whose concentrations may differ with time and place. Water vapor is normally present in air in ranges of 1 to 3% by volume. Air also contains solid or liquid particles that range in size from clusters of a few molecules to diameters measured in microns.

The five most common air pollutants (carbon dioxide not included) in tons emitted annually in the United States are carbon monoxide, sulfur oxides, hydrocarbons, nitrogen oxides, and particles (see Table 2-2). *Carbon dioxide* was omitted from Table 2-2 because it was not considered a contaminant that can be controlled except by replacing the combustion process with another source of energy, such as nuclear power. Carbon dioxide emissions, which far exceed those of the five most common pollutants, may be affecting the temperature of the earth by their effect on the global atmosphere.

TABLE 2-2

NATIONAL AIR POLLUTANT EMISSIONS, 1965
(millions of tons per year)

	Total	Percent of Total	Carbon Monoxide	Sulfur Oxides	Hydro-Carbons	Nitrogen Oxides	Particles
Automobiles	86	60	66	1	12	6	1
Industry	23	17	2	9	4	2	6
Electric power plants	20	14	1	12	1	3	3
Space heating	8	6	2	3	1	1	1
Refuse disposal	5	3	1	1	1	1	1
TOTAL	142		72	26	19	13	12

Source: "The Sources of Air Pollution and Their Control," *Public Health Service Publication No. 1548*, Government Printing Office, Washington, D.C., 1966.

Carbon Dioxide

Although carbon dioxide is not generally thought of as an air pollutant, man generates an enormous amount of it in combustion processes. Most of this carbon dioxide is evolved from the combustion of fossil fuels such as coal, oil, and natural gas. It is a normal constituent of the air, where it takes part in no significant chemical reactions with other substances in the atmosphere. However, its global concentration is rising measurably, and this could affect climate markedly.

The temperature of the earth rises when the amount of carbon dioxide in the atmosphere increases as this gas absorbs and traps heat and prevents some of it from radiating back into space. Visible rays from the sun are little affected, but carbon dioxide absorbs and reflects back portions of the infrared radiations from the earth, particularly in the wavelengths from 12 to 18 micrometers (μm; older term, microns). Water and ozone also have similar characteristics. However, there is such a small amount of ozone in the atmosphere that it can probably be neglected.

Water vapor is concentrated within a few thousand feet of the earth's surface and is highly variable in time and place. On the other hand, carbon dioxide is uniformly distributed up to at least 50 miles (mi). Where water vapor is in the form of clouds, that portion of light energy from the sun that reaches the earth is reduced. Therefore, water vapor can both increase and decrease the earth's temperature, depending upon cloud formation. It is therefore difficult to assess the total effect of water vapor on global temperatures. Nitrogen (78 %), oxygen (21 %), and argon (0.9 %) do not possess the insulating properties of carbon dioxide.[7]

Within a few generations, man is burning the fossil fuels that slowly accumulated in the earth over the past 500 million years. These huge man-made emissions of carbon dioxide have been entering the atmosphere faster than the natural cycle can adjust. About half of the extra carbon dioxide produced by combustion of hydrocarbons is retained by the atmosphere. By the year 2000 the increase in carbon dioxide in the atmosphere will be about 125% of its 1850 level. This may be sufficient to produce marked changes in climate.

The natural cycle of carbon dioxide emissions and withdrawals from the atmosphere can be separated into several parts, each requiring a progressively longer period of time. The exchange of carbon dioxide between air and plant and animal life is the basis for all organic compounds. Animals consume as food the carbon compounds produced by plants. Plants acquire carbon dioxide from the air or dissolved in the water. Animals produce carbon dioxide as a body waste. This may be viewed as a form of biochemical combustion. Carbon dioxide moves through this plant and animal cycle relatively rapidly and with little net effect on the global concentration of the gas in air.

Another part of the natural cycle involves the oceans, which are a huge sink for carbon dioxide. The world's oceans contain about 50 times as much carbon as the atmosphere. It was originally felt that the oceans had almost unlimited capacity to absorb carbon dioxide, but several hundred years will be required before the buffering effect of the oceans becomes very effective. This exchange must take place in two steps: between the air and the top 50–100 m of water; and between the top layer of water and the water beneath. The exchange of carbon dioxide between the top layer of water and the air depends strongly on such factors as the acidity, salinity, and temperature of the water. Atmospheric carbon dioxide is being absorbed in the Arctic regions and released in the tropics. The net global effect is that it takes about 5 years for the air and top layer of seawater to reach equilibrium concentrations of carbon dioxide following a massive new injection of the gas into the air. The upper and lower layers of seawater exchange carbon dioxide in several chemical forms, but this process may take up to 1500 years to adjust to a fresh injection of carbon dioxide into the air.[8]

Still other processes involve the reactions between ocean water and limestone and the weathering of silicate rocks by carbon dioxide in the air. These processes are quite slow and would take 10,000 years or more to reach equilibrium with a new charge of carbon dioxide into the atmosphere.

The carbon cycle is probably the most sensitive of the natural systems that control life on earth. Changes in this cycle cause compensating changes in the hydrologic, nitrogen, and oxygen cycles. Through rapid exploitation of fossil fuels, man may be triggering unprecedented changes in weather and global ecology within the next 50 years.

Carbon Monoxide

Carbon monoxide, one of the most common pollutants in urban air, appears to be almost exclusively a man-made pollutant. Data on the background concentration of carbon monoxide in the atmosphere are very limited. It has a global level in the range of 0.1 part per million (ppm) by volume. Although carbon monoxide appears not to react significantly with any other constituent of the atmosphere, calculations indicate that it will last about 3 years once released into the air. There have been no indications that the amount of carbon monoxide is increasing in the atmosphere.

The automobile is estimated to produce more than 80% of the carbon monoxide emissions globally, with lesser amounts coming from other combustion processes. Some small amounts of carbon monoxide are produced by some plants and animals, especially under stress. This gas kills under high concentrations and retards physical and mental activity at lower levels.

Inside an automobile operating in traffic the concentrations of carbon monoxide can reach high-enough levels to affect the occupants. At approximately 100 ppm most people experience dizziness, headaches, lassitude, and other symptoms. Concentrations higher than this sometimes occur in garages, tunnels, or behind autos in heavy traffic. The concentration of carbon monoxide correlates with traffic volume (see Fig. 2-4).

The toxicity of carbon monoxide is due primarily to its affinity for hemoglobin, the oxygen carrier of the blood. It appears to affect the brain through an interference with the subject's awareness of his environment. It is nontoxic to

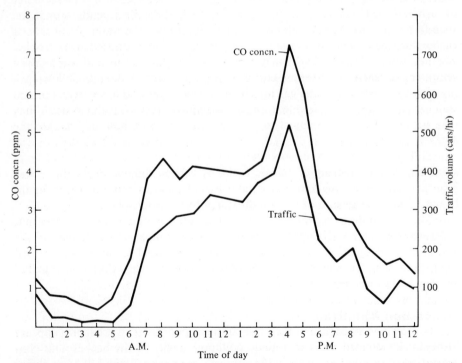

FIGURE 2-4 Carbon monoxide levels in traffic. [Source: 1962–65 monitoring survey in Detroit, Mich.]

insects and other lower forms of life which have no red blood cells. Small amounts of carbon monoxide compete with large amounts of oxygen for hemoglobin binding sites. Its affinity for hemoglobin is approximately 210 times greater than that of oxygen. The main concern about carbon monoxide is the long-term ecological effect from low-level concentrations.

Sulphur Oxides

The coal and oil that are burned so abundantly all over this country to heat buildings and to generate electric power contain elemental sulfur as an impurity. When the fuel is burned, the sulfur also burns, producing *sulfur dioxide* gas and, to a much smaller extent, *sulfur trioxide* gas. The sulfur trioxide in the atmosphere immediately reacts with water to form sulfuric acid. On a global basis, sulfur dioxide is natural to the atmosphere, as about 80% of the molecules of the gas are derived from hydrogen sulfide produced naturally by decaying organic matter on lands and in the oceans.

Sulfur dioxide gas alone can irritate the upper respiratory tract and can be carried deep into the lungs, where it can injure delicate tissues. Many investigations have shown that certain portions of the population are especially vulnerable to sulfur pollution. Studies have demonstrated that the annual incidence of respiratory disease was directly proportional to the annual sulfur oxides level. As the oxides level went up, so did the incidence of disease.

Even though most of the sulfur dioxide in the air on a global basis is of natural origin, it is the contribution by the combustion of coal and oil on a local or regional basis that produces significant effects. This gas is a principal air pollutant and has been extensively studied in urban and industrial areas. These studies have shown distinct seasonal trends in sulfur dioxide emissions that coincide with the increased consumption of coal and oil in winter months. Annual consumption of coal and oil in many cities amounts to millions of tons. The sulfur content for good-quality anthracite coal is around 1% while some bituminous coals are over 5% (Table 2-3). The average sulfur content of coal runs about 2.5% in the United States.

Besides the problems of human health, sulfur compounds in the air are readily absorbed by vegetation, soil, and water surfaces, causing corrosion and other damages. The sulfur compounds also affect wires, metal surfaces, textiles, paints, and building materials. Many green plants are particularly sensitive to sulfur dioxide and are injured by exposure of a few hours to concentrations as low as 0.3 ppm. Rain and snow absorb sulfur compounds and become more acid. In 1958 in Europe, precipitation showed values of less than pH 5 in a few limited areas over the Netherlands. By 1962, values of less than pH 5 could be found over central Europe and values of less than pH 4 could be found over the Netherlands. The area where precipitation of pH 4.5 could be found had reached central Sweden by 1966. This has caused the pH of some lakes and rivers in the affected areas to show a lowering trend. Declining pH is a threat to aquatic life, since most organisms cannot live at less than pH 4, and some fish, such as salmon, are killed at pH 5.5.

TABLE 2-3

SUMMARY OF DEMONSTRATED COAL RESERVE BASE OF
THE UNITED STATES
(billion tons)

Rank of Coal	Underground Mining Reserve	Surface Mining Reserve	Total	Estimated Total Heat Value (quadrillion · Btu)
Bituminous	192	41	233	6,100
Subbituminous	98	67	165	2,800
Lignite	0	28	28	400
Anthracite	7	*a*	7	200
Total	297	137	434	9,500

a Less than 0.5 billion tons.

Source: "Demonstrated Coal Reserve Base of the U.S. on January 1, 1974" (Washington, D.C.: Mineral Industry Survey, U.S. Department of the Interior, Bureau of Mines, Division of Fossil Fuels, Mineral Supply, June 1974).

The global importance of man-made emissions of sulfur compounds to the air is still unknown. Existing data are not sufficient to indicate whether sulfur compounds are accumulating in the atmosphere. Local and regional damages from sulfur compounds released to the air by man are well documented. One question of ecological concern concerns the biogeochemical cycles of oxygen, sulfur, carbon, and nitrogen. All organisms build their protein basically from these four elements and hydrogen. If man were to destroy any of at least half a dozen types of bacteria involved in the nitrogen cycle, life on earth as it now exists could end.

Nitrogen Oxides

The major man-made sources of *nitrogen oxides* are combustion processes. At the temperature commonly reached when fuels are burned for various purposes, nitrogen in the air combines with oxygen to form nitric oxide, which is relatively harmless, but which converts in the atmosphere to nitrogen dioxide.

$$N_2 + O_2 \longrightarrow 2\,NO$$

$$2NO + O_2 \longrightarrow 2NO_2 \qquad\qquad (2\text{-}1)$$

The quality and rate of nitric oxide production is a function of several parameters, such as flame temperature and the amount of excess air used in combustion. It has been estimated that the nitric oxide production in the City of Los Angeles during 1967 amounted to 920 tons/day.

As the nitric oxide is converted to nitrogen dioxide, it creates a problem in polluted air. Nitrogen dioxide is a strong absorber of ultraviolet light from the sun and is the trigger for photochemical reactions that produce smog. *Smog* is a word coined several decades ago from "smoke" and "fog," to describe the characteristic highly polluted fogs of London. The chemical reaction involved is:

$$\text{hydrocarbon} + O_2 + NO_2 + \text{light} \longrightarrow \text{peroxyacyl nitrates} + O_3 \qquad (2\text{-}2)$$

Equation 2-2 is intended to be illustrative and not definitive. The detailed chemistry of the smog-forming process is still under investigation. Ozone and peroxyacyl nitrates (PAN) are oxidizing by nature. They irritate the eyes at about 0.1 ppm, and this level is usually reached several days per year to, in some urban locations, as frequently as one-third of the days. The frequency depends on many factors, including the amount of sunlight, the density of automobile population, and the ability of the local atmosphere to cleanse itself.

Studies have shown that the products of photochemical smog at levels routinely found in our cities make it more difficult for people to breathe, especially people already suffering from respiratory disease. One of the products, ozone, can irritate mucous membranes. At levels commonly found in urban atmosphere, ozone will irritate the nose and throat after only 10 minutes' exposure. At slightly higher levels, ozone produces coughing, choking, and severe fatigue. Research at Wright-Patterson Air Force Base suggests that ozone may impair vision and depress body temperature. The latter could be the result of peripheral vasodilation or an impairment of the brain center that regulates body temperature.

Nitrogen dioxide is present in the air whenever fuels are burned and whether or not photochemical smog has been produced. There is very little information on the effects on man of exposures to the levels of nitrogen dioxide commonly found in polluted air, 1–3 ppm. Experiments with animals have shown that exposure to 10–20 ppm produced persistent pathologic changes in the lungs. Exposure to nitrogen dioxide has shown increased susceptibility in laboratory animals to infection. Animals infected with pneumonia bacteria and also continuously exposed to low levels (0.5–3.5 ppm) of nitrogen dioxide died more frequently than animals who were infected but not exposed to the gas.[11] This relationship between exposure to an irritant gas and increased mortality among infected laboratory animals has profound implications for humans in considering levels of tolerable air pollution.

The adverse influence of air pollution on vegetation was recognized over a century ago. Interest in its effect on animals developed much later. As a result, more knowledge is available relative to plant and tree damage than about harm to man.

Plants sensitive to ozone and PAN include cotton, grapes, tobacco, cereal crops, ornamentals, vegetables, citrus, forage, and salad crops. A plant is held to be damaged when its economic value, quality or quantity of its yield, or external appearance are impaired. Ozone injures primarily the upper surface of plant leaves, while PAN causes a silvering, glazing, or bronzing of their lower surfaces. Economic losses resulting from the effects of air pollution on controlled vegetation are substantial but some of the effects on native vegetation may eventually have a greater impact.

Particles

The atmosphere also receives *particulate matter* of varying sizes. Some particles are large enough (>1 μm) that they settle rapidly to the ground while others are small enough (<1 μm) to remain suspended in the atmosphere until removed by rain and wind. Particles with diameters between 1 and 10 μm are more numerous in the atmosphere and usually include the largest weight fraction. They generally come from industrial processes and include dusts, ash, and the like. Particles in the size range below 1 μm tend to contain more products of condensation, such as ammonium sulfate, and particles produced by combustion along with aerosols formed by photochemical reactions in the atmosphere.

Particles generated in an urban atmosphere normally remain airborne for only a few days, although some, depending upon their size, may remain airborne for many weeks. There is little information on the effects of particulates at the levels found in polluted atmospheres; however, some observations can be made. Arsenic, given off by copper smelters, is suspected of inducing cancer. Asbestos fibers have been associated with chronic lung disease and with lung cancer. The fibers are given off by the brake linings of automobiles and industry, roofing materials, insulation, and shingles. Beryllium, used in the production of metallic alloys and as a fuel for rocket engines, has been indicated with malignant tumors in monkeys. Cadmium is a respiratory poison and may contribute to heart disease. Fluorides are discharged to the atmosphere during manufacture of phosphate fertilizers and have caused severe damage to cattle and vegetation on occasion. Lead is a cumulative poison taken into the body in food and water as well as air. Most lead in the atmosphere is the result of leaded gasoline in automobiles. The brain is protected against foreign materials that enter the blood but heavy metals such as lead are exceptions. Lead appears to interfere with brain function rather than to damage the cells themselves. These are but a few of the specific particulate materials in the air that can effect health.

Air pollution has tremendous economic effect on materials. Steel corrodes two to four times faster in urban and industrial areas than it does in rural areas. In England it has been estimated that one-third of the replacement cost of steel rails is attributable to air pollution. Women in several U.S. cities have had the unfortunate experience of developing runs in their nylons in the time it took them to walk to lunch and back again. Stone statuary and building finishes are deteriorating in many cities. Cleopatra's needle has deteriorated more since its arrival in New York in 1881 than it did during 3000 years in Egypt.

Control of Air Pollution

Many technological and economic factors are concerned with each engineering application in the control of air pollution, but the choice lies fundamentally among five alternatives:

1   Select process inputs, such as fuels, that eliminate the pollutant.
2   Remove the pollutant from the process input.
3   Operate the process, such as combustion temperature, so as to minimize generation of the pollutant.
4   Remove the pollutant from the process effluent.
5   Replace the process with one that does not generate the pollutant.

Alternative 1 is being used to reduce the lead in our air. The lead exhausted by automobiles originates in the antiknock compound added to gasoline. Car engines are now being designed to use unleaded gasoline. Within a short while, all leaded gasoline will be off the market and then the lead problem should cease to exist because lead from other sources such as water and food have decreased over the last 50 years.

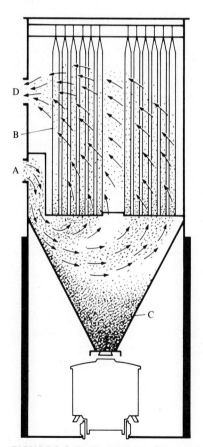

FIGURE 2-5 Baghouse dust collector. A baghouse dust collecting system functions just like a huge vacuum cleaner, passing polluted air through a series of large bags which filter the fine particulate matter, permitting only clean air to the atmosphere. A—dust-laden particles from furnace; B—filter bags; C—dust storage hoppers; D—cleansed gases are discharged through a stack.

Fuel oil for heating and power plant combustion uses alternative 2. Technology is available and used for desulfurizing residual fuel oil. Costs indicate that to reduce the sulfur content of fuel oil from 2.6% to 0.5% would increase the cost to the consumer by about 35%. An increase from the present low sulfur prices would help reduce the cost. Unfortunately, residual fuel oil accounts for only about 10% of electric utility power production, so coal must continue in use for some time to come. Some processes have been developed to desulfur coal, but at present none of them is economically feasible in reducing the sulfur content at the lower levels (e.g., from 2.5 to less than 1% sulfur).

Alternative 3 is used in certain combustion processes to reduce nitrogen oxides in their emissions. The higher the peak combustion temperature, the more nitric oxide is formed. One method recycles part of the combustion exhaust gas back into the combustion chamber, thus reducing combustion temperature and the amount of nitric oxide finally produced.

Alternative 4, remove the pollutant from the process effluent, is widely used in many major industries and combustion processes. One method uses the equivalent of a large vacuum cleaner on the stack gases (Fig. 2-5). This process passes polluted air through a series of large bags which filter the fine particle matter, permitting only clean air to return to the atmosphere. Bag filters using cloth or paper filter material can remove submicron particles with head losses through the filter of 0.5 to 6 in. of water. Wet scrubbers and electrostatic precipitators are also used for this type of removal (see Fig. 2-6).

Alternative 5 is being followed to a limited extent in the electric utility field and being carefully studied in the case of the automobile. Nuclear power reactors

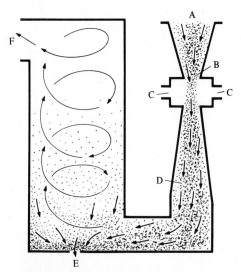

FIGURE 2-6 Wet scrubber. High-efficiency wet scrubbers clean the polluted air as it passes through the system, recovering better than 99% of the fine fume particles generated by the steelmaking furnaces. A—Dust-laden gases from the furnace; B—Venturi throat; C—Scrubber; D—Wetted particles; E—Wash water to treatment facilites; F—Cleansed gases are discharged through a stack.

are being used at an increasing rate in power generation. They are probably the ultimate solution to much of our air pollution problems of the future. Electrically powered vehicles seem likely to become more economically feasible in the next decade. Technological needs include more effective devices for storing and converting electrical energy, and improvements in motors. Should a significant shift take place toward the electric auto using some type of rechargeable system, it would put a tremendous strain on electrical power production.

The movement of people from rural sections to urban areas has led to the rapid evolution of cities into large metropolitan complexes. The result of this movement is an ever-increasing density of industrial activity. Therefore, rigid control of all air pollution is now a necessary way of life (see Fig. 2-7).

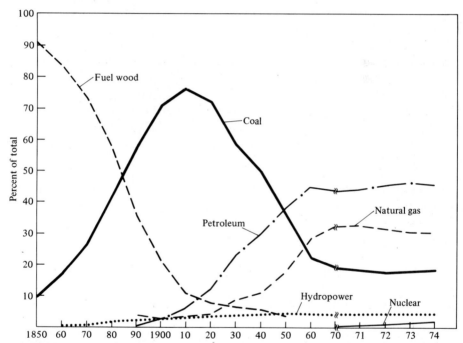

FIGURE 2-7 U.S. energy-consumption patterns, 1850–1974. [Source: *Historical Statistics of the United States* (Washington, D.C.: Bureau of the Census; U.S. Bureau of Mines, 1974).]

2-2

ANIMAL WASTES

Historically, farm animal wastes deposited by animals over pastureland or in concentrated deposits around barns have been returned to cultivated fields. In recent years, the trend has been toward intensive livestock production. The complementary relationship between crop production and livestock production has been further weakened by the lower costs of alternative sources of fertilizers, and the practice of distributing manure on land has become uneconomical. This has resulted in large volumes of animal wastes accumulating around animal production facilities.

As long as animals were widely dispersed over large farms and range lands, their waste products had generally not created problems. The new methods of feedlots for thousands of cattle or chicken facilities for a million or more chickens has greatly intensified the problem. The magnitude of animal waste produced annually in the United States is somewhat difficult to equate to human wastes. The production of animal wastes in this country exceeds the wastes produced by the human population by factors of about 5:1 to 10:1, depending upon the basis of comparison (see Table 2-4).

TABLE 2-4

PRODUCTION OF WASTES BY LIVESTOCK TYPE IN THE UNITED STATES, 1965

Livestock	Population (millions)	Annual Production of Solid Wastes (million tons)	Annual Production of Liquid Wastes (million tons)
Cattle	107	1004.0	390.0
Horses[a]	3	17.5	4.4
Hogs	53	57.3	33.9
Sheep	26	11.8	7.1
Chickens	375	27.4	—
Turkeys	104	19.0	—
Ducks	11	1.6	—
TOTAL		1138.6	435.4

[a] Horses and mules on farms as work stock.

Source: " Wastes in Relation to Agriculture and Forestry," U.S. Department of Agriculture, *Miscellaneous Publication No. 1065* (Washington, D.C.: Government Printing Office, March 1968).

There are over 3 billion chickens produced annually in the United States, and their waste products alone exceed the total annual excrement of the human population. The animal waste problem is growing not only because of our population increase, but it is also due to increased consumption of beef and chickens (see Fig. 2-8). The United States is demonstrating a growing taste for beef and chicken and a slightly decreased taste for pork.

Livestock produce nearly 2 billion tons of manure each year. Dairy cattle produce about 3 pounds (lb) of wet manure for each quart of milk.[14] The enormous quantity of waste must be collected, transported, and disposed of in a nonpollutional manner. The management of these wastes is one of the most severe problems currently facing the livestock industry.

FIGURE 2-8 Number of cattle and chickens in the United States, 1940–1970. [Source: U.S. Department of Agriculture, *Agriculture Statistics—1970* (Washington, D.C.: Government Printing Office).]

The nature of the waste produced is unique for each species of animal and to the conditions by which it is housed and fed. These differences are due to differences in size, diet, and metabolism. Such animals as swine produce feces and urine similar to that produced by humans. The diets consumed by swine and chickens are highly digestible, and therefore the amount of excreta is relatively small. The manure from cattle, on the other hand, is relatively large. This is due to the nature of their diet and method of metabolism. The bacteria that inhabit the stomach of ruminants such as cattle enable these animals to use cellulosic feed. There are other compounds, such as lignin, in feed plants which are difficult to digest in the rumen. This tends to produce a relatively large amount of waste with a different composition than wastes from single-stomached species. Within the same species, the makeup of manure from grass-fed animals is quite different from that of animals liberally fed on concentrates.

The rapid change to raising livestock in close confinement and the increased number of these livestock has increased the volume of waste to be handled. The type of feeds used has also increased the oxidation rates of these wastes. These wastes then have a greater pollutional potential than before. Animals in confinement are fed to cause the most economical weight gain in the shortest period of time. Highly efficient consumption of feed by the animal is subordinate to continuous and rapid weight gain. These animals excrete more of the nutritive material. From the pollutional point of view, it would be highly desirable to reduce the quantity of nondigested material in the waste.

In some areas of the country, livestock operations are a source of serious water pollution. It is estimated that animals produce over 17 billion lb of nitrogen annually. Nitrogen in water can be a problem, owing to the toxicity of nitrates and the stimulation of aquatic plants in receiving waters. Together with nitrogen and phosphorus, most of the salt intake of an animal is excreted in the urine and feces.

The list of infectious disease organisms common to man and animals is lengthy, and many of these can be waterborne. Therefore, improper management of animal wastes could be a serious threat to health when drainage from animal production units reach a water course. The logical solution is to process animal wastes prior to disposal. The magnitude of this problem can be judged from recalling that the wastes from livestock operations are greater in both volume and strength than domestic wastes from the human population, and there is still a large segment of the human population in the United States that is without adequate waste treatment.

2-3

THERMAL POLLUTION

The temperature of many of the nation's rivers is rising as a result of increasing use of water for cooling purposes, particularly by certain industrial operations such as steam electric power, steel mills, petroleum refineries, and paper mills. The principal contributor of this heat is the electric power industry. In 1968, the cooling of generating plants accounted for three-fourths of the 60,000 billion gallons (gal) of water used in the United States for industrial cooling.

During the past 50 years, the power industry has grown by an average rate of 7.2% annually, and has doubled its capacity every 10 years. It appears that this rate will continue or even accelerate. Some estimates are that by 1980, one-sixth of the nation's freshwater runoff will be required by power plants.

According to the Department of the Interior's Bureau of Mines "energy balance sheet," it took 65,645 trillion British thermal units (Btu) to meet the nation's total 1969 needs for heat, light, and all forms of power. This represents a 5.1% increase over 1968 consumption. The record energy demand was met principally through increased use of natural gas and petroleum, plus slight increases in the use of coal, hydropower, and nuclear power. Petroleum continued as the dominant fuel, supplying 43.2% of all U.S. energy requirements. Natural gas, excluding natural gas liquids, supplied 32.1% of energy demand;

bituminous coal and lignite, 20.1 %; water power, 4 %; and nuclear energy, 0.2 %. Nuclear energy, whose use in generating electric power was negligible 10 years ago, jumped 141 trillion Btu in 1969.

As usually defined, thermal pollution means the addition of heat to natural waters to such an extent that it creates adverse conditions for survival and pre- servation of aquatic life. Heat accelerates biological processes in the stream, resulting in reduction of dissolved oxygen content of the water. It also contributes to taste and odor problems by increasing growth of aquatic plants and makes the water less suitable for domestic, recreational, and other industrial uses, including cooling. The federal government has declared that waters above 93°F are essen- tially uninhabitable for most fishes in the country. The problem is that many rivers already reach a temperature of 90°F or more in the summer through natural heating alone. The waste from a single power plant of the size planned for the immediate future [some 1000 megawatts (MW)] could be expected to raise the temperature of a river flowing 3000 cubic feet per second (cfs) by about 10°F. Since a number of industrial and power plants are expected to be built on the banks of a single river, there is a high probability that many American waters will become uninhabitable unless other methods of cooling are used.

In most plants, a once-through cooling process has proved to be most economic (see Fig. 2-9). Because of this, plants have been located where large quantities of water are available for cooling at all times during the year. Such sites in inland areas are limited in number and the increasing density of power plants on rivers will ultimately require power plants to use other means of cool- ing. In areas of short water supply, such as the Southwest, cooling ponds, spray

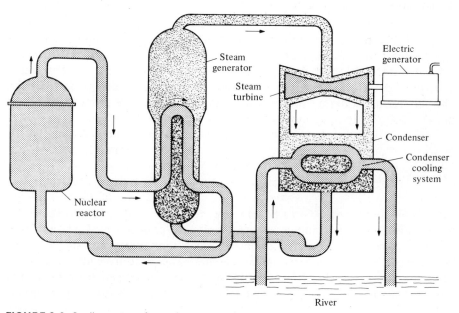

FIGURE 2-9 Cooling system for nuclear power plant.

ponds, and cooling towers have been used to extend the capabilities of water available for power generation. Although such practices can limit the amount of heat dissipated to the natural water environment, they increase the consumptive use of water and may create a saltwater-disposal problem. They may also cause fog and ice problems in their immediate area. Towers for a 1000-MW power plant might eject 20,000–25,000 gal of evaporated water per minute to the atmosphere. This is equivalent to a daily rainfall of 1 in. over an area of 2 mi^2.

Cooling towers are of two basic designs, draft systems using either forced or induced blower systems. Mechanical draft systems have the advantages of relatively low capital cost, less space requirement, and a broad range of loadings. Their principal disadvantages are high operating and maintenance costs. Hyperbolic, natural-draft, evaporation towers have become increasingly popular since their first use in 1963. The hyperbolic towers derive their name from their distinctive profile, which approximates the pattern of air flow within the tower. The principal advantage of natural draft towers is low operating costs, since air flow is induced by convective forces, created by the difference in density between the internal and external air. The disadvantage of these towers is their size. Some units run 450 ft in height and 350 ft in base diameter. Such massive structures require high capital costs, both for the tower and its foundation.

Cooling towers are expensive both in capital cost and annual operation. Vermont Yankee is installing mechanical draft towers to limit effluents to the Connecticut River to a 4°F temperature rise. Capital costs of the installation will be 5.1 % of the total plant cost of $118 million. The operating costs for the towers are estimated at $900,000 per year.

So far there have been few recorded instances of direct fish kills by thermal pollution in the United States. Some people hope to use the heated water for some commercial purpose, and, although dozens of schemes have been advanced no practical use has yet been proven. One scheme that is in the pilot plant stage in both the United States and in Britain is to use the heated water from coastal plants to grow fish and shellfish. The preliminary indications are that better growth of both fish and shellfish can take place in the artificially heated water than in normal waters. Further consideration needs to be given to means of making beneficial use of the waste heat from thermal power plants.

2-4

TOXIC CHEMICALS

Man's ability to manufacture and use new chemical substances has far outstripped his understanding of their biological consequences. The principal interest has centered around chlorinated hydrocarbon insecticides such as DDT, and food additives such as cyclamates. What happens to man or animal when they consume new chemical substances and what happens to these substances are two complicated and partly unanswered questions of key importance to both public health officers and biochemists. These problems become even more complex when several new chemicals are ingested together. Some of this is due to interactions between a new chemical and the liver microsomal enzymes. These enzymes are largely responsible for the oxidation and metabolism of many drugs, pesticides, and other foreign chemicals.

Many substances, such as food additives and commercial drugs, are ingested directly by man, but others, such as pesticides and other toxic substances, are distributed in the plant and animal communities and reach man through the food chain. Biologists define communities broadly to include all species, including man. Over a period of time these communities each develop a complex of organisms that are balanced and self-sustaining. There develops in this community a series of cycles through which they share certain resources, such as mineral nutrients and energy. These nutrients move along pathways called *food chains*. Such chains start with plants which use the sun's energy to synthesize organic matter through photosynthesis. Animals eat these plants and may be eaten by other animals, which may in turn be eaten by other animals. In order that the lower orders in the food chain can survive, they are fed back nutrients through organic decay (mainly microorganisms) that break down organic debris into substances used by plants.

These natural food chains serve as biological amplifiers for trace chemicals spread out in our environment. No matter how widely dispersed certain substances are around us, it is possible to concentrate some of this material to levels many factors above that considered safe. Cesium 137, a fission product of fallout, emits gamma rays, behaves chemically like potassium, and becomes widely distributed once it enters an animal or man. A study in Alaska followed this material through the food chain. The first link in the chain was lichens growing in the forest. The lichens collected cesium 137 from fallout in rain. Caribou live on lichens during the winter; and caribou meat, in turn, is the principal diet of Eskimos in the area. The study indicated that the caribou had accumulated about 15 micromicrocuries ($\mu\mu$Ci) of cesium 137 radioactivity per gram of body weight in their tissue. The Eskimos who fed on the caribou had concentrations of 30 $\mu\mu$Ci per gram of tissue after eating caribou meat in the course of a season. Wolves and foxes that ate caribou contained three times the content of the caribou. It is easy to see that a food chain involving several links could increase the concentrations manyfold. Examples of biological amplification of pesticides in nature have shown increases of several thousand times the original level in the water or air. A major question as yet unanswered is: Will it be safe for people to eat fish and wildlife that are storing up pesticides? Human levels of tolerance to these poisons are unknown.

In the nineteenth century, mercury compounds were used to treat the felt material used to make hats. So many of the people who worked with this material eventually suffered poisoning and brain damage from prolonged exposure to the chemical that the phrase "mad as a hatter" became common. A few years ago, more than 100 Japanese were poisoned, some fatally, after eating fish that had fed on mercury wastes discharged from a plastics factory. In the fall of 1969, seven members of an Alamogordo, New Mexico, family almost died from eating pork that had been fattened on grain treated with a mercury-base fungicide.

The Province of Ontario, Canada, during April 1970, ordered a halt to all fishing in several lakes and rivers. They had earlier imposed a ban on the export and sale of any fish taken from Lake Erie and Lake St. Clair. Most of the fish caught by Canadian fisheries are sold in U.S. markets. The government ordered the export ban after a Canadian Wildlife Service investigation showed dangerously high levels of mercury in fish. Sampled fish contained from 1.3 to 7 ppm of mercury. This is in excess of the Canadian standard of 0.5 ppm, which is also the U.S. standard for mercury.

The mercury came from a few chemical companies that were dumping metal residue from manufacturing processes into the waterways. The Canadian government has ordered firms to eliminate mercury from their effluents or be closed. The point here is that it is not just the new materials on the market that are presenting problems. Many of the waste products are well documented for their environmental effects.

Chemical additives are added to food to improve flavor, color, and texture, and to enhance the keeping qualities. They act as leavening agents, antifoaming agents, clarifying agents, buffering agents, curing agents, and emulsifying agents. Flavoring agents alone amounted to 73 million lb in 1966. While many of these substances are believed not to be harmful to humans, there is little known about their effects from long-term consumption. We may be paying a heavy price for most of these substances used for aesthetic purposes that do not contribute nutritionally to our diet.

Monosodium glutamate is one of the most widely used of all food additives. It is generally present in all frozen food, dry soup mixes, and canned food that contain meat, fish, or their derivatives. It is used to intensify flavor and is one of the chief active ingredients in soy sauces. How monosodium glutamate works is unknown, but one theory is that it acts by stimulating greater saliva formation.

The American public is diet-conscious and has taken avidly to low-calorie foods. Most of these diet foods contain artificial sweeteners, usually saccharin or cyclamates. The use of cyclamates was greatly reduced by a federal order issued during the fall of 1969. About 60 million people are using low-calorie foods, including soft drinks, canned fruit, frozen desserts, and salad dressings among many others. Low-calorie soft drinks constitute the biggest single market for artificial sweeteners.

The artificial sweeteners may provde a sweetness resembling that of sugar, but they do not duplicate a number of sugar's other properties. The soft drink companies have attempted to make up for some of the deficiencies of artificial sweeteners by using other additives, such as sodium carboxymethylcellulose, as bodying agents. Considerable research is being directed toward new compounds to be used as sweeteners. An ideal substance would have properties similar to natural sugar except that of caloric value.

In terms of the amount used, the top-selling food additives are emulsifiers. The United States used about 150 million lb of these materials in foods during 1966. Emulsifiers are used for a variety of reasons in food, but their main function is to disperse an oil in water. They are used to speed the dispersion of such products as artificial cream in coffee and to keep fats from coming to the surface in many confectionary products. These emulsifiers are also used in such products as ice cream, candy, cake, and dessert toppings. Other products are added as stabilizers and thickeners. They are added to ice cream to provide a smooth creamy texture by binding part of the water and thus preventing the water from freezing into large grainy crystals. Still other compounds are added to prevent the escape of moisture from food such as candies and marshmallows.

Coloring agents are added to food to make them more attractive. There are colors added to natural butter to make its color more uniform despite seasonal changes. They are used in cheeses, soft drinks, cakes, cereals, candies, and many other products. The latest thing is simulated foods from imitation bacon to orange juice. These products contain most of the food additives as flavors, flavor enhancers, emulsifiers, thickeners, preservatives, and about every other type of food additive known.

Obviously most of these substances are not harmful at low levels of intake, but the percentage increase of these materials in our food is of such magnitude that they pose a substantial health problem. Because so many of the artificial additives are added to speciality items that have particular appeal to the very young (ice cream, candies, cakes), there may be other problems during these formative years.

2-5

SOLID WASTES

A more complete discussion of solid waste is given in Chapter 12. Public and private agencies collect more than 5.3 lb of solid wastes per person per day, and this figure is expected to rise to more than 8 lb by 1980. Overall, the nation is currently generating an estimated 10 lb per person per day of household, commercial, and industrial wastes. This does not include 7 million vehicles junked annually in this country. Many of these cars are salvaged, but others add to the already large number just lying around. Among the obvious effects of solid wastes are the aesthetics problem and the air and water pollution caused by unsatisfactory means of disposal. Solids can occupy and diminish the value of land and thus contain useful materials and energy in a form that usually makes their recovery uneconomical (see Table 2-5).

One method of solid waste recovery which does not use large land areas or add to air and water pollution is *pyrolysis*. This process can return to the energy cycle much of what is now lost as wastes, according to B. H. Rosen of Cities Service Oil. In a 1970 laboratory-scale pyrolysis project on a typical municipal

TABLE 2-5

RECLAMATION OF NONFERROUS METALS

Metal	U.S. Consumption, 1967		1969 Value[a] (millions)
	Quantity	Units	
Copper	1,243,000	short tons	$1096
Aluminum	885,000	short tons	403
Lead	544,000	short tons	161
Zinc	263,000	short tons	76
Antimony	25,568	short tons	26
Tin	22,790	short tons	71
Mercury	22,150	76-lb flasks	12
Silver	59,000,000	ounces	101
Gold	2,000,000	ounces	70
Total			$2016

[a] Based on prices in *American Metal Market* (May 1969).
Source: U.S. Bureau of Mines.

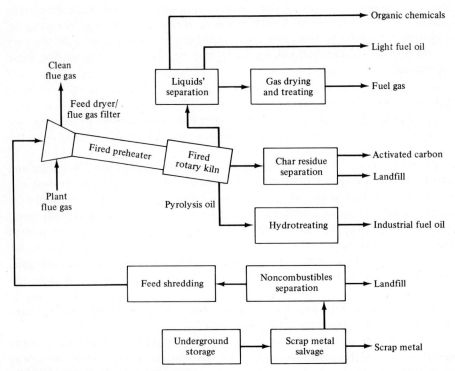

FIGURE 2-10 Pyrolysis of solid wastes.

refuse, excluding inorganic material, Rosen and his associates produced four basic product streams: water, organic liquids, gases, and char.

They concluded that a plant handling 5000 tons/day would require a total investment of about $33 million. Operating costs were estimated at $2.14 per ton input. In return, the plant would yield at least the value of the fuels produced at about $2.80 per ton of refuse fed to the plant. They also pointed out that municipal subsidies are not unreasonable, as the City of New York spends about $8 per ton on incineration, excluding collection costs. A pyrolysis outline is shown in Fig. 2-10. The technology used to handle solid wastes lags well behind that used to control air and water pollution. A more massive scientific effort is needed in this area of environmental control.

2-6

NOISE

We make contact with our surroundings through our senses, which include hearing, sight, touch, taste, and smell. The function of a sense organ is to receive information about our surroundings and to transmit this information to the nervous system in order that we may respond to the stimuli in a coordinated fashion. This process always involves the transfer of energy. In the case of sound, this energy may be transmitted from source to receptor through a region of space which may range from a few inches up to many miles.

Sound can cause both pleasure and pain. The boom of a cannon and the blast of an explosion are unpleasant while the melody and harmony of music are generally admitted to contribute aesthetic enjoyment. The loudness of a sound depends on the rate at which acoustic energy enters the ear. When a sound wave travels through the air, the individual gas molecules oscillate back and forth and energy is exchanged from one molecule to the next. This flow of energy can be measured by physical means. The intensity level is measured in bels. Usually a smaller unit, the *decibel* (dB) is used (1 dB = 0.1 bel). The decibel is a convenient unit because it approximates the threshold of hearing level for the human ear.

Noise pollution is already a serious public problem. No other age has ever thrust so many auditory images so promiscously at the human brain. Much of this sound energy is classified as noise. *Noise* may be defined as any sound that you do not want. It is an obstacle to brainwork and can easily disrupt a vital train of thought. Some noise levels in our environment are shown in Table 2-6. Prolonged exposure to a noise level of about 85 dB usually results in a loss of hearing acuity for sounds in the frequency range most important for understanding human speech.[15] This range is roughly 256 hertz (Hz; old name, cycles per second), the pitch of middle C, to about 2000 Hz, C three octaves higher.

Most people, at least in industrial areas, are obliged to live and work under conditions in which noise often attains an alarming level. The sudden roar of a passing aircraft or the din of a revving automobile disrupts sleep. Older people

TABLE 2-6
NOISE LEVELS (decibels)

130 + Riveting
120 + Threshold of pain
110 + Jet aircraft (at 90 m)
 90 + Sports car
 80 + Motorcycle, motor scooter
 70 + Busy street
 60 + Conversation
 50 + Quiet street
 40 + Quiet room
 30 + Tick of watch (at 1 m)
 20 + Whisper
 10 + Leaves rustling in the wind
  1 + Threshold of audibility

are more sensitive to these disturbances. These people are roused the next morning and have to face another day of work, often in noisy surroundings, which means that more nervous energy is burned up in resisting the stress of noise.

Noise is not just annoying, it is expensive. Hearing loss resulting from exposure to excessive noise has long been a concern of industry, where claims made by workers amount to millions of dollars annually. Loud noise damages hearing mechanisms to a point where the sensory nerve function is depressed. There is also evidence that noise can seriously affect mental well-being.

Unless technical measures are taken, our society will create more noise, not less. With the proliferation of machines, the noise level in urban centers is growing. Much of today's noise is largely the result of people's desire to reach distant places more rapidly and comfortably. As we have developed faster means of transportation, we have created a noise nuisance that is becoming increasingly difficult to live with.

The problem of aircraft noise exists at every major airport in the world. It became aggravated in 1958 with the introduction of the first jets and increased size of planes. There are no short-term or instantaneous areas of substantial improvement. It is generally believed that at most locations, aircraft noise will become more aggravated before the condition starts to improve.

There have been some technical advances in noise reduction in the past 10 years, and most areas are more promising than with aircraft. Automobile noises can be reduced by more effective exhaust mufflers and by nonslam doors and trunk lids. In the construction industry, relatively silent machines can be used in a variety of ways with a little increase in initial cost and some loss of efficiency. The home and office can be more restful through the use of sound-absorbing materials in drapes, ceilings, and walls. Reconstruction and renovation of cities offer opportunities for reducing many noises. Green spaces and protective screens can be incorporated into street and building design.

We will probably have to adapt in some measure to the noises of our civilization. The extent to which adaptation is necessary is largely a matter of economics and convenience. At some moderate cost and at some loss of convenience, our noise levels can be maintained at a reasonable level.

2-7

OIL SPILLS

Destruction of the Torrey Canyon and the associated pollution of part of the English Coast dramatized a problem that has reached a new dimension both in the United States and in most areas of the world. Much of the problem has come about because of the increased size of oil tankers, the density of the water-borne traffic, and offshore petroleum production.

Oil floating on water has disastrous effects on waterfowl. The magnitude of these effects depends on the amount and type of oil released and the concentrations of birds present. The chances of survival of birds that have come into contact with oil are quite small. They die from exhaustion, starvation, and exposure. Small concentrations of many petroleum products and derivatives appear to be highly toxic to fish and shellfish. There is an abundant amount of literature concerning the concentrations of various petroleum products capable of causing direct mortality of fish under laboratory conditions.[16] The effects on plant life are usually of short duration unless the exposure to oil is continuous.

Ships are a prime oil pollution threat. They may waste oil to the water during cargo transfer and handling, in the deballasting of vessels, and in cleaning oil tanks and pumping of bilge water, which usually contains waste oil. Commercial shipping fleets are powered almost exclusively by oil, and about 20% of all vessels are engaged in transporting oil. During 1966, more than 50,000 ships carrying oil visited U.S. ports. A 50,000-ton tanker may have 1200 barrels (bbl) of oil to be cleaned before unloading.

Control of oil pollution is difficult because international law allows ships farther than 50 mi from the shoreline to dump oil wastes into the ocean. Where accidental discharges of oil occur in coastal or inland waters, chemical and physical treatment methods are available. Floating absorbents such as straw and saw dust have been used. These materials are usually gathered from the water surface and burned as a means of final disposal. There are numerous solid absorbents for sinking oil, but because of the weight required per unit of oil removed from the surface, their use has been limited. This method has limitations in shallow waters, where aquatic life is involved. Oil removed from the surface by this method will probably return over a period of time. A large number of chemical dispersants and emulsifiers exist that could be used to remove oil slicks, but these agents are highly toxic to aquatic life and usually do more harm than would have been done by the oil.

The best method to control accidental oil spills is to confine the oil on the water surface and physically to remove the oil. Floating booms and air-bubble barriers have been used effectively in calm waters. Where a vessel is in the process of breaking up, the burning of the oil cargo is probably the best method of control.

The restoration of bathing beaches has proved to be one of the most difficult problems associated with oil pollution. The sand must either be physically removed or plowed under to a considerable depth to be effective. Detergent cleansing has not been satisfactory except for the necessary cleanup work. Plowing under is less desirable, because wave action will ultimately cause resurfacing of the oil.

A major source of accidental oil pollution is offshore drilling. More than 7000 wells have been drilled in the Gulf of Mexico since 1960. These wells require more than 1800 mi of pipline. Well platforms and pipelines have been ruptured by storms, and pipelines have been broken by ships' anchors. The probability of accidental oil spills is increasing as the number of wells, pipelines, and associated storage tanks increase.

In spite of all preventative efforts, oil pollution will occur either because of human error or hazards beyond control. Present cleanup procedures leave much to be desired. More research into the prevention and control of oil pollution is necessary before we can hope to combat this important environmental problem.

2-8

WHAT SOLUTION

The country is experiencing a number of serious environmental pollution problems, whose immediacy and gravity vary from region to region. Some of the public is under the illusion that, by the appropriation of large sums of money, the problem can someday be solved. To view environmental pollution in this manner is totally unrealistic. A feasible solution will require immediate unselfish political action, education, research, planning, construction, and management.

PROBLEMS

2-1 Prepare a summary of the air pollution regulations in your state. How are these regulations administered?

2-2 Investigate a particular state's laws concerning air pollution and outline those provisions that should be considered in the design and construction of an industrial plant.

2-3 Prepare an outline of the administration of an interstate airshed. Is your state a party to any interstate airshed?

2-4 Discuss the procedures involved in bringing a new pesticide to market.

2-5 How are new food additives tested for the market?

2-6 A new nuclear power plant (1000 MW) is planned near your university. What type of cooling should be installed? If all rejected heat is to be discharged to a local river, what temperature rise in the river would be expected at minimum flow?

REFERENCES

1. *Population Estimates and Projections* (Washington, D.C.: U.S. Department of Commerce, Bureau of the Census, October 1975).

2. *The Nation's Water Resources* (Washington, D.C.: Government Printing Office, 1968), pp. 1–2.

3. A. T. Rosanno (ed.), *Air Pollution Control* (Stamford, Conn.: E.R.A., Inc., 1969), p. 26.

4. *U.S. Vital Statistics* (Washington, D.C.: U.S. Public Health Service).

5. "Air Pollution and Health." Statement by the American Thoracic Society. Reprinted from the *Am. Rev. Resp. Dis.* 93, no. 2 (February 1966).

6. "The Effects of Air Pollution," *U.S. Public Health Service Publication No. 1556* (Washington, D.C.: Government Printing Office, 1966), p. 5.

7. E. K. Peterson, "Carbon Dioxide Affects Global Ecology," *Environ. Sci. Technol.* 3, no. 11 (1969): 1162–1167.

8. *Cleansing Our Environment, the Chemical Basis for Action* (Washington, D.C.: American Chemical Society, 1969), pp. 23–92.

9. A Symposium—"The Technical Significance of Air Quality Standards," *Environ. Sci. Technol.* 3, no. 7 (1969): 636.

10. G. D. Friedlander, "Airborne Asphyxia—An International Problem," reprinted from the *IEEE Spectrum* (Washington, D.C.: U.S. Public Health Service, October 1965).

11. *U.S. Public Health Service Publication No. 1556* (Washington, D.C.: Government Printing Office, 1966).

12. "Wastes in Relation to Agriculture and Forestry," *U.S. Department of Agriculture Publication No. 1065* (Washington, D.C.: Government Printing Office).

13. U.S. Department of Agriculture, *Agriculture Statistics—1970* (Washington, D.C.: Government Printing Office).

14. R. C. Lochr, "Animal Wastes—A National Problem," *Proc. Am. Soc. Civil Engrs., J. San. Eng. Div.* SA2 (April 1969): 200.

15. *Time* (August 1968): 47.

16. W. H. Swift, C. J. Tauhill, W. L. Templeton, and D. P. Roseman, "Oil Spillage Prevention, Control and Restoration," *J. Water Poll. Control Fed.* 41, no. 3 (March 1969): 392–412.

3

Development of Water Supplies

Water is located in all regions of the earth. The problem is that the distribution, quality, quantity, and mode of occurrence are highly variable from one locale to another.

    The most voluminous water source is the oceans. It is estimated that they contain about 1060 trillion acre-ft of water.[1] The most valuable water supply (in terms of quality or freshness) is contained within the atmosphere, on the earth's surface, or underground. This supply, however, amounts to only about 3% of that contained in the oceans.

Water Quantity

Water resources vary widely in regional and local patterns of availability. The supply is dependent upon topographic and meteorological conditions as they influence precipitation and evapotranspiration. Quantities of water stored are dependent to a large extent on the physical features of the earth and on the earth's geological structure. Table 3-1 shows the major components of the water resources of the continental United States.

TABLE 3-1

SUMMARY DATA CONCERNING WATER RESOURCES OF THE
CONTINENTAL UNITED STATES

	Square Miles	Acre-ft ($\times 10^6$)
Gross area of continental United States	3,080,809	—
Land area, excluding inland water	2,974,726	—
Volume of average annual precipitation	—	4,750
Volume of average annual runoff (discharge to sea)	—	1,372
Estimated total usable groundwater	—	47,500
Average amount of soil moisture	—	635
Estimated total lake storage	—	13,000
Total reservoir storage (capacity of 5,000 acre-ft or more)	—	365

Source: E. A. Ackerman and G. O. Löf, *Technology in American Water Development* (Baltimore, Md.: The Johns Hopkins Press, 1959).

3-1

SOIL MOISTURE

Soil moisture is the most broadly used water source on the earth's surface. Agriculture and natural plant life are dependent upon it for sustenance. The quantity of water stored as soil moisture at any specified time is small, however. Estimates indicate an equivalent layer about 4.6 in. in thickness distributed over 57 million mi^2 of land surface. This in itself would be insufficient to support adequate plant growth without renewal. It is therefore important that the frequency with which this supply is renewed and the length of time it remains available for use be known. The supply of soil moisture is dependent upon geographical location, climatic conditions, geologic structure, and soil type. Variations may be experienced on a seasonal, weekly, or even daily basis.

It is considered that the natural supply of soil moisture in most of the agricultural areas of this country is less than optimum for crop growth during an average year. It is evident, then, that a greater understanding of optimum water requirements for crops is essential if we are to economically and efficiently supply water artificially to overcome natural soil moisture deficiencies.

3-2

SURFACE WATERS AND GROUNDWATER

Surface waters are found nonuniformly distributed over the earth's surface. It is estimated that only about 4 % of the U.S. land surface is occupied by fresh surface waters. These waters vary widely in quantity, seasonal distribution, quality, and frequency of occurrence.

Groundwater supplies are much more widely distributed than surface supplies. Nevertheless, strong local concentrations are found as a result of the variety of soils, rocks, and geologic structures located underground.

3-3

RUNOFF DISTRIBUTION

Approximately 30% of the average annual rainfall in the United States is estimated to appear as surface runoff. The allocation of this water is directly related to precipitation patterns and thus to meteorologic, geographic, topographic, and geologic conditions. In the West, large regions are devoid of permanent runoff, and some localities, such as Death Valley, California, receive no runoff for years at a time. In contrast, some areas in the Pacific Northwest average about 6 ft of runoff annually. Mountain regions are usually the most productive of runoff, whereas flat areas, especially those experiencing lower precipitation rates, are generally poor runoff producers.

Runoff is distributed in a far-from-uniform pattern over the continental United States. In addition, it is subject to seasonal variations and definite annual periods during which the concentration is significantly greater. For example, about 75% of the runoff in the semiarid and arid regions of the country occurs during a period of a few weeks following the snowmelt on the upper watershed.

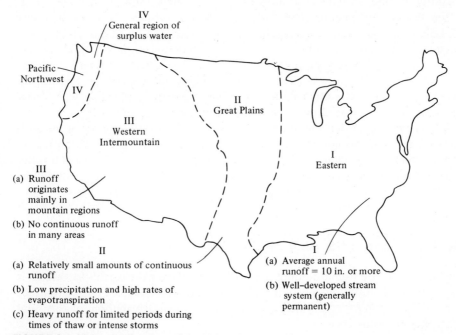

FIGURE 3-1 Major runoff regions of the United States.

Even in the well-watered East, an uneven distribution prevails, which has been one of the factors in the expansion of supplemental irrigation to be found there. Figure 3-1 indicates the four major runoff regions in the United States.

3-4

GROUNDWATER DISTRIBUTION

The usable groundwater storage in the United States is estimated to be about 48 billion acre-ft. This vast reservoir is distributed across the nation in quantities determined primarily by precipitation, evapotranspiration, and geologic structure. There are two components to this supply: one, a part of the hydrologic cycle, the other, water trapped underground in past ages, and no longer naturally circulated in the cycle.

Figure 3-2 shows the principal groundwater areas of the United States as depicted by Thomas.[11] Generally, it is evident that the mountain regions in the

Patterns show areas underlain by aquifers generally capable of yielding to individual wells 50 gpm or more of water containing not more than 2000 ppm of dissolved solids (includes some areas where more highly mineralized water is actually used)

⤲ Watercourses in which groundwater can be replenished by perennial streams

⤲ Buried valleys not now occupied by perennial streams

▨ Unconsolidated and semiconsolidated aquifers

▥ Consolidated rock aquifers

▦ Both unconsolidated and consolidated rock aquifers

☐ Not known to be underlain by aquifers that will generally yield as much as 50 gpm to wells

FIGURE 3-2 Groundwater areas in the United States. [Source: H. E. Thomas, "Water," *The Yearbook of Agriculture* (Washington, D.C.: U.S. Government Printing Office, 1955).]

East and West, the northern Great Plains, and the granitic and metamorphic rock areas of New England and the southern Piedmont do not contain important groundwater supplies.

Aquifers may be generally classified into four categories:

1   Those directly connected to surface supplies which are replenished by gravitational water and which release water to surface flow. Gravels found in floodplains or river valleys are examples.
2   "Regional" aquifers occurring east of the 100th meridian. These aquifers produce some of the largest permanent groundwater yields and have moderate to high rates of recharge. Good examples are found in the Atlantic and Gulf coastal plain areas.
3   Low recharge aquifers between the 100th and 120th meridians. These aquifers have relatively little inflow compared to potential or actual drafts. Although storage volumes are often large, the low rate of replenishment indicates that the water must be considered more of a minable material than a renewable resource. This possible "limited-life" category poses particular problems of development and management.
4   Aquifers subject to saline water intrusion. These are usually found in coastal regions, but inland saline waters also exist, principally in the Western states.

Water Quality

Water-supply development is concerned with both the quantity and quality of water required to meet the needs of man in an efficient and economical manner. Neither factor can be neglected. The usefulness of the maximum available water supply is determined in large part by its quality.

Through the years, quantities of waste have grown until many once-sparkling streams have been degraded to turbid sewers. Such impairment of water quality has resulted in extensive loss of aquatic life and in the destruction of the proper biological balance of streams. In addition, and of utmost importance, pollution has effected an actual decrease in our country's available water resources. If the water is so foul that its treatment for public consumption or industrial use is not economically feasible, other sources of supply must be sought and the polluted body of water can be considered to be as much of a loss as if it were physically unavailable.

The quality of precipitation is usually quite high, but once the precipitated water has penetrated the soil with its component minerals and rocks and flowed in streams contaminated by municipalities and industries, its quality may be seriously degraded by bacteria, organic matter, dissolved salts, acids, and possibly radioisotopes.

Drinking-water regulations prescribed by the EPA recommend that the total dissolved-solids content for human consumption should not exceed 500 milligrams (mg) per liter (l). The U.S. Geological Survey states that waters containing more than about 2000 mg/l dissolved solids are generally unfit for longterm irrigation under average conditions. The usability will depend, however, on the elements present, the soil type, and tolerance of the crop, and may thus vary considerably from the average allowable concentrations. The sodium and potassium balance is particularly important in agricultural waters.[3]

Increased emphasis on the quality of our water resources comes in the face of increasing population, accelerated industrial activity, and large-scale pollution.[5] Waste-abatement operations range from simple technical adjustments to the consideration of exceedingly complex social–political–ecological–psychological problems. A complicating factor is that each pollution-abatement problem is different. In addition, the motivations for abatement programs are shifting from the pure health-hazard base to an inclusion of aesthetic valuations. As stated by Renn, "We undoubtedly have more common interest in pleasant living and are less moved by moralistic views. If we don't like dirty water for any reason, we simply don't like it, and we don't argue that it must necessarily be toxic or bad for health. There is profit in offering a more pleasant future and we know it. We take it for granted that we have a right to some minutes of leisure, irresponsibility, and beauty without reproach. In fact, a very large fraction of our economy rests on this presumption."[10]

3-5

GROUNDWATER

Groundwater quality is influenced considerably by the quality of the source. Changes in source waters or degraded quality of normal supplies may seriously impair the quality of the groundwater supply. Municipal and industrial wastes entering an aquifer are major sources of organic and inorganic pollution. Large-scale organic pollution of groundwaters is infrequent, however, since significant quantities of organic wastes usually cannot be easily introduced underground. The problem is quite different with inorganic solutions, since these move easily through the soil and once introduced are removed only with great difficulty. In addition, the effects of such pollution may continue for indefinite periods since natural dilution is slow and artificial flushing or treatment is generally impractical or too expensive. The number of harmful enteric organisms is generally reduced to tolerable levels by the percolation of water through 6 or 7 ft of fine-grained soil.[4] However, as the water passes through the soil, a significant increase in the amounts of dissolved salts may occur. These salts are added by soluble products of soil weathering and of erosion by rainfall and flowing water. Locations downstream from heavily irrigated areas may find that the water they are receiving is too saline for satisfactory crop production. These saline contaminants are difficult to control because removal methods are exceedingly expensive. A possible solution is to dilute with water of lower salt concentration (wastewater-treatment-plant effluent, for example) so that the average water produced by mixing will be suitable for use.

Considerable care should be exercised to protect groundwater storage capacity from irreparable harm through the disposal of waste materials.

3-6

SURFACE WATER

The primary causes of deterioration of surface-water quality are municipal and domestic wastewater, industrial and agricultural wastes (organic, inorganic, heat), and solid and semisolid refuse. A municipality obtaining its water supply from a surface body may find its source so fouled by wastes and toxic chemicals that it is unsuitable or too costly to treat for use as a water supply. Fortunately, waste products discharged by cities and industry can be controlled at the point of initiation. This has been borne out by successes in cleaning up such watersheds as the Delaware and Susquehanna in the eastern United States.

Objectives of Water Resources Development

Only a small proportion of the total volume of the world's water resources is presently being harnessed and used for human welfare. In general, no stream has has been fully regulated or used to its maximum potential. The degree to which the physical limit of regulation is approached is a function of technological, political, economic, sociological, and ethical factors.

An objective shared by most water-development projects is to meet expected requirements for water by varying the quantity and quality of delivered supplies and by modifying the time and place of use of these supplies as conditions permit.[36] Normally, available supplies are allocated among potential uses on the basis of value, but this approach may be affected to a considerable extent by legal, political, or other considerations. In formulating the project, all possible uses of water should be studied; otherwise a truly comprehensive development plan will not be produced. For example, municipal, industrial, and agricultural supply, hydroelectric power, flood control, navigation, pollution control, recreation, and esthetic considerations may all be motivating factors in the development of river-basin projects.

In order to select the optimum combination of alternative, competitive, and complementary water uses while capitalizing on the economy and resource allocations permitted by multipurpose project development, a coordinated exercising of engineering, economic, ecological, and social principles is a requirement. An efficient balance of purposes should be determined on the basis of the nature and extent of the resources available, relative requirements for various uses, legal restrictions, local and regional policies, and public interest.[36]

3-7

PLANNING FOR WATER RESOURCES DEVELOPMENT

Water and land resources planners are concerned with the preservation, developmental, and use aspects of natural resources. Increased concern about the quality of life demands that future plans focus more sharply on the environment and how it relates to health, welfare, and enjoyment. Plans based on ecologically sound policies for man–environment relationships must be developed. The upsurge of public concern over environmental questions finally reflects a recognition of past planning failures to adequately consider (or consider at all) desirable man–environment challenges and dangers. We can no longer afford complacency with respect to environmental threats. Programs and projects for solving existing and future problems must be designed to give full consideration to the interaction of all key environmental factors.

Water resources planning is at best hazardous, because it has little immunity to failure. Of even greater importance is the fact that such failures often have far-reaching consequences. The difficulty is that water planning, like any other type of planning, is not an exact science. Physical processes which are only partially understood, coupled with even more complex and less-well-known human behavior patterns are the principal ingredients. The failure of poorly conceived plans may be disastrous, in that vast numbers of human beings, animals, birds, aquatic organisms, and other forms of life may be affected adversely and irreversibly.

An exploding population and a rapidly advancing technology are important causative agents in many environmental problems. Such factors not only create problems but make the solution to these problems all the more complex, as well. A fact often overlooked is that the mobility of modern society and changing social values require a vastly modified concept and scope of water resources planning. It is becoming increasingly important that we know how our planning regions function as complex interacting systems. New and powerful analytical techniques are being developed which have great potential applicability to the analysis of sophisticated problems involving cross-relationships between many kinds of components, such as people, land, water, economic factors, technologic factors, and others. These new approaches offer the opportunity for the exploration of many and varied planning alternatives prior to the time planning decisions are required. Various options and strategies for water utilization and development can be concurrently investigated, including interrelationships of conservation, reclamation, water-basin transfers, wastewater disposal, recreation, and others. This will assure the development of more efficient and economic, and less irreversible, plans for the management of vital water resources. Thus it appears that technological progress can be directed to sustain population and other factors of growth (if these are desired) while preserving the quality of the environment. A key objective of water resources planning should be to exploit technological capabilities so that planned development, management, or construction activities will not degenerate into liabilities.

Unfortunately, the planning process must be carried out with incomplete information about phenomena whose occurrence is uncertain. Most of the physical processes of concern are probabilistic in nature. At best, the availability of good historic records permits estimation of the likelihood of an event occurring with some defined frequency in the future—when, or if, the event will occur is entirely uncertain. The planner is therefore faced with not knowing whether previously recorded conditions will ever recur or to predict the occurrence of events of greater severity. In some cases even the probability of an extreme event will be elusive.

Equal in importance to the stochastic nature of planning variables is the time span over which plans must be conceived. Advance planning may extend from a few years to periods of 50 years or more—far beyond the limitations of human foresight. In addition, the planner must face the fact that human activities in the future may not be motivated by current institutional concepts. Social goals and objectives may also be significantly changed.

The greatest difficulty of all, however, is man himself. People are complex and have the potential to make a significant impact on water resources processes. Unfortunately, we understand little about human behavior and preferences as they relate to water resources today and practically nothing in terms of the future. Nevertheless, the planning process must provide recommendations for action which will be compatible with the society they are intended to serve.

In general, water resources planning can be divided into three basic steps: (1) the determination of current and future needs, (2) the appraisal of all possible means to meet these needs, and (3) the selection of the most economic approaches for satisfying the anticipated requirements. While straightforward in concept, the planning process may be exceedingly complex in practice. Numerous detailed investigations and studies are usually required. Examples of the most important of these are as follows[37]:

1 Population, economic land use, and other associated planning studies essential for the prediction of the water quality and quantity requirements for the various anticipated uses for the duration of the planning period chosen.

2 Hydrologic investigations to provide estimates of the quantities of fresh surface water and groundwater entering the basin and of the temporal distribution of these supplies. It is often the case that sufficient data on the meteorologic and hydrologic characteristics of the basin are not available. Additional field measurements and statistical analyses must frequently be relied upon, therefore, to yield the required information.

3 Comprehensive field investigations to evaluate the physical, chemical, radiological, and biological characteristics of the surface waters of the basin. The information determined will indicate (a) the degree of treatment the water will have to be subjected to for various purposes, (b) the ability of the watercourse to assimilate waste discharges while continuing to sustain fish and wildlife, and (c) the recreational and aesthetic values of the waters.

4 Studies of the storage capacities and uses of existing and proposed reservoirs in the basin. In some cases it will also be necessary to make detailed studies of the topography and geology of the basin with the object of determining additional reservoir sites of a satisfactory nature. Storage estimates for these potential storage locations would also be required.

5 Investigations to determine the location and extent of groundwater storage, aquifer characteristics, and the quality of the underground water supplies.

6 Studies to provide information on existing and anticipated sources of wastes which will be introduced into surface or groundwater bodies. A knowledge of the characteristics and volumes of these wastes will be necessary. Information on the optimum treatment that these wastes could be economically expected to undergo will also be required.

7 Estimates of water requirements to be used to augment low flows for quality and quantity control of surface-water bodies. Low-flow augmentation can play an important role in fish and wildlife preservation and in combating pollution.

8 Evaluation of benefits which will result from the proposed water resources program. In arriving at conclusions regarding these benefits, it is important that the optimum economic or socially desirable development of water resources be based on studies of all practical combinations and alternatives of objectives.

In making plans for the future development and allocation of water resources, it should be emphasized that economic, social, and technological advances have an impact on water demand. An attempt should be made to include these effects, insofar as possible, in all planning for future water needs. To counteract the effects of unforeseen modifying factors on water requirements or type of allocation, maximum flexibility should be built into the project plan. Programming events for stage construction is one way of providing the desired flexibility.

3-8

WATER USES

Water resources are developed primarily for urban water supply, irrigation, power generation, flood control, recreation, pollution control, navigation, and fish and wildlife conservation. Because it is important that our water resources be utilized to the fullest extent, a consideration of the multiple use of water is mandatory for all but the smallest projects.

In the design and operation of multipurpose systems, it is important to realize that some form of compromise between uses is essential.[35,39,40] This is because different uses are not always compatible. For example, maintaining an empty storage volume to impound flood waters for short periods during which they are slowly released does not satisfy the objective of obtaining the maximum storage of water for other purposes. The operating rules for the project must thus be devised to permit relatively efficient operation for various uses while recognizing that maximum efficiency might not be obtained for even one purpose.

The Hoover Dam, constructed on the Colorado River, is an outstanding example of the use of a single project unit to impound water for multiple purposes. The dam is a concrete gravity type, 726 ft in height. This structure intercepts the flow of the Colorado River in Black Canyon and provides ample storage and control so that the flow can be utilized downstream for four major purposes. The impounded waters are used for irrigation, residential supplies, and commercial and industrial needs in Southern California. Hydroelectric power is generated and sold in Arizona and California. Approximately 9,500,000 acre-ft of storage volume is reserved for flood-control purposes. In addition, the dam was also designed to aid in maintaining flows in the downstream channel for navigational purposes.[38]

Before a water resource can be developed efficiently, there must be an understanding of the nature of the various water uses. A brief summary of the most important uses follows.

Irrigation

The water requirements for irrigation are in general seasonal. The quantities of water required for irrigation purposes vary with the climate of the region and the type of crops that are to be raised. For purposes of estimation it is sometimes assumed that the growing season lasts for about 3 months during the summer and that the irrigation draft is uniform during that period. The growing period for all crops is not coincident, however, so the overall irrigation season in any region is variable. In humid regions the water requirements for irrigation may not be uniform and may range from about 10% in May to 30% in September. For arid and semiarid regions, the rate of withdrawal will be more nearly uniform. In general, as much reserve storage as possible should be provided to ensure against critical shortages during protracted dry periods. Storage and use for irrigation needs may conflict with many other uses. For example, during the winter months irrigators want to store as much water as possible and minimize releases while maximum releases are desirable for generating power.

Surface waters used for irrigation may be diverted to canals which afford gravity flows to the irrigation project. Various means of controlling the rates of delivery are employed (gates, flumes, weirs, etc.). Where the irrigated lands lie above the river valley, pumping may be required.

In some cases, water requirements might be less than the safe yield of a neighboring stream and flows can be taken directly with no thought of storage. For large-scale irrigation projects, particularly in the arid or semiarid regions, storage reservoirs must be constructed to provide the necessary quantities of water at the right season. The primary engineering works are reservoirs, canals, ditches, pumping plants, and measuring stations.

Hydroelectric Power

The requirements for hydroelectric power are usually seasonal, with the winter months often the critical period. Available hydroelectric capacity should be used on the system load so that the energy derived is employed in the most effective manner. This requires that water be taken from storage during periods of peak demand and that storage be replenished during periods of light load. The principal causes of conflict between water use for power versus other uses result from different seasonal requirements. For example, heavy summer releases for navigation on the Missouri River can conflict with needs for considerable hydropower generation during the winter months. In many cases power production is compatible with other uses. Water passed through the turbines during the summer months can also be used downstream to meet irrigation, navigational, flow-augmentation, or other needs.

There are three categories of hydroelectric plants—run-of-the-river, storage, or pumped-storage. Run-of-the-river plants are characterized by very limited storage capacities. They take water from the river essentially as it comes. Storage plants contain adequate reservoir capacity to impound wet-season flows for use during low-flow periods. They are able to develop a firm flow which is considerably greater than the minimum river flows. Pumped-storage plants are used to produce power during peak-load periods. Releases from the turbines are retained in a tailwater pool and pumped back to the headwater pool for reuse during periods of load.

Power available from a river is directly proportional to the quantity of water which passes through the turbines and the head available for operation of the turbines. When expressed as horsepower (hp), the following relation is obtained:

$$hp = \frac{Q\gamma h}{550} \tag{3-1}$$

where Q = discharge, cfs
γ = specific weight of water, lb/ft^3
h = head, ft

Adjustments must be made in values obtained by Eq. 3-1 to account for efficiency.

The prime feature of water control for power development is the production and maintenance of the maximum possible head at the power plant. Any flow releases from the reservoir will reduce the pool level and thereby affect the power potential. For example, if the water level is reduced by 50%, the power output of the plant will also be cut by 50% for the same rate of flow. Where the storage of water is intended to satisfy uses other than power production, an operating schedule must be designed to provide for maximum efficiency of power production under the circumstances.

Navigation

As in the case of many other water uses, navigational requirements are highly seasonal. The largest releases are generally required during the driest months of the year. Navigation requirements are usually competitive with irrigation demands, hydropower demands, and recreational needs at the reservoir site.

Channel-depth requirements can generally be met by dredging out the channel bottom or by raising the water level by the use of dams or flow releases. During periods of low flow many rivers are actually composed of a series of steps navigationwise, and ships must travel through locks to pass from one slack-water pool to the next.

The control of navigation by a series of low dams generally has little effect on other water uses within the river basin. This is because the purpose of these structures is to maintain a river level close to the minimum required for navigation. Flood flows are not appreciably affected by this form of control, nor are flow rates of the river. Some important advantages may be derived, however, from the maintenance of minimum depths. For example, benefits to fish and wildlife, recreation, pollution control, and aesthetic enhancement may directly result.

The depth of water in a river reach can also be controlled by confining the channel between levees. This approach can produce flood-control benefits at the same time and also serve to provide more usable land along the water course.

Multipurpose reservoirs may have as part of their function the release of flows to maintain minimum navigational depths downstream. Storage volumes required for this purpose can be estimated by considering low-flow sequences and the seasonal nature of the navigational requirements. In operating the reservoir to release the required flows it is necessary to determine the time required for the flow to reach an eventual location. Adequate forecasts of downstream river stages are thus essential in scheduling releases to augment natural flows.

Recreation

Water requirements for recreational purposes are usually greatest in the summer months. The sportsman and vacationer desire to have the maximum storage during this period. This permits optimum conditions for fishing, boating, and other water sports. Obviously, this condition is not compatible with many other summer water requirements.

Flood Control

The utilization of storage works for flood control requires that ample volume be available to impound floodwaters and that means be available to release this volume as rapidly as possible to prepare for the next flood. Of all the potential uses, flood control is the least compatible.

Floods may be defined as abnormally high flows or flows in excess of some base value for a particular stream. In general, any flow that overlaps the normally confining channel can be classified as a flood.

The problems resulting from flood flows are generally associated with the attempt by man to occupy the floodplains of rivers and streams for various purposes. Although these regions may be dry for years at a time, they are in reality a part of the river or stream channel during periods of peak flows. Each year an economic toll is exacted from man by nature for occupancy of these areas. The magnitude of this charge is variable and may range from a simple delay in time to the loss of human life. In dollars, the average annual charge for the United States likely exceeds $95 million in property losses alone, and very few regions in this country are free of the flood hazard. The charge can be avoided or reduced by not using floodplain areas, by the controlled use of these areas, or by the development of protective works.

To alleviate the problems caused by flooding, various methods for controlling floods or minimizing their influence have been employed. Principal among these are dams, levees, channel improvements, diversion works or floodways, zoning laws, land-management practices, floodproofing, and evacuation.

Urban Water Supply

Urban water requirements may be quite variable from region to region. For specific locations use rates can be determined as shown in Chapter 4. In the development of supplies for municipal uses it is important to understand the seasonal variation of these uses and the seasonal and annual fluctuations in supply. In general, the maximum daily municipal use will be about 180% of the annual average, with summer averages higher and winter averages lower. In planning projects to furnish municipal water a curve of demand versus time is important. In addition, storage must often be provided and maintained to carry the community through long periods of drought. The storage of water for drinking purposes may restrict the scope of recreational activities at the reservoir site. The storage and use of a specific volume of water for municipal purposes may also be in direct competition with the storage of water for other uses, such as agriculture.

Water-Quality Control

If our waters are to be effectively and economically used, pollution-control techniques must play a primary role. Indirect reuse practices, fish and wildlife conservation, and recreation are all affected by the quality of our surface and groundwater supplies. Principal control methods include various forms of physical, chemical, and biological treatment and dilution. Major control works are composed of process units and storage works. Where large impoundments are constructed along watercourses, it is conceivable that during dry periods no flow would appear downstream of the dam. Thus any wastes entering the stream below the reservoir would be undiluted and could possibly destroy many desirable aquatic forms and other wildlife. To preclude this type of circumstance it is often the policy to augment the downstream flows by releasing water from storage. In this way a minimum flow having an acceptable quality can be maintained, with the result that downstream pollution problems are alleviated. The low-flow operating policy of the reservoir then becomes a primary form of engineering control.

The Water Budget

In theory, accounting for the water resources of an area is relatively simple. The basic procedure involves the separate evaluation of each factor in the water budget so that a quantitative comparison of the available water resources with the known or anticipated water requirements of the area can be made. In practice, however, the evaluation of the water budget is often quite complex, and extensive and time-consuming investigations are generally required.

Both natural and artificial gains and losses in the water supply must be considered. The primary natural gains to surface bodies are those which result from direct runoff caused by precipitation and effluent seepage of groundwater. Evapotranspiration and unrecovered infiltration are the major natural losses. Dependable dry-season supplies can be increased through diversion from other areas, through low-flow augmentation, through saline-water conversion, and perhaps in the future through induced precipitation. On the other hand, diversions out of the basin will decrease the quantity of water available.

After the gross dependable water supply has been estimated, the net dependable supply may be determined at any point of interest by subtracting the quantity used, detained, or lost as a result of man's activities, from the gross supply. When water is withdrawn from a river, a decrease in flow between the point of withdrawal and the point of return is experienced. As the water is used, part of it will be lost to the atmosphere through various forms of consumption. These consumptive losses are accumulative downstream and effect a permanent decrease in the dependable water supply. Temporary decreases in dependable supply occur along the watercourse between intake and discharge points. The dependable supply in the reach of watercourse between the two points is thus diminished by the amount of the withdrawal.

A water supply may be considered adequate for present needs but inadequate to provide for future requirements. In many cases it is therefore necessary to predict future water needs based on estimated changes in population, industrial development, agricultural practices, and on changes in water policy and technology which will affect the supply and use of water.[2]

3-9

DEFINITION OF TERMS IN THE WATER BUDGET

The evaluation of problems in water supply and demand requires clear-cut definitions of the various components that comprise the water budget. For this reason some of the most troublesome or confused terms will be defined here.

Consumptive Use

Water consumption and *water use* are terms that have frequently been employed indiscriminately and interchangeably. Several definitions of *consumption* or *consumptive use* may be found in the literature. A special committee of the American Water Works Association defines consumptive use as water used in connection with vegetative growth, food processing, or incidental to an industrial process, which is discharged to the atmosphere or incorporated in the products of the process.[12] A slight modification of this definition which includes air-conditioning losses will be adopted here.

Withdrawal Use
The use of water for any purpose which requires that it be physically removed from the source. Depending on the use to which the water is put, a significant percentage may be returned to the original source and then be available for reuse.

Nonwithdrawal Use
The use of water for any purpose which does not require that it be removed from the original source, such as water used for navigation.

Nonconsumptive Losses
Certain water losses, although not "consumptive" in the sense of the definition, may have the effect of reducing the available water supply of an area. For example, dead storage (storage below outlet elevations) in impoundments is unavailable for downstream use. Diversion of water from one drainage basin to another represents an additional form of nonconsumptive loss. An example of this is the use of the Delaware River basin for the municipal supply of New York City. This decreases the total flow in the Delaware River below the point of diversion. New York is required, however, to augment low flows through compensating the downstream interests for diversion losses. Water contaminated or polluted during use which cannot be economically treated for reuse also constitutes a real loss from the total supply.

Surface-Water Supplies

Surface-water supplies may be generally categorized as perennial or continuous unregulated rivers, rivers or streams containing impoundments, or natural lakes. Evaluation of the surface-water resources of an area requires a determination of the general characteristics of the region. The area should be located geographically on the basis of established references. All available data on the climate, hydrology, geology, and topography of the area should be accumulated. The type of industrial, agricultural, and residential development and the predicted growth rates of these factors are necessary information. A description of the principal metropolitan areas and a knowledge of the economy of the region is also important. An evaluation of the natural resources of the area and the impact of their development on the hydrology and economy of the area are additional requirements.

3-10

BASIN CHARACTERISTICS AFFECTING RUNOFF

Important natural characteristics of the basin affecting stream flow are the basin's topographic and geologic features.

Topography determines the slopes and location of drainage channels and the storage capacity of the basin. Channel slope and configuration are directly related to the rate of flow in a basin and the magnitude of the peak flows. A steep watershed will generally indicate a rapid rate of runoff with little storage, whereas relatively flat areas are subject to considerable storage and lower rates of flow.

The soils in a basin affect infiltration capacity and the ability of underground strata to transmit or hold groundwater. A thorough understanding of the characteristics of the underground formations is necessary to properly evaluate these factors.

3-11

NATURAL AND REGULATED RUNOFF

Natural runoff is defined as that runoff which is unaffected by any other than natural influences. Runoff subject to withdrawals by man or artificial storage is defined as *regulated runoff*. If an unregulated stream is to be developed as a primary water source, the safe yield will be the lowest-dry-weather flow of the stream. Under this condition, the user will always have an adequate supply, provided that his maximum requirements do not exceed this minimum flow. If during any time interval the expected demands exceed the lowest dry-weather flow, periods of water shortage can be anticipated unless supplementary supplies are provided.

Regulated runoff is normally the type of runoff for which information is available, although many streams for which runoff records are at hand were unregulated at the time they were first gaged. Most records are seriously affected by artificial regulation from upstream storage works or by the diversion of flows into or out of the stream at points above the gaging station. These withdrawals from the basin or diversions into the basin from outside sources affect the hydrology of the basin. Fortunately, diversions and withdrawals usually lend themselves to accurate estimates, since they are for specific purposes and are gaged or can be measured with little difficulty. Examples of diversion uses are as follows:

1 *Municipal*—the use of the Delaware River for the water supply of New York City.
2 *Agricultural*—the use of the Colorado River to supply irrigation water to California.
3 *Waste disposal*—the use of the Chicago Drainage Canal to remove wastewater from Chicago which originated in Lake Michigan.

The safe yield of a stream which is regulated approaches the average annual flow as storage approaches full development. Economical yields are between the safe yields for unregulated and fully regulated flows. Usually safe yields of 75–90 % of the mean annual flow can be realistically developed. Through regulation, the greatest benefits are normally derived from a stream and, by means of diversified usage, a method of achieving the maximum economy in the utilization of the water is provided.

3-12

STORAGE

Water may be stored for single or multiple uses such as navigation, flood control, hydroelectric power, agriculture, water supply, pollution abatement, recreation, and flow augmentation. Either surface or subsurface storage can be utilized but both necessitate a reservoir, which affects the hydrology of a basin.

Reservoirs regulate stream flow for beneficial use by storing water for later release. The term *regulation* can be defined as the amount of water stored or released from storage in a period of time, usually 1 year. The ability of a reservoir to regulate river flow depends on the ratio of its capacity to the volume of river flow. Evaluation of the regulation provided by existing storage facilities can be determined by studying the records of typical reservoirs. A representative group of reservoirs having detention periods from 0.01 year to 20 years is given by Langbein, with the usable capacity, detention period, and annual regulation of each.[8]

About 190 million acre-ft of water, representing approximately 13 % of the total river flow, has been made available through reservoir storage development in the United States.[8] The degree of storage development is exceedingly variable but is generally greatest in the Colorado River basin and least in the Ohio River basin. Substantial increases in water supply can be attained through the development of additional storage, but water regulation of this type follows a law of diminishing returns. There are limitations on the amount of storage that can be used. The storage development of the Colorado River basin, for example, may be approaching (if not already in excess of) the maximum useful limit.

Reservoir construction seems to be accelerating, with maximum development a definite possibility in the not-too-distant future. Water supply, pollution control, and recreation appear to be the coming primary objectives of water storage.

Reservoirs

Where natural storage in the form of ponds or lakes is not available, artificial impoundments or reservoirs must often be constructed if optimum development of the surface supply is to be obtained.

3-13

DETERMINATION OF REQUIRED RESERVOIR CAPACITY

The amount of storage that must be provided is a function of the expected demands and the inflow to the impoundment. Mathematically this may be stated as follows:

$$\Delta S = I - O \qquad\qquad (3\text{-}2)$$

where ΔS = change in storage volume during a specified time interval
I = total inflow volume during this period
O = total outflow volume during this period

Normally, O will be the draft requirement imposed by the various types of use, but it may also include evaporation and transpiration and flood discharges during periods of high runoff when inflow may greatly exceed draft plus available storage and outflow seepage from the bottom or sides of the reservoir.

Realizing that the natural inflow to any impoundment area is often highly variable from year to year, season to season, or even day to day, it is obvious that the reservoir function must be that of redistributing this inflow with respect to time so that the projected demands are satisfied.

3-14

CLASSICAL METHOD OF COMPUTATION

Several approaches may be taken in the selection of reservoir capacities. Actual or synthetic records of stream flow and a knowledge of the proposed operating rules of the reservoir are fundamental to all solutions. Determination of storage may be accomplished by either graphical or analytical techniques.

Historically, one of the most used methods of determining storage has been the selection of some low-flow period considered to be critical. The most severe drought of record might be selected, for example. Once the critical period is chosen, storage is usually calculated by the mass-curve analysis introduced in 1883 by Rippl.[16] This method evaluates the cumulative deficiency between outflow and inflow $(O - I)$ and selects the maximum cumulative value as the required storage.[43] Examples 3-1 and 3-2 illustrate the procedure.

●EXAMPLE 3-1
Find the storage capacity required to provide a safe yield of 67,000 acre ft/yr for the data given in Fig. 3-3.

○*Solution*
Construct tangents at A, B, and C having slopes equal to 67,000 acre ft/yr. Find the maximum vertical ordinate between the inflow mass curve and the constructed draft rates. From Fig. 3-3 the maximum ordinate is found to be 38,000 acre-ft, which is the required capacity.

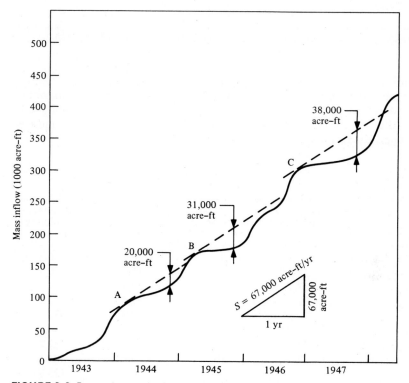

FIGURE 3-3 Reservoir capacity for a specified yield as determined by use of a mass curve.

This example shows that the magnitude of the required storage capacity depends entirely on the time period chosen. Since the period of record given covers only 5.5 years, it is clear that a design storage of 38,000 acre-ft might be totally inadequate for the next 3 years, for example. Unless the frequency of the flow conditions used in the design is known, little can be said regarding the long- or short-term adequacy of the design.

Example 3-1 also illustrates the fact that the period during which storage must be provided is dependent upon hydrologic conditions. Since reservoir yield is defined as the amount of water which can be supplied during a specific time interval, choice of the interval is critical. For distribution reservoirs a period of 1 day is often sufficient. For large impounding reservoirs periods of several months, a year, or several years storage may be required.

●**EXAMPLE 3-2**
Consider an impounding reservoir that is expected to provide for a constant draft of 448 million gallons (mg)/mi^2/yr. The following record of monthly mean inflow values is representative of the critical or design period. Find the storage requirement.

Month	F	M	A	M	J	J	A	S	O	N	D	J	F
Observed Inflow (mg/mi²/month)	31	54	90	10	7	8	2	28	42	108	92	22	50

○*Solution*
Set up the tabulation of values shown in Table 3-2. It can be seen that the maximum cumulative deficiency is 131.5 mg/mi², which occurs in September. The number of months of draft is 131.5/37.3 = 3.53, or stated differently, enough water must be stored to supply the region for about $3\frac{1}{2}$ months.

TABLE 3-2
STORAGE-REQUIREMENT COMPUTATIONS

Month	Inflow, I (mg/mi²/ month)	Draft, O (mg/mi²/ month)	Cumulative Inflow, ΣI	Deficiency, $O - I$	Cumulative Deficiency, $\Sigma (I - O)^a$
F	31	37.3	31	6.3	6.3
M	54	37.3	85	−16.7	0
A	90	37.3	175	−52.7	0
M	10	37.3	185	27.3	27.3
J	7	37.3	192	30.3	57.6
J	8	37.3	200	29.3	86.9
A	2	37.3	202	35.3	122.2
S	28	37.3	230	9.3	131.5
O	42	37.3	272	−4.7	126.8
N	108	37.3	380	−70.7	56.1
D	98	37.3	478	−60.7	0
J	22	37.3	500	15.3	15.3
F	50	37.3	550	−12.7	2.6

a Only positive values of cumulative deficiency are tabulated.

This example gives a numerical answer to the question posed in determining a design storage. It does not, however, give any expression of the probabilities of the shortages or excesses that may result from this design. Past practice has been to use the lowest recorded flow of the stream as the critical period. Obviously this approach overlooks the possibility that a more serious drought might occur, with a resultant yield less than the anticipated safe yield.

3-15

FREQUENCY OF EXTREME EVENTS

If a hydrologic event has a true recurrence interval of T_R years, the probability that this magnitude will be equaled or exceeded in any particular year is

$$P = \frac{1}{T_R} \tag{3-3}$$

where T_R is the recurrence interval of the event. Recurrence interval is defined as the average interval in years between the occurrence of an event of stated magnitude and an equal or more serious event.

Both annual series and partial duration series are used in estimating the recurrence intervals of extreme events.[42] An annual series is composed of one significant event for each year of record. The nature of the event depends on the object of the study. Usually the event will be a maximum or minimum flow. A partial duration series consists of all events exceeding, in significance, a base value. The two series compare favorably at the larger recurrence intervals, but for the smaller recurrence intervals the partial duration series will normally indicate events of greater magnitude.

There are two possibilities regarding an event; it either will or will not occur in a specified year. The probability that at least one event of equal or greater significance than the T_R-year event will occur in any series of N years is shown in Table 3-3. For example, there exists a probability of 0.22 that the 100-year event will occur in a design period of 25 years.

TABLE 3-3
PROBABILITY THAT AN EVENT HAVING A
PRESCRIBED RECURRENCE INTERVAL WILL BE
EQUALED OR EXCEEDED DURING A SPECIFIED
DESIGN PERIOD

T_R (years)	Design Period (years)					
	1	5	10	25	50	100
1	1.0	1.0	1.0	1.0	1.0	1.0
2	0.5	0.97	0.999	a	a	a
5	0.2	0.67	0.89	0.996	a	a
10	0.1	0.41	0.65	0.93	0.995	a
50	0.02	0.10	0.18	0.40	0.64	0.87
100	0.01	0.05	0.10	0.22	0.40	0.63
200	0.005	0.02	0.05	0.12	0.22	0.39

[a] Values are approximately $= 1.0$.

Table 3-3 was derived by means of the binomial distribution, which gives the probability $p(X; N)$ that a particular event will occur X times out of N trials as

$$p(X; N) = \binom{N}{X} P^X (1 - P)^{N-X}$$

$$\binom{N}{X} = \frac{N!}{X!(N - X)!} \tag{3-4}$$

where P is the probability that an event will occur in each individual trial ($P = 1/T_R$ in this case). Now, if we let the number of occurrences equal zero ($X = 0$) in a given period of years N (number of trials) and substitute this value in Eq. 3-4, the result is

$$p(0; N) = (1 - P)^N \qquad (3\text{-}5)$$

This is the probability of zero events equal to or greater than the T_R year event. Then the probability Z that at least one event equal to or greater than the T_R year event will occur in a sequence of N years is given by

$$Z = 1 - \left(1 - \frac{1}{T_R}\right)^N \qquad (3\text{-}6)$$

Solution of Eq. 3-6 for various values of N and T_R provided the data for Table 3-3.

3-16

PROBABILISTIC MASS TYPE OF ANALYSIS

Low-flow information is the basis for reservoir design and for studies of the waste assimilative capacity of streams. For example, a customary critical period used by water quality personnel is the average low flow for 7 consecutive days occurring on the average of once in 10 years. For reservoir design, the critical period will usually be measured in months or even years and return periods are normally on the order of 20, 50, or 100 years.

Information on the probability of occurrence of droughts of various severities during any single year or during any specified period of years can be developed from climatological and hydrologic records. Using such information, the engineer can gain considerable insight into the risks associated with reservoir development. A knowledge of these risks permits an estimate of the numerical odds for any specified yield. The designer may thus evaluate the performance of the reservoir under any set of risks which he chooses.

In designing a reservoir to meet a specific draft, certain basic information must be at hand. First, the duration of the low flow or critical period of design must be known. Second, the magnitude of the critical low flow must be determined. Third, the frequency of occurrence of the critical event must be known.

The question of the critical period can be answered by experimenting with a number of low-flow durations and then selecting, by judgment or by policy, some duration with which to work. From existing or synthetic stream-flow data, a series of magnitudes of critical flows for the specified duration can be obtained. This information answers the question of magnitude. Finally, by assigning recurrence intervals to critical events, the frequency of events can be established.

The method for estimating reservoir capacity which is presented here was developed by Stall and Neill of the Illinois State Water Survey.[14,17] Initially, a historic record or a generated record of stream flows for the watershed in question must be provided. Because historic records are often short (sometimes non-existent), generated stream-flow data must frequently be developed for study purposes. These synthetic data should be developed so that they exhibit the same statistical attributes as the historic data. Various techniques for this have been devised, and detailed information on this subject may be found in Refs. 35 and 40–42.

From the stream-flow record, a sequence of partial-duration series for various low-flow periods is developed. For example, a partial duration series of 18-month low flows for an N-year stream-flow record would be derived as follows:

1 Select a base flow above which all values will be excluded from the series. Stall[17] uses the mean flow at the station of interest. Note that this mean flow should be converted to cumulative flow for the 18-month period.

2 Compute 18-month running totals for the entire period of record.

3 Rank the 18-month running totals with a rank of one assigned to the smallest cumulative 18-month flow. Note that each 18-month period should be independent (i.e., no two periods should overlap). Once the minimum 18-month period is found, the second lowest period should be determined which does not overlap the first and so on.

4 The array obtained by ranking becomes the partial duration series for the 18-month low-flow periods.

Once the partial-duration series for each low-flow period of interest is obtained, the recurrence intervals for the values in the series can be computed as shown in Example 3-3.

●EXAMPLE 3-3
The following partial-duration series of 30-month low flows is available from 49 years of record on the Little Elk River. This river serves as the input to the Little Elk Reservoir. The drainage area equals 330 mi². Find the recurrence intervals for these 30-month low-flow sequences.

30 Month Flow (in./mi²)	4.95	5.40	6.12	8.90	12.70	13.91	15.10	18.30	21.00	24.15
Rank of Events	1	2	3	4	5	6	7	8	9	10

○**Solution**
Determine the recurrence interval in years for these data using the relationship

$$T_R = \frac{N}{m} \tag{3-7}$$

where T_R = recurrence interval, years
   N = number of years of record
   m = rank of the events (arranged in order of magnitude, the tabulation shown in Table 3-4 is obtained)

TABLE 3-4

RECURRENCE INTERVALS FOR A 30-MONTH LOW-FLOW SERIES FOR LITTLE ELK RIVER

Rank of Event	T_R (years)	30-Month Flow (in./mi²)
1	49	4.95
2	24.5	5.40
3	16.3	6.12
4	12.3	8.90
5	9.8	12.70
6	8.2	13.91
7	7.0	15.10
8	6.1	18.30
9	5.4	21.00
10	4.9	24.15

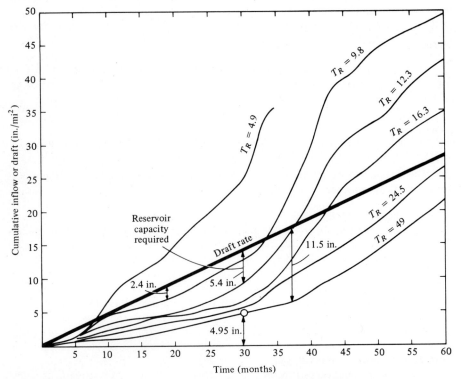

FIGURE 3-4 Example of the determination of reservoir capacity by mass-type analysis using the method of Stall. [Source: J. B. Stall, "Reservoir Mass Analysis by a Low-Flow Series," *Proc. Am. Soc. Civil Engrs., J. San. Eng. Div.* 88, no. SA5 (September 1962).]

Using the recurrence intervals determined for the values in the low-flow arrays, a plot of the kind shown on Fig. 3-4 can be obtained. The mass-type curves shown on this figure are obtained by plotting the cumulative flows for each low-flow period versus the length of period and then connecting all the points having equal recurrence intervals. Thus the curve labeled $T_R = 49$ would give the 49-year low flow for any period of time up to 60 months. For the 30-month period of Example 3-3, the value of 4.95 in. is indicated by a circle.

Figure 3-4 can be used to graphically determine the reservoir capacity required to meet any given draft rate for any critical period and recurrence interval. For example, to meet the draft rate shown on the figure for a return period of 12.3 years, a reservoir capacity of 5.4 in.^{2}/mi^2 of watershed area would be required. This would be associated with a critical period of 30 months.

3-17

RISKS ASSOCIATED WITH RESERVOIR YIELD

For every draft imposed on an impounding reservoir, there is a risk associated with being able to meet this demand over a specified interval (the design period). Such risks can be computed and interpreted by methods developed by Stall.[41] The calculations are based on the concepts of recurrence interval and design period given in Sec. 3-15.

An example will serve to illustrate the application of frequency studies in determining the calculated risks of impounding reservoir yields. The example is based primarily on the work of Stall and Neill.[14,17,41]

●EXAMPLE 3-4
Develop a set of yield probability curves for the Little Elk Reservoir of Example 3-3. Interpret these curves with respect to a design pumping rate of 75 million gallons per day (mgd). The drainage area is 330 mi^2 and the reservoir capacity is 7 in./mi^2.

○*Solution*
a. Develop partial-duration series for various critical periods as outlined in Sec. 3-16. For each value in each series, determine the recurrence interval as shown in Example 3-3 for the 30-month critical period. For each recurrence interval, plot cumulative flow versus critical period. Several such curves are plotted in Fig. 3-4.

b. Impose an ordinate of 7 in. (the reservoir storage) on each T_R curve to determine the draft rate this storage volume could maintain for the given recurrence interval. The procedure is illustrated in Fig. 3-5 for the 49-year return period. Note that the ordinate 7 in. must be located as shown in the figure so that it represents the maximum deviation between the T_R curve and the constructed draft rate line. Several trials may be required to locate the ordinate exactly. The draft rates determined in this manner represent the gross yields of the reservoir for their respective return periods.

c. Plot the values of gross yield versus recurrence interval as shown in Fig. 3-6. From these values construct the line of best fit (the solid line in Fig. 3-6). This is the frequency curve of gross yield. Note that the vertical scale is logarithmic while the

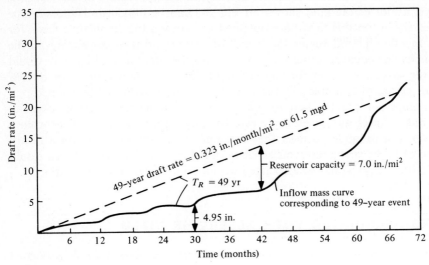

FIGURE 3-5 Determination of 49-year draft rate for Example 3-4.

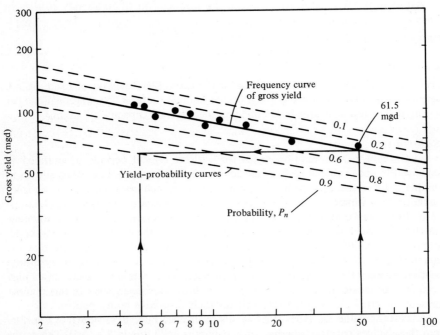

FIGURE 3-6 Frequency and yield-probability curves. Solid Curve—recurrence interval (yr); dashed curves—design period n (yr).

horizontal scale is an extreme-value scale. A simple procedure for constructing a Gumbel extreme-value scale is presented by Linsley and Franzini (Ref. 35, p. 117).

From the frequency curve of gross yield, values of gross yield for any desired frequency can be tabulated. Table 3-5 lists values for several recurrence intervals of interest.

TABLE 3-5

GROSS YIELD FOR A SPECIFIED DROUGHT RECURRENCE INTERVAL FOR THE LITTLE ELK RIVER

Drought Recurrence Interval (years)	Gross Yield (mgd)
2	136
4	110
6	100
8	94
10	88
20	76
40	64

   d. Plot a set of yield probability curves (dashed lines in Fig. 3-6). These curves are constructed in the following manner.

$$p(0; N) = \left(1 - \frac{1}{T_R}\right)^N$$

Use Eq. 3-5 to compute various values of $p(0; N)$ for various values of T_R and N. Note that the values in Table 3-3 may also be used if they are subtracted from 1, the reason being that we are now concerned with the probability $p(0; N)$ that an event will *not* be equaled or exceeded during the design period N. Plot values of $p(0; N)$ as illustrated here for $T_R = 50$, $N = 5$. From Table 3-3 for $T_R = 50$, $N = 5$, $p(0; N)$ is found to be $1.00 - 0.10 = 0.90$. Enter Fig. 3-6 with a value of $T_R = 50$ and project upward to an intersection with the frequency curve of gross yield. Project this point horizontally to an intersection with the vertical extension of $N = 5$. This intersection represents one point on the 0.9 yield-probability curve. The lines with arrows in Fig. 3-6 illustrate this procedure. Repeat using various combinations of T_R and N until enough points of equal probability are plotted to construct each yield-probability curve.

   e. Using Fig. 3-6, a plot of the type shown in Fig. 3-7 can be easily obtained by replotting the yield-probability curves on arithmetic coordinate paper. This plot permits the direct determination of the probability that various yields will be met by the reservoir.

   Consider that the reservoir is to be operated at a design pumpage of 75 mgd. Figure 3-7 will yield the following type of information. There is a probability of 0.8 (or 8 chances out of 10) that the design pumpage will be met for the next 5 years

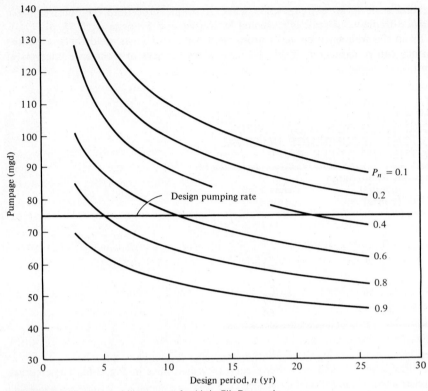

FIGURE 3-7 Yield-probability curves for Little Elk Reservoir.

without failure, a probability of 0.6 that the rate will be met for the next 11 years without failure, and probability of 0.4 that the design pumpage will be met successfully for the next 21 years.

The example illustrates the association of risks with a specified reservoir development. An engineer is thereby offered the opportunity to evaluate the severity of these risks as they relate to a particular draft-storage situation.

In recent years the science of operations research or systems analysis has also been applied successfully to some of the problems related to reservoir design and operation. Langbein, Fiering, and Thomas have made notable contributions to this field.[13,15,18]

3-18

LOSSES FROM STORAGE

The design of an impounding reservoir must include an evaluation of storage losses that may result from natural or artificial phenomena. Natural losses occur through evaporation, seepage, and siltation, while artificial losses are usually the product of withdrawals made to satisfy prior water rights.

Once a dam has been built and the impoundment filled, the exposed water-surface area is increased significantly over that of the natural stream. The resultant effect is that losses are incurred through increased evaporation. In addition, the opportunity for the derivation of runoff from the flooded land is eliminated. On the credit side, gains are made through the catchment of direct precipitation. These phenomena are commonly known as the *water-surface effect*. Net gains, then, usually result in well-watered regions. In arid lands, losses are the typical outcome, since evaporation generally exceeds precipitation.

The magnitude of seepage losses depends mainly on the geology of the region. If porous strata underlie the reservoir valley, considerable losses can occur. On the other hand, where permeability is low, seepage may be negligible. A thorough subsurface exploration is a prerequisite to the adequate evaluation of such losses.

Sedimentation studies are extremely valuable in planning for the development of surface water supplied by impoundment.[35,19] Since the useful life of a reservoir will be materially affected by the deposition of sediment, a knowledge of sedimentation rates is important in reaching a decision regarding the feasibility of its construction.

The rate and characteristic of the sediment inflow can be controlled by using sedimentation basins, providing vegetative screens, and by employing various erosion-control techniques.[9] Dams can be designed so that part of the sediment load can be passed through or over them. A last resort is the physical removal of sediment deposits. Normally this is not economically feasible.

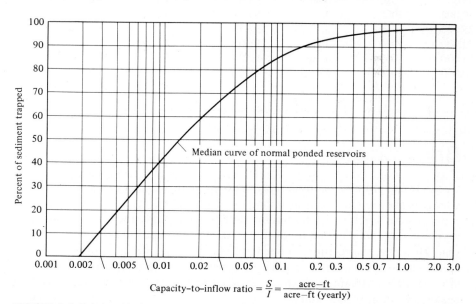

FIGURE 3-8 Relationship between reservoir sediment trap efficiency and capacity/inflow ratio. [Developed from data by G. M. Brune, "Trap Efficiency of Reservoirs," *Trans. Am. Geophys. Union* 34 (June 1953) : 407–418.]

●EXAMPLE 3-5

Determine the expected life of the Lost Valley Reservoir. The initial capacity of the reservoir is 45,000 acre-ft, and the average annual inflow is 76,000 acre-ft. A sediment inflow of 176 acre-ft/yr is reported. Assume that the useful life of the reservoir is exceeded when 77.8% of the original capacity is lost.

○*Solution*

The solution is obtained through application of the data given in Fig. 3-8. The results are tabulated in Table 3-6.

TABLE 3-6
DETERMINATION OF THE PROBABLE LIFE OF THE
LOST VALLEY RESERVOIR

(1)	(2)	(3)	(4)	(5)	(6)	(7)
Reservoir Capacity (acre-ft)	Volume Increment (acre-ft)	Capacity Inflow Ratio: (1)/76,000	Percent Sediment Trapped, from Fig. 3-8	Average Percent Sediment Trapped per Volume Increment	Acre-ft Sediment Trapped Annually: (5) × 176	Number of Years Required to Fill the Volume Increment: (2) ÷ (6)
45,000	5000	0.59	96.5			
40,000	5000	0.52	96.1	96.3	169	30
35,000	5000	0.46	95.8	95.9	169	30
30,000	5000	0.39	95.0	95.4	168	30
25,000	5000	0.33	94.5	94.7	167	30
20,000	5000	0.26	93.0	93.8	165	30
15,000	5000	0.20	92.0	92.5	163	31
10,000	5000	0.13	88.0	90.0	158	32

Total number of years of useful life   213

Groundwater

Groundwater storage is considerably in excess of all artificial and natural surface storage in the United States, including the Great Lakes.[7] This enormous groundwater reserve sustains the continuing outflow of streams and lakes during prolonged periods which often follow the relatively few runoff-producing rains each year. However, the relation between groundwater and surface storage is one of mutual interdependence. For example, groundwater intercepted by wells as it moves toward a stream represents a real diversion just as if the water had actually been taken from the stream.

Increased usage and development of groundwater supplies have stimulated the search for knowledge regarding the occurrence, origin, and movement of these supplies. The study of groundwater is vital. It is also exceedingly complex, owing to the fact that groundwater location and movement are determined primarily by the geology of the area. This geologic role cannot be overemphasized. Groundwater distribution in the United States was discussed in Sec. 3-4.

Practically all groundwater can be considered to be part of the hydrologic cycle, even though small amounts may enter the cycle from other origins. *Connate water*, water that was entrapped in the pores of sedimentary rock at the time of deposition, is an example.

3-19

THE SUBSURFACE DISTRIBUTION OF WATER

Groundwater distribution may be generally categorized into zones of aeration and saturation. The *saturated zone* is one in which all the voids are filled with water under hydrostatic pressure. The *aeration zone*, in which the interstices are filled partly with air and partly with water, may be subdivided into several sub-zones. Todd classifies these as follows:[6]

1   *Soil–water zone:* begins at the ground surface and extends downward through the major root zone. Its total depth is variable and dependent upon soil type and vegetation. The zone is unsaturated except during periods of heavy infiltration. Three categories of water classification may be encountered in this region: *hygroscopic water*, which is adsorbed from the air; *capillary water*, which is held by surface tension; and *gravitational water*, which is excess soil water draining through the soil.
2   *Intermediate zone:* extends from the bottom of the soil–water zone to the top of the capillary fringe and may vary from nonexistence to several hundred feet in thickness. The zone is essentially a connecting link between the near-ground surface region and the near-water table region, through which infiltrating waters must pass.
3   *Capillary zone:* extends from the water table to a height determined by the capillary rise which can be generated in the soil. The capillary-zone thickness is a function of soil texture and may vary not only from region to region but also within a local area.
4   *Groundwater zone:* groundwater fills the pore spaces completely, and porosity is therefore a direct measure of storage volume. Part of this water (specific retention) cannot be removed by pumping or drainage because of molecular and surface-tension forces. The specific retention is the ratio of the volume of water retained against gravity drainage to the gross volume of the soil.

The water that can be drained from a soil by gravity is known as the *specific yield.* It is expressed as the ratio of the volume of water that can be drained by gravity to the gross volume of the soil. Values of specific yield are dependent upon soil particle size, shape and distribution of pores, and degree of compaction of the soil. Average values of specific yield for alluvial aquifers range from 10 to 20%. Meinzer and others have proposed numerous procedures for determining specific yield.[20]

3-20

AQUIFERS

An *aquifer* is a water-bearing stratum or formation capable of transmitting water in quantities sufficient to permit development. Aquifers may be considered as falling into two categories, confined and unconfined, depending upon whether or not a water table or free surface exists under atmospheric pressure. The storage volume within an aquifer is changed whenever water is recharged to or discharged from an aquifer. In the case of an unconfined aquifer, this may be easily determined as

$$\Delta S = S_y \, \Delta V \tag{3-8}$$

where ΔS = change in storage volume

S_y = average specific yield of the aquifer

ΔV = volume of the aquifer lying between the original water table and the water table at a later, specified time

For saturated, confined aquifers, pressure changes produce only slight changes in storage volume. In this case, the weight of the overburden is supported partly by hydrostatic pressure and partly by the solid material in the aquifer. When the hydrostatic pressure in a confined aquifer is reduced by pumping or other means, the load on the aquifer increases, causing its compression, with the result that some water is forced from it. Decreasing the hydrostatic pressure also causes a small expansion, which in turn produces an additional release of water.

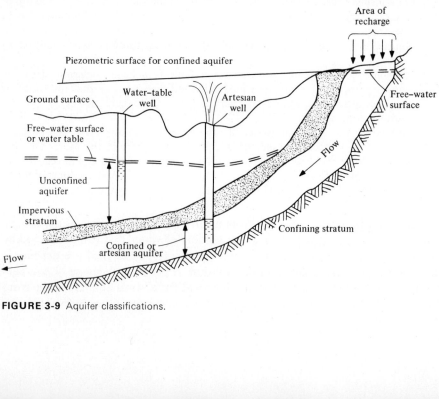

FIGURE 3-9 Aquifer classifications.

For confined aquifers, the water yield is expressed in terms of a storage coefficient S_c. This storage coefficient may be defined as the volume of water that an aquifer takes in or releases per unit surface area of aquifer per unit change in head normal to the surface. Figure 3-9 illustrates the classifications of aquifers.

In addition to water-bearing strata exhibiting satisfactory rates of yield, there are also non-water-bearing and impermeable strata. An *aquiclude* is an impermeable stratum that may contain large quantities of water but whose transmission rates are not high enough to permit effective development. An aquifuge is a formation that is impermeable and devoid of water.

3-21

FLUCTUATIONS IN GROUNDWATER LEVEL

Any circumstance that alters the pressure imposed on underground water will also cause a variation in the groundwater level. Seasonal factors, change in stream and river stages, evapotranspiration, atmospheric pressure changes, winds, tides, external loads, various forms of withdrawal and recharge, and earthquakes all may produce fluctuations in the level of the water table or the piezometric surface, depending upon whether the aquifer is free or confined.[6] It is important that the engineer concerned with the development and utilization of groundwater supplies be aware of these factors. He should also be able to evaluate their importance relative to the operation of a specific groundwater basin.

3-22

SAFE YIELD OF AN AQUIFER

The development engineer must be able to determine the quantity of water that can be produced from a groundwater basin in a given period of time. He must also be able to evaluate the consequences that will result from the imposition of various drafts on the underground supply. A knowledge of the safe yield of an aquifer is therefore exceedingly important. Following are pertinent definitions:

1 *Safe yield:* the quantity of water that can be withdrawn annually without the ultimate depletion of the aquifer.
2 *Maximum sustained yield:* maximum rate at which water can be withdrawn on a continuing basis from a given source.
3 *Permissive sustained yield:* maximum rate at which withdrawals can be made legally and economically on a continuing basis for beneficial use without the development of undesired results.
4 *Maximum mining yield:* total storage volume in a given source which can be withdrawn and used.
5 *Permissive mining yield:* maximum volume of water which can be withdrawn legally and economically, to be used for beneficial purposes, without causing an undesired result.

After studying these definitions the reader should note that extreme caution must be exercised in the development and utilization of groundwater resources. These resources are finite and not inexhaustible. If drafts are imposed such that the various recharge mechanisms will balance them over a period of time, no difficulty will result. On the other hand, excessive drafts may reduce underground storage volumes to a point at which economic development is no longer feasible. The mining of water, as with any other natural resource, will ultimately result in exhaustion of the supply. Figure 3-10 indicates those areas in the United States where perennial overdrafts occur.[11]

Methods for determining the safe yield of an aquifer have been proposed by Hill, Harding, Simpson, and others.[6] The Hill method based on groundwater studies in Southern California and Arizona will be presented here. In this method, the annual change in groundwater table elevation or piezometric surface elevation is plotted against the annual draft. The points can be fitted by a straight line, provided that the water supply to the basin is fairly uniform.

FIGURE 3-10 Groundwater reservoirs with perennial overdraft. [Source: H. E. Thomas, "Water," *The Yearbook of Agriculture* (Washington, D.C.: Government Printing Office, 1955).]

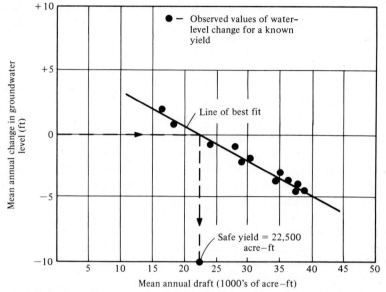

FIGURE 3-11 Example of the determination of safe yield by the Hill method.

That draft which corresponds to zero change in elevation is considered to be the safe yield. The period of record should be such that the supply during this period approximates the long-time average supply. Even though the draft during the period of record may be an overdraft, the safe yield can be determined by extending the line of best fit to an intersection with the zero change in elevation line. An example of this procedure is given in Fig. 3-11.

It should be noted that the safe yield of a groundwater basin may be variable with time. This is because the groundwater-basin conditions under which the safe yield was determined are subject to change, and these changes will be reflected in the modified value of the safe yield.

3-23

GROUNDWATER FLOW

The rate of movement of water through the ground is of an entirely different magnitude than that through natural or artificial channels or conduits. Typical values range from 5 ft/day to a few feet per year. Methods for determining these transmission rates are primarily based on the principles of fluid flow represented by *Darcy's law*. Mathematically, this law may be stated as

$$V = kS \qquad (3-9)$$

where V = velocity of flow
k = coefficient having the same units as velocity
S = slope of the hydraulic gradient

Darcy's law is limited in its applicability to flows in the laminar region. The controlling criterion is the *Reynolds number*.

$$N_R = \frac{Vd}{v} \tag{3-10}$$

where $V =$ flow velocity
$\quad\quad d =$ mean grain diameter
$\quad\quad v =$ kinematic viscosity

For Reynolds numbers of less than unity, groundwater flow may be considered laminar. Departure from laminar conditions develops normally in the range of Reynolds numbers from 1 to 10, depending on grain size and shape. Under most conditions, with the exception of regions in close proximity to collecting devices, the flow of groundwater is laminar and Darcy's law applies.

●**EXAMPLE 3-6**

a. Find the Reynolds number for the portion of an aquifer distant from any collection device where water temperature is 50°F ($v = 1.41 \times 10^{-5}$ ft²/sec), flow velocity is 1.0 ft/day, and mean grain diameter is 0.09 in.

b. Find the Reynolds number for a flow 4 ft from the centerline of a well being pumped at a rate of 3800 gpm if the well completely penetrates a confined aquifer 28 ft thick. Assume a mean grain diameter of 0.10 in., a porosity of 35%, and $v = 1.41 \times 10^{-5}$ ft²/sec.

○**Solution**

a. Using Eq. 3-10, we obtain

$$N_R = \frac{Vd}{v}$$

where

$$V = \left(\frac{1.0}{86,400}\right) \text{ fps} \quad\quad d = \left(\frac{0.09}{12}\right) \text{ ft}$$

$$N_R = \frac{1.0}{86,400} \times \frac{0.09}{12} \times \frac{1}{1.41 \times 10^{-5}}$$

$$= 0.0062 \quad \text{(including laminar flow)}$$

b. Using Eq. 3-10 yields

$$N_R = \frac{Vd}{v} = \frac{Q}{A}\frac{d}{v}$$

$Q = 3800 \times 2.23 \times 10^{-3} = 8.46$ cfs
$V = 8.46/2\,\pi rh \times$ porosity
$\quad = 8.46/8\pi \times 28 \times 0.35$
$\quad = 0.0344$ fps

$$N_R = \frac{0.0344 \times 0.10}{1.41 \times 10^{-5} \times 12} = 20.3 \quad \text{(beyond Darcy's law range)}$$

To compute discharge it is necessary to multiply Eq. 3-9 by the effective cross-sectional area. The equation then becomes

$$Q = pAkS \tag{3-11}$$

where p = porosity or ratio of the void volume to the total volume of the mass

   A = gross cross-sectional area

and the other terms are as previously defined. By combining k and p into a single term, Eq. 3-11 may be written in its most common form,

$$Q = KAS \tag{3-12}$$

where K is known as the *coefficient of permeability*. A number of ways of expressing K may be found in the literature. The U.S. Geological Survey defines the standard coefficient of permeability K_s as the number of gallons of water per day that will flow through a medium of 1-ft^2 cross-sectional area under a hydraulic gradient of unity at 60°F. The field coefficient of permeability is obtained directly from the standard coefficient by correcting for temperature.

$$K_f = K_s \frac{\mu_{60}}{\mu_f} \tag{3-13}$$

where K_f = field coefficient

   μ_{60} = dynamic viscosity at 60°F

   μ_f = dynamic viscosity at field temperature

An additional term which is much used in groundwater computations is the coefficient of transmissibility, T. It is equal to the field coefficient of permeability multiplied by the saturated thickness of the aquifer in feet. Using this terminology, Eq. 3-11 may also be written as

$$Q = T \times \text{section width} \times S \tag{3-14}$$

Table 3-7 gives typical values of the standard coefficient of permeability for a range of sedimentary materials. It should be noted that the permeabilities for specific materials vary widely. Traces of silt and clay can significantly decrease the permeability of an aquifer. Differences in particle orientation and shape can cause striking changes in permeability within aquifers composed of the same geologic material. Careful evaluation of geologic information is absolutely essential if realistic values of permeability are to be used in groundwater-flow computations.

It is of interest to note that Darcy's equation is analogous to the electrical equation known as *Ohm's law*,

$$i = \frac{1}{R} E \tag{3-15}$$

TABLE 3-7

TYPICAL VALUES OF THE STANDARD COEFFICIENT
OF PERMEABILITY FOR VARIOUS MATERIALS

Material	Approximate Range in K_s (gpd/ft^2)
I Clean gravel	10^6–10^4
II Clean sands; mixtures of clean sands and gravels	10^4–10
III Very find sands; silts; mixtures of sand, silt, and clay; stratified clays, etc.	10–10^{-3}
IV Unweathered clays	10^{-3}–10^{-4}

where i = current
R = resistance
E = voltage

The quantities i and Q, K and $1/R$, and E and S are comparable. This equivalency permits the use of electrical models in solving many groundwater-flow problems.[6]

3-24

HYDRAULICS OF WELLS

The collection of groundwater is accomplished primarily through the construction of wells or infiltration galleries. Numerous factors are involved in the numerical estimation of the performance of these collection works. Some cases are amenable to solution through the utilization of relatively simple mathematical expressions. Other cases can be solved only through graphical analysis or the use of various kinds of models.[6] Several of the less difficult cases will be discussed here. The reader is cautioned not to be misled by the simplicity of some of the solutions presented and should observe that many of these are special-case solutions and are not indiscriminately applicable to all groundwater-flow situations. A more thorough treatment of groundwater and seepage problems may be found in numerous references.[6,21,22]

Flow to Wells

A well system may be considered to be composed of three elements—the well structure, the pump, and the discharge piping. The well itself contains an open section through which flow enters and a casing through which the flow is transported to the ground surface. The open section is usually a perforated casing or a slotted metal screen which permits the flow to enter and at the same time prevents collapse of the hole. Occasionally gravel is placed at the bottom of the well casing around the screen. Hoffman states that it is not economical to use anything but a properly sized screen for industrial wells.[23]

When a well is pumped, water is removed from the aquifer immediately adjacent to the screen. Flow then becomes established at locations some distance from the well in order to replenish this withdrawal. Owing to the resistance to flow offered by the soil, a head loss is encountered and the piezometric surface adjacent to the well is depressed. This is known as the *cone of depression* (Fig. 3-12). The cone of depression spreads until a condition of equilibrium is reached and steady-state conditions are established.

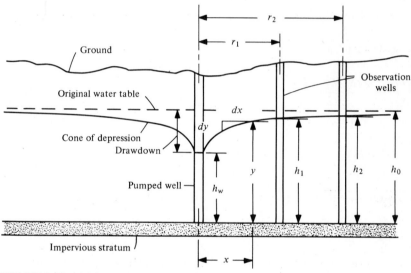

FIGURE 3-12 Well in an unconfined aquifer.

The hydraulic characteristics of an aquifer (which are described by the storage coefficient and the aquifer permeability) may be determined by laboratory or field tests. The three most commonly used field methods are the application of tracers, use of field permeameters, and aquifer performance tests.[6] A discussion of aquifer performance tests will be given here along with the development of flow equations for wells.[23,24,30]

Aquifer performance tests may be classified as either equilibrium or nonequilibrium tests. In the former, the cone of depression must be stabilized for the flow equation to be derived. In the latter, the derivation includes the condition that steady-state conditions have not been reached. Thiem published the first performance tests based on equilibrium conditions in 1906.[27]

The basic equilibrium equation for an unconfined aquifer can be derived using the notation of Fig. 3-12. In this case the flow is assumed to be radial, the original water table is considered to be horizontal, the well is considered to fully penetrate the aquifer of infinite areal extent, and steady-state conditions must prevail. Then the flow toward the well at any location x from the well must equal the product of the cylindrical element of area at that section and the flow velocity. Using Darcy's law, this becomes

$$Q = 2\pi x y K_f \frac{dy}{dx} \tag{3-16}$$

where $2\pi x y$ = area at any section
$K_f \, dy/dx$ = flow velocity
Q = discharge, cfs

Integrating over the limits specified below yields

$$\int_{r_1}^{r_2} Q \frac{dx}{x} = 2\pi K_f \int_{h_1}^{h_2} y \, dy \tag{3-17}$$

$$Q \log_e\left(\frac{r_2}{r_1}\right) = \frac{2\pi K_f (h_2^2 - h_1^2)}{2} \tag{3-18}$$

and

$$Q = \frac{\pi K_f (h_2^2 - h_1^2)}{\log_e(r_2/r_1)} \tag{3-19}$$

This equation may then be solved for K_f, yielding

$$K_f = \frac{1055Q \log_{10}(r_2/r_1)}{h_2^2 - h_1^2} \tag{3-20}$$

where $\log_e$ has been converted to $\log_{10}$, K_f is in gallons per day per square foot, Q is in gallons per minute, and r and h are measured in feet. If the drawdown is small compared with the total aquifer thickness, an approximate formula for the discharge of the pumped well can be obtained by inserting h_w for h_1 and the height of the aquifer for h_2 in Eq. 3-19.

The basic equilibrium equation for a confined aquifer can be obtained in a similar manner, using the notation of Fig. 3-13. The same assumptions apply. Mathematically the flow in cubic feet per second may be determined as follows:

$$Q = 2\pi x m K_f \frac{dy}{dx} \tag{3-21}$$

Integrating, we obtain

$$Q = 2\pi K_f m \frac{h_2 - h_1}{\log_e(r_2/r_1)} \tag{3-22}$$

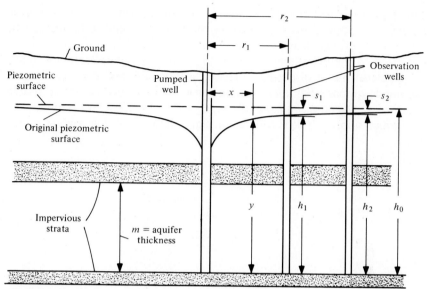

FIGURE 3-13 Radial flow to a well in a confined aquifer.

The coefficient of permeability may be determined by rearranging Eq. 3-22 to the form

$$K_f = \frac{528Q \log_{10}(r_2/r_1)}{m(h_2 - h_1)} \tag{3-23}$$

where Q is in gallons per minute, K_f is the permeability in gallons per day per square foot, and r and h are measured in feet.

●**EXAMPLE 3-7**
Determine the permeability of an artesian aquifer being pumped by a fully penetrating well. The aquifer is composed of medium sand and is 90 ft thick. The steady-state pumping rate is 850 gpm. The drawdown of an observation well 50 ft away is 10 ft, and the drawdown in a second observation well 500 ft away is 1 ft.

○**Solution**

$$K_f = \frac{528Q \log_{10}(r_2/r_1)}{m(h_2 - h_1)}$$

$$= \frac{528 \times 850 \times \log_{10}(10)}{90 \times (10 - 1)}$$

$$= 554 \text{ gpd/ft}^2$$

For a steady-state well in a uniform flow field where the original piezometric surface is not horizontal, a somewhat different situation than that previously assumed prevails. Consider the artesian aquifer shown in Fig. 3-14. The heretofore assumed circular area of influence becomes distorted in this case. This problem may be solved by application of potential theory, or by graphical means; or, if the slope of the piezometric surface is very slight, Eq. 3-22 may be applied without serious error.

Referring to the definition sketch of Fig. 3-14, a graphical solution to this type of problem will be discussed. First, an orthogonal flow net consisting of flow lines and equipotential lines must be constructed. The construction should be performed so that the completed flow net will be composed of a number of elements which approach little squares in shape. Reference 21 is a good source

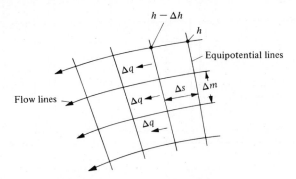

Definition Sketch of a Segment of a Flow Net

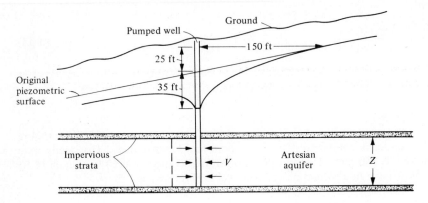

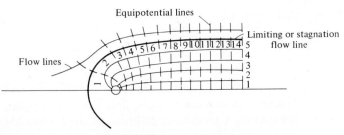

FIGURE 3-14 Well in a uniform flow field and flow-net definition.

of information on this subject for the interested reader. A comprehensive discussion cannot be provided here.

Once the net is complete, it may be analyzed by considering the net geometry and using Darcy's law in the manner of Todd.[6] In the definition sketch of Fig. 3-14, the hydraulic gradient is

$$h_g = \frac{\Delta h}{\Delta s} \tag{3-24}$$

and the flow increment between adjacent flow lines is

$$\Delta q = K \frac{\Delta h}{\Delta s} \Delta m \tag{3-25}$$

where for a unit thickness, Δm represents the cross-sectional area. If the flow net is properly constructed so that it is orthogonal and composed of little square elements,

$$\Delta m \approx \Delta s \tag{3-26}$$

and

$$\Delta q = K \Delta h \tag{3-27}$$

Now considering the entire flow net,

$$\Delta h = \frac{h}{n} \tag{3-28}$$

where n is the number of subdivisions between equipotential lines. If the flow is divided into m sections by the flow lines, the discharge per unit width of the aquifer will be

$$q = \frac{Kmh}{n} \tag{3-29}$$

A knowledge of the aquifer permeability and the flow-net geometry permits solution of Eq. 3-29.

●EXAMPLE 3-8
Find the discharge to the well of Fig. 3-14 by using the applicable flow net. Consider the aquifer to be 35 ft thick, K_f to be 3.65×10^{-4} fps, and the other dimensions as shown.

○**Solution**
Using Eq. 3-29.

$$q = \frac{Kmh}{n}$$

where $h = (35 + 25) = 60$ ft

$\qquad m = 2 \times 5 = 10$

$\qquad n = 14$

$$q = \frac{3.65 \times 10^{-4} \times 60 \times 10}{14}$$

$\qquad = 0.0156$ cfs per unit thickness of the aquifer

The total discharge Q is thus

$$Q = 0.0156 \times 35 = 0.55 \text{ cfs} \quad \text{or} \quad 245 \text{ gpm}$$

When a new well is first pumped, a large portion of the discharge is produced directly from the storage volume released as the cone of depression develops. Under these circumstances the equilibrium equations overestimate permeability and therefore the yield of the well. Where steady-state conditions are not encountered—as is usually the case in practice—a nonequilibrium equation must be used. Two approaches can be taken, the rather rigorous method of Theis, or a simplified procedure such as that proposed by Cooper and Jacob.[28,29]

In 1935 Theis published a nonequilibrium approach which takes into consideration time and the storage characteristic of the aquifer.[28] His method utilizes an analogy between heat transfer described by the Biot–Fourier law, and groundwater flow to a well. Theis states that the drawdown(s) in an observation well located at a distance r from the pumped well is given by

$$s = \frac{114.6Q}{T} \int_u^\infty \frac{e^{-u}}{u} \, du \qquad (3\text{-}30)$$

where T = transmissibility
$\qquad Q$ = discharge, gpm

and u is given by

$$u = \frac{1.87 r^2 S_c}{Tt} \qquad (3\text{-}31)$$

where S_c = storage coefficient
$\qquad T$ = transmissibility
$\qquad t$ = time since the start of pumping, days

The integral in Eq. 3-30 is usually known as the well function of u and is commonly written as $W(u)$. It may be evaluated from the infinite series

$$W(u) = -0.577216 - \log_e u + u - \frac{u^2}{2 \times 2!} + \frac{u^3}{3 \times 3!} \cdots \qquad (3\text{-}32)$$

The basic assumptions employed in the Theis equation are essentially the same as those for the Thiem equation except for the non-steady-state condition.

Equations 3-30 and 3-31 can be solved by comparing a log-log plot of u versus $W(u)$ known as a "type curve," with a log-log plot of the observed data r^2/t versus s. In plotting the type curve, $W(u)$ is the ordinate and u is the abscissa. The two curves are superimposed and moved about until some of their segments coincide. In doing this, the axes must be maintained parallel. A coincident point is then selected on the matched curves and both plots are marked. The type curve then yields values of u and $W(u)$ for the selected point. Corresponding values of s and r^2/t are determined from the plot of the observed data. Inserting these values in Eqs. 3-30 and 3.31 and rearranging, values for the transmissibility T and the storage coefficient S_c may be found.

Often, this procedure can be shortened and simplified. When r is small and t large, Jacob found that values of u are generally small.[29] Thus the terms in the series of Eq. 3-32 beyond the second term become negligible and the expression for T becomes

$$T = \frac{264 Q(\log_{10} t_2 - \log_{10} t_1)}{h_0 - h} \qquad (3\text{-}33)$$

which can be further reduced to

$$T = \frac{264 Q}{\Delta h} \qquad (3\text{-}34)$$

where Δh = drawdown per log cycle of time $[(h_0 - h)/(\log_{10} t_2 - \log_{10} t_1)]$
Q = well discharge, gpm

h_0 and h are defined as shown on Fig. 3-13, and T is the transmissibility in gallons per day per foot. Field data on drawdown $(h_0 - h)$ versus t are plotted on semilogarithmic paper. The drawdown is plotted on the arithmetic scale as shown on Fig. 3-15. This plot forms a straight line, the slope of which permits the determination of the formation constants using Eq. 3-34 and

$$S_c = \frac{0.3 T t_0}{r^2} \qquad (3\text{-}35)$$

where t_0 is the time that corresponds to zero drawdown.

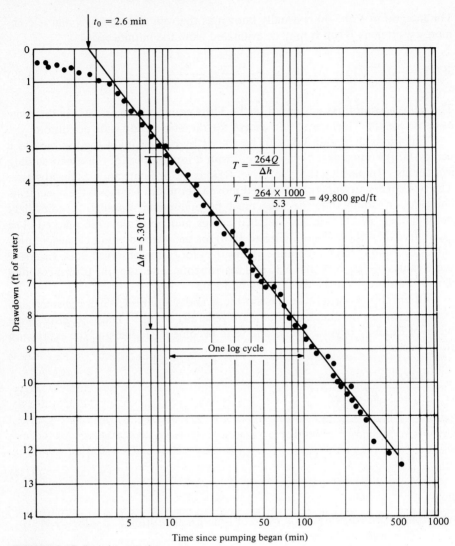

FIGURE 3-15 Pumping test data.

●EXAMPLE 3-9

Using the data given on Fig. 3-15, find the coefficient of transmissibility T and the storage coefficient S_c for the aquifer. Given $Q = 1000$ gpm and $r = 300$ ft.

○*Solution*

Find the value of Δh from the graph. This is 5.3 ft. Then, using Eq. 3-34, we obtain

$$T = \frac{264Q}{\Delta h} = \frac{264 \times 1000}{5.3}$$

$$= 49,800 \text{ gpd/ft}$$

Using Eq. 3-53 yields

$$S_c = \frac{0.3Tt_0}{r^2}$$

Note from Fig. 3-15 that $t_0 = 2.6$ min. When we convert to days, this becomes

$$t_0 = 1.81 \times 10^{-3} \text{ day}$$

and

$$S_c = \frac{0.3 \times 49,800 \times 1.81 \times 10^{-3}}{(300)^2}$$

$$= 0.0003$$

3-25

BOUNDARY EFFECTS

Only the effect of pumping a single well has been previously considered. If more than one well is pumped in an area, however, a composite effect or interference due to the overlap of the cones of depression will result. In this case the drawdown at any location is obtained by summing the individual drawdowns of the various wells involved. An additional problem is that of boundary conditions. The previous derivations have been based on the supposition of a homogeneous aquifer of infinite areal extent. A situation such as this is rarely encountered in practice. Computations based on this assumption are often sufficiently accurate, however, providing field conditions closely approximate the basic hypotheses. Boundary effects may be evaluated by using the theory of image wells proposed by Lord Kelvin, through the use of electrical and membrane analogies and through the use of relaxation procedures. For a detailed discussion of these topics the reader is referred to the many references on groundwater flow.[6,21]

3-26

SALTWATER INTRUSION

Saltwater contamination of freshwater aquifers presents a serious water-quality problem in island locations; in coastal areas; and occasionally inland, as in Arizona, where some aquifers contain highly saline waters. Because fresh water is lighter than salt water (specific gravity of seawater is about 1.025), it will usually float above a layer of salt water. When an aquifer is pumped, the original equilibrium is disturbed and salt water replaces the fresh water. Under equilibrium conditions a drawdown of 1 ft in the freshwater table will result in a rise by the salt water of approximately 40 ft. Pumping rates of wells subject to saltwater intrusion are therefore seriously limited. In coastal areas, recharge wells are sometimes used in an attempt to maintain a sufficient head to prevent seawater intrusion. Injection wells have been used effectively in this manner in Southern California.

A prime example of freshwater contamination by seawater is noted in Long Island, New York.[30] During the first part of this century the rate of pumping far exceeded the natural recharge rate. The problem was further complicated because storm-water runoff from the highly developed land areas was transported directly to the sea. This precluded the opportunity for this water to return to the ground. As pumping continued, the water table dropped well below sea level and saline water entered the aquifer. The result was such a serious impairment of local water quality that Long Island was forced to transport its water supply from upper New York State.

3-27

GROUNDWATER RECHARGE

The volumes of groundwater replaced annually through natural mechanisms are relatively small because of the slow rates of movement of groundwaters and the limited opportunity for surface waters to penetrate the earth's surface. To supplement this natural recharge process, a recent trend toward artificial recharge has been developing. In 1955 over 700 million gallons of water per day were artificially recharged in the United States.[26] This water was derived from natural surface sources, returns from air conditioning, industrial wastes, and municipal water supplies. The total recharge volume was equal, however, to only about 1.5% of the groundwater withdrawn that year. In California, for example, artificial recharge is presently a primary method of water conservation. During the 1957–1958 period a daily recharge volume of about 560 million gallons was reported for 63 projects in that state alone.[26]

Numerous methods are employed in artificial recharge operations. One of the most common plans is the utilization of holding basins. The usual practice is to impound the water in a series of reservoirs arranged so that the overflow of one will enter the next, and so on. These artificial storage works are generally formed by the construction of dikes or levees. A second method is the modified streambed, which makes use of the natural water supply. The stream channel is widened, leveled, scarified, or treated by a combination of methods to increase its recharge capabilities. Ditches and furrows are also used. The basic types of arrangement are the contour type, in which the ditch follows the contour of the ground; the lateral type, in which water is diverted into a number of small furrows from the main canal or channel; and the tree-shaped or branching type, where water is diverted from the primary channel into successively smaller canals and ditches. Where slopes are relatively flat and uniform, flooding provides an economical means of recharge. Normal practice is to spread the recharge water over the ground at relatively small depths so as not to disturb the soil or native vegetation. An additional method is the use of injection wells.

Recharge rates are normally less than pumping rates for the same head conditions, however, because of the clogging that is often encountered in the area adjacent to the well casing.[25] Clogging may result from the entrapment of fine aquifer particles, from suspended material in the recharge water which is subsequently strained out and deposited in the vicinity of the well screen, from air binding, from chemical reactions between recharge and natural waters, and from bacteria. For best results the recharge water should be clear, contain little or no sodium, and be chlorinated.

Richter and Chun have outlined several factors worthy of consideration relative to the selection of artificial recharge project sites.[26] Essentially they are availability and character of local and imported water supply; factors relating to infiltration rates of the native soils; operation and maintenance problems; net benefits; water quality; and legal considerations.

Concurrent Development of Groundwater and Surface-Water Sources

The maximum practical conservation of our water resources is based on the coordinated development of groundwater and surface-water supplies. Geologic, hydrologic, economic, and legal factors must be carefully considered.

Concurrent utilization is primarily founded on the premise of transference of impounded surface water to groundwater storage at optimum rates.[31,32] Annual water requirements are generally met by surface storage while groundwater storage is used to meet cyclic requirements covering periods of dry years. The operational procedure involves a lowering of groundwater levels during periods of below-average precipitation and a subsequent raising of levels during wet years. Transfer rates of surface waters to underground storage must be large enough to assure that surface-water reservoirs will be drawn down sufficiently to permit impounding significant volumes during periods of high runoff. To provide the required maximum transfer capacity, methods of artificial recharge such as spreading, ponding, injecting, returning flows from irrigation, or other techniques must be used.

The coordinated use of groundwater and surface-water sources will result in the provision of larger quantities of water at lower costs. As an example, it has been found that the conjunctive operation of the Folsom Reservoir (California) and its groundwater basin yields a conservation and utilization efficiency of approximately 82% as compared with about 51% efficiency for the operation of the surface reservoir alone.[33] There is little doubt that the inclusion of groundwater resources should be given very careful consideration in future planning for large-scale water development projects.

In general, the analysis of a conjunctive system consisting of a dam and an aquifer requires the solution of three fundamental problems. The first is to establish the design criteria for the dam and the recharge facilities. The second is to determine the service area for the combined system. Finally, a set of operating rules which define the reservoir drafts and pumpages to be taken from the aquifer are required. A mathematical model for an analysis such as this has been proposed by Buras.[31]

PROBLEMS

3-1 An industry uses 8.6×10^6 gal of water per day. Forty percent of this flow is discharged to a nearby stream as a concentrated waste containing 3000 ppm dissolved solids. The stream has a watershed of 80 mi^2 and a mean annual flow rate of 1 cfs/mi^2. Will the water downstream of the refinery meet the U.S. EPA regulations regarding recommended total dissolved solids for: (a) drinking purposes; (b) irrigation purposes? Assume that the stream normally contains 100 ppm dissolved solids.

3-2 Compare the amounts of water required by the various users in your state. What is the relative worth of water in its various uses?

3-3 A mean draft of 115 mgd is to be developed from a 175-mi^2 catchment area. At the flow line the reservoir is estimated to be 4000 acres. The annual rainfall is 36 in., the mean annual runoff is 13 in., and the mean annual evaporation is 48 in. Find the net gain or loss in storage which this represents. Compute the volume of water evaporated. State this figure in a form such as the number of years the volume could supply a given community.

3-4 Discuss how you would go about collecting data for an analysis of the water budget of a region. What agencies would you contact? What other sources of information would you seek out?

3-5 For some area of your choice, make a plot of mean monthly precipitation versus time. Explain how this fits the pattern of seasonal water uses for the area. Will the form of precipitation be an important consideration?

3-6 Given the following 10-year record of annual precipitation, plot a rough precipitation frequency curve. Tabulate the data to be plotted and show the method of computation. Data: annual precipitation in inches: 28, 21, 33, 26, 29, 27, 19, 28, 18, 22. (*Note:* Frequency in percent of years $= 1/T_R \times 100$.)

3-7 Find a maximum reservoir storage requirement if a uniform draft of 726,000 gpd/mi^2 from a specific stream is to be maintained. The following record of average monthly runoff values is given:

Month	A	M	J	J	A	S	O	N	D	J	F	M	A	M	J
Runoff (mgal/mi^2)	97	136	59	14	6	5	3	7	19	13	74	96	37	63	49

3-8 Using the information given in Table 3-3, plot reocurrence interval in years as the ordinate, design period in years as the abscissa, and construct a series of recurrence interval–design period probabilities that an event will not be exceeded during the design period. Use arithmetic coordinate paper. [*Note:* To conform to this, probabilities in the table must be subtracted from 1.0. Where sufficient information is not provided by the table, probabilities may be computed using $P_n = (1 - 1/T_R)^n$, where $n = $ design period in years.]

3-9 The following partial-duration series of 16-month low flows is available from 63 years of record on the Black River. The river supplies a reservoir which has a drainage area of 350 mi². For a 16-month flow (in./mi²) 6.2, 8.5, 4.9, 12.7, 19.3, 14.7, 16.0, 5.1, 7.6, and 8.3. Find the recurrence interval for these events.

3-10 Consider that in Prob. 3-9 the various low flows have been plotted and for a reservoir capacity of 8 in./mi² (350 mi²), the gross-yield values were found to be 183, 179, 160, 148, 149, 130, 126, 98, 92, and 75 mgd. Graphically determine a gross yield–recurrence interval curve and plot yield-probability curves. For a reservoir supply of 70 mgd, what is the probability of satisfying this requirement if the design period is: (a) 10 years; (b) 20 years?

3-11 Given the following 50-month record of mean monthly discharge, find the magnitude of the 20-month low flow. Consecutive average monthly flows (cfs): 14, 17, 19, 21, 18, 16, 18, 25, 29, 32, 34, 33, 30, 28, 20, 23, 16, 14, 12, 13, 16, 13, 12, 12, 13, 14, 16, 13, 12, 11, 10, 12, 10, 9, 8, 7, 6, 4, 6, 7, 8, 9, 11, 9, 8, 6, 7, 9, 13, 17.

3-12 Find the expected life of a reservoir having an initial capacity of 45,000 acre-ft. The average annual inflow is 63,000 acre-ft, and a sediment inflow of 180 acre-ft/yr is reported. Consider the useful life of the reservoir to be exceeded when 80% of the original capacity is lost. Use 5,000-acre-ft volume increments. Obtain values of percent sediment trapped from Fig. 3-8.

3-13 What precautions might be taken to avoid or minimize silting when planning a reservoir?

3-14 Given the following data relating mean annual change in groundwater level to mean annual draft, find the safe yield.

Mean Annual Change in Ground water Level (ft)	+1	+2	−1	−3	−4	+1.5	+1.2	−2.6
Mean Annual Draft (thousands of acre-ft)	23	19	31	42	44	21	19	33

3-15 What is the Reynolds number for flow in a soil when the water temperature is 50°F, velocity is 0.6 ft/day, and mean grain diameter 0.08 in.?

3-16 A 12-in. well fully penetrates a confined aquifer 100 ft thick. The coefficient of permeability is 600 gpd/ft². Two test wells located 45 ft and 120 ft away show a difference in drawdown between them of 8 ft. Find the rate of flow delivered by the well.

3-17 Determine the permeability of an artesian aquifer being pumped by a fully penetrating well. The aquifer is composed of medium sand and is 100 ft thick. The steady-state pumping rate is 1200 gpm. The drawdown in an observation well 75 ft away is 14 ft, and the drawdown in a second observation well 500 ft away is 1.2 ft. Find K_f in gallons per day per square foot.

3-18 Consider a confined aquifer with a coefficient of transmissibility $T = 680$ ft^3/day/ft. At $t = 5$ minutes the drawdown $= 5.6$ ft; at 50 minutes, $s = 23.1$ ft; and at 100 minutes, $s = 28.2$ ft. The observation well is 75 ft away from the pumping well. Find the discharge of the well.

3-19 Assume that an aquifer is being pumped at a rate of 300 gpm. The aquifer is confined and the pumping test data are given below. Find the coefficient of transmissibility T and the storage coefficients S for $r = 60$ ft.

Time Since Pumping Started (minutes)	1.3	2.5	4.2	8.0	11.0	100.0
Drawdown s (ft)	4.6	8.1	9.3	12.0	15.1	29.0

3-20 Given the following data: $Q = 59{,}000$ cfd, $T = 630$ cfd, $t = 30$ days, $r = 1$ ft, and $S_c = 6.4 \times 10^{-4}$. Consider this to be a nonequilibrium problem. Find the drawdown s. Note for

$$u = 8.0 \times 10^{-9} \qquad W(u) = 18.06$$
$$u = 8.2 \times 10^{-9} \qquad W(u) = 18.04$$
$$u = 8.6 \times 10^{-9} \qquad W(u) = 17.99$$

REFERENCES

1. E. A. Ackerman and G. O. Löf, *Technology in American Water Development* (Baltimore, Md.: The Johns Hokpins Press, 1959).
2. H. E. Babbitt, J. J. Doland, and J. L. Cleasby, *Water Supply Engineering* (New York: McGraw-Hill Book Company, 1962).
3. H. F. Dregne and H. J. Maker, "Irrigation Well Waters of New Mexico," *Agricultural Experiment Station, Bull. No. 386*, New Mexico State University, (1954).
4. J. Hirshleifer, J. DeHaven, and J. Milliman, *Water Supply—Economics, Technology, and Policy* (Chicago: University of Chicago Press, 1960).
5. J. M. McKee, "We Need Researchers," *Eng. News-Record* (October 1962).
6. D. K. Todd, *Ground Water Hydrology* (New York: John Wiley and Sons, Inc., 1960).
7. J. G. Ferris, "Ground Water " *Mech. Eng.* (January 1960).
8. W. B. Langbein, "Water Yield and Reservoir Storage in the United States," *U.S. Geol. Survey Circular* (1959).
9. R. K. Linsley, M. A. Kohler, and J. L. H. Paulhus, *Applied Hydrology* (New York: McGraw-Hill Book Company, 1949).
10. C. E. Renn, "Sources of Information and Research in Industrial Waste Treatment," *Proc. 2nd Ann. Symp. Ind. Waste Control* (Baltimore, Md.: The Johns Hopkins Press, 1961).
11. H. E. Thomas, "Underground Sources of Water," *The Yearbook of Agriculture, 1955* (Washington, D.C.: Government Printing Office, 1956).
12. Task Group A4, D1, "Water Conservation in Industry," *J. Am. Water Works Assoc.* 45 (December 1958).
13. W. B. Langbein, "Queuing Theory and Water Storage," *Proc. Am. Soc. Civil Engrs., J. Hydraulics Div.*, no. HY5 (October 1958).
14. J. B. Stall and J. C. Neill, "Calculated Risks of Impounding Reservoir Yield," *Proc. Am. Soc. Civil Engrs., J. Hydraulics Div.* 89, no. HY1 (January 1963).

15. M. B. Fiering, "Queuing Theory and Simulation in Reservoir Design." *Proc. Am. Soc. Civil Engrs., J. Hydraulics Div.* 87, no. HY6 (November 1961).

16. W. Rippl, "The Capacity of Storage Reservoirs for Water Supply," *Proc. Inst. Civil Engrs.* (*London*) 71 (1883): 270.

17. J. B. Stall, "Reservoir Mass Analysis by a Low-Flow Series," *Proc. Am. Soc. Civil Engrs., J. San. Eng. Div.* 88, no. SA5 (September 1962).

18. H. A. Thomas, Jr., "Queuing Theory of Stream Flow Regulation Applied to the Cost–Benefit Analysis of a Multipurpose Reservoir." Private communication to the Harvard Water Resources Program, Harvard University, Cambridge, Mass. (November 20, 1959).

19. G. M. Brune and R. E. Allen, "A Consideration of Factors Influencing Reservoir Sedimentation in the Ohio Valley Region," *Trans. Am. Geophys. Union* 22 (1941): 649–655.

20. O. E. Meinzer, "Outline of Methods for Estimating Ground Water Supplies," *U.S. Geol. Survey Water Supply Paper No. 638C* (1932).

21. M. E. Harr, *Groundwater and Seepage* (New York: McGraw-Hill Book Company, 1962).

22. M. Muskat, *The Flow of Homogeneous Fluids through Porous Media* (Ann Arbor, Mich.: J. W. Edwards, Inc., 1946).

23. J. F. Hoffman, "Field Tests Determine Potential Quantity, Quality of Ground Water Supply," *Heating, Piping, Air. Cond.* (August 1961).

24. J. F. Hoffman, "Well Location and Design," *Heating, Piping, Air Cond.* (August 1963).

25. D. K. Todd, "Ground Water Has To Be Replenished," *Chem. Eng. Progr.* 59, no. 11 (November 1963).

26. R. C. Richter and R. Y. D. Chun, "Artificial Recharge of Ground Water Reservoirs in California," *Proc. Am. Soc. Civil Engrs., J. Irrigation Drainage Div.* 85, no. IR4 (December 1959).

27. G. Thiem, *Hydrologische Methodern* (Leipzig: Gebhart, 1906), p. 56.

28. C. V. Theis, "The Relation Between the Lowering of the Piezometric Surface and the Rate and Duration of Discharge of a Well Using Ground Water Storage," *Trans. Am. Geophys. Union* 16 (1935): 519–524.

29. H. H. Cooper, Jr., and C. E. Jacob, "A Generalized Graphical Method for Evaluating Formation Constants and Summarizing Well-Field History," *Trans. Am. Geophys. Union* 27 (1946): 526–534.

30. J. F. Hoffman, "How Underground Reservoirs Provide Cool Water for Industrial Uses," *Heating, Piping, Air Cond.* (October 1960).

31. N. Buras, "Conjunctive Operation of Dams and Aquifers," *Proc. Am. Soc. Civil Engrs., J. Hydraulics Div.*, 89, no. HY6 (Nobember 1963).

32. F. B. Clendenen, "A Comprehensive Plan for the Conjunctive Utilization of a Surface Reservoir with Underground Storage for Basin-wide Water Supply Development: Solano Project, California." D. Eng. thesis, University of California, Berkeley, Calif. (1959), p. 160.

33. "Ground Water Basin Management," *Manual of Engineering Practice No. 40* (New York: American Society of Civil Engineers, 1961).

34. G. M. Brune, "Trap Efficiency of Reservoirs," *Trans. Am. Geophys. Union* 34 (June 1953): 407–418.

35. R. K. Linsley and J. B. Franzini, *Water Resources Engineering* (New York: McGraw-Hill Book Company, 1964).

36. K. C. Nobe, "Another Look at River Basin Planning," *J. Soil Water Conserv.* 16, no. 5 (September–October 1961).

37. M. E. Scheidt, "Planning for Comprehensive Water Quality Management Programs," presented at Water Resources Workshop of Division of Water Supply and Pollution Control, U.S. Public Health Service, at Sanitary Engineering Center, Cincinnati, Ohio (November 1961).

38. G. F. White, "A Perspective of River Basin Development," *Law and Contemp. Problems*, 22, no. 2 (Spring 1957).

39. A. T. Lenz, "Some Engineering Aspects of River Basin Development," *Law and Contemp. Problems* 22, no. 2 (Spring 1957).

40. A. Maass, et al., *Design of Water Resources Systems* (Cambridge, Mass.: Harvard University Press, 1962).

41. J. B. Stall, "Low Flows of Illinois Streams for Impounding Reservoir Design," *Illinois State Water Survey Bull. No. 51* (1964).

42. V. T. Chow (ed.), *Handbook of Applied Hydrology* (New York: McGraw-Hill Book Company, 1964).

43. G. M. Fair and J. C. Geyer, *Water Supply and Waste Water Disposal* (New York: John Wiley & Sons, Inc., 1954).

4

Water Requirements and Waste Volumes

Every day each of us uses about 5 pints of water for drinking purposes. Other community water uses bring the average total demand to approximately 165 gpcd. Residential, commercial, industrial, and public requirements each represent a component of the demand. Where lawn sprinkling is widespread and prolonged, requirements may be considerably higher than 165 gpcd.[23] Other local conditions may also have a pronounced effect. Caution should therefore be exercised in using representative figures.

In humid regions, municipal water demands are often met with little difficulty. Important exceptions are in evidence, however, where cities have exceedingly large demands combined with complex collection and transportation systems. The arid lands face a considerably different situation. Municipalities must often transport water for considerable distances, and groundwater in many instances becomes the only source of supply.

Demands imposed on our water resources by industry and agriculture are often considerable. The manufacture of a single automobile requires about 100,000 gal of water; some 70,000 gal are used to produce 1 ton of wood pulp for paper manufacturing, and the national average water consumption per ton of steel produced is approximately 65,000 gal. The production of 32 bushels of wheat under optimum conditions may require as much as 650,000 gal of water.

Part of this water is lost to the region through evaporation or transpiration while the remainder is available for various forms of reuse. The reader should understand that these figures are reported here only to provide a guideline to values of water used for various purposes. The quantities given should not be confused with absolute requirements. For example, the Fontana Division of Kaiser Steel in California draws only about 1500 gal/ton of steel or approximately $\frac{1}{4\frac{1}{3}}$ of that drawn by many steel plants in the humid regions.[30] Political decisions, scarcity of water, and other factors can obviously affect the water used for various purposes in any given region.

Population Estimates

The types of population estimates required for the operation and design of water-supply and waste-treatment works are (1) short-term estimates in the range 1–10 years, and (2) long-term estimates of 10–50 years or more. The methods used to make these two kinds of forecasts differ appreciably.

The prediction of future population is at best complex. It should be emphasized that there is no exact solution, even though seemingly sophisticated mathematical equations are often used. War, technological developments, new scientific discoveries, government operations, and a whole host of other factors can drastically disrupt population trends. There is no present way to predict many of these occurrences, and thus their impact cannot be estimated. Nevertheless, population forecasts are exceedingly important and must be made.

4-1

METHODS OF FORECASTING

Both mathematical and graphical methods are used in estimating future populations. Usually the computations or analyses are based on past census records for the area, or on the records of what are considered to be similar communities. These estimates are based primarily on an extension of existing trends. They do not take into consideration factors such as the influx of workers when new industries settle in the area, the loss of residents due to curtailment of military activities, or changes in business or transportation facilities. To optimize the estimate, all possible information regarding anticipated industrial growth, local birth and death rates, government activities, and other related factors should be obtained and used. The local census bureau, the planning commission, the bureau of vital statistics, local utility companies, movers, and the chamber of commerce all are sources of information. An example of the application of utilities information in estimating current populations is the count of residential light meters in Cuyahoga Falls, Ohio. A factor of 3.42 persons per meter was found to give reliable estimates for that area in 1963.

The most widely employed mathematical or graphical methods for extending past municipal population data are (1) arithmetical progression or uniform growth rate, (2) constant percentage growth rate, (3) decreasing rate of increase, (4) graphical extension, (5) graphical comparison with the growth rates of similar and larger cities, (6) the ratio method based on a comparison of the local and national population figures for past census years,[9] and (7) the use of mathematical trends such as the Gompertz and logistic curves.

4-2

SHORT-TERM ESTIMATING

Short-term estimates (1–10 years) are generally made by arithmetic progression, geometric progression, decreasing rate of increase, or graphical extension. Each of the first of these three procedures is based upon a population growth curve of the type shown on Fig. 4-1. The figure shows that the typical S curve can be considered to consist of three segments, which can be described approximately by a geometric increase, an arithmetic increase, and a decreasing rate of increase.

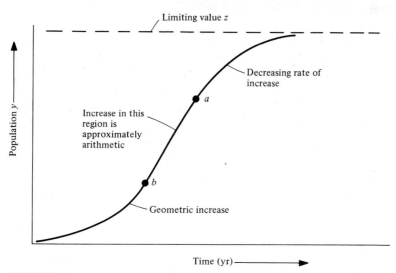

FIGURE 4-1 Population growth curve.

Arithmetical Progression

This method of estimation is based on a constant increment of increase and may be stated as follows:

$$\frac{dY}{dt} = K_u \tag{4-1}$$

where Y = population
t = time (usually years)
K_u = uniform growth-rate constant

If Y_1 represents the population at the census preceding the last census (time t_1), and Y_2 represents the population at the last census (time t_2), then

$$\int_{Y_1}^{Y_2} dY = K_u \int_{t_1}^{t_2} dt$$

Integrating and inserting the limits, we obtain

$$Y_2 - Y_1 = K_u(t_2 - t_1)$$

Therefore,

$$K_u = \frac{Y_2 - Y_1}{t_2 - t_1} \qquad (4\text{-}2)$$

We use Eq. 4-2 to write an expression for short-term arithmetic estimates of population growth:

$$Y = Y_2 + \frac{Y_2 - Y_1}{t_2 - t_1}(t - t_2) \qquad (4\text{-}3)$$

where t represents the end of the forecast period.

Constant-Percentage Growth Rate

For equal periods of time this procedure assumes constant growth percentages. If the population increased from 90,000 to 100,000 in the past 10 years, it would be estimated that the growth in the ensuing decade would be to 100,000 + 0.11 × 100,000 or 111,000. Mathematically this may be formulated as

$$\frac{dY}{dt} = K_p Y \qquad (4\text{-}4)$$

where the variables are defined as before, except that K_p represents a constant percentage increase per unit time. Integrating this expression and setting the limits yields

$$K_p = \frac{\log_e Y_2 - \log_e Y_1}{t_2 - t_1} \qquad (4\text{-}5)$$

A short-term geometric estimate of population growth is thus given by

$$\log_e Y = \log_e Y_2 + K_p(t - t_2) \qquad (4\text{-}6)$$

Base 10 logs may be used in Eqs. 4-5 and 4-6.

Decreasing Rate of Increase

Estimates made on the basis of a decreasing rate of increase assume a variable rate of change. Mathematically, the decreasing rate of increase may be formulated as

$$\frac{dY}{dt} = K(Z - Y) \qquad (4\text{-}7)$$

where Z is the saturation or limiting value which must be estimated, and the other variables are as defined before. Then

$$\int_{Y_1}^{Y_2} \frac{dY}{Z-Y} = K_D \int_{t_1}^{t_2} dt$$

and upon integration,

$$-\log_e \frac{Z-Y_2}{Z-Y_1} = K_D(t_2 - t_1)$$

Rearranging yields

$$Z - Y_2 = (Z - Y_1)e^{-K_D \Delta t}$$

Then, subtracting both sides of the equation from $(Z - Y_1)$,

$$(Z - Y_1) - (Z - Y_2) = (Z - Y_1) - (Z - Y_1)e^{-K_D \Delta t}$$

and

$$Y_2 - Y_1 = (Z - Y_1)(1 - e^{-K_D \Delta t}) \qquad (4\text{-}8)$$

Equation 4-8 may be used to make short-term estimates in the limiting region.

4-3

LONG-TERM POPULATION PREDICTIONS

Long-term predictions are usually made by graphical comparison with growth rates of similar and larger cities, or by selecting a mathematical trend such as the Gompertz or logistic curve and fitting this to the observed data.[1] These predictions are usually less reliable than short-term estimates, since there is considerable opportunity for unpredictable factors to affect the anticipated trend.

Graphical Comparison with Other Cities

The population–time curve of a given community can be extrapolated on the basis of trends experienced by similar and larger communities. Population trends are plotted in such a manner that all the curves are coincident at the present population value of the city being studied (see Fig. 4-2). The cities selected for comparison should not have reached the reference population value too far in the past since the historical periods involved may be considerably different. It should be understood that the future growth of a city may digress significantly from the observed development of communities of similar size. In making the final projection, consideration should be given to conditions which are anticipated for the growth of the community in question. With the exercise of due caution, this method should give reasonable results.

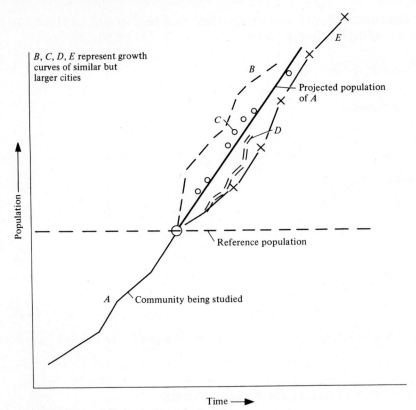

B, C, D, E represent growth curves of similar but larger cities

Projected population of *A*

Reference population

A   Community being studied

Population

Time ⟶

FIGURE 4-2 Graphical prediction of population by comparison.

Mathematical Curve Fitting

Generally, mathematical curve fitting has its greatest utility in the study of large population centers, or nations. The Gompertz curve and the logistic curve are both used in establishing long-term population trends. Both of these curves are S-shaped and have upper and lower asymptotes, with the lower asymptotes being equal to zero.

The logistic curve in its simplest form is[1]

$$Y_c = \frac{K}{1 + 10^{a+bX}} \tag{4-9}$$

where Y_c = ordinate of the curve
X = time period in years (10-year intervals are frequently chosen)
K, a, b = constants

To fit this curve, three years, represented by X_0, X_1, and X_2, each equidistant from the other in succession, must be selected. These years are chosen so that one will be near the earliest recorded population for the area, one near the middle, and one near the end of the available record. The fitted curve will pass through the values of Y_0, Y_1, and Y_2 which are associated with X_0, X_1, and X_2. The origin on the X axis is at the year indicated by X_0. The number of years from X_0, to X_1 or from X_1 to X_2 is designated as n. The constants are then obtained by using the following equations:

$$K = \frac{2 Y_0\, Y_1\, Y_2 - Y_1^2 (Y_0 + Y_2)}{Y_0\, Y_2 - Y_1^2} \tag{4-10}$$

$$a = \log \frac{K - Y_0}{Y_0} \tag{4-11}$$

$$b = \frac{1}{n} \left[\log \frac{Y_0 (K - Y_1)}{Y_1 (K - Y_0)} \right] \tag{4-12}$$

Substitution of these values in Eq. 4-9 permits determination of Y_c for any desired value of X.

4-4

POPULATION DENSITIES

A knowledge of the total population of a region will allow estimates of the total volume of water or wastewater to be considered. In order to design conveyance systems for these flows, additional information regarding the physical distribution of the population to be served must be used. A knowledge of the population density as well as of the total population is important. Population densities may be estimated from data collected on existing areas and from zoning master plans for undeveloped areas. Table 4-1 may be used as a guide if more reliable local data are not available.

TABLE 4-1
GUIDE TO POPULATION DENSITY

Area Type	Number of Persons per Acre
Residential—single-family units	5–35
Residential—multiple-family units	30–100
Apartments	100–1000
Commercial areas	15–30
Industrial areas	5–15

Water Requirements

A reliable estimate of the quantity of water required for municipal, industrial, and agricultural uses in a region must be made before systems can be designed to transport, process, or distribute these flows. It has been estimated that by

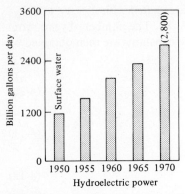
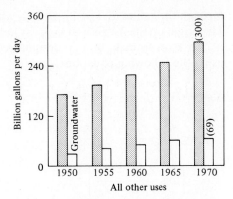

FIGURE 4-3. Graphs showing trends in use of water for hydroelectric power and in all other withdrawal uses combined, 1950–1970. [Source: C. R. Murray and E. B. Reeves, "Estimated Use of Water in the United States in 1970," *U.S. Geological Survey Circular No. 676* (Washington, D.C.: 1972).]

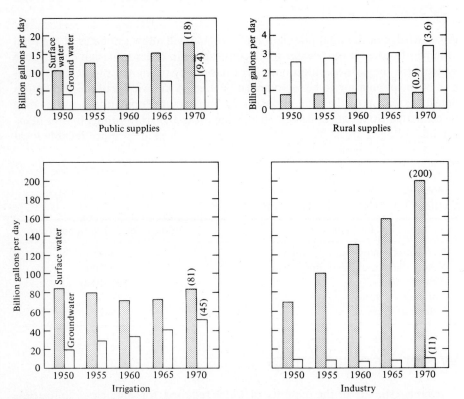

FIGURE 4-4. Graphs showing trends in use of water for public supplies, rural supplies, irrigation, and self-supplied industry, 1950–1970. [Source: C. R. Murray and E. B. Reeves, "Estimated Use of Water in the United States in 1970," *U.S. Geological Survey Circular No. 676* (Washington, D.C.: 1972).]

1980 the demands on the national water resources will nearly double those of 1954, and that by the year 2000 demands will be triple those experienced in 1954.

During 1965, the nation's freshwater withdrawals for all purposes averaged about 269 billion gallons per day (bgd).[32] This included a substantial reuse of flows. Of the total amount used, approximately 77 bgd was consumed through evaporation or incorporation into products. It is estimated that by 1985, the total national withdrawal of fresh water will reach 600 bgd, including reuse. Depending upon the nature of the energy–water picture, between 116 and 154 bgd is estimated to be consumed. Figures 4-3 to 4-5 show historic trends in the use of the nation's water resources from 1950 or 1955 to 1970.[31,32]

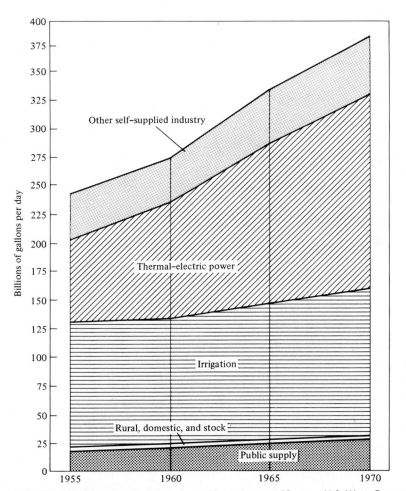

FIGURE 4-5. Historic withdrawal of water for major uses. [Source: U.S. Water Resources Council, "Water Requirements, Availabilities, Constraints, and Recommended Federal Actions—Project Independence" (Washington, D.C.: November 1974).]

TABLE 4-2

ESTIMATED WATER WITHDRAWALS PER CAPITA FOR PUBLIC WATER SUPPLY SYSTEMS (gpcd)

Year	Domestic	Public	Commercial	Industrial	Total
1965	73	20	28	36	157

Source: U.S. Water Resources Council, "The Nation's Water Resources" (Washington, D.C.: Government Printing Office, 1968).

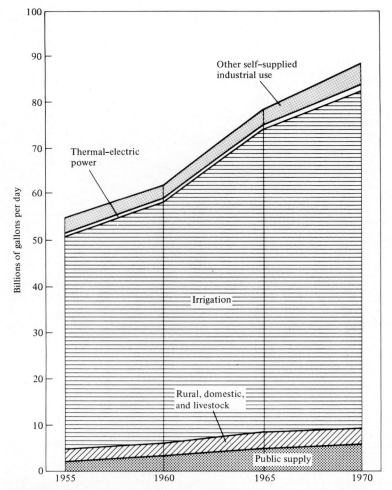

FIGURE 4-6. Historic consumption of water for major uses. [Source: U.S. Water Resources Council, "Water Requirements, Availabilities, Constraints, and Recommended Federal Actions—Project Independence" (Washington, D.C.: November 1974).]

Table 4-2 indicates the estimated amounts of water withdrawn from public water supply systems on a per capita basis for 1965.[23] The national average of the total is expected to increase about 8% during the period 1965–2020.

Surveys of water use by the U.S. Geological Survey show that by 1965, thermal electric plant withdrawals, including reuse, exceeded irrigation withdrawals, to become the leading use of water (Fig. 4-5). This shift is a result of the rapid growth of electric demand and the widespread use of once-through systems for cooling.

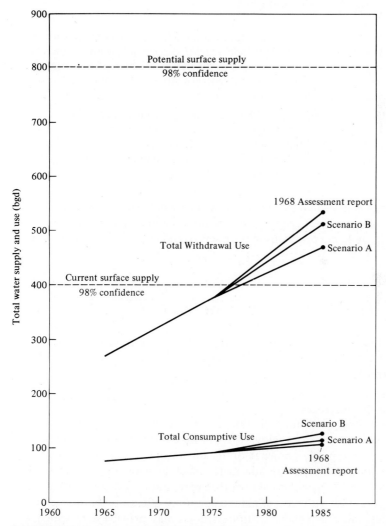

FIGURE 4-7. Current and projected fresh- and saline-water requirements and supplies. [Source: U.S. Water Resources Council, "Water Requirements, Availabilities, Constraints, and Recommended Federal Actions—Project Independence" (Washington D.C.: November 1974).]

While the future indicates an increase in thermal electric facilities, the trend is toward greater use of closed evaporative systems for cooling. This should reduce total withdrawals for this use but result in increased consumption of the water used. In terms of overall consumptive use, however, irrigation stands as the frontrunner (Fig. 4-6).

From a nationwide perspective, there is sufficient water to meet projected needs well beyond 1985. This is illustrated by the Water Resources Council's projections shown on Fig. 4-7. These are based on the 1968 assessment, "The Nation's Water Resources"[23] and on two scenarios for energy development by the Federal Energy Administration in 1974.[32]

> *Scenario A*—A "business as usual" future for 1985, which assumes a world price for oil of $11.00 per barrel (at New York) and that no major conservation actions will be employed to induce reductions in the nation's demand for energy. Also, it assumes that *no* significant additional federal actions will be employed to stimulate domestic production of energy.

> *Scenario B*—An "accelerated" future for 1985, which also assumes a world price for oil of $11.00 per barrel (at New York) and no major conservation actions, but does assume that *some* significant federal actions will be employed to stimulate the domestic production of energy and thereby, by 1985, provide a significant reduction in the dependency upon foreign sources of energy.

The optimisim indicated by Fig. 4-7 should be tempered by a realization that national totals do not reflect geographic or temporal variations. While the overall picture is bright, regional problems may be severe and require detailed consideration.

4-5

FACTORS AFFECTING THE USE OF WATER

Past records and estimated future average water consumption rates indicate a wide range in values across the United States. Some of the factors responsible for the nonuniformity are listed.

Climatic Conditions
Lawn-sprinkling, gardening, bathing, and air-conditioning demands are usually greater in warm, dry regions than in humid areas. Regions subject to extreme cold often report significant drafts to prevent freezing of water lines. Definite correlations between climatic factors such as temperature and rainfall have been reported.[4,8] Brock has shown that correlation of the number of days prior to the last rainfall with days of high temperature offers a potential means for forecasting high water demands.[10] Future studies may provide data which will permit estimating the frequency of peak water demands in much the same way as flood and drought frequencies are now determined.

Data published by a task group of the American Water Works Association indicate that precipitation is the climatic factor having the greatest influence on per capita residential consumption.[15] Where summer precipitation exceeds 1.2 in., the area is termed the (f) climatic classification, where summer precipitation is less than 1.2 in., the area is classified as (s). The (f) area is located generally east of the 100th meridian in the United States and has an average domestic consumption of about 50 gpcd.

Economic Conditions

Yarbrough, Linaweaver, and others have demonstrated that water use is a function of the economic status or living standard of the consumer.[3,4] High-priced residential dwelling units, for example, will normally show rates which are significantly greater than those for medium- and low-priced units.

Howe and Linaweaver have reported that domestic demands are relatively inelastic with respect to price.[27] Their studies also show that lawn-sprinkling demands are elastic with respect to price, but to a lesser extent in arid regions. Maximum day sprinkling demands (very important to system design) are relatively elastic in humid regions but are inelastic in the west.

Composition of the Community or Region

The type and magnitude of residential, commercial, and industrial development in an area will have a pronounced effect on local water-use rates. Industrial requirements often are exceedingly large. The per capita requirements of a region endowed with large-scale industrial development might therefore be strikingly affected, provided that industrial water is supplied by the municipality.

Water Pressure

Rates of water use increase with increases in pressure. This result is due partly to leakage and partly to the increased volumes of flow through fixture units per unit of time. For example, the water-use rate has been known to increase by as much as 30 % for a 20-pounds per square inch (psi) change in line pressure. Pressures in excess of those required for satisfactory service should be avoided whenever possible.

Cost of Water

There is a tendency toward conservation when costs are high. Babbitt, Doland, and Cleasby have indicated the following relationship based on data by Seidel and Baumann[5,6]:

$$C = 21 - 10 \log Q \tag{4-13}$$

where C = cost, dollars/1000 ft^3

Q = rate of water used, thousands of gallons per year

Other factors affecting water-use rates are metering, water quality, presence or absence of sewers, air conditioning, and management.

4-6

RESIDENTIAL WATER USE

The growth and water-use patterns of communities should be reviewed annually for trends if water-supply systems are to keep abreast of continually increasing demands. Evidence in recent years indicates that the average per capita consumption, the peak daily consumption, and the peak hourly consumption are all on the increase.[11] Each of these demands is of special interest to the designer. The first, average consumption, is used in estimating total water requirements and in designing storage works. Peak daily and peak hourly rates are the basis for designing distribution works.

Residential water-use rates are continually fluctuating. There are variations from hour to hour, day to day, and season to season. Average daily winter consumption is only about 80% of the annual daily average, while summer consumption averages are about 25% greater than the annual daily average. Figure 4-8 compares a typical winter day with a typical maximum summer day

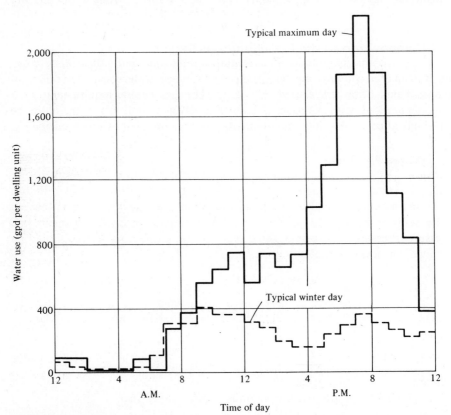

FIGURE 4-8. Daily water-use patterns, maximum day and winter day. (Source: Residential Water-Use Research Project of The Johns Hopkins University and The Office of Technical Studies of the Architectural Standards Division of the Federal Housing Administration.)

in Baltimore, Maryland. Note the hourly fluctuation and the tendency toward two peaks. Studies by Wolff indicate that hydrographs of systems serving predominantly residential communities generally show two peak rates, the first, between the hours of 7 A.M. and 1 P.M., the second in the evening between 5 and 9 P.M.[7] During the summer when sprinkling demands are high, the second peak is usually the greatest, while during the colder months or during periods of high rainfall, the morning peak is commonly the larger of the two.

In Fig. 4-9 the effects of lawn-sprinkling demands are strikingly demonstrated. This figure shows the effects of rainfall on two different days—one a hot,

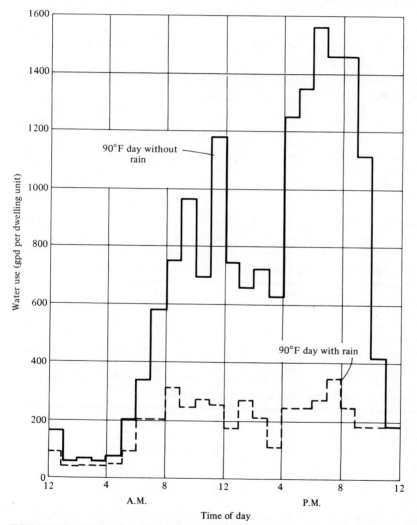

FIGURE 4-9. Daily water-use patterns in R-6 area: maximum day and minimum day. (Source: Residential Water-Use Research Project of the Johns Hopkins University and the Office of Technical Studies of the Architectural Standards Division of the Federal Housing Administration.)

dry day with no antecedent rainfall for 4 days, and the other a day during which rainfall in excess of one inch was recorded. Lawn sprinkling has been found to represent as much as 75% of total daily volumes and as high as 95% of peak hourly demands where large residential lots are involved.[4]

A 1963 report on the residential Water Use Research Project carried on at The Johns Hopkins University has produced some interesting results obtained from gaging programs on four strata of residential areas in Towson, Maryland.[4] This study indicates that an inverse relationship exists between the number of persons per dwelling unit and the average daily per capita use. Values ranging from 85 gpcd for two persons per dwelling to 47 gpcd for five persons per dwelling

TABLE 4-3
WATER-USE DATA

Area Studied	Net Lot Size (ft²)	Consumption for Given Period (gal/day/dwelling unit)			Ratios	
		Average Annual	Maximum Day	Peak Hour	Maximum Day to Average Annual	Peak Hour to Average Annual
Donnybrook Apartments	1,100	156	211	532	1.35	3.42
Country Club Park	7,000	227	657	1,380	2.90	6.07
Pine Valley	7,600	263	1,090	2,380	4.14	9.05
Hampton	28,000	332	1,380	4,100	4.16	12.1

Domestic Water-Use Data (sprinkling use not included)

Area Studied	Net Lot Size (ft²)	Average Annual	Maximum Day	Peak Hour	Maximum Day to Average Annual	Peak Hour to Average Annual
Donnybrook Apartments	1,100	144	183	490	1.27	3.40
Country Club Park	7,000	187	213	414	1.14	2.21
Pine Valley	7,600	214	250	565	1.17	2.64
Hampton	28,000	202	290	860	1.44	4.26

Sprinkling-Use Data (domestic use not included)

Area Studied	Net Lot Size (ft²)	Average Annual	Maximum Day	Peak Hour	Maximum Day to Average Annual	Peak Hour to Average Annual
Donnybrook Apartments	1,100	12	67	303	5.58	25.2
Country Club Park	7,000	40	470	1,120	11.8	28.0
Pine Valley	7,600	49	880	2,080	18.0	42.5
Hampton	28,000	130	1,180	3,900	9.08	30.0

Source: Residential Water Use Research Project of The Johns Hopkins University and the Office of Technical Studies of the Architectural Standards Division of the Federal Housing Administration.

are reported. The relationship is reported to hold regardless of the type of housing or season of the year. The overall average annual domestic use for the study areas was reported as 56 gpcd. This figure does not include sprinkling demands. The generally reported range in domestic requirements is 20–90 gpcd.

Table 4-3 indicates the measured water-use rates for the Towson, Maryland, area. The data clearly illustrate the tremendous impact of lawn sprinkling on total usage. For the Hampton area, the annual sprinkling load was found to be 30% of the total annual load, but during the peak hour this increased to 95% of the total usage. The high 12.1 : 1 ratio of peak hourly rate to average annual rate for this area is due primarily to the sprinkling load. Figures such as these point to the conclusion that where heavy peak-hour consumption is imposed on a distribution system, this flow and not the fire flow plus maximum daily consumption may govern the design.[13] An illustration of this is given in Sec. 4-7.

On the basis of findings resulting from a study of 41 homogeneous residential areas ranging in size from 44 to 410 dwelling units in various climatic regions in the United States, Linaweaver, Geyer, and Wolff have developed relationships for the design maximum daily demand and the design peak hourly demand.[24]

The basic equation for the expected demand in a metered and sewered residential area having a number of consumers is[24]

$$\bar{Q} = \bar{Q}_d + 0.6ca\,\bar{L}_s(\bar{E}_{pot} - \bar{P}_{eff}) \text{ with } \bar{Q} \geq \bar{Q}_d \tag{4-14}$$

where

$\bar{Q}$ = expected average demand for any period, gpd

$\bar{Q}_d$ = expected average domestic (household) use, gpd, which applies for all periods of a day or longer; this can be reliably estimated from a simple function of the average market value of dwellings (Eq. 4-16)

0.6 = coefficient to adjust for the difference between actual and potential evapotranspiration from lawns

c = constant to adjust for units, 2.72×10^4 gal/acre-in. of water

a = number of dwelling units

L_s = average irrigable area in acres per dwelling unit (Eq. 4-17)

$\bar{E}_{pot}$ = estimated average potential evapotranspiration for the period of demand in question, inches of water per day

$\bar{P}_{eff}$ = amount of natural precipitation effective in satisfying evapotranspiration for the period and thereby reducing the requirements for lawn sprinkling, inches of water per day

The first right-hand term of Eq. 4-14 is the estimated nonconsumptive use (water returned to the sewers). The second term is the estimated consumption use.

During high-demand periods, precipitation effects become negligible and the relationship reduces to

$$\bar{Q} = \bar{Q}_d + 0.6caL_s\bar{E}_{pot} \tag{4-15}$$

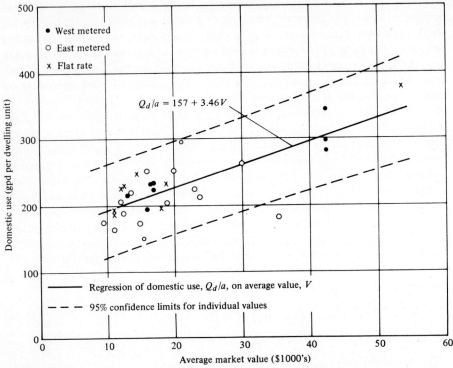

FIGURE 4-10. Domestic (household) water use in areas with public water and public sewers versus average market value of the home. (Source: Residential Water-Use Research Project of The Johns Hopkins University and the Office of Technical Studies of the Architectural Standards Division of the Federal Housing Administration.)

The potential evapotranspiration can be determined from climatological data.[18,19] The overall mean value of $\bar{E}_{pot}$ for the study areas was found to be 0.28 in. of water using Thornthwaite's method.[18,19] The domestic use, $\bar{Q}_d/a$, may be estimated from

$$\frac{\bar{Q}_d}{a} = 157 + 3.46V \tag{4-16}$$

where V is the average market value in $1000 per dwelling unit. A plot of this expression is given in Fig. 4-10. Although the property values given on Fig. 4-10 are for 1963 conditions, the graph can still be expected to yield reasonable values for domestic use if the current property values are deflated to the 1963 base using local indexes. Deflated values would then be used to enter the graph. The average irrigable area, $\bar{L}_s$, is determined by correlation with gross housing density using the relationship

$$\bar{L}_s = 0.803W^{-1.26} \tag{4-17}$$

where W is the gross housing density in dwelling units per acre.

Figures 4-11 and 4-12 relate the number of dwelling units to expected and design values of maximum daily demand and peak hourly demand. These design curves were developed from the basic equation (4-14) and other data obtained from the residential water-use study. The design curves relate to a 95 % confidence interval or a 2.5 % chance of being exceeded. For a complete derivation, the reader is referred to Ref. 24. Examples 4-1 and 4-2 illustrate the use of these curves.

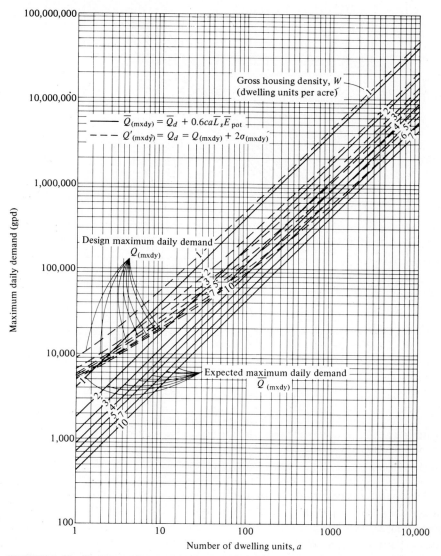

FIGURE 4-11. Design maximum daily demands in relation to the number of dwelling units and the housing density assuming $\bar{E}_{pot} = 0.28$ in. water per day and $\bar{P}_{eff} = 0$. (Source: Residential Water-Use Research Project of The Johns Hopkins University and the Office of Technical Studies of the Architectural Standards Division of the Federal Housing Admininistration.)

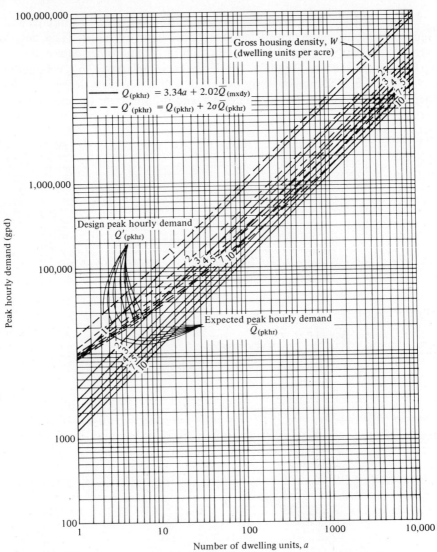

FIGURE 4-12. Design peak hourly demands in relation to the number of dwelling units and the housing density assuming $\bar{E}_{pot} = 0.28$ in. water per day and $\bar{P}_{eff} = 0$. (Source: Residential Water-Use Research Project of The Johns Hopkins University and the Office of Technical Studies of the Architectural Standards Division of the Federal Housing Administration.)

In general, maximum daily use can be considered to be about 180% of the average daily use with values ranging from about 120% to over 400%. Maximum hourly figures have been found to range from about 1.5 to over 12 times the average daily flow. A 1954 study of hourly demands in Baltimore County, Maryland, resulted in the following tabulation of peak-demand ratios shown in Table 4-4.[14]

TABLE 4-4
PEAK-DEMAND RATIOS

Neighborhood	Percent Peak Hour over Average Day
Older neighborhoods, well-settled, small lots	500–600
Newer neighborhoods average-size lots (0.25–0.50 acre)	900
New and old neighborhoods, large lots (0.5 to 3 acres)	1,500

Source: J. B. Wolff and J. F. Loos, "Analysis of Peak Water Demands," *Public Works* (September 1956).

The many factors affecting the use of water preclude any general statement regarding fluctuations in water-use rates on a nationwide basis. General trends and average figures reported here are useful tools, but it should be understood that values from many individual areas vary considerably from the stated mean values. The application of the data presented here must be done with the exercise of qualified judgment and experience by the designer. A careful study must be made of past records of the type and pattern of community water use, the physical and climatic characteristics of the area, expected trends in development, projected population values, and other factors related to water use.

●EXAMPLE 4-1
Consider that an urban area of 400 acres is to be developed with a housing density of 6 houses per acre. Find the design maximum daily demand for this area.

○*Solution*
The number of dwelling units a is equal to $400 \times 6 = 2400$. Follow the curve marked 6.0 Density-dwelling units per acre (Fig. 4-11) and read up from $a = 2400$. Project this intersection over to the ordinate and read a value of 1.4 mgd for the peak hourly demand.

4-7

FIRE DEMANDS

Fire-fighting demands must be considered in any municipal water-system design. Annual volumes required for fire purposes are small, but during periods of need the demand may be exceedingly large and in many cases may govern the design of distribution systems, distribution storage, and pumping equipment. Quantities of water required for fire fighting in high-value community districts are recommended by the National Board of Fire Underwriters (American Insurance Association).[12]

Fire-fighting requirements for residential areas vary from 500 to 3000 gpm, the required rate being a function of population density. Hydrant pressures should generally exceed 20 psi where modern motor pumpers are used; otherwise, pressures in excess of 100 psi might be required. If recommended fire flows cannot be maintained for the indicated time periods, community fire-insurance rates may be adjusted upward.

The coincident draft during fire fighting is usually considered to be equal to the maximum daily demand since the probability of the maximum rate of water usage for community purposes occurring simultaneously with a major conflagration is slight.

4-8

COMMERCIAL WATER USE

The modern community generally is required to serve a variety of commercial water users along with residential and industrial water demands. In considering commercial requirements, it is important that both the magnitude and the time of occurrence of the peak flow be known. Table 4-5 indicates typical requirements and periods of maximum demand for apartments, motels, hotels, office buildings, shopping centers, laundries, washmobiles, and service stations.[4] Generally, it can be stated that commercial water users do not materially affect peak municipal demands. In fact, peak hours for many commercial establishments tend to coincide with the secondary residential peak period. The cessation of numerous commercial activities at about 6 P.M. precludes the imposition of large demands in the early evening when sprinkling demands are often high.

The commercial peak-hour demands given in Table 4-5 are compared to peak-hour demands for a typical individual residence on a 1-acre lot. The data show that maximum commercial needs are considerably less important than peak sprinkling demands in determining peak loads on a distribution system subject to heavy sprinkling loads. An average figure of 20 gpcd is normally considered representative of commercial water consumption. The range is normally reported as 10–130 gpcd.

4-9

INDUSTRIAL WATER USE

As indicated in Fig. 4-4, industrial water use is rapidly increasing. An interesting attribute of this use is the relatively small quantity of water consumed compared to the amount used in the plant operation.[17] Also of interest is the fact that approximately 80% of the industrial water demand is imposed by only about 5% of the industries, whereas nearly 70% of the industrial plants use less than 2% of the total requirement. Major water users are the steel, petroleum products,

TABLE 4-5
COMMERCIAL WATER USE

	Unit	Average Annual Demand (gpd)	Maximum Hourly Demand Rate (gpd)	Hour of Peak Occurrence	Ratio Maximum Hourly to Average Annual	Average Annual Demand per Unit	Ratio Maximum Hourly Demand to R-40 Demand[a]
Miscellaneous residential							
Apartment building	22 units	3,430	11,700	5–6 P.M.	3.41	156 gpd/unit	2.2 : 1
Motel	166 units	11,400	21,600	7–8 A.M.	1.89	69 gpd/unit	4.0 : 1
Hotels							
Belvedere	275 rooms	112,000	156,000	9–10 A.M.	1.39	407 gpd/room	29 : 1
Emerson	410 rooms	126,000				307 gpd/room	
Office buildings							
Commercial Credit	490,000 ft^2	41,400	206,000	10–11 A.M.	4.89	0.084 gpd/ft^2	38 : 1
Internal Revenue	182,000 ft^2	14,900	74,700	11–12 A.M.	5.01	0.082 gpd/ft^2	14 : 1
State Office Building	389,000 ft^2	27,000	71,800	10–11 A.M.	2.58	0.070 gpd/ft^{2b}	13 : 1
Shopping centers							
Towson Plaza	240,000 ft^2	35,500	89,900	2–3 P.M.	2.50	0.15 gpd/ft^2	17 : 1
Hillendale	145,000 ft^2	26,000				0.18 gpd/ft^2	
Miscellaneous commercial							
Laundries							
Laudromat	10 8-lb washers	1,840	12,600	11–12 A.M.	6.85	184 gpd/washer	2.3 : 1
Commercial	Equivalent to 10 8-lb washers	2,510	16,200	10–11 A.M.	6.45	251 gpd/washer equivalent	3.0 : 1
Washmobile	Capacity of 24 cars per hour	7,930	75,000	11–12 A.M.	9.46	330 gpd per car per hour of capacity	14 : 1
Service station	1 lift	472	12,500	6–7 P.M.	26.5	472 gpd/lift	2.3 : 1

[a] Lot type R-40 (1 acre) peak hourly demand for single service is 5400 gpd.
[b] Exclusive of air conditioning.

Source: Residential Water Use Research Project of The Johns Hopkins University and the Office of Technical Studies of the Architectural Standards Division of the Federal Housing Administration.

wood pulp and paper, coke, and beverage industries. Table 4-6 presents some typical industrial water requirements.

As can be seen from the table, the quantities of water used by industry vary widely. They are also affected by many factors, such as cost and availability of water, waste-disposal problems, management, and the type of process employed. Individual studies of the water requirements of a specific industry should therefore be made for each location. The values in Table 4-6 should serve only as approximations.

TABLE 4-6
INDUSTRIAL WATER REQUIREMENTS

Product	Water Requirement
Milk, dairy	340 gal/1000 lb raw milk
Woolens	140,000 gal/ton
Thermoelectricity	80 gal/kWh
Coke	3600 gal/ton
Steel	65,000 gal/ton
Oil refining	770 gal/bbl

Industrial loads in a particular area may impose a draft on the distribution system in excess of that which a domestic population occupying the same area would impose. Accurate information relative to industrial location and requirements is therefore of considerable importance. The time distribution of the load is a further requirement. Future industrial demands for a region may be estimated on the basis of proposed industrial zoning and by considering the type of industries most likely to develop in the area. Accurate projected requirements are not easily secured, however, as there is often little reliable information on land-area requirements for various industries; nor are there consistent data on water requirements per unit of product.

In many instances industries develop their own water-supply systems and under these circumstances impose no demand on the local municipal system except possibly for the source.

4-10

AGRICULTURAL WATER USE

Total annual water requirements for agricultural use are exceedingly large. In many areas of the country, particularly in the arid regions, irrigation-water requirements comprise the bulk of the total developed water supply. This is clearly shown by Fig. 4-6.

Quantities of water required for agriculture vary widely based on the type of crop being irrigated, the distribution of precipitation in the region, and other related factors. In general, irrigation waters are supplied separately from municipal requirements and are therefore of no importance in designing municipal waterworks. They are mentioned here because they may be an important factor in multipurpose water-development projects and cannot be neglected when considering the appropriation of water in an area. For example, in designing combined storage works which will serve municipal, power, and irrigation requirements, reliable estimates of total agricultural requirements and periods of need are a prime requirement. Detailed information on irrigation requirements and practices is readily available in the literature. Table 4-7 indicates the general order of magnitude of quantities of water used for irrigation in 17 Western states.

4-11

WATER REQUIREMENTS FOR ENERGY

Water is used to produce energy directly (hydropower), to process other energy-producing resources, and for restoring lands despoiled during mining operations. Water requirements for extraction of coal, oil shale, uranium, and oil gas are not large, although secondary recovery operations for oil are heavy water users.[33] Significant quantities of water may be used in coal slurry pipelines and are needed in the retorting and disposal of spent oil shale. Conversion of coal to synthetic gas or oil to electric power and electric power generation with nuclear energy all require large quantities of water especially for cooling purposes.

In the energy industry, withdrawals for cooling of thermal electric plants is the largest category of water use, totaling approximately 170 bgd in 1970.[33]

Availability of water will influence the location and design of energy-conversion facilities. This is a particularly important consideration in many water-deficient areas of the West which are rich in coal and oil shale. Tables 8 and 9 display estimated unit water requirements for producing energy.[32] It should be noted that there is a considerable range in some estimates, and these values should be used with caution or supported with other information.

Quantities of Wastes

The design of municipal waste-disposal systems must be based on a knowledge of the expected wastewater flows. The time variation of these flows is also important, since sewers, which normally are gravity-flow systems, must be capable of handling the peak loads and must also be able to transport the minimum loads at velocities sufficient to assure cleansing action. The flow in a sewer consists principally of the wastes of the community and groundwater seepage or infiltration, although in the case of combined sewers or where illegal connections are made, stormwater runoff must also be considered.

TABLE 4-7

WATER USED FOR IRRIGATION, BY STATES, 1970

(partial figures may not add to totals because of independent rounding)

State	Acres Irrigated (1000 acres)	Total Water Withdrawn (1000 acre-feet/yr)						Total Water Withdrawn (mgd)					
		Ground-water	Surface Water	Re-claimed Sewage	All Water	Water Consumed[a] (1000 acre-ft/yr)	Convey-ance Loss (1000 acre-ft/yr)	Ground-water	Surface Water	Re-claimed Sewage	All Water	Water Con-sumed[a] (mgd)	Convey-ance Loss (mgd)
Alabama	27	6.0	14	0	20	20	0	5.4	12	0	18	18	0
Alaska	2.2	0.5	0.5	0	1.0	0.9	0	0.4	0.4	0	0.9	0.8	0
Arizona	1,200	4,300	2,700	0	7,000	5,000	270	3,800	2,400	0	6,300	4,500	240
Arkansas	1,100	1,200	260	0	1,400	1,000	100	1,100	230	0	1,300	890	90
California	8,700	18,000	19,000	140	37,000	23,000	5,500	16,000	17,000	120	33,000	20,000	4,900
Colorado	4,600	2,100	12,000	90	14,000	7,400	1,600	1,900	11,000	80	13,000	6,600	1,400
Connecticut	14	0.7	6.2	0	6.9	6.9	0	0.5	5.4	0	5.9	5.9	0
Delaware	17	2.6	0.7	0	3.3	3.3	0	2.2	0.5	0	2.7	2.7	0
Florida	1,700	1,400	1,100	0	2,500	1,500	160	1,300	970	0	2,200	1,300	150
Georgia	150	7.6	45	0	53	53	0	6.6	40	0	47	47	0
Hawaii	160	610	760	64	1,400	840	250	550	680	57	1,300	750	220
Idaho	3,700	2,300	15,000	2.8	17,000	5,200	4,800	2,100	13,000	2.5	15,000	4,700	4,300
Illinois	36	17	7.2	0	24	24	0	15	6.0	0	21	21	0
Indiana	34	20	8.9	0	29	29	0	18	7.4	0	25	25	0
Iowa	54	26	3.8	0	30	30	0	23	3.1	0	26	26	0
Kansas	1,800	3,100	230	0	3,300	2,600	48	2,800	200	0	3,000	2,300	43
Kentucky	25	0.4	7.8	0	8.2	8.2	0	0.4	6.7	0	7.1	6.8	0
Louisiana	670	870	880	0	1,700	1,300	480	770	780	0	1,600	1,100	430
Maine	22	0.2	9.9	0	10	10	0	0.2	8.7	0	8.9	8.8	0
Maryland	16	2.5	4.9	0.2	7.6	7.5	0	2.1	4.3	0.2	6.6	6.6	0
Massachusetts	34	21	44	0	65	49	0	18	40	0	58	43	0
Michigan	100	25	41	0	66	66	0	22	36	0	58	58	0
Minnesota	50	14	9.5	0	23	23	0	12	8.0	0	20	20	0
Mississippi	200	240	180	0	420	210	42	220	160	0	370	190	37

State												
Missouri	180	79	8.2	0	87	62	6.7	70	6.9	0	77	55	5.5
Montana	2,200	71	8,500	0.1	8,600	6,000	2,500	63	7,600	0.1	7,600	5,400	2,200
Nebraska	4,100	3,000	2,300	0	5,300	3,900	1,400	2,700	2,100	0	4,700	3,500	1,300
Nevada	830	430	2,900	4.8	3,400	1,600	1,800	380	2,600	4.2	3,000	1,400	1,600
New Hampshire	3.3	0	3.3	0	3.3	2.5	0	0	2.8	0	2.8	2.0	0
New Jersey	170	63	22	0	85	78	0	56	20	0	76	70	0
New Mexico	1,100	1,500	1,700	25	3,200	1,500	170	1,300	1,500	22	2,800	1,300	160
New York	75	16	15	0	31	31	0	14	13	0	27	27	0
North Carolina	470	56	36	0	92	92	0	50	32	0	82	82	0
North Dakota	74	30	190	0	210	150	66	26	170	0	190	130	59
Ohio	32	11	25	0	35	32	0	9.0	22	0	31	28	0
Oklahoma	620	810	110	0	920	640	22	720	99	0	820	570	20
Oregon	1,900	710	4,700	3.6	5,400	2,600	1,700	630	4,200	3.1	4,800	2,300	1,500
Pennsylvania	35	0.9	11	0	12	12	0	0.8	9.4	0	10	10	0
Rhode Island	3.8	0.5	4.7	0	5.2	3.9	0	0.4	4.1	0	4.5	3.4	0
South Carolina	42	10	22	0	32	32	88	8.9	20	0	29	29	0
South Dakota	150	35	230	0	260	147	88	31	200	0	230	130	79
Tennessee	9.3	1.8	3.6	0	5.4	4.6	0	1.3	2.9	0	4.2	3.8	0
Texas	8,300	8,800	2,800	17	12,000	9,100	540	7,800	2,500	15	10,000	8,100	480
Utah	1,300	470	3,500	58	4,100	2,200	710	420	3,200	52	3,600	2,000	630
Vermont	0.3	0	0.1	0	0.1	0.1	0	0	0.1	0	0.1	0.1	0
Virginia	80	6.1	34	0	40	38	0	5.2	30	0	35	34	0
Washington	1,400	390	5,900	0	6,300	2,500	1,200	350	5,300	0	5,600	2,200	1,000
West Virginia	2.6	0	1.6	0	1.6	1.6	0	0	1.3	0	1.3	1.3	0
Wisconsin	100	37	22	0	60	45	1.2	33	20	0	52	40	0.7
Wyoming	1,700	140	5,900	8.7	6,000	2,600	1,700	130	5,200	7.7	5,400	2,300	1,500
District of Columbia	0	0	0	0	0	0	0	0	0	0	0	0	0
Puerto Rico	91	75	82	0	160	100	47	67	73	0	140	98	42
United States[b]	50,000	51,000	91,000	410	140,000	82,000	25,000	45,000	81,000	370	130,000	73,000	22,000

[a] Excluding conveyance losses by evapotranspiration.
[b] Including Puerto Rico.
Source: C. R. Murray and E. B. Reeves, "Estimated Use of Water in the United States in 1970," *U.S. Geological Survey Circular No. 676* (Washington, D.C.: 1972).

TABLE 4-8

UNIT WATER REQUIREMENTS FOR PRODUCING ENERGY

Energy Source	Standard Unit	Consumption for Water	Water Needed (gal/10⁶ Btu)	Water Uses Considered
Western coal mining	ton	6–14.7 gal/ton	0.25–0.61	Dust control, coal washing
Eastern surface mining	ton	15.8–18.0 gal/ton	0.66–0.75	Dust control, coal washing
Oil shale	barrel	145.4 gal/bbl	30.1	See Table 4-9
Coal gasification	mscf[a]	72–158 gal/mscf	72–158	Process use, cooking use
Nuclear	kWh	0.80 gal/kWh	234.46	Cooling, uranium mining
Oil and gas production	barrel	17.3 gal	3.05	Well drilling, secondary and tertiary recovery
Refineries	barrel	43 gal/bbl	7.58	Process water, cooling water
Fossil-fuel power plants	kWh	0.41 gal/kWh	120.16	Cooling water
Gas processing plants	mscf	1.67 gal/mscf	1.67	Cooling water

[a] Million standard cubic feet.
Source: U.S. Water Resources Council, "Project Independence" (Washington, D.C.: November 1974).

TABLE 4-9

OIL-SHALE-PRODUCTION WATER REQUIREMENTS FOR A 100,000 bpd SURFACE MINE PLANT

Production Processes	Gallons per Minute	Acre-feet per Year
Processed shale disposal	4,500	7,245
Shale-oil upgrading	2,300	3,703
Power requirements	1,100	1,771
Retorting	800	1,288
Mining and crushing	550	886
Revegetation	220	354
Sanitary use	30	48
Associated urban	900	1,449
TOTAL	10,400	16,744

Source: U.S. Water Resources Council, "Project Independence" (Washington, D.C.: November 1974).

4-12

RESIDENTIAL WASTEWATER FLOWS

Figure 4-13 gives a comparison of water use and wastewater flow on days without lawn sprinkling. The data are from the Pine Valley Subdivision in Baltimore County, Maryland.[16] Domestic sewage flows are highly variable throughout the day and, as in the case of the hydrograph of water use, two distinct peaks have been observed. The primary peak takes place in the morning hours and the secondary peak occurs about dinner time and maintains itself during the evening hours. Extraneous flows resulting from infiltration or storm runoff tend to distort the basic hydrograph shape. Infiltration rates generally

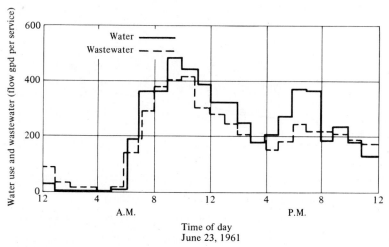

FIGURE 4-13. Comparison of water-use and wastewater flow on days when little sprinkling occurred. (Source: Residential Water-Use Research Project of The Johns Hopkins University and the Office of Technical Studies of the Architectural Standards Division of the Federal Housing Administration.)

tend to gradually increase the total daily volume but do not ordinarily alter the twin-peaked character of the hydrograph. Storm runoff which enters the system may impose almost instantaneous changes; if the quantity is large, the entire characteristic of the hydrograph may be changed. Estimation of the various components of the flow is essential for design purposes. A 1963 study by Lentz of wastewater flows in communities in California, Florida, Missouri, and Maryland provides considerable useful information regarding residential flows, and the components of these flows.[16,25,29]

Average Rates of Flow

Lentz and Linaweaver have shown that when residential water is not being used for consumptive purposes (principally lawn sprinkling) and when infiltration and exfiltration do not produce large flow components, the wastewater flow is essentially equal to the water use.[4,16] Thus average daily water-use rates which do not reflect sprinkling demands can be used to estimate annual average domestic wastewater flows. It is generally reported that about 60–70% of the total water supplied becomes wastewater. Figure 4-14 shows a derived relationship between average per capita waste flow and average assessed valuation of property. This relationship was developed by Lentz, from data on the areas mentioned previously.[16]

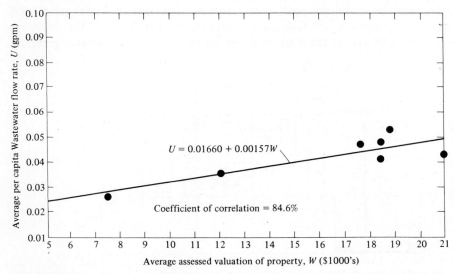

FIGURE 4-14. Average per capita domestic wastewater flow rate versus average assessed valuation of property. (Source: Residential Water-Use Research Project of The Johns Hopkins University and the Office of Technical Studies of the Architectural Standards Division of the Federal Housing Administration.)

Maximum and Minimum Rates of Flow

In 1963, Lentz showed that maximum per capita rates of wastewater flow could be successfully estimated by using an extreme value frequency distribution.[16,25] The method developed for estimating long-term maximum domestic flow rates is carried out in five steps.[16,25] First, an initial estimate of a base period maximum flow is made. This is defined as Gumbel's modal maximum flow and depends on economic stratum of the community and geometric character of the sewer system. Second, an incremental maximum flow is estimated. This is a function of population size and period for which the estimate is made. The third step is to combine the value of the first two steps, and the fourth and fifth steps make adjustments for statistical errors and random occurrence of larger maximums associated with longer return periods.

The analytical solution is complex and is not presented here. Instead, a summary of Lentz and Geyer's graphical procedure for estimating design domestic wastewater flow rates follows.[16,25]

The graphical procedure makes use of Figs. 4-15 through 4-18. The methodology is as described in Appendix B, pp. 97–99, of Ref. 25. The study by Lentz and Geyer was supported by the Federal Housing Administration Technical Studies program and is based on the use of the Extreme value techniques developed in Ref. 16.

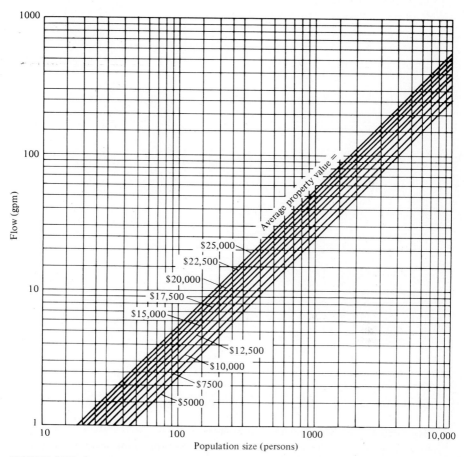

FIGURE 4-15. Long-term average domestic wastewater flow rate versus population size as a function of average assessed valuation of property. (Source: Residential Water-Use Research Project of The Johns Hopkins University and the Office of Technical Studies of the Architectural Standards Division of the Federal Housing Administration.)

Long-Term Daily Average Wastewater Flow Rate

Enter Fig. 4-15 using the known values of population size and property value. Read the long-term daily average domestic wastewater flow as the ordinate. Note that current property values should be deflated to the 1963 level before using the graph. Local real estate indexes should be used for making the adjustment.

Minimum Daily Flushing Flow Rate

Enter Fig. 4-16 with the given population size. Project upward to an intersection with the appropriate *solid* average-property-value line. Project horizontally to determine the minimum daily flushing flow.

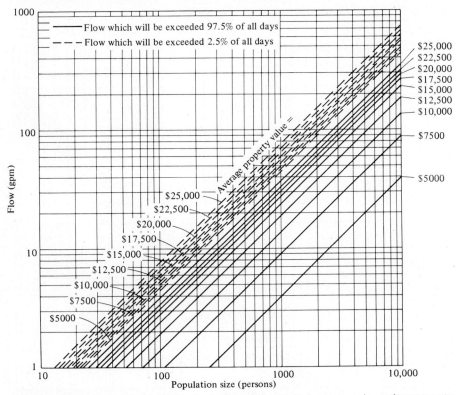

FIGURE 4-16. Upper and lower long-term 2.5% prediction limits for average domestic wastewater flow rates versus population size as a function of average assessed valuation of property. (Source: Residential Water-Use Research Project of The Johns Hopkins University and the Office of Technical Studies of the Architectural Standards Division of the Federal Housing Administration.)

Routine Daily Maximum Flow Rate

Enter Fig. 4-16 in the same manner as for the minimum daily flushing flow rate but project upward to the appropriate *dashed* average property line. The routine daily maximum flow rate is read as the ordinate.

Long-Term Extreme Daily Maximum Flow Rate

1 Using a knowledge of the topography of the neighborhood to be sewered, estimate the average hydraulic gradients of the main sewer channel (percent).
2 Determine a concentration index, C, for the neighborhood to be sewered. Lentz describes this as the ratio of the total length of sewer to length of sewer in the main channel.

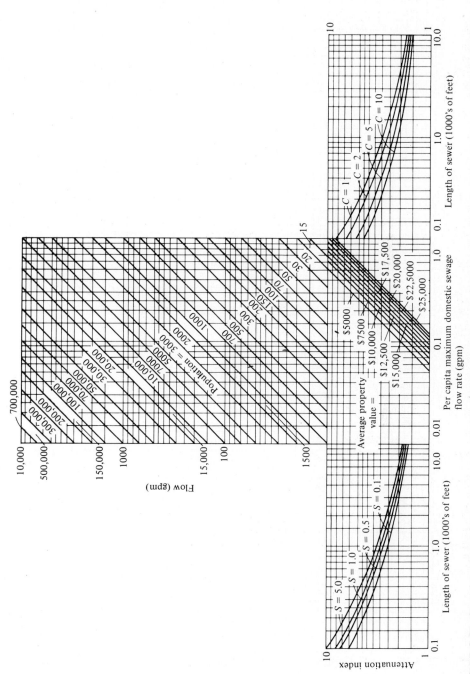

FIGURE 4-17. Chart for graphical determination of Gumbel's modal maximum domestic wastewater flow. (Source: Residential Water-Use Research Project of the Johns Hopkins University and the Office of Technical Studies Division of the Architectural Standards Division of the Federal Housing Administration.)

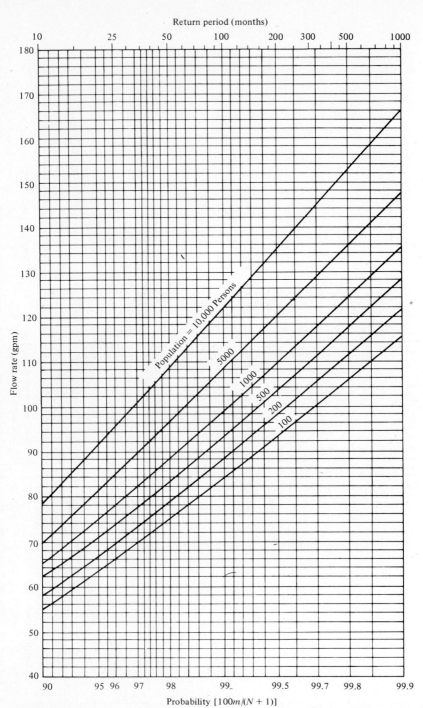

FIGURE 4-18. Chart for graphical determination of incremental maximum domestic wastewater flow as a function of population size and desired return period. (Source: Residential Water-Use Research Project of The Johns Hopkins University and the Office of Technical Studies of the Architectural Standards Division of the Federal Housing Administration.)

3   Next, an attenuation index is estimated. Lentz defines this as the geometric median of a distribution of day-to-day ratios of maximum to average domestic flow.[25] The index is a function of geometric and physical characteristics of the sewer system. Using the estimated values of S and C, enter Fig. 4-17 at the abscissa corresponding to the sewer length. Either the left or right subgraph can be used, depending on the value of C or S which is closest to unity. An attenuation index is now determined by horizontally projecting the intersection of the sewer-length value with the curve for the desired hydraulic gradient (for $C = 1$), or with the curve for the desired value of C (for $S = 1$). The attenuation index is read as the ordinate.

In the event that both C and S differ significantly from unity, the attenuation index must be computed as[25]

$$A = \exp (1.075 \, L^{-0.29} S^{0.06} C^{-0.15}) \tag{4-18}$$

where A is the attenuation index; L is the length of the longest single channel, thousands of feet; S is the average hydraulic gradient of the longest single channel in the system, percent; and C is the concentration index, the quotient of the total length of sewer in the system divided by the length of sewer in the longest single channel.

4   Project the value of the attenuation index horizontally across the lower central section of Fig. 4-17 to an intersection with the line for the appropriate average property value. Read Gumbel's modal maximum daily *per capita* domestic wastewater flow as the abscissa.

5   Extend the modal maximum value vertically to the upper section of Fig. 4-17 to an intersection with the value of total population. Project horizontally and read Gumbel's modal daily maximum domestic flow rate for the entire neighborhood as the ordinate. Record this value.

6   Select the desired return period for estimating a long-term daily extreme maximum daily flow rate.

7   Enter Fig. 4-18 using the selected return period. At the intersection of the return period value with the desired population, read Gumbel's incremental maximum flow rate as the ordinate.

8   Add the value obtained in step 7 to that recorded in step 5. This sum is the desired long-term daily extreme maximum domestic wastewater flow rate. This value has a conditional probability of 85.5% of not being exceeded during the given return period.[25]

Table 4-10 can also be used as a guide for estimating maximum and minimum wastewater flows.

Infiltration and Exfiltration

Infiltration and exfiltration are both functions of height of the groundwater table in the vicinity of the sewer, type and tightness of sewer joints, and soil type. Exfiltration is undesirable since it may tend to pollute local groundwaters, while infiltration has the effect of reducing the capacity of the sewer for conveying the waste flows for which it was designed. If the sewer is well above the ground-

TABLE 4-10

RESIDENTIAL WASTEWATER FLOWS AS RATIOS TO THE AVERAGE

Description of Flow	Ratio to the Average
Maximum daily	2.25 : 1
Maximum hourly	3 : 1
Minimum daily	0.67 : 1
Minimum hourly	0.33 : 1

water table, infiltration will occur only during or after periods of precipitation when water is percolating downward through the soil. Where groundwater tables are high and sewers are not tight, infiltration rates in excess of 60,000 gpd/mi of sewer might be experienced. Rates of 3500–5000 gpd/mi/24 hr for 8-in. pipe, 4500–6000 for 12-in. pipe, and 10,000–12,000 for 24-in. pipe represent the range in which the greater number of specifications fall.[21] Common practice is to design for the peak-design rate of wastewater flow plus 30,000 gpd infiltration per mile of sewer and house connections.[20] This allowance represents average conditions and should be revised by the designer as required on the basis of the physical characteristics of the area and the type of pipe joint to be used.

Storm Runoff

Except for large combined sewers, storm runoff should be excluded from the sewerage system. Storm runoff may enter at manholes or through illicit roof drains connected to the sanitary system. Quantities of flow that enter in this manner vary with the degree of enforcement of regulations and the types of preventive measures that are taken. The American Society of Civil Engineers reports that test on leakage through manhole covers show that 20–70 gpm may enter a manhole cover submerged by 1 in. of water.[20] Rates of this magnitude may be considerably in excess of average wastewater flows. Small sewers can be surcharged easily by a very few roof-drain connections. For example, a rainfall of 1 in./hour on a 1000-ft^2 roof area will contribute flows in excess of 10 gpm. The average domestic sewage flow from a dwelling having this approximate roof area (consider 4 persons) would equal only about 1.5% of this.

●EXAMPLE 4-2

Estimate the maximum hourly, average daily, and minimum hourly residential wastewater flows from an area occupied by 750 people and having an average assessed valuation of $15,000. Consider the length of sewer and house drains equal to 1.3 mi.

○**Solution**

From Fig. 4-14 for $W = 15,000$, find $U = 0.041$ gpm per capita, or $0.041 \times 1440 = 59$ gpcd average flow. Total average daily flow $= 59 \times 750 = 44,200$ gpd.

infiltration $= 30,000 \times 1.3 = 39,000$ gpd

Using Table 4-10, we obtain

$$
\begin{aligned}
\text{maximum hourly flow} &= 39,000 + 44,200 \times 3 = 172,000 \text{ gpd} \\
\text{average daily flow} &= 39,000 + 44,200 \quad\ = 83,200 \text{ gpd} \\
\text{minimum hourly flow} &= 39,000 + 44,200 \times \tfrac{1}{3} = 53,600 \text{ gpd}
\end{aligned}
$$

●**EXAMPLE 4-3**

Estimate the long-term extreme daily flow rate for residential wastewater from a neighborhood of 750 people with an average property value of $15,000. Assume an average gradient of 1 %, a total sewer length of 1.3 mi, and a main channel of 0.8 mi. Consider a return period of 50 months.

○**Solution**

$$
\begin{aligned}
\text{concentration index } C &= 1.3/0.8 = 1.62 \\
\text{length of main channel in feet} &= 0.8 \times 5280 = 4220
\end{aligned}
$$

Using Fig. 4-17, the modal maximum domestic flow is 120 gpm. From Fig. 4-18 for a return period of 50 months, the flow rate is 86 gpm. The combined flow rate is thus $120 + 86 = 206$ gpm. This is the long-term extreme daily maximum flow rate.

4-13

INDUSTRIAL-WASTE VOLUMES

Industrial-waste volumes are highly variable in both quantity and quality, depending principally on the product produced. Since very little water is consumed in industrial processing, large volumes are often returned as waste. These wastes may include toxic metals, chemicals, organic materials, biological contaminants, and radioactive materials. The design of treatment processes for these wastes is a highly specialized operation. Where industrial wastes must be processed in municipal sewage-treatment works, accurate estimates of the time distribution and total volume of the load are necessary, together with a complete analysis of the characteristics of the waste. Under these circumstances, metering and analyzing the industrial waste is normally required and carried out by an engineer when the required information cannot be obtained from the industry or industries involved. For more complete information on volumes as well as characteristics of all types of industrial wastes, the reader is directed to N. L. Nemerow, *Theories and Practices of Industrial Waste Treatment* (Reading, Mass: Addison Wesley Publishing Company, Inc., 1963).

4-14

AGRICULTURAL WASTES

The quantities and character of wastes from agricultural lands are highly variable. The most important pollutants to be found in runoff from agricultural areas are: sediment, animal wastes, wastes from industrial processing of raw agricultural products, plant nutrients, forest and crop residues, inorganic salts and minerals, and pesticides.[22,26,28] Because waste volumes are determined by numerous factors for any given area, it is impossible to indicate any general rules. Nevertheless, the importance of these wastes to any region should not be overlooked. It is clear that large quantities of agricultural drainage are discharged into streams, rivers, and lakes. Regional water quality control will not be a reality unless these wastes are considered along with those of the municipalities and industries. A comprehensive treatment of this topic is beyond the scope of this book, but the student should not minimize the importance of the subject. More detailed information can be found in the references cited.[22,26,28]

PROBLEMS

4-1 Use the following census figures to estimate the 1955 and 1965 populations by assuming arithmetic, geometric, and decreasing rates of increase.

Year	Population (thousands)
1900	6
1910	12
1920	21.1
1930	26.8
1940	30.7
1950	37.1
1960	41.9
1970	43.1

4-2 Obtain census data for your community through 1965. Estimate the 1970 and 1975 population by the methods outlined. Compare the estimated 1965 value with the actual value. Explain the difference.

4-3 The Elephant Butte reservoir has a capacity of 2.64 million acre-ft. How many years would this supply the city of Las Cruces (population 60,000) if evaporation losses are neglected? Assume a use rate of 166 gpcd.

4-4 If the minimum flow of a stream having a 175-mi² watershed is 0.09 cfs/mi², what population can be supplied continuously from the stream? Assume that only distribution storage is provided. Consider a maximum average daily consumption rate of 170 gpcd.

4-5 Estimate the 1970, 1980, and 1990 population of a community for the data of Prob. 4-1 by plotting the data for 1900 onward on arithmetic coordinate paper and extending the curve by eye.

4-6 Of the 18-mgd average consumption in a town of 75,000 population, 2.7 mgd is estimated to be lost through leaks in water mains. Consider these leaks to behave as orifices and determine the amount of water that can be saved by reducing street-main pressure from 60 to 35 psi.

4-7 Estimate the size of water main needed to carry water from a municipal distribution system to a community of 7000 if the fire-fighting requirement is estimated to be 4400 gpm and the coincident draft is 160 gpcd.

4-8 Consider a 620-acre residential area with a proposed housing density of six dwellings per acre. Find the peak-hour requirement.

4-9 Estimate the maximum hourly, average daily, and minimum hourly residential sewage flows from an area serving 1000 persons and having an average assessed property valuation of $20,000. Assume that the length of sewer and house drains equals 2.7 mi and that infiltration occurs.

4-10 Compare the maximum hourly domestic sewage flow from 10 houses (4 persons per house) with the roof drainage from these houses if the roof dimensions are 60 by 35 ft and the rainfall intensity is 2.6 in./hr. Approximately what size of sewer would be required to handle: (a) the domestic flow alone; (b) the combined flow if the pipe is laid on a 1.1% grade?

4-11 What would be the average dilution of a sewage effluent from a community of 60,000 persons if the flow enters a stream having a watershed of 100 mi^2 and a mean annual flow of 0.65 cfs/mi^2?

4-12 Make a comparison between the annual water requirements of a 1000-acre irrigation farm and a city of 50,000 population. Assume an irrigation requirement of 3 acre-ft/yr/acre.

4-13 If 10 acres of farmland was developed for urban housing (four houses per acre), what would be the difference in average annual water requirement after the changeover? Assume that 2.5 acre-ft of water represents the annual agricultural requirement.

REFERENCES

1. F. E. Croxton and D. J. Cowden, *Applied General Statistics* (Englewood Cliffs, N.J.: Prentice-Hall, Inc., 1960).

2. R. H. Marks, "Water: How Industry Can Curb Growing Demands," *Power* (January 1963).

3. K. A. Yarbrough, *J. Am. Water Works Assoc.* (April 1956).

4. F. P. Linaweaver, Jr., "Report on Phase One, Residential Water Use Research Project." The Johns Hopkins University, Department of Sanitary Engineering, Baltimore, Md. (October 1963).

5. H. F. Seidel and E. R. Baumann, *J. Am. Water Works Assoc.* (December 1955): 150.

6. H. E. Babbitt, J. J. Doland, and J. L. Cleasby, *Water Supply Engineering*. 6th ed. (New York: McGraw-Hill Book Company, 1962).

7. J. B. Wolff, "Peak Demands in Residential Areas," *J. Am. Water Works Assoc.* (October 1961).

8. Anonymous, *Water Sewage Works* (September 1958): R116.

9. R. C. Schmitt, "Forecasting Population by the Ratio Method," *J. Am. Water Works Assoc.* 46 (1954): 960.

10. O. A. Brock, "Multiple Regression Analysis of Maximum Day Water Consumption of Dallas, Texas," *J. Am. Water Works Assoc.* 50 (October 1958): 1391.

11. J. B. Wolff, "Forecasting Residential Requirements," presented at the Chesapeake Section, American Water Works Association, Baltimore, Md. (October 1956; mimeograph).

12. "Standard Schedule for Grading Cities and Towns of the United States with Reference to Their Fire Defenses and Physical Conditions" (New York: 1956). National Board Fire Underwriters.

13. K. Carl, Jr., "Extension of Public Service to Suburban Areas—Fire Protection," *J. Am. Water Works Assoc.* 47, no. 10 (October 1955).

14. J. B. Wolff and J. F. Loos, "Analysis of Peak Water Demands," *Public Works* (September 1956).

15. Task Group Report, "Study of Domestic Water Use," *J. Am. Water Works Assoc.* 50 (November 1958): 1408.

16. J. J. Lentz, "Special Report No. 4 of the Residential Sewerage Research Project to the Federal Housing Administration," The Johns Hopkins University, Department of Sanitary Engineering, Baltimore, Md. (May 1963).

17. J. M. Willis, "Forecasting Industrial Requirements," presented at the Chesapeake Section, American Water Works Association, Baltimore, Md. (October 1956; mimeograph).

18. C. W. Thornthwaite and J. R. Mather, "The Water Balance," *Publications in Climatology* VIII, no. 1 (Centerton, N.J.: 1955).

19. C. W. Thornthwaite and J. R. Mather, "Instructions and Tables for Computing Potential Evapotranspiration and the Water Balance," *Publications in Climatology* X, no. 3 (Centerton, N.J.: 1957).

20. "Design and Construction of Sanitary and Storm Sewers," *Manual of Engineering Practice No. 37* (New York: American Society of Civil Engineers, 1960).

21. C. R. Vlezy and J. M. Sprague, *Sewage Ind. Wastes* 27, no. 3 (March 1955).

22. R. C. Loehr, "Animal Wastes—A National Problem," *Proc. Am. Soc. Civil Engrs., J. San. Eng. Div.* 95, no. SA2 (April 1969): 189–221.

23. U.S. Water Resources Council, "The Nation's Water Resources" (Washington, D.C.: Government Printing Office, 1968).

24. F. P. Linaweaver, Jr., J. C. Geyer, and J. B. Wolff, "Residential Water Use," Report V. Phase Two, The Johns Hopkins University, Department of Environmental Engineering Science, Baltimore, Md. (June 1966).

25. J. C. Geyer and J. J. Lentz, "An Evaluation of the Problems of Sanitary Sewer System Design," The Johns Hopkins University, Department of Environmental Engineering Science, Baltimore, Md. (December 1963).

26. "Control of Agriculture-related Pollution." A Report to the President submitted by the Secretary of Agriculture and the Director of the Office of Science and Technology, Washington, D.C. (January 1969).

27. C. W. Howe and F. P. Linaweaver, Jr., "The Impact of Price on Residential Water Demand and Its Relation to System Design and Price Structure," *Water Resources Research* 3, no. 1 (1967).

28. "Agricultural Waste Waters." *Proceedings Symposium on Agricultural Waste Waters, Report No. 10*, Water Resources Center, University of California, Davis, Calif. (April 1966).

29. J. C. Geyer and J. J. Lentz, "An Evaluation of the Problems of Sanitary Sewer System Design," *J. Water Poll. Control Fed.* 38, no. 7 (July 1966).

30. J. Hirshleifer, J. DeHaven, and J. Milliman, *Water Supply—Economics, Technology, and Policy* (Chicago: University of Chicago Press, 1960).

31. C. R. Murray and E. B. Reeves, "Estimated Use of Water in the United States in 1970," *U.S. Geological Survey Circular No. 676* (Washington, D.C.: 1972).

32. U.S. Water Resources Council, "Water Requirements, Availabilities, Constraints, and Recommended Federal Actions—Project Independence" (Washington, D.C.: November 1974).

33. G. H. Davis and L. A. Wood, "Water Demands For Expanding Energy Development," *U.S. Geological Survey Circular No. 703* (Washington, D.C.: 1974).

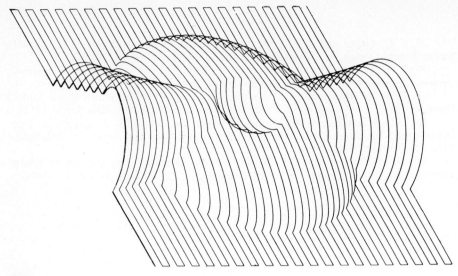

CHAPTER
5

Transportation and Distribution of Water

Transportation is the controlled movement of bulk quantities of water for relatively long distances. Distribution is the delivery of water through complex pipe networks that serve urban centers.

5-1

THE AQUEDUCTS OF ANCIENT ROME

The aqueducts of ancient Rome were usually formed of brick and stone and were covered by an arch to keep the water cool and free of impurity. The invert of the conveyance was normally coated with mortar to seal it. The aqueducts were built above ground where possible, but when necessity required they ran underground, through tunnels, or overhead on arched structures. An idea of the size and quantity of flow carried by several of them is indicated in Table 5-1.

These figures serve to show that large-scale water transportation feats are not new. The quantities of water transported by the Romans are significant even on today's scale. The flow delivered by the aqueduct Appia, for example, would easily serve most present-day American cities of about 30,000 population.

TABLE 5-1

ON DATA SOME OF THE ROMAN AQUEDUCTS DURING THE TIME OF FRONTINUS

Aqueduct	Approximate Date	Length of Conduit (Roman miles ≅ 5000 ft)	Approximate Flow (mgd)	Remarks
Appia	312 B.C.	11	16.7	Mostly underground
Marcia	144 B.C.	62	42.8	54 mi underground
Claudia	A.D. 50	46.5	42.1	36 mi underground
Anio Nova	A.D. 52	58.8	43.3	49 mi underground, passed over arches 109 ft high

[a] Source: P. C. G. Isaac, "Roman Public Works, Engineering," University of Durham, King's College, *Department of Civil Engineering Bull. No. 13* (February 1958).

5-2

THE CALIFORNIA STATE WATER PROJECT

The California State Water Project is probably the best example that can be given to demonstrate man's attempt to solve the problem of unequal distribution of water over the earth's surface. About 70% of California's water supply is generated in the northern part of the state while about 77% of the state's water needs are in the south. In addition, the population of the state is experiencing rapid increases.

The overall objectives of the California water plan are to evaluate the current degree of water development, to estimate ultimate water requirements, and to determine means for providing the necessary water. In 1959 impetus to a segment of this plan was provided when the legislature adopted the Burns–Porter Act to finance the state water project.

This massive transportation undertaking is the largest in the world. It is designed to deliver about 4 million acre-ft of water annually to areas in the Sacramento Valley, San Francisco Bay region, San Joaquin Valley, and Southern California.[3] The project includes 16 reservoirs having a combined storage in excess of 6 million acre-ft, 662 mi of aqueduct, and eight power plants capable of producing more than 5 billion kWh of energy annually. The primary storage facility is the 3.5 million acre-ft Oroville Reservoir, located on the Feather River (see Fig. 5-1). This dam is over 700 ft high, the highest of its type in the United States.

Water released from the Oroville dam flows down the Feather River into the Sacramento River and then on to the Sacramento–San Joaquin delta. From that point it is transported by the California Aqueduct (maximum capacity of 10,000 cfs) to a location south of Bakersfield for pumping across the Tehachapi Mountains. This mountain crossing involves a lift of about 2000 ft, which is

FIGURE 5-1. California's state water project. 1. Upper Feather River Reservoirs; 2. Oroville Facilities; 3. North Bay Aqueduct; 4. Delta Project; 5. South Bay Aqueduct; 6. San Luis Project (joint with United States); 7. Coastal Aqueduct; 8. Castaic Reservoir; 9. Cedar Springs Reservoir; 10. Perris Reservoir. (Courtesy California Department of Water Resources.)

unprecedented in the United States. A series of tunnels 6 mi long and 17–20.5 ft in diameter will be used to convey the water over the higher elevations of the mountains.[4]

South of the Tehachapis the aqueduct follows the south side of Antelope Valley and the Mohave Desert to Cedar Springs Reservoir. It then turns south and passes through a tunnel in the San Bernardino Mountains, down Devils Canyon, and then in an underground pipeline to the Perris Reservoir. About

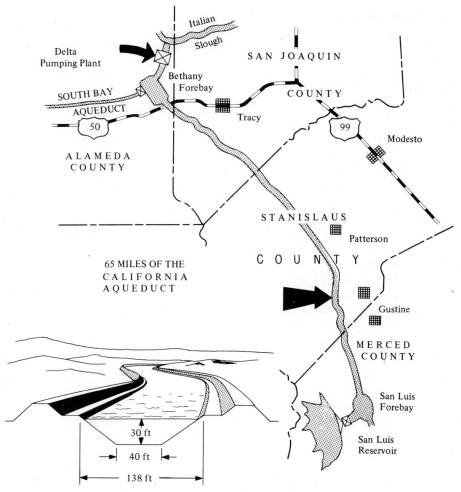

FIGURE 5-2. California Aqueduct. (Courtesy California Department of Water Resources.)

450 mi separate the Delta Pumping Plant from the Perris Reservoir (Fig. 5-1). A cross section of the California Aqueduct is given in Fig. 5-2.

5-3
TYPES OF AQUEDUCTS

Various types of conduit can be used for transporting water.[2] The final selection rests on such factors as topography, head availability, construction practices, economic considerations, and water quality. In addition, the transported water must be safeguarded against pollution by inferior water sources. This is a special problem when open channels or conduits operating at low pressure are used.

Open Channels

Open channels are designed to convey water under conditions of atmospheric pressure. By this definition, the hydraulic gradient and free-water surface are coincident. If the channel is supported on or above the ground, it is classified as a flume. Open channels may be covered or open and may take on a variety of shapes.

The choice of an open channel as the means of conveyance will usually be predicated on suitable topographic conditions which permit gravity flow with minimal excavation or fill. If the channel is unlined, the perviousness of the soil must be considered relative to seepage losses. Other considerations of importance are the potential pollution hazard and evaporative losses.

Many modern open channels are lined with concrete, bituminous materials, butyl rubber, vinyl, synthetic fabrics, or other products to reduce the resistance to flow, minimize seepage, and lower maintenance costs. Flumes are usually constructed of concrete, steel, or timber.

Pipelines

Pipelines are usually built where topographic conditions preclude the use of canals. Pipelines may be laid above or below ground or may be partly buried. Most modern pressure conduits are built of concrete, steel, cast iron, or asbestos cement.

Pipelines used in important transportation systems may require gate valves, check valves, air-release valves, drains, surge control equipment, expansion joints, insulation joints, manholes, and pumping stations. These appurtenances are provided to ensure safe and efficient operation, provide for easy inspection, and facilitate maintenance. Check valves are normally located on the upstream side of pumping equipment and at the beginning of each rise in the pipeline to prevent backflow. Gate valves are often spaced about 1200 ft apart so that the intervening section of line can be drained for inspection or repair and on either side of a check valve to permit its removal for inspection or repair. Air-release valves are needed at the high points in the line to release trapped gases and to vent the line to prevent vacuum formation. Drains are located at low points to permit removal of sediment and allow the conduit to be emptied. Surge tanks or quick-opening valves provide relief from problems of hydraulic surge.

Tunnels

Where it is not practical or economical to lay a pipeline on the surface or provide an open trench for underground installation, a tunnel is selected. Tunnels are well suited to mountain or river crossings. They may be operated under pressure or act as open channels.

5-4

HYDRAULIC CONSIDERATIONS

The analysis of flows in an aqueduct system is carried out through application of basic principles of open-channel and closed-conduit hydraulics. It is assumed that the student already has been exposed to these concepts in his courses in hydraulics or fluid mechanics.

Except for sludges, most flows may be treated hydraulically in the same manner as clean water even though considerable quantities of suspended material are being carried. The Hazen–Williams and Manning formulas are equations used extensively in water transportation problems. The Hazen–Williams formula is used primarily for pressure conduits, while the Manning equation has found its major application in open-channel problems. Both equations are applicable when normal temperatures prevail, a relatively high degree of turbulence is developed, and ordinary commercial materials are used.[5] The Hazen–Williams equation is

$$V = 1.318CR^{0.63}S^{0.54} \qquad (5\text{-}1)$$

where V = velocity, fps
   C = a coefficient, which is a function of the material and age of the conduit
   R = hydraulic radius, ft (flow area divided by the wetted perimeter)
   S = slope of the energy grade line, ft/ft

For circular conduits flowing full, the equation may be restated as

$$Q = 0.279CD^{2.63}S^{0.54} \qquad (5\text{-}2)$$

where Q = flow, mgd
   D = pipe diameter, ft

Some values of C for use in the Hazen–Williams formula are given in Table 5-2. Charts and nomographs which facilitate the solution of the equation are given in most books on hydraulics.

TABLE 5-2

SOME VALUES OF THE HAZEN–WILLIAMS COEFFICIENT

Pipe Material	C
New cast iron	130
5-year-old cast iron	120
20-year-old cast iron	100
Average concrete	130
New welded steel	120
Asbestos cement	140

The Manning equation is stated in the form

$$V = \frac{1.49}{n} R^{2/3} S^{1/2} \tag{5-3}$$

where V = velocity of flow, fps
n = coefficient of roughness
R = hydraulic radius, ft
S = slope of the energy grade line

The equation is applicable as long as S does not materially exceed 0.10. In channels having nonuniform roughness, an average value of n is selected. Where the cross-sectional roughness changes considerably, as in a channel with a paved center section and grassed outer sections, common practice is to compute the flow for each section independently and sum these flows to obtain the total. As in the case of the Hazen–Williams equation, numerous tables, charts, and nomographs are available to permit rapid computations. Values of n for use in Manning's equation are indicated in Table 5-3.

TABLE 5-3
VALUES OF MANNING'S ROUGHNESS COEFFICIENT

Material	n
Concrete	0.013
Cast-iron pipe	0.015
Vitrified clay	0.014
Brick	0.016
Corrugated metal pipe	0.022
Bituminous concrete	0.015
Uniform, firm sodded earth	0.025

Head lost as a result of pipe friction can be computed by solving Eq. 5-1 or 5-3 for S and multiplying by the length of the pipeline. A slightly more direct method is to use the Darcy–Weisbach equation.

$$h_L = f \frac{LV^2}{D2g} \tag{5-4}$$

where h_L = head loss
L = pipe length
D = pipe diameter
f = friction factor
V = flow velocity

The friction factor is related to the Reynolds number and the relative roughness of the pipe. For conditions of complete turbulence, Fig. 5-3 relates the friction factor to pipe geometry and characteristics.

In transportation systems, the pipe friction head loss is usually predominant and other losses can ordinarily be neglected without serious error. In short

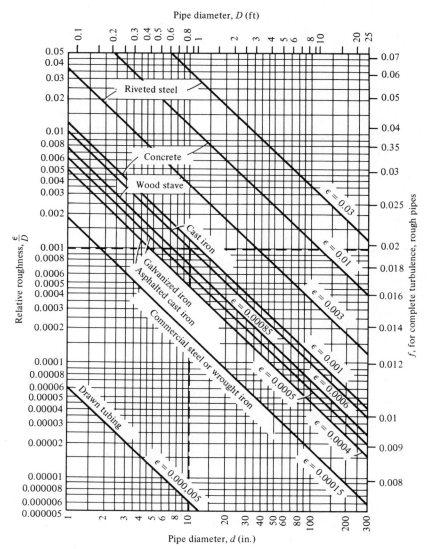

FIGURE 5-3. Relative roughness of pipe materials and friction factors for complete turbulence. (Courtesy Crane Co., Chicago.)

pipelines, such as found in water-treatment plants, the minor losses may be exceedingly important. If there is doubt, it is always best to include them. Minor losses result from valves, fittings, bends, and changes in flow characteristics at inlets and outlets. For turbulent-flow conditions, minor losses are customarily expressed as a function of velocity head. Adequate information is available from manufacturers and other sources on losses in various types of valves and fittings.[5,6]

●**EXAMPLE 5-1**

Consider that water is pumped 10 mi from a reservoir at elevation 100 ft to a second reservoir at elevation 230 ft. The pipeline connecting the reservoirs is 48 in. in diameter and is constructed of concrete with an absolute roughness of 0.003. If the flow is 25 mgd and the efficiency of the pumping station is 80%, what will be the monthly power bill if electricity costs 1 cent/kWh?

○*Solution*

1 Writing the energy equation between a point on the water surface of reservoir A and a point on the water surface of reservoir B, the following equation is obtained:

$$Z_A + \frac{P_A}{W} + \frac{V_A^2}{2g} + H_P = Z_B + \frac{P_B}{W} + \frac{V_B^2}{2g} + H_L$$

2 Letting $Z_A = 0$, and noting that $P_A = P_B =$ atmospheric pressure, and $V_A = V_B = 0$ for a large reservoir, the equation reduces to

$$H_P = Z_B + H_L$$

where $H_P =$ head developed by the pump
$H_L =$ total head lost between A and B, including pipe friction and all minor losses

3 Using Fig. 5-3, determine the value of f as 0.0182.
4 Using Eq. 5-4, find the pipe friction-head loss. Assuming that the minor losses are negligible in this problem, this is equal to H_L.

$$H_L = f \frac{L}{D} \frac{V^2}{2g}$$

V must be determined before Eq. 5-4 can be solved.

$$V = \frac{Q}{A} = \frac{25 \times 10^6 \times 1.55}{\pi \times 4 \times 10^6} = 3.09 \text{ fps}$$

$$H_L = 0.0182 \times \frac{5280 \times 10}{4} \times \frac{(3.09)^2}{64.4}$$

$$= 35.6 \text{ ft}$$

5 $H_p = (230 - 100) + 35.6$

$\qquad = 130 + 35.6 = 165.6$ ft-lb/lb

the energy imparted by the pump to the water.

6 The power requirement may be computed as

$P = Q\gamma H_P$

$\qquad = 25 \times 1.55 \times 62.4 \times 165.6 = 400{,}000$ ft-lb/sec

7 For 80% efficiency, the power requirement is

$\dfrac{400{,}000}{0.80} = 500{,}000$ ft-lb/sec

8 $5.00 \times 10^5 \times 3.766 \times 10^{-7} = 18.8 \times 10^{-2}$ kWh/sec

The number of kWh per 30-day month is then

$18.8 \times 10^{-2} \times 30 \times 864 \times 10^2 = 485{,}000$ kWh/month

9 The monthly power cost is therefore $485{,}000 \times 0.01 = \$4{,}850.00$.

5-5

DESIGN OF TRANSPORTATION SYSTEMS

The design of transportation systems involves primarily a determination of hydraulic adequacy, structural adequacy, and economic efficiency. The required waterway area is a function of the flow to be carried, the head available, the character of the conduit material, and limiting velocities.

Locating the Aqueduct

The location of an aqueduct is based on engineering and economic considerations. Since the terminal locations are dictated by the source of supply and the region to be served, the problem becomes one of finding the most practical and economic route between them. The choice of location has an obvious bearing on the type of aqueduct that can be built. Aqueducts built to grade require topography such that cut-and-cover operations can be closely balanced. Pressure aqueducts, on the other hand, can follow the topography. Pumping and material and construction costs must be given full consideration.

Dimensioning the Conduit

The size and configuration of the aqueduct finally selected will in all probability be variable along the route.

For a given type of aqueduct (pipe, tunnel, flume, canal) the size will usually be determined on the basis of hydraulic, economic, and construction considerations. Occasionally, construction practices dictate a minimum size in excess of that required to handle the flow under the prevailing hydraulic conditions (available head). This condition would most likely be encountered where a tunnel was involved. Hydraulic factors that control the design are the head available and permissible velocities. Available heads are affected by reservoir drawdown and local pressure requirements. Limiting velocities are based on the character of the water to be transported and the need to protect transmission lines against excessive pressures which might be developed through hydraulic surge. Where silt is transported with the water, minimum velocities of about 2.5 fps should be maintained. Maximum velocities must preclude pipe erosion or hydraulic surge problems, and are ordinarily between 10 and 20 fps.[11] The usual range in velocities is from about 4 to 6 fps.

Where power generation is involved, pumping costs and/or the worth of power and conduit costs jointly determine the conduit size. For single gravity-flow pipelines the size should be determined such that all the head available is consumed by friction.

Determining the Most Economical Aqueduct

Hydraulic head has a real economic value. It costs money to produce the head at the upstream end of the system because head can be used for increased flow, for power production, or a combination of these factors. A definite relationship always exists among aqueduct size, hydraulic gradient, and the value of head. In some cases construction costs are related to the elevation of the hydraulic gradient. The elevation of the gradient also affects pumping costs and power-production values, as does the slope of the hydraulic gradient. In long lines composed of different types of conduits passing through varied topography, a means of coordinating conduit types, choosing dam elevations, and selecting pump lifts or power drops is important. This problem can be approached through a joint application of hydaulic and economic principles.[7,10]

In any conduit, sufficient hydraulic slope must exist to obtain the required flow. Steep slopes generate high velocities with smaller conduit requirements. When sufficient fall is available, steep slopes are often economical. On the other hand, if head can be generated only by pumping or through construction of a dam, flatter slopes calling for larger conduits are probably necessary to reduce the cost of the lift. Apparently, then, some combination of lift and slope will yield the optimum economy.

Usually in designing water-transmission lines some controlling feature establishes the elevation of the line at a specified point. Examples of possible governing features are dam heights, tunnel locations, terminal reservoirs, and hilltops. These controls are valuable aids in carrying out the overall system design.

Basic principles of the economic location of a pipeline are given in Fig. 5-4. Water is to be pumped from reservoir A to a second reservoir M. Consider first the use of the pump line AB and the grade tunnel BD. The water must be pumped to a height great enough at B so that the hydraulic gradient in the tunnel BD will permit economic construction. The required tunnel size can be arrived at by using the plot of cost versus slope in Fig. 5-4. Curve AB gives the total cost

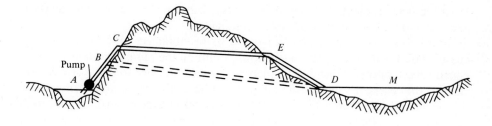

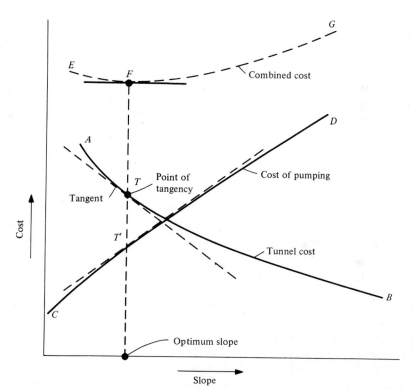

FIGURE 5-4. Economic location of a pipeline.

of tunnel for any specified hydraulic gradient. The curve is obtained by selecting various hydraulic slopes, determining the tunnel size required to transmit the required flow at that slope, and then computing costs for the various sizes. Curve CD represents the capitalized cost of raising water to heights which correspond to the several tunnel slopes. Summing curves AB and CD gives the combined cost curve. Point F indicates the minimum combined cost and thus permits determination of the optimum hydraulic gradient as shown in the figure.

Projecting the gradient upstream from D establishes the height of pumping lift and allows the grade tunnel BD to be designed.

The most economical gradient can also be obtained without the use of curve EFG. To do this, select any arbitrary point on the pumping curve CD and construct a tangent at this point. Then construct a tangent to curve AB whose slope is the reverse of that of the tangent to CD. If the two points of tangency do not lie on approximately the same vertical, select a new starting point and repeat the procedure. At the tangent point T' (Fig. 5-4), the slope of the tangent is numerically equal to the cost of an additional foot of head. When information on this cost is available from other sources, CD need not be drawn since the tangent is determined directly from the cost of a foot of head.

Consider now the alternative aqueduct $ACED$ in Fig. 5-4. A question arises: Is it more economical to pump to C and then carry the flow through the shorter tunnel CE, or to pump to B and use the longer tunnel BD? This problem may be solved by making trial estimates. First, assume the control point to be at E. An estimate is then prepared for line $ACED$ in the manner previously indicated and compared with the one made for ABD where the control was assumed at D. If the total aqueduct cost plus the capitalized cost of pumping is less for $ACED$ than for ABD, the critical point is E. If this is not the case, the control will be at D.

When the head loss between two points in an aqueduct is fixed, it may be most economical to divide this amount unequally between the various types of conduits used. The division will be determined largely by economic considerations. For example, since tunnel costs are often very high, it may be best to build the smallest possible tunnel and thereby consume a disproportionate share of the head available in this particular reach.

If an aqueduct is to be constructed of several different types of conduits and if the total head loss is fixed, application of the principles of Lagrange's method of undetermined multipliers will permit an evaluation of the most economical distribution of head loss.[8] This occurs when the ratios of change in cost to change in head are equal for each type of conduit. The total available head must also equal the sum of the various component losses.

Application of this theory will be illustrated with the aid of Fig. 5-5. First, it is necessary that a set of curves of cost versus head loss be plotted for each conduit type. These curves are derived from data obtained by designing the conduit types for various conditions of head loss and then estimating their cost. Second, a series of parallel tangents drawn to each curve will be constructed by trial and error so that the sum of the individual head losses for each conduit equals the total head. The ratio of change in cost to change in head will be equal for all conduits when the tangents are parallel. These conditions satisfy the requirements for the most economical design.

●EXAMPLE 5-2
Using the data supplied in Fig. 5-5, find the most economical division of the total head of 212 ft and the minimum aqueduct cost.

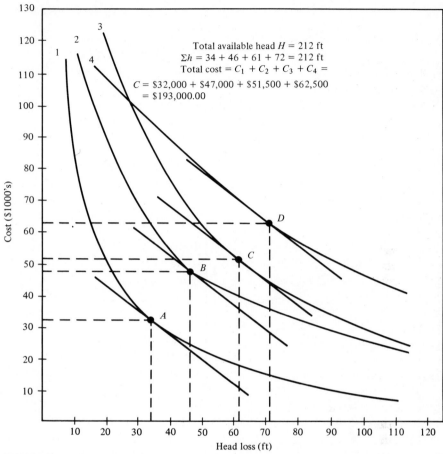

FIGURE 5-5. Graphical determination of the minimum cost of an aqueduct composed of four conduit types.

○*Solution*
1 By repeated trial, find the set of tangents *A, B, C,* and *D* such that $\sum h = H$.
2 Find the related costs *C*1, *C*2, *C*3, and *C*4 by projecting the points of tangency to intersections with the cost axis.
3 Numerical solutions are found on Fig. 5-5.

The method of parallel tangents just discussed will yield the required economic information, provided that the total head loss is known.

Referring to Fig. 5-6, assume that it is necessary to design an aqueduct system between a reservoir at *A* and a lake at *G* where power production is required. An investigation of the route indicates that the best arrangement is a pipeline between *A* and *B*, a canal between *B* and *D*, and a tunnel between *D* and *E*. Assume further that a power drop is necessary at *F*. Determine the conduit gradients and their location for the optimum system.

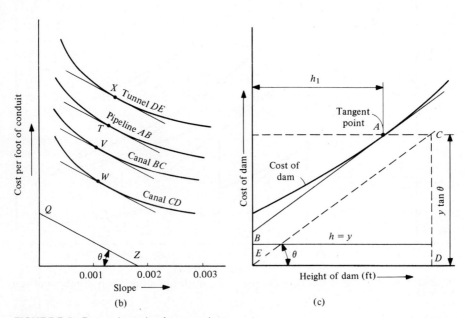

FIGURE 5-6. Economic study of an aqueduct.

Suppose that the tunnel DE is a controlling factor and that point E will fix the vertical location of the aqueduct. The most economical slopes may then be determined in the following manner[10]:

1. Plot a set of curves showing cost per linear foot of conduit versus slope for the various types of conduit (Fig. 5-6b). Note that these curves must correspond to the designated design discharge.

2. Plot a cost curve for the dam (Fig. 5-6c). Assume some trial height of dam h_1 and construct a tangent at the intersection of the ordinate constructed at h_1 with the cost curve (point A). The slope of this tangent represents the cost of an added foot of dam height at h_1.

3   Lay off the line QZ on Fig. 5-6b such that the tan θ equals the cost of an additional foot of dam at the trial height h_1; that is, make the value of tan θ in (b) numerically equal to that in (c). This slope is determined from the inclination of the tangent at point A in Fig. 5-6c but with the slope reversed.

4   Construct tangents to the cost curves AB, BC, CD, and ED in Fig. 5-6b. These tangents are all drawn parallel to QZ. The locations of the tangent points T, V, W, and X permit determination of the optimum economical slopes for each type of conduit.

5   Using the slopes defined by the tangents, and the known or assumed elevation at E, compute the required elevation of the dam at A.

6   Compare the computed dam height with h_1. If there is considerable difference, repeat the procedure outlined until a satisfactory solution is obtained. In this manner the most economical combination of dam height, conduit size, and corresponding hydraulic gradient is established between the dam and the assumed fixed point E.

7   From E to G the optimum system will have to be resolved through a consideration of the value of power developed in the power drop FG. To accomplish this, a procedure similar to that already illustrated is followed. Begin by assuming a trial power drop FG. This tentatively establishes point F. Estimate the value of a foot of head resulting from this drop. A plot similar to that of Fig. 5-6c, or an analytical procedure, may be used.

8   Plot cost curves for the types of conduits required between E and F. On this plot construct a new line similar to QZ which has tan θ equal to the value of a foot of head at F.

9   Draw parallel tangents to the cost curves and find the most economical slopes as before.

10   Using these slopes, find the actual elevation at F. If the power drop found in this manner differs enough to indicate an important change in the value of a foot of head at F, repeat the procedure until satisfactory agreement is obtained.

A method for determining the optimum hydraulic grade line which does not require the assumption of control points (points at which the elevation of the hydraulic gradient is known) has been developed by Jackson and Edmonston for use in designing the Feather River Project in California.[7] The method appears to have particular merit, especially where pumping is involved.

The method is based on the Euler–Lagrange equation[9]

$$\frac{\partial f}{\partial y} - \frac{d}{dx}\left(\frac{\partial f}{\partial y'}\right) = 0 \tag{5-5}$$

which in this case governs the selection of the most economical grade line.

The equation is a perfectly general relationship applicable to any location on any type of conduit structure in the aqueduct. The terms in the equation are defined as follows:[7]

f = cost per foot of the conduit
y = elevation of the hydraulic grade line at a specified location
x = distance along the conduit measured from any origin

$$y' = \frac{dy}{dx}, \text{ hydraulic gradient}$$

$\dfrac{\partial f}{\partial y}$ = rate of change of unit conduit cost with respect to change in elevation of the hydraulic gradient at a specific location

$\dfrac{\partial f}{\partial y'}$ = rate of change of unit conduit cost with respect to change of the hydraulic grade line (numerically, it equals the value of a foot of head at any given point; the value of a foot of head N as defined by Hines is the cost of conveying the design flow through 1 ft of lift[10]; for a pumping plant, the cost of 1 ft of head equals the capitalized cost of pumping the design flow through a foot of lift, including the incremental cost of the additional plant capacity required for the higher lift)

$\dfrac{d}{dx}\left(\dfrac{\partial f}{\partial y'}\right)$ = rate of change of the value of 1 ft of head with respect to change in location along the aqueduct

This equation is the model for establishing the optimum hydraulic gradient. In practice it is not solved analytically but serves as an expression of the basic principle to follow in evaluating the various factors considered in an economical aqueduct design. Jackson and Edmonston give a more complete treatment of the theoretical aspects of this relationship.

In practice, the method can be applied in the following fashion:[7]

1 Determine the relationship between the unit costs of the various types of conduit and the hydraulic gradient. This can be accomplished by plotting cost per foot of conduit as the ordinate versus hydraulic gradient as the abscissa. This gives a plot similar to that of Fig. 5-4.

2 Estimate the value of a foot of head N at a control point, such as a pumping plant, a power plant, or a dam. Figure 5-6c illustrates this approach.

3 Using the information developed in steps 1 and 2, make rough approximations of the optimum hydraulic gradient for each type of conduit. The method of tangents can be used for this.

4 Select a trial starting elevation of the hydraulic gradient after the first pumping lift.

5 Using a topographic map select a tentative route for the aqueduct, and, combining it with field inspections, tentatively select the types of conduits to be utilized.

6 Plot the ground profile of the selected route.

7 Where there is an afterbay reservoir following the initial lift, determine a value of 1 ft of head N, which includes the N of the pumping plant plus the cost of raising the afterbay elevation 1 ft.

8 From the value of N at the afterbay and the conduit cost curves, find the optimum gradient of the first type of conveyance reach. This will be determined by the tangent method as previously illustrated. Project this gradient ahead until N has changed by at least $1000.

9 Using the value of N at the end of the first gradient, set a new gradient for the following reach. Repeat this process until the next pumping station or power plant is reached.

10 At the second pumping or power plant compare the value of N obtained in the preceding manner with the known value of 1 ft of head at that plant. If the computed value differs by more than 5 % from the known value, choose a higher or lower starting elevation for the hydraulic grade line and begin again.

11 The procedure may be repeated several times as mapping and cost data are refined.

The conduit sizes established by the hydraulic gradients determined in this manner are the most economical for a specified set of operating rules over a particular ground profile. Note that not more than two end conditions may be met. These conditions might be known values of N or fixed elevations of the hydraulic gradient at the ends of a line.

Strength Considerations

Water-conveyance structures are required to resist numerous forces such as those resulting from water pressure within the conduit, hydraulic surge (transient internal pressure generated when the velocity of flow is rapidly reduced), external loads, forces at bends or changes in cross section, expansion and contraction, and flexural stresses. The student will find these topics adequately covered by most standard hydraulics textbooks.[5,6,22]

Distribution Systems

Water-distribution systems are ordinarily designed to adequately satisfy the water requirements for a combination of domestic, commercial, industrial, and fire-fighting purposes. The system should be capable of meeting the demands placed on it at all times and at satisfactory pressures. Pipe systems, pumping stations, storage facilities, fire hydrants, house service connections, meters, and other appurtenances are the main elements of the system.

5-6

SYSTEM CONFIGURATIONS

Distribution systems may be generally classified as grid systems, branching systems, or a combination of these. The configuration of the system is dictated primarily by street patterns, topography, degree and type of development of the area, and location of treatment and storage works. Figure 5-7 illustrates the nature of several basic types of systems. A grid system is usually preferred to a branching system, since it can furnish a supply to any point from at least two directions. The branching system does not permit this type of circulation, since it has numerous terminals or dead ends. A grid or combination system can also

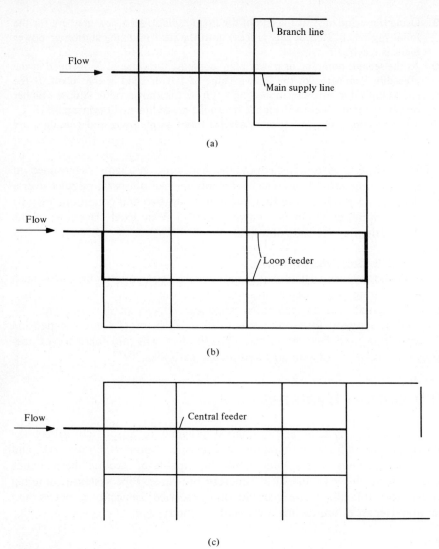

Flow

Branch line

Main supply line

(a)

Flow

Loop feeder

(b)

Flow

Central feeder

(c)

FIGURE 5-7. Types of water-distribution systems. (a) Branching system; (b) grid system. (c) Combination System.

incorporate loop feeders, which act to distribute the flow to an area from several directions.

In locations where sharp changes in topography occur (hilly or mountainous regions) it is common practice to divide the distribution system into two or more service areas or zones. This precludes the difficulty of extremely high pressure in low-lying areas in order to maintain resonable pressures at higher elevations. Usual practice is to interconnect the various systems, with the interconnections closed off by valves during normal operations.

5-7

BASIC SYSTEM REQUIREMENTS

The performance of a distribution system can be judged on the basis of the pressures available in the system for a specific rate of flow.[13,19] Pressures should be great enough to adequately meet consumer and fire-fighting needs. At the same time they should not be excessive, since the development of pressure head is an important cost consideration. In addition, as pressures increase, leakage increases, and money is then spent to transport and process a product that is wasted. Because the investment in a distribution system is exceedingly large, it is important that the design be optimized economically.

For commercial areas, pressures in excess of 60 pounds per square inch, gage (psig) are usually required. Adequate pressures for residential areas usually range from 40 to 50 psig. In tower buildings it is often necessary to provide booster pumps to elevate the water to upper floors. Storage tanks are usually provided at the highest level and distribution is made directly from them.

The capacity of the distribution system is determined on the basis of local water needs plus fire demands as outlined in Chapter 4. Pipe sizes should be selected so that high velocities are avoided. Once the flow has been determined, pipe sizes can be selected by assuming velocities of from 3 to 5 fps. Where fire-fighting requirements are to be met, a minimum diameter of 6 in. is recommended. The National Board of Fire Underwriters recommends 8 in. as a minimum but permits 6-in. pipes in grid systems provided that the length between connections does not exceed 600 ft.

5-8

HYDRAULIC DESIGN

To effect the hydraulic design of a water-distibution system, information must be available on the anticipated local rates of water consumption, the manner in which these design flows are distibuted geographically, and the required pressure gradients for the system. The first is obtained in the manner indicated in Chapter 4. It should be reemphasized that the designer ought to investigate both the maximum day rate plus fire protection and the maximum hourly rate to determine which will govern the design.

The spatial distribution of consumption can be estimated by studying population densities, and commercial and industrial use patterns, which are known or predicted for the region. Consider the design of a feeder to an area composed of residential, commercial, and industrial users. In investigating the peak hour, for example, it will be important to have the predicted hydrograph for each type of user so that the specific hour in which the summation of the three component flows is greatest will be used for the design. Students are cautioned that the regional maximum hour might well coincide with the residential maximum within the region but not coincide with the commercial or industrial peaks, or vice versa. For this reason, information on the hourly variation of water use for all users is extremely valuable. Once the water consumption has been estimated, it is usual practice to consider it to be concentrated at specified points on the feeder–main system. Computations based on the concentration of consumption in this manner normally appear to correlate well with field observations.

Distribution systems are usually designed so that reasonably uniform pressures prevail. A pressure of 30 psi is normally considered to be the minimum desirable in any area, with the exception that during a serious fire it may be permissible to allow the pressure to drop to about 20 psi. Main feeders should be designed for pressures between 40 and 75 psi whenever possible.[13]

The most used methods of pipe-network analysis are the Hardy Cross method, and the equivalent-pipe method. The Hazen–Williams pipe-flow formula is commonly used to compute flows.

Equivalent-Pipe Method
The analysis of a distribution system is often expedited by first skeletonizing the system. This might involve the replacement of a series of pipes of varying diameter with one equivalent pipe or replacing a system of pipes with an equivalent pipe. An equivalent pipe is one in which the loss of head for a specified flow is the same as the loss in head of the system which it replaces. For any system there are theoretically an infinite number of equivalent pipes. An example will illustrate the method of analysis.

●EXAMPLE 5-3
Considering the pipe system shown in Fig. 5-8, replace (a) pipes BC and CD with an equivalent 12-in. pipe and (b) the system from B to D with an equivalent 20-in. pipe.

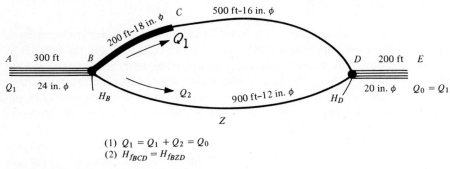

(1) $Q_1 = Q_1 + Q_2 = Q_0$
(2) $H_{fBCD} = H_{fBZD}$

FIGURE 5-8. Diagram for Example 5-3.

○*Solution*
(a) Assume a discharge through BCD of 8 cfs. Using the Hazen–Williams nomograph (Fig. 5-9), find the head loss for $BC = 6.1$ ft/1000 ft and for $CD = 11$ ft/1000 ft.

The total head loss between B and D is therefore

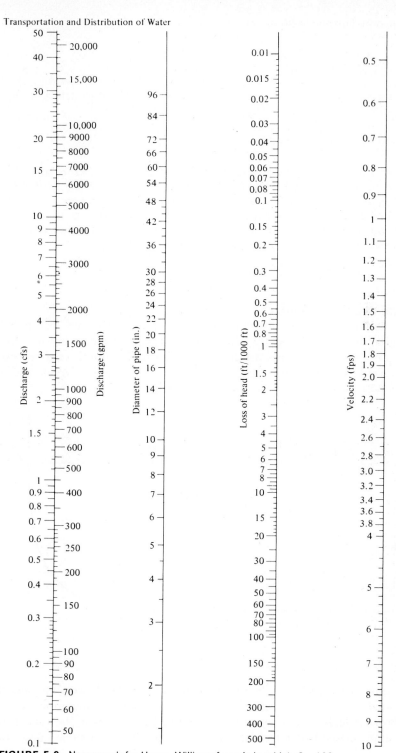

FIGURE 5-9. Nomograph for Hazen–Williams formula in which $C = 100$.

$$6.1 \times \frac{200}{1000} + 11.0 \times \frac{500}{1000} = 6.72 \text{ ft}$$

Using a discharge of 8 cfs, Fig. 5-9 indicates a head loss of 45 ft/1000 ft for a 12-in. pipe. The equivalent length of 12-in. pipe is therefore

$$L_{12} = \frac{6.72 \times 1000}{45} = 149 \text{ ft}$$

(b) Assume a total head loss between B and D of 5.0 ft. For the 12-in. equivalent pipe this is 33.5 ft/1000 ft and for the 900 ft of 12-in. pipe it is 5.5 ft/1000 ft. Using these values and Fig. 5-9, the discharges for the two pipes are found to be 6.8 and 2.6 cfs, respectively. The total flow is thus 9.4 cfs at a head loss of 5 ft. For this discharge, a 20-in. pipe will have a head loss of 4.8 ft/1000 ft. The equivalent 20-in. pipe to replace the whole system will be

$$\frac{5}{4.8} \times 1000 = 1042 \text{ ft long}$$

Hardy Cross Method

The analysis of a simple hydraulic system such as that shown in Fig. 5-8 presents little difficulty. A slightly more complex system is shown in Fig. 5-10. The method of equivalent pipes will fail to yield a solution in this case because there are crossover pipes (pipes that operate in more than one circuit and a number of withdrawal points throughout the system.

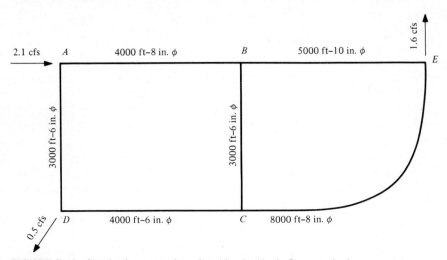

FIGURE 5-10. Simple pipe network analyzed by the Hardy Cross method.

This type of network may be solved by using the Hardy Cross method of network analysis.[15] The procedure permits the accurate computation of the rates of flow through the system and the resulting head losses in the system. It is a relaxation method by which corrections are applied to assumed flows or

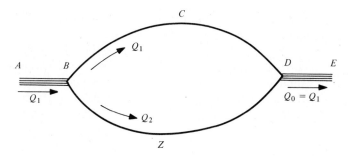

(1) $Q_1 = Q_1 + Q_2 = Q_0$
(2) $H_{f_{BCD}} = H_{f_{BZD}}$

FIGURE 5-11. Derivation of the Hardy Cross method.

assumed heads until an acceptable hydraulic balance of the system is achieved.

The Hardy Cross analysis is based on the principles that (1) in any system continuity must be preserved, and (2) the pressure at any junction of pipes is singled-valued. Referring to the simple network of Fig. 5-11, the elements of the procedure can be explained. First the system must be defined in terms of pipe size, length, and roughness. Then for any inflow Q_I, the system can be balanced hydraulically only if $H_{f_{BCD}} = H_{f_{BZD}}$. This restriction limits the possibilities to only one value of Q_1 and Q_2 which will satisfy the conditions.

Derivation of the basic equation for balancing heads by correcting assumed flows will now be given for the loop of Fig. 5-11.[15] First find the required inflow Q_I. Then arbitrarily divide this flow into components Q_1 and Q_2. The only restriction on the selection of these values is that $Q_1 + Q_2 = Q_I$. An attempt should be made, however, to select realistic values. Since the procedure involves a number of trials, the amount of work involved will be dependent upon the accuracy of the value originally selected. For example, in the network shown, BCD is considerably larger in diameter than BZD. A logical choice would therefore assume that Q_1 will be larger than Q_2. The final solution to the problem will be the same regardless of the original choice, but much more rapid progress results from reasonable initial assumptions.

After Q_1 and Q_2 have been chosen, $H_{f_{BCD}}$ and $H_{f_{BZD}}$ can be computed using the Hazen–Williams or some other pipe-flow formula. Remembering that the Hazen–Williams equation is of the form

$$Q = 0.279CD^{2.63}S^{0.54} \tag{5-2}$$

the equation may be rewritten as

$$Q = K_a S^{0.54} \tag{5-6}$$

where K_a is a constant when we are dealing only with a single pipe of specified size and material. Rearranging this equation and substituting H_f/L for S,

$$H_f = KQ^n \tag{5-7}$$

where $n = 1.85$ in the Hazen–Williams equation. Equation 5-7 is convenient for expressing head loss as a function of flow in network analyses.

If the computed values of $H_{f_{BCD}}$ and $H_{f_{BZD}}$ are not equal (which is usually the case on the first trial) a correction must be applied to the initial values. Call this correction ΔQ. If, for example, $H_{f_{BCD}} > H_{f_{BZD}}$, then the new value for Q_1 will be $Q_1 - \Delta Q = Q_1'$ and the new value for Q_2 must be $Q_2 + \Delta Q = Q_2'$. The corresponding values of head loss will be $H_{f_{BCD}}'$ and $H_{f_{BZD}}'$. If ΔQ is the true correction, then

$$H_{f_{BCD}}' - H_{f_{BZD}}' = 0 = K_1(Q_1 - \Delta Q)^n - K_2(Q_2 + \Delta Q)^n$$

The binomials may be expanded as follows:

$$K_1(Q_1^n - n\,\Delta Q Q_1^{n-1} + \cdots) - K_2(Q_2^n + n\,\Delta Q Q_2^{n-1} + \cdots) = 0$$

If ΔQ is small, the terms in the expansion involving ΔQ to powers greater than unity can be neglected. Therefore,

$$K_1 Q_1^n - nK_1\,\Delta Q Q_1^{n-1} - K_2 Q_2^n - nK_2\,\Delta Q Q_2^{n-1} = 0$$

Substituting $H_{f_{BCD}}$ for $K_1 Q_1^n$, $H_{f_{BZD}}$ for $K_2 Q_2^n$, and rewriting the terms KQ^{n-1} as $K(Q^n/Q)$,

$$H_{f_{BCD}} - \Delta Q nK_1 \frac{Q_1^n}{Q_1} - H_{f_{BZD}} - \Delta Q nK_2 \frac{Q_2^n}{Q_2} = 0$$

$$H_{f_{BCD}} - H_{f_{BZD}} = \Delta Q n \left(\frac{H_{f_{BCD}}}{Q_1} + \frac{H_{f_{BZD}}}{Q_2} \right)$$

and

$$\Delta Q = \frac{H_{f_{BCD}} - H_{f_{BZD}}}{n(H_{f_{BCD}}/Q_1 + H_{f_{BZD}}/Q_2)} \tag{5-8}$$

Expanding this expression to the more general case, the equation for the flow correction ΔQ becomes

$$\Delta Q = - \frac{\sum H}{n \sum (H/Q)} \tag{5-9}$$

Application of this equation involves an initial assumption of discharge and a sign convention for the flow. Either clockwise or counterclockwise flows may be considered positive, and the terms in the numerator will bear the appropriate sign. For example, if the counterclockwise direction is considered positive, all H values for counterclockwise flows will be positive and all H values for clockwise flows will be negative. The denominator, however, is the absolute sum without regard to sign convention. The correction ΔQ has a single direction for all pipes in the loop, and thus the sign convention must also be considered in applying the correction. Example 5-4 illustrates the application of the procedure to a network problem. Example 5-5 illustrates the solution of a similar problem by digital computer.

● **EXAMPLE 5-4**

Given the network, the inflow at A, and outflows at B, C, and D in Fig. 5-12, find the flows in the individual pipes comprising the network.

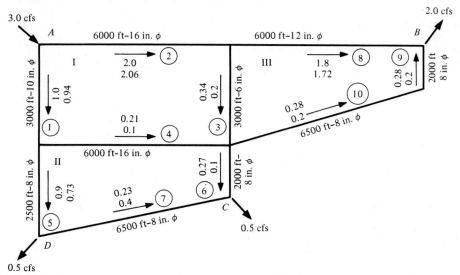

FIGURE 5-12. Pipe network analyzed by the Hardy Cross method. Note: Clockwise direction considered positive. Flows shown are initial assumption and final corrected value.

○ **Solution**

The computational procedure is given in Table 5-4. The final flows are shown on Fig. 5-12.

It should be noted that since pipes 3 and 4 appear in more than one loop, they are subject to the combined correction for loops I and III and I and II, respectively.

A similar procedure to that just discussed is to assume values of H and then balance the flows by correcting the assumed heads. The mechanics of the two methods are the same and the applicable relationship,

$$\Delta H = - \frac{n \sum Q}{\sum (Q/H)} \tag{5-10}$$

can be derived in a manner similar to that for Eq. 5-9. The number of trials required for the satisfactory solution of any problem using Eq. 5-9 or Eq. 5-10 depends to a large extent on the accuracy of the initial set of assumed values and on the desired degree of accuracy of the results.

In using the Hardy Cross method to analyze large distribution systems, it is often useful to reduce the system to a skeleton network of main feeders.[13] Where the main-feeder system has a very large capacity relative to that of the smaller mains, field observations indicate that this type of skeletonizing yields reasonable results. Where no well-defined feeder system is apparent serious

TABLE 5-4

COMPUTATIONS FOR EXAMPLE 5-4

					Trial I			
1	2	3	4	5	6	7	8	9
Loop No.	Pipe No.	Pipe Diam. (in.)	Length (ft)	Q (cfs)	H_L (ft)	$\dfrac{H_L}{Q}$	$n \Sigma \left(\dfrac{H_L}{Q}\right)$	ΣH_L
I	1	10	3000	−1.0	− 6.9	6.9		
	2	16	6000	+2.0	+ 4.92	2.46		
	3	6	3000	+0.2	+ 4.05	20.30		
	4	6	6000	−0.1	− 1.92	19.2	90.3	+ 0.15
II	4	6	6000	+0.1	+ 1.92	19.2		
	6	8	2000	+0.1	+ 0.16	1.6		
	7	8	6500	−0.4	− 7.8	19.5		
	5	8	2500	−0.9	−13.7	15.2	102.5	−19.42
III	3	6	3000	−0.2	− 4.05	20.3		
	8	12	6000	+1.8	+16.2	9.0		
	9	8	2000	−0.2	− 0.7	3.5		
	10	8	6500	−0.2	− 2.28	11.4	81.7	+ 9.8
					Trial III			
I	1			−0.95	− 6.3	6.64		
	2			+2.05	+ 5.1	2.48		
	3			+0.35	+11.85	33.90		
	4			−0.20	− 8.1	40.50	154.8	+ 2.55
II	4			+0.20	+ 8.1	40.50		
	6			+0.25	+ 1.08	4.32		
	7			−0.25	− 3.51	14.1		
	5			−0.75	−10.00	13.33	133.8	− 4.31
III	3			−0.35	−11.85	33.90		
	8			+1.70	+14.38	8.46		
	9			−0.30	− 1.44	4.80		
	10			−0.30	− 4.68	15.61	116.3	− 3.57

errors may result from skeletonizing. Figure 5-13 illustrates a skeletonized distribution network consisting of arterial mains only. Figure 5-14 shows how a portion of the distribution system of Fig. 5-13 (that part lying within the dashed rectangle) looked before skeletonizing. A more complete discussion of such procedures is given by Reh.[13] The analysis of a large network may also be expedited by balancing portions of the system successively instead of analyzing the whole network simultaneously.

Normally, minor losses are neglected in network studies but they can easily be introduced as equivalent lengths of pipe when it is felt that they should

Trial II							
10	**5**	**6**	**7**	**8**	**9**	**10**	**11**
ΔQ	Q_1	H_L	$\dfrac{H_L}{Q}$	$n\,\Sigma\left(\dfrac{H_L}{Q}\right)$	ΣH_L	ΔQ	
−0.002	−1.0	− 6.9	6.9			+0.05	
−0.002	+2.0	+ 4.92	2.46			+0.05	
+0.12	+0.32	+ 9.45	29.5			+0.03	
−0.19	−0.29	−15.6	53.8	171.5	−8.13	+0.09	
+0.19	+0.29	+15.6	53.8			−0.09	
+0.19	+0.29	+ 1.4	4.83			−0.04	
+0.19	−0.21	− 2.4	11.43			−0.04	
+0.19	−0.71	− 9.0	12.69	153.1	+5.6	−0.04	
−0.12	−0.32	− 9.45	29.5			−0.03	
−0.12	+1.68	+14.4	8.58			+0.02	
−0.12	−0.32	− 1.6	5.0			+0.02	
−0.12	−0.32	− 5.2	16.25	109.9	−1.85	+0.02	
Trial IV							
−0.016	−0.966	− 6.59	6.83			+0.027	−0.94
−0.016	+2.034	+ 5.09	2.51			+0.027	+2.06
−0.047	+0.303	+ 9.0	29.70			+0.038	+0.34
−0.048	−0.248	−12.0	48.30	163.5	−4.50	+0.035	−0.21
+0.048	+0.248	+12.0	48.30			−0.035	+0.21
+0.032	+0.282	+ 1.30	4.61			−0.008	+0.27
+0.032	−0.218	− 2.92	13.4			−0.008	−0.23
+0.032	−0.718	− 9.24	12.85	146.3	+1.14	−0.008	−0.73
+0.047	−0.303	− 9.0	29.70			−0.038	−0.34
+0.031	+1.731	+15.25	8.80			−0.011	+1.72
+0.031	−0.269	− 1.2	4.46			−0.011	−0.28
+0.031	−0.269	− 3.89	14.48	106.2	+1.16	−0.011	−0.28

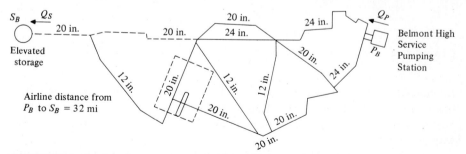

FIGURE 5-13. Arterial piping network of Belmount High Service District, Philadelphia. (Courtesy Civil Engineering Department, University of Illinois, Urbana, Ill.)

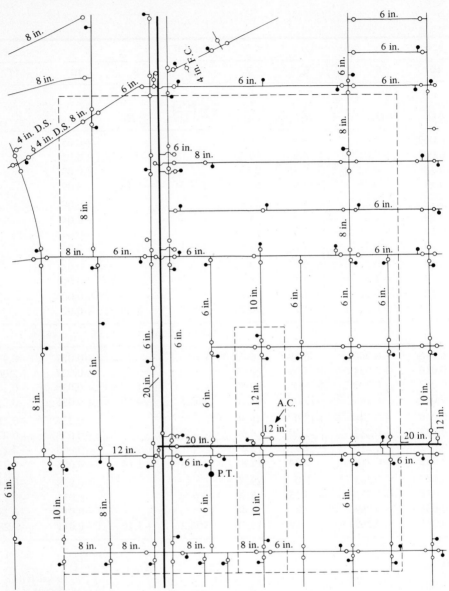

FIGURE 5-14. Intermediate Grid Sector, Belmont High Service District, Philadelphia. (Courtesy Civil Engineering Department, University of Illinois, Urbana, Ill.)

be included. Where *C* values are determined from field measurements, they invariably include a component due to the various minor losses encountered. McPherson gives a good discussion of local losses in water distribution networks.[14]

The construction of pressure contours helps to isolate shortcomings in the hydraulic performance of distribution systems. Contours are often drawn with intervals of 1–5 ft of head loss but may have other intervals depending on local circumstances. For a given set of operating rules applicable to a particular network, the pressure contours indicate the distribution of head loss and are helpful in showing regions where head losses are excessive. Figure 5-15 illustrates contours constructed for a distribution network.

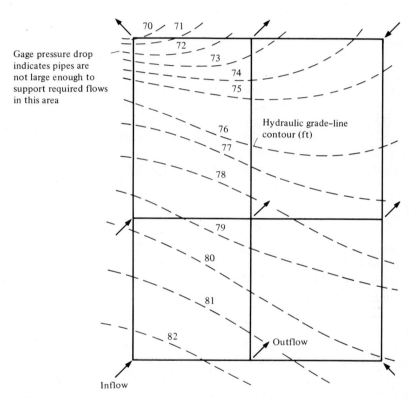

FIGURE 5-15. Pressure contours for a distribution network.

Computations for a given network may be reduced considerably by making use of the proportional-flow method outlined by McPherson when studies of different inputs and outflows are needed.[12] In applying the technique, demands are first concentrated at key points in the network and a solution effected in the normal manner. Then if the inputs and demands on the system are changed in a truly proportional manner, a new set of pipe flows and head losses may be readily obtained. McPherson states that *proportional load* may be defined as the design assumption that each consolidated demand on the system fluctuates about its mean value in direct proportion to the manner in which the total system load fluctuates. Note that the assumption of proportional loading is not always realistic, and in such cases the procedure outlined here does not apply. The significance of this condition can best be illustrated by referring to Fig. 5-11. In the simple network pictured we may state that

(1) $Q_I = Q_O = Q_1 + Q_2$
(2) $H_1 - H_2 = 0$ or $K_1 Q_1^n - K_2 Q_2^n = 0$

Solving these equations simultaneously,

$$Q_1 = AQ_I \quad \text{and} \quad Q_2 = BQ_I \tag{5-11}$$

where A and B are constants for the sytem. Thus it can be seen that if Q_I is doubled, Q_1 and Q_2 are also doubled. Extending this thinking to a slightly more complex system having several outflows in place of the single one, Q_L, as in the network of Fig. 5-11, the same reasoning holds true, provided that the drafts and the inflow are all increased or decreased by the same percentage. New values for Q_1, Q_2, etc., may then be found by using a set of equations such as Eq. 5-11 without the need for making a complete network analysis.

McPherson has shown that under proportional-loading conditions the head loss of a complex network can be reduced to the following equation[12]:

$$\frac{\sum h}{Q_d^m} = \Phi\left(\frac{Q_p}{Q_d}\right)^n \tag{5-12}$$

where h = loss in head between an input and a point in the district (such as an elevated storage location)
Q_d = demand placed on the network
Q_p = input rate at the point from which h is referenced

To determine values for Φ and n, only two network balances (computations) are required. If $Q_p = Q_d$, which is the case when there is no equalizing storage (see Sec. 5-10), Φ can be determined by a single network balance. For more detailed information on proportional loading, the reader should refer to McPherson's work.

Computer Solutions

Both the analog computer and the digital computer are used in analyzing distribution networks. The first deals with continuous physical variables, while the second is concerned only with numerical values. The analog computer acts as a physical model of the system to be studied and produces results which are limited in accuracy only by the physical elements of the model. The digital computer is theoretically limited in accuracy only by the reliability of the original input data.

The pioneering analog computer was the McIlroy Fluid Network Analyzer.

The McIlroy Analyzer uses Fluistors, which are nonlinear resistances. The voltage drop E for a Fluistor with a direct current I is given by

$$E = RI^m \qquad\qquad (5\text{-}13)$$

where R is the resistance. Any value of m between 1.85 and 2.00 can be approximated by the judicious selection of Fluistors. It is evident that Eq. 5-13 is in a form identical to that of Eq. 5-7, $H = KQ^n$. As a result of this analogy, a dc network can be developed to simulate any desired hydraulic network. Once the system is energized, measured voltages and currents, converted into units of discharge and head loss, can be read directly. A detailed discussion of the McIlroy Analyzer can be found in several of the references at the end of the chapter.[14,17,18]

Solutions of the Hardy Cross method by digital computer involve a computation of the flow correction

$$\Delta Q = -\frac{\sum H}{n \sum (H/Q)} \qquad\qquad (5\text{-}9)$$

for each loop of the network after an initial assumption of flows has been made. Repeated sets of corrections are computed and applied until a satisfactorily balanced system is attained. The process of repeated trials is known as the iterative convergence technique. *Feedback* is the term applied to the modification of an initial value by some fraction of the corresponding output. The Hardy Cross procedure can be said to apply a feedback operation to sequential calculations, a procedure extremely well suited to the digital computer.

An early review of merits of various computers used in distribution network balancing was presented by McPherson and Radziul.[19] The McIlroy Analyzer has the advantage of yielding a direct analogy of the system being investigated. The operation of the equipment is such that system inputs (from pumps or storage) are raised from zero to full system capacity, whereas drafts are set initially at full load values. This procedure is particularly valuable in design analysis, since it can indicate system deficiencies at less than design rates. In such a case, revisions in the system piping can be made prior to any detailed test runs. Using the McIlroy Analyzer, trial changes can be made very rapidly and unsatisfactory arrangements eliminated quickly and directly. The analyzer also has great value in studying pumping station and service reservoir performance.

Using the iteration process combined with a convergence technique, the digital computer adjusts an initial set of flow rates to any desired degree of balance. The information obtained from the computer is usually in the form of tabulated head loss and directional flow rate for individual pipe branches. Example 5-5 illustrates the solution of a network problem by digital computer.

●EXAMPLE 5-5

Given the network, inflow at A and outflows at B through M on Fig. 5-16, find the flows in the individual pipes comprising the network.

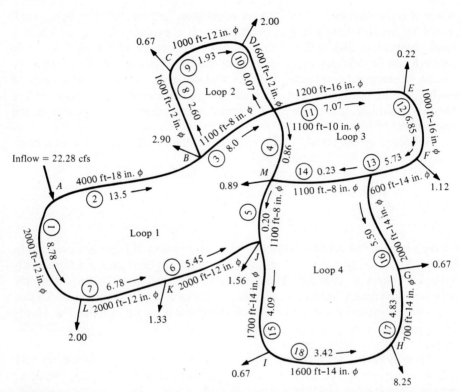

FIGURE 5-16. Pipe network analyzed in Example 5-5. Note: Clockwise direction considered positive. Flows shown are initial assumption. All flows are in cfs, numbers shown, thus ② are pipe numbers.

○Solution

The procedure is outlined in the following steps. Note that the definitions of variables used in the flow diagram and computer program are given in the comment statements at the beginning of the computer program. The printout of input data immediately follows the program, which in turn is followed by a table of final results.

1 Prepare a flow diagram for the computations as shown in Fig. 5-17.
2 Using the flow diagram as a guide, write a program in Fortran IV to repeatedly solve Eq. 5-9 until the flow correction is less than ϵ, a specified small value.
3 Run the program and obtain the required results.

FORTRAN IV PROGRAM FOR HARDY CROSS PIPE NETWORK ANALYSIS

```
C       HARDY CROSS ANALYSIS II
C       THIS PROGRAM ANALYZES A PIPE NETWORK BY THE HARDY CROSS METHOD
C       NUML = NUMBER OF LOOPS
C       NUMPI = TOTAL NUMBER OF PIPES
C       EPS = EPSILON THE ACCEPTABLE ERROR
C       L IS THE LOOP INDEX
C       K IS THE NUMBER OF PIPES IN A LOOP. THE VALUE IS OBTAINED
C       FROM J(L).
C       ID(L,K) IS AN INDEX WHICH ENABLES NUMBERING THE PIPES IN EACH LOOP
C       FROM ONE TO THE NUMBER OF PIPES IN THE LOOP WHILE STILL RETAINING THE
C       ORIGINAL PIPE NUMBERING SCHEME.
C       ID(L,K) IS AN INDEX WHICH INCLUDES THE PIPE NUMBER AND A SIGN H
C       THE SIGN IS POSITIVE FOR ALL PIPES NOT COMMON TO MORE THAN ONE
C       LOOP. FOR PIPES FALLING IN TWO LOOPS, THE SIGN IS POSITIVE FOR THE
C       FIRST LOOP IN WHICH THE PIPE APPEARS AND NEGATIVE FOR THE SECOND
C       LOOP. FOR EXAMPLE, IF PIPE 3 IS COMMON TO LOOPS 1 AND 2, ID(1,3)
C       = +3 AND ID(2,1) = -3 WHERE 3 IS THE THIRD PIPE IN LOOP 1 AND
C       THE FIRST PIPE IN LOOP 2.
C       D IS THE PIPE DIAMETER IN FEET
C       Q(I) IS THE FLOW IN PIPE I (CFS)
C       THE OVERALL PIPE NETWORK IS BROKEN UP INTO L LOOPS
C       PL IS THE PIPE LENGTH IN THOUSANDS OF FEFT. FOR A PIPE LENGTH OF
C       6,000 FEET, PL WOULD BE 6.
C       S IS THE SLOPE IN FEET PER THOUSAND FEET. IT IS COMPUTED USING THE
C       HAZEN WILLIAMS PIPE FORMULA. S = (Q/(0.0103*C*D**2.63))**1.85
C       WHERE C IS THE HAZEN WILLIAMS C.
C       FOR VALUES OF C WHICH DIFFER FOR INDIVIDUAL PIPES A SIMPLE
C       PROGRAMMING CHANGE TO READ THE C VALUES IN AS DATA WOULD BE
C       SUFFICIENT.
C       H IS THE HEAD LOSS IN EACH PIPE IN FEET. H = S*PL
C       HOVQ IS THE RATIO H/Q.
C       SHOVQ IS THE SUM OF THE RATIOS H/Q.
C       SUMH IS THE SUM OF THE HEAD LOSSES H.
C       CORR IS THE CORRECTION TO BE APPLIED TO THE LOOPS.
C       CORR = -SUMH/(1.85*SHOVQ).
C       SCOR IS THE SUM OF THE ABSOLUTE VALUES OF THE CORR(L)'S FOR EACH
C       LOOP. THIS VALUE IS COMPARED TO AN EPSILON. IF THE VALUE OF SCOR
C       IS LESS THAN EPS THE COMPUTATION STOPS. IF SCOR IS GREATER THAN
C       EPS, ANOTHER ITERATION IS MADE.
        DIMENSION Q(18),D(18),PL(18),S(18),H(18),HOVQ(18),SHOVQ(4),SUMH(4)
       1,CORR(4),SQ(18),J(4),ID(4,7)
        C = 110.
        NUML = 4
        NUMPI = 18
        EPS = 0.002
        READ(5,20)(J(L), L = 1,NUML)
     20 FORMAT(4I5)
        WRITE(6,36) (J(L), L = 1,NUML)
     36 FORMAT(1X,4I5)
        READ(5,21) (D(I),Q(I),PL(I),I=1,NUMPI)
     21 FORMAT(9F6.3)
        WRITE(6,29) (D(I),Q(I),PL(I),I=1,NUMPI)
     29 FORMAT(1X,9F6.3)
        DO 23 L = 1,NUML
        M = J(L)
        DO 23 K = 1,M
        READ(5,24) ID(L,K)
     23 WRITE(6,35) ID(L,K)
     24 FORMAT(I5)
     35 FORMAT(1X,I5)
        R = C*0.0103
     17 SCOR = 0.
        DO 1 I = 1,NUMPI
        IF(Q(I).LT.0.) SQ(I) = -Q(I)
        IF(Q(I).GE.0.) SQ(I) = Q(I)
        S(I) = (SQ(I)/(R*D(I)**2.63))**1.85
        H(I) = S(I)*PL(I)
      1 HOVQ(I) = H(I)/SQ(I)
        DO 16 L = 1,NUML
        SUMH(L) = 0.
        SHOVQ(L) = 0.
        M = J(L)
        DO 7 K = 1,M
        IF(ID(L,K))3,3,4
      3 I = -ID(L,K)
        SHOVQ(L) = SHOVQ(L) + HOVQ(I)
        IF(Q(I))5,5,6
      4 I = ID(L,K)
        SHOVQ(L) = SHOVQ(L) + HOVQ(I)
        IF(Q(I))6,5,5
      6 SUMH(L) = SUMH(L) - H(I)
```

```
      GO TO 7
 5  SUMH(L) = SUMH(L) + H(I)
 7  CONTINUE
    CORR(L) = -SUMH(L)/(1.85*SHOVQ(L))
    M = J(L)
    DO 8 K = 1,M
    IF(ID(L,K))10,10,11
10  I = -ID(L,K)
    Q(I) = Q(I) - CORR(L)
    GO TO 8
11  I = ID(L,K)
    Q(I) = Q(I) + CORR(L)
 8  CONTINUE
16  SCOR = SCOR + ABS(CORR(L))
    IF(SCOR - EPS.GT.0.) GO TO 17
    WRITE(6,25)
25  FORMAT(1H1,20X,15HTABLE OF VALUES,///)
    WRITE(6,26)
26  FORMAT(7X,1HI,11X,4HQ(I),13X,4HS(I),8X,7HH(I)ABS,8X,5HPL(I),///)
    DO 27 I = 1,NUMPI
27  WRITE(6,28) I,Q(I),S(I),H(I),PL(I)
28  FORMAT(6X,I2,3X,E14.4,3X,E14.4,3X,E14.4,3X,F6.3)
    STOP
    END
```

INPUT DATA

```
   7     4     5     6
1.000-8.780 2.000 1.50013.500 4.000 1.500 8.000 1.000
0.833 0.860 1.100 0.666 0.200 1.100 1.000-5.450 2.000
1.000-6.780 2.000 1.000 2.600 1.600 1.000 1.930 1.000
1.000-0.070 1.600 1.333 7.070 1.200 1.333 5.850 1.000
1.167 5.730 0.600 0.666 0.230 1.100 1.167-4.090 1.700
1.167 5.500 2.000 1.167 4.930 0.700 1.167-3.420 1.600
   1
   2
   3
   4
   5
   6
   7
   8
   9
  10
  -3
  -4
  11
  12
  13
  14
  15
  -5
 -14
  16
  17
  18
```

COMPUTER OUTPUT FOR SOLUTION OF EXAMPLE 5-5

I	Q(I)	S(I)	H(I)ABS	PL(I)
1	-0.7030E 01	0.2928E 02	0.5856E 02	2.000
2	0.1525E 02	0.1706E 02	0.6823E 02	4.000
3	0.1003E 02	0.7856E 01	0.7856E 01	1.000
4	0.3048E 01	0.1518E 02	0.1669E 02	1.100
5	0.1605E 01	0.1376E 02	0.1514E 02	1.100
6	-0.3700E 01	0.8932E 01	0.1786E 02	2.000
7	-0.5030E 01	0.1576E 02	0.3153E 02	2.000
8	0.2321E 01	0.3768E 01	0.6029E 01	1.600
9	0.1651E 01	0.2007E 01	0.2007E 01	1.000
10	-0.3489E 00	0.1133E 00	0.1812E 00	1.600
11	0.6632E 01	0.6491E 01	0.7789E 01	1.200
12	0.6412E 01	0.6098E 01	0.6098E 01	1.000
13	0.5292E 01	0.8166E 01	0.4899E 01	0.600
14	-0.5530E 00	0.1917E 01	0.2109E 01	1.100
15	-0.3745E 01	0.4308E 01	0.7324E 01	1.700
16	0.5845E 01	0.9814E 01	0.1963E 02	2.000
17	0.5175E 01	0.7835E 01	0.5485E 01	0.700
18	-0.3075E 01	0.2992E 01	0.4787E 01	1.600

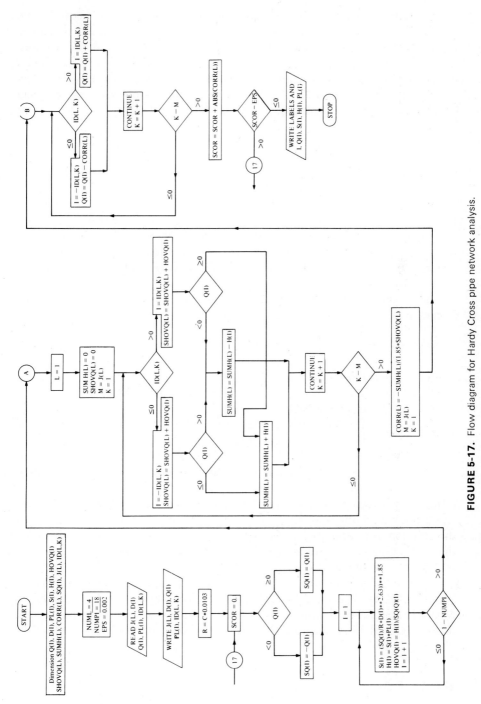

FIGURE 5-17. Flow diagram for Hardy Cross pipe network analysis.

173

A number of computer programs for the solution of pipe networks are available. Many of these are very sophisticated and are designed to handle large systems with pumping stations, storage facilities and other features. Such programs are constantly being improved and the interested reader should consult the literature for detail. [24-26] The elements of network theory presented herein are basic to all of these, however, and should serve as adequate background for the mastering of more complex algorithms.

Linear Theory Method

One additional approach to the analysis of pipe networks will be introduced. It is the *linear theory method* presented by Jeppson. [24,27] This method is relatively simple if external flows to the system are known, but it can be extended to more complex networks containing pumps and reservoirs. It has several advantages over the Hardy Cross method and several other solution algorithms, in that it does not require initialization and converges in a few iterations.[27] The linear theory method is not recommended for solving head-oriented equations or corrective loop equations.

The nonlinear loop equations can be linearized by approximating the head in each pipe by

$$h_{f_i} = (K_i Q_i^{n-1})Q_i = K_i^1 Q_i \tag{5-14}$$

where h_{f_i} = head loss in pipe i

K_i, n_i = coefficients derived from the Hazen–Williams, Manning, or Darcy–Weisbach equation

K_i^1 = coefficient obtained as the product of K_i and Q_i^{n-1}, where Q_i is the estimated flow rate in the pipe

By combining the linearized loop equations with the $j - 1$ continuity equations (j = number of junctions), a system of N linear equations in N unknown results. These equations can then be solved using standard linear methods.

The solution obtained will probably be in error since the Q_i's produced by the solution will probably differ from original estimates. Through iteration the estimates of $Q_i(m)$ will approach closely those of $Q_i(m - 1)$, and a correct solution will be obtained.

The initial assumption made is that $K_i^1 = K_i$, and this avoids estimating initial values for the Q_i's. Noting that successive iterative solutions tend to oscillate about the final solution, Wood suggests that after two iterative solutions, flow rates used in subsequent computations should be the average of values obtained in the past two solutions. This can be stated as

$$Q_i(m) = \frac{Q_i(m - 1) + Q_i(m - 2)}{2} \tag{5-15}$$

To illustrate this procedure, consider the pipe network given in Fig. 5-10. There are five junctions, and thus four independent continuity equations. These are

$$Q_{AB} + Q_{AD} = 2.1$$
$$-Q_{AD} + Q_{BC} = -0.5$$
$$-Q_{AB} - Q_{BC} + Q_{BE} = 0$$
$$-Q_{BE} - Q_{CE} = -1.6$$

The negative sign is used to represent pipe flows into a junction and external flows away from a junction. The two head-loss equations are

$$K^1_{AB} Q_{AB} - K^1_{AD} Q_{AD} - K^1_{DC} Q_{DC} - K^1_{BC} Q_{BC} = 0$$
$$K^1_{BC} Q_{BC} + K^1_{BE} Q_{BE} - K^1_{CE} Q_{CE} = 0$$

where the K^1 values are obtained as indicated in Eq. 5-14.

5-9

SYSTEM LAYOUT AND DESIGN

The design of a water-distribution network involves the selection of a system of pipes so that the predicted design flows can be carried with head losses which do not exceed those deemed necessary for adequate operation of the system. Normally, the design flows should be based on estimated future requirements, since a distribution system is expected to provide effective service for many years (often as long as 100 years). A logical sequence of design and layout operations is as follows:

1   On a development plan of the area to be serviced, sketch the tentative location of all water mains that will be needed to supply the area. The completed drawing should differentiate between proposed feeder mains and smaller service mains. The various pipelines comprising the system should be interconnected at intervals of 1200 ft or less. Looped feeder systems are desirable and should be used wherever possible. Two small feeder mains running parallel several blocks apart are preferable to a single large main with an equal or slightly larger capacity than the two mains combined.

2   Using estimated values of the anticipated design flows, select appropriate pipe sizes by assuming velocities of from 3 to 5 fps.

3   Mark the position of building service connections, fire hydrants, and valves. Service connections form the link between the distribution system and the individual consumer. Normally the practice is, one customer per service pipe. Figure 5-18 illustrates the details of a typical service connection for a private residence. Fire hydrants are located to provide complete protection to the area covered by the distribution system. Recommendations regarding average area per hydrant for various populations and required fire flow are given by the

National Board of Fire Underwriters. Hydrants generally should not be farther than about 500 ft apart. Figure 5-19 illustrates a typical fire-hydrant setting.

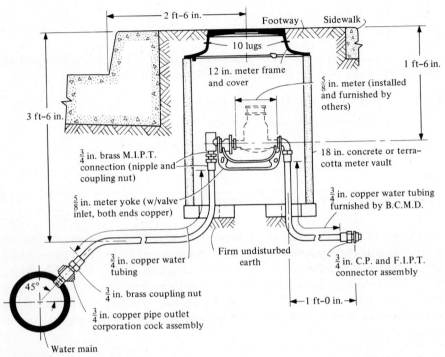

FIGURE 5-18. Typical installation of ¾-in. metered domestic service. (Courtesy Baltimore County Maryland, Department of Public Works.)

4 Using the Hazen–Williams or other appropriate method, compute pressures at key points in the distribution system. From this information, construct a map of pressure contours. Kincaid presents detailed information on requirements prior to a thorough network analysis.[16]

5 Revise the initial trial network as necessary to get adquate operating pressures. The pressure-contour map will help select those parts of the system not adequately sized.

It must be emphasized that although a network analysis might yield very reliable results for the case studied, the results are no better than the initial assumptions of inputs, takeoffs, and pipe roughness.

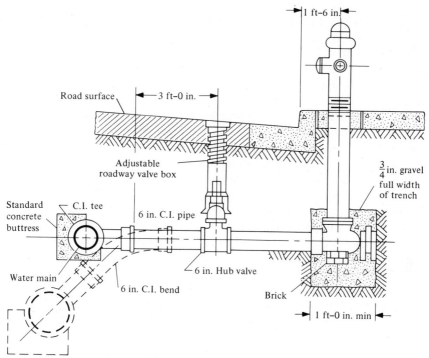

FIGURE 5-19. Typical fire-hydrant setting. (Courtesy Baltimore County, Maryland, Department of Public Works.)

5-10

DISTRIBUTION RESERVOIRS AND SERVICE STORAGE

Distribution reservoirs provide service storage to meet the widely fluctuating demands often imposed on a distribution system, to provide storage for fire fighting and emergencies, and to equalize operating pressures. They are either elevated or at or just below ground level.

The main categories of storage reservoirs are surface reservoirs, standpipes, and elevated tanks. Common practice is to line surface reservoirs with concrete, gunite, asphalt, or an asphaltic membrane. Surface reservoirs may be covered or uncovered. Whenever possible a cover should be considered for the prevention of contamination of the water supply by animals or humans and to prevent the formation of algae.

Standpipes or elevated tanks are normally employed where the construction of a surface reservoir would not provide sufficient head. A standpipe is essentially a tall cylindrical tank whose storage volume includes an upper portion (the useful storage), which is above the entrance to the discharge pipe, and a lower portion (supporting storage), which acts only to support the useful storage and provide the required head. For this reason, standpipes over 50 ft high are usually uneconomical. Steel, concrete, and wood are used in the con-

struction of standpipes and elevated tanks. When it becomes more economical
to build the supporting structure for an elevated tank than to provide for the
supporting storage in a standpipe, the elevated tank is used. A good discussion
of the construction details and characteristics of various kinds of distribution
reservoirs is given by Babbitt, Doland, and Cleasby.[20]

Distribution reservoirs should be located strategically for maximum benefits. Normally the reservoir should be near the center of use, but in large metropolitan areas a number of distribution reservoirs may be located at key points.
Reservoirs providing service storage must be high enough to develop adequate
pressures in the system they are to serve. A central location decreases friction
losses by reducing the distance from supply point to the area served. Positioning
the reservoir so that pressures may be approximately equalized is an additional
consideration of importance. Figure 5-20 will illustrate this point. The location
of the tank as shown in part (a) results in a very large loss of head by the time
the far end of the municipality is reached. Thus pressures too low will prevail
at the far end or excessive pressures will be in evidence at the near end. In part
(b) it is seen that pressures over the whole municipal area are much more uniform for periods of both high and low demand. Note that during periods of
high demand the tank is supplying flow in both directions (being emptied),
while during periods of low demand the pump is supplying the tank and the
municipality.

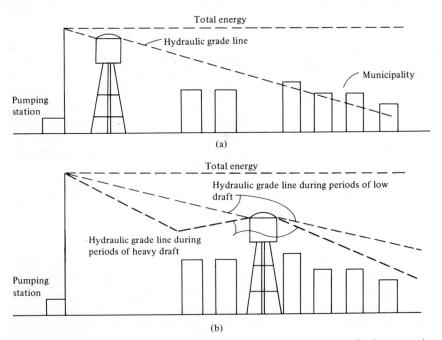

FIGURE 5-20. Pressure distribution as influenced by the location of a distribution reservoir.

The amount of storage to be provided is a function of the capacity of the distribution network, the location of the service storage, and the use to which it is to be put. When water-treatment facilities are required, it is preferable to operate them at a uniform rate such as the maximum daily rate. It is also usually desirable to operate pumping units at constant rates. Demands on the system in excess of these rates must therefore be met by storage, previously defined as operating storage. Requirements for fire-fighting purposes should be sufficient to provide fire flows for 10–12 hr in large communities and for 2 or more hours in smaller ones. Emergency storage is provided to sustain the community's needs during periods when the inflow to the reservoir is shut off, for example, through a failure of the supply work, failure of pumping equipment, or need to take a supply line out of service for maintenance or repair. The length of time the supply system is expected to be out of service dictates the amount of emergency storage to be provided. Emergency storage volumes sufficient to last for several days are common.

The amount of storage required for emergency and fire-fighting purposes is readily computed once the time period over which these flows are to be provided has been selected.[21] An emergency storage of 3 days for a community of 8000 having an average use rate of 150 gpcd is $3 \times 150 \times 8000 = 3.6$ mg. Given that a fire flow of 2750 gpm must be provided for a duration of 10 hours, this means a total fire-fighting storage of 1.65 mg. To the sum of these values, an additional equalizing or operating storage requirement would be added. The determination of this volume is slightly more complex and needs further explanation.

To compute the required equalizing or operating storage, a mass diagram or hydrograph indicating the hourly rate of consumption is required. The procedure to be used in determining the needed storage volume is then:

1   Obtain a hydrograph of hourly demands for the maximum day. This may be obtained through a study of available records, by gaging the existing system during dry periods when lawn-sprinkling demands are high, or by using available design criteria such as presented in Chapter 4 to predict a hydrograph for some future condition of development.
2   Tabulate the hourly demand data for the maximum day as shown in Table 5-5.
3   Find the required operating storage by using mass diagrams such as in Figs. 5-21 and 5-22, the hydrograph of Fig. 5-23, or the values tabulated in column (6) of Table 5-5.

The required operating storage is found by using a mass diagram with the cumulative pumping curve plotted on it. Figure 5-21 illustrates this diagram for a uniform 24-hour pumping rate. Note that the total volume pumped in 24 hours must equal the total 24-hour demand, and thus the mass curve and cumulative pumping curve must be coincident at the origin and at the end of the day. Next, construct a tangent to the mass curve which is parallel to the pumping

TABLE 5-5

HOURLY DEMAND FOR THE MAXIMUM DAY

(1) Time	(2) Average Hourly Demand Rate (gpm)	(3) Hourly Demand (gal)	(4) Cumulative Demand (gal)	(5) Hourly Demand as a Per-cent of Average	(6) Average Hourly Demand Minus Hourly Demand: 286,250 − col. (3) −	+
12 –.M.	0	0	0	0	—	—
1 A.M.	2,170	130,000	130,000	45.4		156.250
2	2,100	126,000	256,000	44.1		160,250
3	2,020	121,000	377,000	42.3		165,250
4	1,970	118,000	495,000	41.3		168,250
5	1,980	119,000	614,000	41.6		167,250
6	2,080	125,000	739,000	43.17		161,250
7	3,630	218,000	957,000	76.2		68,250
8	5,190	312,000	1,269,000	108.9	25,750	
9	5,620	337,000	1,606,000	117.8	50,750	
10	5,900	354,000	1,960,000	123.6	67,750	
11	6,040	363,000	2,323,000	126.7	76,750	
12	6,320	379,000	2,702,000	132.4	92,750	
1 P.M.	6,440	387,000	3,089,000	135.2	100,750	
2	6,370	382,000	3,471,000	133.4	95,750	
3	6,320	379,000	3,850,000	132.4	92,750	
4	6,340	381,000	4,231,000	133.0	94,750	
5	6,640	399,000	4,630,000	139.5	112,750	
6	7,320	439,000	5,069,000	153.3	152,750	
7	9,333	560,000	5,629,000	195.5	273,750	
8	8,320	499,000	6,128,000	174.4	212,750	
9	5,050	303,000	6,431,000	105.8	16,750	
10	2,570	154,000	6,585,000	53.8		132,250
11	2,470	148,000	6,733,000	51.7		138,250
12	2,290	137,000	6,870,000	47.9		149,250
TOTAL		6,870,000			1,466,500	1,466,500

$$\text{average hourly demand} = \frac{6,870,000}{24} = 286,250 \text{ gal}$$

curve at point A in the figure. Then draw a second parallel tangent to the mass curve at point C and drop a vertical from C to an intersection with tangent AB at B. The required storage is equal to the magnitude of the ordinate CB measured on the vertical scale. In the example shown, the necessary storage volume is found to be 1.47 million gal for a 24-hour pumping period.

Readers should recognize that the reservoir is full at A, empty at C, is filling whenever the slope of the pump curve exceeds that of the cumulative

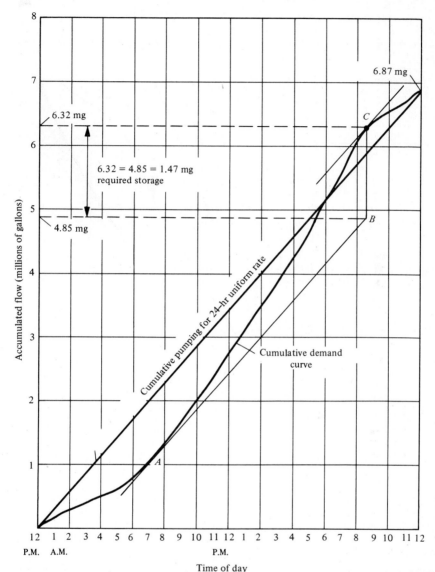

FIGURE 5-21. Operating storage for 24-hour pumping determined by use of the mass diagram.

demand curve, and is being drawn down when the rate of demand exceeds the rate of pumping.

It is often desirable to operate an equalizing reservoir so that pumping will take place at a uniform rate but for a period less than 24 hours. In small communities, for example, it is often advantageous to pump only during the normal working day. It may also be more economical to operate the pumping station at off-peak periods when electric power rates are low.[21]

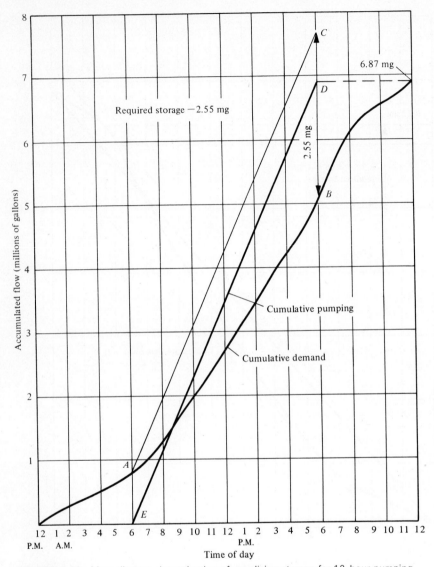

FIGURE 5-22. Mass diagram determination of equalizing storage for 12-hour pumping.

Figure 5-22 illustrates the operation of a storage reservoir where pumping occurs during the period between 6 A.M. and 6 P.M. only. To find the required storage in this case, construct the cumulative pumping curve *ED* so that the total volume of 6.87 mg is pumped uniformly from 6 A.M. to 6 P.M. Then project point *E* vertically upward to an intersection with the cumulative demand curve at *A*. Construct line *AC* parallel to *ED*. Point *C* will be at the intersection of line *AC* with the vertical extended upward from 6 P.M. on the abscissa. The required storage equals the value of the ordinate CBD. Numerically it is 2.55 mg and exceeds the storage requirement for 24-hour pumping.

Another graphical solution to the storage problem may be obtained as outlined in Fig. 5-23. The figure is a plot of the demand hydrograph for the maximum day. For uniform 24-hour pumping, the pumping rate will be equal to the mean hourly demand. This is shown as line PQ. The required storage is then obtained by planimetering or determining in some other manner the area between curve BEC and line PQ. Conversion of this area to units of volume yields the required storage of 1.47 mg. The required storage for 24-hour pumping may also be determined by summing either the plus or minus values of column (6) in Table 5-5.

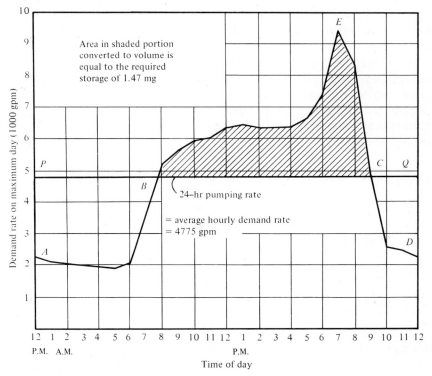

FIGURE 5-23. Graphical determination of equalizing storage.

Unless pumping follows the demand curve or demand hydrograph, storage will be required. Figure 5-23 shows that a maximum pumping rate of about 9400 gpm will be required with no storage, whereas if storage is provided a maximum pumping rate of 4775 gpm (about 50% of that required with no storage) will suffice. This example tends to illustrate the economics of providing operating storage.

Variable-rate pumping is normally not economical. In practice it is common to provide storage and pumping facilities so that pumping at the average rate for the maximum day can be maintained. On days of lesser demand, some pumping units will stand idle. Another operational procedure is to provide enough storage for pumping at the average rate for the average day, with idle reserve capacity, and then to overload all available units on the maximum day. Provision of pumping and storage capacity to meet peak demands experienced for a few hours only every few years has been found to be economically impractical.

Analyses of distribution systems are commonly concerned with the pipe network, topographic conditions, pumping-station performance, and the operating characteristics of the distribution storage. One, all, or any combination of these features may be the object of study. Where multiple sources of supply operate under variable-head conditions, the hydraulic balancing of the system becomes more complex. The simple system of Fig. 5-24 illustrates this point

Considering that the demand for water by the municipal load center fluctuates hourly, it is evident there are essentially two modes of operation of the given distribution system. When municipal requirements are light, such as in the early morning hours, the pumping station will meet these demands and in addition supply the reservoir. The solution of the problem may then be had by writing the equations

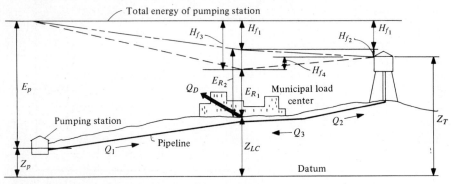

FIGURE 5-24. Modes of operation of a distribution system.

(1) $\quad Q_1 - Q_D = Q_2$

(2) $\quad Z_p + E_p = Z_{LC} + E_{R_2} + H_{f_1}$

(3) $\quad Z_{LC} + E_{R_2} = Z_T + H_{f_2}$

(2 + 3) $\quad H_{f_1} + H_{f_2} = E_p + Z_p - Z_T$

where $\quad Q_1 =$ flow from the pump

$\qquad Q_D =$ municipal demand

$\qquad Q_2 =$ flow to the tank

$Z_p, Z_{LC}, Z_T =$ elevation above the arbitrary datum ($Z_T =$ elevation of water surface in tank)

E_p = energy produced by the pump

E_{R_2} = residual energy of the load center (pressure head plus velocity head)

H_{f_1}, etc. = friction-head losses

If Q_D, Z_p, Z_{LC}, and Z_T are specified, the equations may be solved by selecting values of E_{R_2} and then solving for H_{f_1} and H_{f_2}. When a solution is reached such that equation (2 + 3) is satisfied, Q_1 and Q_2 may be computed.

When demands are high, both the tank and the pump will supply the community. The direction of flow will then be reversed in the line from the tank to the pump and the applicable equation will be

(1) $$Q_1 + Q_3 = Q_D$$

(2) $$Z_p + E_p = Z_{LC} + E_{R_1} + H_{f_3}$$

(3) $$Z_{LC} + E_{R_1} + H_{f_4} = Z_T$$

Again, an assumed value for E_R will be taken and trial solutions carried out until equation (1) is satisfied. Note that the foregoing illustration is a simple case, since Z_T has been specified. Actually, since Z_T fluctuates with time, it is necessary to have information on storage volume available versus water elevation in the tank so that at any specified condition of draft, the actual value for Z_T can be determined and used in the computations.

Water-distribution systems generally are considered to be a composite of four basic constituents: the pipe network, storage, pump performance, and the pumping station and its suction source. These components must be integrated into a functioning system for various schedules of demand. A thorough analysis of each system must be made to ensure that it will operate satisfactorily under all anticipated combinations of demand and hydraulic-component characteristics. The system may work well under one set of conditions but will not necessarily be operable under some other set. A comprehensive system balance requires an hourly simulation of performance for the expected operating schedule.[12,14]

There is an infinite number of possible arrangements of the basic components in a distribution system. The hydraulic analyses applicable to each system are, however, common to all and have been illustrated in detail in this chapter.

Pumping

Pumping equipment forms an important part of the transportation and distribution facilities for water and wastes. Requirements vary from small units used to pump only a few gallons per minute to large units capable of handling several hundred cubic feet per second. The two primary types of pumping equipment of interest here are the centrifugal pump and the displacement pump. Air-lift pumps, jet pumps, and hydraulic rams are also used in special applications. In water and sewage works the centrifugal pump finds the widest use.

Centrifugal pumps have a rotating element (impeller) which imparts energy to the water. *Displacement pumps* are often of the reciprocating type where a piston draws water into a closed chamber and then expels it under pressure. Reciprocating pumps are widely used to handle sludge in sewage-treatment works.

Electrical power is the primary source of energy for driving pumping equipment, but gasoline, steam, and diesel power are also used. Often a standby engine powered by one of these forms is included in primary pumping stations to operate in emergency situations when electric power fails.

5-11

PUMPING HEAD

Anticipated operating conditions of the system must be known to effectively design a pumping station. A knowledge of the *total dynamic head*, (TDH) against which the pump must operate is needed. The TDH is composed of the difference in elevation between the pump center line and the elevation to which the water is to be raised, the difference in elevation between the level of the suction pool and the pump center line, the frictional losses encountered in the pump, pipe, valves and fittings, and the velocity head. Expressed in equation form, this becomes

$$\text{TDH} = H_L + H_F + H_V \qquad (5\text{-}16)$$

where H_L is the total static head or elevation difference between the pumping source and the point of delivery, H_F is the total friction head loss, and H_V is the velocity head $V^2/2g$. Figure 5-25 illustrates the total static head.

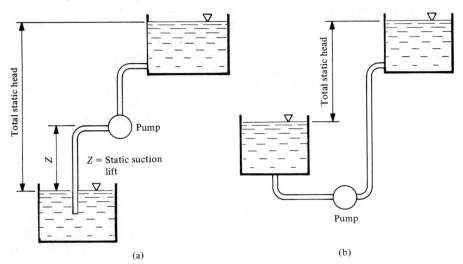

FIGURE 5-25. Total static head. (a) Intake below pump centerline; (b) intake above pump centerline.

5-12

POWER

For a known discharge and total pump lift, the theoretical horsepower required may be found by using

$$\text{hp} = \frac{Q\gamma H}{550} \qquad (5\text{-}17)$$

where Q = discharge, cfs
$\quad\quad \gamma$ = specific weight of water
$\quad\quad H$ = total dynamic head
$\quad\quad 550$ = conversion from foot-pounds per second to horsepower

The actual horsepower required is obtained by dividing the theoretical horse-power by the efficiency of the pump and driving unit.

5-13

CAVITATION

Cavitation is the phenomenon of cavity formation or the formation and col-lapse of cavities.[23] Cavities are considered to develop when the absolute pressure in a liquid reaches the vapor pressure related to the liquid temperature. Under severe conditions, cavitation can result in breakdown of pumping equipment. As the net positive suction head (NPSH) for a pump is reduced, a point is reached where cavitation becomes detrimental. This point is usually referred to as the minimum net positive suction head ($\text{NPSH}_{\min}$) and is a function of the type of pump and the discharge through the pump. NPSH is calculated as

$$\text{NPSH} = \frac{V_1^2}{2g} + \frac{p_1}{\gamma} - \frac{p_v}{\gamma} \qquad (5\text{-}18)$$

where V_1 is the velocity of flow at the centerline of the inlet to the pump, p_1 is the pressure at the centerline of the pump inlet, and p_v is the vapor pressure of the fluid. Referring to Fig. 5-25a and writing the energy equation between the intake pool and the inlet to the pump, we have

$$\frac{p_a}{\gamma} = \frac{V^2}{2g} + \frac{p_1}{\gamma} + Z + H_L$$

or

$$\frac{p_a}{\gamma} - Z - H_L = \frac{V^2}{2g} + \frac{p_1}{\gamma}$$

Subtracting p_v/γ from both sides, we have

$$\frac{p_a}{\gamma} - Z - H_L - \frac{p_v}{\gamma} = \frac{V^2}{2g} + \frac{p_1}{\gamma} - \frac{p_v}{\gamma}$$

This may be written as

$$\frac{p_a}{\gamma} - Z - H_L - \frac{p_v}{\gamma} = \text{NPSH}$$

The minimum value of the static lift is then determined as

$$Z_{min} = \frac{p_a - p_v}{\gamma} - \text{NPSH}_{min} - H_L. \tag{5-19}$$

The required NPSH for any pump must be obtained from the manufacturer. This value can then be checked against the proposed installation using Eqs. 5-18 and 5-19 to assure that the available NPSH is greater than the manufacturer's requirement.

5-14

SYSTEM HEAD

The system head is represented by a plot of TDH versus discharge for the system being studied. Such plots are very useful in selecting pumping units. It should be clear that the system head curve will vary with flow since H_F and H_V are both a function of discharge. In addition, the static head H_L may vary as a result of

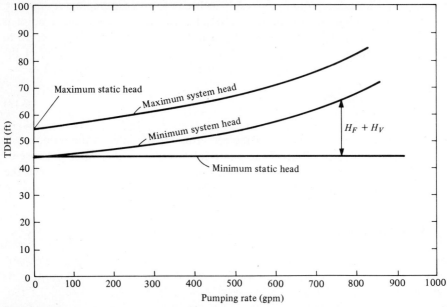

FIGURE 5-26. System-head curves for a fluctuating static pumping head.

fluctuating water levels and similar factors, and it is often necessary to plot system-head curves covering the range of variations in static head. Figure 5-26 illustrates typical system-head curves for a fluctuating static water level.

5-15

PUMP CHARACTERISTICS

Each pump has its own characteristics relative to power requirements, efficiency, and head developed as a function of rate of flow. These relationships are usually given as a set of pump characteristic curves for a specified speed. They are used in conjunction with system-head curves to select correct pumping equipment for a particular installation. A set of characteristic curves are shown in Fig. 5-27.

At no flow, the head is known as the *shutoff head*. The pump head may rise slightly or fall from the shutoff value as discharge increases. Ultimately, however, the head for any centrifugal pump will fall with increase in flow. At maximum efficiency the discharge is known as the *normal* or *rated discharge* of the pump. Varying the pump discharge by throttling will lower the efficiency of the unit. By changing the speed of the pump, discharge can be varied within a

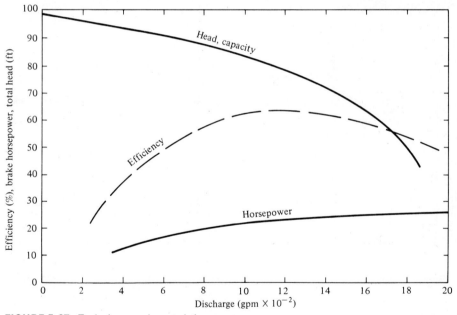

FIGURE 5-27. Typical pump characteristic curves.

certain range without a loss of efficiency. The most practical and efficient approach to a variable-flow problem is to provide two or more pumps in parallel so that the flow may be carried at close to the peak efficiency of the units which are operating.

The normal range of efficiencies for centrifugal pumps is between 50 and 85%, although efficiencies in excess of 90% have been reported. Pump efficiency usually increases with the size and capacity of the pump.[23]

5-16

SELECTION OF PUMPING UNITS

Normally the engineer is given the system-head characteristics and is required to find a pump or pumps to deliver the anticipated flows. This is done by plotting the system-head curve on a sheet with the pump characteristic curves. The operating point is at the intersection of the system-head curve and the pump-head capacity curve. This gives the head and flow at which the pump will be operating. A pump should be selected so that the operating point is also as close as possible to peak efficiency. This procedure is shown on Fig. 5-28.

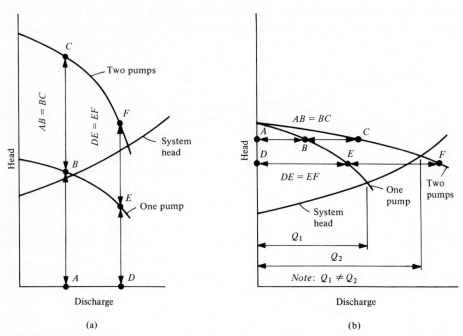

FIGURE 5-28. Characteristic curves for series and parallel pump operations of equal pumps. (a) Series; (b) parallel.

Pumps may be connected in series or in parallel. For series operation at a given capacity, the total head equals the sum of the heads added by each pump. For parallel operation, the total discharge is multiplied by the number of pumps for a given head. It should be noted, however, that when two pumps are used in series or parallel, neither the head or capacity for a given system head curve is doubled (Fig. 5-28).

●EXAMPLE 5-6

A pumping station is to be designed for an ultimate capacity of 1200 gpm at a total head of 80 ft. The present requirements are that the station deliver 750 gpm at a total head of 60 ft. One pump will be required as a standby.

○*Solution*

(a)   The total curve for dynamic head versus discharge is plotted as shown in Fig. 5-29. Values for the curve are obtained as indicated in Sec. 5-11.

(b)   Consider that three pumps will ultimately be needed (one as a standby). Determine the design flows as follows:

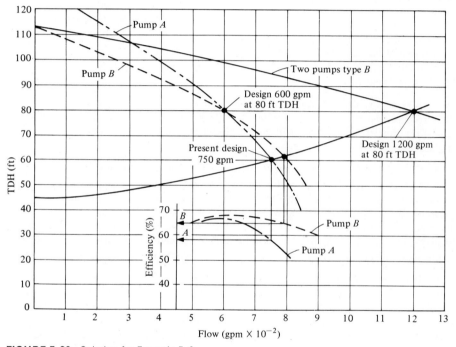

FIGURE 5-29.   Solution for Example 5-6.

1   Two pumps at 1200 gpm at 80 ft of TDH
2   One pump at 1200/2 = 600 gpm at 80 ft of TDH
3   One pump must also be able to meet the present requirements of 750 gpm at 60 ft of TDH.

(c)   From manufacturers' catalogs, two pumps, *A* and *B*, are found that will meet the specifications. The characteristic curves for each pump are shown in Fig. 5-29. The intersection of the characteristic curves with the system-head curve indicates that pump *A* can deliver 750 gpm at a TDH of 60 ft while pump *B* can deliver 790 gpm at a TDH of 62 ft. A check of the efficiency curves for each pump indicates that pump *B* will deliver the present flow at a much greater efficiency than pump *A*. Therefore, select pump *B*.

(d) For the present, select two pumps of type *B* and use one as a standby. For the future add one more pump of type *B*.

Operating characteristics for a wide range of pump sizes and speeds are available from pump manufacturers. Special equipment manufacturered to satisfy the prescribed requirements of the customer must customarily satisfy an acceptance test after it has been installed. Babbitt, Doland, and Cleasby give a detailed discussion of a variety of pump types.[20]

PROBLEMS

5-1 Water is pumped 9 mi, from a reservoir at elevation 100 ft to a second reservoir at elevation 210 ft. The pipeline connecting the reservoirs is 54 in. in diameter. It is concrete and has an absolute roughness of 0.003. If the flow is 25 mgd and pumping-station efficiency is 80%, what will be the monthly power bill if electricity costs 3 cents per kilowatthour?

5-2 A reservoir at elevation 700 ft is to supply a second reservoir at elevation 460 ft. The reservoirs are connected by 1300 ft of 24-in. cast-iron pipe and 2000 ft of 20-in. cast-iron pipe in series. What will be the discharge delivered from the upper reservoir to the lower one?

5-3 Consider that the water needs of a city for the next 40 years may be met by constructing a single 10-in. main for the first 15 years followed by a second 10-in. main for the next 25 years. At this time both mains will be in service. An alternative solution is to build one 18-in. line which will serve the entire 40-year period. Consider that the cost per foot installed is $20.00 for the first 10-in. line, $26.00 for the next 10-in. line, and $31.00 for the 18-in. line. The combined interest and amortization rate is 7.0%/yr. Find the most economical method of meeting the community's water requirements.

5-4 It is necessary to pump 6000 gpm of water from a reservoir at an elevation of 900 ft to a tank whose bottom is at an elevation of 1100 ft. The pumping unit is located at elevation 900. The suction pipe is 24 in. in diameter and very short, so head losses may be neglected. The pipeline from the pump to the upper tank is 410 ft long and is 20 in. in diameter. Consider the minor losses in the line to equal 2.5 ft of water. The maximum depth of water in the tank is 38 ft and the supply lines are cast iron. Find the maximum lift of the pump and the horsepower required for pumping if the pump efficiency is 76%.

5-5 If a flow of 5.0 cfs is to be carried by an 11,000-ft cast-iron pipeline without exceeding a head loss of 137 ft, what must the pipe diameter be?

5-6 A 48-in. water main carries 79 cfs and branches into two pipes at point *A*. The branching pipes are 36 in. and 20 in. in diameter and 2800 and 5000 ft long, respectively. These pipes rejoin at point *B* and again form a single 48-in. pipe. If the friction factor is 0.022 for the 36-in. pipe and 0.024 for the 20-in. pipe, what will be the discharge in each branch?

5-7 Water is pumped from a reservoir whose surface elevation is 1390 ft to a second reservoir whose surface elevation is 1475 ft. The connecting pipeline is 4500 ft long and

12 in. in diameter. If the pressure during pumping is 80 psi at a point midway on the pipe at elevation 1320 ft, find the rate of flow and the power exerted by the pumps. Also, plot the hydraulic grade line. Assume that $f = 0.022$.

5-8 A concrete channel 18 ft wide at the bottom is constructed with side slopes of 2.2 horizontal to 1 vertical. The slope of the energy gradient is 1 in 1400 and the depth of flow is 4.0 ft. Find the velocity and the discharge.

5-9 A rectangular channel is to carry 200 cfs. The mean velocity must be greater than 2.5 fps. The channel bottom should be about twice the channel depth. Find the channel cross section and the required channel slope.

5-10 A rectangular channel carries a flow of 10 cfs/ft of width. Plot a curve of specific energy versus depth. Compute the minimum value of specific energy and the critical depth. What are the alternative depths for $E_s = 5.0$?

5-11 A water-transmission line is to connect point A with points B, C, and D in succession. The line must be capable of carrying 180 mgd with a total head loss not to exceed 40 ft. Use a Hazen–Williams C of 100. Line AB is 25,000 ft, line BC is 28,000 ft, and line CD is 15,000 ft. Find the most economical diameter for each of the three sections of pipe. The dollar cost per foot of pipe installed is as follows:

Pipe	Size of Pipe (in.)					
	84	96	108	120	132	144
AB	73	86	98	121	142	180
BC	96	110	129	160	175	230
CD	129	138	163	190	201	275

5-12 A water-transportation system is to consist of a tunnel, a pipeline, and a canal Given the data on cost (dollars) versus head loss shown in the table, find the most economical division of the total head of 120 ft.

	Head Loss (ft)							
	10	20	30	40	50	60	70	80
Tunnel	—	120	98	80	60	53	49	46
Pipeline	—	105	76	53	47	43	40	39
Canal	110	55	35	29	26	23	20	19

5-13 Determine an equivalent pipe for the system shown in the accompanying diagram.

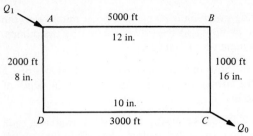

5-14   From the diagram, compute: (a) the total head loss from A to C; (b) P_a if $P_c = 25$ psi; and (c) the flow in each line.

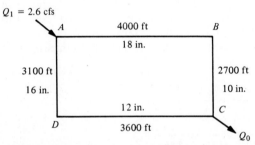

5-15   In the system shown, determine the flow distribution by a Hardy Cross analysis. Find the sizes of pipes EF and GF. Use a Hazen–Williams C of 100.

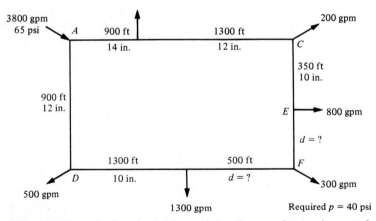

5-16   Compute the flow in the lines of the diagram shown. Assume that the elevation at A is 1500 ft and the elevation at D is 1420 ft. The pressure at D is 38 psi. Find the pressure at A.

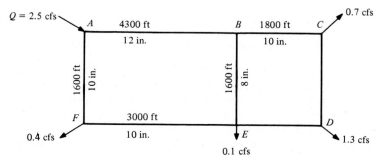

5-17   In the diagram shown, $P_A = 70$ psi and $P_C = 50$ psi. Find the discharge in all lines and determine the size of lines BC and CD.

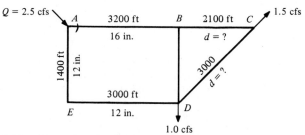

5-18   Given the pipe layout shown, determine the length of an equivalent 18-in. pipe.

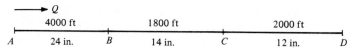

5-19   Given the pipe layout shown, find the diameter of an equivalent 3000-ft pipe.

Q ⟶
	1000 ft		1600 ft		2000 ft	
A	18 in.	B	14 in.	C	12 in.	D

5-20   Given the following average hourly demand rates in gallons per minute, find the uniform 24-hour pumping rate and the required storage.

12 P.M.	0	12 A.M.	6300
1 A.M.	1900	1 P.M.	6500
2	1800	2	6460
3	1795	3	6430
4	1700	4	6500
5	1800	5	6700
6	1910	6	7119
7	3200	7	9000
8	5000	8	8690
9	5650	9	5220
10	6000	10	2200
11	6210	11	2100
		12	2000

5-21 Solve Prob. 5-20 if the period of pumping is from 6 A.M. to 6 P.M. only.

5-22 Solve Prob. 5-20 by the method outlined in Fig. 5-25, assuming 24-hour pumping.

5-23 Use the data given in Prob. 5-20 and add a fire-flow requirement of 4.7 mgd. Refer to Fig. 5-24 and assume that this flow requirement can be delivered to the city by the pumping station–tank system shown. The pipeline from the pumping station to the town is 30 in. in diameter and 5.6 mi long, and the pipeline from the town to the tank is 28 in. in diameter and 3.8 mi long. Elevations are 415 ft at the pump center line, 570 ft at the ground surface in the town, 637 ft at the ground surface at the tank, and 700 ft at the bottom of the tank. The tank is 30 ft high and contains a total of 1.5 mg storage. Find the pressure in town during the peak average hour plus fire flow if the pumping station can maintain a maximum pressure of 130 psi and the tank is half-full.

REFERENCES

1. P. C. G. Isaac, "Roman Public Works, Engineering," University of Durham, King's College, *Department of Civil Engineering Bull. No. 13* (February 1958).
2. Anonymous, "Canadian River Project Moves Ahead," *Public Works* (April 1963).
3. W. E. Warne, "The Big Flow," *California*, magazine of California State Chamber of Commerce (December 1963).
4. A. R. Golze, "The California Water Program in 1964," paper presented before American Right-of-Way Association, Sacramento, Calif. (January 15, 1964).
5. H. M. Morris, *Applied Hydraulics in Engineering.* (New York: The Ronald Press Company, 1963).
6. "Flow of Fluids Through Valves, Fittings, and Pipe," Crane Co., Chicago, Ill., *Tech. Paper No. 410* (1957).
7. J. M. Edmonston and E. E. Jackson, "The Feather River Project and the Establishment of the Optimum Hydraulic Grade Line for the Project Aqueduct," Los Angeles.
8. I. S. Sokolnikoff and E. S. Sokolnikoff, *Higher Mathematics for Engineers and Physicists* (New York: McGraw-Hill Book Company, 1941).
9. R. Weinstock, *Calculus of Variations* (New York: McGraw-Hill Book Company, 1952).
10. J. Hinds, "Economic Water Conduit Size," *Eng. News-Record* (January 1937).
11. N. Joukowsky, "Water Hammer" (translated by O. Simin), *Proc. Am. Water Works Assoc.* 24 (1904): 341–424.
12. M. B. McPherson, "Generalized Distribution Network Head Loss Characteristics," *Proc. Am. Soc. Civil Engrs., J. Hydraulics Div.* 86, no. HY1 (January 1960).
13. C. W. Reh, "Hydraulics of Water Distribution Systems," *University of Illinois Engineering Experiment Station Circular No. 75* (February 1962).
14. M. B. McPherson, "Applications of System Analyzers: A Summary," *University of Illinois Engineering Experiment Station Circular No. 75* (February 1962).
15. H. Cross, "Analysis of Flow in Networks of Conduits or Conductors," *University of Illinois Bull. No. 286* (November 1936).
16. R. G. Kincaid, "Analyzing Your Distribution System," *Water Works Engr.* 97, nos. 2, 6, 10, 16, 21 (January, March, May, August, October 1944).
17. M. S. McIlroy, "Direct Reading Electric Analyzer for Pipeline Networks," *J. Am. Water Works Assoc.* 42, no. 4 (April 1950).

18. V. A. Appleyard and F. P. Linaweaver, Jr., "The McIlroy Fluid Analyzer in Water Works Practice," *J. Am. Water Works Assoc.* 49, no. 1 (January 1957).

19. M. B. McPherson and J. V. Radziul, "Water Distribution Design and the McIlroy Network Analyzer," *Proc. Am. Soc. Civil Engrs.* 84, Paper 1588 (April 1958).

20. H. E. Babbitt, J. J. Doland, and J. L. Cleasby, *Water Supply Engineering* (New York: McGraw-Hill Book Company, 1962).

21. J. E. Kiker, "Design Criteria for Water Distribution Storage," *Public Works* (March 1964).

22. G. E. Russell, *Hydraulics* (New York: Holt, Rinehart and Winston, Inc., 1957).

23. R. M. Olson, *Engineering Fluid Mechanics*, 3rd ed. (New York: IEP Publishers, 1973).

24. R. W. Jeppson, "Steady Flow Analysis of Pipe Networks," Utah State University, Department of Civil Engineering, Logan, Utah (September 1974).

25. J. L. Gerlt and G. F. Haddix, "Distribution System Operation Analysis Model," *J. Am. Water Works Assoc.* (July 1975).

26. R. DeMoyer, Jr., and L. B. Horwitz, "Macroscopic Distribution System Modeling," *J. Am. Water Works Assoc.* (July 1975).

27. D. J. Wood and C. O. A. Charles, "Hydraulic Network Analysis Using Linear Theory," *Proc. Am. Soc. Civil Engrs., Hydraulics Div.* 98, no. HY7 (July 1972).

6
Wastewater Systems

The engineer is as much concerned with the collection and transportation of storm-water runoff and waste flows as he is with the transportation and distribution of water to be processed or used for drinking purposes. The engineering importance of the conveyance systems for these flows is indicated by their costs. In 1963 the Committee on Urban Hydrology of the American Society of Civil Engineers stated that the probable direct capital cost of urban drainage works would be about $25 billion in the next 40 years and that the per capita costs are often in excess of $100.[11]

The collection and movement of surface drainage and waste flows from residential, commercial, and industrial areas pose problems of a different nature than those for water supply. Waste must be transported from the point of collection to the treatment or disposal area as quickly as possible to prevent development of septic conditions. In addition, waste flows are highly variable and contain coarse solids which may be floating or suspended. Storm-water runoff is characterized by exceedingly rapid changes in rate of flow during periods of precipitation. These flows commonly carry various forms of debris from the drainage area.

Hydraulic Considerations

Wastewater systems are usually designed as open channels except where lift stations are required to overcome topographic barriers. The hydraulic problems associated with these flows are complicated in some cases by the quality of the fluid, the highly variable nature of the flows, and the fact that an unconfined or free surface exists. The driving force for open-channel flow is gravity. The forces retarding flow are derived from viscous shear along the channel bed.

6-1

UNIFORM FLOW

For the condition of uniform flow, the velocity in an open channel is usually determined by Manning's equation,

$$V = \frac{1.49}{n} R^{2/3} S^{1/2} \tag{6-1}$$

which has been discussed in Chapter 5. Table 5-3 gives values of the roughness coefficient n for various materials used in open channels. Figure 6-1 is a nomograph which facilitates the solution of the equation when $n = 0.013$. Values obtained from the nomograph may be adjusted to other values of n by making the applicable proportional correction.

Many sanitary sewers and storm drains are circular in cross section, and thus it is cumbersome to compute values for the hydraulic radius and the cross-sectional area for conditions when the pipe is flowing partially full. Figure 6-2 is useful in computing partial-flow values from full-flow conditions. Example 6-1 illustrates this technique.

●EXAMPLE 6-1
Given a discharge flowing full of 16 cfs and a velocity of 8 fps, find the velocity and depth of flow when $Q = 10$ cfs.

○Solution
Enter the chart at $10/16 = 62.5\%$ of value for full section. Obtain a depth of flow of 57.5% of full-flow depth and a velocity of $1.05 \times 8 = 8.4$ fps.

The depth at which uniform flow occurs in an open channel is termed the *normal depth*, d_n. This depth can be computed by using Manning's equation for discharge after the cross-sectional area A and the hydraulic radius R have been translated into functions of depth. The solution for d_n is often obtained through trial and error.

For a specified channel cross section and discharge, there are three possible values for the normal depth, all dependent on the channel slope. Uniform flow for

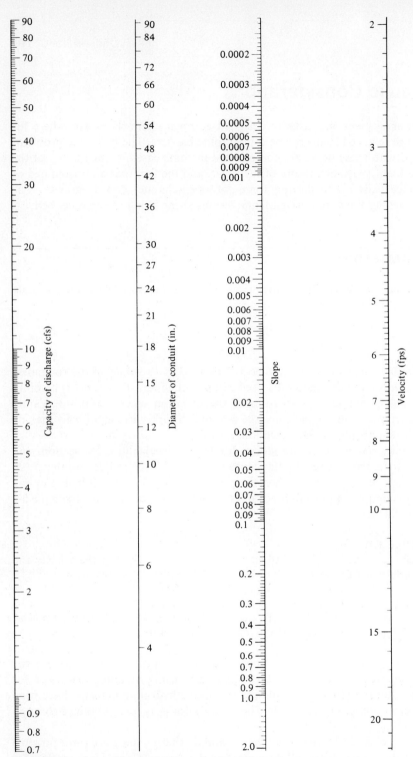

FIGURE 6-1 Nomograph based on Manning's formula for circular pipes flowing full in which $n = 0.013$.

200

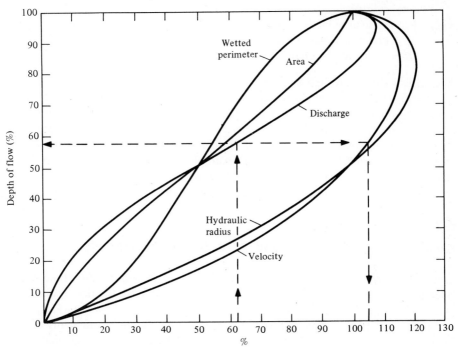

FIGURE 6-2 Hydraulic elements of a circular section.

a given discharge may occur at critical depth, at less than critical depth, or at greater than critical depth. Critical depth occurs when the specific energy is a minimum. Specific energy is defined as

$$E_s = d + \frac{V^2}{2g} \tag{6-2}$$

where d = depth of flow
$\quad\quad\;\, V$ = mean velocity

Flow at critical depth is highly unstable, and designs indicating flow at or near critical depth are to be avoided. For any value of E_s above the minimum, two alternative depths of flow are possible. One is greater than critical depth, the other less than critical depth. The former case indicates subcritical flow, the latter, supercritical flow. Critical depth for a channel can be found by taking the derivative of Eq. 6-2 with respect to depth, setting this equal to zero, and solving for d_c. For mild slopes the normal depth is greater than d_c and subcritical flow prevails. On steep slopes the normal depth is less than the critical depth and flow is supercritical. Once the critical depth has been computed, critical velocity is easily obtained. Critical velocity for a channel of any cross section can be shown to be

$$V_c = \sqrt{g\,\frac{A}{B}} \tag{6-3}$$

where V_c = critical velocity

A = cross-sectional area of the channel

B = width of the channel at the water surface

The critical slope can then be found by using Manning's equation.

In practice, uniform flows are encountered only in long channels after a transition from nonuniform conditions. Nevertheless, a knowledge of uniform-flow hydraulics is extremely important, as numerous varied flow problems are solved through partial applications of uniform-flow theory. A basic assumption in gradually varied flow analyses, for example, is that energy losses are considered to be the same as for uniform flow at the average depth between two sections along the channel which are closely spaced.

6-2

GRADUALLY VARIED FLOW AND SURFACE PROFILES

Gradually varied flow results from gradual changes in depth which take place over relatively long reaches of a channel. Abrupt changes in the flow regime are classified as rapidly varied flow. Problems in gradually varied flow are widespread and represent the majority of flows in natural open channels and many of the flows in man-made channels. They can be caused by such factors as change in channel slope, in cross-sectional area, or in roughness, or by obstructions to flow, such as dams, gates, culverts, and bridges. The pressure distribution in gradually varied flow is hydrostatic and the streamlines are considered to be approximately parallel.

The significance of gradually varied flow problems may be illustrated by considering the following case. Assume that the maximum design flow for a uniform rectangular canal will occur at a depth of 8 ft under uniform-flow conditions. For these circumstances the requisite canal depth, including 1 ft of freeboard, will be 9 ft. Now consider that a gate is placed at the lower end of this canal. Assume that at maximum flow, the depth immediately upstream from the gate will have to be 12 ft in order to produce the required flow through the gate. The depth of flow will then begin decreasing gradually in an upstream direction and approach the uniform depth of 8 ft. Obviously, unless the depth of flow is known all along the channel, the channel depth cannot be designed.

Determination of the water-surface profile for a given discharge in an open channel is of importance in solving many engineering problems. There are 12 classifications of water-surface profiles, or *backwater curves*.[1] Figure 6-3 illustrates several of these and also a typical change which might take place in the transition from one type of flow regime to another. For any channel, the applicable backwater curve will be a function of the relationship between the actual depth of flow and the normal and critical depths for the channel. Often it is helpful to sketch the type curves for a given problem before attempting to effect an actual solution. In doing this, the type curves given in Fig. 6-3 are useful. Additional type curves are presented by Woodward and Posey.[1]

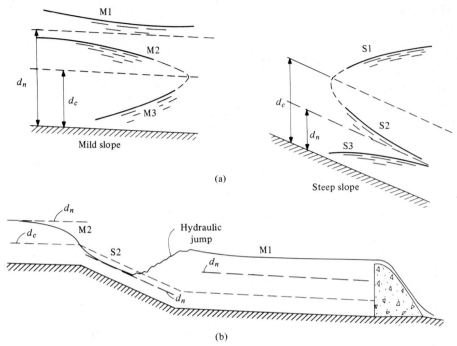

FIGURE 6-3 Gradually varied flow profiles.

Numerous procedures have been proposed for computing backwater curves. The one to be discussed here is known as the direct-step method. Referring to Fig. 6-4, the energy equation may be written

$$Z_1' + d_1 + \frac{V_1^2}{2g} = d_2 + \frac{V_2^2}{2g} + H_f$$

A rearrangement of this equation yields

$$\left(\frac{V_2^2}{2g} + d_2\right) - \left(\frac{V_1^2}{2g} + d_1\right) = Z_1' - H_f$$

or

$$E_2 - E_1 = S_c L - \bar{S}_e L$$

so

$$L = \frac{E_2 - E_1}{S_c - \bar{S}_e} \tag{6-4}$$

where E_2, E_1 = values of specific energy at sections 1 and 2
   S_c = slope of the channel bottom
   $\bar{S}_e$ = slope of the energy gradient

The value of S_e is obtained by assuming that (1) the actual energy gradient is the same as that obtained for uniform flow at a velocity equal to the average of the velocities at sections 1 and 2 (this can be found using Manning's equation) or (2) the slope of the energy gradient is equal to the mean of the slopes of the energy gradients for uniform flow at the two sections. The procedure in using Eq. 6-4 is to select some starting point where the depth of flow is known. A second depth is then selected in an upstream or downstream direction, and the

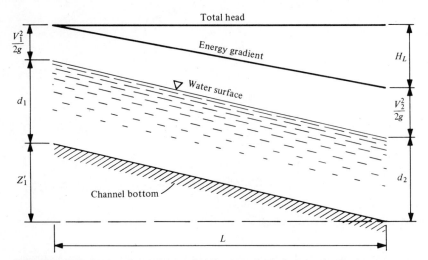

FIGURE 6-4 Definition sketch for the direct-step method of computing backwater curves.

distance from the known depth to this point is computed. Then, using the second depth as a reference, a third is selected and another length increment calculated and so on. The results obtained will be reasonably accurate provided the selected depth increments are small, since the energy-loss assumptions are fairly accurate under these conditions.

●EXAMPLE 6-2
Water flows in a rectangular concrete channel 10 ft wide, 8 ft deep, inclined at a grade of 0.10%. The channel carries a flow of 245 cfs and has a roughness coefficient n of 0.013. At the intersection of this channel with a canal, the depth of flow is 7.5 ft. Find the distance upstream to a point where normal depth prevails and determine the surface profile.

○*Solution*
Using Manning's equation and the given value of discharge, the normal depth is found to be 4 ft. Critical depth y_c is determined by solving Eq. 6-3 after rearranging and substituting by_c for A and b for B, where b equals the width of the rectangular channel.

TABLE 6-1

CALCULATIONS FOR THE SURFACE PROFILE OF EXAMPLE 6-2

(1) y	(2) A	(3) V	(4) $\dfrac{V^2}{2g}$	(5) $E = d + \dfrac{V^2}{2g}$	(6) P	(7) R	(8) $R^{4/3}$	(9) $S_e = \dfrac{V^2}{(1.49)^2 R^{4/3}/n^2}$	(10) $\overline{S}_e$	(11) ΔE	(12) $S_c - \overline{S}_e$	(13) $\Delta L = \dfrac{\Delta E}{S_c - \overline{S}_e}$
7.5	75	3.27	0.166	7.666	25	3.0	4.33	0.000188				
									0.000207	0.475	0.000793	598
7.0	70	3.50	0.191	7.191	24	2.91	4.15	0.000225				
									0.000279	0.932	0.000721	1292
6.0	60	4.08	0.259	6.259	22	2.73	3.82	0.000333				
									0.000436	0.885	0.000564	1570
5.0	50	4.90	0.374	5.374	20	2.50	3.40	0.000539				
									0.000762	0.792	0.000238	3328
4.0	40	6.12	0.582	4.582	18	2.22	2.90	0.000984				

$$L = \Sigma\, \Delta L = 6788$$

Then

$$V_c = \sqrt{g\left(\frac{by_c}{b}\right)}$$

$$= \sqrt{gy_c}$$

$$y_c = \frac{V_c^2}{g} = \frac{Q^2}{A^2 g}$$

$$y_c^3 = \frac{Q^2}{gb^2}$$

$$y_c = \sqrt[3]{\frac{Q^2}{gb^2}} \tag{6-5}$$

Substituting the given values for Q and b.

$$y_c = \sqrt[3]{\frac{(245)^2}{32.2 \times (10)^2}}$$

and

$$y_c = 2.65 \text{ ft}$$

Since $y > y_n > y_c$, an M1 profile will depict the water surface. The calculations then follow the procedure indicated in Table 6-1.

The distance upstream from the junction to the point of normal depth is 6788 ft. The surface profile can be plotted by using the values in columns (1) and (13) of the table. In practice greater accuracy could be obtained by using smaller depth increments, such as 0.25 ft, but the computational procedure would be the same.

6-3

VELOCITY CONSIDERATIONS

Both maximum and minimum velocities are prescribed for waste-transportation systems. Minimum velocities are set to assure that suspended matter does not settle out in the conduit, while maximum velocities are set to prevent erosion of the channel material.

Normally, velocities in excess of 20 fps are to be avoided in concrete or tile sewers, and whenever possible velocities of 10 fps or less should be used. Specially lined inverts are sometimes employed to combat channel erosion. The average value of the channel shearing stress is related to both erosion and sediment deposition.

Using the Chézy equation for velocity, Fair and Geyer have shown that the minimum velocity for self-cleansing V_m can be determined according to [13]

$$V_m = C\sqrt{\frac{K(\gamma_s - \gamma)}{\gamma} d} \tag{6-6}$$

where γ_s is the specific weight of the particles, γ is the specific weight of water, d is the particle diameter, and C is the Chézy coefficient (equal to $1.49R^{1/6}/n$ as evaluated by Manning). The value of C is selected with consideration given to the containment of solids in the flow. The value of K must be found experimentally and appears to range from 0.04 for the initiation of scour to more than 0.8 for effective cleansing.[13]

If nonflocculating particles have mean diameters between 0.05 and 0.5 mm, the minimum velocity may be determined using[9]

$$V_m = \left[25.3 \times 10^{-3} \; g \, \frac{d(\rho_s - \rho)}{\rho} \right]^{0.816} \left(\frac{D\rho_m}{\mu_m} \right)^{0.633} \tag{6-7}$$

where V_m = minimum velocity, fps
   d = mean particle diameter, ft
   D = pipe diameter, ft
   ρ_m = mean mass density of the suspension, pcf
   ρ_s = particle density, pcf
   μ_m = viscosity of the suspension, lb mass/ft-sec
   ρ = liquid mass density, pcf

The value of ρ_m may be determined by using

$$\rho_m = X_v \rho_s + (1 - X_v)\rho \tag{6-8}$$

where X_v is the volumetric fraction of the suspended solids.

Design of Sanitary Sewers

The design of sanitary sewers involves (1) the application of the principles outlined in Chapter 4 to determine the design flows, and (2) the application of hydraulic engineering principles to devise adequate conveyances for transporting these flows.

6-4

HOUSE AND BUILDING CONNECTIONS

Connections from the main sewer to houses or other buildings are commonly constructed of vitrified clay, concrete, or asbestos cement pipe. Building connections are usually made on about a 2% grade with 6-in. or larger pipe. Grades less than 1% are to be avoided; likewise extremely sharp grades.

6-5

COLLECTING SEWERS

Collecting sewers gather flows from individual buildings and transport the material to an interceptor or main sewer. Local standards usually dictate the location of these sewers, but they are commonly located under the street paving on one side of the storm drain, which is usually centered. The collecting sewer should be capable of conveying the flow of the present and anticipated population of the area it is to serve. Design flows are the sum of the peak domestic, commercial, and industrial flows and infiltration. The collecting sewer must transport this design flow when flowing full. Grades requiring minimum excavation while maximum and minimum restrictions on velocity are satisfied are best. Manholes are normally located at changes in direction, grade, pipe size, or at intersections of collecting sewers. For 8-, 10-, or 12-in. sewers, manholes spaced no farther than about 400 ft apart permit inspection and cleaning when necessary. For larger sizes, the maximum spacing can be increased.

6-6

INTERCEPTING SEWERS

Intercepting sewers are expected to carry flows from the collector sewers in the drainage basin to the point of treatment or disposal of the waste water. These sewers normally follow valleys or natural streambeds of the drainage area. When the sewer is built through undeveloped areas, precaution should be taken to ensure the proper location of manholes for future connections. For 15- to 27-in. sewers, manholes are constructed at least every 600 ft; for larger sewers, increased spacing is common. Manholes or other transition structures are usually built at every change in pipe size, grade, or alignment. For large sewers, horizontal or vertical curves are sometimes constructed. Grades should be designed so that the criteria regarding maximum and minimum velocities are satisfied.

6-7

MATERIALS

Collecting and intercepting sewers are constructed of asbestos cement sewer pipe, cast-iron pipe, concrete pipe, vitrified clay pipe, brick, and bituminized fiber pipe. Care should be exercised in designing the system so that permissible structural loadings are not exceeded for the material selected. Information on pipe loading is readily available from the Clay Products Association, the Portland Cement Association, the Cast Iron Pipe Manufacturers, and others.

6-8

SYSTEM LAYOUT

The first step in designing a sewerage system is to establish an overall system layout which includes a plan of the area to be sewered, showing roads, streets, buildings, other utilities, topography, soil type, and the cellar or lowest floor elevation of all buildings to be drained. Where part of the drainage area to be served is undeveloped and proposed development plans are not yet available, care must be taken to provide adequate terminal manholes which can later be connected to the system constructed serving the area. If the proposed sewer connects to an existing one, an accurate location of the existing terminal manhole, giving invert elevation, size, and slope, is essential.

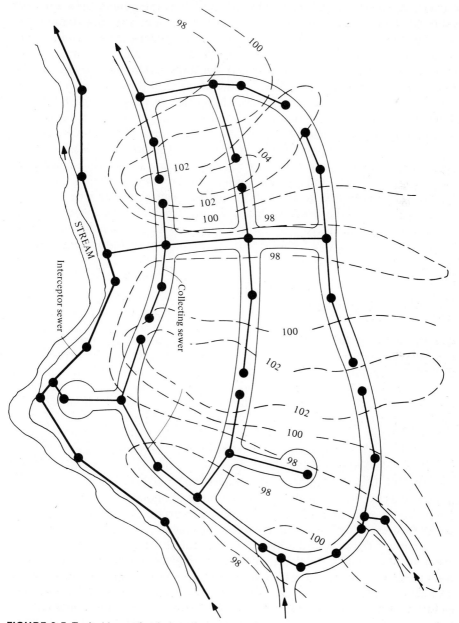

FIGURE 6-5 Typical layout for design of a sewerage system.

On the drainage-area plan just described, a tentative layout of collecting sewers and the intercepting sewer or sewers should be made. When feasible, sewer location should minimize the length required while providing service to the entire area. Length may be sacrificed if the shorter runs would require costlier excavation. Normally, the sewer slope should follow the ground surface so that waste flows can follow the approximate path of the area's surface drainage. In some instances it may be necessary to lay the sewer slope in opposition to the surface slope, or to pump wastes across a drainage divide. This situation can occur when a developer buys land lying in two adjacent drainage basins and, for economic or other reasons, must sewer the whole tract through only one basin.

Intercepting sewers or trunk lines are located to pass through the lowest point in the drainage area (the outlet) and to extend through the entire area to the drainage divide. Normally, they follow major natural drainage ways and are located in a designated right-of-way or street. Land slopes on both sides should be toward the intercepting sewer. Lateral or collecting sewers are connected to the intercepting sewer and proceed upslope to the drainage divide. Collecting sewers should be located in all streets or rights-of-way within the area to provide service throughout the drainage basin. Figure 6-5 illustrates a typical sewer layout. Note that the interceptor sewer is located in the stream valley and collector sewers transport flows from the various tributary areas to the interceptor.

6-9

HYDRAULIC DESIGN

The hydraulic design of a sanitary sewer can be carried out in a systematic manner. Figure 6-6 is typical of the way in which data can be organized to facilitate computations. The essential elements of hydraulic design require that peak design flows be carried by the pipe flowing full at velocities great enough to prevent sedimentation, yet small enough to prevent erosion. To minimize head losses at transitions and eliminate backwater effects, the hydraulic gradient must not change abruptly or slope in a direction adverse to the flow at changes in horizontal direction, pipe size, or quantity of flow. Sewers are usually designed to closely follow the grade of the ground surface or street paving under which they are laid. In addition, the depth of cover should be kept as close to the minimum as effective hydraulic design will permit to minimize excavation costs. The sewer cannot, of course, conflict with other subsurface utilities or structures.

Pipe sizes are determined in the following manner. A profile of the proposed sewer route is drawn and the hydraulic gradient at the downstream end of the sewer noted (normally the elevation of the hydraulic gradient of the sewer being met). Where discharge is to a treatment plant or open body of water, the hydraulic gradient is the elevation of the free water surface at this point. At the beginning elevation of the hydraulic gradient, a tentative gradient approximately following the ground surface is drawn upstream to the next point of control (usually a

JOB NAME _____   J.O. _____ DISTRICT _____ DATE _____

PREPARED BY _____   OF _____ SHEET _____ OF _____

CHECKED BY _____   (FIRM)

Area Symbol	Area Acres	Population Density		Population		Nondomestic Average Flow (mgd)		Average Sewage Flow (mgd)		Peak Flow (mgd)		Infiltration (mgd)	Peak Industrial Waste Flow (mgd)		Design Flow (mgd)		Approx. Grade	Tentative Size	Velocity	
		Pres.	Ult.	Pres.	Ult.	Pres.	Ult.	Pres.	Ult.	Pres.	Ult.		Pres.	Ult.	Pres.	Ult.			+	++
(1)	(2)	(3)	(4)	(5)	(6)	(7)*	(8)*	(9)	(10)	(11)	(12)	(13)	(14)	(15)	(16)	(17)	(18)	(19)	(20)	(21)

Notes:
      Where columns 7, 8, 14, and 15 are used, show calculations on back of form.
  * Includes average sewage flow from institutional, industrial, and commercial establishment.
  + Velocity based upon present average sewage flow.
  ++ Velocity based upon ultimate peak design flow.

BALTIMORE COUNTY DEPARTMENT OF PUBLIC WORKS

APPROVED
‾ ‾ ‾ ‾ ‾ ‾ ‾ ‾
DIRECTOR
‾ ‾ ‾ ‾ ‾ ‾ ‾ ‾
DIVISION OF ENGINEERING
‾ ‾ ‾ ‾ ‾ ‾ ‾ ‾ DATE

REVIEWED

Revised
Date	By

STANDARD SANITARY SEWER DESIGN
FLOW TABULATION FORM

FIGURE 6-6 Typical computation sheet for sanitary sewer design. (Courtesy Baltimore County Department of Public Works, Towson, Maryland.)

manhole). Exceptions to this occur when the surface slope is less than adequate to provide cleansing velocities, where obstructions preclude using this slope, or where adequate cover cannot be maintained. Using the tentative gradient slope, a pipe size is then selected which comes closest to carrying the design flow under full-flow conditions. Usually, a standard-size pipe is not found that will carry the maximum flow exactly at full depth with this gradient. Common practice is then to select the next larger size, modify the slope, or both. The choice will depend upon a comparison of pipe-cost savings versus excavation costs, and on additional local conditions, such as the fact that other utilities may control the vertical location of the sewer being designed.

After the first section of pipe has been designed, the hydraulic gradient at its upstream end is then used as a beginning point for the next section, and so on. Where significant changes in flow, direction, velocity, or any combination of these occur at manholes, it is important that the head lost in the manhole structure be computed and added to the elevation of the hydraulic gradient at the entrance to the outflow pipe. This total value then serves as the initial elevation of the hydraulic gradient for the succeeding section upstream. An example later in the chapter of the design of a storm-drainage system will illustrate the procedure described here.

6-10

PROTECTION AGAINST FLOODWATERS

Because the volume of sanitary wastewater is extremely small compared with flood flows, it is important that sewers be constructed to prevent admittance of large surface-runoff volumes. This will preclude overloads on treatment plants with their resultant reduction in degree of treatment or complete elimination of treatment in some instances.

Where interceptor sewers are built along streambeds, manhole stacks frequently are raised above a design flood level, such as the 50-year level. In addition, the manhole structures are waterproofed. Where stacks cannot be raised, watertight manhole covers are employed. Such measures as these keep surface drainage into the sewer at a minimum.

6-11

INVERTED SIPHONS

An inverted siphon is a section of sewer constructed below the hydraulic gradient due to some obstruction and operates under pressure. The term *depressed sewer* is actually more appropriate, since no real siphon action is involved. Usually, two or more pipes are needed for a siphon in order to handle flow variability. Normally the water-surface elevation at the entrance and exit to a siphon is fixed. Under these conditions the hydraulic design consists of selecting a pipe or pipes that will carry the design flow with a head loss equivalent to the difference in entrance and exit water-surface elevations. A transition structure is generally required at the entrance and exit of the depressed sewer to properly proportion or combine the flows.

The minimum flow in a siphon must be great enough to prevent deposition of suspended solids. Normally, velocities less than 3 fps are unsatisfactory. When the siphon is required to handle flows which vary considerably during any 24-hour period, it is customary to provide two or more pipes. A small pipe is provided to handle the low flows; intermediate flows may be carried by the smallest pipe and a larger pipe. Maximum flows may require the use of three or more pipes. By subdividing the flow in this manner, adequate cleansing

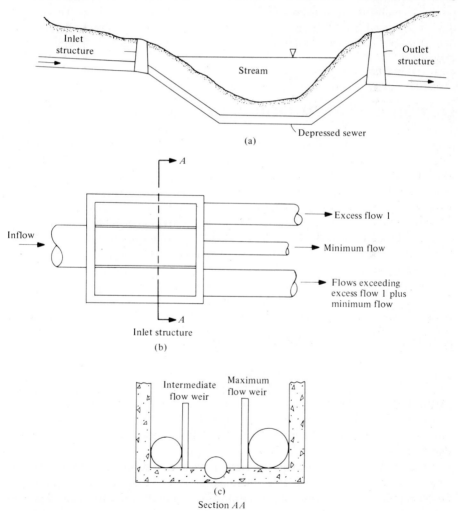

FIGURE 6-7 Inverted siphon or depressed sewer.

velocities are assured for all flow magnitudes. The entrance transition structure is designed to properly channel the flow into the various pipes. Figure 6-7 illustrates a typical transition structure.

6-12

WASTEWATER PUMPING STATIONS

It is often necessary to accumulate wastewater at a low point in the collection system and pump it to treatment works or to a continuation of the system at a higher elevation. Pumping stations consist primarily of a wet well, which intercepts incoming flows and permits equalization of pump loadings, and a bank of pumps, which lift the wastewater from the wet well. In most cases, centrifugal pumps are used and standby equipment is required for emergency purposes. Selection of centrifugal pumps is made according to the principles outlined in Chapter 5.

Design of Storm-Drainage Systems

Rapid and effective removal of storm runoff was a luxury not found in many cities in the early nineteenth century. Today, however, the modern city dweller has come to think of this as an essential service. Urban drainage facilities have progressed from crude ditches and stepping stones to the present intricate coordinated systems of curbs, gutters, inlets, and underground conveyances.

Handling surface runoff in urban drainage areas is a complex and costly undertaking with several primary difficulties—notably quantity and variability. Volumes of surface runoff can be exceedingly large during intense storms, yet such storms may occur only on a very infrequent basis. This poses the problem of building drainage works which perhaps are used for only a short time.

6-13

HYDROLOGIC CONSIDERATIONS

The hydrologic phase of urban drainage design is concerned with determining the magnitude, distribution, and timing of the various runoff events. Maximum events are of utmost importance since they are the basis for design of major structures. In some cases a knowledge of peak flow will suffice, but if storage or routing considerations are important, the volume and time distribution must also be known.

Runoff that occurs on any drainage area is a function of climate and the physical characteristics of the area. Factors that may be pertinent in precipitation–runoff relationships include precipitation type, rainfall intensity, duration, and distribution; storm direction; antecedent precipitation; initial soil-moisture conditions; soil type; evaporation; transpiration; and the size, shape, slope, elevation, directional orientation, and land-use characteristics of the drainage area. Indirect and artificial drainage must also be considered when applicable.

A brief study of these items should be sufficient proof that the hydrologic problem is complex. If urban drainage works are to effectively and economically serve the areas for which they are designed, considerable emphasis must be placed on the determination of accurate and reliable estimates of flow. Unfortunately, it is all too common that months are spent on the structural design of hydraulic conveyances, while only meager computations are made of the flow magnitudes on which these designs are based. Except in the case of very important structures, computations are often founded on formulas of questionable reliability or on the assumption of a past maximum flow plus a factor of safety. The greatest number of hydraulic failures have been caused by faulty determinations of runoff magnitude, not by structural inadequacies.

Peak flows result from excess surface-runoff volumes. The conditions which may generate these excesses are intense storms, snowmelt, and snowmelt combined with rainfall. Maximum flows on urban areas in the United States usually result from high-intensity short-duration rainfall of the thunderstorm type, whereas floods on large drainage basins are habitually caused by a combination of rainfall and snowmelt. The particular factors that produce maximum flow on a specific drainage area must be determined if reasonable reliability is to be accorded the estimated quantities of discharge.

6-14

DESIGN FLOW

Design flow is the maximum flow that a specified structure can pass safely. It is significant that the first question to be answered in relation to the design flow is its probability of occurrence. Should a drainage system be designed to carry the maximum possible flow, the 5%, the 1%, or some other chance discharge? Once this question has been answered, the magnitude of flow having the selected "design" frequency must be determined. Note particularly that there is little relationship between the normal life of the drainage works and the frequency of the design flow.

Table 6-2 shows, for example, that there is a 22% chance that the 100-year storm will occur in any 25-year period. This is good reason for questioning the design of a structure for a life expectancy of 25 years, then using design flows expected on the average of once in 25 years or less. If such a design is used, the chances are good that the structure will be damaged, destroyed, or at least overloaded before it has served its useful life.

Selection of a design frequency must be based on consideration of potential damage to human life, property damage, and inconvenience. Human life cannot be judged in terms of monetary values. If it is apparent that failure of a proposed

TABLE 6-2

PROBABILITY THAT AN EVENT HAVING A PRESCRIBED
RECURRENCE INTERVAL WILL BE EQUALED OR
EXCEEDED DURING A SPECIFIED PERIOD

T_R (years)	Period (years)					
	1	5	10	25	50	100
1	1.0	1.0	1.0	1.0	1.0	1.0
2	0.5	0.97	0.999	[a]	[a]	[a]
5	0.2	0.67	0.89	0.996	[a]	[a]
10	0.1	0.41	0.65	0.93	0.995	[a]
50	0.02	0.10	0.18	0.40	0.64	0.87
100	0.01	0.05	0.10	0.22	0.40	0.63

[a] Values are approximately equal to unity.

drainage system or structure will imperil human lives, the design should be appraised accordingly. Property damage is a purely economic consideration, and design flows can be determined on the basis of the limiting size flow against which it is economically practical to protect. Inconvenience is an intangible quantity for the most part but can be related to economic values under many circumstances and should also be heeded in selecting a design frequency.

6-15

PROCEDURES FOR ESTIMATING RUNOFF

Procedures used in estimating runoff magnitude and frequency can be generally categorized as (1) empirical approaches, (2) statistical or probability methods, and (3) methods relating rainfall to runoff. Historically, numerous empirical equations have been developed for use in the prediction of runoff. Most of them have been based on the correlation of only two or three variables and, at best, have given only rough approximations. Many are applicable only to specific localities—a fact that should be carefully considered before they are used. In most cases the frequency of the computed flow is unknown. Formulas of this type are useful only where a more reliable means is not available—and then only with a complete understanding of the relationship and the exercise of due caution.

Statistical analyses provide good results if sufficient records are available and if no significant changes in stream regimen are experienced or expected in the future. Estimates generally are based on the use of duration or probability curves. In applying these curves, the larger the sample population, the greater the validity of the estimate. For example, a determination of the 10-year peak flow based on records of only 10 years might be seriously in error, whereas the same answer computed from a 100-year record would yield good-results. About 10 independent samples normally can be expected to provide satisfactory estimates of maximum flow magnitudes of any frequency. The need for long-term records emphasizes the limitation inherent in the use of probability methods, since extreme values expected to occur only on the average of once in 50 to 100 years or more are not often obtainable within the available record. They must therefore be determined by extrapolation of the probability curve, an extremely dangerous undertaking. Most streams in the United States do not have reliable gaging records older than 50 years and only a few records exist for urban areas.

Of the methods relating runoff to rainfall, the most used are the unit hydrograph method and the rational method, or modifications of these. The *unit-hydrograph method* is an extremely valuable tool for estimating runoff magnitudes of various frequencies that may occur on a specific stream.[15] To use this approach it is necessary to have continuous records of runoff and precipitation for the particular drainage area. Determinations must be made of infiltration capacity variations throughout the year. The method is limited to areas for which precipitation patterns do not vary markedly. On large drainage basins, hydrographs must be developed for the various reaches and then synthesized into a single design flow at the critical location.

A unit hydrograph represents the runoff volume of 1 in. from a specified drainage area for a particular rainfall duration. A separate unit hydrograph is theoretically required for every possible rainfall length of interest. Ordinarily, however, variations of $\pm 25\%$ from any duration are considered acceptable. In addition, unit hydrographs for short periods can be synthesized into hydrographs for longer durations.

Once a unit hydrograph has been derived for a particular drainage area and rainfall duration, the hydrograph for any other storm of equal duration can be obtained. This new hydrograph is developed by applying the unit-hydrograph theorem, which states that the ordinates of all hydrographs resulting from equal unit time rains are proportional to the total direct runoff from that rain. The condition may be stated mathematically as

$$\frac{Q_s}{V_s} = \frac{Q_u}{1} \tag{6-9}$$

where Q_s = magnitude of a hydrograph ordinate of direct runoff having a volume equal to V_s (in inches) at some instant of time after the start of runoff

Q_u = ordinate of the unit hydrograph having a volume of 1 in. at the same instant of time

Storms of reasonably uniform rainfall intensity, having a duration of about 25% of the drainage area lag time (difference in time between the center of mass of rainfall and center of mass of resulting runoff) and producing a total of 1 in. or more, are most suitable in deriving a unit hydrograph.

Essential steps in the development of a unit hydrograph are as follows:

1 Analysis of the stream-flow hydrograph to permit separation of the surface runoff from the groundwater flow. It can be accomplished by any one of several methods.[14]

2 Measure the total volume of surface runoff (direct runoff) from the storm producing the original hydrograph. The volume is equal to the area under the hydrograph after the groundwater flow has been removed.

3 Divide the ordinates of the direct runoff by the total direct runoff volume in inches. The resulting plot of the answers versus time is a unit graph for the basin.

4 Finally, the effective duration of the runoff-producing rain for this unit graph must be found. The answer can be obtained by a study of the hydrograph of the storm used.

The steps listed are used in deriving the unit hydrograph from an isolated storm. Other procedures are required for complex storms or in developing synthetic unit graphs when few data are available. Unit hydrographs may also be transposed from one basin to another under certain circumstances. An example will serve to illustrate the derivation and application of the unit hydrograph.

●EXAMPLE 6-3

Using the hydrograph given in Fig. 6-8, derive a unit hydrograph for the 3-mi^2 drainage area. From this unit hydrograph, derive a hydrograph of direct runoff for the rainfall sequence given in Table 6-4.

○*Solution*

The steps are as follows:

1 Separate the base or groundwater flow so that the total direct runoff hydrograph may be obtained. As stated previously, a number of procedures are reported in the literature. A common method is to draw a straight line AC which begins when the hydrograph starts an appreciable rise and ends where the recession curve intersects the base-flow curve. The most important point here is to be consistent in methodology from storm to storm.

2 Determine the duration of the effective rainfall (rainfall that actually produces surface runoff). The effective rainfall volume must be equivalent to the volume of direct surface runoff. Usually the unit time of the effective rainfall will be 1 day, 1 hour, 12 hours, or some other interval appropriate for the size of drainage area being studied. As stated before, the unit storm duration should not exceed about 25 % of the drainage-area lag time. The effective portion of the rainstorm for this example is given in Fig. 6-8 together with its duration. The effective volume is 1.4 in.

3 Project the base length of the unit hydrograph down to the abscissa, giving the horizontal projection of the base-flow-separation line AC. It should be noted that the unit-hydrograph theory assumes that for all storms of equal duration, regardless of intensity, the period of surface runoff is the same.

4   Tabulate the ordinates of direct runoff at the peak rate of flow and at sufficient other positions to determine the hydrograph shape. Note that the direct runoff ordinate is the ordinate above the base-flow-separation line. See Table 6-3.

5   Compute the ordinates of the unit hydrograph by using Eq. 6-9. In this example the values are obtained by dividing the direct runoff ordinates by 1.4. Table 6-3 outlines the computation of the unit-hydrograph ordinates.

6   Using the values from Table 6-3, plot the unit hydrograph as shown in Fig. 6-8.

7   Using the derived ordinates of the unit hydrograph, determine the ordinates of the hydrographs for each consecutive rainfall period as given in Table 6-4.

8   Determine the synthesized hydrograph for unit storms 1–3 by plotting the three hydrographs and summing the ordinates. The procedure is indicated in Fig. 6-9.

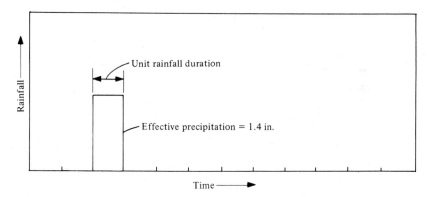

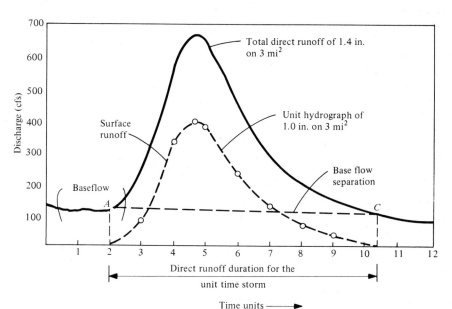

FIGURE 6-8 Derivation of a unit hydrograph from an isolated storm.

TABLE 6-3

**DETERMINATION OF A UNIT HYDROGRAPH FROM AN
ISOLATED STORM**

(1)	(2)	(3)	(4) Total Direct Runoff: (2) — (3) (cfs)	(5) Unit Hydrograph Ordinate: (4) ÷ 1.4 (cfs)
Time Unit	Total Runoff (cfs)	Base Flow (cfs)		
1	110	110	0	0
2	122	122	0	0
3	230	120	110	78.7
4	578	118	460	328
4.7	666	116	550	393
5	645	115	530	379
6	434	114	320	229
7	293	113	180	129
8	202	112	90	64.2
9	160	110	50	35.7
10	117	105	12	8.6
10.5	105	105	0	0
11	90	90	0	0
12	80	80	0	0

TABLE 6-4

UNIT-HYDROGRAPH APPLICATION

(1)	(2)	(3)	(4)[a]		
		Effective	Hydrograph Ordinates for		
Time Unit Sequence	Rain Unit Number	Rainfall (in.)	Rainfall Units 1–3		
			1	2	3
1	1	0.7	55.1	—	—
2	2	1.7	229	134	—
2.7	3	1.2	275	—	—
3	—	—	265	558	94.3
3.7	—	—	—	668	—
4	—	—	161	664	393
4.7	—	—	—	—	472
5	—	—	90.5	389	455
6	—	—	44.9	219	275
7	—	—	25.0	109	155
8	—	—	6.0	60.7	77
9	—	—	—	14.6	42.8
10	—	—	—	—	10.3

[a] Values in column (4) obtained by multiplying values in column (3) by unit-hydrograph ordinates.

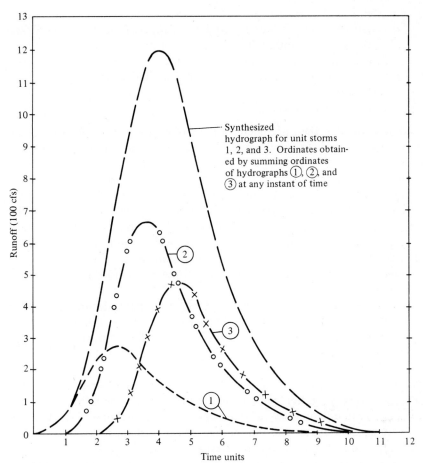

FIGURE 6-9 Synthesized hydrograph derived by the unit-hydrograph method.

The unit-hydrograph method provides the entire hydrograph resulting from a particular storm and offers certain definite advantages over procedures which produce peak flows alone. It has the disadvantage of requiring both rainfall and runoff data for its derivation and of being limited in its application to a particular drainage basin. A number of refinements to the procedure may be found in the literature.[14-17] Procedures for producing synthetic unit hydrographs covering areas where adequate records are not available have also been derived.[14]

The other rainfall–runoff relationship that will be discussed here is known as the *rational method*. It was first proposed in 1889 and is currently the most used method in this country for computing quantities of storm-water runoff.[18] The rational formula relates runoff to rainfall in the following manner

$$Q = cia \qquad\qquad (6\text{-}10)$$

where Q = peak runoff rate, cfs

    c = runoff coefficient, which is actually the ratio of the peak runoff rate to the average rainfall rate for a period known as the time of concentration

    i = average rainfall intensity, in./hr, for a period equal to the time of concentration

    a = drainage area, acres

It should be noted that the assignment of units of cfs to Q is satisfactory for all practical purposes, since 1.008 cfs equals 1 in. of rainfall in 1 hour on an area of 1 acre.

Assumptions basic to the rational method are: (1) the maximum runoff rate to any design location is a function of the average rate of rainfall during the time of concentration, and (2) the maximum rate of rainfall occurs during the time of concentration. The variability of the storm pattern is not taken into consideration. The time of concentration t_c is defined as the flow time from the most remote point in the drainage area to the point in question. Usually it is considered to be composed of an overland flow time or, in most urban areas, an inlet time plus a channel flow time.

The channel flow time can be estimated with reasonable accuracy from the hydraulic characteristics of the sewer. Normally, the average full-flow velocity of the conveyance for the existing or proposed hydraulic gradient is used. The channel flow time is then determined as the flow length divided by the average velocity.

The inlet time consists of the time required for water to reach a defined channel such as a street gutter, plus the gutter flow time to the inlet. Numerous factors, such as rainfall intensity, surface slope, surface roughness, flow distance, infiltration capacity, and depression storage, affect inlet time. Because of this, accurate values are difficult to obtain. Design inlet flow times of from 5 to 30 minutes are used in practice. In highly developed areas with closely spaced inlets, inlet times of 5–15 minutes are common, for similar areas with flat slopes, periods of 10–15 minutes are common, and for very level areas with widely spaced inlets, inlet times of 20–30 minutes are frequently used.[18]

Inlet times are also estimated by breaking the flow path into various components, such as grass, asphalt, and so on, then computing individual times for each surface and adding them. In theory, this would seem desirable, but in practice, so many variables affect the flow that the reliability of computations of this type is questionable. The standard procedure is to use inlet times which have been found through experience to be applicable to the various types of urban areas.

Izzard found the concentration time for small experimental plots without developed channels to be[19]

$$t_c = \frac{41bL_o^{1/3}}{(ki)^{2/3}} \qquad\qquad (6\text{-}11)$$

where t_c = time of concentration, minutes
   b = coefficient
   L_o = overland flow length, ft
   k = rational runoff coefficient (see Table 6-6)
   i = rainfall intensity, in./hr, during time t_c

The equation is valid only for laminar flow conditions where the product iL_o is less than 500. The coefficient b is found by using

$$b = \frac{0.0007i + C_r}{S_o^{1/3}} \qquad (6\text{-}12)$$

where S_o = surface slope
   C_r = coefficient of retardance

Values of C_r are given in Table 6-5.

TABLE 6-5
IZZARD'S RETARDANCE COEFFICIENT, C_r

Surface	C_r
Smooth asphalt	0.007
Concrete paving	0.012
Tar and gravel paving	0.017
Closely clipped sod	0.046
Dense bluegrass turf	0.060

The runoff coefficient c is the component of the rational formula which requires the greatest exercise of judgment by the engineer. It is not amenable to exact determination, since it includes the influence of a number of variables, such as infiltration capacity, interception by vegetation, depression storage, and antecedent conditions. As used in the rational equation, the coefficient c represents a fixed ratio of runoff to rainfall, while in actuality it is not fixed and may vary for a specific drainage basin with time during a particular storm, from storm to storm, and with change in season. Fortunately, the closer the area comes to being impervious, the more reasonable the selection of c becomes. This is true since for highly impervious areas c approaches unity, and for these areas the nature of the surface is much less variable for changing seasonal, meteorological, or antecedent conditions. The rational method therefore is best suited for use on urban areas, where a high percentage of imperviousness is common.

At present there is no precise method for evaluating the runoff coefficient c, although some modern research is being directed toward that end.[20] Common engineering practice is to make use of average values of the coefficient for various surface types which are normally found in urban regions. Table 6-6 lists some values of the runoff coefficient as reported in the American Society of Civil

TABLE 6-6

SOME VALUES OF THE RATIONAL COEFFICIENT C[18]

Surface Type	C Value
Bituminous streets	0.70–0.95
Concrete streets	0.80–0.95
Driveways, walks	0.75–0.85
Roofs	0.75–0.95
Lawns; sandy soil	
Flat, 2%	0.05–0.10
Average, 2–7%	0.10–0.15
Steep, 7%	0.15–0.20
Lawns, heavy soil	
Flat, 2%	0.13–0.17
Average, 2–7%	0.18–0.22
Steep, 7%	0.25–0.35

Engineers' *Manual on the Design and Construction of Sanitary and Storm Sewers.*[18]

Figure 6-10 relates the rational C to imperviousness, soil type, and lawn slope. This graph is used in designing storm drains in Baltimore County, Maryland. Most engineering designers make use of information reported in similar tabular or graphical form, inserting local conditions following their experience and practice.

In applying the rational method, a rainfall intensity i must be used which represents the average intensity of a storm of given frequency for the time of concentration. The frequency chosen is largely a matter of economics. Factors related to the choice of a design frequency have already been discussed. Frequencies of 1–10 years are commonly used where residential areas are to be protected. For higher-value districts, 10- to 20-year or higher return periods often are selected. Local conditions and practice normally dictate the selection of these design criteria.

After t_c and the rainfall frequency have been ascertained, the rainfall intensity i is usually obtained by making use of a set of rainfall intensity–duration–frequency curves such as those shown in Fig. 6-11. Entering the curves on the abscissa with the appropriate value of t_c and then projecting upward to an intersection with the desired frequency curve, i can be found by projecting this intersection point horizontally to an intersection with the ordinate. If an adequate number of years of local rainfall records is available, curves similar to Fig. 6-11 may be developed. Otherwise, data compiled by the Weather Bureau, the Department of Commerce, the Department of Agriculture, and other government agencies, which are available for numerous localities and regions, can be used.

Generally, the rational method should be used only on areas that are smaller than about 2 mi² (approximately 1280 acres) in size. For areas larger than 100 acres, due caution should be exercised. Most urban drainage areas served by storm drains become tributary to natural drainage channels or large conveyances before they reach 100 acres or more in size, however, and for these tributary areas the rational method can be put to reasonable use.

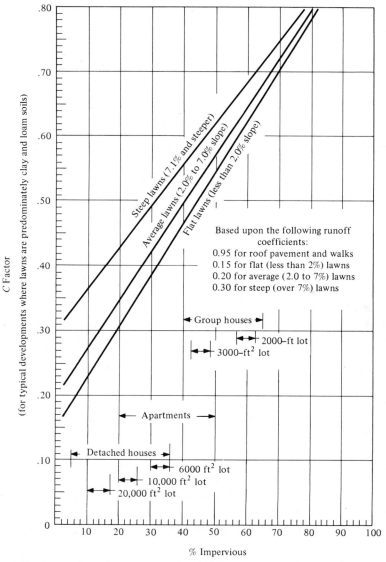

FIGURE 6-10 *C* factors for typical developments with clay soils. (Courtesy Baltimore County Department of Public Works, Towson, Maryland.)

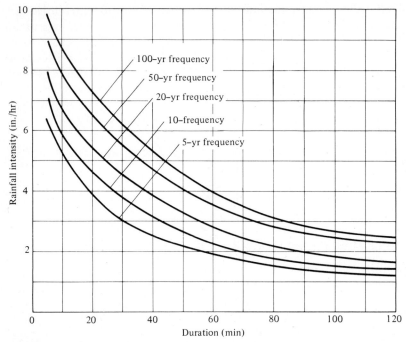

FIGURE 6-11 Typical intensity–duration–frequency curves.

6-16

STORM-WATER INLETS

Storm-water-inlet capacity is a subject which has received little emphasis in the design of storm-drainage systems for highways and streets. It is, nevertheless, of great importance, because regardless of the adequacy of the underground drainage system, proper drainage cannot result unless storm water is quickly and efficiently collected and introduced into the system.

A reliable knowledge of the behavior of storm-water inlets has broad implications: overdesign and inefficient use of the drainage system due to inadequate inlets can be avoided, debris and leaf stoppage of inlets can be reduced, and traffic interference on streets and highways can be minimized. No specific inlet type can be considered best for all conditions of use. Street grade, cross slope, and depression geometry affect the hydraulic efficiency. Eliminating stoppages or minimizing traffic interference often take precedence over hydraulic considerations in design.

Ideally, a simple opening across the flow path is the most effective type of inlet structure. Construction of this type would be impractical and unsafe, however, and the opening must therefore be covered with a grate or located in the curb. Unfortunately, grates obstruct the fall of water and often serve to collect debris, while curb openings are not in the direct path of flow. Nevertheless, a compromise of this type must be made.

Increasing the street cross slope will increase the depth of flow of the gutter, gutter depressions will concentrate flows at the inlet, and curb and gutter openings can be combined. These and other modifications provide increased inlet capacities, although some of them are not compatible with high-volume traffic.

Numerous inlet designs are seen across the United States, most of which have been developed from the practical experience of engineers or by rule-of-thumb procedures. The hydraulic capacity of many of these designs is totally unknown, and estimates frequently are considerably in error. Only in recent years have laboratory studies been conducted to produce efficient designs and develop a better understanding of inlet behavior.[21,23]

Four major types of inlets are being built: curb inlets, gutter inlets, combination inlets, and multiple inlets. A multitude of varieties is possible in each classification. A brief general description of the basic types follows.

a Curb inlet—a vertical opening in the curb through which gutter flow passes.

b Gutter inlet—a depressed or undepressed opening in the gutter section through which the surface drainage falls, covered by one or more grates.

c Combination inlet—an inlet composed of both curb and gutter openings, which acts as an integrated unit. Gutter openings may be placed directly in front of the curb opening (contiguous combination inlet) or upstream or downstream of the gutter opening (offset inlet). Combination inlets may be depressed or undepressed. Figure 6-12 depicts a typical combination inlet. Figure 6-13 relates the capacity of the inlet to the percent gutter slope.

d Multiple inlets—closely spaced interconnected inlets acting as a unit. Identical inlets end to end are called " double inlets."

Several general design recommendations regarding storm-water inlets may be made. The final selection of the optimum inlet type for a specific location will necessarily have to be based on the exercise of engineering judgment relative to the importance of clogging, traffic hazard, safety, and cost. In general, use cross-slopes that are as steep as possible considering traffic safety and comfort. Locate and design the inlets so there will be a 5–10% bypass in gutter flow. This greatly increases the inlet capacity. The bypass flow should not be great enough to inconvenience pedestrians or vehicular traffic, however. On streets where parking is permitted or where vehicles are not expected to travel near the curb, use contiguous combination curb-and-gutter inlets with longitudinal grate bars, or build depressed gutter inlets if clogging is not a problem. Where clogging is important, use depressed curb inlets if the gutter flow is small. For large flows, use depressed combination inlets with the curb opening upstream. On streets having slopes in excess of 5%, where traffic passes close to the curb, use deflector inlets if road dirt will not pack in the grooves. For flat slopes or where dirt is a problem, use undepressed gutter inlets or combination inlets

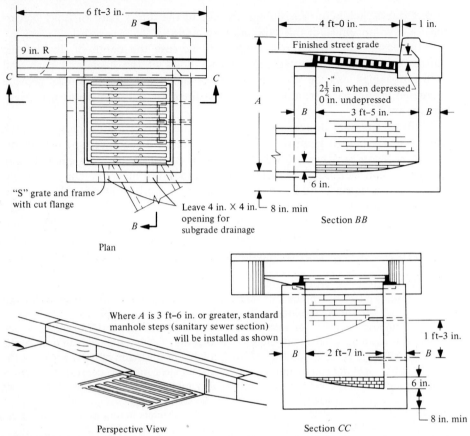

FIGURE 6-12 Baltimore type-S combination inlet. (Courtesy Baltimore County Department of Public Works, Towson, Maryland.)

with longitudinal bars only. For streets having flat grades or sumps (lows), pitch the grade toward the inlet on both ends. This will have the effect of providing a sump at the inlet. For true sump locations, use curb openings or combination inlets. The total open area, not the size and arrangement of bars, is important because the inlet will act as an orifice. Normally, sump inlets should be overdesigned because of the unique clogging problems which develop in depressed areas.

Inlets should be constructed in all sumps and at all street intersections where the quantity of flow is significant or where nuisance conditions warrant such construction. Inlets are required at intermediate points along streets where the curb and gutter capacity would be exceeded without them. Inlet capacities should be equal to or greater than the design flows. As shown in Fig. 6-13, inlet capacity is related to gutter slope and must be taken into account when selecting an inlet for a specific location.

Rapid and efficient removal of surface runoff from streets and highways is exceedingly important for maximum safety and minimum nuisance. Street grade,

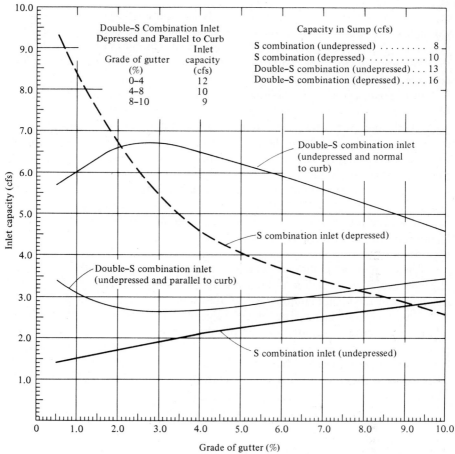

FIGURE 6-13 Inlet-capacity curves for type S combination inlets. Inlet capacities have been reduced for the estimated effect of debris partially clogging the inlet by 20 percent (on slopes). Capacities based on Standard 7-3/16 in. curb and gutter and 1-18 average crown slope for a distance of approximately 5 ft from face of curb. [Courtesy the Baltimore County Department of Public Works, Towson, Maryland. Curves based on data from "The Design of Storm Water Inlets," The Johns Hopkins University, Baltimore Md. (June 1956).]

crown slope, inlet type, grate design, and tolerable bypass flow are all important factors in the selection of inlet structures. Reliable information on the hydraulic performance of various inlet works is essential if effective designs are to be made.

6-17

SYSTEM LAYOUT

Storm-drainage systems may be closed-conduit, open-conduit, or some combination of the two. In most urban areas, the smaller drains frequently are closed conduits, and as the system moves downstream, open channels are often employed. Because quantities of storm-water runoff are usually quite large when contrasted with flows in most water mains and sanitary sewers, the drainage works needed to carry these flows are also quite large and occupy an important position from the standpoint of utilities placement. Common practice is to build the storm drains along the center line of the street, then offset the water main to one side and the sanitary sewer to the other.

In order to produce a workable system layout, a map of the area is needed showing contours, streets, buildings, other existing utilities, natural drainage ways, and areas for future development. As in the case of the sanitary sewer system, the storm drains will lie principally under the streets or in designated drainage rights-of-way. The entire area to the physical divide should be served by the drain or by an extension of it.

In laying out the drainage system, the first step is to tentatively locate all the necessary inlet structures. Once this has been done, a skeleton pipe system connecting all the inlets within a particular drainage area is drawn along with the location of all manholes, wye branches, special bend or junction structures, and the outfall.

Customarily, a manhole or junction structure will be required at all changes in grade, pipe size, direction of flow, and quantity of flow. Figure 6-14 illustrates the details of a typical manhole. Storm drains are usually not built smaller than 12 in. in diameter, because pipes of lesser size tend to clog readily with debris and therefore present serious malfunction problems. Maximum manhole spacing for pipes 27 in. and under should not exceed about 600 ft. For larger pipes, no maximum is prescribed and the normal requirements for structures should provide access to the drain for inspection, cleaning, or maintenance.

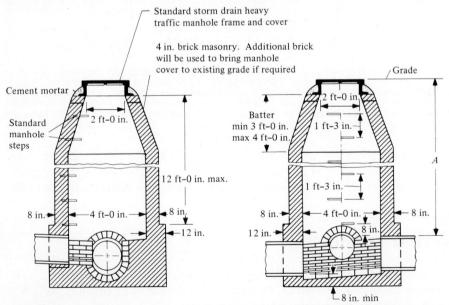

FIGURE 6-14 Typical storm-drainage manhole details. (Courtesy Baltimore County Department of Public Works, Towson, Maryland.)

6-18

HYDRAULIC DESIGN OF URBAN STORM-DRAINAGE SYSTEMS

Basic principles outlined at the beginning of the chapter are sufficient to adequately design an urban drainage system. Calculations for the drainage area shown in Fig. 6-15 are presented in detail to illustrate the mechanical procedure and the rational method as applied to an actual design problem. The overall Mextex area is an urban residential area made up of single-family dwellings and is divided into eight subareas which are tributary to individual storm-water inlets.

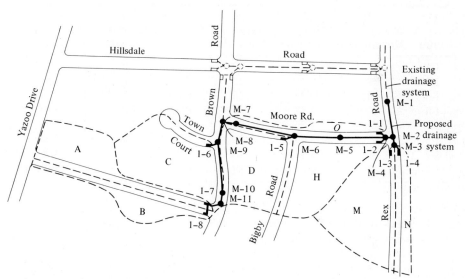

FIGURE 6-15 Plan view of Mextex storm-drainage system. Drainage areas, acres. A, 1.93; B, 1.34; C, 3.19; D, 2.66; H, 2.25; M, 2.75; N, 1.27; O, 0.40.

●EXAMPLE 6-4

Design a storm-drainage system to carry the flows from the eight inlet areas given in Fig. 6-15. It will be assumed that a 10-year frequency rainfall satisfies the local design requirements. Assume clay soil to be predominant in the area, with average lawn slopes.

○*Solution*

The steps are as follows:

1 Prepare a drainage-area map showing drainage limits, streets, impervious areas, and direction of surface flow.
2 Divide the drainage area into subareas tributary to the proposed storm-water inlets (Fig. 6-15).
3 Compute the acreage and imperviousness of each area.

4 Calculate the required capacity of each inlet, using the rational method. Assume a 5-minute inlet time to be appropriate and compute inlet flows for a rainfall intensity of 7.0 in./hr. This is obtained by using the 10-year frequency curve in Fig. 6-11 with a 5-minute concentration time. Appropriate C values are obtained from Fig. 6-10 by entering the graph with the calculated percentage imperviousness (the percent of the inlet area which is covered by streets, sidewalks, drives, roofs, etc.), projecting up to the average lawn-slope curve, and reading C on the ordinate. Computations for the inlet flows are tabulated in Table 6-7.

TABLE 6-7

REQUIRED STORM-WATER INLET CAPACITIES FOR EXAMPLE 6-4

(1) Inlet	(2) Area Designation	(3) Area (acres)	(4) Percent Impervious	(5) C	(6) Rainfall Intensity	(7) (3) × (5) × (6) Q: (cfs)
I-1	O	0.40	49	0.57	7.0	1.59
I-2	H	2.25	20	0.35	7.0	5.52
I-3	M	2.75	26	0.40	7.0	7.70
I-4	N	1.27	26	0.40	7.0	3.55
I-5	D	2.66	26	0.40	7.0	7.43
I-6	C	3.19	24	0.38	7.0	8.48
I-7	A	1.93	23	0.37	7.0	4.99
I-8	B	1.34	29	0.42	7.0	3.93

5 Select the type inlets required to adequately drain the flows in Table 6-7. The choice will be based on a knowledge of the street slopes and their relation to various inlet capacities. Inlet-capacity curves such as those given in Fig. 6-13 would be used. For the purposes of this example, no actual selections will be made, but the reader should recognize that this is the next logical step and an exceedingly important one.

6 Beginning at the upstream end of the system, compute the discharge to be carried by each successive length of pipe, moving downstream. These calculations are summarized in Table 6-8. Note that at each point downstream where a new flow is introduced, a new time of concentration must be determined as well as new values of C and drainage-area size. As the upstream inlet areas are combined to produce a larger tributary area at some design point, a revised C value representing these combined areas must be obtained. Usually the procedure is to take a weighted average of all the individual C values of which the larger area is composed. For example, when computing the flow to be carried by the pipe from M-9 to M-8, the tributary area is $A + B + C = 6.46$ acres, and the composite value of C will be

$$C = \frac{\sum c_i a_i}{\sum a_i} = \frac{0.37 \times 1.93 + 0.42 \times 1.34 + 0.38 \times 3.19}{6.46} = 0.38$$

TABLE 6-8

COMPUTATION OF DESIGN PIPE FLOWS FOR THE STORM-DRAINAGE SYSTEM OF EXAMPLE 6-4

(1) Pipe Section	(2) Tributary Area	(3) Area (acres)	(4) Flow Time (min) Inlet	Pipe	Total	(5) Rainfall Intensity i	(6) C	(7) Q (cfs)	(8) Pipe Slope (%)	Size (in.)	Full-Flow Velocity (fps)	Length (ft)
I-8–I-7	B	1.34	5	0.10	5	7.0	0.42	3.93	1.0	15	5.2	30
I-7–M-11	A+B	3.27	5	0.13	5.10	7.0	0.39	8.93	1.0	18	5.9	46
M-11–M-10	A+B	3.27	—	0.24	5.23	—	0.39	8.93	1.0	18	5.9	85
M-10–M-9	A+B	3.27	—	0.37	5.47	—	0.39	8.93	2.0	18	8.1	178
I-6–M-9	C	3.19	5	—	5	7.0	0.38	8.48	1.0	18	5.9	40
M-9–M-8	A+B+C	6.46	—	0.21	5.80	6.9	0.38	16.90	1.8	21	8.5	110
M-8–M-7	A+B+C	6.46	—	0.11	6.01	—	0.38	16.90	1.8	21	8.5	57
M-7–M-6	A+B+C	6.46	—	0.47	6.12	—	0.38	16.90	1.6	21	8.1	230
I-5–M-6	D	2.66	5	—	5	7.0	0.40	7.43	2.0	15	7.4	19
M-6–M-5	A+B+C+D	9.12	—	0.38	6.59	6.8	0.39	24.20	2.0	24	10.0	230
M-5–M-4	A+B+C+D	9.12	—	0.42	6.97	—	0.39	24.20	1.9	24	9.8	247
I-1–M-4	0	0.40	5	—	5	7.0	0.57	1.59	3.0	15	9.0	19
I-2–M-4	H	2.25	5	—	5	7.0	0.35	5.52	3.0	15	9.0	17
M-4–M-2	A+B+C+D+O+H	11.77	5	0.05	7.39	6.6	0.39	30.3	1.5	27	9.4	29
I-3–M-3	M	2.75	5	—	5	7.0	0.40	7.70	2.0	15	7.4	15
I-4–M-3	N	1.27	5	—	5	7.0	0.40	3.55	2.0	15	7.4	20
M-3–M-2	M+N	4.02	—	—	5	7.0	0.40	11.30	1.8	18	7.8	37
M-2–M-1	A+B+C+D+O+H+M+N	15.79	—	—	7.44	6.6	0.39	40.6	1.4	30	9.8	176

At the design location the value of t_c will be equal to the inlet time at I-8 plus the pipe-flow time from I-8 to M-9 (see Table 6-8), which must be known to permit solving the rainfall intensity to be used in computing the runoff from composite area A + B + C.

7 Using the computed discharge values, select tentative pipe sizes for the approximate slopes given in column (8) of Table 6-8. Once the pipe sizes are known, flow velocities between input locations can be determined. Normally, these velocities are approximated by computing the full-flow velocities for maximum discharge at the specified grade. These velocities are then used to compute channel flow time for estimating the time of concentration. If upon completing the hydraulic design, enough change has been made in any concentration time to alter the design discharge, new values of flow should be computed. Generally, this will not be the case.

8 Using the pipe sizes selected in step 7, draw a profile of the proposed drainage system. Begin the profile at the point farthest downstream, which can be an outfall into a natural channel, an artificial channel, or an existing drain, as in the case of the example. In constructing the profile, be certain that the pipes have at least the minimum required cover. Normally 1.5–2 ft is sufficient. Pipe slopes should conform to the surface slope wherever possible. At all manholes indicate the necessary change in invert elevation. In this example, where there is no change in pipe size through the manhole, a drop of 0.2 ft will be used. Where the size decreases upstream through a manhole, the upstream invert will be set above the downstream invert a distance equal to the difference in the

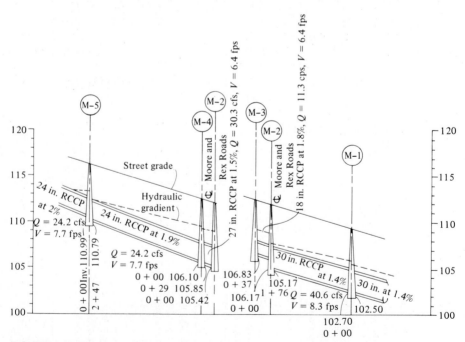

FIGURE 6-16 Profile of part of the Mextex storm drain, showing the hydraulic gradient.

two diameters. In this way the crowns are kept at the same elevation. A part of the profile of the drainage system in the example is given in Fig. 6-16.

9 Compute the position of the hydraulic gradient along the profile of the pipe. If this gradient lies less than 1.5 ft below the ground surface, it must be lowered to preclude the possibility of surcharge during the design flow. Note that the value of 1.5 ft is arbitrarily chosen here. In practice, local standards indicate the limiting value. Hydraulic gradients may be lowered by increasing pipe sizes, decreasing head losses at structures, designing special transitions, lowering the system below ground, or a combination of these means.

Computations for a portion of the hydraulic gradient of the example will now be given. Head losses in the pipes are determined by applying Manning's equation, assuming $n = 0.013$ in this example. Head losses in the structures will be determined by using the relationships defined in Figs. 6-17 and 6-18. These curves were developed for surcharged pipes entering rectangular structures but may be applied to wye branches, manholes, and junction chambers as pictured on the curves.[22] The A curve is used to find entrance and exit losses, the B curve to evaluate the head loss due to an increased velocity in the downstream direction. The loss is designated as the difference between the head losses found for the downstream and upstream pipe ($V_{h-2} - V_{h-1}$). In cases where the greatest velocity occurs upstream, the difference will be negative and may be applied to offset other losses in the structure. The C loss results from a change in direction in a manhole, wye branch, or bend structure. The D loss is related to the effects produced by the entrance of secondary flows into the structure. Several examples of the use of these curves are shown in Figs. 6-17 and 6-18.

Computations for the hydraulic gradient shown in Fig. 6-16 are as follows:

(a) Begin at the elevation of the hydraulic gradient at the upstream end of the existing 30-in. reinforced-concrete culvert pipe (RCCP). This elevation is 105.50. The existing hydraulic gradient is shown in Fig. 6-16.

(b) Compute the head losses in manhole M-1 using Figs. 6-17 and 6-18.

A loss = 0.36 ft ($V = 8.3$ fps $= Q/A$)
B loss = 0; no change in velocity, $V_1 = V_2$
C loss = 0; no change in direction
D loss = 0; no secondary flow
  total = 0.36 ft

The hydraulic gradient therefore rises in the manhole to an elevation of $105.50 + 0.36 = 105.86$ ft, plotted in M-1 in Fig. 6-16.

(c) Compute the head loss due to friction in the 30-in. drain from M-1 to M-2. Assume that $n = 0.013$. Using Manning's equation, the head loss per linear foot of drain is

$$S = \frac{(nV)^2}{2.21\ R^{4/3}} \qquad (6\text{-}13)$$

and from M-1 to M-2,

$$S = \frac{(0.013 \times 8.3)^2}{2.21 \times 0.534} = 0.0166$$

The total frictional head loss is therefore

$$hf = S \times L = 0.0116 \times 176 = 1.73 \text{ ft}$$

Elevation of the hydraulic gradient at the downstream end of M-2 is thus $105.86 + 1.73 = 107.59$ ft. This elevation is plotted in Fig. 6-16, and the hydraulic gradient in this reach is drawn in.

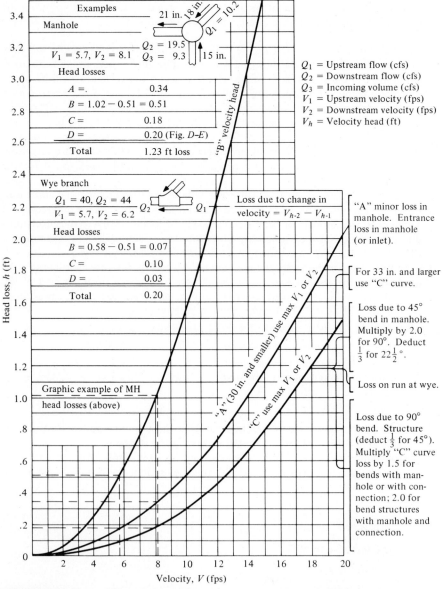

FIGURE 6-17 Types A, B, and C head losses in structures. (Courtesy Baltimore County Department of Public Works, Towson, Maryland.)

(d) Compute the head losses in M-2.

$A = 0.36 \ (V = 8.3 \text{ fps})$

$B = 1.07 - 0.90 = 0.17 \ (V_2 = 8.3, \ V_1 = 7.6 = Q/A \text{ for 27-in. drain})$

$C = 0.20 \times 2.0 \ (\text{multiply by 2 for } 90° \text{ bend in manhole—see Fig. 6-17}) = 0.40$

$D = 0.22 \text{ for } Q_3/Q_1 = 11.3/30.3 = 37\%$

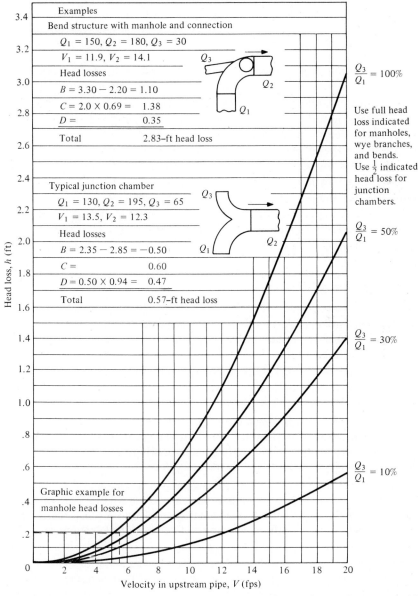

FIGURE 6-18 Type D head loss in structures. (Courtesy Baltimore County Department of Public Works, Towson, Maryland.)

Total head loss in M-2 equals 1.15 ft and the elevation of the hydraulic gradient in M-2 is therefore $107.59 + 1.15 = 108.74$ ft.

(e) Compute the friction head loss in the section of pipe from M-2 to M-3.

$$S = \frac{(0.013 \times 6.4)^2}{2.21 \times 0.272} = 0.0113$$

$hf = 0.0113 \times 37 = 0.42$ ft

Elevation of the hydraulic gradient at the downstream end of M-3 is therefore $108.74 + 0.42 = 109.16$ ft. Plot this point on the profile and draw the gradient from M-2 to M-3.

Students should realize that the hydraulic gradient in this example was computed under the assumption of uniform flow. In closed conduit systems, if the pipes are flowing full or the system is surcharged (the usual design flow conditions), this method will produce good results. Where open conduits are used, or in partial-flow systems, the hydraulic gradient can be determined by computing surface profiles in the manner described in Sec. 6-2.

Computations for the remainder of the hydraulic gradient are identical to those just given and will not be presented. It should be noted that none of the gradients shown in Fig. 6-16 come within 1.5 ft of the surface, so no revisions are needed. If too close to the surface, it would be necessary to modify all or a portion of the system to lower the gradient. This could be accomplished by reducing head losses, by increasing the depth of the system, or both. The choice would depend primarily on cost.

PROBLEMS

6-1   Wastewater flows in a rectangular concrete channel 6.0 ft wide and 3.0 ft deep. The design flow is 30 cfs. Find the critical velocity. Also find the slope of the channel if the flow velocity is to be 2.5 fps.

6-2   A 16-in. sewer pipe flowing full is expected to carry 5.8 cfs. The n value is 0.011. The minimum flow is $\frac{1}{12}$ that of the maximum. Find the depth and velocity at minimum flow.

6-3   The sewage from an area of 400 acres is to be carried by a circular sewer at a velocity not less than 2.5 fps. Manning's $n = 0.013$. The population density is 16 persons per acre. Find the maximum hourly and minimum hourly flows. Determine the pipe size required to handle these flows and the required slope.

6-4   Water flows in a rectangular concrete channel 9.5 ft wide and 8 ft deep. The channel invert has a slope of 0.10% and the n value applicable is 0.013. Flow carried by the channel is 189 cfs. At an intersection of this channel with a canal, the depth of flow is 7.1 ft. Find the distance upstream to a point where normal depth prevails. Plot the surface profile.

6-5   A 54-in. sewer ($n = 0.013$) laid on a 0.17% grade carries a flow of 19 mgpd. At a junction with a second sewer the sewage depth is 36 in. above the invert. Plot the surface profile back to the point of uniform depth.

6-6 Determine the minimum velocity and gradient required to transport $\frac{3}{16}$-in. gravel through a 42-in.-diameter pipe, given $n = 0.013$ and $K = 0.05$.

6-7 An inverted siphon is to carry a minimum dry-weather flow of 1.5 cfs, a maximum dry-weather flow of 3.7 cfs, and a storm flow of 48.00 cfs in three pipes. Select the proper diameters to assure velocities of 3.0 fps in all pipes. Make a detailed sketch of your design. Assume that the siphon goes under a highway with a 3.2-ft drop and is 73 ft long.

6-8 Given the following 25-year record of 24-hour maximum annual stream flows (cfs), plot these values versus the percent of years during which runoff was equal to or less than the indicated value on log-probability paper. Find the peak flow expected on the average of: (a) once every 5 years; (b) once every 15 years.

220	196	89	53	47
200	129	76	50	38
218	142	62	52	36
199	127	67	49	32
180	118	54	47	28

6-9 Given the following unit storm, storm pattern, and unit hydrograph, determine the composite hydrograph.

    Unit storm $= 1$ unit of rainfall for 1 unit of time

Actual storm:	Time units:	1	2	3	4
Pattern:	Rainfall units:	1	2	4	3

    Unit hydrograph: triangular with base length $= 4$ time units, time of rise $= 1$ time unit, and maximum ordinate $= \frac{1}{2}$ rainfall unit height.

6.10 Given a unit rainfall duration of 1 time unit, an effective precipitation of 1.5 in., and the following hydrograph and storm sequence, determine: (a) the unit hydrograph; (b) the hydrograph for the given storm sequence.

STORM SEQUENCE

Time units	1	2	3	4
Precipitation (in.)	0.4	1.1	1.8	0.9

HYDROGRAPH

Time units	1	2	3	4	4.5	5	6	7	8	9	10	11	12	13
Flow, cfs	101	96	218	512	610	580	460	320	200	180	100	86	60	50

6-11 Do Prob. 6-10 if the storm sequence is as follows:

Time units	1	2	3	4
Precipitation (in.)	0.6	1.3	1.5	0.8

6-12   Compute, by the rational method, the design flows for the pipes shown in the accompanying sketch: I-1 to M-4, M-4 to M-3, M-3 to M-2, M-2 to M-1, and M-1 to outfall. A-1 = 2.1 acres, $c_1 = 0.5$; A-2 = 3.0 acres, $c_2 = 0.4$; A-3 = 4.1 acres, $c_3 = 0.7$. Given pipe-flow times are I-1 to M-3, 1 minute, and M-3 to M-1, 1.5 minutes.

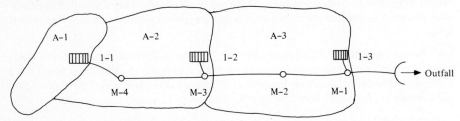

6-13   Given the data below, design a section of sanitary sewer to carry these flows at a minimum velocity of 2.5 fps. Assume that the minimum depth of the sewer below the street must be 8.5 ft and that an n value of 0.013 is applicable. Make a plan and profile drawing. (*Note:* The invert into manhole F must be 937 ft. Assume an invert drop of 0.2 ft across each manhole.)

Manholes	Distance between Manholes (ft)	Flow (gpm)	Street Elevations at Manholes (ft)
A to B	180	1200	A—978
B to C	300	1450	B—972
C to D	450	2200	C—964
D to E	300	2500	D—958
E to F	400	3000	E—954
			F—949

6-14   For an inlet area of 2.0 acres having an imperviousness of 0.50 and a clay soil, find the peak rate of runoff for the 5-, 10-, and 20-year storms.

6-15   Using the manhole spacing–elevation data given in Prob. 6-13 and the data below, design a storm-drainage system such that manhole A is at the upper end of the drainage area and manhole F is replaced by an outfall to a stream. The outfall invert elevation is equal to 940.3 ft. Design for the 5-year storm.

Inflow Point	Incremental Area Contributing to Inflow Point (acres)	Imperviousness of Areas (%)
A to B	3.0	41
B to C	2.7	50
C to D	5.3	55
D to E	3.6	45
Total area = 14.6 acres		

REFERENCES

1. S. M. Woodward and C. J. Posey, *Hydraulics of Steady Flow in Open Channels* (New York: John Wiley & Sons, Inc., 1955).
2. R. M. Olson, *Essentials of Engineering Fluid Mechanics*, 3rd Ed. (New York: IEP Publishers, 1973).
3. C. E. Kindsvater, "The Hydraulic Jump in Sloping Channels," *Trans. Am. Soc. Civil Engrs.* 109 (1944).
4. V. T. Chow, *Open-Channel Hydraulics* (New York: McGraw-Hill Book Company, 1959).
5. J. Hinds, "The Hydraulic Design of Flume and Siphon Transitions," *Trans. Am. Soc. Civil Engrs.* 92 (1928).
6. W. M. Sangster, A. W. Wood, E. T. Smerdon, and H. G. Bossy, "Pressure Changes at Storm Drain Junctions," University of Missouri, *Engineering Experiment Station Bull. No. 41*, Columbia, Missouri.
7. H. Rouse (ed.), *Engineering Hydraulics* (New York: John Wiley & Sons, Inc., 1950).
8. G. M. Fair and J. C. Geyer, *Elements of Water Supply and Waste-Water Disposal* (New York: John Wiley & Sons, Inc., 1958).
9. K. E. Spells, *Trans. Inst. Chem. Engrs.* (*London*) 33 (1955).
10. T. R. Camp, "Design of Sewers to Facilitate Flow," *Sewage Works J.* 18 (1946).
11. Anonymous, *Newsletter Hydraulics Div., Am. Soc. Civil Engrs.* (August 1963).
12. H. M. Morris, *Applied Hydraulics in Engineering* (New York: The Ronald Press Company, 1963).
13. G. M. Fair and J. C. Geyer, *Water Supply and Waste-Water Disposal* (New York: John Wiley & Sons, Inc., 1954).
14. R. K. Linsley, Jr., M. A. Kohler, and J. L. H. Paulhus, *Applied Hydrology* (New York: McGraw-Hill Book Company, 1949), pp. 387–411, 444–464.
15. L. K. Sherman, "Streamflow from Rainfall by the Unit Hydrograph Method," *Eng. News-Record* 108 (1932): 501–505.
16. J. E. Nash, "Systematic Determination of Unit Hydrograph Parameters," *J. Geophys. Res.* 64 (1959): 111–115.
17. B. S. Barnes, "Consistency in Unitgraphs," *Proc. Am. Soc. Civil Engrs., J. Hydraulics Div.* 85 (August 1959): 39–61.
18. "Design and Construction of Sanitary and Storm Sewers," *Manual of Engineering Practice No. 37* (New York: American Society of Civil Engineers, 1960).
19. C. F. Izzard, "Hydraulics of Runoff from Developed Surfaces, *Proc. Highway Res. Bd.* 26 (1946): 129–150.
20. J. C. Schaake, "Report No. XI of the Storm Drainage Research Project," The Johns Hopkins University, Department of Sanitary Engineering, Baltimore, Md. (September 1963).
21. ———, "The Design of Storm Water Inlets," The Johns Hopkins University, Department of Sanitary Engineering and Water Resources, Baltimore, Md. (June 1956).
22. ———, "Baltimore County Design Manual," Baltimore County Department of Public Works, Towson, Md. (1955).
23. R. J. Wasley, "Hydrodynamics of Flow into Curb-Opening Inlets," Stanford University, Department of Civil Engineering, Stanford, California. *Tech. Rept. No. 6* (1960).
24. R. T. Knapp, "Design of Channel Curves for Supercritical Flow," *Trans. Am. Soc. Civil Engrs.* 116 (1951).

CHAPTER
7
Water Quality

Water pollution manifests itself through changes in water quality. Quality water is water whose characteristics make it suited to the needs of the user. Acceptable water quality is therefore dependent upon the requirements of many kinds of water consumers. Characteristics that make water unsuitable to one water user may be unimportant or even desirable to another.

In past years concern for water quality has related primarily to bacterial contaminants, gross hardness, and aesthetic characteristics. There is an increasing concern with microchemical and microphysical substances today entering raw-water supplies. We are, at present, poorly prepared to evaluate the hazards of many of these substances in the water phase of the environment. Water resource planning and policy historically were oriented primarily to the quantity of water in the extensive river basins and rural areas. In the future we will have to relate more directly to issues at the urban–metropolitan level, where most of the population now lives and will concentrate to an even greater extent in the future.

242

7-1

WATER AND HEALTH

The potential of water to spread massive epidemics is a matter of public record. In the early part of the twentieth century, typhoid fever and other enteric diseases were major causes of death. Since about 1920, however, enteric diseases have contributed little to sickness and death in the United States, and this same pattern is shown in many developing countries. This remarkable record is a credit to water resource engineering. Waterborne disease outbreaks still occur from time to time but are usually the result of accidents commonly involving small or private water supplies.

Concern over waterborne viral diseases is a result of increased water reuse by man, intensifying the need to know more about enteric viruses. Although it is believed that all human enteric viruses have the potential to cause illness in man, not all have been etiologically associated with clinical illness. A number of waterborne local outbreaks attributed to virus have occurred in the United States, but no large-scale spread of viral disease by the water route is known to have occurred. Additional information on viruses is given in Section 7-2.

A definite lack of knowledge exists about trace amounts of some potentially toxic chemicals or excessive amounts of some common minerals in drinking water. Many of the possible contaminants are organic compounds. These come from chemicals used as automotive fuel additives, insecticides, detergents, lubricants, and from many other types of industrial production. Toxicological effects of waterborne organics have been observed principally in connection with the chlorinated hydrocarbons and organic phosphorous compounds used as pesticides. These substances may enter the water from runoff, irrigation return flow, air drifting, and by direct application for the control of algae.

Hazards from pesticides in water result both from direct effects, because they tend to persist in their original form over long periods, and from indirect effects because they may be concentrated biologically in man's food chain (see Sec. 2-4).

Inorganic substances that may exert harmful effects on man include nitrate and selenium. The evidence that links nitrates with methemoglobinemia and death in infants is not strong but is widely stated in the literature. Selenium, which is widely distributed in nature, has been linked to bad teeth, gastrointestinal disturbances, and skin discoloration.

The identification and quantification of chemical compounds in water which may have possible effects on human health pose an analytical problem. Instrumental techniques are not capable of operation at nanogram- or microgram- or, in some cases, at milligram-per-liter levels. As water use increases, it will become increasingly important to single out potentially harmful substances and to reduce these concentrations to acceptable levels. This will require improved knowledge of the specific compounds and their possible effects on man.

7-2

DOMESTIC WATER SUPPLY

Drinking-water standards for some public water supplies have been established since 1914. On December 16, 1974, the National Safe Drinking Water Act was signed into law and on December 24, 1975, the National Interim Primary Drinking Water Regulations were issued.[1] The establishment of secondary regulations related to taste, odor, and appearance of drinking water are being developed.

The maximum contaminant levels for inorganic chemicals shown in Table 7-1, with the exception of nitrates, are based upon possible health effects that may occur after a lifetime of exposure of approximately 2 liters of water per day.

TABLE 7-1

MAXIMUM CONTAMINANT LEVELS FOR INORGANIC CHEMICALS OTHER THAN FLUORIDE FOR PUBLIC WATER SUPPLIES

Contaminant	Level (mg/l)
Arsenic	0.05
Barium	1
Cadmium	0.010
Chromium	0.05
Lead	0.05
Mercury	0.002
Nitrate (as nitrogen)	10
Selenium	0.01
Silver	0.05

Source: The National Safe Drinking Water Act, Public Law 93–523, December 16, 1974.

The maximum contaminant levels for pesticides, Table 7-2, have been derived from the recent data on effects of acute and chronic exposure to both organochlorine and chlorophenoxy pesticides. In setting specific limits for chemical constituents, the total lifetime environmental exposure of man to the specific toxicant has been taken into consideration. The limits have been determined with a factor of safety included to minimize the amount of toxicant contributed by water when other sources (milk, food, or air) are known to represent additional sources of exposure to man. On this basis maximum contaminant

TABLE 7-2

MAXIMUM CONTAMINANT LEVELS FOR
ORGANIC CHEMICALS FOR PUBLIC
WATER SUPPLIES

Chlorinated Hydrocarbons	Level (mg/l)
Endrin (1,2,3,4,10,10-Hexachloro-6,7-epoxy-1,4,4a,5,6,7,8a-octahydro-1,4-endo, endo-5,8-dimethanonaphthalene)	0.0002
Lindane (1,2,3,4,5,6-Hexachloro-cyclohexane, gamma isomer)	0.004
Methoxychlor (1,1,1-Trichloro-2,2-bis-[*p*-methoxyphenyl]ethane)	0.1
Toxaphene ($C_{10}H_{10}Cl_8$-Technical chlorinated camphene, 67–69% chlorine)	0.005
Chlorophenoxys	
2,4-D (2,4-Dichlorophenoxyacetic acid)	0.1
2,4,5-TP Silvex (2,4,5-Trichlorophenoxypropionic acid)	0.01

Source: The National Safe Drinking Water Act, Public Law 93–523, December 16, 1974.

levels should not be regarded as fine lines between safe and dangerous concentrations.

Since the maximum contaminant levels were established to protect consumers based on long-term exposure to the water supply, it was clear that these contaminant levels should not apply to transients or intermittent users. Therefore, the final regulations on the maximum contaminant levels for organic chemicals, and for inorganic chemicals other than nitrates, are not applicable to noncommunity systems. Since infants may be adversely affected by nitrates in a short period of time, the maximum contaminant levels for nitrate have been made applicable to noncommunity systems.

The regulations have a maximum contaminant level for turbidity, because turbidity interferes with disinfection efficiency and because high turbidity often signals the presence of other health hazards. The growth of microorgan-

TABLE 7-3

RAW-SURFACE-WATER CRITERIA FOR PUBLIC WATER SUPPLIES

Substance	Surface-Water Criteria	
	Permissive Criteria	Desirable Criteria
Coliforms (MPN)	10,000	<100
Fecal coliforms (MPN)	2,000	<20
Inorganic chemicals (mg/l)		
Ammonia-N	0.5	<0.01
Arsenic[a]	0.05	Absent
Barium[a]	1.0	Absent
Boron[a]	1.0	Absent
Cadmium[a]	0.01	Absent
Chloride[a]	250	250
Chromium[a] (hexavalent)	0.05	Absent
Copper[a]	1.0	Virtually absent
Dissolved oxygen	≥ 4	Near saturation
Iron	0.3	Virtually absent
Lead[a]	0.05	Absent
Manganese[a]	0.05	Absent
Nitrate[a]-N	10	Virtually absent
Selenium[a]	0.01	Absent
Silver[a]	0.05	Absent
Sulfate[a]	250	<50
Total dissolved solids[a]	500	<200
Urany ion[a]	5	Absent
Zinc[a]	5	Virtually absent
Organic chemicals (mg/l)		
ABS		
Carbon chloroform extract[a]	0.15	<0.04
Cyanide[a]	0.20	Absent
Herbicides		
2,4-D + 2,4,5-T + 2,4-TP[a]	0.1	Absent
Oil and gases[a]	Virtually absent	Absent
Pesticides[a]		
Adrian	0.017	Absent
Chlordane	0.003	Absent
DDT	0.042	Absent
Dieldrin	0.017	Absent
Endrin	0.001	Absent
Heptachlor	0.018	Absent
Lindane	0.056	Absent
Methoxychlor	0.035	Absent
Toxaphene	0.005	Absent
Phenols[a]	0.001	Absent

[a] Substances that are not significantly affected by the following treatment process: coagulation (less than about 50 mg/liter of alum, ferric sulfate, or copperas, with alkali addition as necessary but without coagulant aids or activated carbon), sedimentation (6 hours or less), rapid sand filtration (3 gpm/ft^2 or less), and disinfection with chlorine (without consideration to concentration or form of chlorine residual).

Source: "Raw Water Quality Criteria for Public Supplies," National Technical Advisory Committee Report (a report to the U.S. Secretary of the Interior, issued by the Federal Water Pollution Control Administration, April 1, 1968).

isms in a distribution system is often stimulated if excessive particulate or organic matter is present. The maximum contaminant levels for microbiological contaminants are in terms of the surrogate coliform bacteria, although the purpose of the standard is to protect against disease-causing bacteria, viruses, protozoans, worms, and fungi. The analytical procedures for direct detection of these microorganisms are not well enough developed or practicable for widespread application at this time. Total coliform counts have been used for nearly 100 years as indicators. Because the organisms are present in large quantity in the intestinal tracts of humans and other warm-blooded animals, the number remaining in a water supply provides a good correlation with sanitary significance.

The fluoride concentrations excessive for a given water supply depends on climatic conditions, because the amount of water (and consequently the amount of fluoride) ingested is primarily influenced by air temperature (see Sec. 7-13).

The maximum contaminant levels in Tables 7-1 and 7-2 indicate to the engineer water quality that must be achieved in public water treatment, but the quality of water desirable for a raw-water source is not discussed. In the vast majority of cases, public water treatment is "treatment," not water renovation or water reclamation involving complex processes. Therefore, treatment processes commonly employed do not remove all undesirable impurities from raw water. The engineer needs criteria for the selection of water sources such that the raw water selected can be treated to drinking water standards by processes that have been proved effective by demonstration and are within reasonable economic limits. The author does not mean to imply by the previous statement that renovation processes should not be employed or that water-treatment costs are excessive. However, the engineer designing a municipal water plant must be sure of the capabilities of the treatment processes used, and the price of treatment must not create undue hardships to the domestic and industrial users.

Raw-surface-water criteria, listed in Table 7-3, can be used to assist the engineer in the selection of a water source.[2] A significant number of the substances listed are not commonly found in nature, but originate from industrial, agricultural, and domestic pollution. Therefore, these criteria are most applicable in the evaluation of polluted surface waters. A large number of the substances are refractory and contaminants not removed by the most commonly used treatment processes in their simplest form (see the footnote to Table 7-3). It is interesting to note that many of these refractory substances, such as the metal cations and pesticides, may have deleterious physiological effects and are not merely a matter of aesthetics.

The subcommittee that developed the information in Table 7-3 considered establishing criteria on phosphorus concentrations but was unable to establish any generally acceptable limit because of the complexity of the problem.[2] The purpose of such a limit would be twofold: (1) to avoid problems associated with algae and other aquatic plants, and (2) to avoid coagulation problems due particularly to complex phosphates.

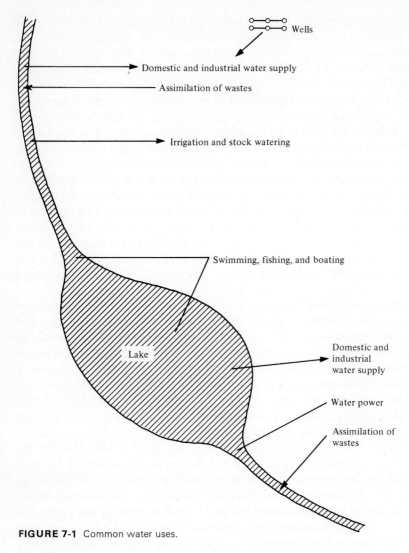

FIGURE 7-1 Common water uses.

The use of water for drinking and other domestic purposes is considered most essential. There are several other beneficial uses for water (Fig. 7-1), and the following tabulation indicates the many purposes served[3]:

1 Domestic water supply.
2 Industrial water supply (including cooling).
3 Agricultural water supply (irrigation).
4 Propagation of aquatic and marine life.
5 Stock and wildlife watering.
6 Water sports.
7 Water power and navigation.
8 Assimilation of wastes.

These uses are not listed in order of economic importance or change of water quality.

The relative importance of these uses for a body of water depends upon the specific geographic area. Many of the uses may be compatible, whereas the use of a stream for gross waste assimilation will be at variance with most other purposes. This is especially true where the body of water is used as a raw-water source for domestic supply.

Raw-water sources have generally decreased in quality because of increased water use, and this trend is expected to continue into the future. This has placed a difficult burden on the water-supply industry, which is caught between two strong trends in water quality. Water users are demanding improved quality and uniformity in the water product while raw-water sources are deteriorating. Fortunately, there are some recent improvements in technology to assist with this difficult task.

7-3

OTHER BENEFICIAL USES

Industrial water supply constitutes the largest category of water use. The magnitude of this demand demonstrates the importance of water to industry. Water is used as a coolant and source of steam, as an ingredient with other materials in the finished product, as a transportation medium, and for cleansing.

The ideal quality of water for industrial use varies widely for the many different applications. There are minimum standards that should be met by all surface waters for all uses.[3] The water should be:

1 Free from substances that will settle to form sludge deposits.
2 Free from floating debris, oil, scum, and other floating materials in amounts sufficient to be unsightly.
3 Free from materials producing color, odor, or other conditions in such a degree as to cause a nuisance.

Industries are generally willing to accept water that meets drinking-water standards. Where higher quality is required, industrial water treatment is employed. It is of primary importance that the concentration of the various constituents of the water remain relatively stable. Short-term variations in water quality require considerable attention and can significantly increase the cost of water treatment. Table 7-4 is a summary of specific quality characteristics that have been used as sources for industrial water supplies.[2] The specific water characteristics are maximums and no water will have all the maximum values shown.

Irrigation water applied to cultivated land may run off as surface flow, evaporate from the surface, or infiltrate the soil. Of the infiltrated water, a part is used by plants, a part is held by the soil for subsequent evapotranspiration, and the remaining surplus, if any, moves downward or laterally through the soil.

TABLE 7-4

SUMMARY OF SPECIFIC QUALITY CHARACTERISTICS OF SURFACE WATERS THAT HAVE BEEN USED AS SOURCES FOR INDUSTRIAL WATER SUPPLIES[a]

Characteristic	Boiler Makeup Water		Cooling Water			
			Fresh		Brackish[b]	
	Industrial 0–1500 psig	Utility 700–5000 psig	Once Through	Makeup Recycle	Once Through	Makeup Recycle
Silica (SiO_2)	150	150	50	150	25	25
Aluminum (Al)	3	3	3	3	—	—
Iron (Fe)	80	80	14	80	1.0	1.0
Manganese (Mn)	10	10	2.5	10	0.02	0.02
Copper (Cu)	—	—	—	—	—	—
Calcium (Ca)	—	—	500	500	1,200	1,200
Magnesium (Mg)	—	—	—	—	—	—
Sodium and potassium (Na + K)	—	—	—	—	—	—
Ammonia (NH_3)	—	—	—	—	—	—
Bicarbonate (HCO_3)	600	600	600	600	180	180
Sulfate (SO_4)	1,400	1,400	680	681	2,700	2,700
Chloride (Cl)	19,000	19,000	600	500	22,000	22,000
Fluoride (F)	—	—	—	—	—	—
Nitrate (NO_3)	—	—	30	30	—	—
Phosphate (PO_4)	—	50	4	4	5	5
Dissolved solids	35,000	35,000	1,000	1,000	35,000	35,000
Suspended solids	15,000	15,000	1,500	15,000	250	250
Hardness ($CaCO_3$)	5,000	5,000	850	850	7,000	7,000
Alkalinity ($CaCO_3$)	500	500	500	500	150	150
Acidity ($CaCO_3$)	1,000	1,000	0	200	0	0
pH (units)	—	—	5.0–8.9	3.5–9.1	5.0–8.4	5.0–8.4
Color (units)	1,200	1,200	—	1,200	—	—
Organics:						
Methylene blue active substances	2[c]	10	1.3	1.3	—	1.3
Carbon tetrachloride extract	100	100	[d]	100	[d]	100
Chemical oxygen demand (O_2)	100	500	—	100	—	200
Hydrogen sulfide (H_2S)	—	—	—	—	4	4
Temperature (°F)	120	120	100	120	100	120

[a] Unless otherwise indicated, units are mg/l and values are maximums. No one water will have all the maximum values shown. Application of these values should be based on Part 23, ASTM book of standards, or APHA standard methods for the examination of water and wastewater.

[b] Water containing in excess of 1000 mg/l dissolved solids.

[c] 1 mg/l for pressures up to 700 psig.

[d] No floating oil.

		Process Water					
Textile Industry, SIC-22	Lumber Industry, SIC-26	Pulp and Paper Industry, SIC-26	Chemical Industry, SIC-28	Petroleum Industry, SIC-29	Prim. Metals Industry, SIC-33	Food and Kindred Products, SIC-20	Leather Industry, SIC-31
—	—	50	—	50	—	For these categories,	
—	—	—	—	—	—	the quality of raw-	
0.3	—	2.6	5	15	—	water surface supply	
1.0	—	—	2	—	—	should be that	
0.5	—	—	—	—	—	prescribed by the	
—	—	—	200	220	—	NTA Subcommittee	
—	—	—	100	85	—	on Water Quality	
						Requirements for	
—	—	—	—	230	—	Public Water Supplies	
—	—	—	—	—	—		
—	—	—	600	480	—		
—	—	—	850	570	—		
—	—	200^e	500	1,600	500		
—	—	—	—	1.2	—		
—	—	—	—	8	—		
—	—	—	—	—	—		
150	—	1,080	2,500	3,500	1,500		
1,000	ƒ	—	10,000	5,000	3,000		
120	—	485	1,000	900	1,000		
—	—	—	500	—	200		
—	—	—	—	—	75		
6.0–8.0	5–9	4.6–9.4	5.5–9.0	6.0–9.0	3–9		
—	—	360	500	25	—		
—	—	—	—	—	—		
—	—	—	—	—	30		
—	—	—	—	—	—		
—	—	—	—	—	—		
—	—	95^g	—	—	100		

^{e} May be ≤1000 for mechanical pulping operations.
^{f} No large particles ≥ 3 mm diameter.
^{g} Applies to bleached chemical pulp and paper only.
Source: "Water Quality Criteria," National Technical Advisory Committee Report, Federal Water Pollution Control Administration (Washington, D.C.: Government Printing Office, 1968).

Soluble salts that occur in soils consist mostly of the cations sodium, calcium, and magnesium, and the anions chloride and sulfate. Other constituents that ordinarily occur only in minor amounts are the cation potassium and the anions bicarbonate, carbonate, and nitrate. Normally, the source of salts in soil is water.

Relatively pure water is used by plants or lost to the soil by evaporation. Therefore, the water retained by the soil becomes more concentrated with dissolved salts. Under humid conditions, soluble materials are carried downward into the groundwater and are transported ultimately to the oceans. In arid regions, leaching and transportation of soluble salts is not complete. This can lead to salinity problems in the soil.

Waters with TDS (total dissolved solids) less than about 500 mg/liter are used for irrigation without salinity problems under normal conditions. Waters with TDS of about 5,000 mg/liter normally have little value for irrigation unless used as an alternative supply. Within these limits, the value of the water generally decreases as the salinity increases.[2] Table 7-5 has a suggested classification as to salinity hazard.

TABLE 7-5

SUGGESTED CLASSIFICATION FOR IRRIGATION WATERS FOR SALINITY HAZARD

	TDS (mg/liter)
Water for which no detrimental effects will usually be noticed	<500
Water that can have detrimental effects on sensitive crops	500–1000
Water that may have adverse effects on many crops and requiring careful management practices	1000–2000
Water that can be used for tolerant plants on permeable soils with careful management practices	2000–5000

Source: "Water Quality Criteria," National Technical Advisory Committee Report, Federal Water Pollution Control Administration (Washington, D.C.: Government Printing Office, 1968).

Propagation of aquatic and marine life is so complex that chemical and physical data alone are insufficient to predict the results of pollution. There are great variations in environmental requirements within the same region. Other factors, such as seasonal changes and daily variations, are essential. Ideally, water-quality criteria should take all such factors into consideration. While probable safe limits of concentrations of various materials can serve as a helpful guide for waste discharges, it is generally a good policy to conduct biological tests upon the organisms involved.

Stock and wildlife consume water in varying amounts, dependent upon such factors as temperature, humidity, water content of diet, degree of exertion

TABLE 7-6
WATER CONSUMPTION BY LIVESTOCK

Animal	Water Consumed (gal/day)
Beef cattle, per head	7–12
Dairy cattle, per head	10–16
Horses, per head	8–12
Swine	3–5
Sheep and goats, per head	1–4
Chickens, per 100 birds	8–10
Turkeys, per 100 birds	10–15

Source: J. E. McKee and H. W. Wolf, "Water Quality Criteria," *California State Water Control Board Publication No. 3-A* (1963).

by the animal, and salinity of the available supply. Table 7-6 indicates some appropriate ranges of water consumption by livestock.

It has been assumed that water safe for human consumption may be used safely by livestock, but it appears that animals can tolerate higher salinities than man. There are very few quantitative data concerning the water-quality tolerances of livestock in the United States. More attention has been given livestock water quality in Australia and South Africa, where available water is frequently highly mineralized. Table 7-7 contains standards developed in Western Australia as safe upper limits for livestock.

TABLE 7-7
PROPOSED SAFE LIMITS OF SALINITY FOR LIVESTOCK

Animal	Threshold Salinity Concentration[a] (TDS, mg/l)
Poultry	2,860
Swine	4,290
Horses	6,435
Dairy cattle	7,150
Beef cattle	10,000
Sheep (adult, dry)	12,000

[a] Total salts, mainly NaCl.
Source: "Water Quality Criteria," National Technical Advisory Committee Report, Federal Water Pollution Control Administration (Washington, D.C.: Government Printing Office, 1968).

Recreational waters, including water-contact sports and aesthetics, must conform to three general conditions: (1) they must be free from obnoxious floating or suspended substances, color, or odors; (2) they must be reasonably free of toxic substances; and (3) they must have a low probability of containing pathogenic organisms.[3]

A report of the National Technical Advisory Committee to the Secretary of the Interior recommends criteria for water-contact recreation[2]:

1 Criteria for mandatory factors:
   a Fecal coliform should be used as the indicator organism for evaluating the microbiological suitability of the water. For a 30-day period the fecal coliform content of the water must not exceed a log mean of 200 per 100 ml, nor may more than 10% of total samples during the period exceed 400 per 100 ml.
   b The pH should be within the range of 6.5–8.3, except when due to natural causes, and in no case may be less than 5.0 or more than 9.0.
2 Criteria for desirable factors:
   a Clarity should be such that a Secchi disc is visible at a minimum depth of 4 ft.
   b Maximum water temperatures shall not exceed 85°F, except where caused by natural conditions.

The remainder of this chapter primarily relates to water quality concerned with public supply.

7-4

PURPOSES OF EXAMINATION

Water and wastes are examined to evaluate their treatability, treatment effectiveness, and quality.

Wastewater is analyzed to determine those constituents that may cause difficulties in treatment or disposal, as an aid in plant operation, and in selecting the correct degree and type of treatment. Plant effluents are investigated to measure their strength and to determine the constituents of the final waste. Receiving waters are tested to evaluate their ability to accept a pollutional load and to indicate the degree of self-purification that occurs in a given reach. The strength of wastewater is usually measured by its nuisance-producing potentialities—odor, solids content, and appearance. The yardstick that measures much of the nuisance potential is the biochemical oxygen demand (BOD). This is the amount of oxygen required by the bacteria to reduce some of the organic matter in a waste under standard conditions (see Sec. 7-22).

Water to be used for a public supply must be potable (drinkable), that is, not contain pollution. Pollution can be defined as the presence of any foreign substance (organic, inorganic, radiological, or biological) which tends to degrade the water quality and constitutes a hazard or impairs the usefulness of the water.[2] Routine analyses of developed water sources are made usually to determine the acceptability of the water for domestic and industrial uses. Results of these analyses also indicate the kind of corrective treatment which should be considered for specific applications of the water. The complete analysis of a potential water source should include a sanitary survey and physical, chemical, and biological analyses.

Methods of collection and analysis must be standardized if results obtained by different laboratories are to be comparable. In the United States, *Standard Methods for the Examination of Water and Wastewater* has been published jointly by the American Public Health Association, the American Water Works Association, and the Water Pollution Control Federation.[4] These methods have also been accepted by the American Chemical Society. No attempt will be made here to describe or explain the various analysis procedures. The reader can acquaint himself with the techniques involved by reading *Standard Methods*.

7-5

SANITARY SURVEY

Sanitary surveys of water and wastewaters include (1) surveys of the conditions under which water is processed, and (2) observations and examinations of certain properties of the water in its natural or artificial field environment. A survey to determine the characteristics of wastewater should furnish information concerning the following: (1) the source, whether domestic or industrial, and if industrial, the types of industries and industrial waste treatment, if any; (2) variations in the rate of flow; (3) the strength and condition of the waste; (4) the amount of infiltration and storm water carried by the sewer, if any; and (5) other factors that affect the characteristics of the wastewater.

Surveys are made of receiving streams for gross qualities such as unsightly floating matter, sludge banks, growths of fungi, and other biological indicators of pollution. Physical properties such as temperature are evaluated in the field. Fixing of chemical constituents—for example, carbon dioxide and dissolved oxygen—is also carried out at the site.

Sanitary surveys are made not only of the source of a water supply but of the water-supply system as well. These surveys uncover environmental conditions that may affect the potability and treatability of the water being considered. Increasing pollution problems point up the need for additional attention to the quality of source waters. Abatement and control of pollution at the source will significantly aid in producing a high-quality water. A periodic survey of water-distribution systems is necessary to control physical defects or other health hazards in the system.

7-6

SAMPLING

Because the environmental engineer is often called upon to collect samples or advise on sampling techniques, a knowledge of correct procedures is important.

A 2-liter portion is adequate for most physical and chemical analyses. Larger volumes may be required for special determinations. Separate portions should be collected for chemical and bacteriological examination, since the methods of collection and handling are quite different. The shorter the time interval between collection of a sample and its analysis, the more reliable the results. Immediate field analysis is required for certain constituents and physical characteristics to assure dependable results, since changes in composition of the sample may occur in transit to the laboratory.

It is difficult to state exactly how much time can be allowed between the collection of a sample and its analysis. A potable water specimen may ordinarily be held for a much longer period than a raw-wastewater sample. The following maximum limits are suggested in *Standard Methods* as reasonable for physical and chemical analyses: unpolluted waters, 72 hours; slightly polluted waters, 48 hours; and polluted waters, 12 hours. The time and place of sampling and analysis should be recorded on the laboratory report. If the portions have been preserved by an additive or deviations made from the procedures outlined in *Standard Methods*, these facts should be recorded on the report.

Certain cations, such as iron and copper, are subject to loss by adsorption on, or ion exchange with, the walls of glass bottles. Also, pH and carbon dioxide are subject to change in transit. Theses changes in pH–alkalinity–carbon dioxide balance can cause calcium carbonate to be precipitated and underestimates of total hardness and calcium may result.

The microbiologic activity of a sample can produce changes in the nitrate–nitrite–ammonia balance in BOD. Color, odor, and turbidity may change significantly between time of sampling and time of laboratory analysis.

It is impossible to prescribe absolute rules for the prevention of all changes that take place in a sample bottle. The sample should be collected and stored consistent with the character of the laboratory examinations to be made. Often much time and trouble can be saved if the person collecting the samples and the laboratory analyst confer in advance on the best technique for collecting and storing the samples.

7-7

REPRESENTATIVE SAMPLES

Care must be taken to obtain a specimen that is truly representative of existing conditions. The sample bottle should be rinsed two or three times with the water to be sampled prior to filling. No general recommendations can be made as to the number or places of sampling, since details of collection vary considerably with local conditions. Care should be exercised to identify every sample bottle, preferably by attaching an appropriately inscribed tag or label. The record needed includes all information pertinent to the purpose of the sample, such as the name of the collector, date, hour, and exact location of the source.

Prior to collecting samples from a water-distribution system, pipes should be flushed for a sufficient period of time to ensure that the sample represents the supply. Samples from wells should be taken only after the well has been pumped for a sufficient period of time to reach equilibrium if the test is to represent the groundwater that feeds the well.

Samples collected from a stream may vary with depth, stream flow, and distance from shore. An "integrated" portion from top to bottom in the middle of the stream is generally representative of stream flow. A grab sample (a single sample) can be collected at middepth in the middle of the stream.

Because the quality of water in lakes and reservoirs is subject to considerable variation, the choice of location and depth of sampling depends upon local conditions and the purpose of the investigation. Lakes and reservoirs are affected by rainfall, runoff, wind, and seasonal stratification.

In testing influents and effluents from waste-treatment plants, a composite made up of several grab samples taken at different times is usually desired. If the character of the waste is constant, only a grab sample is necessary, but normally both the character and rate of flow are variable. Under these conditions the collection of a representative sample is difficult. A survey should obtain data on the average concentrations of pollutants over an 8- to 24-hour period from a composite portion, and the concentrations at high and low rates of flow from grab samples.

Composite specimens are prepared from a series of grab samples or from an automatic sampler. A series of grab samples might be taken every 20 minutes and an amount proportional to the flow rate at the sampling time is placed in the composite container. For example, 1 ml may be taken for each gallon per minute of flow.

It is better to obtain too large a sample rather than one too small as the analyst may wish to make check determinations or run additional tests. Special preservation methods are necessary for portions that are not to be analyzed immediately. Cooling is the most common technique, but special chemicals are added for certain tests. Instructions regarding sampling procedures for specific tests are given in *Standard Methods*.

7-8

STANDARD TESTS

Many of the analyses employed in the examination of samples of water and wastes are identical, but the information sought is used for different purposes. The usual objective of a water analysis is to determine the acceptability of the water for its intended use and as a guide in treatment. The normal purpose of a waste analysis is to learn the composition, concentration, and condition of the waste.

Waters are classified as potable or polluted, safe or unsafe, pure or impure, hard or soft, corrosive or uncorrosive, sweet or sour. Wastewater is classified as strong or weak, fresh or septic, putrescible or nonputrescible, domestic or industrial, to mention but a few.

Standard Methods describes the following tests, which are performed on waters and wastes. Tests that are not normally a part of a sanitary analysis are printed in italics.

Water in the Absence of Gross Pollution
Physical and chemical examination:
>Temperature, turbidity, color, taste, and odor
>Residue, solids after evaporation (total, *filterable*, *nonfilterable*; and for each solid the fixed portion and the loss on burning)
>*Solids by electrolytic conductivity*
>Hardness by calculation from a mineral analysis and by the versenate (EDTA titration) method
>Acidity, including mineral acids, alkalinity (phenolphthalein and total), pH value, carbon dioxide (free and *total*), *bicarbonate ion*, *carbonate ion*, and *hydroxide*
>*Silica*
>*Copper*, *lead*, *aluminum*, iron, *chromium*, *manganese*, and *zinc*
>*Magnesium*, *calcium*, *sodium*, and *potassium*
>Nitrogen: *ammonia*, *albuminoid*, *organic*, *nitrite*, and nitrate
>Chloride, *iodide*, and fluoride
>Phosphate: *orthophosphate*, *total phosphate*, and *polyphosphate*
>*Sulfate*, *sulfide*, and *sulfite*
>*Arsenic*, *boron*, *cyanide*, *selenium*, *barium*, *cadmium*, and *silver*
>*Alkyl benzene sulfonate* (*surfactants*), phenols
>Chlorine: free available, *monochloramine*, *dichloramine*, *nitrogen trichloride*, and chlorine demand
>*Tannin* and *lignin*
>*Oxygen*: *dissolved*, *biochemical demand*, *chemical demand*, and *ozone*
>*Oil* and *grease*

Biological examination:
>*Plate count*
>Coliform group, group density, and *differentiation of coliform Enterococcus group*
>*Examination and enumeration of microscopic organisms*

Radiologic examination:
>*Alpha and gross beta* (*total*, *suspended*, and *dissolved*)
>*Strontium*, *total radioactive*, and *strontium 90*

Examination of Waste Water
Physical and chemical examination:
>Temperature, *turbidity*, color, and odor

Residue: by evaporation, total, dissolved, suspended, and settleable (for each
   solid, the volatile and fixed portion)
Acidity, alkalinity, and pH value
Nitrogen: *ammonia*, *nitrate*, *nitrite*, and *organic*
Chloride and *sulfide*
Chlorine: residual and demand
Grease
Alkyl benzene sulfonate (surfactants)
Oxygen: dissolved, demand chemical, and demand biochemical
Biological examination:
   Examination and identification of some microorganisms common in wastes
Radiologic examination:
   Same as for water

7-9

BACTERIOLOGICAL TESTS

Most bacteria in water are derived from contact with air, soil, living or decaying plants or animals, mineral sources, and fecal excrement. Some of the
most common types of bacteria that may be present in water include the coliform group, fecal streptococcus, fluorescent bacteria, chromogenic bacteria,
proteus group, spore-producing rods, and *Achromobacter*.[15] Many of these
bacteria are without sanitary significance because they die rapidly in water,
come from unknown sources, are widely distributed in air or soil, or have no
known or suspected association with pathogenic organisms.

A variety of procedures have been used in the last 100 years to measure the
bacteriological quality of water. These biological tests and procedures include the
following:

1 The total plate count on gelatin at 20°C and on agar at 37°C.
2 The specific identification of pathogenic bacteria.
3 Use of the coliform group as a sewage-pollution indicator.
4 Checking the fecal streptococcus group to indicate fecal pollution.
5 Examining *Clostridium perfringens* as an indicator of pollution.
6 Employing miscellaneous indicators, including bacteriophage tests, serological
   methods, and identification of other specific bacteria and virus.

The specific identification of pathogenic bacteria as pollution indicators
requires extremely large samples and a wide variety of media and methods to
exclude all the various pathogens that could be present. Neglecting the expense,
the time required for this test procedure would greatly reduce the usefulness of
the results. Such a group of test procedures is not applicable to the frequent
routine sampling of water supples.

The coliform group is considered a reliable indicator of the adequacy of
treatment for bacterial pathogens. The 1974 Safe Drinking Water Act reaffirms

this standard and includes all of the aerobic and facultative anaerobic, gram-negative, non-spore-forming, rod-shaped bacilli which ferment lactose with gas production within 48 hours at 35°C in the coliform group.

This coliform grouping includes organisms that differ in biochemical and serologic characteristics and in their natural sources and habitats. *Escherichia coli* is characteristically an inhabitant of the intestines of man and animals. *Aerobactor aerogenes* and *A. cloacae* are frequently found in various types of vegetation and as deposits in the distribution system. The intermediate-aerogenes-cloacae (IAC) subgroups are found in fecal discharges but normally in smaller numbers than *E. coli*. All of the coliform group except *E. coli* are commonly found in soil and in waters polluted in the past.

Organisms of the IAC group tend to survive longer in water than do *E. coli* and are more resistant to chlorination. The relative survival times of the coliform subgroups may be used to distinguish recent from less recent pollution. Waters that have been newly polluted will show an *E. coli* density greater than the IAC density. Polluted waters that have not lately received fecal material will tend to have a higher IAC group density than *E. coli*. There is considerable confusion and controversy regarding both the advantages and disadvantages claimed for *E. coli* as an indicator of recent pollution. *Standard Methods* lists a standard for differentiation of coliform-group organisms. Recent investigations indicate that the portion of the coliform group which is present in the feces of warm-blooded animals generally includes organisms which are capable of producing gas from a suitable culture medium at 44.5°C. This differentiation of the coliform group yields information relative to the possible source of pollution and to the remoteness of the pollution.

The fecal streptococci are characteristic of fecal pollution and are consistently present in the feces of all warm-blooded animals.[16] They do not multiply in water as sometimes occurs with the coliform group; therefore, the presence of fecal streptococci in water indicates fecal pollution with a density equal to or less than that originally present. Improved methods and media are needed for the routine analysis of the streptococcal group which is not recognized as an official test procedure in the United States, although *Standard Methods* lists a tentative procedure for enterococcus. The fecal streptococci tests have application in stream-pollution investigations.

Clostridium perfringens is widely distributed over the earth's surface and uniformly present in the intestinal tract of warm-blooded animals.[17] The spores have a long survival time in water and are resistant to chemical treatment and natural purification. The presence of spores in deep wells indicates a direct connection between the surface and the underground water source. Methods of isolation and identification are unsatisfactory for routine use. There is no test for *C. perfringens* in *Standard Methods*.

Numerous other biological indicators of pollution have been suggested by various investigators—for example, bacteriophage tests. Phage types in water cannot be interpreted with our current knowledge and therefore have little use as evidence of pollution. Agglutination tests have been propounded for the serological separation of the coliform group. Difficulty has been encountered in duplication of these tests.

7-10

THE COLIFORM GROUP—FERMENTATION TUBE TEST

The coliform group of bacteria possesses the faculty of fermenting lactose or lauryl tryptose broth and producing gas. This offers a simple, visible evidence of a member of the coliform group. Some other bacteria also ferment the broth under certain conditions and combinations and additional growth reactions must be used to confirm the presence of a coliform. The number of coliform organisms in human feces is very great, the daily per capita excretion varying from 125 to 400 billion. The total number of bacteria in fecal matter that can be counted by simple bacteriological techniques is approximately 1000 times greater still. The Safe Drinking Water Act limits the number of coliforms for potable water to less than 1 per 100 ml, or 40 coliforms per gallon. To reach this figure of 40 coliforms per gallon by dilution would require a very large volume of clean water. Assuming 400 billion coliforms per capita-day in summer and 100 gpcd wastewater produced, the approximate dilution factor would be $(400 \times 10^9)/(40 \times 10^2) = 10^8$, or 100 million. This means that every gallon of wastewater would have to be diluted with 100 mgal of clean water to be acceptable as drinking water. Actually the number of coliforms is materially reduced by death from nonnormal environment, by removal and destruction in waste-treatment process, and by their removal and destruction in water treatment. Nevertheless, the large numbers of coliform organisms are a good indicator of bacterial pollution.

   Figure 7-2 shows the anticipated reduction in coliform bacteria in a stream for various types of wastewater treatment. There is some question at the lower end of the activated-sludge-plant curve and the authors suggest that the dashed portion more nearly represents the anticipated reduction.[8] This dashed portion was arrived at by interpolation of data between primary and biological treatment with postchlorination. The figure also includes dashed lines at the 1000 and 5000 coliforms per 100 ml. These values are typical coliform limits used by a number of state and interstate water-pollution-control authorities. The 1000-coliforms-per-100-ml criterion is used for waters where body-contact sports such as swimming and skiing take place. The 5000-coliforms-per-100-ml standard is used for waters that serve as sources for municipal supply and aquatic recreation exclusive of body-contact sports. Using the graph, Fig. 7-2, activated-sludge treatment improves approximately 50 mi of the river as compared to primary treatment for a raw-water-supply source. It requires approximately 60 river miles to reduce the coliform count to 5000 per 100 ml for effluent from an activated-sludge plant and roughly 110 river miles for an effluent from primary treatment.

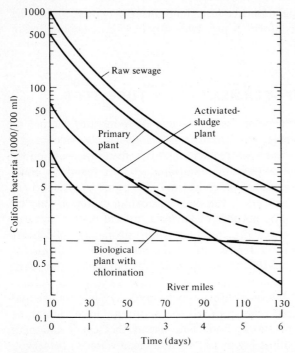

FIGURE 7-2 Anticipated reductions in coliform bacteria in a stream for various types of wastewater treatment. [Source: After F. W. Kittrell and S. A. Furfari, "Observations of Coliform Bacteria in Streams," *Water Poll. Control Fed.* 35 no. 11 (November 1963) : 1379.]

The standard of less than 1 coliform bacteria per 100 ml of sample does not exclude the possibility of intestinal infection, but it is a practical economic degree of acceptability. The test for organisms of the coliform group is simple to perform and can be done by routine laboratory personnel.

Samples for microbiological analysis are to be taken at regular intervals throughout the month proportional to the population served. Samples should be collected from representative locations throughout the system. The minimum number of samples to be collected each month is given in Fig. 7-3 and should be in accordance with the following: for a population of 25 to 1000, one; 1000 to 2,500, 2; 25,001 to 100,000, to the nearest multiple of 5; and over 100,000, to the nearest multiple of 10.

Tests for coliform bacteria are made on raw water and water as it passes through the treatment process and give an indication of its sanitary quality. Coliform bacteria density measurements are made on bathing waters, shellfish waters, and some industrial water. Various state health departments normally divide surface waters suitable for public drinking supplies into two or more classes, depending upon the degree of treatment needed. Since the standard adopted is based upon the number of coliform organisms per unit volume of

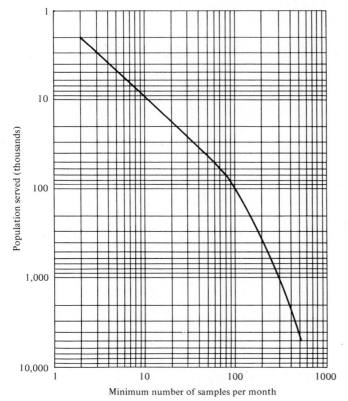

FIGURE 7-3 Minimum number of samples per month required by the National Safe Drinking Water Act. See the text for populations under 2000.

the sample, the coliform test is used over a broad range in water quality. It is important not only to know that coliform organisms are present but also to determine the probable number per unit volume in the water. Polluted waters must be diluted to calculate the probable number of organisms. Figure 7-4 illustrates a scheme for making these dilutions.

Tests for the presence and density of coliform organisms utilize the inoculation of lactose broth or lauryl tryptose broth by multiple portions of a series of decimal dilutions of a water sample. The number of portions planted and the range of dilutions made will depend upon the apparent quality of the water under examination and on the records of previous samples. Practical considerations limit the size of the largest portion to 100 ml.

The test sequence is divided into three parts, the presumptive, the confirmed, and the completed test. All examinations for drinking-water purposes should be carried through the confirmed part and occasionally a run should be carried through the completed portion to check the effectiveness of the testing procedure. Analysis other than for drinking water requires only the presumptive test.

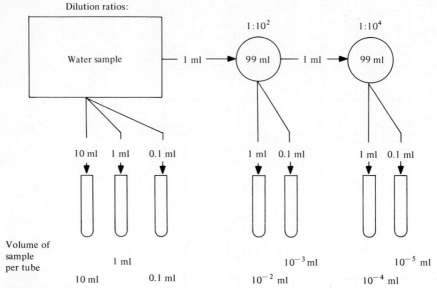

FIGURE 7-4 Preparation of dilutions of a water sample for the coliform-density test.

7-11

MOST PROBABLE NUMBER (MPN)

The most probable number of coliform organisms in a water sample is the density more likely to produce a particular result. A number of portions of different sizes of the water sample are examined for the presence of coliform organisms. It is assumed that variations in the distribution of coliform organisms will follow a probability curve and that the MPN can be determined from the following expression[9,10]:

$$Y = \frac{1}{a}[(1 - e^{-N_1\lambda})^p(e^{-N_1\lambda})^q][(1 - e^{-N_2\lambda})^r(e^{-N_2\lambda})^s][(1 - e^{-N_3\lambda})^t(e^{-N_3\lambda})^u] \qquad (7\text{-}1)$$

where N_1, N_2, N_3 = sizes of portions examined, ml

p, r, t = number of portions of respective sizes giving positive tests for coliforms

q, s, u = number of portions of respective sizes giving negative tests for coliforms

λ = concentrations of coliforms per mililiter

Y = probability of occurrence of a particular result

e = base of Napierian logarithms, 2.718

a = constant for any particular set of conditions and therefore may be omitted in computation of λ

TABLE 7-8

MPN AND 95% CONFIDENCE LIMITS FOR SELECTED COMBINATIONS OF POSITIVE RESULTS IN FERMENTATION-TUBE TESTS

Number of Positive Tubes Out of:				Limits of MPN	
Five 10-ml Tubes	Five 1-ml Tubes	Five 0.1-ml Tubes	MPN	Lower	Upper
0	0	1	2	0.5	7
0	1	0	2	0.5	7
1	0	0	2	0.5	7
1	0	1	4	0.5	11
1	1	1	6	0.5	15
2	1	1	9	2	21
2	2	2	14	4	34
3	0	0	8	1	19
3	2	2	20	6	60
4	0	0	13	3	31
4	3	2	39	13	106
5	0	0	23	7	70
5	1	1	46	16	120
5	2	0	49	17	126
5	3	3	175	44	503
5	4	4	345	117	999

Five 10-ml Tubes	One 1-ml Tube	One 0.1-ml Tube	MPN	Limits of MPN	
				Lower	Upper
0	0	0	0	0	5.9
1	0	0	2.2	0.05	13
2	0	0	5	0.54	19
2	1	0	7.6	1.5	19
3	0	0	8.8	1.6	29
3	1	0	12	3.1	30
4	0	0	15	3.3	46
4	0	1	20	5.9	48
4	1	0	21	6.0	53
5	0	0	38	6.4	330
5	0	1	96	12	370
5	1	0	240	12	3700
5	1	1	—	88	—

●EXAMPLE 7-1

What is the most probable number of coliform organisms per 100 ml as determined by the following series of observations: five 10-ml tubes were used of which four were positive; one 1-ml tube was used and it was negative; and one 0.1-ml tube was used and it was positive?

○*Solution*

From Eq. 7-1, $N_1 = 10$, $N_2 = 1$, $N_3 = 0.1$, $p = 4$, $r = 0$, $t = 1$, $q = 1$, $s = 1$, and $u = 0$.

The MPN of this series is at the mode of the curve. The modal value is determined by selecting the value of λ that will make Y a maximum. The value of a is a constant for any particular set of observations and may be omitted in computation of λ.

In the analytical result cited, let the following trial values be assumed: $\lambda = 0.19, 0.20,$ and 0.21, respectively. Then the detailed solution is as follows:

$$Y = \frac{1}{a}(1 - e^{-10\lambda})^4(e^{-10\lambda})(1 - e^{-\lambda})^0(e^{-\lambda})(1 - e^{-0.1\lambda})(e^{-0.1\lambda})^0$$

The value of aY is maximum where $\lambda = 0.20$ and the MPN per milliliter is 0.20, or 20 per 100 ml.

	Values of λ		
Terms	0.19	0.20	0.21
$(1 - e^{-10\lambda})^4 =$	0.522	0.560	0.594
$e^{-10\lambda} =$	0.150	0.135	0.122
$e^{-\lambda} =$	0.827	0.819	0.811
$1 - e^{-0.1\lambda} =$	0.019	0.020	0.021
$aY =$ products $=$	0.00123	0.00124	0.00123

Table 7-8 gives the MPN and 95% confidence limits for various combinations of positive results in fermentation-tube tests. More complete tables listing all combination of results expected to occur more than 1% of the time are given in *Standard Methods*.

7-12

VIRUS IN WATER

During the past 20 years more than 70 new enteric viruses have been recognized, all of which can be found in human feces. Many of these can be recovered from domestic waste.[11] Some strains of viruses have been found to be more resistant to chlorination than the common bacteria. Kelly and Sanderson[12] have reported this for the polio virus and Clarke and Kabler[13] for the Coxsackie virus in water. The density of enteroviruses in wastewater is low in comparison with that of coliform bacteria. The low densities do not appear to be the result of inactivation of viruses en route to the plant since detention by primary treatment does not affect the density. The presence of minimum densities of virus in the winter months indicates that there is a small reservoir of enterovirus infection in the population at all seasons.[14]

The density of plague-forming enteroviruses in wastewater, shown in the study by Kelly and Sanderson,[14] estimated from the density in swab expressions, was 2–44 per 100 ml. The highest densities were found in August and early September in samples of raw waste. The enteroviruses isolated were identified as strains of Coxsackie, ECHO, and polio viruses.

The mechanism for the removal of enteroviruses from wastewater by activated sludge is not fully understood. Removal of polio virus by activated-sludge experiments in the laboratory has been shown to involve at least two steps: (1) aeration with sludge floc and nutrient, and (2) the settling of the floc.[15] Removal by step 2 depended on the metabolic state of the floc. Aeration with sludge floc was essential and was affected by respiratory metabolites and the absence of air. Aeration itself appeared to have no effect, nor was there any correlation with the oxidation–reduction potential. Viruses were inactivated during the removal process, and the isolation of at least four strains of bacteria with antiviral activity indicates that biological antagonism is a possible third step.[15] The percentage of virus isolations indicated that there was a progressive decrease in virus content of the wastewater as it passed through the plant.[16] This has not been borne out in some other experiments.

Small-scale water-treatment pilot plants have been used to study the removal of polio virus Type I as water was passed through sand under groundwater conditions along with a flocculation–filtration process at rapid rates.[26] Groundwater flow conditions through 2 ft of well-packed sand removed polio virus from water (velocites less than 4 ft/day). Some virus passed through the rapid sand filters but 99% removal was possible with increased alum and conventional flocculators and settling. The results were erratic, and more testing is necessary to resolve design and operating problems with rapid sand filters.

There is a need for improved methods of quantitative estimation of virus concentrations and easier methods of identification. It should be remembered that a possibility always exists that a water may be classified as potable by any of the official bacterial tests and still contain dangerous concentrations of virus.

7-13

FLUORIDES

Small concentrations of fluorides in drinking water reduce the prevalence of dental caries. Excessive amount of fluorides are definitely associated with the mottling of teeth. Fluorides, therefore, must be regarded as both a beneficial and a dangerous mineral.

Fluoridation has become firmly established as an effective and economical public health measure. The process is recommended as a proven scientific procedure and as an accepted adjunct of water-treatment processes.

When the annual average of the maximum daily air temperatures for the location in which the water system is situated, the corresponding maximum levels for fluoride are shown in Table 7-9.

TABLE 7-9

MAXIMUM LEVELS FOR FLUORIDE, 1975 INTERIM PRIMARY DRINKING WATER REGULATIONS

Temperature (°F)	Level (mg/l)
53.7 and below	2.4
53.8–58.3	2.2
58.4–63.8	2.0
63.9–70.6	1.8
70.7–79.2	1.6
79.3–90.5	1.4

Several fluorine compounds are available for water treatment: sodium fluoride, hydrofluoric acid, hydrofluocilicic acid, and sodium silicofluoride. Sodium fluoride and sodium silicofluoride are the most widely used. Water apparently has no fluorine demand, and all fluorine added is indicated by the standard tests.

Most compounds that form fluoride ions in water can be used for adding fluoride to drinking water. However, there are practical considerations in selecting compounds to be used. The material must have sufficient solubility to permit its use in routine plant practices. The cation to which the fluoride ion is attached must not have undesirable water-quality characteristics, and the material should be relatively inexpensive. See Table 7-10 for information on some of the common fluoride compounds used in water supplies.

The courts of 10 states have held that fluoridation of public water supplies does not infringe on the constitutional or legal rights of an individual. These decisions were rendered by the courts of last resort in California, Louisiana, Ohio, Oklahoma, Oregon, Washington, and Wisconsin and by trial courts in Maryland, Pennsylvania, and North Dakota.[18] These decisions are strengthened by the fact that the U.S. Supreme Court has refused to review some of these decisions for the stated reason that no substantial federal constitutional question is involved. Only one court has ever rendered an opinion adverse to fluoridation, and in that case the decision was promptly reversed by the state supreme court, the U.S. Supreme Court refusing to review the case.

TABLE 7-10

CHARACTERISTICS OF FLUORIDE COMPOUNDS

Item	Sodium Fluoride, NaF	Sodium Silicofluoride Na_2SiF_6	Fluosilicic Acid, H_2SiF_6
Form	Powder or crystal	Powder or very fine crystal	Liquid
Molecular weight	42.00	188.05	144.08
Commercial purity (%)	90–98	98–99	22–30
Fluoride ion (%) (100% pure material)	45.25	60.7	79.2
Lb required per mgal for 1.0 ppm F at indicated purity	18.8 (98%)	14.0 (98.5%)	35.2 (30%)
pH of saturated solution	7.6	3.5	1.2 (1% solution)
Sodium ion contributed at 1.0 ppm F (ppm)	1.17	0.40	0.00
Storage space (ft^3/ 1000 lb F ion)	22–34	23–30	54–73
Solubility (g/100 g water at 25°C)	4.05	0.762	Infinite
Weight (pcf)	65–90	55–72	10.5 lb/gal (30%)
Cost			
Cents per lb	18–25	8–10	$2\frac{1}{2}$–15
Cents per lb available F	41–57	13–17	14–63
Shipping containers	100-lb bags; 125–400-lb fiber drums, bulk	100-lb bags; 125–400-lb fiber drums, bulk	13-gal carboys; 55-gal drums, bulk

Opponents of fluoridation have alleged that the procedure violates constitutional rights such as religious freedom and other liberties. They have argued that fluoridation represents the unlicensed practice of medicine and dentistry and is mass medication. Every conceivable legal and constitutional objection to fluoridation has been argued unsuccessfully in the courts.

7-14

RADIOLOGICAL

The effects of radiation on human beings are viewed as harmful, and any unnecessary intake of radioactive material in water should be avoided. Development of the nuclear industry and nuclear weapons testing has caused an increase of radioactivity in the environment. The 1962 Drinking Water Standards established limits for radium 226 and strontium 90. Iodine 131 is not found in significant quantities in public water supplies frequently enough to warrant a limit, and strontium 89 levels are not likely to be significant unless strontium 90 levels are also high.

The Environmental Protection Agency issued proposed maximum levels for radioactivity under the National Safe Drinking Water Act on August 14, 1975. The proposed maximum level for radium is 5 pCi per liter (a picocurie is a measure of radioactivity equivalent to 3.70×10^{-2} atoms disintegrating per minute). The agency also set proposed maximum limits for man-made radioactivity at 4 mrem/yr for a total body or organ dose.

The removal of radioactive materials for either surface waters or groundwater starts before the water enters the treatment process. In surface water, the physical process of sedimentation reduces particulate activity. The extent of this reduction is dependent upon the stream turbulence and size of the particles. Many silts have ion-exchange capabilities and can reduce the level of soluble activity. Biologic processes may be effective in reducing the radiochemical concentration, and some forms of algae have been known to concentrate radio-phosphorus by factors up to 10.[19,20] These benefits by natural processes are only temporary because the radioactivity is merely stored in the stream and not removed. During periods of increased turbulence the bottom sediments are resuspended and plants and animals die, returning the activity they contained to the stream.

Percolating groundwater normally undergoes some natural treatment. Filtration is effective on particulate substances and some ion exchange probably takes place with soluble material. Many minerals have good ion-exchange properties, especially some of the clays. The actual degree of decontamination attained depends upon the chemical and physical properties of the radioactive material and the geology of the area.

The radiologic quality of water is further improved by conventional treatment processes. Radioactive materials are retained in sludges, filter backwash water, and ion-exchange resins. However, special handling problems may be required.

7-15

STREAM POLLUTION

All industrial and domestic wastes affect, in some way, the normal life of a river. When the effect is sufficient to render the stream unacceptable for its best usage, it is said to be polluted. Best usage means any beneficial use, such as drinking, bathing, fish propagation, or irrigation.

The following are common pollutants:

Organic Matter

Biological decomposition of waste organic matter in a stream depletes the dissolved oxygen content of the water. Fish and most aquatic life are stifled by a lack of oxygen, and unpleasant tastes and odors are produced if the oxygen supply is sufficiently reduced.

Suspended Solids

Organic solids create sludge banks and decompose, causing odors and unsightly conditions. Inorganic suspended solids blanket the streambed, affecting benthos organisms.

Floating Solids

These include oils, greases, and other materials that float on the surface and create unsightly conditions and obstruct the passage of light vital to plant growth.

Microorganisms

Certain microorganisms, such as saprophytic bacteria and algae, may be considered as living organic matter. However, pathogenic bacteria and viruses are a health hazard to water users downstream.

Inorganic Solids

The buildup of salts can interfere with water reuse by municipalities, industries manufacturing textiles, paper and food-products, and agriculture for irrigation water. Inorganic nutrient salts, especially nitrogen and phosphorus, induce the growth of algae in surface waters.

Foam-Producing Matter and Color

These are indicators of contamination and lead to an undesirable appearance of the receiving streams.

Acids, Alkalis, and Toxic Chemicals

Chemicals adversely affect the aquatic life in the stream and impair recreational uses, such as fishing, swimming, and boating. Certain chemicals are poisonous to man, as well as aquatic life, rendering the water unsuitable for domestic water supplies.

Heated Water

An increase in water temperature by discharging wastes, such as condenser water, accelerates oxygen depletion, adversely affects fish, and may decrease the value of the water downstream for industrial cooling.

Radioactive Materials

In excessive quantities, these adversely influence the biological life and public use of the stream waters.

Under the provisions of the Federal Water Quality Act of 1965, the states developed (1) water-quality criteria applicable to interstate waters within each state and (2) a plan for implementation and enforcement of the water-quality criteria adopted.[21] Refer to Section 16-6 for additional information on this act. The criteria and associated implementation plan, after adoption by a state, become the water-quality standards applicable to such waters.

The distinction between a standard and a criterion should be noted. A criterion is a standard of judgment used to test the suitability of a water supply for a particular purpose: a quality sought for an intended use, a goal to be achieved. A standard is a definite rule, measure, or limit for a particular water-quality parameter established by a legal authority. This is a limit or quality which should not be exceeded except under the express conditions incorporated in the language of the standard. A criterion would remain the same regardless of the water supply, but a standard would be subject to change.

The fact that a standard has been established by law makes it quite rigid. This authoritative origin does not necessarily mean that the standard is fair or based on sound scientific knowledge. It has been suggested that a better way to describe a decision by a regulatory body is "requirement" instead of standard. Requirement does not give an impression of immutability.

Adoption of water-quality standards grew out of the desire to protect public health. Originally, public health concern was directed to the prevention of specific illnesses, usually pathogenic organisms. Now standards have been set for protection of interests of beneficial uses not directly involving public health. There are also social goals that are difficult to put into perspective.

Small streams should be kept free of pollution, but rivers must be permitted to take on some municipal and industrial waste in a manner that does not interfere with other beneficial uses. Classification of streams and the setting of standards is an important and difficult task and can best be done at the state-government level. Once the stream standards are set, then industries and municipalities must treat their effluents to maintain the established standards.

Treated effluents normally require dilution in streams. The volume of water for dilution is seldom constant, as the stream discharge varies from year to year and day to day. Therefore, the volume of water for dilution is at a minimum during low-flow periods. Low-flow periods are used as the basis of design for treatment capabilities of wastewater treatment plants. The probability of occurrence of low flows can be determined from a study of stream-discharge data contained in U.S. Geological Survey Water Supply Papers and records of state departments of water resources.

Various frequency curves of annual minimum flows have had wide application in estimating the waste assimilative capacity of streams. It is recognized that hydrologic extremes (floods and droughts) are skewed and do not follow a normal symmetrical distribution. Velz and Gannon[22] adapted Gumbel's theory of extreme values for drought flows. However, Gumbel's theory of extreme values is based upon an unlimited distribution and this theory is not true for streams in which the minimum discharge approaches zero as a limit. To account for such situations, Gumbel[24] proposed the use of logarithmic extremal probability paper wherein the data are plotted as the logarithms of the drought flows.

A drought flow is defined as the lowest discharge which occurs for a defined time period, such as the average low flow for 7 consecutive days. For stream-assimilation capacity, one may wish to distinguish between warm- and cold-weather droughts. A study by Ray and Walker showed that this method in Virginia provided a minimum flow that was equal to or larger than 99% of the daily flows.[25]

The procedure for plotting observed low flows on extremal probability paper is as follows (see Example 7-2):

1 Arrange the minimum flows for each year in the order of severity, (i.e., from highest to lowest flow).
2 Assign a serial number m to each of the n values 1, 2, 3, ..., n.
3 Compute the plotting position of each serial value, giving the probability equal to, or less than, for each value as $m/(n + 1)$.
4 Plot the flow on the y axis (logarithmic scale) and the corresponding probability value on the x axis using logarithmic extremal probability paper.
5 Draw the best fit line through the plotted data.

● **EXAMPLE 7-2**

Determine the 10-year low flow for 14 consecutive days given the following information:

Year	Lowest Mean Discharge for 14 Consecutive Days (cfs)	Year	Lowest Mean Discharge for 14 Consecutive Days (cfs)
1941	22.7	1952	64.8
1942	33.9	1953	38.4
1943	20.4	1954	61.7
1944	38.2	1955	30.4
1945	30.4	1956	31.2
1946	38.6	1957	30.5
1947	40.0	1958	33.7
1948	29.6	1959	34.0
1949	35.0	1960	54.8
1950	38.6	1961	37.0
1951	98.1	1962	67.0

○*Solution*

Minimum Flows in Order of Severity (cfs)	Serial Number, m	$\dfrac{m}{n+1}$
98.1	1	$\frac{1}{23} = 0.0435$
64.8	2	$\frac{2}{23} = 0.087$
61.7	3	0.130
60.4	4	0.174
54.8	5	0.217
40.0	6	0.261
38.6	7	0.305
38.6	8	0.347
38.4	9	0.390
38.2	10	0.435
37.0	11	0.477
35.0	12	0.520
34.0	13	0.565
33.9	14	0.605
33.7	15	0.650
31.2	16	0.695
30.5	17	0.740
30.4	18	0.782
30.4	19	0.825
29.6	20	0.870
22.7	21	0.912
20.4	$22 = n$	0.955

This information is plotted on extremal probability paper, and the 10-year low flow is found to be 25.7 cfs (see Fig. 7-5).

In New Mexico and other Western states where streambeds may be dry for many weeks or months of each year, frequency curves have little application for treatment-plant design. For the reach of the Rio Grande below Elephant Butte Reservoir, New Mexico, and the Texas state line, the river bed is dry for about 5 months each year. For the purpose of setting river water standards, the river is only recognized as a river when the flow is above 350 cfs.

Two possibilities exist to provide utilization of stream flow for waste assimilation: (1) Regulate or treat the waste discharge and tailor it to the varying pattern of stream flow, or (2) regulate the stream flow to provide greater dilution druing periods of low or no flow. The first approach is sometimes employed for seasonal industrial wastes where sufficient storage lagoons are available to retain the wastes during periods of low flow. The second approach can only be used where sufficient water is available from dams upstream.

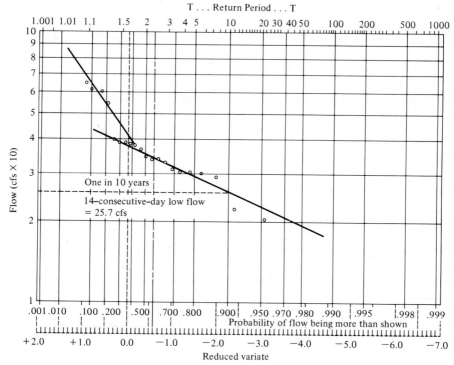

FIGURE 7-5 Plotting low flows from Example 7-2.

Where surface water from a stream is used as a raw-water source for municipal purposes, periods of low flow present special problems. The finished water is much more likely to become unsatisfactory if treatment-plant irregularities occur during such periods. As most water-treatment plants are relatively small and without sophisticated technical controls, marginal raw-water characteristics assume considerable importance to plant operations.

7-16

EUTROPHICATION

Eutrophication of natural waters is one of the most significant water-quality problems today. Nutrient enrichment of lakes, streams, and estuaries arises from pollution associated with population growth, industrial development and intensified agriculture. The process of accelerated fertilization (cultural eutrophication) produces excessive growths of algae and aquatic plants that deteriorate the water environment. Blooms of algae cause tastes and odors in water supplies, reduce transparency, and result in odorous scums. Aquatic plants choke shallow water near the edge of a lake or stream, thus reducing accessibility for recreation. Although most lakes become naturally eutrophic with age by gradually filling

with sediments, man-induced enrichment is far more critical. Cultural eutrophication can occur in a period of a few years, whereas natural aging is a very
slow process, taking place over hundreds to thousands of years.[26]

7-17

BIOLOGICAL FACTORS OF EUTROPHICATION

Lakes and reservoirs may be classified according to their biological productivity.
Oligotrophic surface waters are relatively unproductive and receive only small
amounts of plant nutrients, while eutrophic lakes exhibit abundant plant and
animal life, owing to high nutrient loadings. The former contain small counts
of organims but many different species of aquatic plants and animals. Eutrophic
conditions create large numbers of plants and animals of fewer species. Normally, lakes that show depletion of dissolved oxygen in the hypolimnion (bottom
layer) during thermal stratification are classified as *eutrophic*, while those that
maintain high oxygen levels throughout the year are *oligotrophic*. Reduction
of dissolved oxygen depends on plant productivity in the epilimnion and physical
characteristics of the lake. For most uses, eutrophication diminishes water quality
because dissolved solids increase, transparency decreases, blue-green algae are
more prevalent, rooted aquatics are more abundant, and finer fish are replaced by
more tolerant species. Mesotrophic is often used to classify a moderately productive lake that is suitable for most recreational pursuits but less desirable as a
water-supply source.

A food chain represents levels of metabolism in an aquatic community.
Primary producers are plants capable of photosynthesis, such as chlorophyll-
bearing phytoplankton (algae). The first-level consumers are grazing herbivores
(zooplankton), and the second link is composed of small-sized carnivores. The
latter are eaten by large carnivores (fish). Energy is transferred through the food
chain in a pyramid fashion such that each higher level has less energy than the
previous one. Thus the biomass at each energy level decreases; the mass of
phytoplankton is greater than the mass of zooplankton, and there are more small
fish than large ones.

The pyramid of biomass for an oligotrophic lake is sketched in Fig. 7-6a.
Inorganic nutrients in short supply limit primary production of green plants.
The populations of zooplankton and fish are thus held in check by restricted
food sources. Oligotrophic lakes are typified by a cold-water mountain lake or
sand-bottomed, spring-fed impoundments that have transparent water, little
plant growth, and low fish production. Abundant plant nutrients increase the
standing crop on all levels of the biomass pyramid and unbalance the normal
succession of the aquatic food chain (Fig. 7-6b). Extensive weed beds are not
inhibited by nutrient deficiency, and phytoplankton production exceeds the
consumptive demands of zooplankton. Consequently, frequent algal blooms
increase water turbidity, and floating masses of decaying algae are windblown

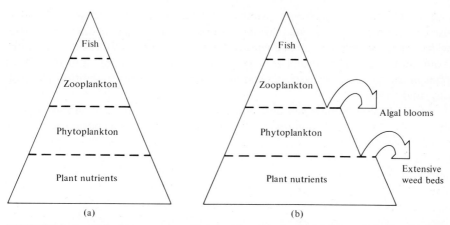

FIGURE 7-6 Pyramids of biomass (standing crop) for : (a) oligotrophic lakes ; (b) eutrophic lakes

to the shore. Diatoms and green algae are replaced by blue-green species that are notorious for causing taste and odor problems in water supplies. Upon decomposition, the excess plant biomass reduces the dissolved oxygen content. Most eutrophic lakes have a greater fish production as a result of the abundant supply of food. However, the preferred food fish cannot survive in the unfavorable environment and are replaced by more tolerant species. For example, in a deep lake, reduction of dissolved oxygen in the hypolimnion results in loss of cold-water fish, such as trout and cisco, and dominance of less desirable warm-water types, such as perch and bass. These are, in turn, succeeded by bullheads and carp, if the environment continues to deteriorate.

7-18

CHEMICAL FACTORS OF EUTROPHICATION

The major aquatic plant nutrients are orthophosphate, inorganic nitrogen in the forms of nitrate and ammonia, and carbon dioxide. Trace nutrients include the elements iron and silica, and organic compounds such as vitamins. The majority of natural waters contain an adequate supply of these elements. Sufficient dissolved carbon dioxide is available in surface waters to support massive growths of algae.[27] In the majority of lakes, either phosphorus or nitrogen appears to be the limiting factor controlling aquatic plant growth. Although not universally accepted, most authorities feel that phosphorus is the key element since several species of blue-green algae can fix atmospheric nitrogen.

Nitrogen and phosphorus are contributed to surface waters from several sources.[28] The most common natural origins are runoff from fertile land, lake sediments, rainfall, and nitrogen fixation. Nevertheless, the majority of plant nutrients entering surface waters come from man-generated wastes and runoff from agricultural land. Domestic waste averages 3.5 lb of phosphorus per capita per year, of which approximately 60% is from the phosphate builders used in synthetic detergents. Primary sources of nitrogen in domestic wastewater are feces, urine, and food wastes. Nitrogen and phosphorus contributions from agricultural land drainage vary from 3–24 lb of nitrogen and 0–15 lb of phosphorus per acre per year, depending on land use, fertilizer additions, hydrology, and topography.

Distribution of nitrogen and phosphorus in an agrarian economy is given in Fig. 7-7. Only small quantities of these nutrients are transported to surface

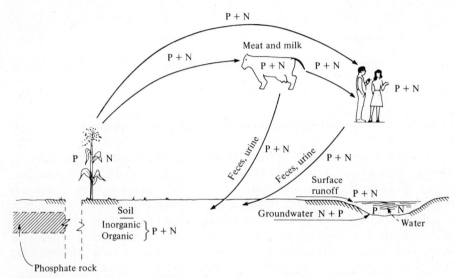

FIGURE 7-7 Phosphorus and nitrogen distribution in an agrarian economy. [Source: C. N. Sawyer, "The Need for Nutrient Control," *J. Water Poll. Control Fed.* 40 no. 3 (1968) : 366.]

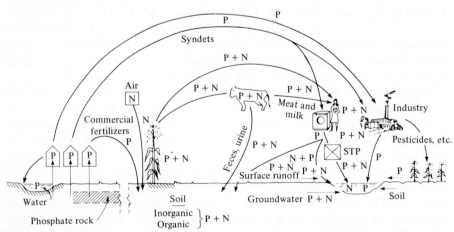

FIGURE 7-8 Phosphorus and nitrogen distribution in a complex urban economy. [Source: C. N. Sawyer, "The Need for Nutrient Control," *J. Water Poll. Control Fed.* 40, no. 3 (1968) : 367.]

waters in runoff from cultivated farmland. The spread of nitrogen and phosphorus in a complex urban economy is illustrated in Fig. 7-8. Phosphate rock is mined and processed into fertilizers, detergents, animal feeds, and other chemicals. A significant amount of this phosphate ultimately reaches surface flows through conventionally treated wastewater discharges and land drainage. Our high standard of living has resulted in the use and disposal of large quantities of phosphorus which had previously been underground. Atmospheric nitrogen is converted into anhydrous ammonia fertilizer, which is widely used in modern agriculture. Thus the primary plant nutrients, not abundant in natural waters, are now supplied by man's activities.

7-19

PHYSICAL CHARACTERISTICS AFFECTING EUTROPHICATION

The major physical factors that influence aquatic-plant production are temperature, light, residence time of the water, and thermal stratification. Although objectionable forms of algae, such as the blue-greens, are generally considered warm-water microorganisms, productivity is not apparently inhibited by cold temperatures. Excessive algal populations are found in lake waters near the freezing point. Inadequate sunlight penetration does restrict plant growth in highly colored or turbid waters. Shallow lakes and reservoirs are often light-limited, owing to soil turbidity held in suspension by wind mixing, even though adequate inorganic nutrients are available.[29] Algal blooms in a highly eutrophic impoundment (e.g., a wastewater stabilization pond) can cause self-shading and reduce the rate of photosynthesis. A very short residence time (detention time) in a watercourse or impoundment appears to reduce the effects of eutrophication. For this reason and because of soil turbidity, many major rivers are not excessively productive even though their nitrogen and phosphorus levels exceed the minimum desirable in a lake. On the other hand, slow-moving clear-water streams are susceptible to overfertilization. Large river and estuary systems, such as the Potomac River, act like long, narrow impoundments during periods of low flow.

Thermal stratification and seasonal mixing of lakes influence productivity. Lakes in the temperate zone thermally stratify during winter and summer and circulate each spring and autumn. In warm latitudes where the water temperature never drops below 4°C, lakes stratify during the summer and circulate continuously during the winter. Typical temperature profiles for a dimictic lake are shown in Fig. 7-9. In the winter, ice near 0°C covers the surface, while the more dense water at 4°C sinks to the bottom. Cold temperature and reduced light penetration inhibits biological productivity. In the spring, after the ice melts, the surface waters warm to 4°C and begin to sink while the bottom waters rise. These convection currents, aided by wind, circulate the lake thoroughly for several weeks while the water temperature gradually increases. As the

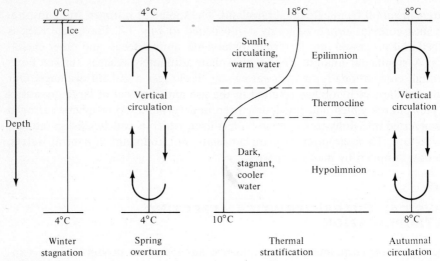

FIGURE 7-9 Temperature profiles of a dimictic lake, showing winter stagnation, thermal stratification, and seasonal overturns.

surface waters warm rapidly and spring winds subside, a less dense surface layer of water forms. This epilimnion mixes continuously during the summer and supports the growth of phytoplankton. The hypolimnion is the dark, stagnant, cooler, bottom water layer. Many fish find the cooler water a more suitable environment than the warm, turbid surface layer. During autumn, the epilimnion is cooled to a higher density than the hypolimnetic water, which leads to vertical currents and the autumnal overturn. This mixing continues until finally the more dense water stays on the bottom and the surface freezes.

Water quality is directly influenced by thermal stratification. The epilimnion supports abundant algal growth, while the hypolimnion in eutrophic lakes decreases in dissolved oxygen content. Hydrogen sulfide, odorous organic compounds, and reduced iron can be released from bottom sediments as a result of anaerobiosis. The water supply of highest quality during thermal stratification is usually found just below the thermocline. During spring and autumn circulations, treatment for taste and odor control may have to be intensified since undesirable matter is distributed throughout the entire water profile.

7-20

TROPHIC LEVEL OF LAKES

Several parameters related to dissolved oxygen content, productivity of algae, and quantity of plant nutrients can be measured to indicate the trophic level of a lake or reservoir. Hypolimnetic oxygen depletion during thermal stratification is probably the most widely used index separating oligotrophic from eutrophic lakes. The major problem with this index is that the rate of oxygen loss depends to a significant degree on the lake's morphology (i.e., the rate of oxygen depletion is less when the limnetic volume is large relative to the surface area).

Productivity of phytoplankton is associated with clarity of the water, amount of chlorophyll, primary productivity, and frequency and number of algal blooms. Transparency is routinely measured using a Secchi disc, which is a small circular plate attached to a metered rope. The depth to which the disc can be lowered and still be visible correlates to the clarity of the water. The extent of green plant growth in the epilimnion is indicated by the amount of chlorophyll extractable from the suspended solids filtered from a water sample. Primary productivity is an *in situ* measurement of the rate of photosynthesis. It is expressed as the quantity of carbon assimilated by the epilimnetic water over a specified time period. A plot of these measurements taken at regular intervals during the growing season can be used to identify a primary production pattern, as well as yielding average values. The number of objectionable blooms of algae (greater than 1 million cells per liter) increases with eutrophication. Also, the types and diversity of algal species present indicate the trophic level of a lake.

Nutrient concentrations and loadings have also been related to the trophic level of lakes. The generally accepted upper concentration limits for lakes free of algal nuisances are 0.3 mg/l of inorganic nitrogen and 0.02 mg/l of ortho-phospate phosphorus at the time of spring overturn. A study of plant growth relative to nutrient concentrations in Wisconsin lakes revealed that annual mean total nitrogen and phosphorus concentrations greater than 0.8 mg/l and 0.1 mg/l, respectively, resulted in algal blooms and nuisance weed growths during most of the growing season. Lakes were essentially free of nuisances if the annual mean for total organic nitrogen was less than 0.2 mg/l and the total average phosphorus concentration was less than 0.03 mg/l.

Information about critical nutrient concentrations cannot be applied directly in an engineering analysis. However, allowable nutrient loadings, such as those developed by Vollenweider,[30] are valuable in the evaluation of lake fertilization. Data correlating annual phosphorus loading, mean lake depth, and degree of enrichment were gathered from lakes throughout the world. Figure 7-10 plots the position of each lake on the basis of annual phosphorus loading and mean water depth. The lakes known to be oligotrophic are separated from the eutrophic lakes by a pair of diagonal lines that represent the mesotrophic transition zone. Table 7-11 lists the nutrient loading levels based on the data plotted in Fig. 7-10. Permissible loadings are defined as the maximum allowable for a lake to remain oligotrophic in the long term; the permissible phosphorus values are points along the line separating oligotrophic and mesotrophic lakes in Fig. 7-10. Dangerous loadings are sufficiently great to cause a lake to become eutrophic; these depth and phosphorus loading values are from the line separating eutrophic from mesotrophic lakes. The nitrogen loadings given in Table 7-11 are based on an allowable nitrogen/phosphorus ratio of 15 : 1 by weight. The

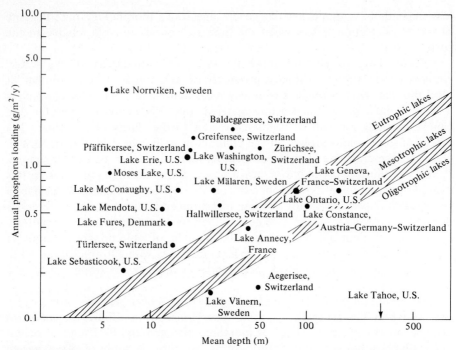

FIGURE 7-10 Graph of annual phosphorus loading versus mean depth for selected oligotrophic and eutrophic lakes. [Source: Based on data of R. A. Vollenweider, *Scientific Fundamentals of the Eutrophication of Lakes and Flowing Waters with Particular Reference to Nitrogen and Phosphorus as Factors in Eutrophication* (Paris, France: Organization for Economic Co-operation and Development, 1970).]

TABLE 7-11

ALLOWABLE NITROGEN AND PHOSPHORUS LOADING LEVELS IN LAKES

Mean Depth (m)	Permissible Loading (g/m² of surface area/yr)		Dangerous Loading (g/m² of surface area/yr)	
	N	P	N	P
5	1.0	0.07	2.0	0.13
10	1.5	0.10	3.0	0.20
50	4.0	0.25	8.0	0.50
100	6.0	0.40	12.0	0.80
150	7.5	0.50	15.0	1.00
200	9.0	0.60	18.0	1.20

Source: R. A. Vollenweider, *Scientific Fundamentals of the Eutrophication of Lakes and Flowing Waters with Particular Reference to Nitrogen and Phosphorus as Factors in Eutrophication* (Paris, France: Organization for Economic Co-operation and Development, 1970).

nutrient loadings were calculated, including all forms of biochemically available phosphorus and nitrogen. Total phosphorus was defined as the content in unfiltered water, including soluble orthophosphates, polyphosphates, absorbed phosphates, and organically bound phosphorus. Total nitrogen included the inorganic forms of nitrite, nitrate, and ammonia, plus organic nitrogen. Example 7-3 illustrates the application of Vollenweider's data.

●EXAMPLE 7-3

A city discharges 5.8 mgd of treated wastewater, containing 7.0 mg/l of phosphorus and 18 mg/l of nitrogen, into a river that enters a lake a short distance downstream. Based on data from a stream gaging and sampling station located immediately upstream from the outfall sewer, the average annual flow is 210 cfs with a mean nitrogen content of 0.20 mg/l and phosphorus concentration of 0.020 mg/l. The lake, used as a municipal water source and for recreation, shows early signs of overfertilization. The surface area is 32,000 acres and the mean depth is 10 m. Developments around the lake contribute an estimated 4000 lb of phosphorus and 25,000 lb of nitrogen annually. What are the permissible and dangerous nutrients loadings recommended by Vollenweider? Calculate the nitrogen and phosphorus loadings resulting from development around the lake and the river, excluding contributions from the wastewater effluent. What are the loadings contributed by the waste discharge? To prevent further eutrophication of the lake, what effluent standards for nitrogen and phosphorus should be required in wastewater treatment?

○*Solution*

From Table 7-11: permissible loadings are 1.5 g of $N/m^2/yr$ and 0.10 g of $P/m^2/yr$; dangerous loadings are 3.0 g of $N/m^2/yr$ and 0.20 g of $P/m^2/yr$. Nutrients loadings from lake developments are:

$$P \text{ loading} = \frac{4000 \text{ lb} \times 454 \text{ g/lb}}{32,000 \text{ acres} \times 4047 \text{ m}^2/\text{acres}} = 0.014 \text{ g/m}^2/\text{yr}$$

$$N \text{ loading} = \frac{25,000 \times 454}{129,500,000} = 0.088 \text{ g/m}^2/\text{yr}$$

Nutrients loadings from river water are:

$$P \text{ loading} = \frac{210 \text{ cfs} \times 31,540,000 \text{ sec/yr} \times 28.3 \text{ l/ft}^3 \times 0.00002 \text{ g/l}}{129,500,000 \text{ m}^2}$$
$$= 0.029 \text{ g/m}^2/\text{yr}$$

$$N \text{ loading} = \frac{18.7 \times 10^{10} \text{ l/yr} \times 0.0002 \text{ g/l}}{129,500,000} = 0.289 \text{ g/m}^2/\text{yr}$$

Total nutrient loadings without wastewater are:

$P \text{ loading} = 0.014 + 0.029 = 0.043 \text{ g/m}^2/\text{yr}$
$N \text{ loading} = 0.088 + 0.289 = 0.38 \text{ g/m}^2/\text{yr}$

Nutrients loadings from the wastewater discharge are:

$$\text{P loading} = \frac{5,800,000 \text{ gpd} \times 365 \text{ days/yr} \times 3.78 \text{ l/gal} \times 0.007 \text{ g/l}}{129,500,000 \text{ sq m}}$$

$$= 0.433 \text{ g/m}^2\text{/yr}$$

$$\text{N loading} = \frac{8.00 \times 10^9 \text{ l/yr} \times 0.018 \text{ g/l}}{129,500,000} = 1.11 \text{ g/m}^2\text{/yr}$$

The total phosphorus loading exceeds the permissible loading:

total P loading $= 0.043 + 0.433 = 0.476 \text{ g/m}^2\text{/yr}$ > 0.20 permissible

whereas the total nitrogen loading equals the permissible loading:

total N loading $= 0.38 + 1.11 = 1.49 \text{ g/m}^2\text{/hr} = 1.5$ permissible

For a phosphorus loading equal to the permissible loading, the P loading from the wastewater cannot exceed

$0.100 - 0.043 = 0.057$ g of $\text{P/m}^2\text{/yr}$

This is equivalent to a phosphorus concentration in the effluent equal to

$$\frac{0.057 \times 129,500,000}{5,800,000 \times 365 \times 3.78} = 0.00092 \text{ g/l} = 0.92 \text{ mg/l}$$

7-21

CONTROL OF EUTROPHICATION

Cultural eutrophication of a lake can be controlled, or its effects minimized, by reducing nutrient input.[31] Prior to taking corrective measures, a quantitative survey of nitrogen and phosphorus sources and limnological study of the lake are essential. These data can reveal the trophic level and determine whether the majority of nutrients are from point or diffuse sources. If the latter dominate, there is no ready solution for reduction of input other than land management to control soil erosion and loss of fertilizers. The success of soil- and water-conservation practices is often dictated by local weather conditions and topography. However, point sources, such as municipal wastewater discharges, can be controlled by alternative disposal on land, diversion around the lake, or advanced waste treatment prior to disposal by dilution in surface waters.

Land disposal by irrigation is practiced by many towns and small cities in semiarid regions. Reclaimed wastewater is also being used for recreational lakes (Santee, California), impounding in a reservoir for recreation and irrigation (Lake Tahoe, California), groundwater recharge and aquifer injection for control of seawater intrusion (Los Angeles), and industrial water supplies.

Muskegon, Michigan, is providing tertiary treatment by a crop-irrigation and soil-filtration system. Biologically treated wastewater is sprayed over a natural filter of sandy soil. A network of underground pipes collects the percolated water, which is then pumped to a river for disposal. Nevertheless, land disposal is not often feasible for major cities because of the large land area required, unsuitable soils, climatic conditions, or other factors.

Nutrient loadings caused by point sources can be eliminated by diverting wastewaters around or away from lakes. While reducing the problem of eutrophication in the lake, diversion may be transferring it to the new receiving watercourse. Madison, Wisconsin, was one of the first metropolitan areas to employ diversion of municipal wastewaters to protect a chain of five lakes. Effluent from the metropolitan treatment plant was piped in 1958 to a creek that flows into the river downstream from the lakes. Although these lakes still receive nutrients form agricultural runoff and urban drainage, the condition of the lakes has stabilized and early indications show a reversal of the eutrophication process. Natural flushing has reduced the soluble phosphorus content, with the best response occurring in the lakes with the shortest liquid detention times. An increase in the species diversity of algae and reduced frequency of blue-green algal blooms have been observed as indicators of improved water quality. The creek receiving the diverted waste has been distressed even though the treatment plant is producing a high-quality effluent. It does not provide significant dilution since about 75% of the flow is treated effluent. A measurable portion of the phosphorus present in downstream rivers is attributed to the Madison wastewater effluent.

Lake Washington in Seattle is another case history in diversion of wastewaters to reduce fertilization. Rapid eutrophication during the 1950s correlated with population growth in the watershed. Increased quantities of municipal wastes accounted for the dramatic rise in nitrogen and phosphorus contributions to Lake Washington. A multimillion-dollar interceptor sewer system was constructed to collect all wastewaters entering the lake and pipe them to two major wastewater treatment plants on Puget Sound. Since completion in 1968, significant improvement in water quality has already occurred, as a result of the flushing action of the nutrient-poor river water entering the lake. The conventionally processed wastewater does not appear to be polluting Puget Sound since tidal action provides adequate mixing and dilution.

Advanced-waste-treatment processes (Chapter 13) are being employed to remove nutrients from discharges that must be disposed of in bodies of water subject to eutrophication. The latter include all the Great Lakes, the Potomac River and estuary, San Francisco Bay and its tributaries, and numerous other lakes. Conventional secondary treatment removes only about 30% of the phosphorus and 50% of the nitrogen; thus a typical biologically treated wastewater contains approximately 7 mg/l of P and 20 mg/l of N. At present, emphasis is being placed on phosphorus removal since phosphate is believed to be the controlling factor in the enrichment of most lakes. For example, approximately two-thirds of the phosphorus loading on the Great Lakes is from municipal and industrial waste effluents, as opposed to about one-third for nitrogen. Furthermore, efficient and relatively inexpensive processes are available for removal of phosphorus up to 90%, whereas comparable elimination of nitrogen is not yet feasible. Effluent standards for phosphorus have been established by several states for wastewater discharges that flow into lakes and estuaries. Effluent limits range from 0.1 to 2.0 mg/l of P, with 1.0 mg/l of P being the most commonly established value. Treatment efficiency requirements for phosphorus removal range from 80 to 95%.

Several temporary controls can be used to arrest or reduce nuisance effects in eutrophic lakes and reservoirs. The most popular are chemical control of plant growths, harvesting of aquatic weeds, destratification by mechanical mixing, and aeration of the hypolimnion. Algicides are used extensively for algal control in water-supply reservoirs.[32] Filamentous green algae can form thick, repugnant scums that clog water plant filters. Blue-green blooms also impart tastes and odors to water, even when their algal concentration is relatively low. Copper sulfate is the main algicide of choice, even though some organic mercury chemicals are more effective. The dosage and frequency of treatment are based on observations of the agal populations, with the personnel in charge using past knowledge as to which species cause trouble. Herbicides, such as 2,4-D and sodium arsenite, are effective in controlling rooted aquatic plants. They are often applied at boat docks, swimming areas, and fishing points of lakes and reservoirs to reduce weed beds. The application of algicides and herbicides is expensive, requires close supervision, and must be done at frequent intervals during the growing season.

Harvesting of aquatic plants can be performed by boat-mounted underwater weed cutters with optional barge-loading conveyors. Cutting weeds below the water line improves the aesthetics of a lake, and clearance for motor boat propellors. Harvesting of plants is not considered a feasible means of removing nutrients from a lake. Two tons of aquatic plants by wet weight contain only about 1 lb of phosphorus and 10 lb of nitrogen.

Mechanical destratification by pumping cold water from the bottom and discharging it at the surface has been effective in improving quality in water-supply reservoirs. This mixing lowers the temperature of the epilimnetic water and may add dissolved oxygen to the hypolimnion. The reduction in surface-water temperature appears to shift algal populations from blue-greens to green algae, which are less troublesome in water treatment. In small lakes and reservoirs where the loss of dissolved oxygen in the hypolimnion is a serious problem during thermal stratification, surface waters may be pumped to the bottom, or deep-water diffused aeration can be used to supply oxygen. The latter is preferred by some authorities since oxygenation can be performed without destratification. Destruction of the thermocline may circulate nutrients, stored in the hypolimnion, to the overlying waters where photosynthesis can take place.

7-22

BIOCHEMICAL OXYGEN DEMAND

The *biochemical oxygen demand* (BOD) of domestic and industrial waste-waters is the amount of molecular oxygen required to stabilize the decomposable matter present in a water by aerobic biochemical action. Oxygen demand is exerted by three classes of matter—carbonaceous material, oxidizable nitrogen, and certain chemical reducing compounds. The classes of matter present and the manner in which oxygen demand is exerted are a function of the type and history of the waste. Complete stabilization of a given waste may require too long a time for practical purposes; therefore, the standard laboratory BOD test is incubated for a period of 5 days at 20°C.

The BOD test is among the most important made in sanitary analyses to determine the strength of wastewater. It is used as a parameter in process design and loading and as a measure of plant efficiency and operation.

The presently accepted definition of biochemical oxygen demand is the fruition of over a century of research on a natural phenomenon that is relatively simple in principle yet exceedingly complex in evaluation and instrumentation.[33] Frankland was the first to use a form of the BOD test in attempting to evaluate the pollutional load of the Thames River.[34] In 1868 he used an excess-oxygen technique involving the incubation of river-water samples in completely filled and tightly stoppered bottles. The dissolved oxygen remaining in the samples at 24-hour intervals was determined by boiling off the dissolved gases in vacuo and performing a gas analysis on the product. The 7-day curve produced in this experiment closely resembles present-day BOD curves.

The first chemical oxygen demand test (COD) using potassium permanganate was developed by Forchamer in 1849.[35] He did not recognize that the decomposition of organic matter is largely a biological process but did attempt to determine the oxidizable matter present in a water quantitatively. The possibility that microorganisms in water had the ability to consume the oxygen dissolved in water for their own metabolic processes was first proposed by Dupré in 1884.[36] He also made comparisons between the BOD and the COD tests, concluding that there was a difference between the amount of oxygen taken up by the microorganisms and by the amount of oxygen which the same water would take up from a solution of permanganate.

Investigators continued active interest in BOD related tests of various types, but the most significant single contribution was reported by the Royal Commission in 1913. The dilution-bottle method proposed by Adeney consisted essentially of the standard method used today. The determination of the biochemical oxygen demand by the standard dilution bottle method is accomplished by diluting suitable portions of the sample with water saturated with oxygen and measuring the dissolved oxygen both immediately and after a period of incubation, usually 5 days at 20°C. The acceptance of the dilution-bottle technique for determining BOD was not extended until the seventh edition of *Standard Methods*, published in 1933. One of the chief reasons for this acceptance was the research done by Theriault. His work was strongly in favor of the dilution method and his "The Oxygen Demand of Polluted Waters," published as *U.S. Public Health Service Bulletin No. 173*, is one of the classics of the field.

Since its introduction, the dilution method has been the object of considerable research to improve its accuracy as an analytical tool. One of the chief areas of study has been toward the development of a satisfactory dilution water. As shown in Fig. 7-11, the progressive exertion of the BOD of freshly polluted water normally breaks down into two stages: a first stage, in which the carbonaceous material is largely oxidized, and a second stage, in which significant amounts of nitrification may take place.

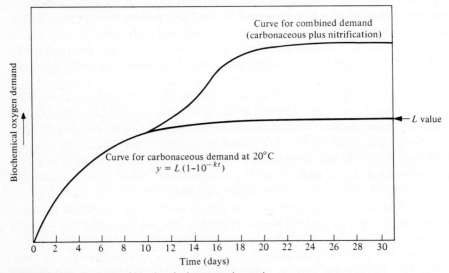

FIGURE 7-11 Progress of biochemical oxygen demand.

Population Dynamics

In the BOD test, the wastewater supplies the biological food and the dilution water the dissolved oxygen. The general biological reaction that takes place is

$$\text{waste organics} + \text{DO} \xrightarrow{\text{microorganisms}} CO_2 + \text{new cell growth} \qquad (7\text{-}2)$$

This equation is an oversimplification of the extremely complex biochemical reactions which take place. Three major factors in population dynamics of a mixed culture are competition for the same food, secondary bacteria predomination, and the predator–prey relationship. In order to grow, a microorganism must be able to derive nutrients from the system. Thus, any mixed population

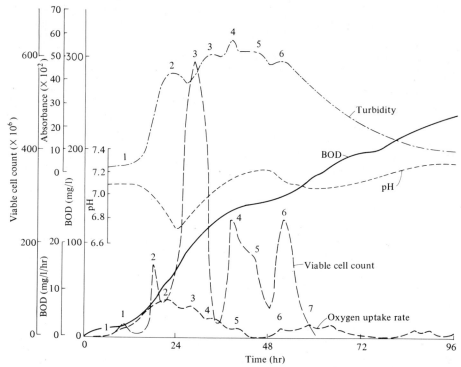

FIGURE 7-12 Variation of dominant species in a 5-day BOD, showing ecological succession [Source: J. C. Young and J. W. Clark, "Growth of Mixed Bacterial Populations at 20°C and 35°C," *Water and Sewage Works* 112 (July 1965): 251–256.]

of microorganisms contains those microorganisms which have been successful in the competition for food. Two types of competition exist, competition for the same food and use of one organism by another organism for food. Primary and secondary bacterial predomination, showing ecological succession, is illustrated in Fig. 7-12. The six peaks on the viable cell count represent different species of bacteria. The predator–prey relationship between protozoans and bacteria, resulting in the diphasic exertion of BOD, is illustrated in Fig. 7-13. The BOD exerted using the protozoan-rich sewage seed is greater than the BOD using pasteurized seed, as a result of the second-stage oxygen uptake attributed to the metabolism of the bacteria by the protozoa.

Nitrification

Nitrifying bacteria can exert an oxygen demand in the BOD test by performing the following reactions:

$$NH_3 + DO \xrightarrow[\text{bacteria}]{\text{nitroso}} NO_2 + DO \xrightarrow[\text{bacteria}]{\text{nitro}} NO_3 \qquad (7\text{-}3)$$

Fortunately, the growth of nitrifying bacteria lags behind the growth of microorganisms which perform the carbonaceous reaction (Eq. 7-2). If the seed

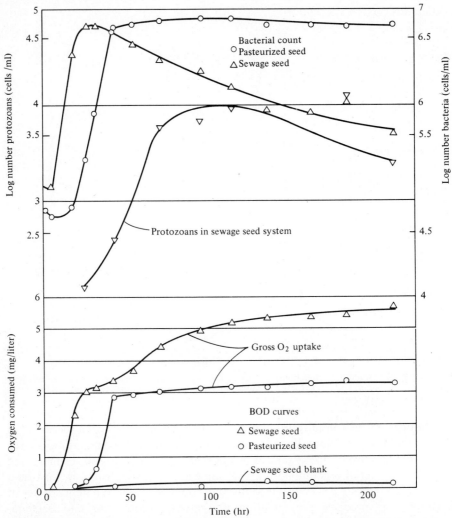

FIGURE 7-13 BOD exertion, bacteria and protozoan counts in a BOD bottle using glucose. [Source: M. N. Bhatla and A. F. Gaudy, "Role of Protozoa in the Diphasic Exertion of BOD," *Proc. Am. Soc. Engrs., J. San. Eng. Div.* 91, no. SA3 (June 1965) : 63–87.]

material used in the BOD test has a low population of nitrifying bacteria, nitrification will generally not occur until several days after the standard 5-day incubation period of the BOD test. Currently, there is no standard method for preventing nitrification in a BOD test.

Nutrients

The principal nutrients required for growth of microorganisms in the metabolism of organic matter are nitrogen and phosphorus. In the BOD test these are provided in the phosphate buffer solution, which contains inorganic phosphate and ammonium chloride. Based on the maximum usable DO in a BOD bottle, the ratio of maximum BOD/nitrogen/phosphorus provided in the dilution water is equal to $60 : 3.2 : 0.73$. The generally accepted BOD/N/P ratio necessary in the BOD test is $60 : 3 : 1$.

Trace nutrients required for biological growth in the BOD test are assumed to be provided by dilution water, wastewater, and seed material.

Mathematical Calculations
The computation for BOD is

$$\text{mg/l BOD} = \frac{(D_1 - D_2) - (B_1 - B_2)f}{P} \qquad (7\text{-}4)$$

where D_1 = DO of diluted sample 15 minutes after preparation

D_2 = DO of diluted sample after incubation

B_1 = DO of dilution of seed control before incubation

B_2 = DO of dilution of seed control after incubation

f = ratio of seed in sample to seed in control

$$= \frac{\text{percent seed in } D_1}{\text{percent seed in } B_1} \quad = \quad \frac{V_{Dilution} \,(test)}{V_D \,(blank)}$$

P = decimal fraction of sample used $= \dfrac{V_{sample}}{V_{Bottle}}$

The first-order monomolecular equation for the carbonaceous phase of biochemical oxygen demand is after Theriault,

$$Y = L(1 - 10^{-kt}) \quad \text{or} \quad L\left(1 - e^{-k't}\right) \qquad Y \uparrow \vdash L \qquad (7\text{-}5)$$

where Y = BOD at any time t, days

L = ultimate or first-stage BOD

k = rate constant

t = time, days

The value of k can be determined graphically from daily laboratory readings using the Thomas method. With the values of the cube root of time over biochemical oxygen demand as ordinates and time as abscissas to natural scale, the best-fit line through the data can be used to determine value of k as follows:

$$k = 2.61\frac{B}{A} \quad \text{or} \quad k' = 6\frac{B}{A} \qquad \left(\frac{t}{Y}\right)^{\frac{1}{3}} \qquad (7\text{-}6)$$

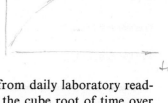

where A = intercept of the line on the ordinate axis

B = slope of the line

Precision and Accuracy
Precision is a measure of the reproducibility of a method when repeated on a homogeneous sample under controlled conditions. Precision can be expressed by standard deviation. Based on a normal curve of frequencies, a range of 1 standard deviation about the mean includes 68 % of the observations. Accuracy

is a measure of the error of a method and may be expressed as a comparison of the amount of compound determined by the test method and the amount actually present.

Precision of the BOD test, according to *Standard Methods*, is based on the precision of the DO determinations. Data presented indicate that the standard deviation of the BOD test on wastewaters range from 0.07 to 0.11 ml of oxygen demand titrated. Therefore, depending on how many milliliters of 0.025 N thiosulfate are used in titrating the remaining DO, the precision of the BOD test for 1 standard deviation is between ± 1 and $\pm 5\%$.

Standard Methods states that there is no standard against which the accuracy of the BOD test can be determined, because of the unavailability of standard substances that can be added in known quantities on which percentage recovery can be based.

Ballinger and Lishka[40] sent prepared samples of a synthetic waste to 34 agencies for BOD determinations to evaluate the precision and reliability of the BOD test when performed by different laboratories. One of the conclusions of this study was: The expected standard deviation of the BOD test using this synthetic waste is shown to be approximately $\pm 20\%$ of the mean value. Although this conclusion is stated in the mathematical terms of precision, the test procedure used (i.e., using a synthetic medium of glucose–glutamic acid and sending prepared samples to 34 different laboratories) indicates something about the accuracy of the BOD test.

In summary, one might conclude that the precision of the BOD test is approximately $\pm 5\%$ of the mean for 68% of the observations and the reliability is approximately $\pm 20\%$.

●EXAMPLE 7-4

The 5-day BOD (y) of a waste is 200 mg/l. Assuming k to be 0.17, what is the ultimate demand, L?

○*Solution*

The data are

$$Y = 200 \quad t = 5 \quad k = 0.17$$
$$Y = L(1 - 10^{-kt})$$
$$200 = L(1 - 10^{-0.17 \times 5}) = L(1 - 10^{-0.85})$$
$$\text{antilog of } (-0.85) = 0.141$$
$$200 = L(1 - 0.141) = L(0.859)$$
$$L = \frac{200}{0.859} = 233 \text{ mg/l of BOD}$$

$$\text{fraction BOD remaining} = 10^{-kt}$$
$$\log = -kt$$
$$= -0.17 \times 5$$
$$= -0.85$$

antilog of $-0.85 = .141$

% BOD oxidized $= 100 - 14.1 = 85.9$

$$L = \frac{200}{0.859} = 233 \text{ mg/l}$$

●EXAMPLE 7-5

The 5-day BOD of a waste is 276 mg/l. The ultimate demand is reported to be 380 mg/l. At what rate is the waste being oxidized?

○Solution

% BOD oxidized $= \dfrac{276}{380} \times 100 = 72.6\%$ $= \dfrac{Y}{L} \times 100 \%$

% BOD remaining $= 100 - 72.6 = 27.4\%$ $= 100\% - \% \text{ BOD oxidized}$

$\log 27.4 = 2 - kt = 2 - (k \times 5)$ $\quad \% \text{ BOD remaining} = 10^{-kt} \times 100 = 10^{2-kt}$

$k \times 5 = 2 - \log 27.4 = 2 - 1.438 = 0.562$

$\quad k = 0.11$

●EXAMPLE 7-6

Theriault tables[37] have been used to simplify calculations and eliminate the need for log tables. The function $1 - 10^{-kt}$ has been calculated by Theriault for a range of values of k and t. In use, the known values of k and t are used to find the fraction oxidized. The problems as stated in Examples 7-4 and 7-5 are used to illustrate this method.

○Solution

As in Example 7-4,

$Y = 200 \qquad t = 5 \qquad k = 0.17$

The value of $1 - 10^{-0.17 \times 5}$ is found in the table to be 0.859. Then

$200 = L(0.859)$

$$L = \frac{200}{0.859} = 233 \text{ mg/liter}$$

As in Example 7-5,

fraction oxidized $= 1 - 10^{-kt} = \dfrac{276}{380} = 0.726$

From the Theriault tables,

fraction oxidized $= 0.726 \quad (t = 5,\ k = 0.11^+)$

BOD's by Electrolysis

The principle of measuring BOD's by electrolysis was developed by Clark in 1959.[41] The method involves using a flask into which 1 liter of undiluted

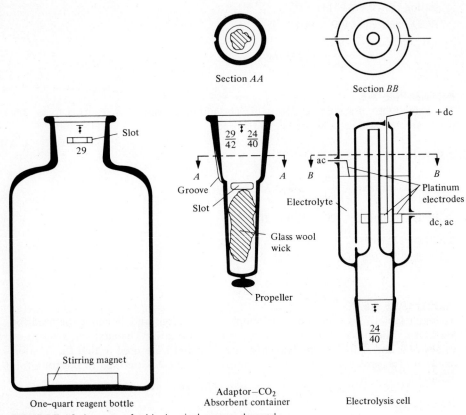

Section *AA*

Section *BB*

Slot

29

+dc

$\frac{29}{42}$ $\frac{24}{40}$

ac

A

Groove

Slot

A

B

Electrolyte

Platinum electrodes

B

dc, ac

Glass wool wick

Propeller

Stirring magnet

$\frac{24}{40}$

One–quart reagent bottle

Adaptor–CO₂ Absorbent container

Electrolysis cell

FIGURE 7-14 Apparatus for biochemical oxygen demand.

waste is placed. An electrolysis cell and pressure switch are mounted on the flask (Fig. 7-14). The flask is then closed to the atmosphere. A magnetic stirrer is rotated inside the flask to continually renew the surface of the liquid. A vial containing potassium hydroxide is suspended from the top of the flask to take up the carbon dioxide.

Bacteria use the oxygen inside the flask to oxidize the organic matter present in the waste. The slight vacuum remaining in the flask actuates a pressure switch which turns on a dc power supply. A small dc current produces oxygen in the electrolysis cell, and this oxygen is piped into the reaction vessel, replacing the oxygen used by the bacteria. The oxygen is produced until pressure equilibrium is established. The periods of oxygen demand are recorded electrically for later calculation of the shape of the BOD curve. The BOD at any point in time is read on a meter.

Oxygen is produced from the electrolysis of water, according to Faraday's law. The faraday is that number of ampere-seconds required to decompose 1 equivalent weight of a substance. In the case of water, 1 faraday separates 8.00 g

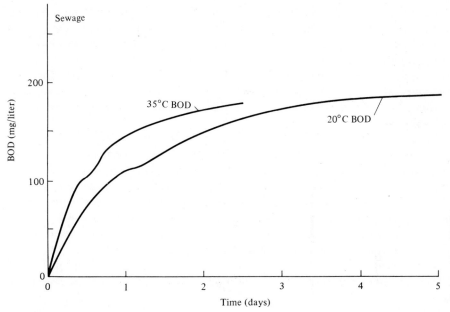

FIGURE 7-15 Biochemical-oxygen-demand curves at 20°C and 35°C recorded continuously by the electrolysis method.

of oxygen per 96,519 amp-sec or coulombs. Then

$$\frac{96{,}519 \text{ A-sec} \times 1 \text{ hr/3600 sec}}{8.00 \text{ mg of oxygen}} = \frac{X \text{ A-hr}}{30.0 \text{ mg of oxygen}}$$

$$0.1007 \text{ A-hr} = 30.0 \text{ mg of oxygen} \qquad (7\text{-}7)$$

or 100.7 mA flowing for 1 hour will produce 30.0 mg of oxygen at the positive electrode of the electrolysis cell. Multiplying the number of hours recorded on the elapsed time meter of the power supply by 30.0 will yield the BOD in milligrams per liter. Larger currents may be used for strong industrial wastes.

The electrolysis method has the advantage over the standard dilution technique in that a short-term BOD is possible at elevated temperatures without loss of precision (Fig. 7-15). The gaseous oxygen diffusion through the solution and the motility of the microorgansims may be factors which have contributed to variations in BOD data at elevated temperatures in tests by the dilution technique.

7-23

ORGANIC LOADS ON STREAMS

The discharge of wastes into a body of water presents a problem of primary importance in the field of water-pollution control. The reduction of this organic matter by bacteria results in the utilization of dissolved oxygen. The primary replacement of this dissolved oxygen occurs through the water surface exposed to the atmosphere. An increase in the pollutional load stimulates the growth of bacteria, and oxidation proceeds at an accelerated rate. The concentration of the organic load can be so great that all the dissolved oxygen in a receiving water is utilized by the bacteria. This lack of oxygen inhibits the higher forms of biological life, and conditions set in that are detrimental to man. The concentration of dissolved oxygen is one of the most significant criteria in stream sanitation.

Every stream is limited in its capacity to assimilate organic wastes. As long as this limit is not exceeded, the disposal of organic wastes in streams represents the most economical method of waste disposal. The evaluation of the natural purification capacity of a stream is of fundamental engineering value. Streams are used as natural treatment plants, and it is necessary to determine their capacity in order not to destroy their usage for other purposes.

The simultaneous action of deoxygenation and reaeration produces a pattern in the dissolved-oxygen concentration of river water. This pattern, known as "the dissolved-oxygen sag," was first described by Streeter and Phelps in 1925.[42] The equation describing the simultaneous action of deoxygenation and reaeration is

$$\frac{dD}{dt} = k_1'L - k_2'D \tag{7-8}$$

where D = dissolved oxygen deficit
$\quad L$ = concentration of the organic matter   <u>or</u>   amt. of BOD remaining)
$\quad k_1'$ = coefficient of deoxygenation
$\quad k_2'$ = coefficient of reaeration

The rate of change in the dissolved oxygen deficit D is the result of oxygen utilization in the oxidation of organic matter and the reaeration which replenishes oxygen from the atmosphere. The concentration of the organic matter L must be expressed in terms of the initial concentration L_0 at the point of waste discharge before integrating (Fig. 7-16).

$$L = L_0 e^{-k_1't} \tag{7-9}$$

where L_0 is the initial concentration of the organic matter in the stream (BOD in mg/l.) The substitution of Eq. 7-9 for the value L in Eq. 7-8 and integration gives

$$D = \frac{k_1'L_0}{k_2' - k_1'}(e^{-k_1't} - e^{-k_2't}) + D_0 e^{-k_2't} \tag{7-10}$$

where D = oxygen deficit in time t, mg/l
$\quad D_0$ = initial oxygen deficit at the point of waste discharge, mg/l

Equation 7-10 is normally used with common logarithms.

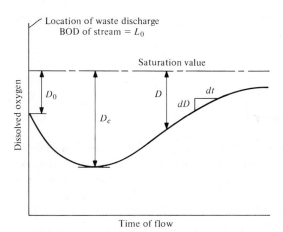

FIGURE 7-16   Dissolved-oxygen-sag curve.

Since $e^{-k't} = 10^{-kt}$, where $k = 0.434k'$,

$$D = \frac{k_1 L_0}{k_2 - k_1}(10^{-k_1 t} - 10^{-k_2 t}) + D_0(10^{-k_2 t}) \qquad (7\text{-}11)$$

The proportionality factor k_1 *or* k'_1 is a temperature function. The proportionality factor k_2 is also a temperature function but, more important, it is a function of the turbulence of the stream.

A general approximate formula for the reaeration coefficient of natural rivers is given by O'Connor and Dobbins.

$$k'_2 = \frac{(D_L U)^{1/2}}{H^{3/2}} \qquad (7\text{-}12)$$

where k'_2 = reaeration coefficient (base e) per hour
  D_L = diffusivity of oxygen in water = 0.000081 ft^2/hr at 20°C
  U = velocity of flow, ft/hr
  H = depth of flow, ft

The effect of temperature on the reaeration coefficient k_2 is as follows:[43]

$$k_{2T} = k_{2\text{-}20} \times 1.047^{T\text{-}20} \qquad (7\text{-}13)$$

where k_{2T} = reaeration coefficient at the temperature T
  $k_{2\text{-}20}$ = reaeration coefficient at 20°C

The value of k'_2 ranges from 0.20 to 10.0 per day, the lower values representing deep slow-moving rivers, and the higher values, shallow streams with steep slopes.

From an engineering design viewpoint, the dissolved-oxygen sag curve indicates the point of minimum DO. This critical point is the place in the stream where the rate of change of the deficit is zero and the demand rate equals the reaeration rate.

$$k_2' D_c = k_1' L = k_1' L_0 e^{-k_1' t_c} \tag{7-14}$$

Solving for the critical time t_c,

$$t_c = \frac{1}{k_2' - k_1'} \ln \left[\frac{k_2'}{k_1'} \left(1 - D_0 \frac{k_2' - k_1'}{k_1' L_0} \right) \right] \tag{7-15}$$

These equations have constants which must be carefully evaluated. The k_1' term reflects the rate at which bacteria demand oxygen and is calculated from the BOD test by running BOD determinations. The k_2 term is the reaeration characteristic of the stream and varies from reach to reach in most streams. Constant k_1 can be evaluated in the laboratory while k_2 must be determined from field studies. In the development of these equations, it is assumed that k_1 and k_2 are constant and that only one source of pollution exists and that the only oxygen demand is the BOD. Variations from these assumptions may be taken into account in any practical case. Some of the following processes, in addition, may be taking place in any given river stretch:

1   Removal of BOD by adsorption or sedimentation.
2   The addition of BOD along the river stretch by tributary inflow.
3   The addition of BOD or the removal of oxygen from the water by the benthal layer.
4   The addition of oxygen by the photosynthetic action of plankton.
5   The removal of oxygen by plankton respiration.

●EXAMPLE 7-7
A city of 200,000 population produces sewage at the rate of 120 gpcd and the sewage plant effluent has a BOD of 28 mg/l. The temperature of the sewage is 25.5°C, and there is 1.8 mg/l of DO in the plant effluent. The stream flow is 250 cfs at 1.2 fps and the average depth is 8 ft. The temperature of the water is 24°C before the sewage is mixed with the stream. The stream is 90% saturated with oxygen and has a BOD of 3.6 mg/l. The deoxygenation coefficient k_1' is equal to 0.50 at 20°C. Determine the following:

1   Sewage flow, cfs.
2   The DO of the mixture of water and sewage-plant effluent.
3   The temperature of the mixture of water and sewage-plant effluent.
4   The value of the initial oxygen deficit for the river just below the plant discharge.
5   The distance downstream to the point of minimum DO.
6   The minimum DO in the stream below the sewage plant.

○Solution

1   The sewage flow, cfs:

$$(120 \text{ gpcd})(2 \times 10^5) = 240 \times 10^5 \text{ gpd} = 24 \text{ mgpd}$$
$$1 \text{ mgpd} = 1.547 \text{ cfs}$$
$$\text{sewage flow} = (24)(1.547) \text{ cfs} = 37.1 \text{ cfs}$$

2   The DO of the mixture of water and sewage-plant effluent. Assuming a pressure of 1 atmosphere and a chloride concentration of zero, the solubility of oxygen in water at 24°C is 8.5 mg/l. The DO of the river water is (8.5 mg/l)(0.90) = 7.65 mg/l. The DO of the mixture is

$$DO_m = \frac{Q_r(DO_r) + Q_s(DO_s)}{Q_r + Q_s} \qquad - \text{ weighted avg.}$$

$$= \frac{(250 \text{ cfs})(7.65 \text{ mg/l}) + (37.1 \text{ cfs})(1.8 \text{ mg/l})}{287.1 \text{ cfs}}$$

$$= 6.89 \text{ mg/l}$$

3   Temperature of the mixture of water and sewage-plant effluent. As in (2).

$$T_m = \frac{(250 \text{ cfs})(24°C) + (37.1 \text{ cfs})(25.5°C)}{287.1 \text{ cfs}} \qquad - \text{ weighted avg.}$$

$$= 24.2°C$$

4   Value of the initial oxygen deficit. Assuming pure water at 24.2°C. The saturation value of O_2 is 8.48 mg/l but we have 6.89 mg/l.

deficit $= 8.48 - 6.89 = 1.59$ mg/l

5–6   The distance downstream to the minimum point of DO and the value of the minimum DO. Constant k_2' may be computed from Eq. 7-12.

$$k_{2\text{-}20}' = \frac{(D_L U)^{1/2}}{H^{3/2}}$$

$$= \frac{(81 \times 10^{-6} \text{ ft}^2/\text{hr} \times 1.2 \text{ ft/sec} \times 3600 \text{ sec/hr})^{1/2} \times 24 \text{ hr/day}}{(8 \text{ ft})^{3/2}}$$

$$= 0.627$$

The values of k_2' and k_1' at 24.2°C are found by Eq. 7-13.

$$k_{2T} = k_{2\text{-}20} \times 1.047^{T-20}$$
$$k_{2\text{-}24.2}' = 0.627 \times 1.047^{24.2-20}$$
$$= 0.76$$
$$k_{1\text{-}24.2}' = 0.50 \times 1.047^{4.2}$$
$$= 0.607$$

The value of the initial BOD, L_0, of the mixture of river water and plant effluent is

$$L_0 = \frac{(250 \text{ cfs})(3.6 \text{ mg/l of BOD}) + (37.1 \text{ cfs})(28 \text{ mg/l of BOD})}{287.1 \text{ cfs}} \qquad - \text{ weighted avg.}$$

$$= 6.75 \text{ mg/l of BOD}$$

The time t_c to the point of minimum DO is found by Eq. 7-15.

$$t_c = \frac{1}{k_2' - k_1'} \ln \left[\frac{k_2'}{k_1'} \left(1 - \frac{D_0(k_2' - k_1')}{k_1' L_0} \right) \right]$$

$$= 1.08 \text{ days}$$

The value of the minimum DO is found by Eq. 7-10 or Eq. 7-11.

$$D_c = \frac{k_1' L_0}{k_2' - k_1'} (e^{-k_1' t_c} - e^{-k_1' t_c}) + D_0 e^{-k_2' t_c}$$

$$= 2.80 \text{ mg/l}$$

The distance downstream at which the critical DO occurs is calculated from the value of t_c and the velocity of flow.

distance $= 1.2$ ft/sec $\times$ 3600 sec/hr $\times$ 24 hr/day $\times$ 1.08 days
$\qquad = 112,000$ ft

Thus, under the given conditions the oxygen-sag curve takes the form shown in Fig. 7-17.

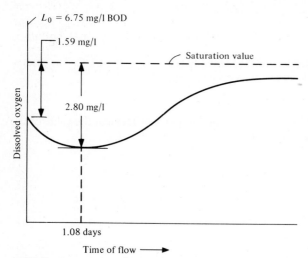

FIGURE 7-17 Dissolved-oxygen-sag curve for Example 7-7.

7-24

HEAT WASTES

As knowledge of the effects of heat additions on lakes and streams and their ecology increases, concern over thermal pollution heightens. The projected increases in water use for industrial and power production cooling indicate that thermal pollution is one of the most serious long-range water resource management problems. (Review Sec. 2-3.) Clearly, there is an upper limit to the amount of heat that can be introduced into a body of water without harmful results.

Projections of total energy release to the environment from the conterminous United States is expected to exceed 190,000 trillion Btu/yr by the year 2000.[44] The Boston–Washington megalopolis area is expected to exceed 30% of the incident solar energy by the same year. A growing portion of this total heat release to the environment will originate from electrical power generation. Currently about 20% of waste heat comes from this source. Thermal efficiency in present power production averages about 35% and is not expected to rise above 50% in the near future.

Uneven geographical distribution of projected heat rejection presents a major problem. The East coast and Middle Atlantic states are projected at faster rates than any other comparable region. The heat loss from energy use in the Atlantic metropolitan area is projected at about 240 Btu/ft^2/day by the year 2000. The signiffcance of this can be gaged by comparing it with average daily total solar radiation for New York City. This amounts to 1690 Btu/ft^2/day in July and 480 Btu/ft^2/day in January. Knowledge concerning climatological effects from releases of this magnitude is far from complete, but there is speculation that unique microclimates may result unrelated to anything now experienced.

With projections such as these, electrical-power generation may become the most severe test for environmental pollution control. Thermal pollution from this source is such that conceived solutions seem hardly adequate. Cooling ponds, for example, require vast acreages of land, and cooling towers pose secondary pollution effects through formation of fog and ice (under some

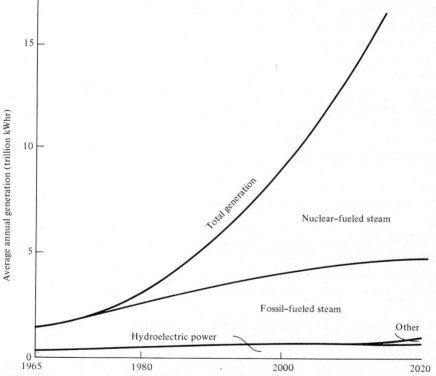

FIGURE 7-18 Projected electric generation by types, 1965–2020. [Source: *The Nation's Water Resources* (Washington, D.C.: Government Printing Office, 1968), pp. 1–11.]

climatic conditions) and dispersal of water-treatment chemicals into air and water.

Elevated temperatures generally make water less satisfactory for other uses. Fish kills may result from a sudden rise in temperature, from a short period of lethal temperature, or from a prolonged period of sublethal temperature. Dissolved oxygen in water decreases at higher temperature, while the requirement for oxygen increases. For each 10°C temperature rise, the oxygen consumption of aquatic fauna nearly doubles, while the minimum critical concentration required by fish increases. In addition, adverse effects of compounds toxic to fish also increase with rise in temperature.

Increases in heat loadings which are expected to accompany future industrial growth could cause serious deterioration in the usefulness of many waters (see Fig. 7-18). Predicted cooling water usage by the electric power industry provides a rough measure of the order of magnitude of future heat discharges. Heat dissipation by industry may become a controlling factor for many industries in selecting new plant sites.

PROBLEMS

7-1 In a community of 70,000 population, what is the minimum number of samples of water for bacteriological tests that should be collected monthly?

7-2 One milliliter of a $1:10^4$ dilution of a water sample was found to contain 35 microorganisms. What was the probable concentration of microorganisms in the original sample?

7-3 What is the most probable number of coliform organisms per 100 ml as determined by the following series of observations? Five 10-ml tubes were used, of which three were positive; five 1-ml tubes were used, of which three were positive; and five 0.1-ml tubes were used, of which one was positive.

7-4 What is the most probable number of coliform organisms per 100 ml as determined by the following series of observations? Three 10-ml tubes were used, of which three were positive; three 1-ml tubes were used, of which one was positive; and three 0.1-ml tubes were used, of which none were positive.

7-5 A test for coliform organisms produced the following results: five of five 100-ml portions were positive, three of five 10-ml portions were positive, and none of five 1-ml portions were positive. Find the MPN per 100 ml.

7-6 How many grams of oxygen are required for the complete combustion of carbon when 700 kcal are liberated? The pressure is constant at 1 atmosphere and the temperature is 25°C.

7-7 How many pounds of urea (N_2H_4CO) are required in the synthesis of 1 lb of dextrose ($C_6H_{12}O_2$) to cell material?

7-8 One thousand grams of dead bacterial cell material is available in an oxidation system. If 55% of the dead cell material is synthesized to new cells, and these new cells live for 7 days and the process is repeated, how long will it be before there is only 1 g of the original cell material remaining?

7-9 The 5-day BOD of a waste is 250 mg/l. If K is equal to 0.15, what is the ultimate BOD of the waste?

7-10 The 5-day BOD of a waste is 280 mg/l. The ultimate BOD is reported to be 410 mg/l. At what rate is the waste being oxidized?

7-11 The 5-day, 20°C BOD of a waste is 250 mg/l and its temperature is 85°F. The 5-day 20°C BOD of the water in a stream is 5 mg/l and its temperature is 55°F. What will be the 5-day, 20°C BOD of the stream below the point of waste discharge when the total flow is 150 cfs and the waste flow is 0.5 mgpd? Assume that the BOD rate is constant and the K at 20°C is 0.35 for both the stream and the waste.

7-12 The sewage flow from a city is 25×10^6 gpd. If the average 5-day 20°C BOD is 265 mg/l, compute the total daily oxygen demand in pounds and the population equivalent of the sewage.

7-13 A city of 150,000 population produces sewage at the rate of 150 gpcd and the waste-treatment-plant effluent has a BOD of 19 mg/l. The temperature of the waste is 26.1°C and there is 1.4 mg/l of DO in the plant effluent. The stream flow is 161 cfs at 1.3 fps and an average depth of 5 ft. The temperature of the stream before the waste flow is introduced is 21.2°C. The stream is 85% saturated with oxygen and has a BOD of 1.8 mg/l. The deoxygenation coefficient k_1 is equal to 0.45 at 20°C. Graph the oxygen-sag curve.

7-14 If all of the 7 mg/l of DO in a clean turbulent stream can be counted on for the oxidation of organic matter in a domestic waste (the reaeration is a factor of safety against nuisance), what should be the dilution factor in cfs per 1000 population if the waste flow is 150 gpd and its BOD is 230 mg/l?

7-15 In a BOD determination 6 ml of sewage are mixed with 294 ml of diluting water containing 8.16 mg/l of dissolved oxygen. After incubation the dissolved oxygen content of the mixture is 5.4 mg/l. Calculate the BOD of the sewage. Assume that the initial DO of the sewage is zero.

7-16 A city discharges 4 mgpd of 15°C sewage with a 5-day, 20°C BOD of 250 mg/l into a stream whose discharge is 78 cfs and the water temperature is 17°C. If k_2 is equal to 0.28 at 20°C, what is the critical oxygen deficit and the time at which it occurs? Assume that the stream is 100% saturated with oxygen before the sewage is added.

7-17 A stream with k_2 equal to 0.31, temperature of 28°C, and minimum flow of 800 cfs receives 10 mgd of sewage from a city. The river water is saturated with DO above the city. What is the maximum permissible BOD of the sewage if the dissolved oxygen content of the stream is never to be below 4.0 mg/l?

REFERENCES

1. The National Safe Drinking Water Act, Public Law 93–523 (December 16, 1974).

2. "Water Quality Criteria," National Technical Advisory Committee Report, Federal Water Pollution Control Administration (Washington, D.C.: Government Printing Office, 1968).

3. J. E. McKee and H. W. Wolf, "Water Quality Criteria," *California State Water Control Board Publication No. 3-A* (1963).

4. *Standard Methods for the Examination of Water and Wastewater*, 13th ed. (New York: American Public Health Association, American Water Works Association, and Water Pollution Control Federation, 1971).

5. E. O. Jordan, "The Kinds of Bacteria Found in River Water," *J. Hygiene* 3 (1903): 1.

6. L. A. Allen, M. A. Pierce, and H. M. Smith, "Enumeration of Streptococcus Faicalis with Particular Reference to Polluted Waters," *J. Hygiene* 51 (1953): 458.

7. L. D. S. Smith and M. V. Gardner, "Vegetative Cells of Clostridium Perfringens in Soil," *J. Bacteriol.* 58 (1949): 407.

8. F. W. Kittrell and S. A. Furfari, "Observations of Coliform Bacteria in Streams," *J. Water Poll. Control Fed.* 35, no. 11 (November 1963): 1361–1385.

9. M. Greenwood and G. U. Yule, "On the Statistical Interpretation of Some Bacteriological Methods Employed in Water Analysis," *J. Hygiene* (1917): 1636.

10. J. K. Hoskins, "Most Probable Numbers for Evaluation of Coliaerogenes Tests by Fermentation Tube Method," *U.S. Public Health Rept.* 49, reprint 1621 (1934): 393.

11. N. A. Clarke, G. Berg, P. W. Kabler, and S. L. Chang, "Human Enteric Viruses: Source, Survival, and Removability in Waste-water," London Program, Section 11, Intern. Conf. Water Pollution Research (1962).

12. S. Kelly and W. W. Sanderson, "The Effect of Chlorine in Water on Enteric Viruses," *Am. J. Public Health* 48 (1958): 1323.

13. N. A. Clarke and P. W. Kabler, "The Inactivation of Purified Coxsackie Virus in Water by Chlorine," *Am. J. Hygiene* 59 (1954): 119.

14. S. Kelly and W. W. Sanderson, "Density of Enteroviruses in Sewage," *J. Water Poll. Control Fed.* 32, no. 12 (December 1960): 1269–1275.

15. S. Kelly, W. W. Sanderson, and C. Neial, "Removal of Enteroviruses from Sewage by Activated Sludge," *J. Water Poll. Control Fed.* 33, no. 10 (October 1961): 1056–1062.

16. W. N. Mack, J. R. Frey, F. J. Riegle, and W. L. Mallman, "Enterovirus Removal by Activated Sludge Treatment," *J. Water Poll. Control Fed.* 34, no. 11 (November 1952): 1133–1139.

17. G. C. Robeck, N. A. Clarke, and K. A. Dostal, "Effectiveness of Water Treatment Processes in Virus Removal," *J. Am. Water Works Assoc.* 54, no. 10 (October 1962): 1275–1292.

18. B. J. Conway, "Legal Aspects of Municipal Fluoridation," *J. Am. Water Works Assoc.* 50, no 10 (October 1958): 1330–1336.

19. R. A. Baker, "Threshold Odors of Organic Chemicals," *J. Am. Water Works Assoc.* 55, no. 7 (July 1963): 913–916.

20. H. A. Bevis, "Significance of Radioactivity in Water Supply and Treatment," *J. Am. Water Works Assoc.* 52, no. 7 (July 1960): 841–846.

21. Federal Water Pollution Control Act, Public Law 660, as amended by the Water Quality Act of 1965 (PL 89–234).

22. C. J. Velz and J. T. Grannon, "Drought Flow Characteristics of Michigan Streams," Water Resources Commission, State of Michigan, Lansing (1960).

23. E. J. Gumbel, "The Return Period of Flood Flows," *Ann. Math. Statistics* 12 (1941).

24. E. J. Gumbel, "Statistical Theory of Droughts," *Proc. Am. Soc. Civil Engrs.* 80, Separate No. 439 (May 1954).

25. W. C. Ray and W. R. Walker, "Low-Flow Criteria for Stream Standards," *Proc. Am. Soc. Civil Engrs., J. San. Eng. Div.* 94, no. SA 3 (June 1968).

26. G. F. Lee, *Eutrophication*, The University of Wisconsin, Water Resources Center, Eutrophication Information Program, Madison, Wis., (1970).

27. J. C. Goldman, D. B. Porcella, E. J. Middlebrooks, and D. F. Toerien, *The Effect of Carbon on Algal Growth—Its Relationship to Eutrophication*, Utah Water Research Laboratory Bulletin, Logan, (1971).

28. Task Group Report, "Sources of Nitrogen and Phosphorus in Water Supplies," *J. Am. Water Works Assoc.* 59 (1967): 344.

29. G. L. Hergenrader and M. J. Hammer, "Eutrophication of Small Reservoirs in the Great Plains," *Geophys. Monograph Ser. No. 17*, Man-Made Lakes: Their Problems and Environmental Effects (Washington, D.C.: American Geophysical Union, 1973), pp. 560–566.

30. R. A. Vollenweider, *Scientific Fundamentals of the Eutrophication of Lakes and Flowing Waters with Particular Reference to Nitrogen and Phosphorus as Factors in Eutrophication* (Paris, France: Organization for Economic Co-operation and Development, 1970).

31. A. D. Hasler, "Cultural Eutrophication Is Reversible," *BioScience* 19, no. 5 (1969): 425–431.

32. G. P. Fitzgerald, *Algicides*, The University of Wisconsin, Water Resources Center, Eutrophication Information Program, Madison, Wis. (1971).

33. E. Frankland, "First Report of the Rivers Pollution Commission of 1868," extracted in Royal Commission Fifth Report, Appendix VI (1908), p. 10.

34. J. B. Weems, "The Evolution of the Oxygen Absorption Test in Water Analysis," Thirteenth Annual Meeting Iowa Engineering Society, Davenport, Iowa, January 1901, abs., *Rev. Am. Chem. Res.* 8 (1902): 323.

35. A. Dupré, "Report on Changes in the Aeration of Water, as Indicating the Nature of the Impurities Present in It," Fourteenth Annual Report of the Local Government's Board, 1883–1884, Appendix B, no. 11 (1884), p. 304.

36. W. E. Adeney, "Studies in the Chemical Analysis of Salt and Fresh Waters, I. Applications of the Aeration Method of Analysis to the Study of River Water," Royal Commission, Fifth Report, Appendix VII (1908), p. 95.

37. E. J. Theriault, "The Oxygen Demand of Polluted Waters," *U.S. Public Health Service Bull. No. 173* (1927); *Chem. Abst.* 25 (1931): 4644.

38. J. C. Young and J. W. Clark, "Growth of Mixed Bacterial Populations at 20°C and 35°C," *Water and Sewage Works* 112 (July 1965): 251–256.

39. M. N. Bhatla and A. F. Gaudy, "Role of Protozoa in the Diphasic Exertion of BOD," *Proc. Am. Soc. Engrs., J. San. Eng. Div.* 91, no. SA3 (June 1965): 63–87.

40. D. G. Ballinger and R. J. Lishka, "Reliability and Precision of BOD and COD Determinations," *J. Water Poll. Control Fed. 34*, no. 5 (May 1962): 470–474.

41. J. W. Clark, "New Method for Biochemical Oxygen Demand," *New Mexico State University Engineering Experiment Station Bull. 11* (December 1959).

42. N. W. Streeter and E. B. Phelps, *U.S. Public Health Service Bull. No. 146* (1925).

43. J. D. O'Connor and W. E. Dobbins, "Mechanism of Reaeration in Natural Streams," *Trans. Am. Soc. Civil Engrs.*, 123 (1958): 655.

44. *Chem. Eng. News* (March 2, 1970): 34.

45. *The Nation's Water Resources* (Washington, D.C.: Government Printing Office, 1968, pp. 1–11.

8
Systems for Treating Water and Wastes

Chapter 7 presented discussions pertaining to stream pollution, eutrophication, surface-water quality, and groundwater quality. The objective of this chapter is to provide students with an overview of water and wastewater treatment systems relative to water pollution and water quality. Subsequent chapters will cover physical, chemical, and biological treatment processes.

The sections on selection of wastewater and water-treatment processes are introductory to subsequent chapters that deal separately with physical, chemical, biological, and sludge-treatment methods. They also act as summaries for subsequent chapters by drawing the unit processes into integrated treatment systems. It is suggested that the student frequently refer to Sections 8-5 and 8-8, while studying Chapters 9 through 12.

Waste-Treatment Systems

The purpose of municipal wastewater treatment is to prevent pollution of the receiving watercourse. Characteristics of a municipal wastewater depend to a considerable extent on the type of sewer collection system and industrial wastes entering the sewers. The degree of treatment required is determined by the bene-

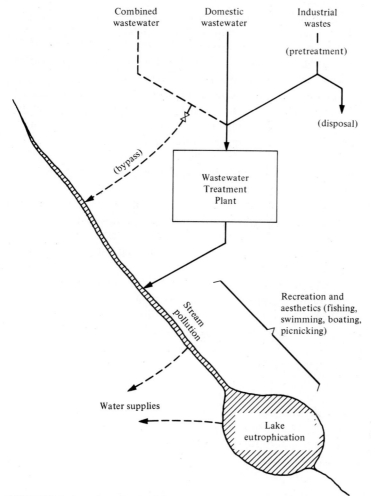

FIGURE 8-1 Location of a municipal wastewater treatment plant.

ficial uses of the receiving stream or lake. Stream pollution and lake eutrophication resulting from municipal wastes are particularly troublesome in water
reuse for water supplies and recreation. Figure 8-1 shows the position of a
typical wastewater-treatment plant relative to the waste sources and receiving
waters.

8-1

INDUSTRIAL WASTES

An industry has three possibilities for disposal of process wastewaters: (1) they
may be treated separately in an industrial waste-treatment plant prior to discharge to a watercourse; (2) raw wastewaters may be discharged to the municipal

treatment plant for complete treatment; or (3) industrial wastes can be pretreated at the industrial site prior to discharge in the municipal sewerage system. A careful study must be performed to determine the most feasible method for disposal. Discharge to a municipal plant relieves the industry of waste treatment, and the municipality accepts the responsibility for its disposal.

Joint treatment of industrial and domestic wastewater has several advantages. A municipality can apply for federal aid for plant construction, whereas private industry is not eligible for such cost-sharing monies. Only one major treatment plant is required, resulting in more economical operation. The responsibility for operation is placed with one owner. The majority of industrial wastes are more amenable to biological treatment after dilution with domestic wastewater.[1]

Three categories of wastes should be excluded from the municipal sewers, those which (1) create a fire or explosion hazard (e.g., gasoline or cleaning solvents); (2) impair hydraulic capacity (e.g., paunch manure or sand); and (3) create a hazard to people, the sewer system, or the biological-treatment system (e.g., toxic metal ions or petroleum wastes).[2]

Large volumes of high-strength wastes must be considered in the design capacity of municipal plants since overloaded treatment plants resulting from industrial expansion become a serious problem. Facilities constructed on funds provided from a 20-year bond issue frequently reach design loading in 5–10 years. A municipal sewer code, user fees, and separate contracts between an industry and the city can provide adequate control and sound financial planning, while accommodating industry by joint waste treatment. A municipality should enact a sewer ordinance and establish charges for industry to protect the public investment in its waste-treatment facilities.

Pretreatment should be considered for industrial discharges which have strengths or characteristics differing significantly from domestic wastewater. Only a fine line of difference may exist between an untreatable trade waste and one which can be jointly treated after certain pretreatment by considering the industrial process design, segregation of wastes, equalization of the wastewaters, and strength reduction.

Process Modifications

Facilities for pretreatment of industrial wastes are frequently an integral part of plant design. Process changes, equipment modifications, by-product recovery, and wastewater reuse can have an economic advantage. Industries with a large volume of high-strength wastewater should consider recovery processes to reduce the pollution load on the municipal treatment plant. For example, a tannery in the dehairing process can use a proprietary unhairing compound with low sulfides and realize significant BOD reductions in comparison with the usual high-sulfide process. Many industries have installed countercurrent washing or spray-rinsing systems to reduce water consumption. Collecting and processing blood is a good example of by-product recovery by the meat industry of a salable commodity. Reuse of water is practiced extensively by closed systems in paper production. Thus, a paper mill can recycle white water (that passing through a wire screen upon which paper is formed), thus saving costs of both supply and treatment.

Segregation of Wastes

In many older industrial plants, process, cooling, and sanitary waste-waters are mixed in one sewer line. Modern industrial treatment dictates segrega-tion of these waste streams for separate disposal, individual pretreatment, or controlled mixing. Cooling water, although slightly contaminated by leakage, corrosion, and heat, generally contains little or no pollutional organic matter. Therefore, it can be disposed of in a storm sewer or directly in a watercourse provided that thermal pollution is not a problem. Segregated waste streams containing coarse solids (i.e., paunch manure, feathers, corn, onion peelings, and hair) can be screened readily. Combined waste streams having coarse solids mixed with grease and sanitary wastes are more difficult to screen. In treating tannery wastes, the chrome dump waste (the spent chrome tanning liquor) can be proportioned into the tannery wastewater to improve primary sedimentation.

Equalization

Industries using a diversity of processes may be required to equalize wastes by holding them in a basin for a certain period of time to get a stable effluent easier to treat in a municipal plant. Neutralization of alkaline and acid waste streams, stabilization of BOD, and settling of heavy metals are some objectives of equalization. Textile mill wastes discharged to a municipal plant must be equalized to prevent fluctuations in pH and BOD which upset the efficiency of a biological-treatment system. Unequalized wastewaters high in alkalinity or acidity commonly require neutralization by chemical addition to prevent upset-ting the system.

Waste Strength Reduction

Certain industrial wastes can be pretreated to reduce the organic or inor-ganic solids prior to disposal in a municipal plant. Tannery or textile wastes may require equalization and sedimentation to decrease settleable solids. Dairy wastes from a bottling plant, a creamery, or a cheese factory, located on the sewer line a considerable distance from the municipal plant, may require pre-treatment by aeration to prevent acid, malodorous influent wastewater. Reduc-tion in drippage and spillage, and by-product recovery in milk processing can reduce the strength of dairy wastes significantly. Whole milk has a BOD of approximately 100,000 mg/l, and whey a BOD of 35,000 mg/l.

Metal-plating wastes containing cyanide, chromium, zinc, copper, and other heavy metals may require treatment in addition to equalization if the toxic ions are not sufficiently diluted with domestic wastewater. Chemical treatment by oxidation or coagulation are commonly used for removal of these inorganic pollutants. Waste streams with high concentrations of refractory pollutants (i.e., salt or ammonia nitrogen) must be regulated at the plant site if the municipal waste-disposal system does not have sufficient assimilative capacity.

8-2

INFILTRATION AND INFLOW

Infiltration is groundwater entering sewers and building connections through defective joints and cracks in pipes and manholes. Inflow is water discharged into service connections and sewer pipes from foundation and roof drains, outdoor paved areas, cooling water from air conditioners, and unpolluted discharges from businesses and industries. Excessive infiltration and inflow cause surcharging of sewer lines with possible backup of sanitary wastes into basements, hydraulic overloading of treatment facilities, and bypassing of pumping stations or processing plants.[3]

The amount of infiltration depends on condition of the sewer system, groundwater levels, and porosity of the soil profile. Proper design, selection of sewer pipe with watertight joints, and supervision during construction can limit the quantity of seepage. Specifications for new construction permit a maximum infiltration rate of 200–500 gpd/mi/in. of pipe diameter. This quantity of flow is equal to about 5% of the peak hourly, or 10% of the average, domestic wastewater flow rate. Evaluation of an existing sewer system is initiated by measuring both the wet- and dry-weather flows to determine sources and quantities of excessive infiltration. Possible corrective measures include grouting or sealing soils, relining pipe, and sewer replacement.

Inflow is the result of deliberately connecting sources of extraneous water to sanitary sewer systems. Storm water and drainage should be disposed of in storm sewers; however, sanitary lines are often more convenient because of their greater depth of burial and convenient location. Establishing and enforcing a sewer-use regulation that excludes unpolluted waters from separate sanitary collectors is an effective method of preventing excess inflow. A municipal sewer code should require disposal of surface runoff, foundation drainage, air-conditioning effluents, industrial cooling waters, and other clean waters, to storm lines or natural drainage courses. Where problems already exist, surveys can be conducted to locate illegal connections and institute corrective measures.

8-3

COMBINED SEWERS

Combined wastewater-collection systems have only one sewer pipe network to collect domestic wastewater, industrial wastes, and storm runoff water. The dry-weather flow (domestic and industrial wastewater, plus street drainage from washing operations and lawn sprinkling) is intercepted for treatment. During storms when the combined waste and runoff water exceed the capacity of the treatment plant, the overflow is bypassed directly to the receiving watercourse without treatment. This excess may result in significant pollution, even during high river stages, and create a health hazard for the downstream water user. Combined sewers are no longer acceptable except in cities having combined systems, where separate systems are not feasible due to a lack of storm sewer outfalls.

In antiquity, sewers were constructed primarily as drains to carry away storm water from urbanized areas. As late as 1850, disposal of sanitary wastes into the sewers of London was prohibited. Development of the water carriage waste-disposal system led to the use of storm drains as combined sewers.

In the United States, many cities along rivers built combined sewers to convey both storm runoff and sanitary wastewater directly to the river through a series of lines oriented perpendicular to the river. Required treatment of sanitary wastewater resulted in construction of a collector sewer along the river to inter- cept the dry-weather flow in combined sewers, and convey it to a treatment plant. The flow entering a collector line from a combined sewer is commonly regulated by a gate or weir arrangement. The dry-weather flow from an overlying combined sewer falls into the interceptor below, but the greater wet-weather flows, with their higher velocities, leap across the opening and flow directly to the river.

Solutions for the problem of overflows from sewer systems are receiving national attention.[4] Most large cities in the United States are served by com- bined sewers and thereby major sources of pollution. One obvious solution is to replace the combined collectors with separate sewers. Unfortunately, the cost of individual sewers based on current construction methods is usually pro- hibitive. Furthermore, it may be desirable to treat urban runoff and prevent pollution in certain cases. Another method being considered is use of stabiliza- tion basins for collection and storage of the combined wastewater overflow, followed by treatment prior to disposal.

In many municipalities, only the older sections are served by combined sewers. In other cities, the sanitary sewers in aging neighborhoods are cracked and the mortar joints deteriorated, so infiltration of groundwater is excessive during wet seasons. Foundation drains and roof downspouts are sometimes connected to the sanitary sewer; hence the design of a municipal plant must consider these hydraulic loads, or the high flows eliminated by reconstruction. If drainage from streets is involved, special consideration should be given to the design of bar screens and grit-removal units to prevent carryover of solids damaging to subsequent treatment equipment.

8-4

PURPOSES OF WASTEWATER TREATMENT

Location of a typical municipal wastewater-treatment plant is illustrated in Fig. 8-1. Wastewaters from households, industries, and combined sewers are collected and transported to the treatment plant with the effluent commonly disposed of by dilution in rivers, lakes, or estuaries. This is normally the only feasible method of disposal and, for several communities on rivers, the only system which ensures adequate water resources for downstream users during drought flows. Other means include irrigation, infiltration, evaporation from lagoons, and submarine outfalls extending into the ocean.

Water-quality criteria[5] have been established for receiving waters to assist in defining the degree of treatment required for disposal by dilution. Effluent standards prescribe the extent of processing for each municipal or industrial wastewater discharge. In general, the minimum specified is secondary treatment. Still, many municipalities and industries on major rivers provide a lesser quality of purification. On the other hand, other cities have been requested to install advanced waste-treatment systems that remove phosphorus and nitrogen to retard eutrophication of receiving lakes. Stream-classification documents, published by each state as required by federal law, categorize surface waters according to their most beneficial present or future use (i.e., for drinking-water supplies, body-contact recreation, etc.). These publications also incorporate stream standards that establish maximum allowable pollutant concentrations for a given stream under defined flow conditions. Effluent standards are used for regulatory purposes to achieve compliance with these stream standards.

Rules and regulations of the Environmental Protection Agency in 1973 defined the minimum level of effluent quality that must be attained by secondary treatment of municipal wastewaters.[6] Acceptable secondary effluent is defined in terms of biochemical oxygen demand, suspended solids, fecal coliform bacteria, and pH. The arithmetic mean of BOD and suspended-solids concentrations for effluent samples collected in a period of 30 consecutive days must not exceed 30 mg/l; and, during any 7-consecutive-day period, the average must not exceed 45 mg/l. Furthermore, removal efficiencies shall not be less than 85% (i.e., if an influent concentration is less than 200 mg/l the effluent cannot exceed 15% of this value). The geometric mean of fecal coliform counts for effluent samples collected in a period of 30 consecutive days must not exceed 200 per 100 ml; and the geometric mean of fecal coliform bacteria for any 7-consecutive-day period must not exceed 400 per 100 ml. The effluent values for pH shall remain within the limits 6.0–9.0. Based on these standards, the purposes of conventional wastewater treatment are reduction of BOD, suspended solids, and pathogens.

Meeting the 30 mg/l BOD and suspended-solids effluent requirement requires a well-designed facility that is properly operated. Moderately loaded activated-sludge processes, biological towers, and two-stage trickling filter plants are capable of achieving this degree of treatment. However, some designs that were popular during the past 20 years are not efficient enough (e.g., single-stage trickling filter plants and high-rate activated-sludge systems). Reduction of fecal coliforms to a level of 200 per 100 ml requires disinfection with chlorine. Facultative stabilization ponds handling raw wastewater are not able to meet discharge standards without further treatment. The main problem is the high suspended-solids content of the lagoon water during the summer due to growth of algae. The best solution in most cases appears to be disposal by land irrigation, or evaporation of the liquid from complete retention lagoons to avoid effluent discharge to surface waters.

Refractory contaminants, both inorganic and organic materials, are pollutants resistant to, or totally unaffected, by conventional treatment processes. The mineral quality of wastewater depends largely on the same property of the municipal water supply but, during water use, numerous substances are added— common salt (sodium chloride) and other dissolved solids. Phosphates, which occur in low concentrations in most natural waters, are increased during domestic use, principally by synthetic detergents. Organic nitrogeneous compounds decompose to ammonia and to a variable extent oxidize to nitrates in waste treatment. Certain high-molecular-weight materials (e.g., tannins and lignins) and aromatic compounds (e.g., surfactants and dyes) are resistant to biological degradation. Rivers are also polluted by waste effluents from industries and agricultural chemicals in land runoff. The latter includes refractory pesticides such as DDT, aldrin, orthonitrochlorobenzene, and others.

If a refractory pollutant originates from an industrial waste which can be segregated, the pollutant may be removed by pretreatment, or by eliminating the isolated waste stream using land burial, or incineration rather than disposal in a sewer. Highly colored industrial waste streams can be handled by chemical treatment or separate disposal. Common plant nutrients are derived from organic matter and phosphate builders of detergents found in domestic wastewaters. Phosphorus and nitrogen removals in conventional waste treatment are only 30 to 60%, depending on the types of processes and concentrations in the raw wastewater. Advanced waste-treatment methods are required for greater nitrogen and phosphorus removals.

Eutrophication of a lake induced by municipal wastes can be retarded by removing the source of plant nutrients, notably phosphorus and nitrogen. This is accomplished by diversion of the treated effluent around the lake, or by treatment of the wastewater employing advanced treatment processes. There is no ready solution for removing nutrients from runoff, either from natural land or cultivated fields. Before instituting phosphorus and nitrogen removal from a municipal wastewater, therefore, a survey should determine existing levels and identify sources. Although in many cases municipal wastes have been the major origin of phosphorus, various receiving waters have concentrations well above the desirable limit originating from land drainage.

8-5

SELECTION OF WASTE-TREATMENT PROCESSES

Conventional wastewater treatment consists of preliminary processes (pumping, screening, and grit removal), primary settling to remove heavy solids and floatable materials, and secondary biological aeration to metabolize and flocculate

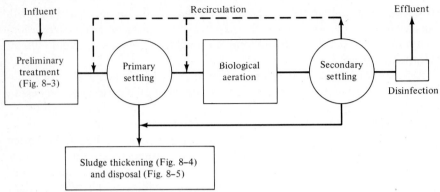

FIGURE 8-2 Schematic diagram of conventional wastewater treatment.

colloidal and dissolved organics. Waste sludge drawn from these unit operations is thickened and processed for ultimate disposal (Fig. 8-2).

Preliminary Treatment Units

The following preliminary processes can be used in municipal wastewater treatment: coarse screening (bar racks), medium screening, comminution, flow measuring, pumping, grit removal, preaeration, flotation, flocculation, and chemical treatment. The latter three are not common in domestic wastewater treatment. These practices are sometimes dictated by the industrial pollutants in the municipal wastewater. Flotation is used to remove fine suspensions, grease, and fats, and performed either in a separate unit or in a preaeration tank also used for grit removal. If adequate pretreatment is provided by petroleum industries and meat-processing plants, flotation units are not required at a municipal facility. Flocculation with or without chemical additions may be practiced on high-strength municipal wastes to provide increased primary removal and prevent excessive loads on the secondary-treatment processes. Chlorination of raw wastewater is sometimes used for odor control and to improve settling characteristics of the wastes.

The arrangement of preliminary-treatment units varies depending on raw-wastewater characteristics, subsequent treatment processes, and preliminary steps employed. A few general rules always apply in arrangement of units. Screens are used to protect pumps and prevent solids from fouling grit-removal units and flumes. In small plants a Parshall flume is normally placed ahead of constant-speed lift pumps, but may be located after them in large plants or where variable-speed pumps are used. Grit removal should be placed ahead of the pumps when heavy loads are anticipated, although the grit chamber follows the lift pumps in most separate sanitary wastewater plants. Three possible arrangements for preliminary units are shown in Fig. 8-3.

Primary Treatment Units

Primary treatment is sedimentation. In common usage, though, the term includes the preliminary treatment processes. Sedimentation of raw wastewater

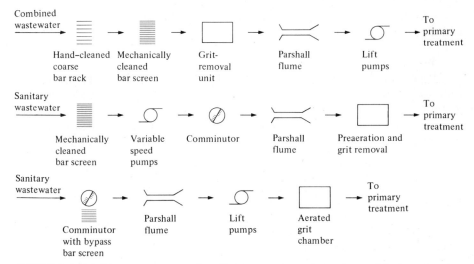

FIGURE 8-3 Possible arrangements of preliminary treatment units in municipal wastewater processes.

is practiced in all large municipal plants and must precede conventional trickling filtration. Complete-mixing activated-sludge processes can be used to treat unsettled raw wastewater; however, this method is restricted to small municipalities because of the costs involved in sludge disposal and operation.

Secondary Treatment Units

Primary sedimentation removes 30–50% of the suspended solids in raw municipal wastewater. Remaining organic matter is extracted in biological secondary treatment to the allowable effluent residual using activated-sludge processes, trickling filters, or biological towers. In the activated-sludge method, wastewater is fed continuously into an aerated tank, where microorganisms synthesize the organics. The resulting microbial floc (activated sludge) is settled from the aerated mixed liquor under quiescent conditions in a final clarifier and returned to the aeration tank. The plant effluent is clear supernatant from secondary settling. Advantages of liquid aeration are high-BOD removals, ability to treat high-strength wastewater, and adaptability for future use in plant conversion to advanced treatment. On the other hand, a high degree of operational control is needed, shock loads may upset the stability of the biological process, and sustained hydraulic or organic overloading results in process failure.

Biological beds, either trickling filters or towers, contain rock or synthetic media to support microbial films. These slime growths extract organics from the wastewater as it trickles through the bed. Oxygen is supplied from air moving through voids in the medium. Excessive biological growth washes out with the underflow and is collected in the secondary clarifier. In northern climates, two-stage rock-filled trickling filters are needed to achieve efficient treatment. Biological towers may be either single- or two-stage systems. Advantages of filtration are ease of operation and capacity to accept shock loads and overloading without causing complete failure.

Sludge Disposal

Primary sedimentation and secondary biological flocculation processes concentrate the waste organics into a volume of sludge significantly less than the quantity of wastewater treated. But disposal of the accumulated waste sludge is a major economic factor in wastewater treatment. Construction cost of a sludge processing facility is approximately one-third that of a treatment plant.

Flow schemes for withdrawal, holding, and thickening raw waste sludge from sedimentation tanks are illustrated in Fig. 8-4. The settled solids from clarification of trickling filter effluent are frequently returned to the plant head for removal with the primary sludge (top diagram Fig. 8-4). Raw settlings may

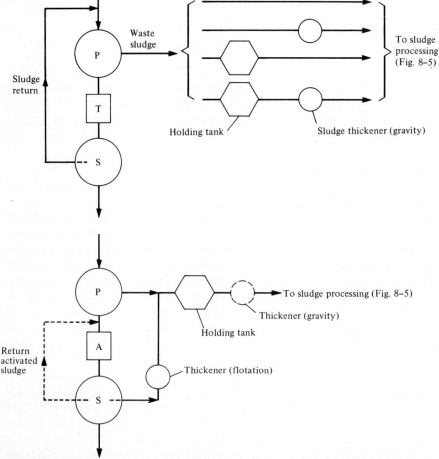

FIGURE 8-4 Flow schemes for withdrawal, holding, and thickening of waste sludge. P—primary settling; T—trickling filtration; A—activated-sludge aeration; S—secondary settling.

be stored in the primary tank bottom until processed, or pumped into a holding tank for storage. The withdrawn sludge may be concentrated in a thickener, usually a gravity-type unit, prior to processing. In the bottom diagram of Fig. 8-4, waste activated sludge is mixed with primary residue after withdrawal. A holding tank is commonly used in this system arrangement, along with a thickener. The waste activated may be thickened separately or the combined sludges thickened prior to processing.

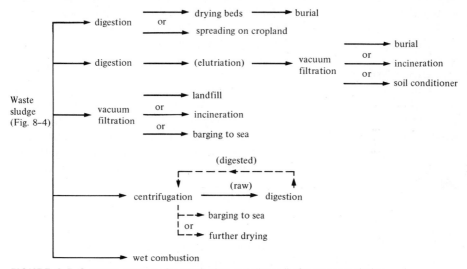

FIGURE 8-5 Common methods for processing and disposal of raw waste sludge.

Alternatives for processing and disposal of raw sludge are shown in Fig. 8-5. Common methods of processing are anaerobic digestion and vacuum filtration, with centrifugation and wet combustion also currently used. Conventional methods of disposal include burial in landfill, incineration, production of soil conditioner, and barging to sea. For coastal cities, ocean dumping is often least expensive, whereas burial is regularly practiced if landfill area is available. Incineration, although more costly, is frequently the only feasible method of disposal in urbanized areas.

All possible sludge-disposal processes for a municipal treatment plant must be given careful consideration. The method selected should be the most economical process, if it is best, with due regard to environmental conditions. Attention must be given to such factors as trucking sludge through residential areas, future use of landfill areas, groundwater pollution, air pollution, other potential public health hazards, and aesthetics.

Chlorination

Disinfection of secondary-treatment-plant effluents is commonly practiced where the receiving watercourse is used for recreation or water supply. Present

Environmental Protection Agency rules and regulations require chlorination of effluents throughout the entire year, whereas previously it had been specified only during the water-recreation season.

Phosphorus and Nitrogen Removal

In recent years, feasible methods for phosphorus removal in wastewater-treatment plants have been developed, and several full-scale facilities are operational. Studies have also been made to determine the best methods for nitrogen removal and complete water reclamation. However, only a few large-scale plants are reclaiming water to its original quality. Advanced waste-water treatment systems are discussed in Chapter 13.

Size of Municipality

Operational management and control, and the necessity for sludge hand-ling, dictates the selection of wastewater-treatment processes for small communities. Methods that do not require sludge disposal (stabilization ponds), or only occasional sludge withdrawals (extended aeration) are preferable for small villages and subdivisions. Towns large enough to employ a part-time operator frequently use systems which require more operational control and maintenance (i.e., contact stabilization and oxidation ditch plants). Cities with trickling filter and activated-sludge-treatment plants employ several workers.

Listed in Table 8-1 are the common types of wastewater-treatment plants built in municipalities of various sizes. Many existing ones do not conform to the listing in Table 8-1. Some are no longer popular (i.e., Imhoff tanks) and others are selected on the basis of unique local conditions.

TABLE 8-1

COMMON TYPES OF WASTEWATER-TREATMENT PLANTS FOR MUNICIPALITIES OF VARIOUS SIZES

Subdivisions, schools, etc.
   Extended aeration (factory-built)
   Stabilization ponds
Towns (population less than 2000)
   Extended aeration
   Contact stabilization (field-erected, factory-built)
   Oxidation ditch
   Stabilization ponds
Small cities (2000–8000 population)
   Contact stabilization
   Oxidation ditch
   Complete mixing activated sludge without primary
   Primary plus trickling filters
Cities (population greater than 10,000)
   Primary plus trickling filters
   Primary plus activated sludge

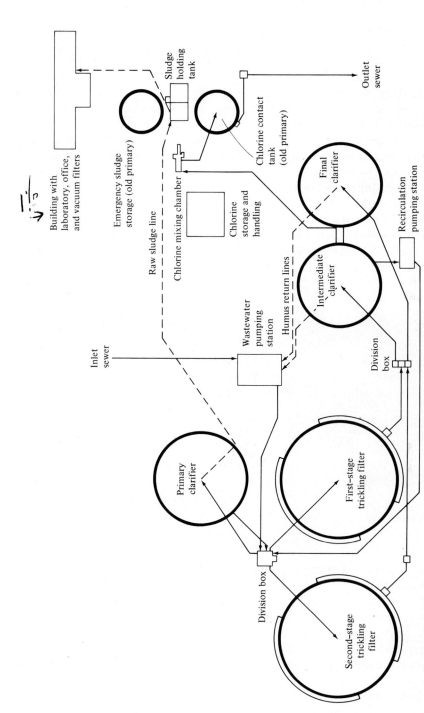

FIGURE 8-6 Plant layout for a two-stage trickling filter wastewater-treatment plant, Kearney, Nebraska. Average design flow 3.0 mgd with 230 mg/l of BOD and 250 mg/l of suspended solids, and maximum wet weather flow of 7.5 mgd. (Courtesy Henningson, Durham & Richardson.)

319

Layout of Treatment Plants

The flow diagram for a two-stage trickling filter plant is shown in Fig. 8-6; an aerial view of the same facility is given in Fig. 8-7. The sequence of treatment units is: raw-wastewater pumping station with a mechanically cleaned bar screen, primary settling tank, first-stage trickling filter, intermediate clarifier, second-stage filter, final settling basin, and chlorine contact tank. The rock-filled biological filters are covered with fiberglass domes to prevent ice formation and maintain an adequate wastewater temperature during the winter for biological activity. Wastewater is recirculated through the trickling filters to maintain optimum hydraulic loading and increase the removal of organic matter. Humus washed off the filter media collects in the intermediate and final clarifiers. Underflow from the clarifiers returns this material to the lift station, where it is pumped to the primary settling basin with the raw wastewater. Primary sludge containing raw organics and filter humus is withdrawn from the bottom of the clarifier and discharged to a holding tank. From here it is pumped to vacuum filters for dewatering prior to disposal in a sanitary landfill.

FIGURE 8-7 Aerial view of the two-stage filter plant shown in Figure 8-6. (Courtesy Henningson, Durham & Richardson, and City of Kearney, Neb.)

The sequence of treatment units in the activated-sludge plant shown in Fig. 8-8 is: mechanically cleaned bar screen, and a grinder that returns shredded solids to the raw wastewater; constant- and variable-speed lift pumps, with standby gas engines; Parshall flume; clarifier-type aerated grit chambers, with separate grit washer; primary settling tanks; complete mixing activated-sludge secondary, and excess sludge gravity-flow return line to wet well; vacuum filtration of raw waste slurry from holding tanks; and filter-cake disposal by burial. The filtrate from the vacuum filter returns to the wet well. Raw wastewater cannot be bypassed ahead of the wet well by gravity flow because of sewer depth. Standby gas engines are provided for two lift pumps used in case of electrical power outage.

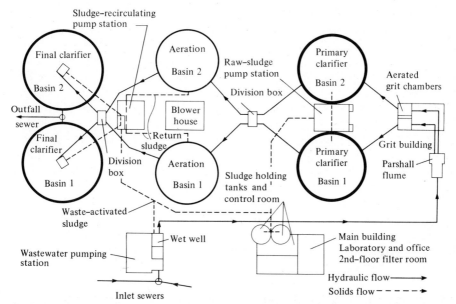

FIGURE 8-8 Plant layout for complete-mixing activated-sludge wastewater-treatment plant, design flow = 11 mgd, Grand Island, Nebraska. (Courtesy Black and Veatch, Consulting Engineers.)

Water Treatment Systems

8-6

WATER SOURCES

Typical water sources for municipal supplies, illustrated in Fig. 8-9, are deep wells, shallow wells, rivers, natural lakes, and impounding reservoirs. No two sources of supply are alike, and the same origin may produce water of varying quality at different times. Water-treatment processes selected must consider the raw-water quality and differences in quality for each particular water source.

Municipal water-quality factors of safety, temperature, appearance, taste and odor, and chemical balance are most easily and frequently satisfied by a deep well source. Furthermore, the treatment processes employed are simplest because of the relatively uniform quality of such origin. Excessive concentrations of iron, manganese, and hardness habitually exist in well waters. Some deep-well supplies may contain hydrogen sulfide, others may have excessive concentrations of chlorides, sulfates, and carbonate. Hydrogen sulfide can be removed by aeration or other oxidation processes, but the prevalent sodium and potassium salts of anions bicarbonate, chloride, and sulfate are refractory inorganics. Calcium and magnesium can be removed by precipitation softening. Excessive concentration of flouride, responsible for mottled enamel of the teeth, has in rare cases caused a municipality to abandon a well-supply source. The mineral constituents

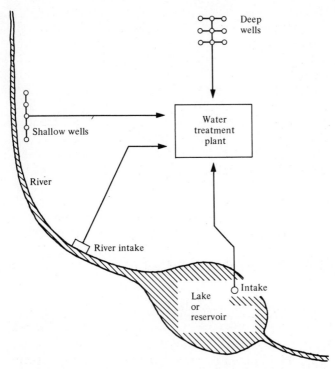

FIGURE 8-9 Common municipal water sources.

of a potential groundwater source should be carefully examined in selecting a deep-well water supply.

Shallow wells recharged by a nearby surface watercourse may have quality characteristics similar to deep wells or may relate more closely to the watercourse quality. A sand aquifer adjacent to a river may act as an effective filter for removal of organic matter and as a heat exchanger for leveling out temperature changes of the recharge water seeping into it. A case in point is the municipal water supply for Lincoln, Nebraska, located in the Platte River Valley near Ashland. The wells are located in a sand aquifer adjacent to the river channel. Chemical composition of the well water and river water are nearly identical. River-water temperature varies from 80°F in the summer to 33°F in the winter while the well-water temperature remains within 3° of 56°F year around. None of the bacterial or organic pollution existing in the river has been identified in the well water. Another shallower well field located adjacent to the Platte River in a coarser sand and gravel aquifer withdraws water of significantly poorer quality after peak pumping periods. The well-water temperature increases to nearly 70°F in late summer and bacterial counts in the raw water rise. Clogging of the well screens, attributed in part to bacterial slime growths, has occurred in some wells.

To predict water quality from shallow wells, careful studies of the aquifer and nature of recharge water are necessary. Quality of the well water depends upon such things as aquifer permeability, well spacing and depth, seasonal changes in river flow, and pumping rates. The best way to evaluate these variables is by full-scale field pumping tests, or by checking similar existing well fields.

Pollution and eutrophication are major concerns in selecting and treating surface-water supplies. Where these are highly contaminated and difficult to treat, municipalities seek either groundwater supplies or alternate less-polluted surface sources within a feasible pumping distance. However, in many regions of the United States, adequate groundwater resources are not available and high-quality surface supplies are not within economic reach. Although the majority of municipal water-supply systems are from underground origins, only about one-fourth of the nation's population is served by these sources. In general, larger cities are dependent upon surface supplies. Treatment of polluted waters is perhaps the greatest challenge in water-supply engineering, and the importance of pollution control is realized most vividly when one ponders the problems of providing safe, palatable water from contaminated sources.

Water quality in rivers depends upon the character of the watershed; pollution caused by municipalities, industries, and agricultural practices; river development such as dams; season of the year; and climatic conditions. During spring runoff or periods of high flows, river water may be muddy and high in taste- and odor-producing compounds. During drought flows, pollutants are often present in higher concentrations and odorous conditions can occur. During late summer, algal blooms frequently create problems. River-temperature variations depend on latitude and the location of the stream headwaters. In Northern states, river water is warm in the summer and cold in the winter. Water-quality conditions vary from one stream to another and, in addition, each has its own peculiar characteristics. River-water quality is usually deteriorating if the watershed is under development by man.

Water supplies from rivers normally require the most extensive treatment facilities, and greatest operational flexibility of any source. A river-water-treatment plant must be capable of handling day-to-day variations, and anticipated quality changes likely to occur within its useful life.

The quality of water in a lake or reservoir depends upon the physical, chemical, and biological characteristics (limnology) of the body. Size, depth, climate, watershed, degree of eutrophication, and other factors influence the nature of an impoundment. The relatively quiescent retention of river water produces marked changes in quality brought about by self-purification. Common benefits derived from impounding a river water include reduction of turbidity, coliform counts and usually color, and elimination of the day-to-day variations in quality.

Lakes and reservoirs are subject to seasonal changes which are particularly noticeable in eutrophic waters. In summer and winter during stratification, the hypolmnion (unmixed bottom water layer) in an eutrophic lake may contain dissolved iron and manganese, and taste and odor compounds. The decrease of oxygen by bacterial activity on the bottom results in dissolution of iron and manganese, and production of hydrogen sulfide and other metabolic intermediates. Algal blooms frequently occur in the epilimnion (warmer mixed surface-water layer) of fertile lakes in early spring and late summer. Heavy growths of algae, particularly certain species of blue-greens produce difficult-to-remove tastes and odors. Normally, the best quality of water is found near middepth, below the epilimnion and above the bottom. The intake for a fertile lake or reservoir should preferably be built so that water can be drawn from selected depths. Lakes which stratify experience spring and fall overturns, which occur when the temperature profile is nearly uniform and wind action stirs the lake from top to bottom. If the bottom water has become anaerobic during stratification, water drawn from all depths during the overturn will contain taste- and odor-producing compounds.

The limnology of a lake or reservoir and the present and/or future level of eutrophication should be thoroughly understood before beginning design of intake structures or treatment systems.

8-7

PURPOSE OF WATER TREATMENT

The purpose of a municipal water-supply system is to provide potable water which is chemically and bacteriologically safe for human consumption and has adequate quality for industrial users. For domestic uses, water should be free from unpleasant tastes and odors, and improved for human health (i.e., by fluoridation). Since the quality of public supplies is based primarily on drinking-water standards, the special temperature and other needs of some industries are not always met by public supplies. Boiler feedwater, water used in food processing, and process water in the manufacture of textiles and paper have special quality tolerances which may require additional treatment of the municipal water at the industrial site.

8-8

SELECTION OF WATER-TREATMENT PROCESSES

Watershed management should be considered part of the operation of a water-supply system. Stream flows may be augmented by controlled releases from upstream storage, or water-supply reservoirs can be used for natural purification of river waters. In the latter case, algicides such as copper sulfate can be introduced to control algal blooms which would interfere with coagulation and filtration processes, and produce taste and odor problems.

Current pretreatment processes in municipal water treatment are screening, presedimentation or desilting, chemical addition, and aeration.[7] Screening is practiced in pretreating surface waters. Presedimentation is regularly used to remove suspended matter from river water. Chemical treatment, in advance of in-plant coagulation, is most frequently applied to improve presedimentation, to pretreat hard-to-remove substances, such as taste and odor compounds and color, and to reduce high bacterial concentrations. Conventional chemicals used with presedimentation are chlorine, potassium permanganate, polyectrolytes, and alum. Aeration is customarily the first step in treatment for removal of iron and manganese from well waters, and a standard way to separate dissolved gases such as hydrogen sulfide and carbon dioxide.

Treatment processes used in water plants depend upon the raw-water source and quality of finished water desired. The specific chemicals selected for treatment depend upon their effectiveness to perform the desired reaction and cost. For example, activated carbon, chlorine, chlorine dioxide, and potassium permanganate are all used for taste and odor control. Superchlorination is generally the least expensive, but for many taste and odor compounds, activated carbon is more effective. In surface-water-treatment plants, equipment for feeding two or three taste- and odor-removal chemicals is usually provided, so the operator can select the most effective and economic chemical applications. There is no fixed rule for color removal applicable to all waters. Alum coagulation with adequate pretreatment, and applying oxidizing chemicals or activated carbon, may provide satisfactory removal. On the other hand, a more expensive coagulant might prove to be more effective and reduce overall chemical costs (i.e., copperas or ferric salts can be substituted for alum in the coagulation process).

Perhaps the most important consideration in designing water-treatment processes is to provide flexibility. The operator should have the means to change the point of application of certain chemicals. For example, chlorine feedlines are normally provided for pre-, intermediate-, and postchlorination. Multiple chemical feeders and storage tanks should be supplied so that various chemicals can be employed in the treatment process. Degradation of the raw water quality, or changes in costs of chemicals, may dictate a change in the type of coagulant or auxiliary chemicals used in coagulation. In the case of surface-water-treatment plants, it is desirable to provide space for the construction of additional pretreatment facilities. The flow in rivers may change due to construction of dams, channel improvements, or upstream water use. The quality of water changes due to man's alteration and occupation of the watershed. Concentrations of pollutants from disposal of municipal and industrial wastes and agricultural land runoff may increase. Lakes become more eutrophic.

The treatment plant described in Fig. 8-10 was built in the 1930s with provisions for iron and manganese removal and lime softening. Since then plant capacity has been expanded and the treatment process converted to iron and manganese removal only by increasing the aerator capacity, adding chlorine feed after the aerators, and using the settling basin as a detention tank. At present this plant is only capable of performing oxidation by aeration and chlorination for iron and manganese removal and disinfection. Finished water quality is excellent all year around because of the stable character of the well supply.

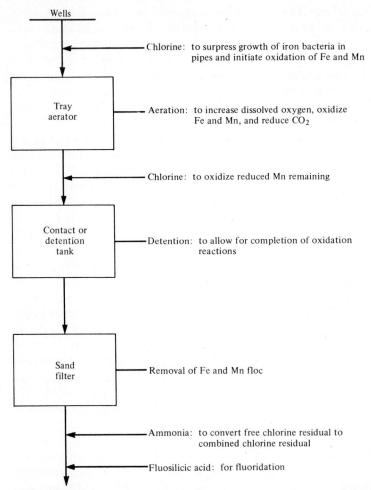

Wells

Chlorine: to surpress growth of iron bacteria in
pipes and initiate oxidation of Fe and Mn

Tray
aerator

Aeration: to increase dissolved oxygen, oxidize
Fe and Mn, and reduce CO_2

Chlorine: to oxidize reduced Mn remaining

Contact or
detention
tank

Detention: to allow for completion of oxidation
reactions

Sand
filter

Removal of Fe and Mn floc

Ammonia: to convert free chlorine residual to
combined chlorine residual

Fluosilicic acid: for fluoridation

FIGURE 8-10 Iron and manganese removal plant using aeration and chlorine for oxidation.

The treatment plant shown in Fig. 8-11 is a 1968 installation for partial softening and iron and manganese removal. Flocculator-clarifiers can be operated in series or parallel for flexibility in processing. The flow diagram illustrates split-treatment operation. An alternative method is single step coagulation using selective calcium softening by applying lime and alum. Both dry and liquid chemical feeders are provided with lines to each flocculator-clarifier. No pretreatment facilities are required in the treatment of the existing water quality; however, land area is available for expansion of the plant facilities (Fig. 8-12).

The treatment plant in Fig. 8-13 was constructed about 1950 as a conventional treatment plant for coagulation, sedimentation, and filtration. Since then the lake-water source has become increasingly eutrophic, and additional facilities

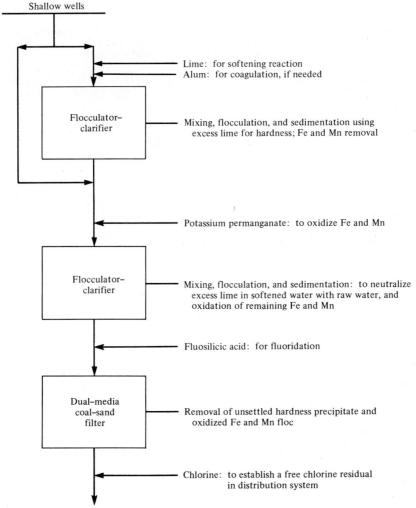

Shallow wells

Lime: for softening reaction
Alum: for coagulation, if needed

Flocculator–
clarifier

Mixing, flocculation, and sedimentation using
excess lime for hardness; Fe and Mn removal

Potassium permanganate: to oxidize Fe and Mn

Flocculator–
clarifier

Mixing, flocculation, and sedimentation: to neutralize
excess lime in softened water with raw water, and
oxidation of remaining Fe and Mn

Fluosilicic acid: for fluoridation

Dual–media
coal–sand
filter

Removal of unsettled hardness precipitate and
oxidized Fe and Mn floc

Chlorine: to establish a free chlorine residual
in distribution system

FIGURE 8-11. Plant using split treatment for partial softening and iron and manganese removal.

for taste and odor control have been provided. Prechlorination, activated carbon, chlorine dioxide, and various auxiliary chemicals for improved chemical treatment are now available to the operator during critical periods of poorer raw-water quality. During most of the year, the finished water is very palatable, but tastes and odors cannot be completely removed during spring and fall lake overturns.

The river-water processing scheme shown in Fig. 8-14 is a very complex, flexible treatment system for a turbid and polluted stream with highly variable quality. The plant has a complex of uncovered and covered mixing and settling basins with a variety of chemicals that can be fed at several points during water

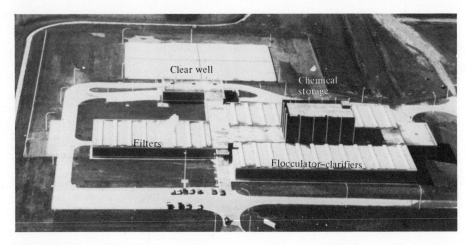

FIGURE 8-12 Aerial view of a water-treatment plant. The flow diagram for this water utility is given in Figure 8-11. (Courtesy Metropolitan Utilities District, Omaha, Neb.)

processing. The general treatment scheme consists of plain sedimentation for desilting; mixing followed by sedimentation, using chlorine regularly and coagulants as necessary; split treatment for partial softening and coagulation in flocculator-clarifiers; blending of the split flows followed by sedimentation; rapid sand filtration; and chemical additions for chlorine residual, pH adjustment, and control of scaling. Operation of this plant is varied from day to day, and season to season depending upon the raw-water quality. In general, the most troublesome time of the year is during spring runoff. Little is known about the degree to which refractory inorganic or organic substances can be transported through a complex treatment system. A river-water-treatment plant should have process depth to prevent any possibility of short-circuiting in pollution emergencies. The plant illustrated in Fig. 8-14 has a capability in emergencies to provide four coagulation–sedimentation steps in series prior to filtration.

8-9

WATER-PROCESSING SLUDGES

Chemical residues from water treatment historically were discharged to surface watercourses without processing. Water-quality regulations now require handling of these wastes to minimize environmental degradation. The method of eliminating residues is unique to each water utility because of the individuality of each treatment scheme and differing characteristics of the waste slurries generated. The two primary wastes are sludge from the settling basins following chemical coagulation, or precipitation softening, and wash water from back-

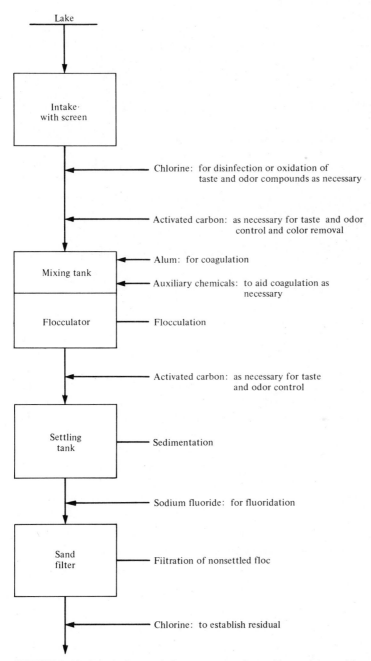

FIGURE 8-13 Chemical-coagulation treatment plant with special provisions for taste and odor control.

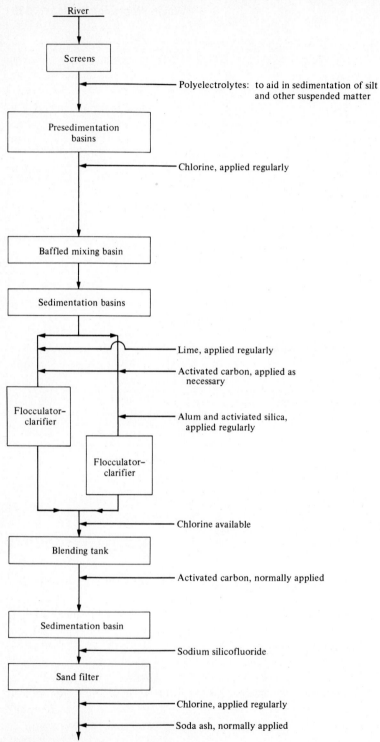

FIGURE 8-14 Chemical-coagulation and partial-softening treatment plant with provisions for handling high turbidity, tastes and odors, and color.

washing the filters. These residues, containing matter removed from the raw water and chemicals added during processing, are relatively nonputrescible and high in mineral content. Figure 8-15 is a generalized scheme of alternatives for disposing of water-processing wastes. Filter wash water is a very dilute waste discharged for only a few minutes once or twice a day for each filter bed. Therefore, a holding tank is necessary for flow equalization (i.e., to accept the intermittent peak discharges and release a relatively uniform effluent flow). Backwash water is settled in a clarifier–thickener with the overflow recycled to the plant influent; the underflow is withdrawn for further processing. Waste sludge from settling basins may also be concentrated further in a clarifier–thickener.

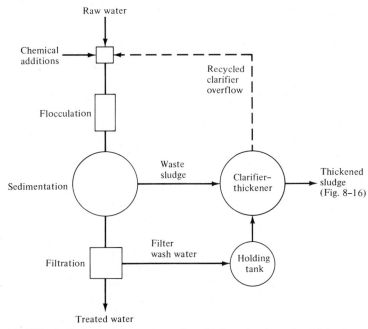

FIGURE 8-15 General flow scheme for withdrawal and gravity thickening of water-processing wastes.

Disposal alternatives for gravity-thickened precipitates are diagrammed in Fig. 8-16. In some municipalities, slurries can be drained into a sanitary sewer for processing with the municipal wastewater. Air drying is accomplished either in lagoons or on sand drying beds. Ponding is a popular method for dewatering, thickening, and temporary storage of waste solids. Where suitable land area is available, this technique is both inexpensive and efficient when compared to other methods. Drying beds are more costly and applicable only to small water plants. Centrifugation and pressure filtration are the two most successful mechanical dewatering processes. Recovery of alum and recalcination for lime are feasible in some locations. Often, economics do not warrant processing of

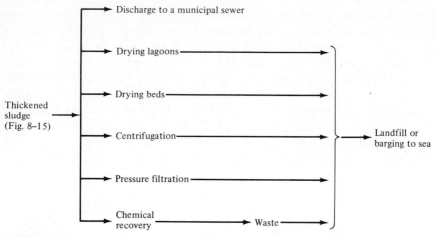

FIGURE 8-16 Common methods for dewatering and disposal of water-processing sludges.

sludge for chemical recovery. Residue from air drying, concentrated cake from dewatering, and waste after chemical recovery may be disposed of by either landfill or barging to sea.

REFERENCES

1. N. L. Nemerow, *Liquid Waste of Industry, Theories, Practices, and Treatment* (Reading, Mass.: Addison-Wesley Publishing Company, Inc., 1971).

2. "Regulation of Sewer Use," *WPCF Manual of Practice No. 3* (Washington, D.C.: Water Pollution Control Federation, 1975).

3. "Prevention and Correction of Excessive Infiltration and Inflow into Sewer Systems" (Washington, D.C.: Environmental Protection Agency, Water Quality Office, 1971).

4. R. Field and A. N. Tafuri, "Stormflow Pollution Control in the U.S.," in "Combined Sewer Overflow Seminar Papers," EPA/New York State Department of Conservation (Washington, D.C.: Environmental Protection Agency-670/2–73–077, 1973).

5. *Report of the Committee on Water Quality Criteria* (Washington, D.C.: Federal Water Pollution Control Administration, U.S. Department of the Interior, 1968; reprinted by EPA, 1972).

6. "Secondary Treatment Information," Environmental Protection Agency, *Federal Register* 38, no. 159 (August 17, 1973).

7. R. Eliassen and D. F. Coburn, "Versatility and Expandability of Pretreatment," *Proc. Am. Soc. Civil Engrs., J. Sanitary Eng. Div.* 95, no. SA2 (1969): 299–310.

8. "Disposal of Wastes from Water Treatment Plants," *AWWA Handbook No. 20101* (Denver, Colo.: American Water Works Association, 1970).

9
Physical-Treatment Processes

Physical-treatment methods are used in both water treatment and wastewater disposal. Except for preliminary steps, most physical processes are associated directly with chemical and biological operations. In water treatment, rapid sand filtration (a physical method) must be preceded by chemical coagulation. In wastewater processing, the physical procedures of mixing and sedimentation in activated sludge are related directly to the biology of the system.

Flow-Measuring Devices

Measurement of flow is essential for operation, process control, and record keeping of water- and wastewater-treatment plants.

9-1

MEASUREMENT OF WATER FLOW

Flow of water through pipes under pressure can be measured by mechanical, or differential head meters, such as a venturi meter, flow nozzle, or orifice meter. Drop in piezometric head between the undisturbed flow and the constriction in a

333

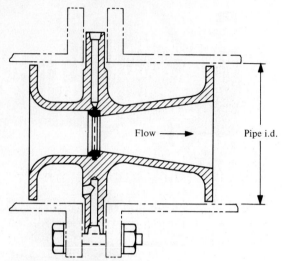

FIGURE 9-1 Cross section of a twin-throat venturi insert. (Courtesy Infilco Degremont Inc.)

differential head meter is a function of the flow rate. A venturi meter (Fig. 9-1), although more expensive than a nozzle or orifice meter, is preferred because of its lower head loss.

Piping changes upstream from a venturi meter produce nonuniform flow, causing inaccuracies in metering. In general, a straight length of pipe equal to 10–20 pipe diameters is recommended to minimize error from flow disturbances created by pipe fittings.

Flow in a venturi meter can be calculated using the formula

$$Q = CA_2\left[2g\left(\frac{P_1 - P_2}{w}\right)\right]^{1/2} \tag{9-1}$$

where Q = flow, cfs
C = discharge coefficient, commonly in the range of 0.90–0.98
A_2 = cross-sectional area of throat, ft^2
g = acceleration of gravity, ft/sec^2
$P_1 - P_2$ = differential pressure, psf
w = specific weight of water, pcf

9-2

MEASUREMENT OF WASTEWATER FLOW

Flow of wastewater through an open channel can be measured by a weir or a venturi-type flume, such as the Parshall flume[1] (see Fig. 9-2). Parshall flumes have the advantages of a lower head loss than a weir and a smooth hydraulic flow preventing deposition of solids.

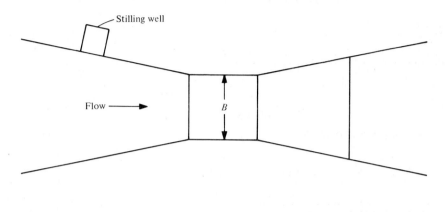

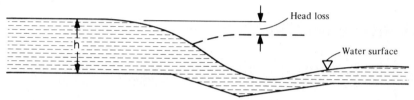

FIGURE 9-2 Parshall measuring flume.

Flow through a Parshall flume, with a throat width of at least 1 ft but less than 8 ft under free-flow conditions, can be calculated by the formula

$$Q = 4Bh^{1.522B^{0.026}}$$ (9-2)

where Q = flow, cfs
$\quad\quad B$ = throat width, ft
$\quad\quad h$ = upper head, ft

● EXAMPLE 9-1

What is the calculated water flow through a venturi meter with $5\frac{1}{16}$-in. throat diameter when the measured pressure differential is 160 in. of water? Assume $C = 0.95$ and a water temperature of 50°F.

○ Solution

Substituting in Eq. 9-1, we obtain

$$Q = 0.95 \times \frac{\pi(0.422)^2}{4} \left(64.4\,\frac{160 \times 62.4}{12 \times 62.4}\right)^{1/2}$$

$$= 3.9 \text{ cfs} = 1750 \text{ gpm}$$

● EXAMPLE 9-2

What is the calculated wastewater flow through a Parshall flume with a throat width of 2.0 ft at the maximum free-flow head of 2.5 ft?

○*Solution*
  By Eq. 9-2,

$$Q = 4 \times 2.0 \times 2.5^{1.522 \times 2.0^{0.026}}$$
$$= 33.4 \text{ cfs}$$

Screening Devices

River water and wastewater in sewers frequently contain large suspended floating debris varying in size from logs to small rags. These solids can clog and damage pumps or impede the hydraulic flow in open channels and pipes. Screening is the first step in treating water containing large solids.

9-3

WATER-INTAKE SCREENS

River-water intakes are commonly located in a protected area along the shore to minimize collection of floating debris. Lake water is withdrawn below the surface to preclude interference from floating materials.

Coarse screens of vertical steel bars having openings of 1–3 in. are employed to exclude large materials. The clear openings should have sufficient total area so that the velocity through them is less than 3 fps. These screens are available with mechanical rakes to take accumulated material from the bars. A coarse screen can be installed ahead of a finer one used to remove leaves, twigs, small fish, and so on. The travelling screen shown in Fig. 9-3 serves this purpose. Trays, sections constructed of wire mesh or slotted metal plates, generally have $\frac{3}{8}$-in. openings. As water passes through the upstream side of the screen, the solids are retained and elevated by the trays. As the trays rise into the head enclosure, the solids are removed by means of water sprays. The operation of a traveling screen is intermittent, being controlled by the water head differential across it resulting from clogging.

9-4

SCREENS IN WASTE TREATMENT

Coarse screens (bar racks) constructed of steel bars with clear openings not exceeding 2.5 in. are normally used to protect wastewater lift pumps. (Centrifugal pumps used for wastewater works, in sizes larger than 100 gpm, are capable of passing spheres at least 3 in. in diameter.) To facilitate cleaning, the bars are usually set in a channel inclined 30–45° with the horizontal.

FIGURE 9-3 Traveling water-intake screen. (Courtesy Water Quality Control Div., Envirex, a Rexnord Co.)

Mechanically cleaned medium screens (Fig. 9-4) with clear bar openings of $\frac{5}{8}$ in. to $1\frac{3}{4}$ in. are commonly used instead of a manually cleaned coarse screen. The maximum velocity through the openings should not exceed 2.5 fps.

Collected screenings are generally hauled away for disposal by land burial or incineration, although they can be shredded in a comminutor and returned to the wastewater flow for removal in a treatment plant.

Fine screens with openings as small as $\frac{1}{32}$ in. have been used in wastewater treatment, but because of their high installation and operation cost are rarely employed in handling municipal waste. Fine screens are used to take out suspended and settleable solids in pretreating certain industrial waste streams. Typical applications are the removal of coarse solids from cannery and packing-house wastewaters.

Cleaning rake

FIGURE 9-4 Mechanically cleaned bar screen for wastewater treatment. (Courtesy Water Quality Control Div., Envirex, a Rexnord Co.)

9-5

COMMINUTORS

Comminutors or shredding devices (Figs. 9-5 and 9-6) chop solids passing through bar screens to about $\frac{1}{4}$ to $\frac{3}{8}$ in. in size. They are installed directly in the wastewater flow channel and provided with a bypass so that the length containing the unit can be isolated and drained for machine maintenance. In small waste-treatment plants, comminutors may be installed without being preceded by medium screens. The bypass channel around the comminutor contains a hand-cleaned medium screen (Fig. 9-5).

Mixing and Agitation

Mixing is blending constituents to a specified state of uniformity. *Agitation* has as a major function the production of fluid turbulence. A primary example of agitation is the promotion of floc growth in suspensions (flocculation). Mixing processes are employed to disperse a variety of chemicals and gases in fluids.

FIGURE 9-5 The Griductor comminutor has a stationary semicircular screen grid and rotating circular cutter discs, and is installed in a wastewater flow channel provided with an emergency bypass containing a fixed bar screen. (Courtesy Infilco Degremont Inc.)

9-6

CONTINUOUS MIXING

Continuous-flow systems are employed almost exclusively in water-processing operations. Mixing characteristics of these systems are particularly important. Unfortunately, their nature is often complex and difficult to describe in exact mathematical terms. Simplified mixing models are therefore used to aid in approximating their characteristics. Knowledge of three basic mixing models will help readers to understand the design of continuous reactors.[2,3]

Plug Flow

Plug flow is defined as flow in which individual particles of feed pass through the reaction vessel in the same sequence they entered.[2] In addition, there is no intermixing or interaction between the particles. Each particle is retained in the reaction vessel for a period equal to the theoretical retention time:

$$t_0 = \frac{V}{Q}$$

(9-3)

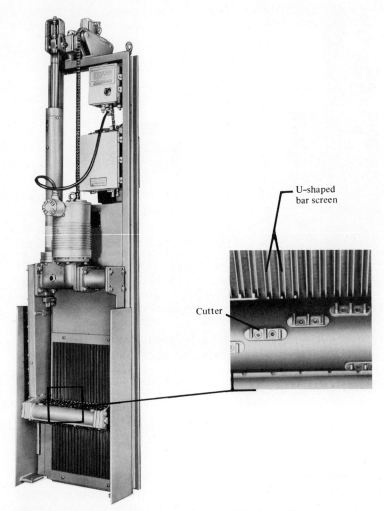

FIGURE 9-6 The Barminutor screening and comminuting machine has a high-speed rotating cutter that travels up and down on a bar screen to shred solids that accumulate on the U-shaped bars. (CHICAGO PUMP Products, courtesy Environmental Equipment Div., FMC Corp.)

where t_0 = theoretical retention time
$\quad\quad V$ = volume of the reactor
$\quad\quad Q$ = flow rate

Long and narrow reaction vessels (tubular reactors, for example) are designed on the basis of the plug-flow model. In this case, the conversion (degree of completion of the reaction) can be estimated from batch data provided such factors as temperature change or volume changes are known. For any reactor, maximum conversion will be indicated if the plug-flow model is used, because it is assumed that all feed particles are retained for exactly the theoretical retention time. In reality, flow increments are usually held for variable amounts of time—some for less and some for more than the theoretical time.

Complete Mixing

A second basic model assumes *complete mixing,* defined as a flow in which the feed particles become completely intermixed in the reactor and lose their identity.[2,3] The effluent and content of the reaction vessel are exactly alike and uniform. This condition is approximately equivalent to that encountered in a rapid mixing tank. Since the contents are uniform, reaction can only proceed at a rate corresponding to the constituent concentrations.

A mathematical statement describing complete mixing is derived by considering a mixing reactor having an influent Q containing a reactant A at concentration c_0, an effluent Q containing the reactant at final concentration c_t, and a holding volume equal to V. For this steady continuous-flow system, the quantity of material converted per unit time is

$$Q(c_0 - c_t) \tag{9-4}$$

If a single, first-order, homogeneous reaction of the type $A \rightarrow$ products is considered, an expression for rate takes the form

$$\frac{dc_A}{dt} = Kc_A \tag{9-5}$$

where c_A = concentration of the reactant
K = rate constant

For complete mixing, the concentration is uniform and equal to c_t, and the amount of material converted per unit time must be VKc_A, and

$$Q(c_0 - c_t) = V\left(\frac{dc}{dt}\right)_{c_t}$$

The theoretical retention time is therefore

$$t_0 = \frac{V}{Q} = \frac{c_0 - c_t}{(dc/dt)_{c_t}} \tag{9-6}$$

Equation 9-6 is basic for the complete-mixing model. The term dc/dt can be evaluated by taking the slope of the batch curve at a point corresponding to c_t. Figure 9-7 illustrates a typical batch curve. The theoretical retention time t_0 is found by substituting for c_t in Eq. 9-6 as a fraction of c_0 and then substituting for dc/dt the value of $1/c_0(dc/dt)$ found from Fig. 9-7 multiplied by c_0. Considerable information is given in the literature on graphical and analytical methods applicable to problems in complete mixing.[4,5] An example will illustrate the determination of retention time for this model.

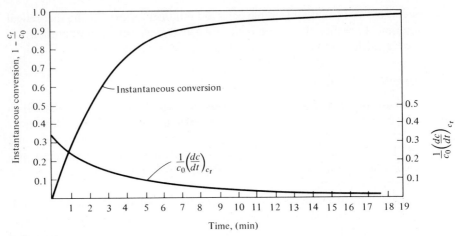

FIGURE 9-7 Typical batch reaction data.

●EXAMPLE 9-3

For an instantaneous conversion of 80% with a reaction whose rate is indicated by Fig. 9-7, find the required holding time.

○**Solution**

For $1 - (c_t/c_0) = 0.80$, $c_t/c_0 = 0.2$ and $c_t = 0.2c_0$. From Fig. 9-7,

$$\frac{1}{c_0}\left(\frac{dc}{dt}\right) = 0.11$$

$$t_0 = \frac{c_0(1 - 0.2)}{0.11c_0} = 7.3 \text{ minutes}$$

Complete Mixing with Zero Intermixing

Complete mixing with zero intermixing model is essentially a combination of the two previous models. In this case feed particles are immediately and uniformly displaced in the reactor but do not intermix or interact. Feed particles leave the reactor vessel in a purely random fashion.

Because the particles do not interact, elements of the flow behave similarly to individual batch reactors.[3] The difference between this condition and ideal plug flow is that each particle remains in the reactor for a different period of time; thus each batch reaction reaches a different degree of completion before leaving the vessel. The effluent is then composed of a series of increments having varied degrees of reaction completion.

The relative probability P that one feed particle will remain in the reactor longer than another if the particles are uniformly distributed is a known mathematical function. Figure 9-8 indicates the probability that a portion of the inflow will be retained in the system at time t and leave before time $t + dt$ for various values of t/t_0 and for differing numbers of vessels in series. Area under the curves represents the feed slug.

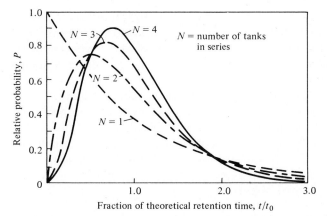

FIGURE 9-8 Probability curves for zero intermixing model. [Source: Adapted from Figure 2 T. R. Camp, "Flocculation," *Trans. Am. Soc. Civil Engrs.* 120 (1955) : 8.]

Considering that each feed element acts as an individual batch reactor, its conversion depends on its residence time. Gilliland and Mason[4] have shown that average conversion of the effluent is equal to the sum of the conversions in each increment of feed. Mathematically, this is stated as

$$\text{average conversion} = \left(1 - \frac{c_t}{c_0}\right) = \frac{1}{t_0}\int_0^\infty P\left(1 - \frac{c_t}{c_0}\right)dt \qquad (9\text{-}7)$$

where the terms are as previously defined. Average conversion may be calculated by integrating the batch-conversion data with the relative probability function.[5] An approximate solution can be had by determining values of $P(1 - c_t/c_0)\,\Delta t$ at various intervals of time and then summing. In doing this, values of P and $(1 - c_t/c_0)$ are obtained from data such as presented on Figs. 9-7 and 9-8.

When data are available on flow rates and residence-time distribution of the feed particles or can be assumed, upper and lower limits of conversion are predictable. For simple homogeneous reactions, individual well-stirred tanks will yield conversions in close agreement with those predicted by the complete mixing model. In long tubular reactors, conversions approach those indicated by the plug-flow model. The zero-intermixing model is theoretically limited in application to zero-order reactions and to first-order single homogeneous reactions.[2]

9-7

RAPID-MIXING DEVICES

Various devices have been employed in rapid-mixing operations. Common equipment includes pumps, venturi flumes, air jets, and rotating impellers (paddles, turbines, propellers). Of those listed, the vertical shaft, turbine-type, impeller is the most widely used rapid-mixing device.[7] It is composed of a vertical shaft driven by a motor on which one or more straight or curved turbine blades are mounted to provide both horizontal and vertical mixing. Flow through mixing tanks is usually from bottom to top. Mixing-tank diameters commonly range from 3 to 10 ft. Detention periods are usually less than 1 minute, although longer times have been used.

Baffles are used in tanks with rotating impellers to reduce vortexing about the impeller shaft. This vortexing action hinders mixing operation and complicates calculation of impeller power consumption. Effective baffling is accomplished by placing vertical strips along the periphery of the tank, usually no more than four being required. These strips tend to disrupt the rotational flow pattern by deflecting the liquid inward toward the impeller shaft. Considerable information on baffling is offered in several references at the end of the chapter.[5,7]

The flow pattern generated by an impeller affects the results of the mixing operation being performed. This pattern is a function of the size, shape, and speed of the impeller, baffling, and nature of the mixing tank.[3] Turbulent flows readily promote mixing through the erratic displacement of fluid particles. The turbulence produced in mixing vessels is principally derived from contact between low- and high-velocity flow streams.

There is scant information available to predict mixing process performance on the basis of specific operations. It has therefore been necessary to conduct model or pilot-plant studies of processes of interest to find the optimum conditions for carrying out these processes. Once this has been done, large-scale or prototype operation can be designed using principles of similitude.[3]

9-8

FLOCCULATION

Flocculation is essentially an operation designed to force agitation in the fluid and induce coagulation. In this manner, very small suspended particles collide and agglomerate into larger heavier particles or flocs, and settle out. Flocculation is a principal mechanism in removing turbidity from water. Floc growth depends on two factors—intermolecular chemical forces and physical action induced by agitation. Chemical reactions are considered in Chapter 10.

Various processes have been employed to accomplish flocculation. Foremost of these are diffused air, baffles, transverse or parallel shaft mixers, vertical turbine mixers, and walking-beam-type mixers. Diffused-air flocculators are costly to operate and do not produce results equal in quality to those obtained with mechanical devices. For this reason they are not extensively used today. Baffled flocculators were very common at one time but their use is decreasing because of inflexibility, high head loss, and construction cost. The vertical-turbine mixer has been used successfully in water-softening operations but appears to have an adverse affect on the more fragile alum floc.

The most common type of flocculator used today is the paddle flocculator. Paddle wheels can be mounted on vertical or horizontal shafts, although for a series of units, vertically mounted paddle wheels are more expensive.[7] The horizontally mounted units may be located transverse to or parallel with the flow. Orientation does not seem to have any appreciable effect on results. A paddle flocculator consists essentially of a shaft with protruding steel arms on which are mounted a number of wooden or metal blades. The shaft slowly rotates (on the order of 60–100 revolutions per hour), causing a gentle agitation of the fluid and collision of the floc particles with one another and with suspended matter in the raw water. The end result is promotion of floc growth so that finely divided suspended solids and colloidal particles can be removed by sedimentation. Figure 9-9 illustrates a plant-sequence rapid mixing of chemicals, promotion of floc growth, and sedimentation of the flocculent particles.

As previously stated, flocculation is the formation of suspended flocs which lend themselves to removal by sedimentation or filtration. Essentially the process begins by dissolving chemicals in a rapid-mix operation such as indicated in Fig. 9-9. Initial collisions between colloidal particles occur from Brownian motion and contacts brought about due to differences in settling velocities of heavier and lighter particles. In a quiescent vessel, floc growth is extremely slow. A period of gentle agitation or turbulent mixing is therefore a necessary adjunct to rapid floc growth and is extremely important in the design of coagulation systems.

Flocculation is directly proportional to the velocity gradients established in the water being treated. The stirring action derived from flocculating mechanisms is responsible for establishing these gradients and is therefore fundamental to the process. The number of contacts between particles in a unit time has been shown to vary with the number and size of particles.[8]

The absolute velocity gradient at a point in a moving fluid du/dy is equal to the square root of the power loss through fluid shear per unit volume of fluid divided by the dynamic viscosity of the fluid. Mean velocity gradient $du/dy = G$ is given by

$$G = \sqrt{\frac{W}{\mu}} \tag{9-8}$$

where μ = dynamic viscosity

W = dissipation function or rate of power dissipation per unit of volume P/V (P is the power input and V the tank volume)

Considering that rate of floc formation is directly proportional to the mean velocity gradient, the time of floc formation should decrease with increasing values of G. There is, however, a maximum size of floc particle associated with each velocity gradient and the velocity gradient should therefore be selected with this in mind. In other words, the desired floc size determines the velocity gradient to be used.

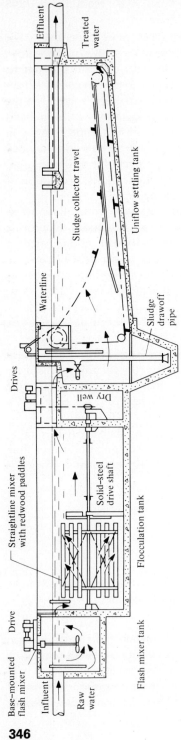

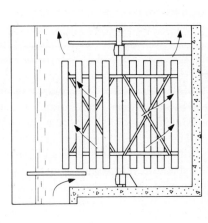

Enlarged view of paddle flocculator

FIGURE 9-9 Rapid mixing, flocculation, and sedimentation. (LINK-BELT Product, courtesy Environmental Equipment Div., FMC Corp.)

At a given point within a fluid, the shearing stress τ is equal to velocity gradient G multiplied by the dynamic viscosity μ, or

$$\tau = \mu G \tag{9-9}$$

It is obvious, therefore, that as velocity gradient increases, shear stress also increases. As the floc particles grow in size they become weaker and subject to being torn apart. One can therefore state in general that if small floc particles are desired, high-velocity gradients should be used, whereas to form large floc particles, lower-velocity gradients are required. Values of G between 10 fps/ft and 75 fps/ft have been found to promote floc growth without destruction of the floc particle.[9] Detention periods less than 10 minutes should be avoided. For small velocity gradients, detention times of 0.5 hour or more are required. Practice has shown optimum conditions are obtained for G values ranging from 30 to 60 fps/ft. Fair and Geyer have stated that, for satisfactory performance, the dimensionless product of velocity gradient and detention time may range within the limits 10^4 to 10^5.[9]

Mechanical mixers are usually the revolving-paddle type. When rotor paddles alone are used, the resistance to rotation of water with the paddles is provided by drag on the tank walls. Where stator blades are also used, these further resist rotation of the fluid. In some instances, very wide paddles have been used and water is moved along with the paddle at a velocity approximately equal to it. Mixing occurs only along the walls and outside edges of the paddles in such cases. Beam[10] recommends that in basins where stator blades are not used, paddles should not exceed 15–20% of the cross-sectional area of the basin to prevent rolling of the water. At paddle areas equal to 25% of the cross-sectional area, significant rotation of the water is to be expected. Drobny[11] indicates the manner in which paddle characteristics affect flocculation for a given input of power.

For paddle flocculators, the useful power input is directly related to the drag of the paddles.[9] The general equation for drag can be written as

$$F = \frac{C_D A \rho v^2}{2} \tag{9-10}$$

where F = drag force, lb
  C_D = dimensionless drag coefficient for plates moving with faces normal to direction of motion
  A = cross-sectional area of the paddles, ft^2
  v = relative velocity between paddles and fluid, fps

The power input may be computed as the product of drag force and velocity, stated as

$$P = Fv = \frac{C_D A \rho v^3}{2} \tag{9-11}$$

Substituting in Eq. 9-8, the mean velocity gradient G becomes[9]

$$G = \sqrt{\frac{C_D A \rho v^3}{2V\mu}} \tag{9-12}$$

where V is the volume of the flocculating basin in cubic feet. Common peripheral speeds of paddles are within the range 3–0.3 fps.

● **EXAMPLE 9-4**

A water-treatment plant is designed to process 25 mgd. The flocculator is 100 ft long, 50 ft wide, and 16 ft deep. Revolving paddles are attached to four horizontal shafts which rotate at 1.5 rpm. Each shaft supports four paddles which are 8 in. wide, 48 ft long, and centered 6 ft from the shaft. Assume the mean water velocity is 30% of the paddle velocity and that C_D equals 1.9. Find: (a) the difference in velocity between paddles and water; (b) the value of G; and (c) the retention time.

○ **Solution**

    a Rotational speed

$$V_p = \frac{2\pi r n}{60}$$

    where V_p = velocity of paddle blades, fps
              n = number of revolutions per minute
              r = distance from shaft to center of the paddle, ft

$$V_p = \frac{2\pi \times 6 \times 1.5}{60} = 0.94 \text{ fps}$$

    The velocity differential between paddles and fluid is therefore

    $0.70 \times 0.94 = 0.66$ fps

    b The value of G may be found by using Eqs. 9-11 and 9-12. Total power input is first determined as

$$P = C_D A \frac{\gamma}{g} \frac{v^3}{2}$$

    where $C_D = 1.9$
            A = paddle area, $4 \times 4 \times 48 \times \frac{8}{12} = 512 \text{ ft}^2$
            v = velocity differential of 0.66 fps

    Then

$$P = 1.9 \times 512 \times \frac{62.4}{64.4} \times (0.66)^3$$

$$= 271 \text{ ft-lb/sec}$$

    and

$$G = \sqrt{\frac{P}{V\mu}}$$

where V = tank volume = $100 \times 50 \times 16 = 80,000$ ft^3
μ = dynamic viscosity

For an assumed temperature of 50°F, $\mu = 2.73 \times 10^{-5}$ (lb-force)(sec)/ft^2 (see Table 9-1). Then

$$G = \sqrt{\frac{271 \times 10^5}{8 \times 10^4 \times 2.73}}$$

$$= 11.1 \text{ fps/ft}$$

TABLE 9-1

VALUES OF DYNAMIC AND KINETIC VISCOSITY OF WATER
(based on data from Smithsonian tables)

Temperature (°F)	μ (centipoises)	μ (lb-sec/ft^2 $\times 10^{-4}$)	ν (ft^2/sec $\times 10^{-5}$)	ν (cm^2/sec)
32	1.792	0.374	1.93	0.0179
39.2	1.567	0.327	1.69	0.0157
40	1.546	0.323	1.67	0.0155
50	1.308	0.273	1.41	0.0131
60	1.124	0.235	1.21	0.0113
70	1.003	0.209	1.08	0.0100
80	0.861	0.180	0.929	0.00863
90	0.766	0.160	0.828	0.00769
100	0.684	0.143	0.741	0.00688
110	0.617	0.129	0.670	0.00623
120	0.560	0.117	0.610	0.00567
130	0.511	0.107	0.559	0.00519
140	0.469	0.0979	0.513	0.00477
150	0.432	0.0905	0.475	0.00442
160	0.400	0.0835	0.440	0.00409
170	0.372	0.0777	0.411	0.00382
180	0.347	0.0725	0.385	0.00358
190	0.325	0.0679	0.362	0.00336
200	0.305	0.0637	0.341	0.00317

c Time of flocculation is found by dividing the tank volume in gallons by the flow value:

$$t = \frac{80,000 \times 7.48 \times 24 \times 60}{25 \times 10^6}$$

$$= 34.5 \text{ minutes}$$

Dimensionless product Gt therefore $= 11.1 \times 34.5 \times 60 = 2.3 \times 10^4$, which is within the range of satisfactory performance.

A problem inherent in the design of flocculation basins, reaction tanks, and other vessels designed to hold a fluid for a specified theoretical retention time is that of short-circuiting. This occurs wherever a part of the influent is passed through a given system without remaining for a satisfactory period of time. (See also Sec. 9-14.)

Short-circuiting may be measured using tracers such as dyes, salts, or radioactive materials. The tracer can be injected at the entrance and measured at the exit at different time intervals. If the mixing disperses the tracer instantaneously, concentration of the tracer at the outlet may be computed as[6]

$$\frac{C}{C_0} = e^{-t/T} \qquad\qquad (9\text{-}13)$$

where C = concentration of tracer at the outlet at time t

$\quad\quad C_0$ = initial concentration of the tracer if it is instantaneously dispersed

$\quad\quad T$ = retention period of the tank

$\quad\quad e$ = base of Napierian logarithms

Figure 9-10 shows a plot of Eq. 9-13 and an experimental curve from a model of a cubical tank with a moderate degree of mixing. In the theoretical curve, 40% of the tracer slug leaves the tank in less than one-half the retention time, and 22% in less than one-fourth the retention time.

To minimize the effects of short-circuiting, several tanks can be placed in series to give the fluid which passes rapidly through the initial tank the opportunity to remain in the succeeding tank or tanks for a longer period. The follow-

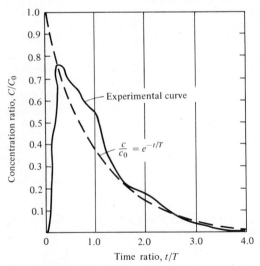

FIGURE 9-10 Tracer solution concentration. [Source: Adapted from Figure 1, T. R. Camp, "Flocculation," *Trans. Am. Soc. Civil Engrs.* 120 (1955) : 7.]

ing equation for the instantaneous dispersion curve for n tanks of equal size in series has been developed by Camp and Stein[8]:

$$\frac{C}{C_0} = \frac{n^n}{(n-1)!}\left(\frac{t}{T}\right)^{n-1} e^{-nt/T} \tag{9-14}$$

where C_0 and T are based on the total volume of n tanks in series.

9-9

RECOMMENDED STANDARDS FOR WATER WORKS

The 1968 *Recommended Standards for Water Works, Great Lakes–Upper Mississippi River Board of State Sanitary Engineers*[12] (Ill., Ind., Iowa, Mich., Minn., Mo., N.Y., Ohio, Pa., and Wis.) recommends that flash or quick mixing for rapid dispersion of chemicals throughout the water to be treated be performed with mechanical mixing devices. The mixing detention period should not be less than 30 seconds.

The following recommendations are presented in the *Standards* with regard to flocculation:

1 Inlet and outlet design should prevent short-circuiting and destruction of floc.
2 Minimum flow-through velocity must not be less than 0.5 or greater than 1.5 fpm, with a detention time for floc formation of at least 30 minutes.
3 Agitators should be driven by variable-speed drives with the peripheral speed of paddles ranging from 0.5 to 2.0 fps.
4 Flocculation and sedimentation basins must be placed as close together as possible. The velocity of flocculated water through pipes or conduits to settling basins should not be less than 0.5 or greater than 1.5 fps. Allowances must be made to minimize turbulence at bends and changes in direction.
5 Baffling is permissible to provide for flocculation in small plants only after consultation with the reviewing authority.

Sedimentation

Sedimentation is the removal of solid particles from a suspension through gravity settling. Other terms used to describe this process are *clarification* and *thickening*.

In water treatment, sedimentation is used to remove particulate matter, flocculated impurities and precipitates which are formed in operations such as water softening or iron removal. In wastewater-treatment operations, sedimentation is used to remove both inorganic and organic materials which are settleable or which have been converted to settleable solids. Most modern sedimentation basins are operated on a continuous-flow basis.

Settling operations may be classified approximately as falling into four separate categories: type I, type II, zone, and compression, all dependent on the concentration of the suspension and the character of the particles.[3] Types I and II clarifications both deal with dilute suspensions, the difference being that *type I* consists of essentially discrete particles while *type II* deals with flocculent materials. *Zone settling* describes a mass-settling process in intermediate-concentration suspensions of flocculent materials, and *compression* results when the concentration increases to the point where particles are in physical contact with one another and supported partly by the compacting mass. The following section presents theoretical aspects of various types of sedimentation and indicates how this knowledge can be used in design.

9-10

TYPE I SEDIMENTATION

Type I sedimentation is concerned with removal of nonflocculating discrete particles in a dilute suspension. Under such circumstances the settling may be said to be unhindered and a function only of fluid properties and characteristics of the particle. Settling of heavy, inert materials would be an example of type I sedimentation.

A discrete particle (one that retains its individual characteristics) placed in a quiescent fluid will accelerate until fluid drag reaches equilibrium with the driving force acting on the particle. At the moment of equilibrium the particle begins to settle at a uniform velocity. Since this condition of equilibrium is rapidly reached for the general conditions encountered in practice, terminal settling velocity is of particular importance in sedimentation studies.

The driving force, which is the net effect of the particle weight acting downward and the buoyant force of the fluid acting upward, is given by

$$F = (\gamma_s - \gamma)V \tag{9-15}$$

where F = driving force
γ_s, γ = specific weights of the particle and fluid, respectively
V = volume of the particle

The drag force acting on a particle is a function of the fluid density and viscosity, settling velocity of the particle v_s, and a characteristic dimension of the particle d. A dimensionally derived relationship for fluid drag can be shown to take the form

$$F_D = \frac{C_D A \rho v_s^2}{2} \tag{9-16}$$

where F_D = drag force
C_D = Newton's drag coefficient
A = projected area of the particle in the direction of motion[13]

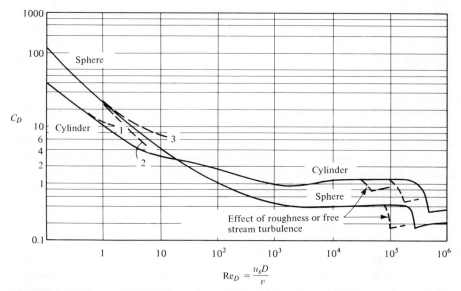

FIGURE 9-11 Drag coefficients for spheres and infinite circular cylinders: curve 1: Lamb's solution for cylinder: curve II: Stokes' solution for sphere; curve III: Oscen's solution for sphere. [Source: R. M. Olson, *Essentials of Engineering Fluid Mechanics* (New York: IEP, A Dun-Donnelley Publisher, 1961), p. 261.]

The drag coefficient is a function of Reynolds number as shown in Fig. 9-11. For Reynolds numbers less than about 1, $C_D = 24/\text{Re}$; for Reynolds numbers between 1 and 10^4, C_D can be approximated by[9]

$$C_D = \frac{24}{\text{Re}} + \frac{3}{\sqrt{\text{Re}}} + 0.34 \qquad (9\text{-}17)$$

Equating driving and drag forces, which must be equivalent for equilibrium conditions,

$$(\gamma_s - \gamma)V = \frac{C_D A \rho v_s^2}{2}$$

A rearrangement of this equation yields

$$v_s = \sqrt{\frac{2(\gamma_s - \gamma)V}{C_D A \rho}} \qquad (9\text{-}18)$$

which is the expression for settling velocity of a discrete particle. If the particles are assumed to be spherical in shape, $V = \pi d^3/6$ and $A = \pi d^2/4$. Making these substitutions and setting $\gamma = \rho g$ and $\gamma_s = \rho_s g$, we obtain

$$v_s = \sqrt{\frac{4}{3}\frac{g}{C_D}\frac{\rho_s - \rho}{\rho}d} \qquad (9\text{-}19)$$

when the Reynolds number is less than 1, substitution of 24/Re for C_D transforms Eq. 9-19 to the form

$$v_s = \frac{g}{18} \frac{\rho_s - \rho}{\mu} d^2 \qquad (9\text{-}20)$$

which is known as *Stokes' law*.

Discrete particles settle at a constant velocity, as indicated, provided that the fluid temperature does not change. A knowledge of this velocity is fundamental in the evaluation of the performance of a sedimentation basin. To illustrate this, the operation of an ideal sedimentation basin will be discussed.[14] Suppose that Fig. 9-12 represents an idealized rectangular continuous horizontal-flow basin which may be divided into four general zones[14]: inlet, settling, outlet,

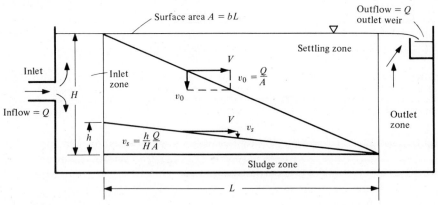

FIGURE 9-12 Sketch of an ideal rectangular sedimentation basin, indicating the settling of discrete particles.

and sludge zone. The *inlet zone* is a region in which it is assumed that the incoming flow is uniformly distributed over the cross section of the tank in such a manner that flow through the settling zone follows horizontal paths. In the *settling zone* it is considered that settling occurs in exactly the same manner as it would under tranquil conditions. It is assumed that flow through this region is steady and that the concentration of each size particle is uniform throughout the cross section normal to the flow direction. In the *outlet zone*, clarified effluent is collected and discharged through an outlet weir. The *sludge zone* is provided for the collection of sludge below the settling zone. It is assumed that particles reaching this zone are effectively removed from the suspension. Where mechanical equipment continually removes the sludge, this zone may be neglected for all practical purposes.

Consider first a uniform dilute suspension of discrete particles which have settling velocities equal to v_s. The path taken by a particle originally at height h is described by the vector sum of flow velocity V and settling velocity of the particle

v_s, as shown on Fig. 9-12. It is apparent that a particle initially at height h will just be removed by the time it traverses the settling zone. Particles initially at heights less than h will also be removed, while those initially at a greater height will not reach the bottom before they reach the outlet zone. If we now consider a particle having a settling velocity v_0, it can be seen from the figure that 100% removal would be effected if all particles in the suspension had settling velocities at least equal to v_0. The velocity is of particular interest as a reference, since it can be said that all particles in a suspension with settling velocities $\geq v_0$ will be removed, while only part of those having settling velocities $< v_0$ will be removed.

From the geometry of Fig. 9-12 it can be seen that if the area of the triangle having legs H and L represents 100% removal of particles, then the removal ratio of particles having a settling velocity equal to v_s will be h/H. Since depth equals the product of settling velocity and retention time t_0,

$$\frac{h}{H} = \frac{v_s t_0}{v_0 t_0} = \frac{v_s}{v_0}$$

Thus the proportion of particles of a given size that are removed in a horizontal flow tank is[14]

$$\frac{v_s}{v_0} = \frac{v_s}{Q/A} \tag{9-21}$$

where Q = rate of flow
   A = surface area of the settling zone

Consider now modification of the tank in Fig. 9-12 by the addition of a tray or false bottom at depth $H/2$ of the tank. The conditions of flow through the tank (Q and V remain as before) and other dimensions are unchanged. Settling velocities also remain the same, but it can be seen that the removal ratio v_s/v_0 is now doubled. The maximum depth through which the particles must settle is reduced by one-half, but the effective floor area of the tank is doubled.

It can be demonstrated that this increased removal is not a function of the depth change but only of the change in floor area. Consider now a tank of depth $H/2$ with a flow Q. For these conditions to prevail, the horizontal flow velocity must now be equal to $2V$, therefore, the particle trajectories will be one-half those shown in Fig. 9-12, since the settling velocities are unchanged. The removal ratio for particles having a settling velocity v_s will however still remain equal to v_s/v_0. This indicates that the rate of removal for a given discharge in an ideal sedimentation tank is entirely independent of depth but is directly related to its floor area. Thus the tank depth may be varied with no effect on removal ratio. Small depths are economically desirable. Other governing factors are space requirements for sludge-removal equipment, and control of the horizontal flow velocity to avoid scouring deposited sludge.

Overflow rate of a settling tank is defined as the settling velocity v_0 of particles which are just removed in an ideal basin if they enter at the surface, or as the discharge per unit surface or floor area of the tank. Overflow rates ranging from 500 to 1000 gpd/ft^2 are common for sedimentation of wastewater. Where silt and clay are to be removed by plain sedimentation, maximum surface loadings as low as 150 gpd/ft^2 of tank surface are not unusual.

The efficiency of a sedimentation basin indicates the overall percentage removal of suspended matter at a given overflow rate v_0. For an ideal basin, this may be formulated by using Eq. 9-21 and referring to Fig. 9-13. For a particular

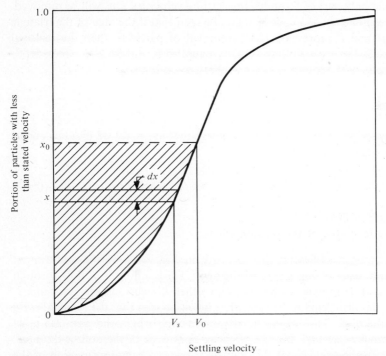

Settling velocity

FIGURE 9-13 Discrete particle velocity cumulative frequency distribution. [Source: After Figure 9, T. R. Camp, "Sedimentation and the Design of Settling Tanks," *Trans. Am. Soc. Civil Engrs.* 111 (1946) : 909.]

clarification rate v_0, it has already been shown that those particles having settling velocities $\geq v_0$ will be completely removed. Of the total number of particles in the suspension, the percentage removed can then be stated as $1 - x_0$, where x_0 represents the portion of particles with a settling velocity $\leq v_0$. For each size particle with a settling velocity $v_s \leq v_0$ it has been indicated that the portion removed would be equal to v_s/v_0. Thus when considering various particle sizes in this category, the percentage removal of these particles is given by

$$\int_0^{x_0} \frac{v_s}{v_0}\, dx \qquad (9\text{-}22)$$

and the overall removal will therefore be

$$R = (1 - x_0) + \frac{1}{v_0} \int_0^{x_o} v_s \, dx \tag{9-23}$$

The second term in the equation can be determined by graphical integration of a settling-analysis curve, such as the shaded portion of Fig. 9-13.

● **EXAMPLE 9-5**
In lime-soda water softening, crystals of calcium carbonate formed in the chemical reaction have a specific gravity of 2.7 and particle size of 15–20 μm. These crystals agglomerate and settle in clusters possessing a specific gravity of about 1.2. Assuming clusters with an average diameter of 0.05 mm, calculate their settling velocity using Stokes' law and express the answer in terms of mm/sec, fps, and gpd/ft^2 (overflow rate).

○**Solution**
   At 50°F, $\rho = 1.00$ and $\mu = 1.308$ (Table 9-1). Substituting in Eq. 9-20, we obtain

$$v_s = \frac{32.2 \times 305}{18} \frac{1.2 - 1.0}{1.308} (0.05)^2$$

$$= 0.21 \text{ mm/sec} = 0.00068 \text{ fps}$$

$$= \frac{Q}{A} = 0.00068 \text{ ft/sec} \times \text{ft}^2/\text{ft}^2 \times 7.48 \text{ gal/ft}^3 \times 86{,}400 \text{ sec/day}$$

$$= 400 \text{ gpd/ft}^2$$

9-11

TYPE II SEDIMENTATION

The settling properties of dilute suspensions of flocculating particles differ from those of nonflocculating particles in that the flocculating properties of the suspension must be considered along with the settling characteristics of the particles.[3] In this case, heavier particles having large settling velocities overtake and coalesce with smaller, lighter particles to form still larger particles with increased rates of subsidence. The opportunity for particle contact increases as the depth of the settling vessel increases. As a result, removal of suspended matter depends not only on clarification rate but on depth as well. This is the important difference between type I and type II clarification. Settling of wastewater solids in the upper levels of a primary clarifier is typical of type II sedimentation.

   Unfortunately, there is no adequate mathematical relationship to determine the effect of flocculation on sedimentation. Settling-column analyses are required to evaluate this effect.[15] A standard method for settling-column analysis is to place the test suspension in a column, which includes sampling ports at various depths, and allow it to settle in a quiescent manner. Samples are withdrawn at various selected time intervals from different depths. The concentration of

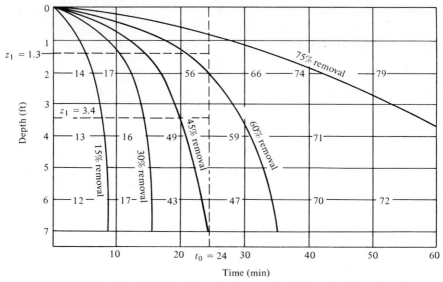

FIGURE 9-14 Settling trajectories for a flocculent suspension. (Note: Numbers plotted represent fraction removed based on settling column analysis. Percentage removal curves are interpolated using these data.)

particles is determined from this portion and the information used to compute the percentage of material removed or settled out. The actual value of the fraction removed is then plotted in the manner shown in Fig. 9-14. A particular value is plotted at the proper coordinates of depth and time. After sufficient data have been plotted, lines of equal concentration or isoconcentration lines can be drawn. These lines represent equal fractions of removal. They also trace the maximum trajectories of particle settling paths for specific concentrations in a flocculent suspension. For example, 45% of the particles in suspension will have average velocities not greater than $4/21 = 0.19$ ft/min by the time the 4-ft depth has been reached.

To determine the overall removal in a specific basin, an approach similar to that outlined for discrete particles is used. An example will illustrate the procedure.

●**EXAMPLE 9-6**

Using settling trajectories determined by the settling column analysis shown in Fig. 9-14, find the overall removal for a settling basin 7 ft deep with an overflow rate v_0 of $\frac{7}{24}$ ft/min.

○**Solution**

From the figure, 45% of the particles will have settling velocities greater than $\frac{7}{24}$ ft/min.

Using 15% increments, it can be seen that an additional $60 - 45 = 15\%$ of the particles will have an average settling velocity $v_s \geq 3.4/24$ ft/min and a second

$75 - 60 = 15\%$ will have an average settling velocity $\geq 1.3/24$ ft/min. The next increment would have a negligible velocity and will therefore be neglected. (*Note*: These values are determined as shown on Fig. 9-14.)

The removal ratio of each percentage of particles considered to have a certain characteristic is v_s/v_0, as stated previously.

Overall removal in percent is therefore

$$R = 45 + \frac{v_{s1}}{v_0} C_1 + \frac{v_{s2}}{v_0} C_2 + \cdots + \frac{v_{sn}}{v_0} C_n \tag{9-24}$$

where in the example $C_1 = C_2 = C_3 = 15\%$. Also, since $v_{s1} = z_1/t_0$ and $v_0 = z_0/t_0$, $v_{s1}/v_0 = z_1/z_0$, etc., where $z_1, z_2, \ldots$ are measured down from the top of the tank and $z_0 = 7$ ft. Thus

$$R = 45 + \frac{3.4}{7}(60 - 45) + \frac{1.3}{7}(75 - 60) + \cdots$$

$$= 45 + 7.3 + 2.8 + \cdots$$

and the overall removal is approximately 55.1%.

9-12

ZONE SETTLING

In concentrated suspensions, the velocity fields of the closely spaced particles are obstructed, causing an upward displacement of the fluid and hindering settlement. In a dilute suspension, particles settle freely at their terminal velocity until they approach the sludge zone at the bottom of the settling tank. In this region the particles decelerate and finally become a part of the sludge blanket. For more concentrated suspensions, a zone will be formed in which more rapidly settling particles act as a group, with a reduction in the velocity of subsidence.[3] With further increases in concentration, formation of this zone occurs at progressively earlier periods. Finally, complete collective settling takes place. When particle concentration reaches this point, a definite interface is formed between settling particles and clarified fluid. Figure 9-15 illustrates the manner in which the location of the interface varies with respect to time. From A to B there is a hindered settling of the particle–liquid interface, and from B to C, a deceleration as the transition into the compression zone is accomplished. Settling in the region from C to D depends upon compression of the sludge blanket.

The manner in which settling of a thick flocculent suspension occurs may be further clarified by considering a batch process carried out in a cylindrical settling column. Initially the column contains a suspension which has a uniform concentration. After $t = 0$, an interface is formed at some depth below the surface and a zone of clarified fluid develops directly above the interface. Below the interface a constant uniform suspension having some specific settling velocity remains. The velocity of subsidence in this zone is defined as

$$v_s = f(C) \tag{9-25}$$

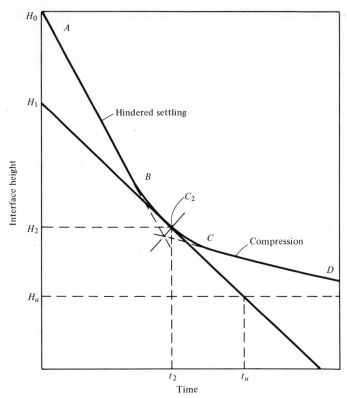

FIGURE 9-15 Graphical representation of interface height versus time for a batch-settling test. [Source: After W. P. Talmadge and E. B. Fitch, "Determining Thickener Unit Areas," *Ind. Eng. Chem.*, 47 (1955).]

where v_s = settling velocity for conditions of hindered settling
   C = initial concentration of particles in the suspension

Two other zones are also formed along with that of clarified liquid. One of these is directly below the zone of uniform suspension and it represents a transition region in which settling velocity is reduced due to an increase in solids concentration. Under this zone, a compression develops and the solids are supported mechanically by those beneath them. Thus there is an upper zone of clarified liquid, a second zone of hindered settling at uniform concentration, a transition zone, and a compression zone. As time passes, all zones except those containing the clarified liquid and sludge will disappear as no new solids are introduced into the column. In a continuous-flow reactor, all zones would be maintained. Zone settling occurs for activated sludges in final clarifiers and for raw wastewaters when the particles approach the bottom of the sedimentation tank.

The surface area required in a continuous-flow system designed to handle concentrated suspensions depends upon the clarification and thickening capacities of the system. Batch-settling tests of the nature shown in Fig. 9-15 can be used to estimate both these factors.

The initial rate at which interface subsidence takes place permits an estimate of the clarification capacity of the system. Essentially, the design must assure a surface area large enough to preclude the rate of liquid rise exceeding the velocity of subsidence. Rate of liquid rise is based on quantity of liquid overflow only (the portion passing out with the sludge is excluded). The area required for clarification can be stated as

$$A = \frac{Q_0}{v_s} \tag{9-26}$$

where A = surface area of the settling zone
Q_0 = overflow rate
v_s = subsidence rate in the zone of hindered settling

A value of v_s is determined from batch data (Fig. 9-15) by computing the slope of the hindered settling portion of the interface height versus time curve. The clarification area found by Eq. 9-26 must then be compared with that computed on the basis of thickening capacity and the larger of the two accepted as the design surface area.

Thickening capacity is obtained from the batch-sedimentation characteristics of thick suspensions. The required cross-sectional area for adequate thickening is given by[3,16]

$$A = \frac{Qt_u}{H_0} \tag{9-27}$$

where A = required surface area of the settling zone
Q = volumetric rate of influent to the settling basin
t_u = time it takes to reach a desired underflow or sludge concentration C_u
H_0 = depth of fluid in the settling column (the initial interface height)

In settling analyses of this type, there is a critical concentration layer which will yield the largest value for A to be used as the basis for design. Usually this concentration occurs in the transition zone at the inception of compression. An approximation of this critical concentration C_2 is had by bisecting the tangents to the hindered settling and compression portions of the interface height versus time curve. This procedure is illustrated in Fig. 9-15. The point at which the bisector intersects the curve is considered to be representative of the actual compression point.

The value of t_u can also be determined from the batch settling curve by first plotting a line parallel to the time axis which passes through the desired underflow concentration C_u. Next, a tangent is constructed at C_2. The time-scale intercept of the tangent with the underflow–interface height line is the required time t_u (Fig. 9-15).

Proper design of a final clarifier must satisfy three criteria: (1) the clarification capacity required by the hindered settling rate, (2) the thickening capacity needed to remove material in the transition and sludge zone, and the (3) detention period, which cannot be excessive (otherwise, the bottom sludge may become anaerobic).

●EXAMPLE 9-7

Use the batch-settling analysis given in Fig. 9-16 to find the minimum surface area of a sedimentation basin. The influent has a solids concentration of 0.175 pcf and is to be settled in a continuous-flow unit operated at a rate of 1.3 cfs. An underflow concentration of 0.823 pcf is considered desirable.

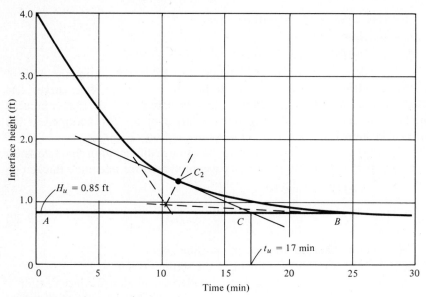

FIGURE 9-16 Batch-settling data for Example 9-7.

○*Solution*

The height of the particle–liquid interface must first be determined for the underflow concentration C_u. Since the total weight of solids in the system must equal $C_0 H_0 A$ or $C_u H_u A$,

$$H_u = \frac{C_0 H_0}{C_u}$$

$$= \frac{0.175 \times 4.0}{0.823}$$

$$= 0.85 \text{ ft}$$

Construct a horizontal line AB through H_u. Then draw tangents to the hindered settling and compression portions of the curve. Bisect the angle formed by these tangents and, using the bisector, locate the compression point C_2. Construct a tangent to the curve at C_2 and project the tangent to an intersection at C with line AB. Project C vertically downward to the time scale and find $t_u = 17$ min.

The thickening area requirement can then be calculated by Eq. 9-27:

$$A = \frac{Qt_u}{H_0}$$

$$= \frac{1.3 \times 17 \times 60}{4}$$

$$= 331 \text{ ft}^2$$

Compute the subsidence velocity v_s in the hindered settling portion of the curve.

$$v_s = \frac{4.0 - 2.25}{6 \times 60}$$

$$= 4.86 \times 10^{-3} \text{ fps}$$

The area required for clarification is given by

$$A = \frac{Q_0}{v_s}$$

where Q_0, the overflow rate, is proportional to the volume above the sludge zone.

$$Q_0 = 1.3 \times \frac{3.15}{4.0}$$

$$= 1.02 \text{ cfs}$$

and

$$A = \frac{Q_0}{v_s} = \frac{1.02}{4.86 \times 10^{-3}}$$

$$= 211 \text{ ft}^2$$

The controlling requirement is therefore the thickening area of 331 ft², since it exceeds that required for clarification.

9-13

COMPRESSION

Consolidation of sediment at the bottom of the basin is extremely time-consuming because the fluid displaced must flow through an ever-decreasing pore space between particles. The rate of settlement decreases with time, owing to increased resistance to flow of the fluid. The porosity of deposited sediment is at a maximum in the lowest portion of the sludge blanket, owing to compression resulting from the weight of supported particles above and because the consolidation time for this lowest portion is also greatest. The consolidation rate in the zone of compression is approximated by[17]

$$-\frac{dH}{dt} = i(H - H_\infty) \tag{9-28}$$

where H = sludge line height at time t
  H_∞ = final sediment depth
    i = constant for a given suspension

The time required for the sludge line to drop from critical height at compression point H_c to a height H can be gotten by integrating Eq. 9-28. The resulting expression is

$$i(t - t_c) = \ln (H_c - H_\infty) - \ln (H - H_\infty) \qquad (9\text{-}29)$$

where t_c = time at which the sludge line is at the compression point
  H_c = interface height at the compression point

and the other variables are as previously defined. A plot of log $(H - H_\infty)$ versus $(t - t_c)$ is a straight line having the slope $-i$. The final sludge height H_∞ is primarily dependent upon the liquid surface film which adheres to the particles. Gentle agitation facilitates compaction.

9-14

SHORT-CIRCUITING

Short-circuiting can be defined as the deviation from plug flow which is exhibited by fluid particles passing through a reaction vessel. For ideal plug flow, all fluid particles are retained for a theoretical time $t_0 = V/Q$, whereas if short-circuiting is apparent, part of a slug will be retained for a period less than t_0 and part for a time greater than t_0. Short-circuiting produces clarification efficiencies less than expected, owing to the nonuniform times of passage. Major causes of this phenomenon are influent velocity, density, and thermal-, wind-, and effluent-structure-induced currents. In general, relative importance of the various causes of short-circuiting is related to the characteristics of the suspended solids being settled.

The short-circuiting characteristics of tanks are usually measured by putting a slug of dye, salt, or radioactive tracer in the inlet of a tank and then measuring the tracer concentration when it appears at the outlet of the vessel at specific intervals of time. If the entire slug traverses the system at nominal tank velocity, the slug passes out of the system in its entirety at the nominal detention time $t_0 = V/Q$. In reality, however, parts of the tracer remain for different periods of time, but the average will approach t_0. Typical concentration–time curves for different types of tanks are given in Fig. 9-17.[14]

The vertical scale of Fig. 9-17 is the ratio of the actual concentration to that which would be obtained if the slug of tracer material were mixed instantaneously with the entire tank contents. The horizontal scale is the ratio of the actual time at which a certain concentration appears at the outlet to the nominal retention time t_0.

Curve A in Fig. 9-17 represents conditions that would prevail in an ideal dispersion tank where the slug mixes immediately with the entire contents of the tank. Characteristics of actual vessels designed for violent mixing closely approximate curve *A*. With such containers, however, a short time is recorded before the tracer first reaches the outlet. Curve *B* is typical of stable flow conditions for radial-flow circular tanks. The curve indicates that the largest concentration of suspended matter passes the outlet in about 50% of the nominal

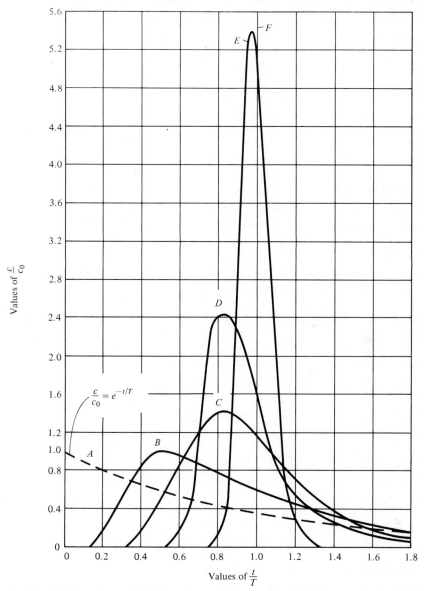

FIGURE 9-17 Typical dispersion curves for tanks. [Source: After T. R. Camp, "Studies of Sedimentation Basin Design," *Sewage and Ind. Wastes* (January 1953).]

detention period. Curve C characterizes wide, shallow, rectangular tanks, while curve D exemplifies long, narrow, rectangular tanks. The largest concentrations of suspended matter for curve D are seen to be retained for about 80% of the detention period. Curve E pictures round-the-end baffled mixing chambers, and curve F represents the condition of ideal plug flow.

It is extremely difficult to reproduce the short-circuiting curves for square tanks and radial-flow circular tanks because flow patterns in these basins are not stable and are readily affected by density and convection currents. Short-circuiting curves in long, narrow, rectangular basins can be repeated quite well, however, owing to the particular stability of such basins.

On a qualitative basis it is generally considered that dye fronts which take long periods to arrive at the outlet, and which are not greatly dispersed, indicate the more nearly optimum settling conditions. This premise is however not valid in all cases. Removal of particles which settle out in an unhindered manner is not affected by detention time but is solely a function of overflow rate. In this case, density currents and vertical short-circuiting do not affect the removal of particles. Dye-test curves do not reflect the difference between density-current short-circuiting and mixing effects or that resulting from the nonuniform lateral distribution of flows. Consequently, there may be no direct correspondence between the dye-front curve and the sedimentation efficiency in type I operations.

For type II particles, detention time may be of greater significance than overflow rate.[18] Under these circumstances, the detention characteristics of the basin are important and the dye test serves as an indicator of removal efficiency. Thus all the possible short-circuiting mechanisms must be considered in design.

Morrill[19] has proposed that volumetric basin efficiency be related to the ratio of the times of occurrence of the 10 and 90 percentiles of a tracer-front curve. Fiedler and Fitch[20] have suggested the use of standard detention efficiency as a functional measure of basin efficiency where detention time is required in a clarification basin. The standard detention efficiency (SDE) is given by

$$\text{SDE} = \frac{\sum\limits_{0}^{\infty}(D - \frac{1}{2}\,\Delta D)\,\Delta t \Big/ \sum\limits_{0}^{\infty}[(D - \frac{1}{2}\,\Delta D)\,\Delta t/(t - \frac{1}{2}\,\Delta t)]}{V/Q} \qquad (9\text{-}30)$$

where D = dye concentration
t = time
Q = influent rate of flow
V = tank volume

For a more complete discussion, the reader is referred to the work of Fiedler and Fitch.[20]

Tracer studies suggest that, in general, long, narrow, rectangular basins are superior to radial-flow circular tanks since unstable, poorly shaped flow waves are normally exhibited by the latter. It has been indicated, however, that unstable flow wave associated with the radial-flow tank may not be a valid index of the efficiency of that type of sedimentation vessel.[21]

Short-circuiting effects in sedimentation basins may be minimized by covering the basin to eliminate effects related to the sun or wind. Large horizontal currents can be induced by the wind, whereas vertical convection currents can be caused by the sun. Influent and density-current short-circuiting may be offset by utilizing of stream-deflecting baffles, feed-stream dividing mechanisms, impingement of opposed jets or streams, radial-type velocity dispersing feedwells, vertical-flow feedwells, and tangentially fed velocity-dispersing feedwells. Short-circuiting due to effluent structures can be minimized by providing an effluent weir length as large as possible.

9-15

SEDIMENTATION BASINS

Sedimentation units often occupy a position of considerable importance in the overall process of treating a specific water supply. Poor design or operation of these basins results in the passage of inadequately conditioned water to the next unit process. Such a happening may adversely affect the outcome of the entire remaining treatment sequence.

The physical construction of a sedimentation basin may vary from an excavation in the ground (often for presedimentation of extremely turbid waters) to a structure of concrete or steel. Sludge-removal equipment may be provided or hand-cleaning methods may be employed. Most modern sedimentation basins are circular concrete tanks, either open or covered, equipped with mechanical scrapers for sludge removal.

Sedimentation basins are constructed in a variety of sizes and shapes. Depths range from about 7 to 15 ft, with 8–12 ft most common. Circular tanks run from about 35 to 150 ft in diameter, although some are as large as 200 ft. Square tanks usually have sides less than 100 ft, while rectangular basins are normally 100 ft or more in length. The length/width ratio of rectangular tanks varies from about 3:1 to 5:1, the width being controlled in many instances by the size of the sludge-removal apparatus. Bottom slopes are from approximately 1% in rectangular tanks to about 8% for circular or square basins.

To minimize the effects of short-circuiting and turbulent flow, considerable attention has been directed toward effective hydraulic design of inlet and outlet structures. Inlet structures are expected to (1) uniformly distribute the influent over the cross section of the settling zone; (2) initiate parallel or radial flow; (3) minimize large-scale turbulence; and (4) preclude excessive velocities near the sludge zone. Generally, the influent is dispersed across the width or radially from the center of the tank through entrance ports or pipes. Various methods of employing baffles or deflectors to dissipate the velocity of influent jets have been devised.[23] Considerable large-scale turbulence is often developed by inlet structures; however, by controlling overall tank dimensions, inflow distribution can be confined to a relatively minor volume of the tank.

Part of the volume in a sedimentation tank is ineffective for settling purposes because particles entering this zone become entrained in the effluent. To minimize this effect, a relatively long flow path is desirable. Long, narrow tanks are especially effective under such circumstances. Outflows are normally controlled by weirs placed along the sides of outlet troughs located beside the walls of rectangular tanks or extending toward the center. In circular basins, weirs are conventionally located on the periphery of the tank. Various types of outlets are of little importance compared with inlet types for control of dispersion characteristics.[24]

9-16

SEDIMENTATION IN WATER TREATMENT

Surface water containing high turbidity (e.g., from a muddy river) may require sedimentation prior to chemical treatment. Presedimentation basins can have hopper bottoms or be equipped with continuous mechanical sludge-removal apparatus. The minimum recommended detention period for presedimentation is 3 hours, although in many cases this is not adequate to settle out fine suspensions which occur at certain times of the year. Chemical feeding equipment for prechlorination or partial coagulation is frequently provided ahead of presedimentation basins for periods when the raw water is too turbid to clarify adequately by plain sedimentation. Sludge withdrawn from the bottom of presedimentation basins is generally discharged back to the river.

Sedimentation following flocculation depends upon settling characteristics of the floc formed in the coagulation process. A general range for the settling velocities of floc from chemical coagulation is 2–6 ft/hr. Detention periods used in floc sedimentation range from 2 to 8 hours, and overflow rates (surface settling rates) vary from 500 to 1000 gpd/ft^2.

The 1968 *Recommended Standards for Water Works, Great Lakes–Upper Mississippi River Board of State Sanitary Engineers*[12] (GLUMRB) recommends the following standards for sedimentation design following flocculation: minimum settling time of 4 hours, maximum horizontal velocity through settling basins of 0.5 ft/min, and maximum flow rate over the outlet weir of 20,000 gpd/ft of weir length.

A typical rectangular sedimentation tank for clarification of flocculated water is shown in Fig. 9-9. Figure 9-18 shows a square clarifier having a sludge scraper equipped with special corner blades. The method of operation for the unit shown in Fig. 9-18 is referred to as cross-flow. Feed is introduced through submerged ports along one side of the tank and flows toward the effluent weir positioned along the opposite side of the clarifier. An alternative method of feed is to pipe influent to the center column of the clarifier, distribute it radially, and

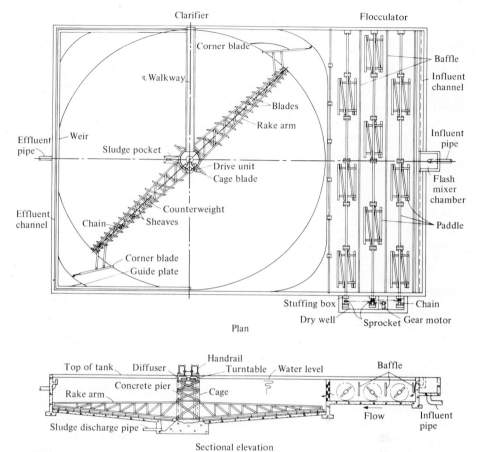

FIGURE 9-18 Flocculator and square sedimentation tank for water clarification, illustrating cross flow operation. (Courtesy Dorr-Oliver.)

collect the effluent over a peripheral weir extending around four sides of the sedimentation tank.

Where raw-water quality does not differ greatly, such as that from a well supply, flocculator–clarifiers have proved to be an efficient method for chemical treatment. One such unit, shown in Fig. 9-19, combines the processes of mixing, flocculation, and sedimentation in a single-compartmented tank. Raw water and added chemicals are mixed with the slurry of previously precipitated solids to promote growth of larger crystals and agglomerated clusters which settle more readily.

The GLUMRB *Standards*[12] state that flocculator–clarifiers " are acceptable for combined softening and clarification where water characteristics are not variable and flow rates are uniform. Before such units are considered as clarifiers without softening, specific approval of the reviewing authority shall be obtained. Clarifiers should be designed for the maximum uniform rate and should be

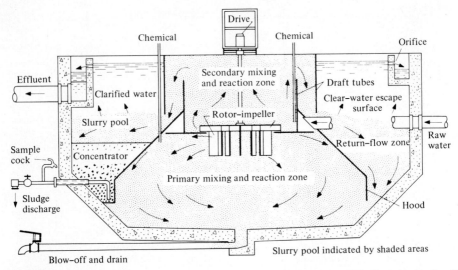

FIGURE 9-19 Vertical-flow sedimentation basin used in water treatment. The "accelerator" unit includes a basin having a raw-water inlet and distribution duct; a primary mixing and reaction zone; two concentric draft tubes which form the secondary mixing and reaction zone; a rotor-impeller for mixing and pumping, driven by a motorized reducer; an effluent channel system; and concentrators to accumulate and remove excess slurry. (Courtesy Infilco Degremont Inc.)

adjustable to changes in flow which are less than the design rate and for changes in water characteristics." Design criteria recommended are: flocculation and mixing period not less than 30 minutes; minimum detention time of 2 hours for chemical coagulation in flocculator–clarifiers and 1 hour for lime–precipitation softeners; weir loadings not exceeding 20 gpm/ft for softener units and 10 gpm/ft for clarifier units; and upflow rates not to exceed 1.75 gpm/ft^2 for softeners and 1.0 gpm/ft^2 for clarifiers.

The volume of sludge removed from clarifiers plus the quantity of backwash water from cleaning sand filters generally runs from 4 to 8% of the processed water. Exact amount of wastewater produced depends upon chemical treatment processes employed, type of sludge-collection apparatus, and kind of filter backwash equipment. Methods of waste disposal are governed by local and state pollution-control-agency requirements. Possible procedures include: direct discharge to a receiving stream or drainage system; lagooning or sludge drying beds; hauling away and spreading on land; discharge to a municipal sewer; and dewatering of sludge by pressure filtration or centrifugation.

●EXAMPLE 9-8

Two rectangular clarifiers each 90 ft long, 16 ft wide, and 12 ft deep are used to settle 0.50 mg of water in an 8-hour operational period. Calculate the detention period, horizontal velocity, overflow (surface settling) rate, and weir loadings, assuming multiple effluent weirs with a total length equal to three tank widths.

○*Solution*

flow $= 0.50$ mg/8 hours $= 8356$ ft^3/hr $= 139$ cfm

$$\text{detention time} = \frac{2 \times 90 \times 16 \times 12}{8356} = 4.1 \text{ hours}$$

$$\text{horizontal velocity} = \frac{140}{2 \times 16 \times 12} = 0.36 \text{ ft/min}$$

$$\text{overflow rate} = \frac{3 \times 500,000}{2 \times 90 \times 16} = 520 \text{ gpd/ft}^2$$

$$\text{weir loading} = \frac{3 \times 500,000}{3 \times 2 \times 16} = 15,600 \text{ gpd/ft}$$

9-17

SEDIMENTATION IN WASTEWATER TREATMENT

Clarifiers used to settle raw wastewater are referred to as *primary tanks.* Sedimentation tanks used between trickling filters in a two-stage secondary-treatment system are called *intermediate clarifiers. Final clarifiers* are settling tanks following secondary aerobic treatment units. Placement of sedimentation basins in single-stage trickling filter plants, two-stage filter plants, and activated-sludge plants are illustrated in Figs. 11-25, 11-26, and 11-42, respectively.

Sedimentation basins in wastewater treatment have overflow rates in the range 300–1000 gpd/ft^2 and detention times between 1 and 3 hours. Side water depths depend upon settling characteristics of the waste and need for sludge storage volume in the tank bottom.

The 1971 *Recommended Standards for Sewage Works, Great Lakes–Upper Mississippi River Board of State Sanitary Engineers*[25] (GLUMRB) recommend the following overflow rates: primary settling tanks not followed by secondary treatment not to exceed 600 gpd/ft^2 (24 m^3/m^2-day) for plants having a design flow of 1.0 mgd or less (higher rates may be permitted for larger plants); intermediate settling tanks not to surpass 1000 gpd/ft^2 (41 m^3/m^2-day); and final settling tanks not to exceed 800 gpd/ft^2 (33 m^3/m^2-day), except for activated-sludge final clarifiers. Detention time, surface settling rate and weir overflow rate must be adjusted for various activated-sludge processes to minimize the problems with sludge loadings, density currents, inlet hydraulic turbulence, and occasional poor sludge settleability. The design parameters listed in Table 9-2 should be observed in the design of final settling tanks following activated-sludge processes. GLUMRB *Standards* recommend that the liquid depth of mechanically cleaned settling tanks be as shallow as practicable but not less than 7 ft (2.1 m), and final clarifiers for activated sludge be not less than 8 ft (2.4 m). Weir loadings are not to exceed 10,000 gpd/ft (124 m^3/m-day) for plants of 1 mgd or smaller, and preferably not to go over 15,000 gpd/ft for design flows above 1.0 mgd.

TABLE 9-2

SUGGESTED DESIGN PARAMETERS FOR FINAL SETTLING TANKS FOLLOWING ACTIVATED-SLUDGE PROCESSES

Type of Process	Average Design Flow (mgd)[a]	Minimum Detention Time (hr)	Maximum Overflow Rate (gpd/ft²)
Conventional,	to 0.5	3.0	600
high rate and	0.5–1.5	2.5	700
step aeration	1.5 up	2.0	800
Contact	to 0.5	3.6	500
stabilization	0.5–1.5	3.0	600
	1.5 up	2.5	700
Extended	to 0.05	4.0	300
aeration	0.05–0.15	3.6	300
	0.15 up	3.0	600

[a] 1.0 mgd = 3790 m³/day.
[b] 100 gpd/ft² = 4.08 m³/m²-day.
Source: 1971 *Recommended Standards for Sewage Works Great Lakes–Upper Mississippi River Board of Sanitary Engineers.*

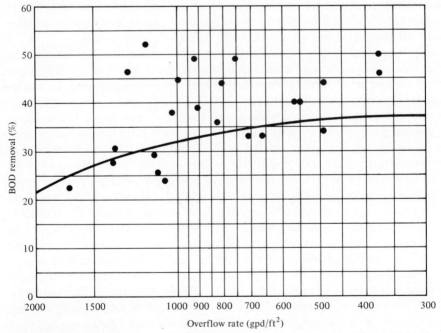

FIGURE 9-20 BOD removal in plain sedimentation of raw wastewater. [The solid line is the suggested value for design from the *Recommended Standards for Sewage Works, Great Lakes–Upper Mississippi River Board of State Sanitary Engineers*. Plotted points are operational data from "Sewage Treatment Plant Design," *WPCF Manual No. 8* or *ASCE Manual No. 36* (1959), p. 91.]

BOD removals from raw domestic wastewater by primary sedimentation permitted for design by GLUMRB are shown by the solid line in Fig. 9-20. BOD removals for wastewater containing appreciable quantities of industrial wastes should be determined by laboratory tests and consideration of the quantity and characteristics of the waste. The plotted points in Fig. 9-20 are BOD removal data for circular primary settling tanks from actual municipal-treatment-plant operation.[26]

(a)

(b)

FIGURE 9-21 Typical primary settling tank for wastewater treatment. (a) Circular settling tank using inboard weir trough. (b) Stilling well and sludge collecting mechanism in circular clarifier. (Courtesy Walker Process Division of Chicago Bridge & Iron Company.)

Figure 9-21 pictures a typical primary settling tank. Wastewater enters at the center behind a stilling baffle and travels outward toward effluent weirs located on the periphery of the tank. The inlet line usually terminates near the surface but the wastewater must travel down behind the stilling baffle before entering the actual settling zone. A stilling well reduces the velocity and imparts a downward motion to the solids, which drop to the tank floor.

The clarifier in Fig. 9-22 is designed as a final clarifier for an activated-sludge secondary. The liquid flow pattern is the same as that of a primary clarifier, but the sludge-collection system is unique. Sludge uptake pipes are attached to and spaced along the scraper mechanism to convey the less dense sludge to a stationary collecting channel constructed around the stilling baffle on the center column. This continuously withdrawn sludge is activated sludge returned to the aeration tank. Heavy sludge plowed to the hopper near the center column is excess activated sludge wasted from the system.

●**EXAMPLE 9-9**
Determine the dimensions for a primary settling tank treating 0.90 mgd of domestic wastewater based on GLUMRB *Standards*. Calculate the detention time and BOD removal efficiency for the tank capacity selected.

○*Solution*

From the GLUMRB *Standards*:

$$\text{maximum overflow rate} = 600 \text{ gpd/ft}^2$$
$$\text{minimum depth} = 7 \text{ ft}$$
$$\text{maximum weir loading} = 10{,}000 \text{ gpd/ft}$$

$$\text{area of settling tank required} = \frac{900{,}000}{600} = 1500 \text{ ft}^2$$

Use a 44-ft-diameter tank ($A = 1520 \text{ ft}^2$):

$$\text{overflow rate} = \frac{900{,}000}{1520} = 592 \text{ gpd/ft}^2$$

Use a liquid-side-wall depth of 7.0 ft plus 2-ft freeboard for wind protection:

$$\text{detention time} = \frac{1520 \times 7.0 \times 7.48}{900{,}000} \times 24 = 2.1 \text{ hours}$$

Use an inboard weir trough on a diameter of 38 ft:

$$\text{weir loading} = \frac{900{,}000}{\pi \times 38} = 7540 \text{ gpd/ft}$$

The BOD removal efficiency from Fig. 9-20 for an overflow rate of 590 gpd/ft^2 = 36%.

●**EXAMPLE 9-10**
Two final clarifiers of the type shown in Fig. 9-22, with 100-ft diameter and 9-ft liquid-side-wall depth, are provided for a plant designed to treat 11 mgd. (a) Calculate the

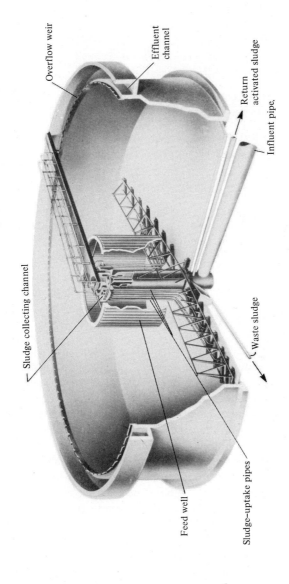

Overflow weir

Effluent channel

Return activated sludge

Influent pipe,

Sludge collecting channel

Waste sludge

Feed well

Sludge-uptake pipes

FIGURE 9-22 Final clarifier for an activated-sludge secondary with rapid-sludge-removal apparatus. (Courtesy Dorr-Oliver.)

overflow rate and detention time based on design flow, and compare these answers with the values given in Table 9-2. (b) If the aeration tank is operated at a mixed liquor–suspended solids (SS) concentration of 4000 mg/l and a recirculation ratio of 0.5, calculate the detention time, including recirculation flow and the solids loading on the clarifier.

○***Solution***

Surface area of clarifiers $= 2 \times \pi(50)^2 = 15,700 \text{ ft}^2$

volume of clarifiers $= 15,700 \times 9 \times 7.48 = 1,060,000 \text{ gal}$

a   Overflow rate $= \dfrac{11,000,000}{15,700} = 700 \text{ gpd/ft}^2 \; (<800 \text{ OK})$

detention time $= \dfrac{1,060,000}{11,000,000} \times 24 = 2.3 \text{ hours} \; (> 2.0 \text{ OK})$

b   Flow from aeration tank to clarifier with a recirculation ratio of $0.5 = 1.5 \times 11.0 = 16.5 \text{ mgd}$.

detention time based on $Q + Q_R = \dfrac{1.06}{16.5} \times 24 = 1.5 \text{ hours}$

solids loading $= \dfrac{16.5 \times 4000 \times 8.34}{15,700} = 35 \text{ lb of SS/day/ft}^2$

9-18

GRIT CHAMBERS IN WASTEWATER TREATMENT

Grit includes gravel, sand, and heavy particulate matter such as corn kernels, bone chips, and coffee grounds. Grit removal in municipal waste treatment protects mechanical equipment and pumps from abnormal abrasive wear, prevents pipe clogging by its deposition, and reduces accumulation in settling tanks and digesters. Grit chambers are commonly placed between lift pumps and primary settling tanks. Wear on the centrifugal pumps is tolerated to obtain the convenience of ground-level construction of grit chambers.

Several types of grit-removal units are used in wastewater treatment. The kind selected depends upon the amount of grit in the wastewater, size of the plant, convenience of operation and maintenance, and costs of installation and operation. Standard types are channel-shaped settling tanks, aerated tanks of various shapes with hopper bottoms, clarifier-type tanks with mechanical scraper arms, and cyclone grit separators with screw-type grit washers.

A channel-type grit-removal unit is depicted in Fig. 9-23. Although most plants now have mechanical collectors, many plants have been built with two or more hand-cleaned channels with grit storage volume provided in the bottom. Flight-type grit collectors stir and scrape the material into a receptacle for convenience of handling and disposal. A proportional weir placed on the discharge end of the channel provides velocity control. The weir opening is shaped to keep the horizontal velocity relatively constant while depth of flow varies.

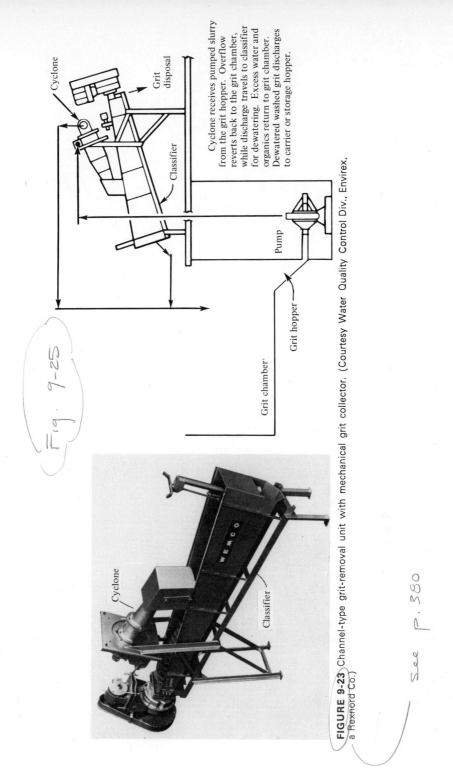

Cyclone

Grit disposal

Classifier

Cyclone receives pumped slurry from the grit hopper. Overflow reverts back to the grit chamber, while discharge travels to classifier for dewatering. Excess water and organics return to grit chamber. Dewatered washed grit discharges to carrier or storage hopper.

Pump

Grit hopper

Grit chamber

Fig. 9-25

Cyclone

Classifier

WEMCO

FIGURE 9-23 Channel-type grit-removal unit with mechanical grit collector. (Courtesy Water Quality Control Div., Envirex, a Rexnord Co.)

See P. 380

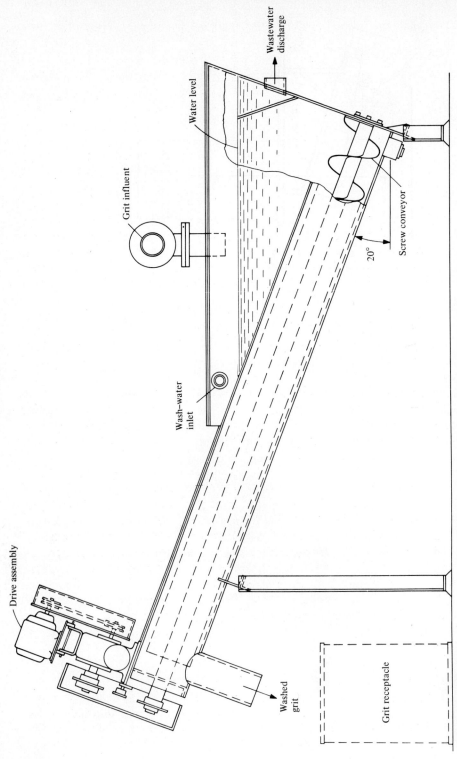

Wastewater discharge

Water level

Grit influent

Wash-water inlet

Screw conveyor

20°

Drive assembly

Washed grit

Grit receptacle

FIGURE 9-24 Counterflow grit washer. (Courtesy Walker Process Division of Chicago Bridge & Iron Company.)

For design purposes grit is defined as fine sand, 0.2-mm-diameter particles with a specific gravity of 2.65 and a settling velocity of 0.075 fps. The scouring velocity for the same size and density of particle is 0.75 fps. A channel type of grit-removal unit is normally designed to provide a controlled horizontal velocity of approximately 1 fps.

Square and rectangular hopper bottom tanks with influent and effluent weirs on opposite sides of the tank are generally used in smaller plants. These units can be mixed by diffused aeration to keep the organics in suspension while the grit settles out. Small tanks are baffled to prevent short-circuiting. Settled substances are removed from the hopper bottom by air-lift pump, screw conveyor, bucket elevator, or, if possible, by gravity flow. Grit taken from this type of unit is sometimes relatively high in organic content. A counterflow grit washer (Fig. 9-24) which functions like a screw conveyor can wash the grit, returning waste organics to the plant influent. Design of an aerated hopper-bottom type of grit tank is based on a detention time of approximately 1 minute at peak hourly flow.

Clarifier-type tanks are generally square, with influent and effluent weirs on opposite sides. A centrally driven collector arm scrapes the settled grit into an end hopper. The grit is then elevated into a receptacle with a chain and bucket, or screw conveyor. The clarifier may be a shallow tank or a deeper aerated chamber.

A cyclone grit separator with screw-type washer is shown in Fig. 9-25. A special centrifugal pump lifts the slurry from a grit sump to a centrifugal cyclone. The cyclone separates the grit from the organic material and discharges it to a classifier for washing and draining. Wash water and washed-out organics from the cyclone and classifier return to the wastewater.

Preaeration of raw wastes prior to primary sedimentation is practiced to restore wastewater freshness, scrub out entrained gases, and improve subsequent settling. The process is similar to flocculation if the detention period is not less than 45 minutes without chemical addition, or about 30 minutes with chemicals. Normally, the purpose of preaeration is freshening wastewater, not flocculation, and little or no BOD reduction occurs. The detention period in preaeration basins is generally less than 20 minutes. Air supplied for proper agitation ranges from 0.05 to 0.20 ft^3/gal of applied wastewater.

Processes of grit removal and preaeration can be performed in the same basin—a hopper bottom or clarifier-type tank provided with grit removal and washing equipment. Aeration mixing improves grit separation while freshening the wastewater.

●EXAMPLE 9-11

Design wastewater flows for a treatment plant are an average daily flow of 280 gpm and a peak hourly rate of 450 gpm. Estimate tank volumes and dimensions for the following grit-removal units: (a) channel-type unit, (b) hopper bottom tank with grit washer, and (c) clarifier-type unit for grit removal and preaeration.

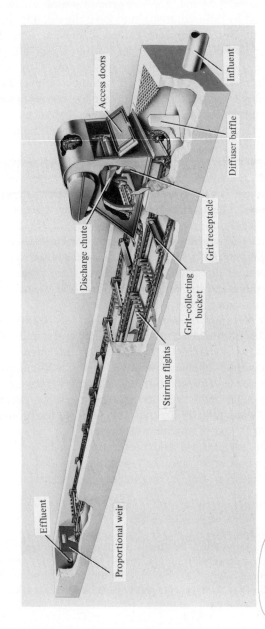

FIGURE 9-25 Cyclone separator and grit washer. (Courtesy WEMCO, Division of Envirotech Corp.)

Influent

Diffuser baffle

Access doors

Grit receptacle

Discharge chute

Grit-collecting bucket

Stirring flights

Effluent

Proportional weir

See p. 377.

○*Solution*

a   Design criteria for channel-type unit:

$v_s = 0.075$ fps $V = 1.0$ fps

Length/depth ratio:

$$\frac{V}{v_s} = \frac{1.0}{0.075} = \frac{13.3}{1}$$

Cross-sectional area required at average flow:

$$A = \frac{280}{449} \times \frac{1}{1.0} = 0.62 \text{ ft}^2$$

and at peak flow:

$$A = \frac{450}{449} \times \frac{1}{1.0} = 1.0 \text{ ft}^2$$

Use a channel 1.0 ft wide, 1.0 ft deep + freeboard, and a free-flowing length of 13.5 ft. Provide a proportional weir at the effluent end, and modify the cross section and length to accommodate a mechanical grit collector (Fig. 9-23).

b   Design criteria for a hopper-bottom grit-removal tank:
  detention period = 1 minute at peak flow
volume of tank required = $450 \times 1.0 = 450$ gal = 60 ft^3

c   Design criteria for preaeration basin with grit removal:
  detention period = 20 minutes
volume of basin required = $280 \times 20 = 5600$ gal = 750 ft^3

Use a 12 ft × 12 ft basin 6 ft deep + freeboard. Provide a revolving clarifier mechanism, a separate grit washer, and diffused aeration along one side of the tank.

Filtration

Filtration is an operation to separate suspended matter from water by passing it through a porous material. This medium may be sand, anthracite, diatomaceous earth, or a finely woven fabric.

9-19

MICROSTRAINING

Microstraining is a form of filtration whose primary objective is the removal of microorganisms (various forms of phytoplankton, zooplankton, and microscopic debris) and other suspended solids. A filtering medium consisting of a finely woven stainless-steel fabric is normally used. This fabric supports a thin layer or mat of removed materials, which further improves filtrability. Micro-strainers usually have high flow ratings, corresponding to low hydraulic resis-tance. Fabrics become matted rapidly and for this reason must be backwashed almost continuously. Most microstrainers are of the rotating-drum type, where the fabric is mounted on the periphery and raw water passes from the inside to the outside of the drum. Backwashing is usually accomplished by wash-water jets.

Microstrainers have been successfully employed in primary clarification of water preceding filtration, in preparing water prior to its use for groundwater recharge, in final clarification of wastewater effluents, in treatment of industrial waters and wastes, and in other applications.[27,28]

9-20

DIATOMACEOUS-EARTH FILTERS

Diatomaceous-earth filters were first used extensively during World War II to provide water to military units in the field within a short period of time after occupancy of the source. The filter medium is supported on a fine metal screen, a porous ceramic material, or a synthetic fabric known as a septum. There are three basic steps in the diatomite filtration cycle[29]: (1) precoat, (2) filtration and body-feed addition, and (3) removal of the filter cake. During the precoat operation a thin layer of diatomite is deposited on the outside of the septum by a flow of water forming a filter. Once completed, raw water combined with a small amount of diatomaceous earth (body feed) is supplied the filter. The body feed continually builds the filter itself, so clogging at the surface is minimized, per-mitting significantly longer filter runs. When the pressure drop or filtration rate reaches an economic limit it becomes necessary to wash the filter, reversing the flow through the septum and discharging dirty filter cake to wash. Once the septum has been cleaned of the original precoat, added body feed, and particles removed by filtration, it is ready for a new precoat and continued use. Common rates of filtration range from about 1 to 5 gpm /ft^2, depending on the quality of raw water and results to be obtained. Use of diatomaceous-earth filters have been generally limited to swimming pools, military field units, and installations for small communities.

9-21

PRESSURE FILTERS

Pressure filters have media (sand, anthracite, or calcite) and underdrains con-tained in a steel tank (Fig. 9-26). Water is pumped through the filter under pressure and the media washed by reversing flow through the bed, flushing out the impurities. Pressure filters have the same general characteristics as rapid sand filters, with rates ranging from 2 to 4 gpm/ft^2.

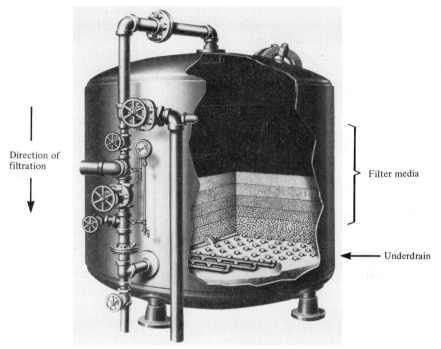

Direction of filtration

Filter media

Underdrain

FIGURE 9-26 Pressure filter. (Courtesy Infilco Degremont Inc.)

Pressure filters are not generally employed in large treatment works because of size limitations; however, they have been used successfully in small municipal softening and iron-removal treatment plants. Their most extensive application has been in treating water for industrial purposes.

9-22

SLOW SAND FILTERS

Slow sand filters, although historically significant, have been abandoned by most cities since development of rapid sand filtration. They are limited to low-turbidity waters not requiring chemical pretreatment. Filtration rates run from 3 to 6 mgd/acre (0.05–0.10 gpm/ft^2). Beds are cleaned by scraping off the top layer (0.5–2 in.) of sand.

Filtering action of a slow sand filter is a combination of straining, adsorption, and biological flocculation. Gelatinous slimes of bacterial growths form on the surface and in the upper sand layer. Effective bacterial and turbidity reductions, and reasonable color removal are realized in slow sand filters, provided that the raw-water turbidity does not exceed the upper practical limit of 30–50 mg/l.

Use of slow sand filters has declined because of their high construction cost, large filter area needed, and unsuitability for treating highly turbid and polluted waters requiring chemical coagulation.

9-23

RAPID SAND FILTRATION

The most common type of device for treating municipal water supplies is the rapid sand filter, which removes nonsettleable floc and impurities remaining after chemical coagulation and sedimentation of the raw water. Water passes downward through the filter media by a combination of positive head and suction from the bottom. Filters are cleaned by reversing flow, a process known as *backwashing*.

Action of a rapid sand filter is extremely complex, consisting of straining, flocculation, and sedimentation. The behavior that takes place in the media depends upon water quality and previous chemical treatment. The straining process occurs principally at the interface between filter media and water. Initially, materials larger than the pore openings at the interface are strained. During the filtration process, material deposited as a mat on the surface builds up and further enhances the straining process; however, it also restricts passage of water through the filter bed. A bed plugs quickly from poor pretreatment of raw water, either by improper coagulation or inadequate sedimentation.

Microscopic particulate matter in raw water not chemically treated will pass through a rapid-sand-filter bed. On the other hand, particulate matter fed to a filter with excess coagulant carryover from chemical treatment produces a heavy mat of flocculated matter clogging the bed surface. Optimum filtration occurs when impurities in the water and coagulant concentration cause "in-depth" filtration. The impurities neither pass through the bed nor are they all strained out on the surface, but rather a significant amount of coagulated impurities is removed and trapped in the bed interstices.

Hazen has stated that a filter medium acts somewhat similar to a sedimentation basin with a large number of trays or false bottoms.[30] When particles smaller than the pore spaces are introduced to a filter, they have an opportunity to settle out on the surface of the medium while passing through the bed. In this respect, each pore space acts as a tiny sedimentation basin. Based on spherical sand grains of 5×10^{-2} cm in diameter, Fair and Geyer indicate that the settling velocity of removable particles is approximately $\frac{1}{400}$ that of particles which can be removed effectively in a sedimentation basin of equal loading.[9] Floc growth is dependent upon the chances for particle contacts to be made. In the filtration process, conditions within pores of a filter bed promote flocculation. Floc thus grow in size and become trapped in the interstices. Figure 9-27 is a schematic diagram illustrating the action in a rapid sand filter. Filtration rates for rapid sand beds range from 2 to 5 gpm/ft^2 (120–290 m^3/m^2-day), depending on medium gradation and character of applied water.

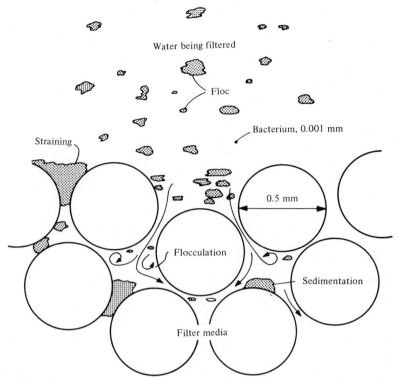

FIGURE 9-27 Schematic diagram illustrating straining, flocculation, and sedimentation actions in a rapid sand filter.

9-24

DESCRIPTION AND OPERATION OF A RAPID SAND FILTER

A typical rapid-sand-filter system is shown in Fig. 9-28. Filters are generally placed on both sides of a pipe gallery which contains inlet and outlet piping, wash-water inlet lines, and backwash drains. The pipe gallery is decked by an operating floor where control consoles are placed near the filters. A clear well for filtered water storage is located under a portion of the filter-bed area.

Figure 9-29 is a cross section of a typical rapid-sand-filter bed which is placed in a box structure a minimum of 8.5 ft deep. The gravel layer containing underdrains is 15–24 in. thick, depth of filter sand is 24–30 in., and top elevation of wash-water troughs is not more than 30 in. above the filter surface. Wash-water trough spacing limits to 3 ft the maximum horizontal travel distance of suspended particles to reach a trough.

A step-by-step description of rapid-sand-filter operation follows the valve numbering illustrated in Fig. 9-30. Initially, valves 1 and 4 are opened, and 2, 3, and 5 are closed, permitting filtration to proceed. Overflow from the settling

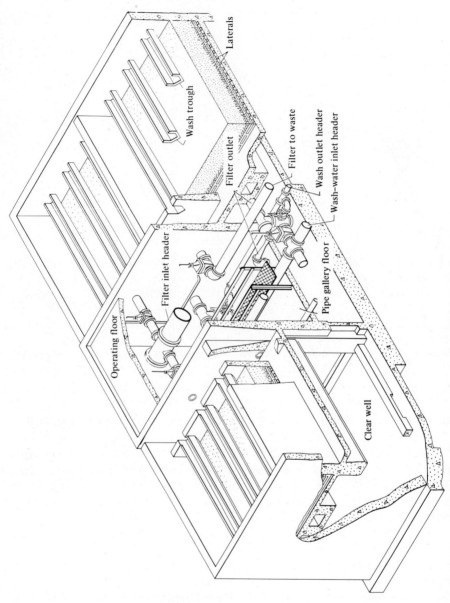

Laterals

Wash trough

Filter outlet

Filter to waste

Wash outlet header

Wash-water inlet header

Filter inlet header

Pipe gallery floor

Operating floor

Clear well

FIGURE 9-28 Typical rapid-sand-filter system. (Courtesy the Permutit Co.)

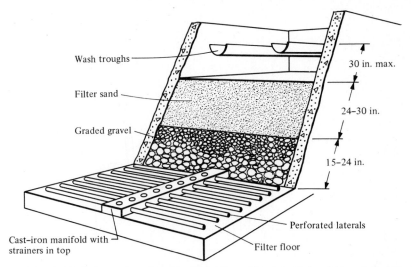

Wash troughs

Filter sand

Graded gravel

30 in. max.

24–30 in.

15–24 in.

Perforated laterals

Cast–iron manifold with
strainers in top

Filter floor

FIGURE 9-29 Cross section of a typical sand filter. (Courtesy National Lime Association.)

basin is applied to the filter, and water passes through the bed into the clear well. After operating head loss becomes excessive, valves 1 and 4 are closed (3 remains closed), and 2 and 5 are opened to permit backwashing. Clear water from the wash-water tank, or pumps, flows into the filter underdrainage system, where it is distributed upward through the sand filter. Dirty wash water is collected by troughs and flows into a channel connected to the drain. Normally, some water at the beginning of a filter run is wasted to flush out through the drain the wash water remaining in the bed. This initial wasting is accomplished by opening valve 3 when valve 1 is opened to start filtration (2, 4, and 5 are shut). Then valve 4 is opened while valve 3 is being closed. The sequence is then repeated.

During filtration, depth of water above the sand surface is a minimum of 3 ft, usually between 3 and 4.5 ft. The filter effluent pipe is trapped in the clear well to provide a connection to water above the sand and to prevent backflow of air to the bottom of the filters. Total head available for filtration is equal to the difference between elevation of the water surface above the filter and the liquid level in the clear well, commonly 9–12 ft. A rate of flow controller (a valve controlled by a venturi meter) in the filter effluent pipe regulates the rate of flow through a clean filter. As the interstices of the filter plug up and a mat forms on the sand surface, head loss in the sand bed increases until the flow controller valve is wide open. When the measured head loss through the filter bed is 8–9 ft, the filter is cleaned by backwashing. Schematic piezometric diagrams in Fig. 9-31 illustrate the approximate hydraulic profiles occurring in a filter system.

The control console for each filter unit is provided with a head-loss gage, flow meter, and rate controller. A run is normally terminated when head loss through the filter reaches a prescribed value between 6 and 9 ft. Filtration may

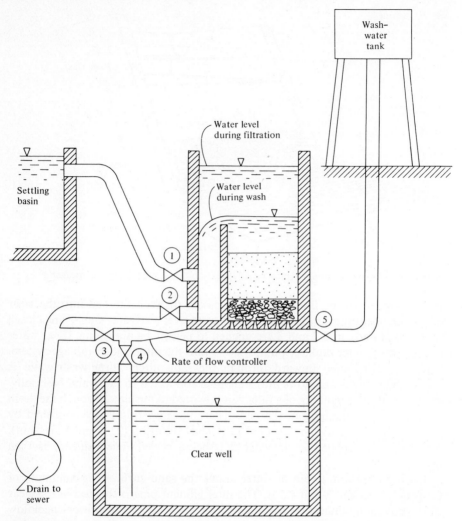

FIGURE 9-30   Schematic diagram indicating operation of a rapid sand filter.

be stopped because of low rate of filtration, passage of excess turbidity through the bed, or " air binding." As head loss increases across the bed, the lower portion of the filter is under a partial vacuum (Fig. 9-31). This negative head permits release of dissolved gases which tend to fill the pores of the filter, causing air binding and reducing the rate of filtration.

Under average operating conditions sand filters are backwashed about once in 24 hours at a rate of 15 gpm/ft^2 for a period of 5–10 minutes. Initial filtered water is wasted for 3–5 minutes. A bed is out of operation for 10–15 minutes to complete the cleaning process. The amount of water used in backwashing varies, but it is usually about 4% of the filtered water.

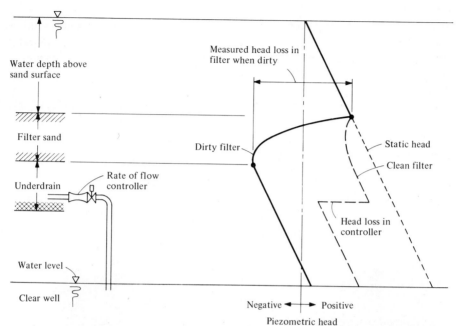

FIGURE 9-31 Schematic diagrams of piezometric head variations with depth through a rapid sand filter.

During backwashing the sand layer expands hydraulically about 50%. Sand grains are cleaned by loosening attached material in the turbulent backwash flow, and impurities released conveyed to the wash troughs. Several devices have been developed to improve backwashing by increasing the scrubbing action in an expanded bed and decreasing the quantity of water involved. The majority of new installations provide surface wash facilities. A system of fixed nozzles or a revolving agitator driven by nozzles can be installed just above the sand surface so that during backwashing they are submerged under the expanded sand surface. The nozzles spray water, under a pressure of 45–75 psi, into the expanded bed increasing turbulence without materially raising the flow rate of wash water.

During backwashing, grains of sand are mixed in the turbulent flow of the expanded bed. When the upward flow of water is stopped, suspended media settle down, forming a stratified bed with the finest grains on top. In a mixed-media bed, the medium of lowest density settles on top.

9-25

FILTERING MATERIALS

The underdrainage system supporting a filter sand is frequently graded layers of gravel containing perforated pipe laterals (Fig. 9-29). Gravel supporting media consist of four or five layers ranging in size from 2.5–1.5 in. in the bottom layer to $\frac{3}{16}$ to $\frac{3}{32}$ in. in the top layer. Several other types of filter bottoms and strainer systems are available to replace the lower coarse gravel layers and perforated lateral underdrains. These include tile filter block, porous plates, depressions filled with glazed earthenware spheres, and various types of manufactured strainers.

The filter underdrain illustrated in Fig. 9-32 consists of vitrified clay blocks fitted with thermoplastic resin nozzles. Nozzles, placed on 8-in. centers, have 0.25-mm-aperture slots to retain the filter sand. No graded gravel or coarse filter layers are needed. This underdrain system is used with a dual-media system employing a combination air and water backwash.

Broadly speaking, filter media should possess the following qualities: (1) coarse enough to retain large quantities of floc, (2) sufficiently fine to prevent passage of suspended solids, (3) adequate depth to allow relatively long filter runs, and (4) graded to permit backwash cleaning. These attributes are not all

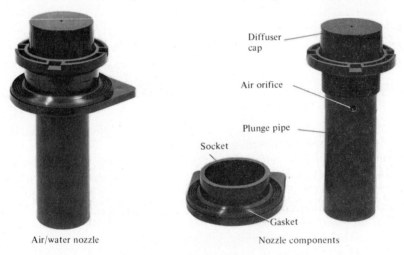

Air/water nozzle   Nozzle components

FIGURE 9-32 Camp filter underdrains consisting of vitrified clay blocks fitted with thermoplastic resin nozzles. (Courtesy Walker Process Division of Chicago Bridge & Iron Company.)

compatible. For example, a very fine sand retains floc, which also tends to shorten the filter run, while for a coarse sand the opposite would be true. Recent trends are toward coarser sands so that higher rates of filtration can be obtained. Care must be exercised, however, because the efficiency of bacterial removal can be reduced if the sand is too coarse. In general, coarse filter beds have greater depth than fine ones.

Filter sand is classed primarily by its effective size and uniformity coefficient. Effective size is the 10 percentile diameter; that is, 10% by weight of the filter sand is less than this diameter. Uniformity coefficient is the ratio of the 60 percentile size to the 10 percentile. Common ranges in effective size and uniformity coefficient are 0.4–0.55 mm and 1.35–1.75, respectively.[31] The GLUMRB *Standards*[12] state that filter sand must have an effective size of 0.45–0.55 mm, depending upon the quality of the raw water, and a uniformity coefficient not greater than 1.65. A 3-in. layer of torpedo sand used to support the filter sand must have an effective size of 0.8–2.0 mm and a uniformity coefficient not greater than 1.7. Clean crushed anthracite or a combination of sand and anthracite must have an effective size from 0.45–0.8 mm and a uniformity coefficient not greater than 1.7.

In a sand filter it is desirable to have a relatively uniform sand so that effective filtration occurs throughout its depth. After the unit has been backwashed, the sand stratifies; if the size variation is too great, effective filtering is confined to the upper few inches of sand.

The problem of surface plugging of sand filters has led to development of dual-media filters.[32] A dual-media filter consists of a sand [2.65 specific gravity (sp gr)] layer topped with a bed of anthracite coal media (1.4 to 1.6 sp gr), both generally 12 in. deep. The coarser anthracite top media layer has intersticial voids about 20% larger than the sand media. These openings are capable of adsorbing and trapping particles so that floc carried over in clarified water does not accumulate prematurely on the filter surface and plug and the sand filter.

Problems in backwashing of dual-media filters have resulted when common backwash and underdrain systems were employed. Nonuniform fluidizing and poor scouring of dual-media filters can result in mud balls dropping through the coarser media and lodging on top of the sand layer. The combination air and water backwash system illustrated in Fig. 9-33 is an effective method for cleaning dual filters.[33] The steps used for performing air and water backwashing of the filter are:

1 Lower the filter water level well below the wash-water troughs but not below the 20% expansion level.
2 Introduce air only at 4 cfm/ft² and continue air addition for 3 minutes to mix and scour media.
3 Admit wash water for about 3–5 minutes to purge and restratify the media. The wash rate should be sufficiently high to provide a bed expansion above 10% (fluidizing threshold) but not so high as to give over 25% expansion.
4 Rewash filter, if required.

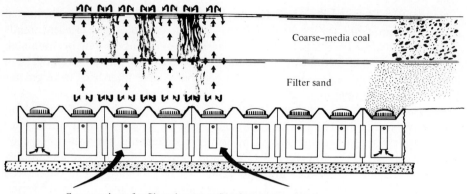

Common ducts for filtered water and backwash air and water

FIGURE 9-33 Dual-media filter, with Camp filter underdrain system, showing air-washing action. (Courtesy Walker Process Division of Chicago Bridge & Iron Company.)

9-26

FILTRATION HYDRAULICS

In a hydraulic sense, granular filters operate much like other porous media. The driving force is the head developed in passing water through the filter. This loss of head can be expressed as

$$H_f = F(e, L, d, v, \mu, \rho, g) \tag{9-31}$$

where H_f = head lost in a depth of filter L
e = porosity of the bed
d = diameter which characterizes the filter-media grains
v = velocity of flow moving toward the filter media
μ = dynamic viscosity
ρ = mass density of the fluid
g = acceleration due to gravity

The range of head loss is an important design parameter. Initial head losses through clean filters are on the order of 1.5–2.5 ft. Filters are normally backwashed when head losses exceed about 8–10 ft. It is common practice to have gravity flow from the filters to the next plant unit, which is generally a clear well. The elevation of the clear well is thus controlled to a great extent by the loss of head through the filter and the connecting transportation system.

Two general equations, one proposed by Rose and one proposed by Carmen and Kozeny, are used to compute the head loss resulting from the passage of the water through the filter media.[34,35] The results obtained using either equation are essentially equivalent.

Carmen–Kozeny Equation

The Carmen–Kozeny equation is derived from the fundamental Darcy–Weisbach relationship for head losses in circular pipes,[3,35]

$$h_L = f \frac{Lv^2}{2gD} \tag{5-4}$$

where h_L = frictional head loss
$\quad f$ = dimensionless friction factor
$\quad L$ = length over which head loss occurs
$\quad D$ = pipe diameter
$\quad v$ = mean pipe-flow velocity
$\quad g$ = acceleration due to gravity

To make this equation representative of conditions in a granular filter, we redefine L as the depth of filter and replace D by an equivalent term $4R$, where R is the hydraulic radius of the flow passage. Equation 5-4 thus becomes

$$h_L = f \frac{Lv^2}{8gR} \tag{9-32}$$

If a unit volume of the filter medium is considered, it can be seen that the channel volume or volume available for flow is essentially equal to the porosity of the bed. For the entire filter bed, then, the channel volume is obtained by multiplying the porosity by the total volume occupied by the bed. Now for N particles of volume V_p, the volume of the solids is NV_p, and the total volume of the bed is $NV_p/(1 - e)$. The total channel volume is therefore

$$\frac{e}{1 - e} NV_p \tag{9-33}$$

where e is the porosity of the bed. The total wetted surface area is considered to be the combined surface area of the particles or the product of the number of particles and the surface area (A_p) of an individual particle NA_p. Remembering that hydraulic radius is defined as area divided by wetted perimeter or volume divided by surface area, we can now state it as

$$R = \frac{e}{1 - e} \frac{V_p}{A_p} \tag{9-34}$$

For spherical particles of diameter d, $V_p = \pi d^3/6$ and $A_p = \pi d^2$. Substituting these values, one obtains

$$\frac{V_p}{A_p} = \frac{d}{6} \tag{9-35}$$

Considering that the granular materials used in practice are usually not spherical, it is necessary to correct Eq. 9-35 by inserting a dimensionless particle shape factor ϕ.[3,9] The general relationship then becomes

$$\frac{V_p}{A_p} = \phi \frac{d}{6} \tag{9-36}$$

where ϕ equals unity for spherical particles. Carman has reported shape factors of 0.73 for pulverized coal, 0.95 for Ottawa sand, 0.82 for rounded sand, and 0.73 for angular sand.[36] These values can be used as a guide, but for reliable results, an analysis of the material to be used should be made. Settling velocity determinations afford one means for evaluating the particle-shape factor.

The velocity of flow downward just above the filter bed is given by Q/A, where A is the surface area of the filter. When the fluid enters the filter, the cross-sectional area is reduced as a result of the space occupied by the filter media. Consequently, the velocity through the interstices of the filter exceeds the face or approach velocity. The face velocity is thus

$$V_s = ev \tag{9-37}$$

where V_s = face or approach velocity
v = mean velocity through the filter
e = bed porosity

By substituting V_s/e for v, and $\phi ed/6(1-e)$ for R in Eq. 9-32, the following relationship for head lost through the filter is obtained:

$$H_L = f_1 \left(\frac{L}{\phi d} \right) \left(\frac{1-e}{e^3} \right) \left(\frac{V_s^2}{g} \right) \tag{9-38}$$

This is known as the *Carmen–Kozeny relationship*.[35,37] The dimensionless friction factor f_1 can be determined in the following manner[3,38]:

$$f_1 = 150 \frac{1-e}{\text{Re}} + 1.75 \tag{9-39}$$

where

$$\text{Re} = \phi \frac{\rho V_s d}{\mu} \tag{9-40}$$

Equation 9-38 is applicable to the determination of head loss in a filter bed consisting entirely of particles of a specific size. The equation can be used for beds of mixed particles and for stratified beds, however, by making a slight modification. By solving Eq. 9-36 for d, substituting this relationship for d in Eq. 9-38, and substituting the total volume V and surface area A of all the particles in the bed for the volume V_p and surface area of an individual particle A_p, a relationship applicable to mixed beds can be obtained. This relationship is of the form

$$h_L = f_1 \frac{L}{6} \frac{1-e}{e^3} \frac{V_s^2}{g} \frac{A}{V} \qquad (9\text{-}41)$$

For particles having a uniform shape which are packed homogeneously, the average area-volume ratio for the bed, based on Eq. 9-36, is[9]

$$\left(\frac{A}{V}\right)_{av} = \frac{6}{\phi} \int_{x=0}^{x=1} \frac{dx}{d} \qquad (9\text{-}42)$$

where dx is the proportion of particles of a specific size d. The value of $(A/V)_{av}$ is customarily determined on the basis of a sieve analysis by using

$$\left(\frac{A}{V}\right)_{av} = \frac{6}{\phi} \sum \frac{x}{d} \qquad (9\text{-}43)$$

where x represents the weight fraction of particles retained between adjacent sieve sizes and d is the geometric mean size of the adjacent sieve openings. Substitution of the value of $(A/V)_{av}$ determined in this manner in Eq. 9-41 will permit computation of the head loss in a homogeneously packed bed of uniformly shaped particles. This type of packing is found in slow sand filters.

For stratified beds, an additional consideration must be made. In the case of the homogeneous bed, one value of f_1 can be used to represent the entire bed. For a stratified bed, each layer will have a different value of the friction factor f_1, since the representative particle size in each layer will be different.

If the porosity of the stratified bed is uniform and the particles are of uniform shape, Eq. 9-38 can be written as

$$\frac{dh_L}{dL} = Kf_1 \frac{1}{d} \qquad (9\text{-}44)$$

for a particular stratum under some set of operating conditions. To obtain the total head lost through the filter depth L, Eq. 9-44 must be integrated:

$$h_L = \int_0^{h_L} dh_L = K \int_0^L \frac{f_1}{d} dL \qquad (9\text{-}45)$$

Noting that $dL = L\, dx$, where dx represents the proportion of particles of a size d, Eq. 9-45 becomes

$$h_L = KL \int_{x=0}^{x=1} f_1 \frac{dx}{d} \qquad (9\text{-}46)$$

If the particles between adjacent sieve sizes are considered uniform, Eq. 9-46 takes the form

$$h_L = LK \sum \frac{f_1 x}{d} \qquad (9\text{-}47)$$

where

$$K = \frac{1}{\phi} \frac{1-e}{e^3} \frac{V_s^2}{g} \tag{9-48}$$

Equation 9-47 can be used to compute the head loss in stratified filter beds. It is therefore applicable to conditions encountered in a rapid-sand-filter plant.

Rose Equation

A second relationship for determining the head lost through filter beds was derived experimentally by Rose.[34] It is applicable to filters having uniform spherical or nearly spherical particles. This equation has been widely used in hydraulic computations for rapid sand filters. The equation is of the form

$$h_L = \frac{1.067}{\phi} \frac{C_D}{g} L \frac{V_s^2}{e^4} \frac{1}{d} \tag{9-49}$$

where C_D is a coefficient of drag and the other variables are as defined previously. Figure 9-11 indicates a relationship between C_D and the Reynolds number for spherical particles.

By relating the particle diameter d to the area/volume ratio, Eq. 9-49 can be used for homogeneous or stratified beds in a manner similar to that indicated for the Carmen–Kozeny equation.[9] For homogeneous mixed beds the equation takes the form

$$h_L = \frac{1.067}{\phi} \frac{C_D}{g} L \frac{V_s^2}{e^4} \sum \frac{x}{d} \tag{9-50}$$

For stratified beds with uniform porosity, the equation takes the form

$$h_L = \frac{1.067}{\phi} \frac{L}{g} \frac{V_s^2}{e^4} \sum \frac{C_D x}{d} \tag{9-51}$$

In both equations the summation terms can be evaluated by making use of a sieve analysis in a manner similar to that already indicated. An example will serve to illustrate the manner in which the equations for head loss can be used.

●EXAMPLE 9-12

Determine the initial head loss in a rapid sand filter which is 24 in. deep and has a porosity of 0.40. Assume a water temperature of 50°F and and a filtration rate of 2.5 gpm/ft². A sieve analysis of the sand is given in Table 9-3. Consider that the sand has a particle-shape factor ϕ of 0.95.

○Solution

  1 Columns (1), (2), and (3) are given data related to the sieve analysis and standard sieve sizes.

TABLE 9-3

SIEVE ANALYSIS AND COMPUTATION OF HEAD LOSS FOR EXAMPLE 9-12

(1) Sieve Number	(2) Percentage of Sand Retained ($\times 100$)	(3) Geometric Mean Size, d (ft $\times 10^3$)	(4) Reynolds Number $\dfrac{\rho V_s d}{\mu}$	(5) C_D	(6) $C_D \dfrac{x}{d}$
14–20	1.10	3.28	1.30	18.5	62.0
20–28	6.60	2.29	0.91	26.4	761.0
28–32	15.94	1.77	0.70	34.3	3,090.0
32–35	18.60	1.51	0.60	40.0	4,930.0
35–42	19.10	1.25	0.49	49.0	7,480.0
42–48	17.60	1.05	0.42	57.2	9,600.0
48–60	14.30	0.88	0.35	68.6	11,150.0
60–65	5.10	0.75	0.30	80.0	5,450.0
65–100	1.66	0.59	0.23	104.4	2,940.0
Summations	100.00				$\sum \dfrac{C_D x}{d} = 45{,}463.0$

2 Column (4) is determined from

$$\text{Re} = \frac{\rho V_s d}{\mu}$$

where $\mu = 8.78 \times 10^{-4}$ lb-mass/ft-sec for water at 50°F
V_s = face velocity of the water, fps, $= 2.5$ gpm/ft$^2 \times 2.228 \times 10^{-3}$ fps
ρ = mass density of water $= 62.4$ lb-mass/ft^3
d = geometric mean particle size from column (3)

3 For Reynolds numbers less than about 1.9, $C_D = 24/\text{Re}$.[5]

4 The head loss h_L is computed by using Eq. 9-51 with $\phi = 0.95$, $L = 2$ ft, and $e = 0.40$.

$$h_L = \frac{1.067}{\phi} \frac{L}{g} \frac{V_s^2}{e^4} \sum \frac{C_D x}{d}$$

$$= \frac{1.067}{0.95} \times \frac{2.0}{32.2} \times \frac{(5.57 \times 10^{-3})^2}{(0.40)^4} \times 45{,}463.0$$

$$= 3.84 \text{ ft}$$

9-27

HYDRAULICS OF EXPANDED BEDS

After a filter has been in operation for a period of time, the operating-head loss increases as a result of the suspended matter which has been collected. When the head loss has reached the point (usually 8–10 ft) where the flow controller from the filter to the clear well is wide open, an additional loss of head would produce a reduction in flow. To prevent this, it becomes necessary to remove the particles which have been trapped in the filter bed. This removal may be accomplished by scraping off the clogged portion of the bed, or by reversing the flow through the bed so that it is expanded and the trapped particles can be washed out. This latter procedure, backwashing, is the method used to cleanse rapid sand filters.

The backwashing process is carried out by reversing the flow in the under-drainage system so that it moves upward through the filter bed. For very low upflow velocities the bed remains fixed, but as the velocity is increased, the lighter particles begin to be moved upward. The velocity at which a given size particle is suspended or "fluidized" is known as the critical velocity. As the velocity is further increased, the particles become more widely separated and behave in an unhindered manner. The lighter soiled materials which have been previously trapped are then freed and pass to waste with the wash water. During the expansion of the bed the trapped particles are dislodged by the shearing action of the water or by the abrasive action resulting from contacts made between the rising bed particles. This scouring action is an important phase of the cleansing operation, and it can be enhanced by agitating the expanded filter bed with hydraulic jets, mechanical rakes, and compressed air have also been employed.

Uniform Beds

When a bed of uniform particles is subjected to backwashing, the bed just begins to open up when the backwash velocity equals the critical velocity of the particles. At this time the effective weight of the particles in the water is exactly balanced by the upward drag on the particles resulting from the upflow velocity. If the velocity is increased above the critical value, the bed opens further. This does not improve the cleansing action of the upflow but may be important in allowing sufficient open space for the trapped suspended matter to be washed away. A study of Fig. 9-34 helps to clarify the discussion of filter-bed expansion which follows.

Consider a filter bed composed of uniform particles of sand. During normal filter operation, these particles occupy the depth L in Fig. 9-34. During backwash the bed expands to depth L_e. As the critical velocity V_c is reached, there is a balance between the frictional resistance of the particles and the head loss of the fluid in expanding the bed.[3,9] Stated mathematically,

$$h_L \gamma = (\gamma_s - \gamma)(1 - e_e)L_e \qquad (9\text{-}52)$$

where h_L = head loss during expansion

γ_s, γ = specific weights of the particles and fluid, respectively

L_e = expanded depth

e_e = porosity of the expanded bed

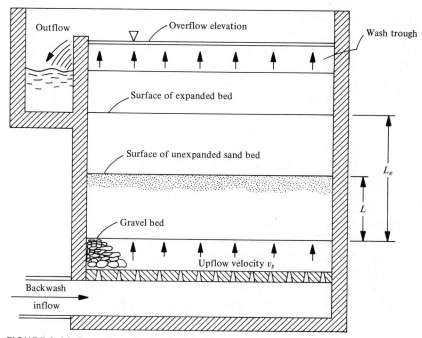

FIGURE 9-34 Expansion of a filter bed during backwashing.

The frictional pressure drop $h_L \gamma$ is just balanced by the weight of the particles in the fluid. Rearranging,

$$h_L = L_e \frac{\gamma_s - \gamma}{\gamma} (1 - e_e)$$ (9-53)

The solution of this equation depends upon a knowledge of the porosity of the expanded bed e_e. For a bed of uniform sand grains, Fair and Geyer have indicated that e_e can be determined by using[9]

$$e_e = \left(\frac{V_s}{v_s}\right)^{0.22}$$ (9-54)

where v_s = terminal settling velocity of the particles
V_s = face or upflow velocity of the fluid

Thus a uniform bed of particles will expand when

$$V_s = v_s e_e^{4.5}$$ (9-55)

The relative expansion of a bed of uniform sand grains can be determined by considering that the volume of sand in the unexpanded bed is exactly equal to that of the expanded bed. Thus it can be written that

$$L(1 - e)A_b = L_e(1 - e_e)A_b$$ (9-56)

where A_b equals the surface area of the bed. Therefore,

$$L_e = L \left(\frac{1 - e}{1 - e_e} \right) \tag{9-57}$$

or

$$L_e = \frac{L(1 - e)}{1 - (V_s/v_s)^{0.22}} \tag{9-58}$$

From this it can be seen that the required flow rate of the wash water to obtain a given expansion depends upon the settling velocity of the particles at the operating temperature.

Stratified Beds

For stratified beds, expansion takes place successively for the different layers. The surface strata are expanded at lower rates of backwash than the deeper ones. The bed is fully expanded when Eq. 9-55 is satisfied for the largest particles.

The relative expansion of the bed can be determined by

$$L_e = L(1 - e) \sum \frac{x}{1 - e_e} \tag{9-59}$$

where x is the fraction by weight of particles having a particular expanded porosity e_e.[9] The value of e_e is obtained by using Eq. 9-54.

●EXAMPLE 9-13

Using the data given in Example 9-12, find: (a) the backwash rate required to expand the bed; (b) the head loss; (c) the depth of the expanded bed. For this filter, $e = 0.40$, $L = 24$ in., and $T = 50°F$. A sieve analysis of the sand is given in Table 9-4. Assume that the specific gravity of the sand grains is 2.65.

○*Solution*

1　Columns (1)–(3) are data provided on the sand filter.
2　To determine the backwash rate, it is necessary to first compute the settling velocity of the largest particle. This is accomplished by using the equation

$$v_s = \left(\frac{4}{3} \frac{g}{C_D} \frac{\rho_s - \rho}{\rho} d \right)^{1/2} \tag{9-19}$$

and a relationship between C_D and Re for Reynolds numbers between 1.9 and 500 (the normal range for this type of problem). The relationship is[3]

$$C_D = \frac{18.5}{Re^{0.6}} \tag{9-60}$$

TABLE 9-4

SIEVE ANALYSIS AND COMPUTATIONS FOR EXAMPLE 9-13

(1) Sieve Number	(2) Percentage of Sand Retained ($\times 100$)	(3) Geometric Mean Size, d (ft $\times 10^3$)	(4) Settling Velocity v_s (fps $\times 10^2$)	(5) Porosity of Expanded Bed, e_e	(6) $\dfrac{x}{1 - e_e}$
14–20	1.10	3.28	46.3	0.378	0.0177
20–28	6.60	2.29	30.6	0.414	0.1125
28–32	15.94	1.77	22.7	0.443	0.287
32–35	18.60	1.51	18.9	0.460	0.344
35–42	19.10	1.25	15.3	0.483	0.370
42–48	17.60	1.05	12.8	0.502	0.353
48–60	14.30	0.88	10.2	0.521	0.299
60–65	5.10	0.75	8.1	0.555	0.1145
65–100	1.66	0.59	6.4	0.585	0.040

$$\sum \frac{x}{1 - e_e} = 1.9377$$

By setting $(\rho_s - \rho)/\rho = S_g - 1$, where S_g is the specific gravity of the particles, and by substituting Eq. 9-60 in Eq. 9-19, the following relationship between v_s and d is obtained:

$$v_s = 313 d^{1.14}$$

For the maximum size-particle, this yields

$$v_s = 313(3.28 \times 10^{-3})^{1.14} \qquad \text{and} \qquad v_s = 46.3 \times 10^{-2} \text{ fps}$$

The backwash rate is then obtained by using

$$V_s = v_s e^{4.5} \tag{9-55}$$

for the largest particle. Thus

$$V_s = 0.463(0.40)^{4.5}$$
$$= 7.42 \times 10^{-3} \text{ fps}$$
$$= 0.45 \text{ ft/min}$$

3 The head loss is computed using Eq. 9-53 after substituting $(1 - e)L$ for $(1 - e_e)L_e$, an equivalent expression as shown in Eq. 9-57.

$$h_L = L \frac{\rho_s - \rho}{\rho} (1 - e) \tag{9-53}$$

$$= L(S_g - 1)(1 - e)$$
$$= 2.0(1.65)(1 - 0.40)$$
$$= 1.98 \text{ ft}$$

4   Column (4) values are computed using the expression $v_s = 313d^{1.14}$.
5   Compute column (5) values using Eq. 9-54.
6   The depth of the expanded bed is found by using

$$L_e = L(1 - e) \sum \frac{x}{1 - e_e} \tag{9-59}$$

$$= 2.0(1 - 0.4) \times 1.9377$$
$$= 2.32 \text{ ft}$$

9-28

DESIGN OF FILTER UNITS

Design of a filter unit must involve capacity of the filter unit, volumetric dimensions of the unit, placement of wash-water troughs, depth of filter medium and gravel layer, the underdrainage system, and various associated equipment or filter appurtenances.

Filter Capacity
Operating rates for rapid-sand-filter plants have been essentially duplicated from plant to plant in the last 50 years, owing largely to a lack of knowledge of filter performance over a wider range of loads. In addition, where public health is concerned, there is a hesitancy to break with tradition. Normally, gravity filters are designed to operate at some multiple or fraction of 1 mgd so that accessories (meters, flow controllers, etc.), which are generally standardized for these rates, can be used. Past practice has been to design rapid sand filters to operate at a load of about 2 gpm/ft.^{2} More recent trends have been toward higher rates up to 5 gpm/ft^2, and future designs will probably follow this trend.

Length, Width, and Depth of a Filter Unit
Rapid filter units are usually rectangular in shape with an average length/width ratio of about 1.25. The surface area of a filter unit commonly ranges from about 450 to 4500 ft^2. Choice of a rectangular section is based primarily on optimum utilization of space and the most efficient system of piping between filter units. These criteria must be evaluated individually for each specific design; hence only a general guide can be given. In the final analysis, economic considerations (dependent largely on piping) dictate the choice.

Depth of a filter unit should be as small as possible for maximum economy and is controlled by the minimum permissible distance from filter bottom to freeboard required above the wash-water trough or by the maximum operating head of the filter. Usually, the overall depth of a filter unit is in excess of 8 ft.

Depth of Filter Medium
Theoretically, thickness of the filter medium should be based on the depth to which impurities will penetrate the filter during a prescribed mode of operation. This depth is generally a function of the characteristics of the materials to be removed by filtration and the manner in which these materials react to removal

by filtration. Rate of filtration, grain size and kind of stratification, porosity of the filter medium, water temperature, and final head loss all affect the removal reaction.

Water applied to rapid sand filters has been coagulated and settled previously. Under these circumstances the load applied can be expressed reasonably in terms of the percentage of iron or aluminum present. Hudson, Stanley, and others have developed empirical relationships for determining depth of penetration of various flocs for several conditions of operation.[39] These relationships are limited in applicability, however, and additional studies are required. A paucity of information on the reaction of filters to various load conditions explains to a large extent why depths of filter materials have not varied significantly in past years.

Wash-Water Troughs

Wash-water troughs are placed above the sand to remove soiled water released from the filter during the backwash operation. These troughs are constructed in a variety of cross sections. Materials commonly used include concrete, steel, cast iron, aluminum, and asbestos cement.

Arrangements of wash-water gutters vary, but it is considered good practice to restrict the horizontal flow distance to 3 ft or less for any one gutter. Edge-to-edge distances of parallel gutters should therefore not exceed 6 ft. An important consideration is that the overflow weirs along gutter edges be level and all troughs be set at the same elevation. The weir along the trough edge should be placed far enough above the sand to preclude loss of fine sand during the backwash operation. At the same time, it should not be set so high that a considerable quantity of dirty water is left in the filter after washing. Normally, the weir should be slightly above the maximum expanded sand depth. The bottom of the wash-water trough should be at least 12 in. above the unexpanded bed. An approximate method for setting the wash-water trough is to place the weir above the surface of the unexpanded bed a distance equal to the rise of the wash water in 1 minute.

The cross-sectional area of the gutter is determined by considering the weirs along the edges of the troughs to act as side-channel spillways. Theoretical and empirical approaches to this problem are available in the literature.[5,9]

Filter Underdrainage

Underdrainage systems serve the dual purpose of furnishing an outlet for filtered water and providing a means for distributing wash water during cleaning operations. It is very important that the rate of removal of filtered water be uniform over the entire filter bottom and that the backwash water be distributed uniformly as well. Since backwash rates are considerably greater than filtration rates, they govern the design of underdrainage systems. Several types of underdrainage systems are in use today. They may be categorized generally as perforated-pipe, pipe and strainer, vitrified-tile block with orifices, porous plates, and precast-concrete underdrains. Detailed information on the various types of underdrainage systems may be found in manufacturers' literature.

Uniform distribution of wash water is obtained by maintaining high head losses in orifices or by using porous plates or clay blocks to form a double bottom. If pipe lateral systems are used, about 15 psi pressure should be maintained at the orifices or strainers, and flow velocities should not exceed about 8 fps. In general, 6- to 12-in. spacings of laterals yield satisfactory results. Laterals should not be longer than about 60 diameters if uniform pressure and equal distribution of flows are to be maintained. Normally, total orifice area should be approximately 0.2–0.3 % of filter surface area. Orifices in perforated pipes are commonly 0.25–0.5 in. in diameter and spaced from about 3 to 8 in. apart. For 0.25-in. orifices the lateral cross-sectional area should total about twice the total orifice area. For 0.5-in. orifices the lateral cross-sectional area should be approximately 4 times the orifice area.

Headers supplying the underdrainage system should deliver the required wash-water flow at a velocity of about 6–8 fps. Cross-sectional area of the header is usually about 1.5 times the total orifice area.

Other Equipment

Other equipment associated with water filtration includes gate valves, metering devices, flow-rate controllers, surface washing equipment, loss-of-head gages, and wash-water controllers. Figure 9-28 is a typical rapid-sand-filter unit showing piping layout and placement of some filter appurtenances. Information on most items is readily available from numerous manufacturers.

PROBLEMS

9-1 A venturi meter with a 6.25-in. throat diameter shows a pressure differential of 147 in. of water. Calculate the water flow using a discharge coefficient of 0.93 and a unit weight of 62.4 pcf.

9-2 Compute the upper head in a Parshall flume with a throat width of 1.5 ft for a flow of 5.56 cfs.

9-3 A wastewater bar screen is constructed using 0.25-in. wide bars spaced 2 in. apart center to center. If the approach velocity in the channel is 2.0 fps, what is the velocity through the screen openings?

9-4 A flocculation basin equipped with revolving paddles is 60 ft long (the direction of flow), 45 ft wide, and 14 ft deep, and treats 10 mgd. The power input to provide paddle-blade velocities of 1.0 and 1.4 fps, for the inner and outer blades, respectively, is 930 ft-lb/sec. Calculate the detention time, horizontal flow-through velocity, and G (the mean velocity gradient).

9-5 A water-treatment plant is designed to process 20 mgd. The flocculation tank must be 15 ft deep to accommodate four horizontal-shaft paddle flocculators. Assume the mean flow velocity to be 28 % of the paddle velocity and $C_D = 1.9$. Design the flocculation tank and paddles. Assume a water temperature of 50°F.

9-6 A fine sand particle 0.10 mm in diameter settles 1 ft in 38 seconds at a water temperature of 50°F. Compute the Reynolds number and find out if Stokes' law is applicable.

9-7 A settling-column analysis is performed on a dilute suspension of discrete particles. Data collected from samples taken at the 5-ft depth are as follows:

Time Required to Settle 5 ft (min)	0.7	1.2	2.3	4.6	6.3	8.8
Portion of Particles with Velocities Less Than Those Indicated	0.58	0.49	0.34	0.16	0.07	0.03

Find the overall removal if the overflow (clarification rate) of the basin is 0.09 fps.

9-8 Use the data below on interface height versus time to find the minimum acceptable surface area of the clarifier. The influent solids concentration is 0.168 pcf and is settled in a continuous flow unit operated at a rate of 1.2 cfs. The desired underflow concentration is 0.80 pcf.

Interface Height (ft)	5.0	4.0	3.0	2.3	1.8	1.4	1.15	1.0	0.95
Time (min)	0	2.6	5.0	5 4	10.0	15.0	20.0	25.0	30.0

9-9 A 100-ft-diameter 10-ft-deep circular clarifier has an influent flow of 7 mgd. Compute the overflow rate and detention time.

9-10 A wastewater-treatment plant has two primary clarifiers, each 20 m in diameter with a 2-m side-water depth. The effluent weirs are inboard channels set on a diameter of 18 m. For a flow of 12,900 m^3/day, calculate overflow rate, detention time, and weir loading. Express answers in units of m^3/m^2-day, hours, and m^3/m-day, respectively.

9-11 Calculate diameter and depth of a circular sedimentation basin for a design flow of 3800 m^3/day based on an overflow rate of 0.000,24 m/sec and a detention time of 3 hours.

9-12 A rectangular sedimentation basin is to be designed for a flow of 1.0 mgd using a 2:1 length/width ratio, an overflow rate of 0.000,77 fps, and a detention time of 3.0 hours. What are the dimensions of the basin?

9-13 A wastewater-treatment plant has two rectangular primary settling tanks, each 40 ft long, 12 ft wide, and 7 ft deep. The effluent weir length in each tank is 45 ft. The average daily wastewater flow is 387,000 gal. Calculate the overflow rate and effluent weir loading, and compare these values to the GLUMRB *Standards*. What is the estimated BOD removal?

9-14 Settling velocity of alum floc is approximately 0.0014 fps in water at 10°C. Compute the equivalent overflow rate in gpd/ft². What is the minimum detention time in hours to settle out alum floc in an ideal basin with a depth of 10 ft?

9-15 List the GLUMRB design criteria for a final clarifier following a step-aeration, activated-sludge basin with a design capacity of 5 mgd. Express these parameters in metric units.

9-16   The design flow for a contact stabilization wastewater-treatment plant is 0.80 mgd. Determine the diameter and side-wall depth required for a circular final clarifier based on the GLUMRB *Standards* (Table 9-2).

9-17   Calculate the dimensions for a channel-type grit-removal unit based on an average wastewater flow of 2 mgd and a peak flow of 4 mgd. Use a square cross-sectional area at peak flow, a horizontal velocity of 1.0 fps, and a settling velocity of 0.075 fps.

9-18   An aerated clarifier-type unit for grit removal and preaeration is 12 ft square with an 8-ft liquid depth. The wastewater flow is 0.80 mgd with an estimated grit volume of 3 ft^3/mg. A separate hopper-bottomed grit storage tank has a usable volume of 3 yd^3. Compute the detention time in the aerated unit and the estimated length of time required to fill the storage tank with grit.

9-19   Why must rapid sand filtration be preceded by chemical coagulation and clarification?

9-20   In a gravity-flow rapid sand filter, how can the head-loss gage record a 9-ft loss when the water depth above the sand filter surface is only $3\frac{1}{2}$ ft?

9-21   Water is to be filtered through 18 in. of uniform sand having a porosity of 0.42 and a grain diameter of 0.9×10^{-3} ft. Assume spherical particles and a water temperature of 55°F. The surface loading is 2.5 gpm/ft^2. Find the head loss using the Carmen–Kozeny equation.

9-22   Solve Prob. 9-17 using the Rose equation.

9-23   What will be the head loss of Prob. 9-17 if a particle-shape factor of 0.82 is introduced?

9-24   Find the initial head loss in a rapid sand filter which is 20 in. deep and has a porosity of 0.42. Assume a water temperature of 55°F and a filtration rate of 2 gpm/ft^2. Consider spherical particles. The sieve analysis is given below. Use the Carmen–Kozeny equation.

SIEVE ANALYSIS FOR PROB. 9-24

Sieve Number	Percentage of Sand Retained	Geometric Mean Size, d (ft $\times$ 10^3)
14–20	1.05	3.28
20–28	6.65	2.29
28–32	15.70	1.77
32–35	18.84	1.51
35–42	18.98	1.25
42–48	17.72	1.05
48–60	14.25	0.88
60–65	5.15	0.75
65–100	1.66	0.59
	100.00	

9-25 Solve Prob. 9-24 using the Rose equation.

9-26 Solve Prob. 9-24 if the water temperature is 45°F and $\phi = 0.95$.

9-27 Solve Prob. 9-24 if the water temperature is 60°F and $\phi = 0.95$.

9-28 Consider the expansion of a bed of uniform particles 20 in. thick. The porosity of the bed is 0.40 and water temperature is 50°F. The specific gravity of particles is 2.65. Find: (a) the upflow velocity needed to just open the bed; (b) the head lost in expanding the bed.

9-29 Use the sieve analysis given in Prob. 9-24 and find: (a) the backwash rate required to expand the bed; (b) the head loss; (c) the depth of the expanded bed. Assume porosity $= 0.43$, $L = 24$ in., and $T = 50°F$. Specific gravity of the particles $= 2.65$.

9-30 Solve Prob. 9-29 if the depth of the unexpanded bed is 20 in. and $T = 55°F$.

REFERENCES

1. R. L. Parshall, "The Parshall Measuring Flume," *Colorado Experimental Station Bulletin No. 423* (1936).
2. R. E. Greenhalgh, R. L. Johnson, and H. D. Nott, "Mixing in Continuous Reactors," *Chem. Eng. Progr.* 55 (February 1959); 44–48.
3. L. G. Rich, *Unit Operations of Sanitary Engineering* (New York: John Wiley & Sons, Inc., 1961).
4. E. R. Gilliland and E. A. Mason, "Gas Mixing in Beds of Fluidized Solids," *Ind. Eng. Chem.* 44 (1952).
5. J. H. Rushton and J. Y. Oldshue, "Mixing of Liquids," *Chem. Eng. Progr. Symp. Series* 55, no. 25 (1959).
6. T. R. Camp, "Flocculation and Flocculation Basins," *Trans. Am. Soc. Civil Engrs.* 120 (1955).
7. S. L. Tolman, "The Mechanics of Mixing and Flocculation," *Public Works* (December 1962).
8. T. R. Camp and P. C. Stein, "Velocity Gradients and Internal Work in Fluid Motion," *J. Boston Soc. Civil Engrs.* 30 (1943).
9. G. M. Fair and J. C. Geyer, *Water Supply and Waste-Water Disposal* (New York: John Wiley & Sons, Inc., 1961).
10. E. L. Beam, "Study of Physical Factors Affecting Flocculation," *Water Works Eng.* (January 1953).
11. N. Drobny, "Effect of Paddle Design on Flocculation," *Proc. Am. Soc. Civil Engrs., J. San. Eng. Div.* 89, no. SA2, part I (1963).
12. Recommended Standards for Water Works, Great Lakes–Upper Mississippi River Board of State Sanitary Engineers, 1968 Edition (Albany, N.Y.: Health Education Service).
13. R. M. Olson, *Essentials of Engineering Fluid Mechanics*, 3rd Ed. (New York: IEP Publishers, 1973).
14. T. R. Camp, "Studies of Sedimentation Basin Design," *Sewage Ind. Wastes* 25, no. 1 (January 1953).
15. W. W. Eckenfelder, Jr., *Industrial Water Pollution Control* (New York: McGraw-Hill Book Company, 1966), pp. 28–51.
16. W. P. Talmadge and E. B. Fitch, "Determining Thickener Unit Areas," *Ind. Eng. Chem.* 47 (January 1955).

17. J. M. Coulson and J. F. Richardson, *Chemical Engineering*, vol. II (New York: McGraw-Hill Book Company, 1955).

18. E. B. Fitch, "The Significance of Detention in Sedimentation," *Sewage Ind. Wastes* 29, no. 10 (1957).

19. A. B. Morrill, "Sedimentation Basin Research and Design," *J. Am. Water Works Assoc.* 24 (1932).

20. R. A. Fiedler and E. B. Fitch, "Appraising Basin Performance from Dye Test Results," *Sewage Ind. Wastes* 31, no. 9 (1959).

21. E. B. Fitch, "Flow Path Effect on Sedimentation," *Sewage Ind. Wastes* 28, no. 1 (January 1956).

22. S. L. Tolman, "Sedimentation Basin Design and Operation," *Public Works* (June 1963).

23. E. B. Fitch and W. A. Lutz, "Feedwells for Density Stabilization," *J. Water Poll. Control Fed.* 32, no. 2 (1960).

24. A. C. Ingersoll, J. E. McKee, and N. H. Brooks, "Fundamental Concepts of Rectangular Settling Tanks," *Trans. Am. Soc. Civil Engrs.* 121 (1956).

25. Recommended Standards for Sewage Works, Great Lakes–Upper Mississippi River Board of State Sanitary Engineers, 1971 Edition (Albany, N.Y.: Health Education Service).

26. "Sewage Treatment Plant Design," prepared by a joint committee of the Water Pollution Control Federation and the American Society of Civil Engineers (*WPCF Manual No. 8* or *ASCE Manual No. 36*, 1959).

27. Anonymous, *Microstraining* (New York: Glenfield and Kennedy, Inc., 1956).

28. P. L. Boucher and G. R. Evans, "Microstraining Description and Application," *Water Sewage Works* (November 1962).

29. Anonymous, "The Filtration of Water" (New York: Johns-Manville Co., 1961) *Publ. FA-74A 11–61.*

30. A. Hazen "On Sedimentation," *Trans. Am. Soc. Civil Engrs.* 53 (1904).

31. Committee Report, "Filter Sand for Water Purification Practice," *J. Am. Water Works Assoc.* (August 1953).

32. T. R. Camp. "Theory of Water Filtration," *Am. Soc. Civil Engrs. J. San. Eng. Div.* 90, no. SA4, part I (August 1964).

33. J. D. Walker, "High Rate Filtration," Walker Process Equipment, 1967 (preprint of a paper presented Sept. 13, 1967 at Georgia Water and Pollution Control Assoc. Annual Conf., Atlanta, Ga.)

34. H. E. Rose, "On the Resistance Coefficient–Reynolds Number Relationship for Fluid Flow through a Bed of Granular Material," *Proc. Inst. Mech. Engrs.* (*London*) (1945): 153, and (1949); 154, 160.

35. G. Kozeny, *Sitzber, Akad. Wiss. Wien, Math-Naturw. Kl. Abt.* IIa, vol. 136 (1927).

36. P. C. Carmen, "Fluid Flow through Granular Beds," *Trans. Inst. Chem. Engrs.* (*London*) 15 (1937).

37. P. C. Carmen, *Trans. Inst. Chem. Engrs.* (*London*) 15 (1937).

38. Sabri Ergun, *Chem. Eng. Progr.*, vol. 48 (1952).

39. H. E. Hudson, "Factors Affecting Filtration Rates," *J. Am. Water Works Assoc.* 48 (1956).

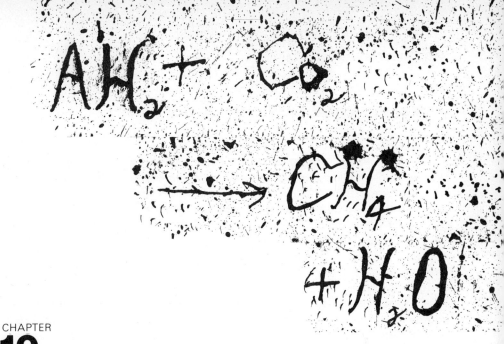

10
Chemical-
Treatment
Processes

Chemical treatment is the most important step in processing public water supplies. Surface water normally requires chemical coagulation to eliminate turbidity, color, and taste and odor-producing compounds, while well water supplies are commonly treated to remove dissolved minerals, such as hardness, iron, and manganese. Through a variety of saline-water conversion processes, saltwater can be converted to fresh water. Chlorine is applied for disinfection and to establish a protective residual in the distribution system, and most utilities fluoridate their water as a health benefit.

Chemical Considerations

The purpose of these initial sections is to refresh students on selected fundamental concepts, including definitions, compounds, units of expression, the bicarbonate–carbonate system, and chemical equilibria. Basic chemistry is also presented as introductory material for each specific unit process discussed in the chapter. These brief comments are not intended to replace formal chemistry ✗ courses prerequisite to sanitary engineering.

409

✗ What formal chemistry?

10-1

INORGANIC CHEMICALS AND COMPOUNDS

Definitions

A table of chemical elements and their atomic weights is given in the Appendix. Atomic weight is the weight of an element relative to that of carbon 12, which has an atomic weight of 12. *Valence* is the combining power of an element relative to that of the hydrogen atom, with an assigned value of 1. Thus an element with a valence of $2+$ can replace two hydrogen atoms in a compound or, in the case of a $2-$ valence, can react with two hydrogen atoms. Equivalent or combining weight of an element is equal to its atomic weight divided by the valence. For example, the equivalent weight of calcium (Ca^{2+}) equals 40.0 divided by 2, or 20.0.

The *molecular weight* of a compound equals the sum of the weights of the combining elements and is conventionally expressed in grams. *Equivalent weight* is the molecular weight divided by the number of positive or negative electrical charges resulting from dissolution of the compound. Consider sulfuric acid, with a molecular weight of 98.1 g. *Ionization* releases two H^+ ions and one SO_4^{2-} radical; therefore, the equivalent weight of sulfuric acid is 98.1 divided by 2, or 49.0

Chemicals Applied in Treatment

The common inorganic compounds used in water and wastewater processing are listed in Table 10-1. Given are the name, formula, common usage, molecular weight, and equivalent weight when appropriate. Common names and purity of commercial-grade chemicals are presented in the sections dealing with specific chemical treatment processes.

Units of Expression

The concentration of ions or chemicals in solution is normally expressed as weight of the element or compound in milligrams per liter of water, abbreviated as mg/l. Occasionally, the term "parts per million" (ppm) is used rather than mg/l. These are identical in meaning, since 1 mg/1,000,000 ml is essentially the same as 1 part by weight to 1 million parts for low concentrations. Chemical dosages may be expressed in units of pounds per million gallons or grains per gallon. To convert milligrams per liter to pounds per million gallons, multiply by 8.34, which is the weight of 1 gal of water. In other words,

$$1.0 \text{ mg/l} = \frac{1 \text{ gal by weight}}{1,000,000 \text{ gal}} = 8.34 \text{ lb/mg}$$

One pound contains 7000 grains, and 1.0 grain per gallon (gpg) equals 17.1 mg/l.

Elemental concentrations expressed in units of mg/l can usually be interpreted to mean that the solution contains the stated number of milligrams of

TABLE 10-1

COMMON CHEMICALS IN WATER AND WASTEWATER PROCESSING

Name	Formula	Common Application	Molecular Weight	Equivalent Weight[a]
Activated carbon	C	Taste and odor control	12.0	n.a.[a]
Aluminum sulfate	$Al_2(SO_4)_3 \cdot 14 \cdot 3H_2O$	Coagulation	600	100
Ammonia	NH_3	Chloramine disinfection	17.0	n.a.
Ammonium sulfate	$(NH_4)_2SO_4$	Coagulation	132	66.1
Calcium hydroxide	$Ca(OH)_2$	Softening	74.1	37.0
Calcium hypochlorite	$Ca(ClO)_2 \cdot 2H_2O$	Disinfection	179	n.a.
Calcium oxide	CaO	Softening	56.1	28.0
Carbon dioxide	CO_2	Recarbonation	44.0	22.0
Chlorine	Cl_2	Disinfection	71.0	n.a.
Chlorine dioxide	ClO_2	Taste and odor control	67.0	n.a.
Copper sulfate	$CuSO_4$	Algae control	160	79.8
Ferric chloride	$FeCl_3$	Coagulation	162	54.1
Ferric sulfate	$Fe_2(SO_4)_3$	Coagulation	400	66.7
Ferrous sulfate	$FeSO_4 \cdot 7H_2O$	Coagulation	278	139
Fluosilicic acid	H_2SiF_6	Fluoridation	144	n.a.
Magnesium hydroxide	$Mg(OH)_2$	Defluoridation	58.3	29.2
Oxygen	O_2	Aeration	32.0	16.0
Potassium permanganate	$KMnO_4$	Oxidation	158	n.a.
Sodium aluminate	$NaAlO_2$	Coagulation	82.0	n.a.
Sodium bicarbonate	$NaHCO_3$	pH adjustment	84.0	84.0
Sodium carbonate	Na_2CO_3	Softening	106	53.0
Sodium chloride	NaCl	Ion-exchanger regeneration	58.4	58.4
Sodium fluoride	NaF	Fluoridation	42.0	n.a.
Sodium fluosilicate	Na_2SiF_6	Fluoridation	188	n.a.
Sodium hexametaphosphate	$(NaPO_3)_n$	Corrosion control	n.a.	n.a.
Sodium hydroxide	NaOH	pH adjustment	40.0	40.0
Sodium hypochlorite	NaClO	Disinfection	74.4	n.a.
Sodium silicate	Na_4SiO_4	Coagulation aid	184	n.a.
Sodium thiosulfate	$Na_2S_2O_3$	Dechlorination	158	n.a.
Sulfur dioxide	SO_2	Dechlorination	64.1	n.a.
Sulfuric acid	H_2SO_4	pH adjustment	98.1	49.0

[a] n.a., not applicable.

that particular element. For example, a water containing 1.0 mg/l of fluoride means that there is 1.0 mg of F ion by weight per liter. However, in some cases, the concentration given in milligrams of weight does not relate to the specific element whose concentration is being expressed. For example, hardness, which is a measure of the calcium ion and magnesium ion content of a water, is given in weight units of calcium carbonate. This facilitates treating hardness as a single value rather than two concentrations expressed in different weight units, one for Ca^{2+} and the other for Mg^{2+}. The alkalinity of a water may consist of one or more of the following ionic forms: OH^-, CO_3^{2-}, and HCO_3^-. For commonality, the concentrations of these various radicals are given as mg/l as $CaCO_3$. According to *Standard Methods*[1], all nitrogen compounds—ammonia, nitrate, and organic nitrogen—are expressed in units of mg/l as nitrogen and phosphates are given as mg/l as phosphorus.

The term milliequivalents per liter (meq/l) expresses the concentration of dissolved substance in terms of its combining weight. Milliequivalents are calculated from milligrams per liter for elemental ions by Eq. 10-1 and for radicals or compounds by Eq. 10–2:

$$meq/l = mg/l \times \frac{valence}{atomic\ weight} = \frac{mg/l}{equivalent\ weight} \tag{10-1}$$

$$meq/l = mg/l \times \frac{electrical\ charge}{molecular\ weight} = \frac{mg/l}{equivalent\ weight} \tag{10-2}$$

Milliequivalents-per-Liter Bar Graph ▷ *completeness of analysis*

Results of a water analysis are normally expressed in milligrams per liter and reported in tabular form. For better visualization of the chemical composition, these data can be expressed in milliequivalents per liter to permit graphical presentation, as illustrated in Fig. 10-1. The top row of the bar graph consists of major cations arranged in the order of calcium, magnesium, sodium, and potassium. Anions in the bottom row are aligned in the sequence of carbonate

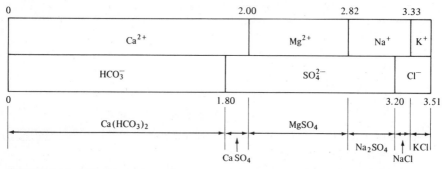

FIGURE 10-1 Milliequivalents-per-liter bar graph for water analysis given in Example 10-1.

(if present), bicarbonate, sulfate, and chloride. The sum of the positive milli-equivalents per liter must equal the sum of the negative values for a water in equilibrium. Hypothetical combinations of positive and negative ions can be written from a bar graph. These combinations are useful in evaluating a water for lime-soda ash softening.

Table 10-2 lists basic data for selected elements, radicals, and compounds. Included are equivalent weights for calculating milliequivalents per liter for use in bar graph presentations and chemical equations.

TABLE 10-2

DATA ON SELECTED ELEMENTS, RADICALS, AND COMPOUNDS

Name	Symbol or Formula	Atomic or Molecular Weight	Equivalent Weight
Aluminum	Al^{3+}	27.0	9.0
Calcium	Ca^{2+}	40.1	20.0
Carbon	C	12.0	
Hydrogen	H^+	1.0	1.0
Magnesium	Mg^{2+}	24.3	12.2
Manganese	Mn^{2+}	54.9	27.5
Nitrogen	N	14.0	
Oxygen	O	16.0	
Phosphorus	P	31.0	
Sodium	Na^+	23.0	23.0
Ammonium	NH_4^+	18.0	18.0
Bicarbonate	HCO_3^-	61.0	61.0
Carbonate	CO_3^{2-}	60.0	30.0
Hydroxyl	OH^-	17.0	17.0
Hypochlorite	OCl^-	51.5	51.5
Nitrate	NO_3^-	62.0	62.0
Orthophosphate	PO_4^{3-}	95.0	31.7
Sulfate	SO_4^{2-}	96.0	48.0
Aluminum hydroxide	$Al(OH)_3$	78.0	26.0
Calcium bicarbonate	$Ca(HCO_3)_2$	162	81.0
Calcium carbonate	$CaCO_3$	100	50.0
Calcium sulfate	$CaSO_4$	136	68.0
Carbon dioxide	CO_2	44.0	22.0
Ferric hydroxide	$Fe(OH)_3$	107	35.6
Hydrochloric acid	HCl	36.5	36.5
Magnesium carbonate	$MgCO_3$	84.3	42.1
Magnesium hydroxide	$Mg(OH)_2$	58.3	29.1
Magnesium sulfate	$MgSO_4$	120	60.1
Sodium sulfate	Na_2SO_4	142	71.0

●EXAMPLE 10-1

The results of a water analysis are: calcium 40.0 mg/l, magnesium 10.0 mg/l, sodium 11.7 mg/l, potassium 7.0 mg/l, bicarbonate 110 mg/l, sulfate 67.2 mg/l, and chloride 11.0 mg/l. Draw a milliequivalents-per-liter bar graph and list the hypothetical combinations. Express the hardness and alkalinity in units of mg/l as $CaCO_3$.

○**Solution**
   Using Eqs. 10-1 and 10-2,

Component	mg/l	Equivalent Weight	meq/l
Ca^{2+}	40.0	20.0	2.00
Mg^{2+}	10.0	12.2	0.82
Na^+	11.7	23.0	0.51
K^+	7.0	39.1	0.18
		Total cations $= 3.51$	
HCO_3^-	110	61.0	1.80
SO_4^{2-}	67.2	48.0	1.40
Cl^-	11.0	35.5	0.31
		Total anions $= 3.51$	

From the bar graph of these data in Fig. 10-1, the hypothetical chemical combinations are:

1.80 meq/l of $Ca(HCO_3)_2$	0.38 meq/l of Na_2SO_4
0.20 meq/l of $CaSO_4$	0.13 meq/l of $NaCl$
0.82 meq/l of $MgSO_4$	0.18 meq/l of KCl

Hardness is the sum of the Ca^{2+} and Mg^{2+} concentrations expressed in mg/l as $CaCO_3$, and alkalinity equals the bicarbonate content. Thus

$$\text{hardness} = 2.82 \text{ meq/l} \times 50 \, \frac{\text{mg/l of } CaCO_3}{\text{meq/l}} = 141 \text{ mg/l}$$

$$\text{alkalinity} = 1.80 \times 50 = 90 \text{ mg/l}$$

10-2

HYDROGEN-ION CONCENTRATION

The hydrogen-ion activity (i.e., intensity of the acid or alkaline condition of a solution) is expressed by the term pH, which is defined as follows:

$$pH = \log \frac{1}{[H^+]} \tag{10-3}$$

Water dissociates to only a slight degree, yielding hydrogen ions equal to 10^{-7} mole/l; thus pure water has a pH of 7. It is also neutral since 10^{-7} mole/l of hydroxyl ion is produced simultaneously,

$$H_2O \rightleftharpoons H^+ + OH^- \qquad (10\text{-}4)$$

When an acid is added to water, the hydrogen-ion concentration increases resulting in a lower pH value. Addition of an alkali reduces the number of free hydrogen ions, causing an increase in pH, because OH^- ions unite with H^+ ions. The pH scale is acid from 0 to 7 and basic from 7 to 14.

The chemical equilibrium of water can be shifted by changing the hydrogen-ion activity in solution. Thus pH adjustment is used to optimize coagulation, softening, and disinfection reactions, and for corrosion control. In wastewater treatment, pH must be maintained in a range favorable for biological activity.

10-3

ALKALINITY AND pH RELATIONSHIPS

Alkalinity is a measure of water's capacity to absorb hydrogen ions without significant pH change (i.e., to neutralize acids). It is determined in the laboratory by titrating a water sample with a standardized sulfuric acid solution. The three chemical forms that contribute to alkalinity are bicarbonates, carbonates, and hydroxides that originate from the salts of weak acids and strong bases. Bicarbonates represent the major form since they originate naturally from reactions of carbon dioxide in water.

Below pH 4.5, dissolved carbon dioxide is in equilibrium with carbonic acid in solution; thus no alkalinity exists. Between pH 4.5 and 8.3, the balance shown in Eq. 10-5 shifts to the right, reducing the CO_2 and creating HCO_3^- ions. Above 8.3 the bicarbonates are converted to carbonate ions. Hydroxide appears at a pH greater than 9.5 and reacts with carbon dioxide to yield both bicarbonates and carbonates (Eq.10-6). The maximum CO_3^{2-} concentration for dilute solutions is in the pH range 10–11. Figure 10-2 shows the relationship between carbon dioxide and the three forms of alkalinity with respect to pH calculated for water having a total alkalinity of 100 mg/l at 25°C. Temperature, total alkalinity and the presence of other ionic species influence alkalinity–pH relationships; nevertheless, Fig. 10-2 is a realistic representation of most natural waters.

$$CO_2 + H_2O \rightleftharpoons H_2CO_3 \rightleftharpoons H^+ + HCO_3^- \qquad (10\text{-}5)$$

$$CO_2 + OH^- \rightleftharpoons HCO_3^- \rightleftharpoons H^+ + CO_3^{2-} \qquad (10\text{-}6)$$

Substances that offer resistance to change in pH as acids or bases are added to a solution are referred to as *buffers*. Since the pH falls between 6 and 9 for most natural waters and wastewaters, the primary buffer is the bicarbonate–carbonate system. When acid is added, a portion of the H^+ ions are combined

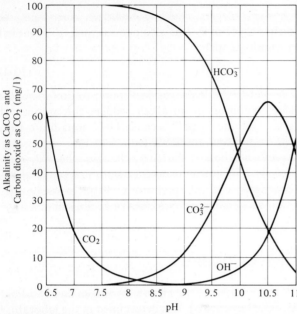

FIGURE 10-2 Carbon dioxide and various forms of alkalinity relative to pH in water at 25°C.

with HCO_3^- to form un-ionized H_2CO_3; only the H^+ remaining free affect pH. If a base is added, the OH^- ions react with free H^+, increasing the pH. However, some of the latter are replaced by a shift of HCO_3^- to CO_3^{2-}, attenuating the change in hydrogen-ion concentration. Both chemical reactions and biological processes depend on this natural buffering action to control pH changes. Sodium carbonate or calcium hydroxide can be added if naturally existing alkalinity is insufficient, such as in the coagulation of water where the chemicals added react with and destroy alkalinity.

Precipitation softening is easily understood by referring to the pH–alkalinity relationship illustrated in Fig. 10-2. Calcium and magnesium ions are soluble when associated with bicarbonate anions. But, if the pH of a hard water is increased, insoluble precipitates of $CaCO_3$ and $Mg(OH)_2$ are formed. This is accomplished in water treatment by adding lime to raise the pH level. At a value of about 10, hydroxyl ions convert bicarbonates to carbonates to allow formation of calcium carbonate precipitate as follows:

$$Ca(HCO_3)_2 + Ca(OH)_2 \longrightarrow 2CaCO_3 + 2H_2O \qquad (10\text{-}7)$$

bicarbonate   lime   solid
hardness .   slurry   precipitate

10-4

CHEMICAL EQUILIBRIA

Most chemical reactions are reversible to some degree, and the concentrations of reactants and products determine the final state of equilibrium. For the general reaction expressed by Eq. 10-8, increase in either A or B shifts the equilibrium to the right, while a larger concentration of either C or D drives it to the left. A reaction in true equilibrium can be expressed by the *mass-action formula*, Eq. 10-9.

$$aA + bB \rightleftharpoons cC + dD \tag{10-8}$$

$$\frac{[C]^c[D]^d}{[A]^a[B]^b} = K \tag{10-9}$$

where A and B = reactants
C and D = products
[ ] = molar concentrations
K = equilibrium constant

Strong acids and bases in dilute solutions approach 100% ionization, while weak acids and bases are poorly ionized. The degree of ionization of the latter is expressed by the mass-action equation. For example, Eqs. 10-10 through 10-13 are, for carbonic acid,

$$H_2CO_3 \rightleftharpoons H^+ + HCO_3^- \tag{10-10}$$

$$\frac{[H^+][HCO_3^-]}{[H_2CO_3]} = K_1 = 4.45 \times 10^{-7} \text{ at } 25°C \tag{10-11}$$

$$HCO_3^- \rightleftharpoons H^+ + CO_3^{2-} \tag{10-12}$$

$$\frac{[H^+][CO_3^{2-}]}{[HCO_3^-]} = K_2 = 4.69 \times 10^{-11} \text{ at } 25°C \tag{10-13}$$

The foregoing characterizes homogeneous chemical equilibria where all reactants and products occur in the same physical state. Heterogeneous equilibrium exists between a substance in two or more physical states. For example, at greater than pH 10, solid calcium carbonate in water reaches a stability with the calcium and carbonate ions in solution,

$$CaCO_{3(s)} \rightleftharpoons Ca^{2+} + CO_3^{2-} \tag{10-14}$$

Equilibrium between crystals of a compound in the solid state and its ions in solution can be treated mathematically as if the equilibrium is homogeneous in nature. For Eq. 10-14 the expression is as follows:

$$\frac{[Ca^{2+}][CO_3^{2-}]}{[CaCO_3]} = K \tag{10-15}$$

Concentration of a solid substance can be treated as a constant K_s in mass-action equilibrium; therefore, $[CaCO_3]$ can be assumed equal to KK_s, and then

$$[Ca^{2+}][CO_3^{2-}] = KK_s = K_{sp} = 5 \times 10^{-9} \text{ at } 25°C \tag{10-16}$$

The constant K_{sp} is called the *solubility-product constant*.

If the product of the ionic molar concentrations is less than the solubility-product constant, the solution is unsaturated. Conversely, a supersaturated solution contains a $[A^+][B^-]$ value greater than K_{sp}. In this case, crystals form and precipitation progresses until the ionic concentrations are reduced equal to those of a saturated solution. Based on Eq. 10-16, the theoretical solubility of calcium carbonate is approximately 7 mg/l.

10-5

WAYS OF SHIFTING CHEMICAL EQUILIBRIA

Chemical reactions in water and wastewater treatment rely on shifting of homogeneous or heterogeneous equilibria to achieve the desired results. The most common methods for completing reactions are by formation of insoluble substances, weakly ionized compounds, gaseous end products, and by oxidation and reduction.

The best example of shifting equilibrium to form precipitates is lime-soda ash softening. Calcium is removed from solution by adding lime, as shown in Eq. 10-7. If insufficient alkalinity is available to complete this reaction, sodium carbonate (soda ash) is also applied. However, magnesium hardness must be removed by forming $Mg(OH)_2$ since $MgCO_3$ is relatively soluble. This is affected by addition of excess lime to increase the value of $[Mg^{2+}][OH^-]^2$ above the solubility product of magnesium hydroxide, which equals 9×10^{-12}.

$$MgCO_3 + Ca(OH)_2 \xrightarrow{\text{excess OH}^-} CaCO_3 \downarrow + Mg(OH)_2 \downarrow \qquad (10\text{-}17)$$

A common example of destroying equilibrium by forming a poorly ionized compound is neutralization of acid or caustic wastes. Here, the combining of hydrogen ions and hydroxyl ions forms poorly ionized water and a soluble salt. For example,

$$2H^+ + SO_4^{2-} + 2Na^+ + 2OH^- \longrightarrow 2H_2O + 2Na^+ + SO_4^{2-} \qquad (10\text{-}18)$$

Reactions involving production of a gaseous product go to practical completion if the gas escapes from solution. One illustration is breakpoint chlorination, which oxidizes ammonia to nitrogen and nitrous oxide gases.

Oxidation and reduction is a very positive method of sending reactions to completion, since one or more of the ions involved in the equilibrium is destroyed. A practical example in water treatment is removal of soluble iron from solution by oxidation using potassium permanganate. In this reaction (Eq. 10-19), the iron gains one positive charge while the manganese in the permanganate ion is reduced from a valence of $7+$ to a valence of $4+$, forming manganese dioxide.

$$Fe(HCO_3)_2 + KMnO_4 \longrightarrow Fe(OH)_3 \downarrow + MnO_2 \downarrow \qquad (10\text{-}19)$$

10-6

COLLOIDAL DISPERSIONS

Colloidal dispersions in water consist of discrete particles held in suspension by their extremely small size (1–200 nm), state of hydration (chemical combination with water), and surface electrical charge. Size of particles is the most significant property responsible for the stability of a *sol* (a colloidal dispersion in a liquid). With larger particles the ratio of surface area to mass is low and mass effects, such as sedimentation by gravity forces, predominate. For colloids, the surface area to mass ratio is high and surface phenomena, for instance, electrostatic repulsion and hydration, become important.

There are two types of colloids—hydrophilic and hydrophobic. *Hydrophilic* colloids are readily dispersed in water, and their stability (lack of tendency to agglomerate) depends upon a marked affinity for water rather than upon the slight charge (usually negative) that they possess. Examples of hydrophilic colloidal materials are soaps, soluble starch, soluble proteins, and synthetic detergents.

Hydrophobic colloids possess no affinity for water and owe their stability to the electric charge they possess. Metal oxide colloids, most of which are positively charged, are samples of hydrophobic sols. A charge on the colloid is gained by adsorbing positive ions from the water solution. Electrostatic repulsion between the charged colloidal particles produces a stable sol.

The concept of *zeta potential* is derived from the diffuse double-layer theory applied to hydrophobic colloids (Fig. 10-3). A fixed covering of positive ions is attracted to the negatively charged particle by electrostatic attraction. This stationary zone of positive ions is referred to as the *Stern layer*, which is surrounded by a movable, diffuse layer of counterions. The concentration of these positive ions in the diffuse zone decreases as it extends into the surrounding bulk of electroneutral solution. Zeta potential is the magnitude of the charge at the surface of shear. The boundary surface between the fixed ion layer and the solution serves as a shear plane when the particle undergoes movement relative to the solution. The zeta potential magnitude can be estimated from electrophoretic measurement of particle mobility in an electric field.

A colloidal suspension is defined as stable when the dispersion shows little or no tendency to aggregate. The repulsive force of the charged double layer disperses particles and prevents aggregation, thus particles with a high zeta potential produce a stable sol.

Factors tending to destabilize a sol are van der Waals' forces of attraction and Brownian movement. *Van der Waals' forces* are the molecular cohesive forces of attraction that increase in intensity as particles approach each other. These forces are negligible when the particles are slightly separated but become dominant when particles contact. *Brownian movement* is the random motion of colloids caused by their bombardment by molecules of the dispersion medium. This movement has a destabilizing effect on a sol because aggregation may result.

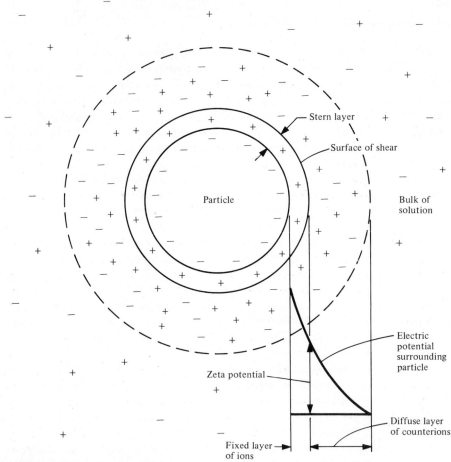

FIGURE 10-3 Concept of zeta potential derived from the diffuse double-layer theory.

Destabilization of hydrophobic colloids can be accomplished by adding electrolytes to the solution. Counterions of the electrolyte suppress the double-layer charge of the colloids sufficiently to permit particles to contact. Upon meeting, van der Waals' forces of attraction become dominant and aggregation results. Electrolytes found to be most effective are multivalent ions of opposite charge to that of the colloidal particles. Sols can also be destabilized by long-chain polyelectrolytes. Although cationic polymers act much the same as neutral salts in suppressing the diffuse double layer, the primary mechanism of organic polyelectrolytes appears to be polymer bridging. The long polymer molecule attaches to absorbent surfaces of colloidal particles by chemical or physical interactions, resulting in aggregation. The destabilizing action of hydrolyzed metal ions (i.e., aluminum and iron salts) appears to fall into an intermediate category between simple ions and polyelectrolytes. Highly charged, soluble hydrolysis products of these metal salts reduce the repulsive forces between colloids by compressing the double-layer charge, bringing on coagulation. Hydrolyzed metal ions are also adsorbed on the colloids, creating bridges between the particles.

In contrast to the electrostatic nature of hydrophobic colloids, stability of hydrophilic colloids is related to their state of hydration (i.e., their marked affinity for water). Chemical coagulation does not materially affect the degree of hydration of colloids. Therefore, hydrophilic colloids are extremely difficult to coagulate and heavy doses of coagulant salts, often 10–20 times the amount used in conventional water treatment, are needed for destabilization.

10-7

COAGULATION AND FLOCCULATION

Removal of turbidity by coagulation depends upon the nature and concentration of the colloidal contaminants; type and dosage of chemical coagulant; use of coagulant aids; and chemical characteristics of the water, such as pH, temperature, and ionic character. Because of the complex nature of coagulation reactions, chemical treatment of water supplies is based primarily upon empirical data derived from laboratory and field studies. However, in recent years a considerable amount of research has been directed toward gaining a better understanding of coagulation mechanisms.

In destabilizing colloids, two basic mechanisms have been described as helping form sufficiently large aggregates to facilitate settling from suspension. The first, referred to as *coagulation*, reduces the net electrical repulsive forces at particle surfaces by electrolytes in solution. The second mechanism, known as *flocculation*, is aggregation by chemical bridging between particles.

In water-treatment practice, chemical coagulation and flocculation are also considered to be dependent upon physical processes. Choice of coagulant dosage, pH, and coagulant aids are related to the mixing process promoting aggregation of the destabilized colloids. And efficiency of the coagulation–flocculation system depends upon subsequent settling and filtration. Trace quantities of impurities frequently present in natural waters (e.g., color and silica) can have significant effects on both the chemical and physical properties of flocs formed during coagulation–flocculation, and thereby alter their settling and filtering characteristics.

Traditionally, sanitary engineers have not restricted the use of the terms "coagulation" and "flocculation" to describing chemical mechanisms only. Common use of these terms refers to both chemical and physical processes in treatment, possibly because the complex reactions that take place in chemical coagulation–flocculation are only partially understood. More important is the fact that engineers tend to tie coagulation to the operational units (mixing devices and flocculation chambers) used in chemical treatment. The physical processes of mixing and flocculation are described in Sections 9-7 and 9-8. Coagulation concerns the series of chemical and mechanical operations by which coagulants are applied and made effective. These operations are customarily considered to comprise two distinct phases: (1) mixing, wherein the dissolved coagulant is rapidly dispersed throughout the water being treated, usually by violent agitation; and (2) flocculation, involving agitation of the water at lower velocities for a much longer period, during which very small particles grow and agglomerate into well-defined flocs of sufficient size to settle readily.

Hydrolyzing Metals

Early investigators assumed that aluminum and iron coagulants behaved in solution as simple salts with Al^{3+} and Fe^{3+} acting as the principle coagulating species. Now it has been demonstrated that salts of these metals hydrolyze in solution, forming species with charges other than the trivalent ions.

Sullivan and Singley[3] experimentally determined the hydrolysis of aluminum ion in dilute, aqueous solution, in the absence of complexing anions other than hydroxide. They found that the predominant species was Al^{3+} up to pH 4.5, $Al(OH)_3$ from pH 4.5 to 8, and $Al(OH)_4^-$ above pH 8. Species present as a function of pH for 0.0001 molar aluminum perchlorate are shown in Fig. 10-4.

Singley and Black[4] experimentally determined distribution of the hydrolysis species of iron(III) in dilute solutions in the absence of extraneous complexing ligands, that is, low ionic strength. Species present as a function of pH for a 0.0001 molar concentration of iron are given in Fig. 10-5.

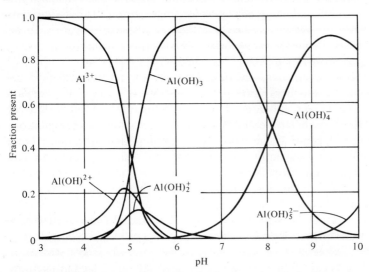

FIGURE 10-4 Distribution diagram for the fraction of total aluminum present in each form as a function of pH for 1×10^{-4} molar aluminum perchlorate. [Source: J. H. Sullivan, Jr., and J. E. Singley, "Reactions of Metal Ions in Dilute Aqueous Solution: Hydrolysis of Aluminum," *J. Am. Water Works Assoc.* 60, no. 11 (1968) : 1286. Copyright 1968 by the American Water Works Association, Inc.]

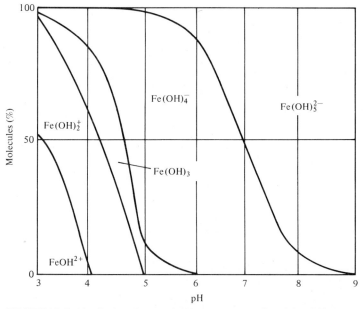

FIGURE 10-5 Distribution diagram for the percentage of total iron(III) present as various species as a function of pH for 1×10^{-4} molar iron(III). [Source: J. E. Singley and A. P. Black, "Hydrolysis Products of Iron(III)," *J. Am. Water Works Assoc.* 59, no. 12 (1967) : 1557. Copyright 1967 by the American Water Works Association, Inc.]

The pH dependence of destabilization by a hydrolyzing metal is illustrated schematically in Fig. 10-6. Laboratory analyses can be performed to find the concentration of metal salt just sufficient to cause coagulation or restabilization in various pH solutions. These critical values can then be plotted as boundaries between stable and unstable zones, thus establishing domains of stability. An uncoagulated dispersion zone indicates that insufficient coagulant has been applied and destabilization does not occur. A restabilization zone is generally attributed to charge reversal at higher concentrations of coagulant by adsorption of coagulant ions onto the surface of colloids. The coagulation region defines the conditions of pH and coagulant concentrations which produce rapid clarification. Turbidity removal in water treatment practice is accomplished by applying the coagulant dosage within the region where it is most efficient.

Stumm and O'Melia,[5] after a review of chemical factors effective in destabilizing colloids, conclude that: "Coagulation is a time-dependent process including several reaction steps: (1) hydrolysis of multivalent metal ions and subsequent polymerization to multinuclear hydrolysis species; (2) adsorption of hydrolysis species at the solid–solution interface to accomplish destabilization of the colloid; (3) aggregation of destabilized particles by interparticle bridging involving particle transport and chemical interactions; (4) aggregation of destabilized particles by particle transport and van der Waals' forces; (5) "aging" of flocs, accompanied by chemical changes in the structure of metal—OH—metal

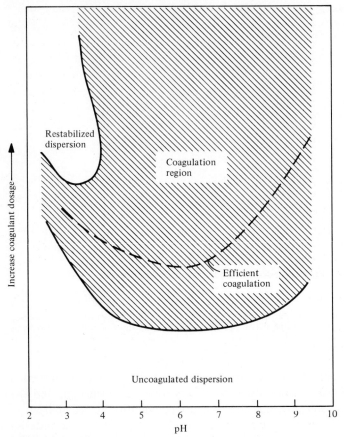

FIGURE 10-6 Schematic diagram of coagulant dosage–pH domains for coagulation and restabilization.

linkages, concurrent change in floc sorbability and in extent of floc hydration; and (6) precipitation of metal hydroxide." The authors further state that some of these steps occur sequentially, some overlap (steps 1 and 2), and several may occur concurrently under certain conditions (steps 3 and 4, or step 6 with steps 1–5).

Polyelectrolytes

Synthetic polyelectrolytes are water-soluble, high-molecular-weight, organic polymers containing chemical groups that undergo electrolytic dissociation in solution resulting in a long chain having highly charged ions.

After studying destabilization of dilute clay suspensions by polymers, Black, Birkner, and Morgan[6] proposed that the destabilization mechanism of cationic polymer-clay systems was a combined partial coagulation–flocculation reaction with the latter serving as the rate-controlling step. The cationic polymer molecule serves first as a coagulant to reduce the forces of repulsion between clay particles, and then as a flocculant to bridge adjacent particles via extended segments of adsorbed polymer molecules. In the case of anionic polymers in dilute clay suspensions, the principal mode of action appears to be flocculation via an interparticle bridging mechanism. It was found that a sufficient concentration of counter ions must be present initially in, or added to, the suspension to reduce particle–particle, polymer–particle, and adsorbed polymer–polymer repulsive forces so that interparticle bridging can occur.[6]

Coagulation

Chemical coagulation, a key step in the overall system, removes turbidity-producing substances (colloidal solids, clay particles, organics, bacteria, and algae) and color in surface water, which result from decaying vegetation or industrial wastes (e.g., dye waste). Taste- and odor-producing compounds from algae, decaying organics, or contaminants from wastewaters are reduced by breakpoint chlorination and activated carbon, as well as coagulation. Chlorine is applied for disinfection and a fluoride compound for fluoridation.

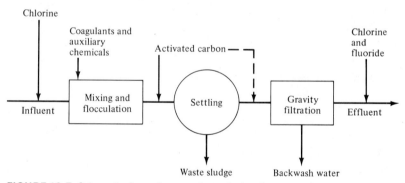

FIGURE 10-7 Schematic flow diagram of a typical surface-water treatment system.

Figure 10-7 is a flow diagram of a typical surface-water treatment system for processing a lake or reservoir supply. Highly turbid river waters are generally held in presedimentation basins to reduce silt and settleable organic matter prior to coagulation. If the source water contains excessive hardness, lime is applied along with chemical coagulants for precipitation softening with turbidity removal.

10-8

COAGULANTS

The most widely used coagulants for water and wastewater treatment are aluminum and iron salts. The common metal salt is aluminum sulfate, which is a good coagulant for water containing appreciable organic matter. Iron coagulants operate over a wider pH range and are generally more effective in removing color from water; however, they are usually more costly. Cationic polyelectrolytes can serve as prime coagulants,[7] but their typical application is as a coagulant aid. Conventional aids adopted for metal coagulants include lime, soda ash, activated silica, and polyelectrolytes.

The final choice of coagulant and chemical aids for a particular water should be based on coagulation control tests, past experience in treatment of water of similar quality, and on overall economics involved.

The following paragraphs discuss various coagulants and present traditionally theoretical chemical reactions. It is apparent from the previous section that these coagulation reactions yield only approximate results. Molecular and equivalent weights of chemical compounds listed in Tables 10-1 and 10-2 are useful in solving numerical problems.

Aluminum Sulfate (Filter Alum)

Aluminum sulfate is the standard coagulant used in water treatment. The commercial product strength ranges from 15 to 22% as Al_2O_3 with a hydration of about 14 moles of water. A formula used for filter alum is $Al_2(SO_4)_3 \cdot 14.3H_2O$ with a molecular weight of 600. The material is shipped and fed in a dry granular form, although available as a powder or alum syrup.

Aluminum sulfate reacts with natural alkalinity in water to form aluminum hydroxide floc.

$$Al_2(SO_4)_3 \cdot 14.3H_2O + 3Ca(HCO_3)_2 \longrightarrow$$
$$2Al(OH)_3 \downarrow + 3CaSO_4 + 14.3H_2O + 6CO_2 \qquad (10\text{-}20)$$

Each mg/l of alum decreases water alkalinity by 0.50 mg/l (as $CaCO_3$) and produces 0.44 mg/l of carbon dioxide. Production of carbon dioxide is undesirable since this increases the corrosiveness of water.

If water does not contain sufficient alkalinity to react with the alum, lime or soda ash is fed to provide the necessary alkalinity.

$$Al_2(SO_4)_3 \cdot 14.3H_2O + 3Ca(OH)_2 \xrightarrow{\text{lime}}$$
$$2Al(OH)_3 \downarrow + 3CaSO_4 + 14.3H_2O \qquad (10\text{-}21)$$
$$Al_2(SO_4)_3 \cdot 14.3H_2O + 3Na_2CO_3 + 3H_2O \xrightarrow{\text{soda ash}}$$
$$2Al(OH)_3 \downarrow + 3Na_2SO_4 + 3CO_2 + 14.3H_2O \qquad (10\text{-}22)$$

An advantage of using sodium carbonate (soda ash) is that unlike lime it does not increase water hardness, only corrosiveness. Lime, more popular, is less expensive than soda ash.

The dosage of alum used in water treatment is in the range 5–50 mg/l. Effective pH range for alum coagulation is 5.5 to 8.0. Alum is preferred in treating relatively high quality surface waters because it is the only chemical needed for coagulation.

Ferrous Sulfate (Copperas)

Commercial ferrous sulfate has a strength of 55% $FeSO_4$ and is supplied as green crystal or granule for dry feeding. Ferrous sulfate reacts with natural alkalinity (Eq. 10-23), but the response is much slower than that between alum and natural alkalinity. Lime is generally added to raise the pH to the point where ferrous ions are precipitated as ferric hydroxide by the caustic alkalinity (Eq. 10-24).

$$2FeSO_4 \cdot 7H_2O + 2Ca(HCO_3)_2 + 0.5O_2 \longrightarrow$$
$$2Fe(OH)_3 \downarrow + 2CaSO_4 + 4CO_2 + 13H_2O \qquad (10\text{-}23)$$

$$2FeSO_4 \cdot 7H_2O + 2Ca(OH)_2 + 0.5O_2 \longrightarrow$$
$$2Fe(OH)_3 \downarrow + 2CaSO_4 + 13H_2O \qquad (10\text{-}24)$$

Treatment using copperas and lime adds some hardness but no corrosiveness to the water. This process is usually cheaper than alum coagulation, but the dosing operation with two chemicals is more difficult. If excess lime is used the water may require treatment for stabilization.

Chlorinated copperas treatment is a second method of using ferrous sulfate. In this process chlorine is used to oxidize the ferrous sulfate to ferric sulfate.

$$3FeSO_4 \cdot 7H_2O + 1.5Cl_2 \longrightarrow Fe_2(SO_4)_3 + FeCl_3 + 21H_2O \qquad (10\text{-}25)$$

Oxidation is generally performed by adding ferrous sulfate to the discharge from a solution feed chlorinator. Theoretically, 1.0 lb of chlorine oxidizes 7.8 lb of copperas. In practice, a chlorine feed slightly in excess of the theoretical amount produces good results.

Ferric sulfate and the ferric chloride react with natural alkalinity or lime, as illustrated by the following reactions with ferric chloride:

$$2FeCl_3 + 3Ca(HCO_3)_2 \longrightarrow 2Fe(OH)_3 \downarrow + 3CaCl_2 + 6CO_2 \qquad (10\text{-}26)$$

$$2FeCl_3 + 3Ca(OH)_2 \longrightarrow 2Fe(OH)_3 \downarrow + 3CaCl_2 \qquad (10\text{-}27)$$

Color in water is generally not affected by copperas and lime treatment, whereas chlorinated copperas is effective in the removal of color.

Ferric Salts

Ferric sulfate and ferric chloride are available as coagulants under a variety of trade names. The reactions of these salts with natural alkalinity and with lime are noted in Eqs. 10-26 and 10-27. Advantages of the ferric coagulants are: (1) coagulation is possible over a wider pH range, generally pH 4–9 for most waters; (2) the precipitate produced is a heavy quick-settling floc; and (3) they are more effective in the removal of color, taste, and odor compounds.

Ferric sulfate is available in crystalline form and may be fed using dry or liquid feeders. Ferric sulfate, although not as aggressive as ferric chloride or chlorinated copperas, must be handled by corrosive resistant equipment.

Ferric chloride is supplied in either crystalline or liquid form. Although ferric chloride can be used in water treatment, its most frequent application is in wastewater treatment (e.g., as a waste sludge conditioning chemical in combination with lime prior to mechanical dewatering).

Other Coagulants

Sodium aluminate is essentially alumina dissolved in sodium hydroxide. The principal use of sodium aluminate is as an additional coagulant with aluminum sulfate, generally, in the treatment of boiler water. The limited employment of sodium aluminate is dictated by its high cost. Sodium aluminate is alkaline in its reactions, instead of acid like other coagulants. Reactions of sodium aluminate with aluminum sulfate and carbon dioxide are:

$$6NaAlO_2 + Al_2(SO_4)_3 \cdot 14.3H_2O \longrightarrow$$
$$8Al(OH)_3 \downarrow + 3Na_2SO_4 + 2.3H_2O \qquad (10\text{-}28)$$

$$2NaAlO_2 + CO_2 + 3H_2O \longrightarrow 2Al(OH)_3 \downarrow + Na_2CO_3 \qquad (10\text{-}29)$$

Ammonia alum, $Al_2(SO_4)_3 \cdot (NH_4)_2SO_4 \cdot 24H_2O$, and potash alum, $Al_2(SO_4)_3 \cdot K_2SO_4 \cdot 24H_2O$, are restricted to small installations (i.e., industrial plants and swimming pools). Alum is placed in a pot-type feeder where water is run through under pressure to dissolve some of it. The solution is then injected in the water ahead of a pressure filter.

10-9

COAGULANT AIDS

Acids and Alkalies

Acids and alkalies are added to water to adjust the pH for optimum coagulation. Typical acids used to lower the pH are sulfuric and phosphoric.

Alkalies used to raise the pH are lime and soda ash. Hydrated lime, with about 70% available CaO, is suitable for dry feeding but costs more than quicklime, which is 90% CaO. The latter must be slaked (combined with water) and fed as a lime slurry. Soda ash is 98% sodium carbonate and can be fed dry but is more expensive than lime.

Activated Silica

Activated silica is sodium silicate which has been activated with sulfuric acid, aluminum sulfate, carbon dioxide, or chlorine. One method of preparing activated silica is to dilute a sodium silicate solution to a level of 1.5% SiO_2 and then add enough sulfuric acid to neutralize 85% of the alkalinity. This solution is aged 2 hours before use. Another procedure is to add 1 part of a 1% silicate solution to 4 parts of a 1% alum solution applying the mixture immediately. These and other methods of preparation require proper equipment and close operational control to successfully produce activated silica.

When activated silica is placed in water, it produces a stable sol having a negative charge. This colloidal dispersion of activated silica, as an aid to coagulation, has several advantages. The activated silica unites with positively charged metal hydroxides of the coagulant, resulting in a larger, tougher, denser floc and causes more rapid, improved settling, which in turn reduces turbidity carryover to the sand filter. Activated silica may lower the coagulant dosage required for good coagulation and thus reduce overall chemical costs.

Polyelectrolytes

Three classifications of polyelectrolytes based on their ionic character are: negatively charged compounds are called anionic polyelectrolytes, positively charged are cationic polymers, and compounds with both positive and negative charges referred to as polyamphotytes. Commercially available polyelectrolytes are manufactured by a number of companies under a variety of trade names.

Polyelectrolytes are effective as coagulant aids; however, the type of polyelectrolyte, dosage, and point of addition must be determined for each water. As coagulant aids they increase the rate and degree of flocculation (aggregation) by adsorption, charge neutralization, and interparticular bridging. The latter appears to be the main mechanism by which polyelectrolytes aid coagulation.

Clays

Clay improves coagulation by adding weight to the floc, thus increasing the settling rate and exerts adsorptive action that aids in the floc formation. Bentonite and other clays are available in different degrees of fineness as coagulant aids.

●EXAMPLE 10-2

A surface water is coagulated with a dosage of 2.0 grains/gal of ferrous sulfate and an equivalent dosage of lime. How many pounds of (a) ferrous sulfate are needed per mg of water treated; (b) hydrated lime are required assuming a purity of 70% CaO; (c) $Fe(OH)_3$ sludge produced per mg of water treated? (*Note*: 1 grain/gal = 17.1 mg/l and 1 lb = 7000 grains.)

○*Solution*

Converting the dosage of ferrous sulfate from grains/gal to lb/mg,

$$2.0 \times 17.1 \times 8.34 = 286 \text{ lb/mg}$$

$$\text{or } 2.0 \times \frac{1,000,000}{7000} = 286 \text{ lb/mg}$$

By Eq. 10-24 and molecular weights from Tables 10-1 and 10-2,

$$\underset{\text{FeSO}_4 \cdot 7\text{H}_2\text{O}}{2 \times 278 \text{ lb}} \quad + \quad \underset{\text{Ca(OH)}_2}{2 \times 74 \text{ lb}} \quad \longrightarrow \quad \underset{\text{Fe(OH)}_3}{2 \times 107 \text{ lb}}$$

Therefore,

$$\frac{556}{286 \text{ lb}} = \frac{148}{Y \text{ lb Ca(OH)}_2} = \frac{214}{Z \text{ lb Fe(OH)}_3}$$

Solving for the lime dosage,

$$Y \text{ lb Ca(OH)}_2 = \frac{286 \times 148}{556} = 75.8 \text{ lb/mg}$$

$$\text{lb } 70\% \text{ CaO} = 75.8 \times \frac{56}{74} \times \frac{1}{0.70} = 82 \text{ lb/mg}$$

The $Fe(OH)_3$ sludge production

$$Z \text{ lb Fe(OH)}_3 = \frac{286 \times 214}{556} = 110 \text{ lb/mg}$$

Alternative solution for lime dosage:
   One equivalent weight of ferrous sulfate (139) reacts with 1 equivalent weight of 70% CaO $(28 \div 0.70 = 40)$. Therefore,

$$\text{lime dosage} = 2.0 \times \frac{40}{139} \times \frac{1,000,000}{7000} = 82 \text{ lb/mg}$$

●EXAMPLE 10-3
Dosage of alum with the alum-lime coagulation of a water is 50 mg/l. It is desired to react only 10 mg/l (as $CaCO_3$) of the natural alkalinity with the alum. Based on the theoretical Eqs. 10-20 and 10-21, what dosage of lime is required, in addition to 10 mg/l of natural alkalinity, to react with the alum dosage?

○*Solution*
   Using equivalent weights, the alum that reacts with 10 mg/l of natural alkalinity = $10 \times (100/50) = 20$ mg/l. Amount of alum remaining to react with lime = $50 - 20 = 30$ mg/l. Lime dosage required to react with 30 mg/l of alum = $30 \times (28/100) = 8.4$ mg/l as CaO.

10-10

JAR TEST

A multiple stirring apparatus with variable-speed drive is used for the jar test. A typical unit consists of six agitator paddles mechanically coupled to operate at the same speed which can be from 10 to 100 rpm. The coagulation containers are 1- or 2-liter beakers or battery jars.

   The jar-test apparatus permits laboratory studies on chemical coagulation and flocculation. Experiments may be conducted to determine the effectiveness of various coagulants, optimum dosage for coagulation, optimum pH for coagulation, concentration of coagulant aid, and most effective order in which to add various chemicals.

General procedure for conducting a jar test is outlined:

1   Fill six 1- or 2-liter beakers with a measured amount of the water to be treated.
2   Add the coagulant and/or other chemicals to each sample.
3   Flash-mix the samples by agitating at maximum speed (100 rpm) for 1 minute.
4   Flocculate the samples at a stirring rate of 20–70 rpm for 10–30 minutes. Record the time of floc appearance for each beaker.
5   Stop the agitation and record the nature of the floc, clarity of supernatant, and settling characteristics of the floc.

Water Softening

Hardness in water is caused by the ions of calcium and magnesium. Although ions of iron, manganese, strontium, and aluminum also produce hardness, they are not present in significant quantities in natural waters.

A singular criterion for maximum hardness in public water supplies is not possible. Water hardness is largely the result of geological formations of the water source. Public acceptance of hardness varies from community to community, consumer sensitivity being related to the degree to which he or she is accustomed.

Hardness of more than 300–500 mg/l as $CaCO_3$ is considered excessive for a public water supply and results in high soap consumption as well as objectional scale in heating vessels and pipes. Many consumers object to water harder than 150 mg/l, a moderate figure being 60–120 mg/l.

10-11

CHEMISTRY OF LIME-SODA ASH PROCESS

The lime-soda water-softening process uses lime, $Ca(OH)_2$, and soda ash, Na_2CO_3, to precipitate hardness from solution. Carbon dioxide and carbonate hardness (calcium and magnesium bicarbonate) are complexed by lime. Non-carbonate hardness (calcium and magnesium sulfates or chlorides) requires addition of soda ash for precipitation.

Chemical reactions in the lime-soda process are:

$$CO_2 + Ca(OH)_2 = CaCO_3 \downarrow + H_2O \tag{10-30}$$

$$Ca(HCO_3)_2 + Ca(OH)_2 = 2CaCO_3 \downarrow + 2H_2O \tag{10-31}$$

$$Mg(HCO_3)_2 + Ca(OH)_2 = CaCO_3 \downarrow + MgCO_3 + 2H_2O \tag{10-32}$$

$$MgCO_3 + Ca(OH)_2 = Mg(OH)_2 \downarrow + CaCO_3 \downarrow \tag{10-33}$$

$$MgSO_4 + Ca(OH)_2 = Mg(OH)_2 \downarrow + CaSO_4 \tag{10-34}$$

$$CaSO_4 + Na_2CO_3 = CaCO_3 \downarrow + Na_2SO_4 \tag{10-35}$$

These equations give all the reactions taking place in softening water containing both carbonate and noncarbonate hardness, by additions of both lime and soda ash. The carbon dioxide in Eq. 10-30 is not hardness as such, but it consumes lime and must therefore be considered in calculating the amount required. Equations 10-31, 10-32, and 10-33 demonstrate removal of carbonate hardness by lime. Note that only 1 mole of lime is needed for each mole of calcium alkalinity, whereas 2 moles are required for each mole of magnesium alkalinity (Eqs. 10-32 and 10-33). Equation 10-34 shows the removal of magnesium noncarbonate hardness by lime. No softening results from this reaction because 1 mole of calcium noncarbonate hardness is formed for each mole of magnesium salt present. Equation 10-35 is for removal of calcium sulfate originally present in the water and also that formed as stated in Eq. 10-34.

Precipitation softening cannot produce a water completely free of hardness because of the solubility of calcium carbonate and magnesium hydroxide. Furthermore, completion of the chemical reactions is limited by physical considerations, such as adequate mixing and limited detention time in settling basins. Therefore, the minimum practical limits of precipitation softening are 30 mg/l of $CaCO_3$ and 10 mg/l of $Mg(OH)_2$ expressed as $CaCO_3$. Hardness levels of 80–100 mg/l are generally considered acceptable for a public water supply, but the magnesium content should not exceed 40 mg/l as $CaCO_3$ in a softened municipal water.

There are several advantages of lime softening in water treatment. The most obvious is that the total dissolved solids are dramatically reduced, hardness is taken out of solution, and the lime added is also removed. When soda ash is applied, sodium ions remain in the finished water; however, noncarbonate hardness requiring the addition of soda ash is generally a small portion of the total hardness. Lime also precipitates soluble iron and manganese often found in groundwaters. In processing surface waters, excess lime treatment provides disinfection and aids in coagulation for removal of turbidity.

10-12

PROCESS VARIATIONS IN LIME-SODA ASH SOFTENING

Three different basic schemes are used to provide a finished water with the desired hardness: excess lime treatment, selective calcium removal, and split treatment.

Excess Lime Treatment

Carbonate hardness associated with the calcium ion can be effectively removed to the practical limit of $CaCO_3$ solubility by stoichiometric additions of lime (Eq. 10-31). Precipitation of the magnesium ion (Eqs. 10-33 and 10-34) calls for a surplus of approximately 35 mg/l of CaO (1.25 meq/l) above stoichiometric requirements. The practice of excess lime treatment reduces total hardness to about 40 mg/l (i.e., 30 mg/l of $CaCO_3$ and 10 mg/l of magnesium hardness).

After excess lime treatment, the water is scale-forming and must be neutralized to remove caustic alkalinity. Recarbonation and soda ash are regularly used to stabilize the water. Carbon dioxide neutralizes excess lime as follows:

$$Ca(OH)_2 + CO_2 = CaCO_3 \downarrow + H_2O \tag{10-36}$$

This reaction precipitates calcium hardness and reduces the pH from near 11 to about 10.2. Further recarbonation of the clarified water converts a portion of the remaining carbonate ions to bicarbonate by the reaction

$$CaCO_3 + CO_2 + H_2O = Ca(HCO_3)_2 \tag{10-37}$$

Final pH is in the range 9.5–8.5, depending on the desired carbonate to bicarbonate ratio (Fig. 10-2).

Equipment required for generating carbon dioxide gas consists of a furnace to burn the fuel (coke, coal, gas, or oil), a scrubber to remove soot and other impurities from the gas, and a compressor for forcing the gas into the water.

A two-stage system is preferred for excess lime treatment (Fig. 10-8). Lime is applied in first-stage mixing and sedimentation to precipitate both calcium and magnesium. Then carbon dioxide is applied to neutralize the excess lime (Eq. 10-36), and soda ash is added to reduce noncarbonate hardness. Solids formed in these reactions are removed by secondary settling and subsequent filtration. Recarbonation immediately ahead of the filters may be used to prevent scaling of the media (Eq. 10-37).

●EXAMPLE 10-4

Water defined by the following analysis is to be softened by excess lime treatment in a two-stage system (Fig. 10-8).

$CO_2 = 8.8$ mg/l as CO_2 Alk(HCO_3^-) = 115 mg/l as $CaCO_3$
$Ca^{2+} = 70$ mg/l $SO_4^{2-} = 96$ mg/l
$Mg^{2+} = 9.7$ mg/l $Cl^- = 10.6$ mg/l
$Na^+ = 6.9$ mg/l

The practical limits of removal can be assumed to be 30 mg/l of $CaCO_3$ and 10 mg/l of $Mg(OH)_2$, expressed as $CaCO_3$. Sketch a meq/l bar graph and list the hypothetical combinations of chemical compounds in the raw water. Calculate the quantity of softening chemicals required in pounds per million gallons of water treated, and the theoretical quantity of carbon dioxide needed to provide a finished water with one-half of the alkalinity converted to bicarbonate ion. Draw a bar graph for the softened water after recarbonation and filtration.

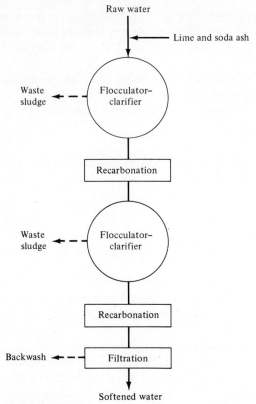

Raw water

Lime and soda ash

Waste sludge

Flocculator-clarifier

Recarbonation

Waste sludge

Flocculator-clarifier

Recarbonation

Backwash

Filtration

Softened water

FIGURE 10-8 Schematic flow diagram for a two-stage excess-lime water-softening plant.

○*Solution*

Components	mg/l	Equivalent Weight	meq/l
CO_2	8.8	22.0	0.40
Ca^{2+}	70	20.0	3.50
Mg^{2+}	9.7	12.2	0.80
Na^+	6.9	23.0	0.30
Alk	115	50.0	2.30
SO_4^{2-}	96	48.0	2.00
Cl^-	10.6	35.5	0.30

The meq/l bar graph of the raw water is shown in Fig. 10-9a, and the hypothetical combinations are listed.

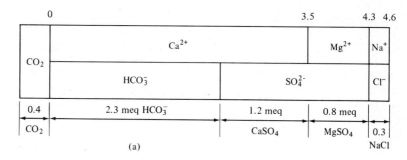

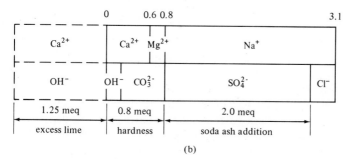

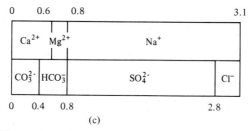

FIGURE 10-9 Milliequivalent bar graphs for Example 10-4. (a) Bar graph and hypothetical chemical combinations in the raw water. (b) Bar graph of the water after lime and soda ash additions and settling, but before recarbonation. (c) Bar graph of the water after two-stage recarbonation and final filtration.

Component	meq/l	Lime	Soda Ash
CO_2	0.4	0.4	0
$Ca(HCO_3)_2$	2.3	2.3	0
$CaSO_4$	1.2	0	1.2
$MgSO_4$	0.8	0.8	0.8
		—	—
		3.5	2.0

$$\text{lime required} = \text{stoichiometric quantity} + \text{excess lime}$$
$$= 3.5 \times 28 + 35 = 133 \text{ mg/l of } CaO = 1100 \text{ lb/mg}$$
$$\text{soda ash required} = 2.0 \times 53 = 106 \text{ mg/l of } Na_2CO_3 = 900 \text{ lb/mg}$$

A hypothetical bar graph for the water after addition of softening chemicals and first-stage sedimentation is shown in Fig. 10-9b. The dashed box is the excess lime addition, 35 mg/l of CaO = 1.25 meq/l. The 0.6 meq/l of Ca^{2+} (30 mg/l as $CaCO_3$) and 0.20 meq/l of Mg^{2+} (10 mg/l as $CaCO_3$) are the practical limits of hardness reduction. The 2.0 meq/l of Na_2SO_4 results from the addition of soda ash. Alkalinity consists of 0.20 meq/l of OH^- associated with $Mg(OH)_2$ and 0.60 meq/l of CO_3^{2-} related to $CaCO_3$.

Recarbonation converts the excess hydroxyl ion to carbonate ion; using the relationship in Eq. 10-36 and 22.0 as the equivalent weight of carbon dioxide = $(1.25 + 0.2)22.0 = 31.9$ mg/l of CO_2. After second-stage processing, final recarbonation converts one-half of the remaining alkalinity to bicarbonate ion by Eq. 10-37, $= 0.5 \times 0.8 \times 22.0 = 8.8$ mg/l of CO_2. Therefore total carbon dioxide reacted

$$= (31.9 + 8.8)8.34 = 340 \text{ lb/mg of } CO_2.$$

The bar graph of the finished water is shown in Fig. 10-9 c.

Selective Calcium Removal

Waters with a magnesium hardness of less than 40 mg/l as $CaCO_3$ can be softened by removing only a portion of the calcium hardness. The processing scheme can be a single-stage system of mixing, sedimentation, recarbonation, and filtration. Enough lime is added to the raw water to precipitate calcium hardness without providing any excess for magnesium removal. Soda ash may be required depending on the amount of noncarbonate hardness. Recarbonation is usually practiced to reduce scaling of the filter sand and produce a stable effluent.

●EXAMPLE 10-5

Determine the chemical dosages needed for selective calcium softening of the water described in Example 10-4. Draw a bar graph of the processed water.

○Solution

The hypothetical combinations of concern in calcium precipitation are CO_2, $Ca(HCO_3)_2$, and $CaSO_4$ (Fig. 10-11a).

Component	meq/l	Lime	Soda Ash
CO_2	0.4	0.4	0
$Ca(HCO_3)_2$	2.3	2.3	0
$CaSO_4$	1.2	0	1.2
		2.7	1.2

lime required $= 2.7 \times 28 = 76$ mg/l of CaO $= 630$ lb/mg
soda ash required $= 1.2 \times 53 = 64$ mg/l of $Na_2CO_3 = 530$ lb/mg

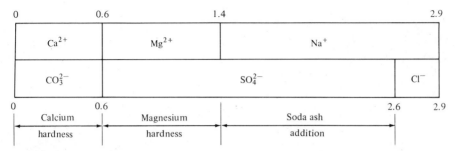

FIGURE 10-10 Bar graph of the softened water after selective calcium removal, Example 10-5.

Final hardness is all the Mg^{2+} in the raw water plus the practical limit of $CaCO_3$ removal (30 mg/l or 0.60 meq/l) $= (0.8 + 0.6)50 = 70$ mg/l as $CaCO_3$. The bar graph of the softened water is shown in Fig. 10-10. Recarbonation would be desirable to stabilize the water by converting a portion of the carbonate alkalinity to bicarbonate.

Split Treatment

Split treatment consists of treating by excess lime a portion of the raw water and then neutralizing the excess lime in the treated flow with the remaining portion of raw water. When split treatment is used, any desired hardness level above 40 mg/l is attainable. Since hardness levels of 80–100 mg/l are generally considered acceptable, split treatment can result in considerable chemical savings. Recarbonation after the first stage is not customarily required; however, it may be needed after the second stage before filtration. Split treatment is particularly advantageous on well waters. In softening surface waters, where taste, odor, and color may be problems, two stages of processing for the total flow are usually preferred over split treatment.

The flow pattern of a typical two-stage split treatment plant, shown in Fig. 10-11, and the following description are from Cleasby and Dillingham.[8] If X is the ratio of the bypassed flow to the total quantity Q, then bypassed flow is XQ, and that through the first stage is $Q - XQ$ or $(1 - X)Q$. The magnesium content leaving the first stage (designated Mg_1) will be less than 10 mg/l (as $CaCO_3$). Magnesium in the bypass will be the same as that in the raw water (designated as Mg_r). Finshed water total hardness is dictated by what is considered acceptable to the consumer, about 80–100 mg/l. Calcium in the finished water should not exceed 30–40 mg/l (as $CaCO_3$) for good operation. Therefore, permissible magnesium in finished water (designated Mg_f) is about 50 mg/l (as $CaCO_3$). Some cities produce a water of 40 mg/l of Mg to reduce problems with hard magnesium silicate in high temperature (180°F) services. The bypass flow fraction can be calculated for any desired level of magnesium using the formula

$$X = \frac{Mg_f - Mg_1}{Mg_r - Mg_1}$$

(10-38)

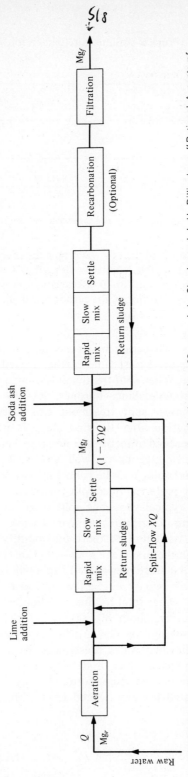

FIGURE 10-11 Flow diagram for a typical split-treatment lime-soda ash softening plant. [Source: J. L. Cleasby and J. H. Dillingham, "Rational Aspects of Split Treatment," *Proc. Am. Soc. Civil Engrs., J. San. Eng. Div.* 92, no. SA2 (1966): 1–7.]

Coagulation and Softening

Addition of a coagulant or a coagulant aid may result in more efficient removal of the hardness precipitates formed in lime-soda softening. Alum is the prevalent coagulant. Activated silica has been effective when applied in softening waters high in calcium.

Lime softening is often used to treat surface waters (see Fig. 8-14). Improved coagulation and turbidity removal usually result from lime-soda softening, as compared to simple coagulation, because of the greater quantity of precipitate formed in chemical treatment processes.

●**EXAMPLE 10-6**

Consider split-treatment softening of the water described by the following analysis. Criteria for the finished water are a maximum permissible magnesium hardness of 40 mg/l as $CaCO_3$ and hardness in the calcium range 30–40 mg/l.

$CO_2 = 0.5$ meq/l

$Ca^{2+} = 3.5$ meq/l $HCO_3^- = 3.2$ meq/l

$Mg^{2+} = 1.8$ meq/l $SO_4^{2-} = 2.2$ meq/l

$Na^+ = 0.5$ meq/l $Cl^- = 0.4$ meq/l

○**Solution**

Dosages for treating flow through the first stage, based on softening equations, are:

Component	meq/l	Lime	Soda Ash
CO_2	0.5	0.5	0
$Ca(HCO_3)_2$	3.2	3.2	0
$CaSO_4$	0.3	0	0.3
$MgSO_4$	1.8	1.8	1.8
		5.5 meq/l	2.1 meq/l

Additional chemicals needed to react with the carbon dioxide and calcium in the bypass flow for precipitation in the second stage are:

lime $= 0.5 + 3.2 = 3.7$ meq/l

soda ash $= 0.3$ meq/l

Solving for the fraction of bypassed flow using Eq. 10-38,

$$X = \frac{40 - 10}{(1.8)(50) - 10} = \frac{30}{80} = 0.375$$

Total chemical applications based on flow of raw water entering plant are:

lime $= [0.625(5.5) + 0.375(3.7)]28 = 135$ mg/l of CaO

soda ash $= [0.625(2.1) + 0.375(0.3)]53 = 75$ mg/l of Na_2CO_3

If all of the required lime is added to the first stage, the concentration of excess lime is

$$\frac{0.375 \times 3.7 \times 28}{0.625} = 62 \text{ mg/l of CaO} \quad (>35 \text{ mg/l OK})$$

The finished water would have a total hardness of 70–80 mg/l, 40 mg/l of magnesium and 30–40 mg/l of calcium.

10-13

WATER STABILIZATION

Water is considered to be stable when it neither dissolves nor deposits calcium carbonate [i.e., the calcium carbonate is in equilibrium with the hydrogen-ion concentration (Eq. 10-39)]. If the pH is raised from the equilibrium point, water becomes scale-forming, depositing calcium carbonate. Water turns corrosive if the pH is lowered.

$$CaCO_3 + H^+ \rightleftharpoons Ca^{2+} + HCO_3^- \tag{10-39}$$

A thin coating of calcium carbonate on the pipe interior protects the metal against excessive corrosion. Such a covering can be maintained permanently if the water is held at a proper level of calcium carbonate saturation.

Saturation Index

Langelier[9] developed an index, in the pH range 6.5–9.5, that makes it possible to predict whether a given water will deposit or dissolve calcium carbonate. The saturation index is calculated by the following equation:

$$SI = pH - pH_s = pH - [(pK_2' - pK_s') + pCa^{2+} + pAlk] \tag{10-40}$$

where
- pH = actual pH of the water
- pH_s = pH at saturation
- $(pK_2' - pK_s')$ = empirical constants
- pCa^{2+} = negative logarithm of the calcium-ion concentration, moles/liter
- $pAlk$ = negative logarithm of the total alkalinity equivalents/liter

A positive value for the index signifies the water is oversaturated and will precipitate calcium carbonate. A negative number indicates that the water is corrosive. The saturation index serves as a measure of the water's tendency to dissolve or precipitate calcium carbonate but does not give either the rate at which stability is attained or the capacity.

The value of $(pK_2' - pK_s')$ based on temperature and ionic strength can be determined from Table 10-3 by Larson and Buswell.[10] Ionic strength is calculated using the following equation:

$$\text{ionic strength} = \frac{1}{2}(C_1 Z_1^2 + C_2 Z_2^2 + \cdots + C_n Z_n^2) \tag{10-41}$$

where C = concentration, moles/1000 g of water
Z = valence of the individual ions

TABLE 10-3

VALUES OF pK'_2 AND pK'_s AT 25°C FOR VARIOUS IONIC STRENGTHS AND THE DIFFERENCE $(pK'_2 - pK'_s)$ FOR VARIOUS TEMPERATURES

Ionic Strength	Total Dissolved Solids (mg/l)	25°C pK'_2	25°C pK'_s	25°C $pK'_2 - pK'_s$	$pK'_2 - pK'_s$ 0°C	10°C	20°C	30°C	40°C	50°C	60°C	70°C	80°C	90°C
0.000		10.26	8.32	1.94	2.45	2.23	2.02	1.86	1.68	1.52	1.36	1.23	1.08	0.95
0.001	40	10.26	8.19	2.07	2.58	2.36	2.15	1.99	1.81	1.65	1.49	1.36	1.21	1.08
0.002	80	10.25	8.14	2.11	2.62	2.40	2.19	2.03	1.85	1.69	1.53	1.40	1.25	1.12
0.003	120	10.25	8.10	2.15	2.66	2.44	2.23	2.07	1.89	1.73	1.57	1.44	1.29	1.16
0.004	160	10.24	8.07	2.17	2.68	2.46	2.25	2.09	1.91	1.75	1.59	1.46	1.31	1.18
0.005	200	10.24	8.04	2.20	2.71	2.49	2.28	2.12	1.94	1.78	1.62	1.49	1.34	1.21
0.006	240	10.24	8.01	2.23	2.74	2.52	2.31	2.15	1.97	1.81	1.65	1.52	1.37	1.24
0.007	280	10.23	7.98	2.25	2.76	2.54	2.33	2.17	1.99	1.83	1.67	1.54	1.39	1.26
0.008	320	10.23	7.96	2.27	2.78	2.56	2.35	2.19	2.01	1.85	1.69	1.56	1.41	1.28
0.009	360	10.22	7.94	2.28	2.79	2.57	2.36	2.20	2.02	1.86	1.70	1.57	1.42	1.29
0.010	400	10.22	7.92	2.30	2.81	2.59	2.38	2.22	2.04	1.88	1.72	1.59	1.44	1.31
0.011	440	10.22	7.90	2.32	2.83	2.61	2.40	2.24	2.06	1.90	1.74	1.61	1.46	1.33
0.012	480	10.21	7.88	2.33	2.84	2.62	2.41	2.25	2.07	1.91	1.75	1.62	1.47	1.34
0.013	520	10.21	7.86	2.35	2.86	2.64	2.43	2.27	2.09	1.93	1.77	1.64	1.49	1.36
0.014	560	10.20	7.85	2.36	2.87	2.65	2.44	2.28	2.10	1.94	1.78	1.65	1.50	1.37
0.015	600	10.20	7.83	2.37	2.88	2.66	2.45	2.29	2.11	1.95	1.79	1.66	1.51	1.38
0.016	640	10.20	7.81	2.39	2.90	2.68	2.47	2.31	2.13	1.97	1.81	1.68	1.53	1.40
0.017	680	10.19	7.80	2.40	2.91	2.69	2.48	2.32	2.14	1.98	1.82	1.69	1.54	1.41
0.018	720	10.19	7.78	2.41	2.92	2.70	2.49	2.33	2.15	1.99	1.83	1.70	1.55	1.42
0.019	760	10.18	7.77	2.41	2.92	2.70	2.49	2.33	2.15	1.99	1.83	1.70	1.55	1.42
0.020	800	10.18	7.76	2.42	2.93	2.71	2.50	2.34	2.16	2.00	1.84	1.71	1.56	1.43

Source: T. E. Larson and A. M. Buswell, "Calcium Carbonate Saturation Index and Alkalinity Interpretations," *J. Am. Water Works Assoc.* 34 (1942): 1667. Copyright 1942 by the American Water Works Association, Inc.

●EXAMPLE 10-7

Calculate the saturation index (SI) for water based on the following information:

Component	mg/l	Molecular Weight	Moles/l
Ca^{2+}	63.3	40.1	0.00158
Mg^{2+}	14.8	24.3	0.00061
Na^+	19.5	23.0	0.00085
K^+	10.1	39.1	0.00026
CO_3^{2-}	7.8	60.0	0.00013
HCO_3^-	94.4	61.0	0.00155
SO_4^{2-}	80.0	96.0	0.00083
Cl^-	17.0	35.5	0.00048

pH = 7.9; Temperature = 15°C.

○**Solution**

Using Eq. 10-41,

$$
\begin{aligned}
\text{Ionic strength } (Ca^{2+}) &= 0.5 \times 0.00158 \times 4 = 0.00316 \\
(Mg^{2+}) &= 0.5 \times 0.00061 \times 4 = 0.00122 \\
(Na^+) &= 0.5 \times 0.00085 \times 1 = 0.00042 \\
(K^+) &= 0.5 \times 0.00026 \times 1 = 0.00013 \\
(CO_3^{2-}) &= 0.5 \times 0.00013 \times 4 = 0.00026 \\
(HCO_3^-) &= 0.5 \times 0.00155 \times 1 = 0.00155 \\
(SO_4^{2-}) &= 0.5 \times 0.00083 \times 4 = 0.00166 \\
(Cl^-) &= 0.5 \times 0.00048 \times 1 = \underline{0.00024} \\
& 0.00864
\end{aligned}
$$

Using Table 10-3, with an ionic strength of 0.009 at a temperature of 15°C:

$$(pK_2' - pK_s') = 2.46$$

$$pCa^{2+} = -\log 0.00158 = \log \frac{1}{0.00158} = \log 634 = 2.80$$

$$pAlk = p[CO_3^{2-} + HCO_3^-]$$

$$= -\log[0.00013 \times 2 + 0.00155] = 2.74$$

Substitution in Eq. 10-5 yields

$$SI = 7.90 - 2.46 - 2.80 - 2.74 = -0.10$$

The negative value of the saturation index indicates that the water is slightly corrosive and will dissolve calcium carbonate.

pH Adjustment

The stability of softened water is established by controlled recarbonation. If the water processing does not include softening, the method most commonly used in corrosion control is upward adjustment of pH, and addition of meta-phosphates. Bringing the water pH above its calcium carbonate saturation value preserves a thin protective coating on the pipe interior. Metaphosphates sequester the slight excess of calcium and carbonate ions, preventing crystal formation of calcium carbonate scale.

Either lime or soda ash can be applied to treat corrosive waters having a hardness exceeding 35 mg/l. The former is preferred in treating soft water to provide the needed calcium as well as pH adjustment, although lime for calcium and soda ash to provide carbonate may be needed.

10-14

ION-EXCHANGE SOFTENING

Ion exchangers are used in water and waste treatment to soften water, selectively remove specific impurities, and recover valuable chemicals otherwise lost in industrial waste discharges. Ion exchangers are also employed to demineralize water completely for laboratory and industrial purposes. In dilute solutions, the ion-exchange process can produce ion-free water.

In cation-exchange softening, the hardness-producing elements of calcium and magnesium are removed and replaced with sodium by a cation resin. Ion-exchange reactions for softening may be written

$$Na_2R + \begin{Bmatrix} Ca \\ Mg \end{Bmatrix} \begin{matrix} (HCO_3)_2 \\ SO_4 \\ Cl_2 \end{matrix} \rightarrow \begin{Bmatrix} Ca \\ Mg \end{Bmatrix}R + \begin{cases} 2NaHCO_3 \\ Na_2SO_4 \\ 2NaCl \end{cases} \tag{10-42}$$

where R represents the exchange resin. They show that if a water containing calcium and magnesium is passed through an ion exchanger, these metals are taken up by the resin, which simultaneously gives up sodium in exchange.

After the ability of the bed to produce soft water has been exhausted, the unit is removed from service and backwashed with a solution of sodium chloride. This removes the calcium and magnesium in the form of their soluble chlorides and at the same time restores the resin to its original sodium condition. The bed is rinsed free of undesirable salts and returned to service. The governing reaction may be written

$$\begin{Bmatrix} Ca \\ Mg \end{Bmatrix}R + 2NaCl \longrightarrow Na_2R + \begin{Bmatrix} Ca \\ Mg \end{Bmatrix}Cl_2 \tag{10-43}$$

A majority of ion-exchange softeners are the pressure type, with either manual or automatic controls. They normally operate at rates of 6–8 gpm/ft^2 of surface filter area. A water meter is usually employed on the inlet or outlet side. For manual-type operations, this meter can be set to turn on a light or sound an alarm at the end of the softening run.

About 8.5 lb of salt is required to regenerate 1 ft^3 of resin and remove approximately 4 lb of hardness in a commercial unit. The reduction is directly related to the amount of cations present in the raw water and the amount of salt used to regenerate the resin bed.

Iron and Manganese Removal

Iron and manganese in concentrations greater than 0.3 mg/l of iron and 0.05 mg/l of manganese stain plumbing fixtures and laundered clothes. Although discoloration from precipitates is the most serious problem associated with water supplies having excessive iron and manganese, foul tastes and odors can be produced by growth of iron bacteria in water distribution mains. These filamentous bacteria, using reduced iron as an energy source, precipitate it, causing pipe incrustations. Decay of the accumulated bacterial slimes creates offensive tastes and odors.

Dissolved iron and manganese are often found in groundwater from wells located in shale, sandstone, and alluvial deposits. Impounded surface-water supplies may also have troubles with iron and manganese. An anaerobic hypolimnion (stagnant bottom-water layer) in a reservoir dissolves precipitated iron and manganese from the bottom muds, and during periods of overturn, these minerals are dispersed throughout the entire depth.

10-15

CHEMISTRY OF IRON AND MANGANESE

Iron(II)(Fe^{2+}) and manganese(II)(Mn^{2+}) are chemically reduced, soluble forms which exist in a reducing environment (absence of dissolved oxygen and low pH). These conditions exist in groundwater and anaerobic reservoir water. When it is pumped from underground or an anaerobic hypolimnion, carbon dioxide and hydrogen sulfide are released, raising the pH. In addition, the water is exposed to air creating an oxidizing environment. The reduced iron and manganese start transforming to their stable, oxidized, insoluble forms of iron(III)(Fe^{3+}) and manganese (IV) (Mn^{4+}).

The rate of oxidation of iron and manganese depends upon the type and concentration of the oxidizing agent, pH, alkalinity, organic content, and presence of catalysts.

Oxygen, chlorine, and potassium permanganate are the most frequent oxidizing agents. The natural reaction by oxygen is enhanced in water treatment by using spray nozzles or waterfall-type aerators. Chlorine and potassium permanganate ($KMnO_4$) are the chemicals commonly used in iron and manganese removal plants. Oxidation reactions using potassium permanganate are

$$3Fe^{2+} + MnO_4^- \longrightarrow 3Fe^{3+} + MnO_2 \tag{10-44}$$

$$3Mn^{2+} + 2MnO_4^- \longrightarrow 5MnO_2 \tag{10-45}$$

Rates of oxidation of the ions are dependent upon pH and bicarbonate-ion concentration. The pH for oxidation of iron should be 7.5 or higher: manganese oxidizes readily at pH 9.5 or higher. Organic substances (i.e, humic or tannic acids) can create complexes with iron(II) and manganese(II) ions holding them in the soluble state to higher pH levels. If a large concentration of organic matter is present, iron can be held in solution at pH levels up to 9.5.

Copper ions and silica have a catalytic effect on the oxidation of iron and manganese.[11] The presence of about 0.1 mg/l of copper increases the rate of iron oxidation 5–6 times. Silica increases oxidation rates of both metals. Manganese oxides are catalytic in the oxidation of manganese. Tray aerators frequently contain coke or stone contact beds through which the water percolates. These media develop and support a catalytic coating of manganese oxides.

10-16

PREVENTATIVE TREATMENT

When an industry or city is confronted with iron and manganese problems, solutions are difficult and probably costly. Treatment of the water supply is the only permanent answer. Control and preventative measures can be employed with expectation of reasonable success.

Sodium hexametaphosphate has been found to be effective in sequestering iron and manganese in some supplies. When applied at the proper dosage, before oxidation of the iron and manganese occurs, metaphosphate tends to hold iron and manganese in solution. Suggested dosage of sodium hexametaphosphate is a minimum of 2 mg/l for each mg/l of iron and manganese present.[12] Metaphosphate does not prevent oxidation of iron and manganese but stops agglomeration of the individual tiny particles of iron and manganese oxides. Thus the sequestered oxides pass through the distribution system without creating accumulations which periodically cause badly discolored water.

Growth of iron bacteria in water mains can be controlled by chlorine and copper sulfate.[13] The required dosage depends upon quality of the water and extent of the bacterial growth. When iron bacteria persist in chlorinated water-distribution systems, heavy chlorination (50–100 mg/l) for 48 hours followed by flushing of the water main may be necessary. A copper sulfate dosage of 03.–0.5 mg/l applied continously for several weeks can be effective.

10-17

IRON AND MANGANESE REMOVAL PROCESSES

Aeration–Sedimentation–Filtration

The simplest form of oxidation treatment uses plain aeration. The units most commonly employed are the tray type, where a vertical riser pipe distributes the water on top of a series of trays, from which it then drips and spatters down through a stack of 3 or 4 of them. Soluble iron is readily oxidized by the following reaction:

$$2Fe(HCO_3)_2 + 0.5O_2 + H_2O = 2Fe(OH)_3 + 4CO_2 \qquad (10\text{-}46)$$

Manganese cannot be oxidized as easily as iron, and aeration alone is generally not effective. If, however, the pH is increased to 8.5 or higher (by the addition of lime, soda ash, or caustic soda), and if aeration is accompanied by contact with coke beds coated with oxides in the aerator, catalytic oxidation of the manganese occurs.

In plants using the aeration–sedimentation–filtration process, most of the oxidized iron and manganese is removed by a sand filter.[14] Flocculant metal oxides are not heavy enough to settle out in the settling basin. It appears that the main function of the basin is to allow sufficient reaction time for oxidation to proceed to near completion.

Aeration–Chemical Oxidation–Sedimentation–Filtration

This sequence of processes is the usual method for removing iron and manganese from well water without softening treatment. Contact tray aeration is designed to displace dissolved gases, (i.e., carbon dioxide) and initiate oxidation of the reduced iron and manganese.

Either chlorine or potassium permanganate can chemically oxidize the manganese. When chlorine is utilized, a free available chlorine residual is maintained throughout the treatment process. The rate of manganese(II) oxidation by chlorine is dependent upon pH, chlorine dosage, mixing conditions, and other factors. Copper sulfate is a catalyst in the oxidation of manganese.

Theoretically, 1 mg/l of potassium permanganate will oxidize 1.06 mg/l of iron or 0.52 mg/l of manganese (Eqs. 10-41 and 10-42). In actual practice, however, the permanganate necessary for oxidation of the soluble manganese is less than the theoretical requirement. One main advantage of potassium permanganate oxidation is the high rate of the reaction, many times faster than for chlorine. Also, the rate of reaction is relatively independent of the hydrogen-ion concentration within a pH range of 5–9.

Filtration following chemical oxidation and sedimentation is very important. Practice has shown that filters pass manganese unless grains of the media are coated with metal oxides. This covering develops naturally during filtration of manganese-bearing water. The coating serves as a catalyst for the oxidation and removal of manganese.

Water Softening

Lime-soda softening will take out iron and manganese. If split treatment is employed, potassium permanganate can oxidize the iron and manganese in water bypassing the first-stage excess lime treatment. Lime-soda softening should be given careful consideration as a possible process for treating a hard water requiring iron and manganese elimination.

Lime treatment has been used to remove organically bound iron and manganese from surface water. The process scheme aeration–coagulation–lime treatment–sedimentation–filtration can treat surface waters containing color, turbidity, and organically bound iron and manganese.

Manganese Zeolite Process

Manganese zeolite is made by coating natural greensand (glauconite) zeolite with oxides.[15] Manganese dioxide removes soluble iron and manganese until it becomes degenerated (Eq. 10-47). The filter is regenerated using potassium permanganate (Eq. 10-48).

$$Z-MnO_2 + \begin{matrix} Fe^{2+} \\ Mn^{2+} \end{matrix} \longrightarrow Z-Mn_2O_3 + \begin{matrix} Fe^{3+} \\ Mn^{3+} \\ Mn^{4+} \end{matrix} \qquad (10\text{-}47)$$

$$Z-Mn_2O_3 + KMnO_4 \longrightarrow Z-MnO_2 \qquad (10\text{-}48)$$

Manganese zeolite filters are generally pressure types (Sec. 9-21).

Disadvantages of the regenerative-batch process are the possibility of soluble manganese leakage when the bed is nearly degenerated, and the waste of excess potassium permanganate needed to regenerate the greensand. These two drawbacks have been substantially overcome by continuously supplying a feed of potassuim permanganate solution ahead of a dual-media filter of anthracite and manganese zeolite. The anthracite filter media remove most insolubles, thereby reducing the problem of plugging the greensand. A continuous feed of permanganate reduces the frequency of greensand regeneration. When the permanganate feed is less than the reduced iron and manganese in the water, excess iron and manganese is oxidized by the greensand. If a surplus is applied, it regenerates the greensand.

●EXAMPLE 10-8

A well-water supply contains 3.2 mg/l of iron and 0.8 mg/l of manganese at pH 7.8. Estimate the dosage of potassium permanganate required for iron and manganese oxidation.

○Solution

Thetheoretical potassium permanganate dosages given in Section 10-17 are 1.0 mg/l per 1.06 mg/l of iron and 1.0 mg/l per 0.52 mg/l of manganese. Therefore,

$$KMnO_4 \text{ required} = \frac{3.2 \times 1.0}{1.06} + \frac{0.8 \times 1.0}{0.52} = 4.6 \text{ mg/l}$$

Disinfection

The purpose of disinfecting drinking water is to destroy organisms that cause diseases in man. Most pathogenic bacteria and many other microorganisms are removed from water in varying degrees by conventional treatment processes as follows: physical elimination through coagulation, sedimentation, and filtration; natural die-away of organisms in an unfavorable environment during storage; and inactivation by chemicals introduced for treatment purposes other than disinfection. However, chlorination is used in water treatment to ensure satisfactory disinfection of potable water supplies.

10-18

CHEMISTRY OF CHLORINATION

Chlorine gas is soluble in water (7160 mg/l at 20°C and 1 atm), and hydrolizes rapidly to form hypochlorous acid.

$$Cl_2 + H_2O \rightleftharpoons HOCl + H^+ + Cl^- \tag{10-49}$$

Hydrolysis goes virtually to completion at pH values and concentrations normally experienced in water- and waste-treatment operations.

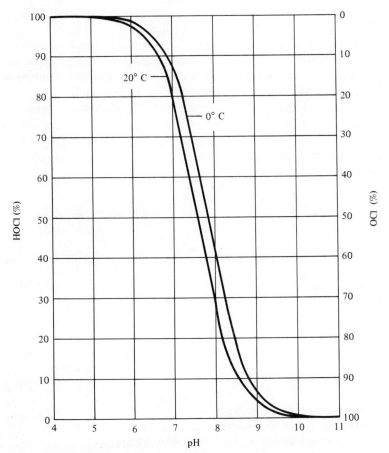

FIGURE 10-12 Relationship between HOCl, OCl⁻, and pH.

Figure 10-12 shows the relationship between HOCl and OCl⁻ at various pH levels. Hypochlorous acid ionizes according to the following equation:

$$HOCl \rightleftharpoons H^+ + OCl^- \tag{10-50}$$

$$\frac{[H^+][OCl^-]}{[HOCl]} = K_i \tag{10-51}$$

The dissociation rate from hypochlorous acid to hypochlorite ion is sufficiently rapid, so equilibrium is maintained even though the former is being continuously consumed. If a reducing agent is put into water containing free available chlorine, the unconsumed residual redistributes itself between HOCl and OCl⁻.

Chlorine reacts with ammonia in water to form chloramines as follows:

$$HOCl + NH_3 \longrightarrow H_2O + NH_2Cl \quad \text{(monochloramine)} \tag{10-52}$$

$$HOCl + NH_2Cl \longrightarrow H_2O + NHCl_2 \quad \text{(dichloramine)} \tag{10-53}$$

$$HOCl + NHCl_2 \longrightarrow H_2O + NCl_3 \quad \text{(trichloramine)} \tag{10-54}$$

The chloramines formed depend upon pH of the water, amount of ammonia available and temperature. In the pH range 4.5–8.5, monochloramine and dichloramine are formed. At room temperature, monochloramine exists alone above pH 8.5 and dichloramine occurs alone at pH 4.5. Below pH 4.4 trichloramine is produced.

Free available residual chlorine is that residual chlorine existing in water as hypochlorous acid or hypochlorite ion. *Combined available residual chlorine* is that residual existing in chemical combination with ammonia (chloramines) or organic nitrogen compounds. *Chlorine demand* is the difference between the amount added to a water and the quantity of free and combined available chlorine remaining at the end of a specified contact period.

When chlorine is added to water containing reducing agents and ammonia, residuals develop which yield a curve similar to Fig. 10-13. Chlorine reacts first with reducing agents present and develops no measurable residual as shown by the portion of the curve extending from *A* to *B*. The chlorine dosage at *B* is the amount required to meet the demand exerted by the reducing agents (those common to water and waste include nitrites, ferrous ions, and hydrogen sulfide).

The addition of chlorine in excess of that required up to point *B* results in forming chloramines. Monochloramines and dichloramines are usually considered together because there is little control over which will be formed. The quantities of each are determined primarily by pH. Chloramines thus established show an available chlorine residual and are effective as disinfectants. When all the ammonia has been reacted with, a free available chlorine residual begins to develop (point *C* on the curve). As the free available chlorine residual increases, the previously produced chloramines are oxidized. This results in the creation of oxidized nitrogen compounds, such as nitrous oxide, nitrogen, and nitrogen trichloride, which in turn reduce the chlorine residual, as seen on the curve between *C* and *D*.

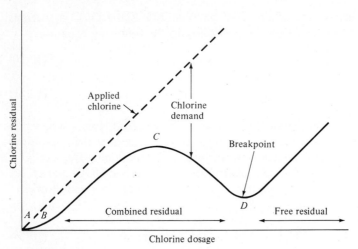

FIGURE 10-13 Chlorine-residual curve for breakpoint chlorination.

Upon completion of oxidation of all chloramines, additional chlorine added to the water creates an equal residual, indicated by point D on the curve. Point D is generally referred to as the "breakpoint," that limit beyond which all the residual is free available chlorine. Some resistant chloramines may still be present beyond D, but their relative importance is small.

Hypochlorites (salts of hypochlorous acid) may be used for chlorination at small installations such as swimming pools, and in emergencies. Since hypochlorites are more expensive, liquid chlorine is applied in most water-treatment plants in the United States. Calcium hypochlorite, $Ca(OCl)_2$, is available commercially in granular and powdered forms that contain about 70% available chlorine. Sodium hypochlorite ($NaOCl$) is handled in liquid form at concentrations between 5 and 15% available chlorine. These salts in water solution yield the hypochlorite ion directly.

10-19

DISINFECTION OF DRINKING WATER

The bactericidal action of chlorine results from its strong oxidizing power on the bacterial cell's chemical structure, destroying the enzymatic processes required for life. The rate of microbial inactivation depends upon the concentration and form of the available chlorine residual, pH and temperature of the water, and time of contact.

Hypochlorous acid is the primary disinfectant. Hypochlorite ions are somewhat less effective. Therefore, the power of free chlorine residual decreases with increasing pH. Bactericidal action of combined available chlorine is significantly less than that of free chlorine residual. Recommended minimum chlorine residuals for disinfection based on studies reported by the U.S. Public Health Service[16] are listed in Table 10-4.

Protozoal cysts and enteroviruses are much more resistant to chlorine than are coliforms and other enteric bacteria. Although there is little evidence to indicate that current water-treatment practices are not adequate, considerable concern remains regarding the resistance of these potential pathogens. The only documented cases involve infectious hepatitus transmitted by untreated water supplies. Thus far, no epideminological surveys in the United States have conclusively demonstrated transmission of protozoal or viral diseases by processed public water supplies. Required free chlorine residuals after 30 minutes' contact are in a broad range from 3 to greater than 20 mg/l; the destructive levels necessary increase with rising pH and lower water temperatures. For virus inactivation, only undissociated HOCl is considered effective. Studies have indicated that a hypochlorous acid residual of 0.5 to 1.0 mg/l is effective within 30 minutes while more than 100 mg/l of hypochlorite is required for an equivalent effect. Therefore, the best disinfection for surface waters is breakpoint chlorination to establish a free chlorine residual in a pH range yielding HOCl.[17]

TABLE 10-4

RECOMMENDED MINIMUM CHLORINE RESIDUALS FOR BACTERICIDAL DISINFECTION OF WATER

pH Value	Minimum Free Available Chlorine Residual After 10 Minutes of Contact (mg/l)	Minimum Combined Available Chlorine Residual After 60 Minutes of Contact (mg/l)
6.0	0.2	1.0
7.0	0.2	1.5
8.0	0.4	1.8
9.0	0.8	Not applicable
10.0	0.8	Not applicable

Well-water supplies obtained from aquifers beneath impervious strata are often not disinfected. Since contamination can occur in the distribution system, it is good practice to retain a chlorine residual to ensure a safe water supply at the consumer's faucet. The chlorine residual may be free or combined residual established by feeding anhydrous ammonia with the chlorine.

Surface-water supplies from polluted rivers or impoundments are normally purified with heavy chlorine dosages. Breakpoint chlorination prior to chemical coagulation may be practiced at certain times of the year for taste and odor control and for disinfection. Intermediate and postchlorination retain the desired residual through the treatment plant and in the distribution system.

10-20

DISINFECTION OF WASTEWATER EFFLUENTS

Wastewater discharges are chlorinated to inactivate enteric bacteria and protect public health, particularly in body-contact recreation and for municipal water supplies. Disinfection is accomplished by chloramines formed when the chlorine reacts with ammonia present in the wastewater (Eqs. 10-52 and 10-53). Prior to establishing specific biological requirements, disinfection of wastewater was empirically defined as the addition of sufficient chlorine so that a residual of 0.5 mg/l existed after 15 minutes. Now, adequate disinfection of a secondary effluent is defined by an average fecal coliform count of less than 200 per 100 ml. Since biologically treated wastewater contains approximately 1,000,000 coliforms per 100 ml, oxidative reduction of fecal coliforms from this large number to only 200 per 100 ml destroys the majority of organisms in the waste.

An efficient chlorination system provides rapid initial mixing of the chlorine solution in the wastewater and contact time in a plug-flow basin for a minimum of 30 minutes at the peak hourly flow rate. Rapid blending can be accomplished by applying the chlorine in a pressure conduit under conditions of highly turbulent flow, or in a channel immediately upstream from a mechanical mixer. Adequate plug flow can be achieved by a baffled contact chamber; circular and rectangular tanks that allow backmixing are not as effective.[18] A well designed unit provides adequate disinfection of a secondary effluent with a dosage of 8–15 mg/l.

Control of chlorine dosage is extremely important for proper operation. Automatic residual monitoring and feedback control is necessary to prevent both inadequate disinfection and excessive chlorination, resulting in discharge of an effluent toxic to aquatic life.[19] To protect receiving streams, some regulatory agencies have specified maximum chlorine residuals in undiluted effluents of 0.1–0.5 mg/l. Dechlorination may be required to detoxify a discharge after chlorination if it is poorly designed. This may be accomplished by adding sulfur dioxide or sodium metabisulfite.

Chlorination reduces the K rate and 5-day BOD of treated wastewater to a limited extent. The unit BOD reduction per mg/l of chlorine absorbed is greatest at the dosage producing the least residual and decreases with increasing chlorine dosage. BOD reductions up to 2 mg/l per mg/l of chlorine absorbed have been observed.[20] Chlorination for lowering BOD of wastewater effluents have been practiced to control nuisance conditions in a receiving stream during drought flows.

10-21

CHLORINE HANDLING AND DOSING

Chlorine is a poisonous, yellow-green gas at room temperature and atmospheric pressure. Liquid chlorine is shipped in pressurized steel cylinders ranging in size from 100 lb to 1 ton. One volume of liquid chlorine yields about 450 volumes of vapor. Moist chlorine gas is extremely corrosive, and therefore piping and dosing equipment are either nonmetal or made of special alloys. The vapor causes respiratory and eye irritation, and high concentrations can cause physiological damage. Chlorine feeding rooms and storage areas should be kept cool and well ventilated. Gas is drawn from a pressurized cylinder through a solution feeder to control the rate of application. Direct feed of chlorine gas into a pipe or channel is not practiced for safety reasons (e.g., the danger of piping gas to points of application). The injector in a solution feed chlorinator dissolves the gas in water, and then this concentrated solution is applied to the water being treated. The automatic proportional-control system illustrated in Fig. 10-14 adjusts the feed rate to maintain a constant preset

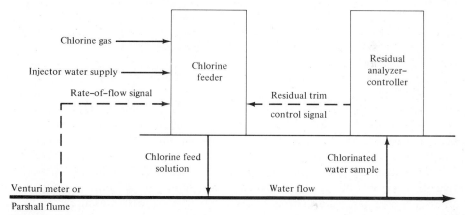

FIGURE 10-14 Automatic proportional control system for feeding to a constant, preestablished chlorine residual.

dosage for all rates of flow. The chlorine feeder is responsive to signals from both the flow-meter transmitter and the chlorine residual analyzer. In this manner rate of flow measurement is the primary feed regulator, and residual monitoring trims the dosage. At small installations, it may be satisfactory to proportion feed to flow and thus apply a constant preset dosage without residual monitoring. However, this type of regulator is only satisfactory where chlorine demand and flow are reasonably constant and an operator is available to make adjustments as necessary.

Fluoridation

The incidence of dental caries relative to the concentration of fluoride in drinking water is well established. Low levels of fluoride result in increasing incidence of caries, while excessive fluoride results in mottled tooth enamel. At the

TABLE 10-5

RECOMMENDED AND APPROVAL FLUORIDE LIMITS FOR
PUBLIC DRINKING-WATER SUPPLIES

| Annual Average of Maximum Daily Air Temperatures Based on Temperature Data Obtained for a Minimum of 5 Years (°F) | Fluoride-Ion Concentrations (mg/l) | | | |
| | Recommended Limits | | | Approval Limit |
	Lower	Optimum	Upper	
50.0–53.7	0.9	1.2	1.7	1.8
53.8–58.3	0.8	1.1	1.5	1.7
58.4–63.8	0.8	1.0	1.3	1.5
63.9–70.6	0.7	0.9	1.2	1.4
70.7–79.2	0.7	0.8	1.0	1.2
79.3–90.5	0.6	0.7	0.8	1.1

Source: *1974 Drinking Water Standards and Guidelines, Water Supply Division, Environmental Protection Agency.*

optimum concentration of approximately 1.0 mg/l, consumers realize maximum reduction in tooth decay with no esthetically significant dental fluorosis. Recommended limits given in Table 10-5 are based on air temperature, since this influences the amount of water ingested by people. Medical studies have also indicated that fluoride benefits older persons by reducing the prevalence of osteoporosis and hardening of the arteries. Most public water supplies add fluoride ion to achieve an optimum level as a public health measure.

10-22

FLUORIDATION

The fluoride compounds most commonly applied in water treatment are sodium fluoride, sodium silicofluoride, and fluosilicic acid (Table 10-6). Sodium silicofluoride, the most widely used compound, is commercially availabe in various gradations for application by dry feeders. Sodium fluoride is also popular, particularly the crystaline type where manual handling is involved. Fluosilicic acid is a strong corrosive acid that must be handled with care. It is preferred in large waterworks where it can be applied by liquid feeders without prior dilution.

Design of a fluoridation system varies with size and type of water facility, chemical selection, and availability of operating personnel. Small utilities often choose liquid feeders to apply solutions of NaF or Na_2SiF_6 that are prepared in batches. The solution tank may be placed on a platform scale for the convenience of weighing during preparation and feed. Saturator tanks containing

TABLE 10-6

COMMON CHEMICALS USED IN THE FLUORIDATION OF
DRINKING WATER

	Sodium Fluoride	Sodium Silicofluoride	Fluosilicic Acid
Formula	NaF	Na_2SiF_6	H_2SiF_6
Fluoride ion (%)	45	61	79
Molecular weight	42	188	144
Commercial purity (%)	90–98	98–99	22–30
Commercial form	Powder or crystal	Powder or fine crystal	Liquid

a bed of sodium fluoride crystals yield a solution of about 4.0% (18,000 mg/l of F). Larger water plants use either gravimetric dry feeders to apply chemicals, or solution feeders to inject full-strength H_2SiF_6 directly from the shipping drum. Automatic control systems use flow meters and recorders to adjust feed rate.

Application of fluoride is best in a channel or water main coming from the filters, or directly to the clear well. If applied prior to filtration, losses could occur due to reactions with other chemicals, such as, coagulation with heavy alum doses or lime softening. If no treatment plant exists, fluoride can be injected into mains carrying water to the distribution system. This may be a single point, or several separate fluoride feeding installations where wells supply water at different points in the piping network.

●**EXAMPLE 10-9**

The fluoride-ion concentration in a water supply is increased from 0.30 mg/l to 1.00 mg/l by applying 98% pure sodium silicofluoride. How many pounds of chemical is required per million gallons of water?

○*Solution*

From Table 10-6, Na_2SiF_6 is 61% F.

$$\text{dosage} = \frac{(1.00 - 0.30)8.34}{0.98 \times 0.61} = 9.77 \text{ lb/mg}$$

10-23

DEFLUORIDATION

The upper limits of fluoride-ion concentrations for approval of public water supplies are listed in Table 10-5. Consumption of water containing excess fluorides over a period of several years results in dental fluorosis, identified by a permanent gray-to-black discoloration of tooth enamel. Studies have indicated that children consuming water containing 5 mg/l develop mottling to the extent of pitted enamel, resulting in loss of teeth. Although expensive, defluoridation is cost-effective when the expenses of dental care and loss of teeth are considered. A few communities have been able to find alternative water sources to solve their fluoride problem without special treatment.

Current methods for defluoridation use either activated alumina or bone char.[21] Fluoride ion is removed by filtering water through these insoluble, granular media. After saturation with fluoride ion, bone char is regenerated by backwashing with a 1 % solution of caustic soda and then rinsing the bed. Reactivation of alumina also involves rinsing the bed with a caustic solution.

Taste and Odor

Surface waters contain tastes and odors associated with decaying organic matter, biological growths, and chemicals originating from industrial waste discharges. Land drainage from snowmelt and spring rains contains pollutants that are difficult to remove, especially at the low water temperatures that occur in northern climates. During the summer, lake and reservoir supplies may be plagued by blooms of algae that impart compounds producing fishy, grassy, or other foul odors. Actinomycetes are moldlike bacteria that create an earthy odor. Well waters may contain dissolved gases, such as hydrogen sulfide, and inorganic salts or metal ions that flavor the water. Tastes in groundwater can be identified, but defining specific odor-bearing substances in surface waters is usually an impossible task. Therefore, the best practical treatment for each water supply is determined by experiment and experience.

10-24

CONTROL OF TASTE AND ODOR

Water-treatment-plant designs should allow maximum flexibility of operations for control of tastes and odors. Problems may vary considerably during different seasons of the year, and future adverse changes in water quality of a supply are difficult to predict. Breakpoint chlorination and treatment with activated carbon are common control techniques for surface-water supplies, while aeration is applied most frequently in groundwater processing. Preventative measures, such as selection of a source with the best water quality, should be given primary consideration. Regulatory controls should be used to reduce contaminants from waste discharges that enter surface and underground waters. Algal blooms may be suppressed in reservoirs by regular application of copper sulfate.

Oxidative Methods

Chlorine, potassium permanganate, and ozone are strong oxidants capable of destroying many odorous compounds. The most popular process is breakpoint chlorination as the first step in processing surface waters. Sufficient chlorine must be applied to destroy ammonia–nitrogen and establish a free residual (Fig. 10-13). If a destructive dosage is not applied, the odors of some organic compounds are intensified rather than reduced. Heavy preliminary chlorination and retention of a free residual through the plant is also recommended for bacterial and viral disinfection. Occasionally, potassium permanganate is more effective than chlorine as an oxidizing agent, and both chemicals can be used to achieve maximum control. The use of ozone is extremely limited in the United States.

Activated Carbon

Adsorption on activated carbon is the most effective means of taste and odor removal. It is primarily a surface phenomenon, where one substance is attracted to the surface of another; the larger the surface area of an adsorber, the greater its power. Carbon for water treatment is rated in terms of square meters of surface area per gram. One pound of activated carbon has an estimated surface area of 100 acres. Besides controlling tastes and odors, powdered carbon aids in sludge stabilization, improved floc formation, and reduction in non-odor-causing organics.

Carbon is fed to water either as a dry powder or as a wet slurry. The latter has the advantage of being cleaner to handle and assures complete effectiveness by thoroughly wetting the carbon. Although granular carbon adsorption beds are used extensively to purify product water in the food and beverage industries, their application in municipal water processing is limited because of economic considerations.

Activated carbon can be introduced at any stage of water processing ahead of filtration. Although adsorption is nearly instantaneous, a contact time of 15 minutes or more is desirable before sedimentation or filtration. The best point of application is generally determined by trial and error based on previous experience. Since carbon adsorbs chlorine, these two chemicals should not be applied simultaneously or in close sequence. Carbon is often fed during flocculation and/or just prior to filtration when the raw water receives breakpoint chlorination.

Aeration

Air stripping is effective for removing dissolved gases and highly volatile odorous compounds. Aeration as a first step in processing well water may achieve any of the following: removal of hydrogen sulfide, reduction of dissolved carbon dioxide, and addition of dissolved oxygen for oxidation of iron and manganese. Aeration is rarely effective in processing surface waters, since the odor-producing substances are generally nonvolatile.

Corrosion and Corrosion Control

10-25

CORROSION OF WATER MAINS

All metal in contact with water is subject to chemical corrosion, a complex and as yet a poorly understood process. In its simplest form, corrosion occurs when positive ions enter a solution and combine with negative ions of the water to form a hydroxide.

Anodic reaction:

$$4Fe + 2O_2 + 8H_2O \rightleftharpoons 4Fe(OH)_3 + 4H^+ + 4e^- \tag{10-55}$$

Cathodic reaction:

$$O_2 + 4H^+ + 4e^- \longrightarrow 2H_2O \tag{10-56}$$

There is a complex mutual interaction of corrosion stimulating and inhibiting factors such as pH, buffer capacity, $CaCO_3$ deposition, alkalinity, and the activity of electrolytic cells.

The mechanism of corrosion is primarily electrochemical in nature, with reactions occurring between metal surfaces and chemicals in the soil or water in contact with the metal surface. An elementary example of this theory is the "corrosion battery," which illustrates the principles involved (Fig. 10-15).

Corroding metal surfaces contain innumerable local galvanic cells or "corrosion batteries" wherein the decay takes place. These "batteries" result from areas of differing electrical potential over the metallic surface which form the anodes and cathodes while the water or soil in contact with the metal surface

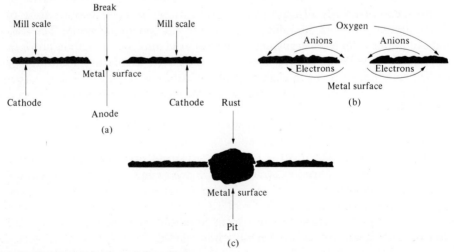

FIGURE 10-15 One form of corrosion battery.

acts as an electrolyte. As the current flows from anodic to cathodic area, metallic ions are released into solution at the anode, resulting in rust and pitting. Figure 10-15 shows the cause, effect, and result.

Dissolved oxygen has a dual influence on the corrosion process. Metallic destruction results primarily from electrolytic action. Hydrogen ions in solution are reduced by electrons to atomic hydrogen at the cathode. If left undisturbed, atomic hydrogen would form a protective film on the surface of the cathode and materially reduce the rate of corrosion. The accumulation of hydrogen is prevented by any dissolved oxygen present, and the oxygen and hydrogen react to form water. High concentrations of dissolved oxygen lower the probability of corrosion by improving the anodic films. It is considered that dissolved oxygen is adsorbed on the surface of the iron and takes part in the formation of protective films. Thus before the effects of dissolved oxygen can be fully evaluated in the corrosion process, it must be determined whether the chemical reaction is governed by cathodic or anodic control. It appears that cathodic reactions control the early stages of rusting on water pipes but that the control over a period of years is largely anodic. In the absence of dissolved oxygen or some other oxidizing agent, such as chlorine, nitrate, nitrite, or dichromate, the corrosion process cannot proceed very far in the early stage.

Metallic corrosion can be retarded or prevented by (1) cathodic protection, (2) application of coatings or linings, (3) regulation of the outside pipe environment, (4) the addition of inhibitors to the water, and (5) production of films by chemically treating the water.

Cathodic Protection

Some forms of electrolytic corrosion can be controlled by cathodic protection. This type of safeguard is simply the original "corrosion battery" with an auxiliary anode acting as a substitute for the local anode on the metal surface to be preserved.

The auxiliary anodes may be either electrolytic (similar metal energized by an external source of direct current) or galvanic (composed of a metal higher in the electromotive series than the metal to be saved). In either case, electrical energy supplied by the auxiliary anodes forms a block to prevent flow of current in the local cells on the protected surface. With the flow of current blocked, escape of metallic ions from the local cell anodes also ceases and corrosion is prevented (see Figs. 10-16 and 10-17).

Where the current demand of the cathodes is satisfied by local anodes on the metal surface to be protected in the "corrosion battery," it is furnished by a power supply and the corrosion processes transferred to the auxiliary anode. The auxiliary anodes are eventually expended but can easily be replaced.

Coatings and Linings

Coatings and linings prevent corrosion and increase or maintain smoothness of pipe walls. Materials used to guard steel and concrete surfaces include asphaltics, resins, rubber bases, galvanizing, and plastics. Several types of plastic

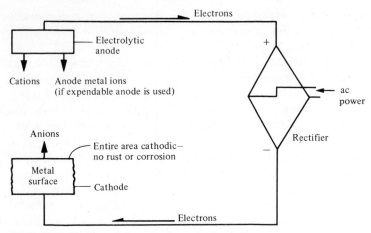

FIGURE 10-16 Electrolytic cathodic-protection battery.

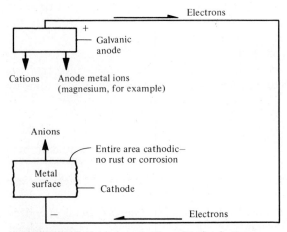

FIGURE 10-17 Galvanic cathodic-protection battery.

coatings appear to be the most promising since they resist corrosion and are chemically inert to acids and alkalies. Although plastic coatings are not very resistant to sunlight, this property presents no problem in pipe-corrosion control.

Control of the Outside Pipe Environment

Corrosion-control measures for the pipe's exterior are similar to those for the inside. The galvanic anode principle is widely employed for pipelines and metal in contact with the ground. Outside coatings are sometimes necessary for the protection of metal and concrete pipes because of the chemical nature of the soil.

The Addition of Inhibitors to Water

Certain compounds, when added to water in small amounts, reduce the rate of corrosion. Such compounds include polyphosphates, widely used in water treatment to form soluble complexes with dissolved iron. These iron-phosphate complexes prevent the formation of iron hydroxide tubercules which reduce the hydraulic efficiency of pipes. "Red water" formation is also elimi-nated. Polyphosphates function as sequestering agents to prevent precipitation of calcium, magnesium, and iron. Sodium silicate apparently functions in a manner similar to polyphosphates, but this compound is not widely used in municipal water treatment.

Films by Chemical Treatment

Deposition of coatings of calcium carbonate on pipes and metal tanks is discussed in Section 10-13. Silicate films are reportedly successful in protecting the walls of pipes. They are dense, but slightly permeable, and corrosion can take place if conditions are favorable after the film is deposited.

10-26

CORROSION OF SEWER PIPES

Corrosion is the destruction (eating away) of pipe materials by chemical action. Sewer corrosion can result from biological production of sulfuric acid, and is caused by strong industrial wastes unless they are neutralized prior to disposal. Most municipal sewer ordinances prohibit industrial wastes having a pH less than 5.5, higher than 9.0, or having other corrosive effects.

Crown corrosion in sanitary sewers is most prevalent in warm climates where they are laid on flat grades, or where the sulfur content of the wastewater is high. Biological activity in wastewater in a sewer creates anaerobic condi-tions, producing hydrogen sulfide. Condensation moisture on the crown and walls of the sewer pipe absorbs hydrogen sulfide and oxygen from the atmos-phere in the sewer. The sulfur-oxidizing bacteria *Thiobacillus* form sulfuric acid in the moisture of condensation.

$$H_2S + O_2 \xrightarrow{\textit{Thiobacillus}} H_2SO_4 \qquad (10\text{-}57)$$

In concrete sewers, sulfuric acid reacts with lime to form calcium sulfate, which lacks structural strength. If the concrete is sufficiently weakened, the pipe might collapse under heavy overburden loads.

The best protection for sanitary sewers is a corrosion-resistant pipe material such as vitrified clay or plastic. In large sewers, where size and economics dictate concrete pipe, crown corrosion can be retarded by ventilation or by chlorinat-ing the wastewater to control hydrogen sulfide generation. Recent advances for protection of concrete pipe include development of synthetic coatings and linings.

Saline-Water Conversion

There are approximately 32×10^7 mi^3 of seawater and large amounts of inland saline water on the earth. Seawater is available in almost unlimited quantities. Figure 10-18 shows the distribution of known inland brackish-water resources of the United States. The conversion of saline water to fresh water is not a new idea. Man has known how to accomplish this for centuries by boiling water and condensing the resulting vapors. In Julius Caesar's time his legionnaries devised a means of desalting seawater by using solar heat to supply fresh water for drinking in order to survive the siege of Alexandria.[23] The first widespread practical application of seawater conversion units came with the advent of the steamship and its requirement of fresh water for boilers. At present, water is rapidly becoming a limiting factor on economic growth in many areas of the United States and the world.

Water quality is usually referred to as fresh, brackish, seawater, or brine.

1 Fresh water normally contains less than 1000 mg/l of dissolved salts.
2 Brackish water ranges from 1000 to 35,000 mg/l of dissolved salts.
3 Seawater generally contains 35,000 mg/l of dissolved salts.
4 Brine is water containing more dissolved salts than seawater, such as the Great Salt Lake or the Dead Sea.

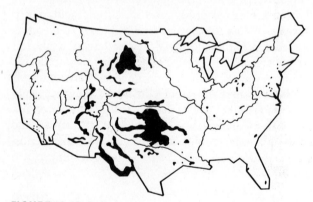

FIGURE 10-18 Preliminary survey of the brackish-water resources of the United States. [Source: *Saline Water Conversion*, Office of Saline Water (Washington, D.C.: Government Printing Office, 1962).]

The problem of obtaining fresh water from brackish water or seawater is essentially one of engineering economics. If costs were of no concern, we could boil the ocean and condense the steam. Costs very definitely are of concern, however, and this gives rise to the technological problems of seawater conversion. Conversion costs result directly from land and materials, labor, and energy. The engineering problem is to design combinations of these resources that result in a minimum cost of fresh water for each given locality.

10-27

ENERGY REQUIREMENTS

Energy is an important consideration in saline-water conversion. The minimum energy requirement can be calculated as the lowest possible energy level at which conversion can take place without violating the second law of thermodynamics. The minimum work required for a completely reversible process is given by the relation

$$W_{min} = \Delta H - T_0 \, \Delta S \tag{10-58}$$

where W_{min} = minimum work
H = enthalpy
S = entropy
T_0 = absolute temperature

The value of work W_{min} does not depend on how the process is carried out but only on the initial state and the final state achieved. Equation 10-58 can be stated in a more convenient way for an isothermal process:

$$-W_{min} = \int_{n_1}^{n_2} RT \ln \frac{P}{P_0} \, dn \tag{10-59}$$

where P = vapor pressure of the salt solution
P_0 = vapor pressure of pure water
T = absolute temperature
R = gas constant
n = number of moles of water removed

For the case where an infinitely small amount of water is reclaimed from a very large amount of saline water (zero recovery), a special case can be developed. The process is assumed to be carried out in three infinitesimally slow steps. In this way reversibility can be achieved.

1 One mole of water is evaporated from a large volume of the salt solution (seawater), isothermally:

$$W_1 = P(V_{vapor} - V_{liquid}) \tag{10-60}$$

where W = work
P = pressure
V = volume

For practical purposes, V_{liquid} can be neglected, in comparison with V_{vapor}. Also, the vapor can be approximated closely as an ideal gas, for which $PV = RT$. Therefore,

$$W_1 = P \frac{RT}{P} = RT \tag{10-61}$$

2   The vapor is compressed reversibly and isothermally from P to P_0.

$$W_2 = \int_P^{P_0} P \, dv = -RT \int_P^{P_0} \frac{dP}{P} = -RT \ln \frac{P_0}{P} \tag{10-62}$$

3   The vapor at pressure P_0 is condensed isothermally.

$$W_3 = P_0(V_{\text{liquid}} - V_{\text{vapor}}) = -P_0 V_{\text{vapor}}$$

$$= -P_0 \frac{RT}{P_0} = -RT \tag{10-63}$$

The total work is the sum of the individual work terms in 1, 2, and 3:

$$W_{\min} = W_{\text{total}} = RT - RT \ln \frac{P_0}{P} - RT = -RT \ln \frac{P_0}{P} \tag{10-64}$$

The negative sign indicates that work is done on the system.

This analysis does not consider specifically the heat absorbed in vaporizing the liquid or the approximately equivalent heat given up in condensation. Infinite-heat reservoirs may be assumed for this purpose.

Compression distillation is based on the principle of solvent transfer given above. Good heat contact is provided between the boiler and condenser, so basically no heat of vaporization needs to be supplied from an external source and no heat needs to be removed.

●EXAMPLE 10-10
What is the minimum energy required to recover 1000 gal of fresh water from the ocean? Assume that the temperature is 25°C and that the salt concentration is 35,000 mg/l as sodium chloride.

○*Solution*
Using Eq. 10-64.
1   Absolute temperature $T = 25 + 273 = 298°K$. The gas constant $= 1.987$ cal/deg/mole. The vapor pressure for pure water is $= 23.76$ mm Hg. The vapor pressure for water with 35,000 mg/l as $NaCl = 23.51$ mm Hg.
2   The number of moles of water in 1000 gal is

$$\frac{1000 \times 8.34 \times 453.6}{18} = 21 \times 10^4 \text{ moles/1000 gal}$$

3   Therefore,

$$-W_{\min} = 1.987 \text{ cal/deg/mole} \times 298°K \times \ln \frac{23.76}{23.51} \times 21 \times 10^4 \text{ moles/1000 gal}$$

$$= 2.98 \text{ kWh/1000 gal}$$

This figure (2.98 kWh/1000 gal) is normally cited as the minimum possible work, and the thermodynamic efficiency of an actual process is often based on it.

Actual processes must have larger energy requirements than this for two reasons: (1) zero recovery is wholly impractical, as it would require infinite work to pump the infinite supply of saline water necessary for the process; and (2) the equation is based upon a reversible process which is an ideal with zero driving forces. The driving forces, which must have finite values for any real process, are temperature difference across heat exchangers, pressure difference for fluid flow, concentration differences, and electromotive-force (emf) differences.

The minimum work for a reversible process for various percentage yields of pure water using Eq. 10-59 is plotted in Fig. 10-19. The work required for pumping the feedwater is based upon a 50-ft head and 70% pump efficiency, and the total of the two is also plotted on the same graph. The minimum separation work increases with increased recovery and pumping work normally decreases, resulting in a minimum total work at about 40% recovery. If Eq. 10-58 is used to calculate the minimum work, higher values will be obtained for all values except zero recovery. This is because the equation represents ideal work for a single-stage process which is not completely reversible.[24]

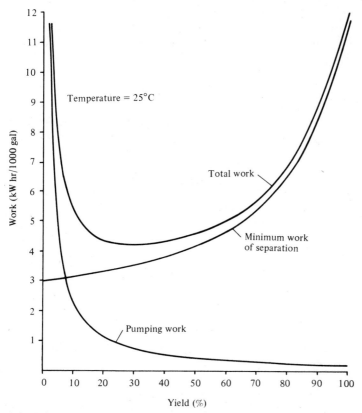

FIGURE 10-19 Minimum work of separation and pumping work as a function of percent yield of fresh water from seawater.

The minimum work is considerably less for lower salt contents. For a 50% recovery from a feed of 5000 mg/l of salt, the minimum reversible work is 0.71 kWh/1000 gal of water produced, as compared to 4.15 kWh for seawater. All practical processes have a relatively low thermodynamic efficiency based upon the minimum work for zero recovery. These efficiencies are of the order of 2–5%. Some improvements are being made and the best efficiency to be achieved in the foreseeable future, according to Dodge, is in the area of 10%.[24] This corresponds to about 30 kWh/1000 gal with seawater.

The reason for this low efficiency is that any decrease in the driving force of an operation such as heat transfer or gas compression results in an increase in the size and cost of the equipment. In the case of a heat exchanger, if the temperature difference is halved, the required surface area will be twice as great for the same conditions.

10-28

COSTS

The "optimum-energy" technique has been used to design plants. This technique employs thermodynamics, information theory, and economics. It has been shown that thermodynamic considerations alone can fix the minimum energy requirements. Figure 10-20 shows the relationship between minimum and opti-

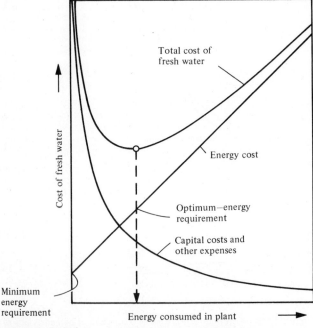

FIGURE 10-20 Minimum and optimum energy requirements.

mum energy. The optimum-energy requirement is the amount of energy consumption that results in the minimum total cost of freshwater production. The cost of fresh water from saline water is the amount the consumer must be charged to pay for the total cost of producing and distributing the fresh water.

The economics of saline-water conversion have been widely discussed. Many of the conclusions drawn indicate that conversion costs are decreasing while other water costs are increasing and that the conversion of saline water will soon be cheaper than methods of supply currently in use. This is not supported by fact. Many conversion costs made public by the federal government and others do not include the cost of delivery to the consumer or administrative costs involved in the sale. Another cost associated with inland saline-water conversion is the disposal of the waste brine. Viessman points out that consideration of the enormous volumes of waste which might be produced by future saline-water-conversion plants is cause for concern (Fig. 10-21).[26] These disposal costs may amount to more than the cost of conversion.

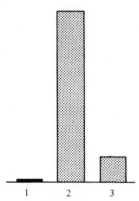

FIGURE 10-21 Relative annual per capita volumes of solid wastes produced by wastewater treatment and saline-water-conversion operations of an inland city with 40,000 population. These values are estimated on the basis of brine waste composition and saline-water-conversion capabilities at Roswell, New Mexico, as of July 1963. Changes in technology may alter these figures but the diagram is considered to be indicative of the general order of magnitude of wastes produced by these operations. (1) Solid waste produced by wastewater treatment (sludge digestion and dewatering are assumed). (2) Saline-water conversion waste estimated on the premise that total community water supply is derived in this manner. (3) Saline-water conversion waste estimated on the assumption that only 15 percent of the community water supply is derived in this manner, with 85 percent furnished by other means. [Source: *Public Works* (December 1963)].

The current role of saline-water conversion in supplementing national water resources may be considered as insignificant, and it is not anticipated that saline-water conversion will contribute significant quantities of fresh water in the near future. It should not, however, be concluded that saline-water conversion has no role in water supply. It is important in isolated areas of critical water shortage, in certain industrial areas where a constant high-quality process water is required, and in removing trace substances for potable water, especially in small communities. There are about 1000 land-based desalting plants larger than 25,000 gpd with a combined capacity of approximately 526 mgd in use worldwide.

On the basis of available operating data and changing energy costs, it is impossible to predict the future cost of saline-water conversion with any degree of accuracy. A United Nations survey shows costs from less than $1 to more than $40 per 1000 gal.[27] The costs, less than $1/1000 gal, may not represent a true cost, because of the favorable fuel supplies in the Middle East.

10-29

CONVERSION PROCESSES

Implementation of the National Safe Drinking Water Act with the proposed interim Primary Drinking Water Regulations will have a great impact on many communities and those business establishments having private water supplies. Many public water supplies currently do not meet the proposed regulations.

The various treatment methods which may be used to improve water quality can be divided into two basic classifications: (1) processes which

TABLE 10-7

CLASSIFICATION OF SALINE-WATER-CONVERSION PROCESSES

A. Processes that separate water from the solution
   1. Distillation or evaporation
      a. Multiple-effect long-tube vertical
      b. Multistage flash
      c. Vapor compression
      d. Humidification (solar)
   2. Crystallization or freezing
      a. Direct freezing
      b. Indirect freezing
      c. Hydrates
   3. Reverse osmosis
   4. Solvent extraction

B. Processes that separate salt from the solution
   1. Electrodialysis
   2. Osmionisis
   3. Absorption
   4. Liquid extraction
   5. Ion exchange
   6. Controlled diffusion
   7. Biological systems

separate water from solution, and (2) processes which separate dissolved species from solution. Further subdivisions of these two classifications into specific unit processes are shown in Table 10-7. Distillation or evaporation, crystallization, and freezing involve phase changes to effect the separation of water from solution. Reverse osmosis utilizes a semipermeable membrane to separate the water from solution. In electrodialysis a semipermeable membrane separation of salt from solution is achieved. All these processes have been investigated for desalination applications. Ion exchange has been used for water treatment in membrane form, but it is most commonly used as a resin.

Of the processes listed in Table 10-7; ion exchange, electrodialysis, and reverse osmosis would appear to have the most promise for economically treating public drinking-water supplies in small communities. Ion exchange is used quite extensively for water softening, and both electrodialysis and reverse osmosis have been evaluated extensively for desalination applications. Electrodialysis and reverse osmosis are very different processes; electrodialysis removes dissolved salts from solution under an electric potential gradient while reverse osmosis removes water from solution using a pressure gradient. Reverse osmosis also produces water essentially free of dissolved organic material and microbial species. These contaminants remain in the waste brine generated. Of these processes, reverse osmosis is probably the most practical basic unit for application to small quantity public drinking-water treatment needs. However, any basic system will probably require pretreatment steps and possibly polishing steps, which would involve others of the processes listed in Table 10-7.

A total treatment-system evaluation should include selection of a basic unit operation and the necessary pretreatment, polishing, and ultimate disposal process of rejected brine. In addition, consideration should be given to alternative drinking-water supplies.

Reverse-osmosis processing equipment is comparatively simple in design; operation can be automated, resulting in minimal labor requirements; and energy consumption is comparatively low. Dissolved-ionic-species removal is high, particularly for the troublesome species found in many raw waters.

Ion-exchange techniques may provide adequate treatment because of the ion-removal selectivity preferences for divalent ions. The major difficulty in treating many public water supplies is the removal of trace divalent heavy metal contaminants in the presence of large quantities of other divalent ionic species. In the ion-exchange treatment process, unless resins are carefully evaluated for specific application, regeneration of the resin could require substantial quantities of regenerant chemicals and the need for high-capacity resins or large volumes of resins.

The possible key to the development of low-cost demineralization processes might lie in the use and reduction of energy required in a separation process. This may require the development of new or little used sources of energy. Table 10-8 lists some presently used and possible energy sources.

TABLE 10-8

POTENTIAL ENERGY SOURCES FOR
DEMINERALIZATION

1. Combustion of fuels
2. Hydropower
3. Utilization of waste heat
   a. Waste heat from conventional industrial processes
   b. Waste heat from nuclear fission
4. Nuclear fission
5. Solar energy
6. Marine energy
   a. Thermal
   b. Waves
   c. Tides
7. Wind power
8. Chemical energy
9. Atmospheric heat
10. Geothermal energy

10-30

EVAPORATION

Distillation is the oldest demineralization process. It consists of evaporating a part or all of the water from a saline solution and subsequent condensation of the mineral-free vapor. The products of distillation are pure water and either a concentrated salt solution or a mixture of crystalline salts. This is a process in which water is removed from the salts. The heat and power requirements are relatively independent of the amount of salt in solution. At present, most of the fresh water produced from the sea is by one of the evaporation methods.

A modification of the distillation process called *vapor compression* is used where the salt water is evaporated at atmospheric pressure. The vapor is compressed to raise the pressure of the steam to about 3 psig and the corresponding temperature to about 222°F. The compressed steam is returned to the heating side of the evaporator to heat more of the brine from which the original vapor was formed (Fig. 10-22). There is a temperature difference of about 9°F between the compressed steam and the boiling brine, which permits heat transfer back to the brine. Substantially all the latent heat of the compressed steam is used in maintaining evaporation of the saline solution. The condensation of the compressed steam occurs directly in the steam chest forming the distilled water product. Therefore, no separate condenser or cooling water is required. From an engineering viewpoint the fact that no cooling water is required with the vapor-compression unit is an important difference between it and multiple-effect

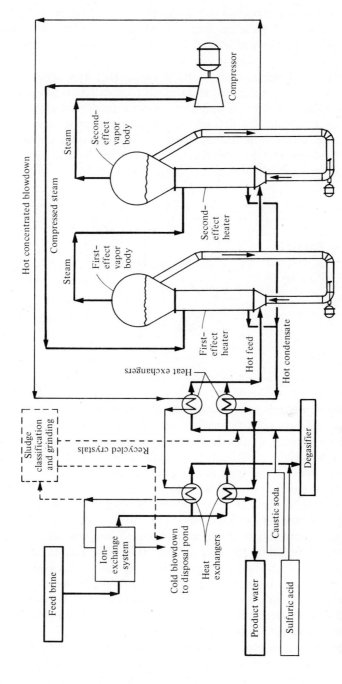

FIGURE 10-22 Forced-circulation vapor-compression distillation.

evaporators. One pound of fuel oil will evaporate about 250 lb of water in a clean vapor-compression unit. Scale formation is a major problem, as it decreases the efficiency of brine evaporators and periodic shutdowns are required for its removal.

In multiple-effect evaporation, water is evaporated at a given pressure in the first stage. The vapor is then fed to a second compartment, where additional water is evaporated at a lower temperature under a small vacuum produced by the condensed steam from the first effect. The evaporation of water is carried out in successive stages (Fig. 10-23). This method has been under extensive commercial development for many years. Costs for this process for triple-effect evaporation have been estimated at $1.20–1.60/1000 gal.[28]

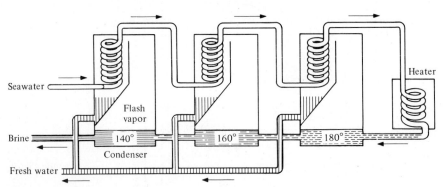

FIGURE 10-23 Flash distillation.

Solar distillation has the advantage of "free" energy. There are vast quantities of solar energy, particularly in areas of the United States and other countries where other fuels and energy sources are not available. Since there is little energy cost, the cost of water produced by solar evaporation depends almost entirely upon capital cost and maintenance. These costs are approximately $1.60/1000 gal with present equipment.

The quantity of solar energy striking the earth per year is approximately 25×10^{20} Btu. Of this total, about 5.5×10^{20} Btu is absorbed by the oceans, with the resulting evaporation of seawater. About 200 billion acre-feet of water returns to the land annually. Another additional 40 billion acre-feet evaporates from land and plant life. The solar energy striking the earth's surface is many times the present energy consumption in all other forms from fuels and hydro-power. The average solar energy in the continental United States is about 40 billion Btu/mi^2 on an average day. This is equivalent to 5 million gal of water/day when compared to natural evaporation processes.

Although the figures concerning solar energy are large, the concentration in comparison with conventional energy-transfer rates used industrially is low. Average solar energies are equivalent to approximately 1500 Btu/ft^2/day. This figure may be compared to common industrial heat-transfer rates, as used in steam boilers, ranging from 100,000 to 3 million $Btu/ft^2/day$. It is apparent that the utilization of solar energy will involve large collectors in relation to the energy delivered. This requirement results in a relatively large capital cost per unit of energy developed.

The simplest solar-evaporation process is the evaporation of water from a shallow blackened pan exposed to the sun and covered with sloping glass sheets (Fig. 10-24). The evaporated water condenses on the sloping glass sheets and runs down to collecting channels at the base of the covers. About 0.63 lb of water can be produced per day per square foot of evaporator pan. This is operating with an efficiency of approximately 35%, which has been achieved in practice with large units.

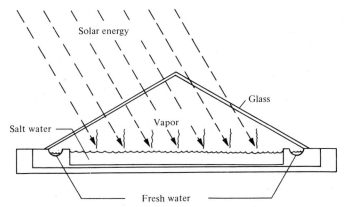

FIGURE 10-24 Simple solar still.

10-31

DESIGN PROBLEMS OF A VAPOR-COMPRESSION PLANT

The saline-water conversion plant located at Roswell, New Mexico, is based on the vapor-compression principle. This plant is designed to produce 1 million gallons of fresh water per day and is the fourth of the demonstration plants authorized by the Office of Saline Water. The forced-circulation vapor-compression process is used. The plant was completed in 1963.

The Roswell plant was designed for a difficult brackish-water feed in which calcium sulfate concentrations can reach very high levels. Table 10-9 gives an analysis of the brine. The water is drawn from a well at the plant site. Sodium chloride is the major constituent of the dissolved salts, accounting for about 80% of the total salinity. The bulk of the remainder of the solids consists of calcium and magnesium sulfates. There are also some calcium and magnesium bicarbonates present. Bicarbonates decompose thermally to form a scale of calcium carbonate and magnesium hydroxide on heat-exchanger surfaces. The Roswell water has only one-half the salinity of seawater but is richer in scale-formers.

TABLE 10-9
ANALYSIS OF ROSWELL BRINE

	mg/l
Total hardness as $CaCO_3$	2,145
Calcium as $CaCO_3$	1,370
Magnesium as $CaCO_3$	775
Alkalinity as $CaCO_3$	202
Chlorides as Cl	8,064
Sulfate as SO_4	1,528
Iron as Fe	Trace
Silica as SiO_2	Trace
Calcium as Ca	549
Magnesium as Mg	303
Sodium as Na	5,000
Total dissolved solids	15,860
pH	7.4

The water is saturated with respect to calcium sulfate and scaling can take place upon evaporation of only a small amount of water. This type of water is more difficult to process than seawater and it occurs widely in the Southwest and in a broad belt extending up into the Dakotas.

A seawater plant has available an unlimited supply of cooling water, while an inland plant normally has only its feedwater. The design must be based upon an efficient exchange of heat between the hot products and the cold feed. With a brine feed rich in salts, an upper limit on concentration is imposed by the boiling-point elevation. This represents a thermodynamic inefficiency which decreases the percent of recoverable energy supplied to the process. The maximum concentration of feed brine is also controlled by scaling considerations.

In the first step of the forced-circulation vapor-compression process (Fig. 10-22) the feedwater exchanges a major portion of its calcium for sodium in a conventional ion-exchange unit. The feed water is then heated to about 145°F by heat exchange with the product water and waste. This is followed by acidification to break down bicarbonates. The carbon dioxide and dissolved gases are then removed under vacuum in the degasifier. This water is then neutralized and further heated by exchange with product water and waste.

The heated water is introduced into the evaporator system. The flow is from the vapor body down to the pump and then up through the tubes of the evaporator heat exchanger at the rate of 90,000 gpm. The water emerges again in the vapor body where a high water level is maintained. This water level develops a sufficient hydrostatic head that boiling cannot occur in the tubes. Boiling occurs as the heated water rises up in the vapor body (Fig. 10-22). About 250 lb of water is recirculated for every pound vaporized. The water temperature is raised 4°F as it passes through the heater tubes. After leaving the tubes, the heat represented by the 4°F temperature rise is used in the heat of vaporization. The remaining liquid is cooled to the original temperature by the vaporization process before starting upon another recirculation through the pump and heater. Therefore, the heat required to vaporize 1 lb of water is supplied by cooling 250 lb of water 4°F. The 250 lb of water is then reheated 4°F and pumped through the evaporator again. The heat taken up by the water is obtained from the condensation of steam on the outside of the tubes of the evaporator heater. The heating and boiling operation in the second-effect vapor body is identical to that in the first except that the temperatures and pressures are lower.

In order to recover the major portion of the heat energy of the steam generated in the second effect, the steam pressure and temperature are raised by the compressor. This operation compresses 75,000 cfm of steam from 2 psi to 8.5 psi. This higher-temperature steam is then condensed in the heat exchanger of the first evaporator heater, which furnishes the heat of evaporation for the first effect.

A portion of the brine feed is withdrawn at a regulated rate from the recirculating water. Much of the heat is recovered from the wastewater by heat exchangers. The wastewater concentrated brine is termed *blowdown*. The cooled blowdown water is used to regenerate the ion-exchange resin in the initial softening system. The condensed and cooled steam is the product water.

If the evaporators were rated as boilers, the rating would be about 1,320,000 hp. The power consumption of the vapor compressor is of the order of 2300 hp or about 2% of the work accomplished in the evaporation system. The compressor represents 75% of the total power consumption of the plant. These values are based on clean heat-exchange surfaces. The Roswell plant represents a synthesis of a number of operations into a design having a capacity never before achieved.

10-32

FREEZING PROCESS

A salt solution, upon cooling, will eventually deposit ice crystals unless the brine concentration is very high, in which case the dissolved salt will crystallize out of solution. Seawater freezes with the formation of ice and a higher-concentation brine solution (Fig. 10-25). The residual brine solution is of a concentration greater than the initial brine solution because the ice crystals are pure water.

The temperature at which ice crystals begin to form is a function of the salt concentration of the brine. A lower temperature is required as the brine concentration is increased. A curve showing the relationship between the freezing temperature and salt concentration is given in Fig. 10-26. Upon cooling, the first solid phase to separate from seawater is ice crystals. Continued cooling will

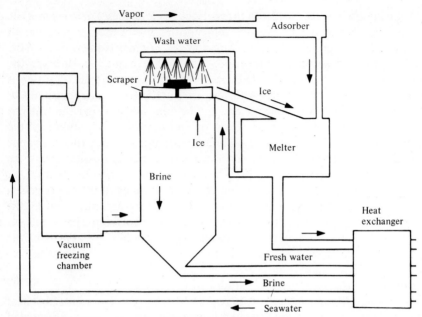

FIGURE 10-25 Direct-freezing saline-water-conversion process.

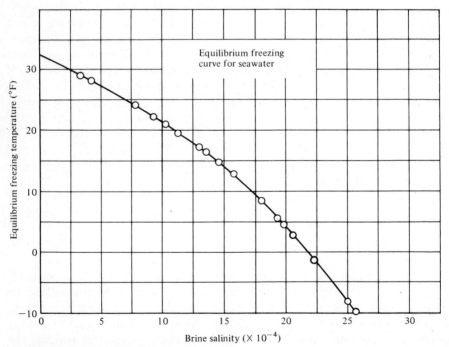

FIGURE 10-26 Equilibrium freezing curve for seawater. [Source: "Research and Development of Processes for Desalting Water by Freezing," *Office of Saline Water Report No.* 10 (Washington, D.C.: U.S. Government Printing Office, 1956), p. 9.]

result in additional ice being formed and the remaining brine being concentrated. The next solid phase to separate upon continued cooling is sodium sulfate decahydrade at a temperature below 17°F. The second salt to separate is sodium chloride dihydrate, at a temperature of about −9°F. Therefore, the freezing process should stay above 17°F for the formation of pure ice from seawater. Even though the temperature is maintained above 17°F, the ice crystals are coated with the brine solution and must be washed with high-quality water to remove this salt. At the higher freezing temperatures, a small variation in the equilibrium freezing temperature corresponds to a large change in the brine concentration. The ice particles formed by freezing salt water consist of practically pure water. Even if the freezing process is carried out above the temperature at which the first solid salt phase is separated out, it is not simple to separate the brine from the ice. A bed of ice particles, when drained by gravity of adhering liquid, contains about equal weights of ice and brine held to the ice by capillary and viscous forces. New processes are being investigated that might provide more efficient ice brine separation.

10-33

REVERSE OSMOSIS

Reverse osmosis is a membrane-separation technique in which a semipermeable membrane allows water permeation while acting as a highly selective barrier to the passage of dissolved, colloidal, and particulate matter. Inorganic, organic, and microbial species are all included in the membrane rejection. *Osmosis* is a term indicating the natural passage of water through a membrane without the application of external forces. For aqueous solutions having different osmotic pressures and separated by a semipermeable membrane, water will spontaneously pass through the membrane so as to equalize the chemical potentials of the water in the membrane-separated solutions. *Osmotic pressure* has been the term applied to the driving force causing osmosis to occur. The greater the osmotic pressure difference between the solutions, the greater the tendency for osmosis to occur. In reverse osmosis an external pressure difference is applied to the solution, causing water to flow against the natural direction through the membrane and thus producing water purer than the original solution. This is illustrated in Fig. 10-27.

Reverse osmosis application to water purification is dependent upon the production of membranes which will reject most of the ions present while permitting acceptable rates of water passage. Many natural osmotic membranes are known, inferring the potential of producing synthetic membranes possessing osmotic properties. Indeed, a large variety of membranes have been developed, but cellulose acetate is currently the most widely used membrane material. Cellulose acetate is cast in a film having an asymmetric structure consisting of a thin dense skin on a porous structure. Typical membranes are approximately 100 μm thick, having a surface skin of about 0.2 μm thickness which serves as

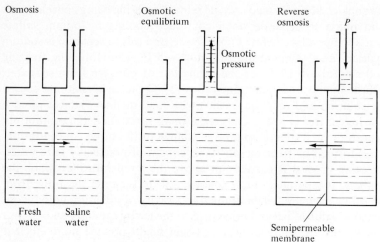

FIGURE 10-27 Principle of reverse osmosis.

the rejecting surface. The remainder of the film is spongy and porous, containing about two-thirds water by weight. Modified cellulose acetate membranes have relatively high sodium chloride rejection, about 95%, while maintaining good water fluxes of 10 gal/day/ft² at 600 psi. Basic limitations in the application of cellulose acetate membranes are:

1. Feedwater pH must be on the acid side to prevent membrane hydrolysis. For best operation the pH should be between 5 and 7.
2. Microbial populations must be limited to prevent microbial degradation of the membrane.
3. Process temperatures should not exceed 75–80°F, if long membrane life is expected.
4. Projected membrane life is 2–3 years of continuous service.

A second commercially important membrane is polyamide (nylon) material. It has high chemical and physical stability, leading to long life expectancies. Polyamide membranes have lower water-permeation coefficients than the cellulose acetate membranes, and therefore to maximize the transfer surface area, these membranes are usually formed into hollow fine fibers having outer and inner diameters of approximately 50 and 25 μm, respectively. The feedwater is fed to the inner passages of the fibers, thus greatly limiting the size of suspended solids which may be permitted to enter. Scaling and chemical precipitation must also be carefully controlled because of the large potential for plugging.

Reverse-osmosis modules suitable for water treatment involve the arrangement of membranes and their supporting structures so that feed water under high pressure can pass along the membrane surface while the product water is easily collected from the opposite surface without brine contamination. In designing reverse-osmosis modules, a number of items require consideration:

1 Support of a fragile membrane to withstand differential pressures of 300–1500 psig.
2 High packaging density to minimize high-pressure-vessel cost.
3 Design of feed channels to minimize concentration polarization and fouling.
4 Avoidance of parasitic pressure drops in the feed, brine, and product streams.
5 Minimization of membrane replacement costs.
6 Avoidance of low-pressure product water contamination by contact with the high-pressure feed and brine waters.

Four different types of module designs have been developed to solve these problems: plate and frame, large tubes, spiral wound, and hollow fine fiber.

The plate and frame assembly was designed to use the concept of the plate and frame filter press. Such a unit may consist of a series of porous support plates with a membrane placed on each side of the plate. Spacer washers, or frames, are placed between the membranes to serve as brine channels. The feedwater thus flows into the first frame and the brine discharge becomes feed into the next frame, and so on through the module until the concentrated brine leaves the module. The product water flows through the membrane and either radially outward or inward, depending upon the design, through the porous plate to collection ports, where the water is discharged from the module. Designs may average 50–100 ft² of membrane surface per cubic foot of pressure vessel. Advantages of this configuration are simple, rugged equipment; practical experience with plate and frame filter-press operations; and single-sheet membrane replacement. Disadvantages are difficult-to-analyze brine-flow patterns, appreciable manual labor in membrane assembly and replacement, and high equipment costs.

Porous-wall tubes serve as the membrane support, with the membrane cast on the inner wall. The feedwater is pumped through the tube and the product water collected in a shell around the outer surface of the porous tube. Some newer designs incorporate cylindrical packed porous inert material as the tube center with the membrane cast on the cylinder. In these designs the brine flows through the shell side with the product water collected from the porous packing. Advantages of the large tube modules include: the geometry of brine-flow passage is well defined; the prefiltration requirements for the feedwater are minimal; and the porous support wall may serve as the pressure vessel for outward product flow tubes. Disadvantages include difficulty in sealing brine chambers from product water at tube sheets; tube costs; handling large number of tubes; and lack of membrane replacement technology. Packing densities of 10–100 ft²/ft³ have been obtained.

Spiral-wound modules consist of a tube for collecting the product water and two membranes, separated by a porous mesh, for receiving and conducting the product water to holes in the tube. The two membranes are sealed together on three sides, with the fourth side encasing the tube perforations and glued to the tube. The membrane blanket is then wrapped around the pipe with a screen on one side to separate the layers of the membrane blanket spiral, providing a passageway for the brine across the membrane surfaces. The entire assembly is inserted into a pipe with the center tube extending through the end seals of the pipe. The spiral assembly fits snugly so the brine or feedwater passes axially through the brine passageways while the product water flows into the spiral toward the collection tube. Advantages are a relatively high packing density of between 250 and 500 ft^2/ft^3, controlled brine side spacing, factory assembly, and easy field replacement of units. Disadvantages are that good prefiltration of feedwater is required to prevent blockage of brine side passages and that total membrane element replacement is necessary.

When membrane tubes are made sufficiently small in diameter, they can be strong enough to withstand considerable pressure. Tube diameters are built in the range 5–100 μm having diameter/wall ratios of 4 : 5. The tubes are arranged in bundles with the ends sealed in headers by materials such as epoxy. Brine flow is inside the tubes with product water flowing through the walls for collection in the shell surrounding the tube bundle. Extremely large packing densities may be obtained, ranging from 10,000 to 20,000 ft^2/ft^3. Advantages are large packing densities, elimination of membrane support material, and potential for achieving high water fluxes before boundary-layer problems become limiting. Disadvantages include extremely good feedwater filtration, need for short tubes

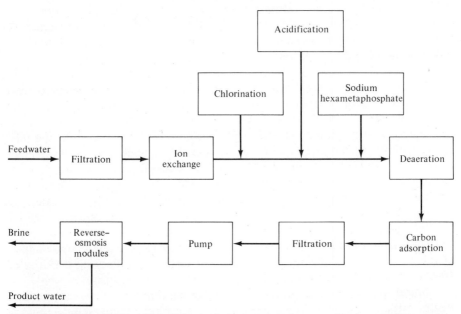

FIGURE 10-28 Basic flow diagram for a single-stage reverse-osmosis plant. SHMP—sodium hexametaphosphate.

to obtain high flux membranes, the enormous number of tubes, need for detection of defective tubes during or prior to assembly, and system pressure limitations of 500 psig or less.

A representative process flow diagram for a single-stage reverse-osmosis treatment facility is shown in Fig. 10-28. Feedwater is chlorinated, acidified, filtered, dechlorinated, and deaerated prior to entering the reverse-osmosis module in a pretreatment phase. The pretreated water is then pumped to the operating pressure and passed into the reverse-osmosis modules with product water polishing, if necessary, and reject brine disposal.

The configuration of the reverse osmosis modules may be in series, parallel, or series–parallel. A common arrangement is parallel units for the first separation stages and as the quantity of brine to be treated is reduced, series of modules are used as shown in Fig. 10-29.

The operating variables to be considered in a reverse-osmosis system are product recovery rate, product water quality, brine flow rate, operating pressures, flux maintenance procedures, pretreatment of feedwater, and brine disposal. These variables greatly affect the capital and operating costs of reverse-osmosis processing.

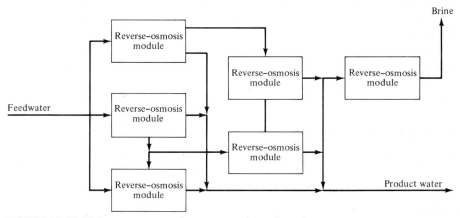

FIGURE 10-29 High-water-recovery reverse-osmosis configuration.

10-34

ION EXCHANGE

Ion exchange is a separation process in which ions, held by electrostatic forces to charged functional groups on the surface of an insoluble solid, are replaced by ions of similar charge in a solution in contact with the solids. Ion exchange is a sorption process, since the exchanging ions must undergo a phase transfer from a solution to a solid surface phase. Unlike simple physical adsorption phenomena, ion exchange is a stoichiometric process in which every ion removed from solution is replaced by an equivalent amount of another ionic species of the same sign. Ion exchange is, in general, a reversible process and is selective in the removal of dissolved ionic species. Although many naturally occurring materials exhibit ion-exchange properties, synthetic ion-exchange resins having a wide range or properties for specific applications have been developed.

The characteristic properties of ion-exchange materials are due primarily to their structure. These materials consist of a framework held together by chemical bonds. Attached to this framework are soluble ionic functional groups containing ions which are relatively free to move and exchange with similarly charged ions in solution.

Ion-exchange materials must possess the following characteristics:

1　Ion-active sites throughout the entire structure, possessing very uniform distribution of activity.
2　High total capacity; that is, a high degree of ion substitution or low equivalent weight.
3　Good degree of selectivity for ionic species but capable of being regenerated.
4　Extremely low solubility.
5　Good structural chemical stability.
6　Good structural physical stability.

Naturally occurring ion-exchange materials are solids, cellulose, wool, protein, coal, metallic oxides, and living cells such as algae and bacteria. Natural aluminosilicates, zeolites, of crystalline structure are commonly used ion exchangers. These materials have relatively open three-dimensional lattice structures with channels and interconnecting cavities available for ion movement. The lattice carries a negative electric charge balanced by alkali or alkaline-earth cations which can be replaced by other cations. Some aluminosilicates also act as anion exchangers; that is, they exchange anions.

Synthetic ion exchangers were developed to impart better characteristics into the material than those exhibited by naturally occurring zeolites. Of the synthetic materials, the organic ion-exchange resins are most important. They are typical gels consisting of a matrix of irregular, macromolecular, three-dimensional network of hydrocarbon chains. The matrix carries ionic groups capable of being exchanged. Cross-linking of the macromolecule is accomplished by carbon–carbon bonding, giving the resin good chemical, thermal, and mechanical stability. In synthetic resins, the ion selectivity can be controlled by the nature of the fixed ion groups attached to the matrix.

Purification by ion exchange is a process for removing ionized species from slightly ionized water. Most ion-exchange operations are carried out in columns, as illustrated in Fig. 10-30. As the solution passes down through the ion-exchange bed, its composition changes. Ion-exchange processing may be considered batch operation for any single column with the operating cycle consisting of four distinct phases: service period, backwash, regeneration, and rinse. During

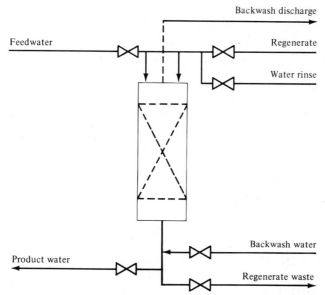

FIGURE 10-30 Typical single column ion-exchanger flow diagram.

the service period, feedwater is passed through the bed with removal of ionized contaminant. As the solution is first fed to the column, it will exchange all its exchangeable ions in a narrow zone of resin near the feed point. As the capacity of the resin in this zone is approached, the exchange region will move through the column until the entire resin bed is exhausted. In actual operation the service period is ended before complete exhaustion and a fresh column is brought into service. The column is placed in a backwash mode in which water is passed up through the column to remove any foreign matter which may have entered and to reclassify the resin particles.

During the regeneration period, regenerant solution is pumped down through the bed, displacing undesirable ions from the resin and restoring it to its original condition. In this period, a waste stream is generated which requires disposal. In the rinse period, feedwater is used to displace spent regenerant remaining in the bed. This water effluent is discarded as long as any contamination with spent regenerant is evident.

Where continuous product water supply is required, an ion-exchange processing unit would consist of several columns. The columns would be cycled in different operational phases, ensuring that at least one column would always be in the service mode.

The economics of ion exchange require high-capacity resins, a high degree of column utilization, high selectivity for exchange ions, efficient regeneration, rapid exchange rates, reasonable production rates, and inexpensive spent-brine disposal methods.

10-35

ELECTRODIALYSIS

The *electrodialysis process* is one of the most practical and widely used methods to treat brackish waters. At present it is not very suitable for the treatment of seawater because of the high energy requirements. The energy requirement is directly proportional to the salt concentration of the water being treated.

The ingenious principle of this process is illustrated in Fig. 10-31. Plastic membranes are used that will not pass water but are ion-selective. Some membranes will pass cations while others will permit only anions to go through. An electrodialysis cell consists of alternate cation- and anion-permeable membranes arranged in a stack with alternately charged electrodes on each side. In Fig. 10-31 the cation-permeable membranes C contain negatively charged ionic groups as in anion-exchange resins and therefore tend to repel the anions but allow the cations to permeate through. The anion-permeable membranes A have positively charged ionic groups that repel the cations but permit the anions to go through.

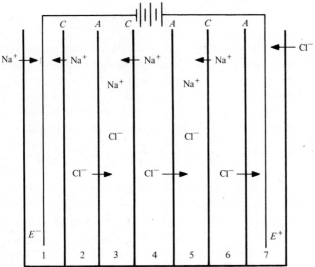

FIGURE 10-31 Schematic cross section of the electrodialysis process. C cation-permeable membrane; A anion-permeable membrane; E electrode.

An electromotive force is imposed across the assembly by electrodes E and all positive ions move toward the anode while all negative ions move toward the cathode. This movement is effected by the membranes in the following way. Na^+ ions can move out of compartment 2 through membrane C into compartment 1, but those in section 3 cannot move out of this compartment because they cannot penetrate an A membrane. In a similar manner, Cl^- ions can move from 2 to 3, but those in 3 are prevented from moving out by the C membrane.

The net result is the depletion of ions in the even-numbered compartments and concentration of ions in the odd-numbered compartments. The even-number sections produce relatively pure water and the odd numbers a concentrated waste brine.

The energy requirements for this process are made up of the electrical energy to provide the emf across the electrodes and the mechanical energy required to move the liquid through the cells. The total energy is affected by the nature of the membrane, the number of membranes in a stack, the spacing, the fluid velocity in the cells, the salt concentration, and the current density. The electrical energy may be calculated from the formula

$$E = \frac{KiAqRFuN}{ne} \tag{10-65}$$

where E = kWh/1000 gal
  K = constant (dependent on units)
  i = current density
  A = transfer area per cell pair
  q = total rate of flow through unit
  R = resistance of unit
  u = fractional removal efficiency
  F = Faraday constant
  N = normality of feed solution
  n = number of cell pairs
  e = current efficiency

Equation 10-65 shows that the electrical energy required for a unit is directly proportional to current density and salt concentration. The electrical energy can be reduced by operating at lower current densities, but the capacity is directly proportional to the current used. Accordingly, a balance must be designed between cost of energy and fixed charges for minimum cost. Equation 10-65 also shows that the electrical energy required is proportional to the salt concentration of the water being treated. In addition to the electrical-energy requirement, there is need for considerable pumping energy. For brackish waters of concentration less than 10,000 mg/l, the energy requirement in an actual installation is of the order of 10–30 kWh/1000 gal of product water.[24] This may be less than the energy required for distillation processes. The energy requirement for seawater for the electrodialysis process is on the order of 125–150 kWh/1000 gal. This is much higher than for several vaporization processes.

●EXAMPLE 10-11

An electrodialysis stack is made up of 400 membranes each 20 × 24 in. This unit will be used to partially demineralize 100,000 gal of saline water per day. The water contains dissolved solids of 2500 mg/l as sodium chloride. The resistance through the unit has been determined to be 7 ohms (Ω) and the current efficiency, 92%. The maximum current density/normality ratio that may be used without serious polarization is 500. Estimate the removal efficiency and the power consumption per 1000 gal of product water.

○**Solution**

1 Normality of the feedwater:

$$N = \frac{2500 \text{ mg/l}}{58,500 \text{ mg/l/mole}} = 0.0427 \text{ mole}$$

2 Maximum current density allowable:

$$CD = 500 \, N = 500 \times 0.0427 = 21.3 \text{ mA/cm}^2$$

3 Current required:

$$I = \frac{21.3 \text{ mA/cm}^2 \times 20 \text{ in.} \times 24 \text{ in.} \times (2.54)^2}{1000 \text{ mA/A}} = 66.2 \text{ A}$$

4 Power consumption:

$$P = I^2 R = (66.2 \text{ A})^2 (7 \, \Omega) = 30,600 \text{ W}$$

$$E = \frac{30,600 \text{ W} \times 24 \text{ hours/day} \times 1000}{1000 \text{ W/kW} \times 100,000 \text{ gal/day}} = 7.34 \text{ kWh/1000 gal}$$

5 The removal efficiency from Eq. 10-65:

$$u = \frac{Ene}{KiARqNF}$$

$$= \frac{7.34 \times 400/2 \times 0.92}{0.001 \times 21.3 \times 20 \times 24 \times 2.54^2 \times 7 \times 100,000/22,800 \times 0.0427 \times 96,500}$$

$$= 0.674 \text{ or } 67.4\%$$

The total energy requirement is the electrical energy E plus the mechanical energy necessary to pump the water through the system. The mechanical energy of pumping must pass the water uniformly between the closely packed membranes. This energy requirement is independent of the dissolved solids and is a constant for a particular flow rate. The pressure drop across a stack handling 100,000 gpd would be in the range 15–40 psi.

Radioactivity in Water and Waste Treatment

10-36

RADIOACTIVITY

The classical picture of an atom consists of a central nucleus about which rotate a number of electrons in orbit. The nucleus is made up of relatively massive protons and neutrons. The proton carries a positive electrostatic charge, whereas the neutron has no charge.

Certain atoms are radioactive—that is, they have nuclei which are unstable. These nuclei emit an energetic particle, a pulse of energy, or both. This process is radioactive decay. The radiations are known as α (alpha) and β (beta) particles, and γ (gamma) rays. Particles and rays are readily distinguished by their behavior in a magnetic field. The α and β particles are deflected by the field in such a way as to indicate that the former carries a positive and the latter a negative electrical charge. The γ rays are undeflected and therefore presumably uncharged. A strong magnetic field will separate a mixed beam of radioactive emanations into the three types.

The penetrating powers of the α and β particles and γ rays increase in the order listed. Alpha particles are stopped by a few centimeters of air or a sheet of paper. Beta particles have about 100 times this penetrating power, while γ rays exceed ordinary X rays in their ability to penetrate materials. The γ rays from 30 mg of radium can conveniently be detected through 30 cm of iron.

An α particle is a nucleus of a helium atom. In α decay the parent nucleus loses two protons and two neutrons. The resulting daughter nucleus, therefore, is reduced by four mass units, and its positive charge is lowered by two units. For example, radium (atomic number 88, mass number 226) decays by α emission to radon (atomic number 86, mass number 222). The α particle picks up two electrons from the environment and becomes an atom of helium gas.

With β decay, an electron is given off by the nucleus. Because electrons have very little mass, the mass number of the daughter nucleus is unchanged but its atomic number is increased by one unit, since the electron carries one negative unit of charge. Radiophosphorus (atomic number 15, mass number 32) decays to sulfur (atomic number 16, mass number 32) by giving off an electron.

Gamma decay is the emission of energy; no new nucleus is formed, but the nucleus is left in a lower energy state.

The scheme of decay is specific for each radioactive isotope. This scheme consists of the type of radiation, its energy level, and its half-life. The half-life is the period of time during which 50% of the number of radioactive atoms initially present will decay. This decay scheme is the basis for many analytical methods by which specific radioisotopes can be identified.

The quantity of radioactivity is measured by the number of emissions that take place in a unit of time. The standard unit, called a *curie*, is 3.7×10^{10} emissions per second. This is the rate of disintegration of 1 g of radium.

10-37

RADIOACTIVE POLLUTANTS

Radioactive materials may contaminate water naturally or may come from such sources as tests of nuclear weapons, wastes from industrial use of atomic energy, or from the use of radioactive substances in research or medicine.

The nature and amount of radioactive fallout from a weapons test depends on the type and energy yield of the nuclear device and the conditions under which it is exploded. Fallout from tests taking place above the ground is readily carried into the atmosphere. There is the possibility that certain volatile products may be vented to the atmosphere from underground tests. From high yield explosions, a substantial portion of the fallout penetrates the stratosphere. This fallout is dispersed to every part of the world by upper-air-currents. Much of this radioactive debris reaches the earth by precipitation.

The refining of uranium ores is an important source of radioactive waste. Raw ore from the mine is processed at the mill to concentrate the uranium. The processing steps include crushing and grinding, washing, and several chemical treatments. The product from the mill is fused uranium salts. These salts are shipped to other installations for further processing, refining, and final use. Wastes from the mill include wash waters, process liquors, and solids. These liquid wastes are normally stored in ponds where the volume is reduced by evaporation and seepage.

Uranium is the mother of a chain of naturally occurring radioisotopes. It decays to thorium, which in turn reduces to isotopes of radium, bismuth, etc. The end product is a stable form of lead. Uranium is normally the only isotope recovered from the ore, and the process wastes include all the radioactive daughters in varying amounts. Radium is the most significant waste product. It is considered to be a hazard in drinking water. It is a bone-seeking α emitter with a half-life of 1620 years.

Nuclear reactors for various industrial and research purposes represent a potential source of radioactive waste. The operation of reactors induces radiation in structural elements and coolants. The production of radioactivity in the materials discharged to the environment is minimized by design, but there is always the possibility of rupture or failure.

The disposal of excessive quantities of radioactive material to streams would interfere with other uses of the water. The maximum allowable concentrations of radioactivity in drinking water are specified in the Environmental Protection Agency Standards. The control of radioactive-waste discharges to water is a problem no different in principle from the control of any other industrial waste. It involves a very direct public health hazard and must be rigidly controlled.

10-38

REMOVAL OF RADIOACTIVITY FROM WATER

Normal water-treatment processes have been found to remove some radioactivity but not significant amounts of cesium and strontium. Filtration will normally remove only those wastes associated with suspended solids. The lime-soda ash process of softening has been found effective in removing most radioactive wastes, including strontium, but not cesium. Ion-exchange resins have been found effective for removing certain radionuclides. Distillation is the most effective method of radioactive-waste removal from water, but it is expensive.

These processes result in a concentrated sludge which must be disposed of by some form of isolation. The disposal of concentrated wastes includes disposal in wells, placement in salt mines, and solids fixation.

Each source of radioactive water pollution represents an individual problem to be solved in terms of a specific location. In each case, the type and quantity of water involved, environmental factors, and planned water uses must be taken into consideration. All radiation exposure should be reduced and minimized to whatever extent is reasonable.

PROBLEMS

10-1 Using atomic weights from the table of elements given as Table A-1 of the Appendix, calculate the molecular and equivalent weights of alum (aluminum sulfate), ferric sulfate, and soda ash (sodium carbonate). The formulas of these compounds are given in Table 10-1.

10-2 Using atomic weights, compute the equivalent weights of the ammonium ion, bicarbonate ion, calcium carbonate, and carbon dioxide. Values are given in Table 10-2.

10-3 Water contains 35 mg/l of calcium ion and 15 mg/l of magnesium ion. Express the hardness as mg/l of $CaCO_3$.

10-4 Alkalinity in water consists of 120 mg/l of bicarbonate ion and 15 mg/l of carbonate ion. Express the alkalinity in units of mg/l of $CaCO_3$.

10-5 Draw a milliequivalent per liter bar graph and list the hypothetical combinations of compounds for the following water analysis:

$Ca^{2+} = 42$ mg/l	$HCO_3^- = 190$ mg/l
$Mg^{2+} = 19$ mg/l	$SO_4^{2-} = 28$ mg/l
$Na^+ = 8$ mg/l	$Cl^- = 14$ mg/l
$K^+ = 3$ mg/l	

10-6 Draw a milliequivalent per liter bar graph for the following water analysis:

$Ca^{2+} = 63$ mg/l	$CO_3^{2-} = 16$ mg/l
$Mg^{2+} = 15$ mg/l	$HCO_3^- = 189$ mg/l
$Na^+ = 20$ mg/l	$SO_4^{2-} = 80$ mg/l
$K^+ = 10$ mg/l	$Cl^- = 10$ mg/l

10-7 What is the dominant form of alkalinity in a natural water at pH 7? What are the forms present at a pH of 10.5?

10-8 Theoretically, what concentrations of alkalinity in mg/l as $CaCO_3$ are required to react with alum feeds of: (a) 30 mg/l ; (b) 300 lb/mg?

10-9 Calculate the dosage of lime as CaO required to react with an alum dosage of 4.0 gpg.

10-10 A ferrous sulfate dosage of 2.5 gpg and an equivalent dosage of lime are used to coagulate a water. (a) How many pounds of ferrous sulfate per million gallons are used?

(b) How many pounds of hydrated lime per million gallons are used, assuming a purity of 70% CaO? (c) How many pounds of ferric hydroxide are theoretically produced per million gallons of water treated?

10-11 In chlorinated copperas coagulation, how many milligrams of ferrous sulfate are oxidized by 1.0 mg of chlorine?

10-12 Treatment of a water supply requires 4.0 gpg of ferric chloride as a coagulant. The natural alkalinity of the water is 40 mg/l. Based on theoretical chemical reactions, what dosage of lime as CaO is required to react with the ferric chloride after the natural alkalinity is exhausted?

10-13 Presedimentation reduces the turbidity of a raw river water from 1500 mg/l suspended solids to 200 mg/l. How many pounds of dry solids are removed per million gallons? If the settled sludge has a concentration of 8% solids and a specific gravity of 1.03, calculate the sludge volume produced per million gallons of river water processed.

10-14 The results from a jar test for coagulation of a turbid alkaline raw water are given in the table. Each jar contained 1000 ml of water. The aluminum sulfate solution used for chemical addition had such strength that each milliliter of the solution added to a jar of water produced a concentration of 0.5 gpg of aluminum sulfate.

Jar Number	Aluminum Sulfate Solution (ml)	Floc Formation
1	1	None
2	2	Smoky
3	3	Fair
4	4	Good
5	5	Good
6	6	Very heavy

(a) What is the strength in mg/l of the aluminum sulfate solution used for chemical dosage of the jars?
(b) Based on the jar-test results, what is the most economical dosage of aluminum sulfate in gpg?

10-15 Groundwater described by the analysis given below is to be softened by excess lime treatment. (a) Sketch a meq/l bar graph. (b) Express the alkalinity and hardness in mg/l as $CaCO_3$. (c) Which chemical equations apply to excess lime softening of this water? (d) Calculate the quantity of softening chemicals required in mg/l and lb/mg.

$Ca^{2+} = 2.7$ meq/l $HCO_3^- = 3.9$ meq/l
$Mg^{2+} = 1.2$ meq/l $SO_4^{2-} = 0.9$ meq/l
$Na^+ = 1.1$ meq/l $Cl^- = 0.2$ meq/l

10-16 The water defined by the analysis given below is to be softened by excess lime treatment. (a) Sketch a meq/l bar graph. (b) Calculate the softening chemicals required. (c) Draw a bar graph for the softened water after recarbonation and filtration, assuming that one-half of the alkalinity is in the bicarbonate form.

$$CO_2 = 8.8 \text{ mg/l} \qquad Alk(HCO_3^-) = 135 \text{ mg/l}$$
$$Ca^{2+} = 40.0 \text{ mg/l} \qquad SO_4^{2-} = 29.0 \text{ mg/l}$$
$$Mg^{2+} = 14.7 \text{ mg/l} \qquad Cl^- = 17.8 \text{ mg/l}$$
$$Na^+ = 13.7 \text{ mg/l}$$

10-17   A settled water after excess lime treatment, before recarbonation and filtration, contains 35 mg/l of CaO excess lime in the form of hydroxyl ion, 30 mg/l of $CaCO_3$ as carbonate ion, and 10 mg/l as $CaCO_3$ of $Mg(OH)_2$ in the form of hydroxyl ion. First-stage recarbonation precipitates the excess lime as $CaCO_3$ for removal by filtration, and second-stage recarbonation converts a portion of the remaining alkalinity to bicarbonate ion. Calculate the carbon dioxide needed to neutralize the excess lime and convert one-half of the alkalinity in the finished water to the bicarbonate form.

10-18   Compute the lime dosage needed for selective calcium-removal softening of the water described in Prob. 10-6. What is the finished water hardness?

10-19   Consider split-treatment softening of the water described by the analysis below. Use the same criteria for the finished water as given in Example 10-6.

$$CO_2 = 15 \text{ mg/l as } CO_2 \qquad HCO_3^- = 200 \text{ mg/l as } CaCO_3$$
$$Ca^{2+} = 60 \text{ mg/l} \qquad SO_4^{2-} = 96 \text{ mg/l}$$
$$Mg^{2+} = 24 \text{ mg/l} \qquad Cl^- = 35 \text{ mg/l}$$
$$Na^+ = 46 \text{ mg/l}$$

10-20   The water described by the analysis given in Prob. 10-16 is to be softened by split treatment. The selected bypass flow fraction X is 40%; 60% is processed in the first stage. Compute the softening chemicals required and the hardness of the finished water.

10-21   Calculate the saturation index of the water described in Prob. 10-5 for pH 7.1 and a temperature of 20°C.

10-22   Calculate the saturation index from the water analysis given in Prob. 10-6 for pH 8.0 and a temperature equal to 50°F.

10-23   Sketch a meq/l bar graph of the water described in Prob. 10-16 after it is softened to zero hardness by cation-exchange softening.

10-24   Consider ion-exchange softening of the water described in Example 10-4. If 0.3 lb of NaCl is required to regenerate the resin bed per 1000 grains of hardness removed, calculate the salt required per million gallons of water softened. Sketch a meq/l bar graph for the ion-exchange-softened water. How does finished water from ion-exchange softening differ from finished water produced in lime-soda softening?

10-25   A small community has used an unchlorinated well-water supply containing approximately 0.3 mg/l of iron and manganese for several years without any apparent iron and manganese problems. A health official suggested that the town install chlorination equipment to disinfect the water and provide a chlorine residual in the distribution system. After initiating chlorination, consumers complained about water staining washed clothes and bathroom fixtures. Explain what is occurring due to chlorination.

10-26   Untreated well water contains 1.2 mg/l of iron and 0.8 mg/l of manganese at a pH of 7.5. Calculate the theoretical dosage of potassium permanganate required for iron and manganese oxidation.

10-27   Results of a chlorine demand test on a raw water at 20°C are given in the table.

Sample Number	Chlorine Dosage (mg/l)	Residual Chlorine after 10 Minutes' Contact (mg/l)
1	0.20	0.19
2	0.40	0.37
3	0.60	0.51
4	0.80	0.50
5	1.00	0.20
6	1.20	0.40
7	1.40	0.60
8	1.60	0.80

(a) Sketch the chlorine demand curve.
(b) What is the "breakpoint" chlorine dosage?
(c) What is the chlorine demand at a chlorine dosage of 1.20 mg/l?

10-28   What is the recommended minimum free chlorine residual to ensure disinfection of a water at pH 8? Assuming that a chlorine dosage of 1.0 mg/l is required to maintain this minimum residual in remote sections of the distribution system, what is the chlorine dosage in lb/mg of water chlorinated?

10-29   Describe the chlorination practice considered necessary to provide virus inactivation.

10-30   The practice of combined residual chlorination involves feeding both chlorine and anhydrous ammonia. Calculate the stoichiometric ratio of chlorine feed to ammonia feed for combined chlorination.

10-31   Shallow wells located in a sand stratum along a river are used for a municipal water supply. The distribution system in the municipality consists of old and new mains varying in size from 4 in. to 12 in. in diameter. During periods of high water consumption, residual water pressure drops very low in some areas of the distribution system. Why should this community chlorinate its water supply? Include comments with regard to sources of contamination and chlorination practice.

10-32   What dosage of commercial fluosilicic acid is needed to increase the fluoride-ion concentration from 0.3 to 1.0 mg/l? Use fluosilicic acid data from Table 10-6 and express the answers as mg/l and lb/mg.

10-33   A 4.0% sodium fluoride solution is applied to increase the fluoride concentration from 0.4 mg/l to 1.0 mg/l in a municipal water supply. (a) What is the application rate of NaF solution in gal/mg? (b) How many pounds of commercial grade sodium fluoride are needed per million gallons?

10-34   What are the common chemicals used for taste and odor control in surface-water treatment? Where are they usually applied in water processing?

10-35   How does a cement-mortar lining on the inside of a cast-iron pipe prevent corrosion?

10-36   What is cathodic protection?

10-37   List three possible methods for controlling crown corrosion in a large concrete sanitary sewer.

10-38   Outlined below is the sequence of unit operations and chemical additions used in the treatment of a well-water supply. Briefly state the function or purpose of each unit process and the reason for each chemical addition.

1   Mixing and flocculation with the addition of lime and soda ash.
2   Sedimentation.
3   Recarbonation.
4   Rapid sand filtration.
5   Postchlorination.

10-39   Outlined below is the sequence of unit operations and chemical additions used in the treatment of a well-water supply. Briefly state the function or purpose of each unit process and the reason for each chemical addition.

1   Prechlorination at the wells.
2   Aeration over coke trays.
3   Rechlorination.
4   Detention in a settling basin.
5   Rapid sand filtration.
6   Addition of anhydrous ammonia.

10-40   Outlined below is the sequence of unit operations and chemical additions used in the treatment of a well-water supply. Briefly state the function or purpose of each unit process and the reason for each chemical addition.

1   Prechlorination at the wells.
2   Mixing–flocculation–sedimentation in solids-contact units using split treatment with lime and alum added to one leg and potassium permanganate to the other leg.
3   Rapid sand filtration.
4   Postchlorination.

10-41   Outlined below is the sequence of unit operations and chemical additions used in the treatment of a river-water supply. Briefly state the function or purpose of each unit process and the reason for each chemical addition.

1   Presedimentation.
2   Chlorination.
3   Mixing and flocculation with the addition of alum and activated silica.
4   Sedimentation.
5   Addition of activated carbon.
6   Rapid sand filtration.
7   Postchlorination.

10-42 Outlined below is the sequence of unit operations and chemical additions used in the treatment of a reservoir-water supply. Briefly state the function or purpose of each unit process and the reason for each chemical addition.

1 Intermittent applications of copper sulfate to reservoir during summer and fall.
2 Prechlorination.
3 Mixing and flocculation with the addition of alum and polyelectrolyte.
4 Sedimentation.
5 Addition of activated carbon.
6 Rapid sand filtration.
7 Postchlorination.
8 Addition of anhydrous ammonia to the effluent of the clear well.

10-43 What is the minimum energy required to recover 1000 gal of fresh water from a brackish water containing 4500 mg/l as sodium chloride? The temperature is 25°C. What would be the probable energy requirements with a present process?

10-44 What is the theoretical minimum work required in a saline-water-conversion process where the feedwater is pumped against a 110-ft head and 35% of the water is recovered as fresh water? Give the answer in hp/1000 gal.

10-45 With energy costing 0.1 cent/kWh and with the use of Fig. 10-19, what would be the necessary efficiency of a separation process in order to compete with the 5 cents/1000 gal for in-plant conventional water treatment?

10-46 An inland city of 300,000 population using water at the rate of 150 gpcd plans to use a brackish raw-water supply. The brackish water contains 35,000 mg/l as sodium chloride. The proposed process will result in a 35% freshwater recovery. What is the annual volume of waste brine? What is the weight of the sodium chloride in this annual volume of waste brine?

10-47 The waste brine in Prob. 10-46 has an average annual evaporation rate in excess of rainfall of 11 in./yr for the particular location. Solar evaporation is being studied as a means of disposal of the liquid brine waste. What is the surface area of the evaporation pond necessary to dispose of this waste?

10-48 An electrodialysis stack is made up of 500 membranes 48 × 48 in. This unit will be used to partially demineralize 300,000 gal of brackish water per day. The water contains 3500 mg/l of dissolved solids as sodium chloride. The resistance through the unit has been determined to be 8 Ω and the current efficiency, 89%. The maximum current density/normality ratio that may be used without serious polarization is 600 Ω. Estimate the removal efficiency and the power consumption per 1000 gal of product water.

REFERENCES

1. American Public Health Association, American Water Works Association, and Water Pollution Control Federation, *Standard Methods for the Examination of Water and Wastewater*, 14th ed. (Washington, D.C.: American Public Health Association, 1975).

2. C. N. Sawyer and P. L. McCarty, *Chemistry for Sanitary Engineers*, 2nd ed. (New York: McGraw-Hill Book Company, 1967).

3. J. H. Sullivan, Jr., and J. E. Singley, "Reactions of Metal Ions in Dilute Aqueous Solution: Hydrolysis of Aluminum," *J. Am. Water Works Assoc.* 60, no. 11 (1968): 1280–1287.

4. J. E. Singley and A. P. Black, "Hydrolysis Products of Iron (III)," *J. Am. Water Works Assoc.* 59, no. 12 (1967): 1549–1564.

5. W. Stumm and C. R. O'Melia, "Stoichiometry of Coagulation," *J. Am. Water Works Assoc.* 60, no. 5 (1968): 514–539.

6. A. P. Black, F. B. Birkner, and J. J. Morgan, "Destabilization of Dilute Clay Suspensions with Labeled Polymers," *J. Am. Water Works Assoc.* 57, no. 12 (1965): 1547–1560.

7. M. Pressman, "Cationic Polyelectrolytes as Prime Coagulants in Natural-Water Treatment," *J. Am. Water Works Assoc.* 59, no. 2 (1967): 169–182.

8. J. L. Cleasby and J. H. Dillingham, "Rational Aspects of Split Treatment," *Proc. Am. Soc. Civil Engrs., J. San. Eng. Div.*, 92, no. SA2 (1966): 1–7.

9. W. F. Langelier, "The Analytical Control of Anticorrosion Water Treatment," *J. Am Water Works Assoc.* 28 (1936): 1500.

10. T. E. Larson and A. M. Buswell, "Calcium Carbonate Saturation Index and Alkalinity Interpretations," *J. Am. Water Works Assoc.* 34 (1942): 1667.

11. H. H. Chambers and R. S. Ingols, "Copper Sulfate Aids in Manganese Removal," *Water and Sewage Works* 103 (1956): 248.

12. George L. Illig, Jr., "Use of Sodium Hexametaphosphate in Manganese Stabilization," *J. Am. Water Works Assoc.* 52, no. 7 (1960): 867.

13. American Water Works Association, *Water Quality and Treatment*, 3rd. ed. (New York: McGraw Hill Book Company, 1971).

14. J. M. Longley, R. S. Engelbrecht, and G. E. Margrave, "Laboratory and Field Studies on the Treatment of Iron-Bearing Waters," *J. Am. Water Works Assoc.*, 54, no. 6 (1962): 731.

15. W. A. Welch, "Potassium Permanganate in Water Treatment," *J. Am. Water Works Assoc.* 55, no. 6 (1963): 735.

16. C. T. Butterfield, "Bactericidal Properties of Chloramines and Free Chlorine in Water," *Pub. Health Repts.* 63 (1948): 934; *J. Am. Water Works Assoc.* 40 (1948): 1305.

17. American Water Works Association, "Water Chlorination, Principles and Practices," *AWWA Manual No. M20* (Denver, Colo.: American Water Works Association, 1973).

18. H. F. Collins, R. E. Selleck, and G. C. White, "Problems in Obtaining Adequate Sewage Disinfection," *Proc. Am. Soc. Civil Engrs., J. San. Eng. Div.* 97 no. SA5 (1971): 549–562.

19. H. F. Collins and D. G. Deaner, "Sewage Chlorination versus Toxicity—A Dilemma?" *Proc. Am. Soc. Civil Engrs., J. Environ. Eng. Div.* 99, no. EE6 (1973): 761–772.

20. R. H. Susag, "BOD Reduction by Chlorination," *J. Water Poll. Control Fed.* 40, no. 11 (1968): R434.

21. E. Bellack, *Fluoridation Engineering Manual*, EPA, Office of Water Programs, Water Supply Programs Division (Environmental Protection Agency, 1972).

22. "Saline Water Conversion Program," *U.S. Office Saline Water Pamphlet No. 649411 0-69-2* (Washington, D.C.: Government Printing Office, 1962).

23. York Sampson, "The Nation's Water Crisis," *Am. Engr.* (October 1963): 29.

24. B. F. Dodge, "Fresh Water from Saline Waters," *Am. Scientist* 48, no. 4 (December 1960): 476–513.

25. J. C. Lamb, III, "Economic Aspects of Saline-Water Conversion," *J. Am. Water Works Assoc.* 54 (July 1962): 781–788.

26. W. Viessman, Jr., "Desalination Brine Waste Disposal," *Public Works* 94, no. 12 (December 1963): 117–118.

27. United Nations Publication, *Second United Nations Desalting Plant Operation Survey*, No. E.73.11.A.10 (New York: United Nations, 1973).

28. O. G. George, "Demineralization of Saline Water with Solar Energy," in *U.S. Office Saline Water, Saline Water Research and Development Progress Report No. 4* (Washington, D.C.: Government Printing Office, 1954).

29. Aerojet-General Corporation, "The Mechanism of Desalination by Reverse Osmosis," in *U.S. Office Saline Water, Saline Water Research and Development Progress Report No. 84* (Washington, D.C.: Government Printing Office, November 1963).

11
Biological-Treatment Processes

Biological treatment is the most important step in processing municipal waste-waters. Physical treatment of raw wastewater by sedimentation removes only about 35% of the BOD, owing to the high percentage of nonsettleable solids (colloidal and dissolved) in domestic wastes. Chemical treatment, although it has been used in domestic wastewater treatment, is not favored because of high chemical costs and inefficiency of dissolved BOD removal by chemical coagulation. A modern treatment plant uses a variety of physical, chemical, and biological processes to provide the best, most economical treatment.

Biological Considerations

Biological treatment systems are "living" systems which rely on mixed biological cultures to break down waste organics and remove organic matter from solution. Domestic wastewater supplies the biological food, growth nutrients, and inoculum. A treatment unit provides a controlled environment for the desired biological process. Civil engineers have traditionally designed treatment systems on the basis of empirical rules. This practice has lead to failures in

497

sanitary design—not unsuccessful in the sense of collapse of a structure but deficient in that the biological process did not function properly. Understanding the biological processes involved in wastewater treatment is essential to a designing engineer.

The following abbreviated discussion of biological considerations is, at best, an overview of the subject matter. It may raise more questions than it answers. Yet, the author believes it is a step in the right direction toward teaching the fundamentals of sanitary engineering to undergraduate civil engineering students. For additional information in the biological science areas, a variety of books on microbiology are available, but a limited number deal with microbiology as applied to wastewater systems. Two books currently available are *Microbiology for Sanitary Engineers* by R. E. McKinney,[1] and *The Ecology of Waste Water Treatment* by H. A. Hawkes.[2]

11-1

BACTERIA AND FUNGI

Bacteria (singular, bacterium) are the simplest forms of plant life which use soluble food and are capable of self-reproduction. Bacteria are fundamental microorganisms in the stabilization of organic wastes and therefore of basic importance in biological treatment. Individual bacterial cells range in size from approximately 0.5 to 5 μm in rod, sphere, and spiral shapes, and occur in a variety of forms: individual, pairs, packets, and chains (Fig. 11-1).

Bacteria reproduce by binary fission (the mature cell divides into two new cells). In most species, the process of reproduction—growth, maturation, and fission—occurs in 20–30 minutes under ideal environmental conditions. Certain bacterial species form spores as a means of survival under adverse environmental conditions. Their tough coating is resistant to heat, lack of moisture, and loss of food supply. Fortunately, only one spore-forming bacterium, *Bacillus anthracis*,* is pathogenic to man. As the result of stringent public health measures, incidents of anthrax in man are rare.

Based on nutritive requirements, bacteria are classified as heterotrophic or autotrophic bacteria, although several species may function both heterotrophically and autotrophically.

Heterotrophic bacteria use organic compounds as an energy and carbon source for synthesis. A term commonly used instead of heterotroph is "saprophyte," which refers to an organism that lives on dead or decaying organic matter. The heterotrophic bacteria are grouped into three classifications, depending upon their action toward free oxygen. *Aerobes* require free dissolved oxygen to live and multiply. *Anaerobes* oxidize organic matter in the complete absence of dissolved oxygen. Pasteur referred to anaerobiosis as "life without

---

*Bacteria are named using a binomial system in which each species is given a name consisting of two words. The first word is the genus and the second the name of the species.

Escherichia coli

Vegetative
cells

Formation
of spores

Bacillus subtilis

Streptococcus lactis

Sheath

Free
cells

Sphaerotilus natans

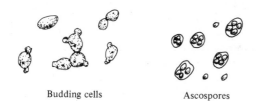

Budding cells

Ascospores

Saccharomyces

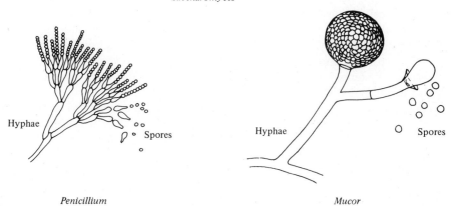

Hyphae

Spores

Hyphae

Spores

Penicillium

Mucor

FIGURE 11-1 Examples of bacteria and fungi forms.

air." *Facultative bacteria* are a class of bacteria which use free dissolved oxygen when available but can also respire and multiply in its absence. *Escherichia coli,* a common coliform, is a facultative bacterium.

Autotrophic bacteria use carbon dioxide as a carbon source and oxidize inorganic compounds for energy. Autotrophs of greatest significance in sanitary engineering are the nitrifying, sulfur, and iron bacteria. Nitrifying bacteria perform the following reactions:

$$NH_3 \text{ (ammonia)} + \text{oxygen} \xrightarrow{\text{\textit{Nitrosomonas}}} NO_2^- \text{ (nitrite)} + \text{energy} \qquad (11\text{-}1)$$

$$NO_2^- \text{ (nitrite)} + \text{oxygen} \xrightarrow{\text{\textit{Nitrobacter}}} NO_3^- \text{ (nitrate)} + \text{energy} \qquad (11\text{-}2)$$

Autotrophic sulfur bacteria, *Thiobacillus,* perform the reaction

$$H_2S \text{ (hydrogen sulfide)} + \text{oxygen} \rightarrow H_2SO_4 + \text{energy} \qquad (11\text{-}3)$$

This bacterial production of sulfuric acid occurs in the moisture of condensation on side walls and crowns of sewers conveying septic wastewater. Since thiobacilli can tolerate pH levels less than 1.0, sanitary sewers constructed on flat grades in warm climates should be built using corrosion-resistant materials.

True iron bacteria are autotrophs which oxidize inorganic ferrous iron as a source of energy. These filamentous bacteria occur in iron-bearing waters and deposit the oxidized iron, $Fe(OH)_3$, in their sheath. All species of the iron bacteria *Leptothrix* and *Crenothrix* may not be strictly autotrophic; however, they are truly iron-accumulating bacteria and thrive in water pipes conveying water containing dissolved iron and form yellow or reddish-colored slimes. When mature bacteria die, they may decompose, imparting foul tastes and odors to water.

$$Fe^{2+} \text{ (ferrous)} + \text{oxygen} \rightarrow Fe^{3+} \text{ (ferric)} + \text{energy} \qquad (11\text{-}4)$$

Bacteriophages are viruses of bacteria, obligate parasites, and dependent upon the bacteria for growth. The submicroscopic bacteriophage attaches to the bacterium, penetrates the cell wall, reproduces within the cell, and finally, when the cell ruptures, new bacteriophages release into solution to infect other bacterial cells. When this occurs, the bacterium is said to have undergone *lysis.* The whole process cycle takes about 30–40 minutes. Bacteriophages normally occur in nature wherever hosts are found. They are certainly present in biological-treatment systems but their significance, other than the knowledge that they perform cell lysis, is not known.

Fungi (singular, fungus) is a common term used to refer to microscopic nonphotosynthetic plants, including yeasts, molds, and bacteria. Because of their importance, bacteria are frequently excluded from the fungi classification. The most important group of yeasts for industrial fermentations are the genus *Saccharomyces* (Fig. 11-1). *Saccharomyces cerevisiae* is the common yeast used by bakers, distillers, and brewers. *Saccharomyces cerevisiae* is single-celled, commonly 5–10 μm in size, and reproduces by budding, in which large, mature cells divide, each producing one or more daughter cells. Under anaerobic conditions, this yeast produces alcohol as an end product. *Saccharomyces cerevisiae* is facultative and performs the following reactions:

$$\text{\textit{Anaerobic:} sugar} \longrightarrow \text{alcohol} + CO_2 + \text{energy} \qquad (11\text{-}5)$$

$$\text{\textit{Aerobic:} sugar} + \text{oxygen} \longrightarrow CO_2 + \text{energy} \qquad (11\text{-}6)$$

Energy yield in the aerobic reaction is much greater than in the anaerobic fermentation.

Molds are saprophytic or parasitic filamentous fungi which resemble higher plants in structure, composed of branched, filamentous, threadlike growths called hyphae (Fig. 11-1). Molds are nonphotosynthetic, multicellular, heterotrophic, aerobic, reproduce by spore formation, and grow best in low-pH solutions (pH 2–5) high in sugar content. Molds are undesirable growths in activated sludge and can be created by low-pH conditions. The operation of an activated-sludge wastewater-treatment system relies on gravity separation of microorganisms from the wastewater effluent. A large growth of molds creates a filamentous activated sludge which does not settle easily.

11-2

ALGAE

Algae (singular, alga) are microscopic photosynthetic plants. The process of photosynthesis is illustrated by the equation

$$CO_2 + 2H_2\dot{O} \underset{\text{dark reaction}}{\overset{\text{sunlight}}{\rightleftharpoons}} \text{new cell tissue} + \dot{O}_2 + H_2O \qquad (11\text{-}7)$$

The overall effect of this reaction is to produce new plant life, thereby increasing the number of algae. By-product oxygen results from the biochemical conversion of water.

Algae are autotrophic, using carbon dioxide (or bicarbonates in solution) as a carbon source. The nutrients of phosphorus (as phosphate) and nitrogen (as ammonia, nitrite, or nitrate) are necessary for growth. Certain species of blue-green algae are able to fix atmospheric nitrogen. In addition, certain trace nutrients are required, such as magnesium, sulfur, boron, cobalt, molybdenum, calcium, potassium, iron, manganese, zinc, and copper. In natural waters the nutrients most frequently limiting algal growth are inorganic phosphorus and nitrogen.

Energy for photosynthesis is derived from sunlight. Photosynthetic pigments biochemically convert the energy in sun's rays to useful energy for plant synthesis. The most common pigment is chlorophyll, which is green in color. Other pigments or combinations of pigments result in algae of a variety of colors, such as blue-green, yellowish green, brown, and red. In the prolonged

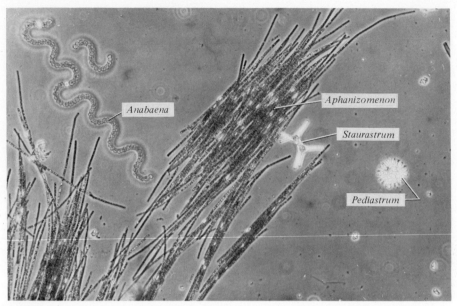

FIGURE 11-2 Photomicrograph of algae (100×).

absence of sunlight the algae perform a dark reaction—for practical purposes the reverse of synthesis. In the dark reaction, the algae degrade stored food or their own protoplasm for energy to perform essential biochemical reactions for survival. The rate of this endogeneous reaction is significantly slower than photosynthetic reaction.

Algal growth in rivers and lakes is not something mysterious or unknown but a simple natural process. Given a suitable environment (temperature, pH, and sunlight) and a proper nutrient supply (phosphates, nitrogen, and trace nutrients) algae will grow and multiply in abundance. Algae which grow un-attached in the water are referred to as "phytoplankton"* (Fig. 11-2). If the supply of nutrients is not a limiting factor in natural waters, excessive growth of phytoplankton creates a "pea-soup" condition referred to as a "bloom," during which the algal populations multiply rapidly to several thousand per ml. The alga *Oscillatoria rubescens* (Fig. 11-3) has been indentified as one which creates blooms in eutrophic lakes. Filamentous algae, such as *Anabaena*, float on the water surface, forming mats which are unsightly and wash onto bathing beaches. Water supplies from eutrophic rivers and lakes can have periodic serious problems, clogging filters, and producing tastes and odors caused by excessive algal growth.

Algae grow in abundance in stabilization ponds rich in inorganic nutrients and carbon dioxide released from bacterial decomposition of waste organics.

---

* The term *plankton* refers to unattached organisms that are dispersed individually or in colonies in water. *Phytoplankton* are plant plankton. *Zooplankton* are planktonic animals.

Anabaena
Blue–green algae associated with taste and
odor problems

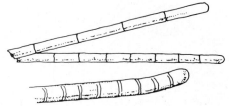

Oscillatoria
Blue–green algae associated with filter clogging

Chlorella
Green phytoflagellate found in
polluted water

Navicula
Clean–water diatom (brown–colored)

FIGURE 11-3 Examples of algae forms.

Green algae *Chlorella* are commonly found in oxidation ponds. Certain genera of algae are identified with clean water, such as *Navicula*. A rather complete treatise on algae in water supplies has been prepared by Palmer.[3]

11-3
PROTOZOANS AND HIGHER ANIMALS

Protozoans are single-celled animals that reproduce by binary fission. The protozoans of significance in biological-treatment systems are strict aerobics found in activated sludge, trickling filters, and oxidation ponds. These microscopic animals have complex digestive systems and use solid organic matter as an energy and carbon source. Protozoans are a vital link in the aquatic chain since they ingest bacteria and algae.

A few common protozoans are illustrated in Fig. 11-4. The species with hairlike cilia are the most prevalent forms found in activated sludge. Protozoans with cilia may be categorized as free-swimming and stalked. Free-swimming forms move rapidly in the water, ingesting organic matter at a very high rate. The stalked forms attach by a stalk to particles of matter and use cilia to propel their head about and bring in food. Another group of protozoans move by flagella. Long hairlike strands (flagella) move with a whiplike action, providing motility. *Amoeba* move and ingest food through the action of a mobile protoplasm. Only the pathogen *Entamoeba histolytica*, which produces amoebic dysentery in tropical climates (e.g., South America and southwestern United States), is of significance in sanitation.

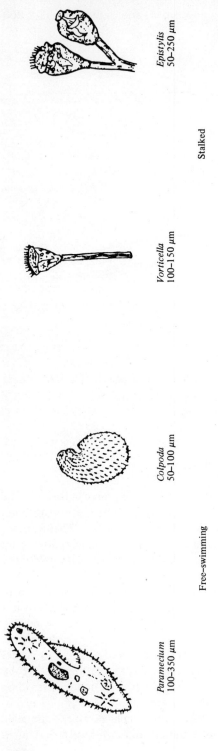

Epistylis
50–250 μm

Stalked

Vorticella
100–150 μm

Colpoda
50–100 μm

Free-swimming

Paramecium
100–350 μm

Amoeba
to 500 μm

Motile protoplasm

Euglena
30–50 μm

Monas
10–15 μm

Motility by flagella

Bodo
10–15 μm

FIGURE 11-4 Some common forms of protozoans.

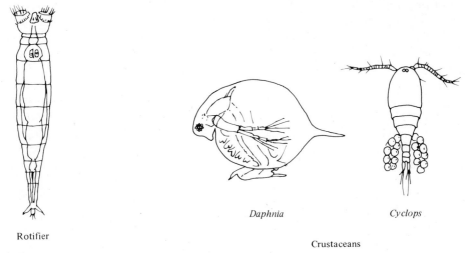

Daphnia   *Cyclops*

Rotifier

Crustaceans

FIGURE 11-5 Typical rotifer and crustaceans.

Rotifers are the simplest multicelled animals. They are strict aerobes and metabolize solid food. A typical rotifer, shown in Fig. 11-5, uses the cilia around its head for catching food. The name *rotifer* is derived from the apparent rotating motion of the cilia on its head. Rotifers are indicators of low pollutional waters and are regularly found in streams and lakes.

Crustaceans (Fig. 11-5) are multicellular animals with branched swimming feet or a shell-like covering, with a variety of appendages (antennae). The two most common crustaceans of interest are *Daphnia* and *Cyclops*. Crustaceans are strict aerobes and ingest microscopic plants. Zooplankton population in a lake includes a wide selection of crustaceans which serve as food for fishes.

11-4

METABOLISM, ENERGY, AND SYNTHESIS

Metabolism (catabolism) is the biochemical process (a series of biochemical oxidation–reduction reactions) performed by living organisms to yield energy for synthesis, motility, and respiration to remain viable. In standard usage, metabolism implies both catabolism and anabolism, that is, both degradation and assimilative reactions.

The metabolism of autotrophic bacteria is illustrated in Eqs. 11-1, 11-2, 11-3, and 11-4. In these reactions, the reduced inorganic compounds are oxidized, yielding energy for synthesis of carbon from carbon dioxide, producing organic cell tissue. (In the case of algae, Eq. 11-7, the carbon source is carbon dioxide but the energy is from sunlight.)

In heterotrophic metabolism, organic matter is the substrate (food) used as an energy source. However, the majority of organic matter in wastewater is in the form of large molecules which cannot penetrate the bacterial cell membrane. The bacteria, in order to metabolize high-molecular-weight substances, must be capable of hydrolyzing the large molecules into diffusible fractions for assimilation into their cells. Therefore, the first biochemical reactions are hydrolysis* of complex carbohydrates into soluble sugar units, protein into amino acids, and insoluble fats into fatty acids. Under aerobic conditions the reduced soluble organic compounds are oxidized to end products of carbon dioxide and water (Eq. 11-8). Under anaerobic conditions, soluble organics are decomposed to intermediate end products, such as organic acids and alcohols, along with the production of carbon dioxide and water (Eq. 11-9). Many intermediates, such as butyric acid, mercaptons (organic compounds with —SH radicals), and hydrogen sulfide have foul odors.

Under anaerobic conditions, if excess organic acids are produced, the pH of the solution will drop sufficiently to "pickle" the fermentation process. This is the principle used for preservation of silage. Bacteria produce an overabundance of organic acids in the anaerobic decomposition of the green fodder stored in the silo, inhibit further bacterial decomposition, and preserve the food value of the fodder. However, if proper environmental conditions exist to prevent excess acidity from the production of organic acid intermediates, populations of acid-splitting methane-forming bacteria will develop and use the organic acids as substrate (Eq. 11-10). The combined biological processes of anaerobic decomposition of raw organic matter to soluble organic intermediates and the gasification of the intermediates to carbon dioxide and methane is referred to as digestion.

$$\textit{Aerobic: } \text{organics} + \text{oxygen} \longrightarrow CO_2 + H_2O + \text{energy} \qquad (11\text{-}8)$$

$$\textit{Anaerobic: } \text{organics} \longrightarrow \text{intermediates} + CO_2 + H_2O + \text{energy} \qquad (11\text{-}9)$$

$$\text{organic acid intermediates} \longrightarrow CH_4 + CO_2 + \text{energy} \qquad (11\text{-}10)$$

Growth and survival of nonphotosynthetic microorganisms is dependent upon their ability to obtain energy from the metabolism of substrate. Biochemical metabolic processes of heterotrophs are energy-yielding oxidation–reduction reactions in which reduced organic compounds serve as hydrogen donors and oxidized organic or inorganic compounds act as hydrogen acceptors. *Oxidation* is the addition of oxygen, removal of hydrogen, or the removal of electrons. *Reduction* is the removal of oxygen, addition of hydrogen, or addition of electrons.

The simplified diagram of substrate dehydrogenation shown in Fig. 11-6 is intended to illustrate the general relationship between energy yields of aerobic and anaerobic metabolism. Enzymatic processes of hydrogen transfer and methods of biologically conserving energy released are beyond the scope of this

* Hydrolysis is the addition of water to split a bond between chemical units.

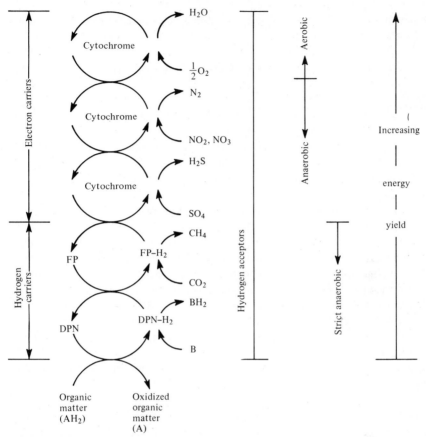

FIGURE 11-6   General scheme of substrate dehydrogenation for energy yield. * FP=flavo-protein. † DPN=diphosphopyridine nucleotide.

discussion. For students to fully understand the mechanisms illustrated in Fig. 11-6, a knowledge of the biochemistry of microorganisms is necessary.[4]

Energy stored in organic matter (AH_2) is released in the process of biological oxidation by dehydrogenation of substrate followed by transfer of hydrogen, or electrons, to an ultimate acceptor. The higher the ultimate hydrogen acceptor is on the energy (electromotive) scale, the greater will be the energy yield from oxidation of 1 mole of a given substrate. Aerobic metabolism using oxygen as the ultimate hydrogen acceptor yields the greatest amount of energy. Aerobic respiration can be traced on Fig. 11-6 from reduced organic matter (AH_2) at the bottom, through the hydrogen and electron carriers to oxygen. Facultative respiration, using oxygen bound in nitrates and sulfates, yields less energy than aerobic metabolism. The least energy yield results from strict anaerobic respiration, where the oxidation of AH_2 is coupled with reduction of B (an oxidized organic compound) to BH_2 (a reduced organic compound). The preferential use of hydrogen acceptors based on energy yield in a mixed bacterial culture is illustrated by the following equations:

Aerobic	$AH_2 + O_2$	$\longrightarrow$	$CO_2 + H_2O + \text{energy}$	(11-11)
$\vert$	$AH_2 + NO_3^-$	$\longrightarrow$	$N_2 + H_2O +$	(11-12)
(Facultative)	$AH_2 + SO_4^{2-}$	$\longrightarrow$	$H_2S + H_2O +$	(11-13)
$\downarrow$	$AH_2 + CO_2$	$\longrightarrow$	$CH_4 + H_2O +$	(11-14)
Anaerobic	$AH_2 + B$	$\longrightarrow$	$BH_2 + A + \text{energy}$	(11-15)

(to the right of equations 11-12 through 11-14: "decreasing energy yield" ↓)

Hydrogen acceptors are used in the sequence of dissolved oxygen first, followed by nitrates, sulfates, and oxidized organic compounds, in this general order. Thus hydrogen sulfide odor formation follows nitrate reduction and precedes methane formation.

The biochemical reactions in Fig. 11-6 are performed by oxidation–reduction enzymes. Enzymes are organic catalysts which perform biochemical reactions at temperatures and chemical conditions compatible with biological life. The coenzyme component of the enzyme determines what chemical reaction will occur. Coenzymes diphosphopyridine nucleotide (DPN) and flavoproteins (FP) are responsible for hydrogen transfer. Cytochromes are respiratory pigments that can undergo oxidation and reduction and serve as hydrogen carriers.

Synthesis (anabolism) is the biochemical process of substrate utilization to form new protoplasm for growth and reproduction. Microorganisms process organic matter to create new cells. The cellular protoplasm formed is a combination of hundreds of complex organic compounds, including proteins, carbohydrates, nucleic acids, and lipids. Major elements in biological cells are carbon, hydrogen, oxygen, nitrogen, and phosphorus. On a dry-weight basis, protoplasm is 10–12% nitrogen and approximately 2.5% phosphorus, the remainder is carbon, hydrogen, oxygen, and trace elements.

Relationships between metabolism, energy, and synthesis are important in understanding biological-treatment systems. The primary product of metabolism is energy, and the chief use of this energy is for synthesis. Energy release and synthesis are coupled biochemical processes which cannot be separated. The maximum rate of synthesis occurs simultaneously with the maximum rate of energy yield (maximum rate of metabolism). Therefore, in heterotrophic metabolism of wastewater organics, maximum rate of removal of organic matter, for a given population of microorganisms, occurs during maximum biological growth. Conversely, the lowest rate of removal of organic matter occurs when growth ceases.

The major limitation of anaerobic growth is energy, owing to the fact that in anaerobic decomposition a low energy yield per unit of substrate results from incomplete breakdown (Fig. 11-7). In other words, the limiting factor in anaerobic metabolism is a lack of hydrogen acceptors. When the supply of biologically available energy is exhausted, the processes of metabolism and synthesis cease.

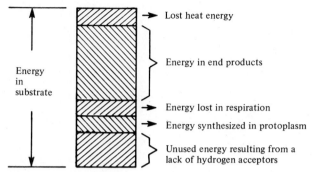

FIGURE 11-7 Energy conversions in anaerobic metabolism.

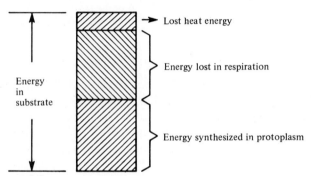

FIGURE 11-8 Energy conversions in aerobic metabolism.

Aerobic metabolism is the antithesis of anaerobiosis, biologically available carbon being the limiting factor (Fig. 11-8). Abundance of oxygen creates no shortage of hydrogen acceptors. But the supply of substrate carbon is rapidly exhausted through respiration of carbon dioxide and synthesis into new cells.

The energy-conversion diagrams shown schematically in Figs. 11-7 and 11-8 illustrate the major features of anaerobic and aerobic metabolism. An anaerobic process has the following characteristics: incomplete metabolism, small quantity of biological growth and production of high-energy products, such as acetic acid and methane. An aerobic process results in complete metabolism and synthesis of the substrate, ending in a large quantity of biological growth.

11-5

GROWTH

The general growth pattern for fission-reproducing microorganisms in a batch culture is sketched in Fig. 11-9. Within a short period of time after a substrate in liquid suspension is inoculated with a healthy population of bacteria or

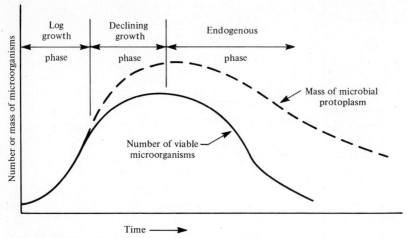

FIGURE 11-9 Microbial growth curves.

protozoans, microorganisms start to reproduce by binary fission. This logarithmic growth results in a rapid increase in the number and mass of microorganisms in the culture medium. The existence of excess food promotes a maximum rate of growth along with a maximum rate of substrate removal. Rate of metabolism in the log growth phase is limited only by the microorganism's ability to process the substrate.

The declining growth phase is caused by an increasing shortage of food. Rate of reproduction decreases until the number of viable microorganisms is stationary (rate of reproduction is equal to rate of death). The total mass of microbial protoplasm exceeds the mass of viable cells, since many of the microorganisms stopped reproducing owing to substrate limiting conditions.

In the endogeneous growth phase, viable microorganisms are competing for the small amount of substrate still in solution. The rate of metabolism is decreasing at an increasing rate resulting in a rapid decrease in number of viable microorganisms. Starvation occurs such that rate of death exceeds rate of reproduction. Total mass of microbial protoplasm decreases as cells utilize their own protoplasm as an energy source. Cells become old, die and lyse, releasing nutrients back into solution. The action of cell lysis decreases both the number and mass of microorganisms.

The batch-culture growth pattern shown in Fig. 11-9 is not directly applicable to biological-treatment processes which are continuous-flow systems. For example, an activated-sludge system is fed continuously, and excess microorganisms are withdrawn, either continuously or intermittently, to maintain the desired mass of microorganisms for metabolizing incoming organic wastes. A schematic diagram, Fig. 11-10, illustrates the flow pattern for food (organic matter) and microorganisms in an activated-sludge system. Food (influent wastewater) is aerated with a mixed culture of microorganisms for a sufficient

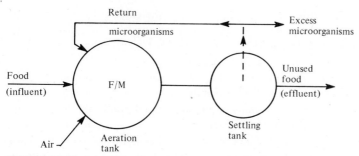

FIGURE 11-10 Schematic diagram of a continuous-flow activated-sludge process.

period of time to permit synthesis of the waste organics into biological cells. The microorganisms are then settled out of solution, removed from the bottom of the settling tank, and returned to the aeration tank to metabolize additional waste organics. Unused food, the nonsettleable fraction of the aeration tank effluent, passes out in the system effluent. Metabolism of the organic matter results in an increased mass of microorganisms in the system. Excess microorganisms are removed (wasted) from the system to maintain proper balance between food supply and mass of microorganisms in the aeration tank. This balance is referred to as the *food to microorganism ratio* (F/M).

The F/M ratio maintained in the aeration tank defines the operation of an activated-sludge system. At a high F/M ratio, microorganisms are in the log growth phase, characterized by excess food and maximum rate of metabolism (Fig. 11-11). Although the log growth phase is desirable for maximum rate of organic matter removal, distinct disadvantages make it undesirable for operation of an activated-sludge system. The microorganisms are in dispersed growth such that they do not settle out of solution by gravity. Consequently, the settling tank

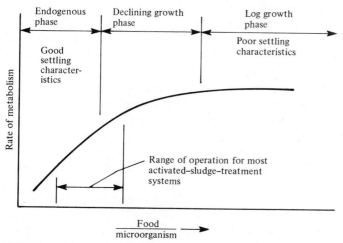

FIGURE 11-11 Rate of metabolism versus increasing food/microorganism ratio.

is not effective in separating microorganisms from the effluent for return to the aeration tank. Second, there is excess unused organic matter in solution which cannot be removed by sedimentation and passes out in the effluent. Operation at a high F/M ratio results in poor BOD removal efficiency.

At a low F/M ratio, overall metabolic activity in the aeration tank may be considered as endogeneous. Although initially there is rapid growth when the influent food and return microorganisms are mixed, competition for the small amount of food made available to the large mass of microorganisms results in near-starvation conditions for the majority of microorganisms within a short period of time. Under these conditions, continued aeration results in auto-oxidation of the microbial mass through cell lysis and resynthesis, and through the predator–prey activity where bacteria are eaten by the protozoans. Although the rate of metabolism is relatively low in the endogeneous phase, metabolism of the organics is nearly complete and the microorganisms flocculate rapidly and settle out of solution by gravity. Good settling characteristics exhibited by activated sludge in the endogeneous phase makes operation in this growth period desirable where a high BOD removal efficiency is desired. Figure 11-11 summarizes the previous discussion and shows the range of operation for most activated-sludge-treatment systems as being between the declining growth phase and the endogenous phase.

11-6

FACTORS AFFECTING GROWTH

Several factors affect the growth of microorganisms. The most important are temperature, pH, availability of nutrients, oxygen supply, presence of toxins, types of substrate, and, in the case of photosynthetic plants, sunlight. Growth, with respect to aerobic and anaerobic conditions and the need for essential nutrients, has been discussed.

Bacteria are classified as psychrophilic, mesophilic, or thermophilic, depending upon their optimum temperature range for growth. Of least significance to sanitary engineers are the *psychrophilic* (cold-loving) bacteria, which grow best at temperatures slightly above freezing (4–10°C).

Thermophilic (heat-loving) bacteria like an optimum temperature range of 50–55°C. They hold sanitary significance in food preservation, and attempts have been made to use a thermophilic temperature range for the anaerobic digestion of waste sludge. Thermophilic digestion has not been successful in practice because thermophilic bacteria are sensitive to small temperature changes, and it is difficult to maintain the required high operating temperature in a digestion tank.

The *mesophilic* (moderation-loving) bacteria grow best in the temperature range 20–40°C. Most bacterial pathogens are mesophilic and thrive at human body temperature, 37°C. Pasteurization of milk [heating to 62°C (143°F) for a period of at least 30 minutes] destroys the vegetative cells of mesophilic pathogens. The vast majority of biological-treatment systems operate in the mesophilic temperature range. Anaerobic digestion tanks are normally heated to near the optimum level of 35°C(95°F). Aeration tanks and trickling filters operate at the temperature of the wastewater as modified by that of the air. Generally, this is within the 15–25°C range. A high wastewater temperature increases biological activity in the treatment process but rarely causes any severe operating problems. At high temperatures odor problems may be more pronounced at a wastewater plant, and, in one case observed by the author, the increased metabolism rate during high loading periods of an activated-sludge system at elevated temperature resulted in serious dissolved-oxygen depletions in the aeration tanks.

Cold wastewater can reduce BOD removal efficiency of biological processes. The efficiency of trickling filters is definitely decreased during cold weather and increased during warm periods. Trickling filters opeating at 5–10°C in mountain parks were observed to have poor BOD removals. However, low-loaded extended aeration systems operating at the same temperatures showed good efficiencies. Extended aeration at a reduced BOD loading and resultant long aeration time compensates for the low metabolism rate of microorganisms. The biological-treatment system most affected by cold winter temperature is the stabilization pond. Heat in the wastewater is not adequate to prevent formation of an ice cover in northern climates during winter.

As a general rule, the rate of biological activity doubles for every 10–15°C temperature rise within the range 5–35°C (Fig. 11-12). Above 40°C mesophilic bacterial metabolism drops off sharply and thermophilic growth starts. Thermophilic bacteria have a range of approximately 45–75°C, with an optimum near 55°C.

The hydrogen-ion concentration of the culture medium has a direct influence on microbial growth. Most biological-treatment systems operate best in

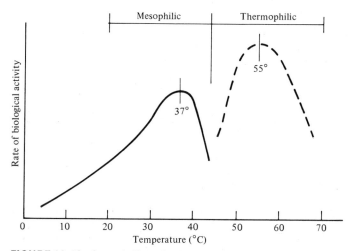

FIGURE 11-12 General effect of temperature on biological activity.

a neutral environment. The general range for operation of activated-sludge systems is between pH 6.5 and 8.5. At pH 9.0 and above, microbial activity is inhibited. Below 6.5, fungi are favored over bacteria in the competition for food. The methane-forming bacteria in anaerobic digestion have a much smaller pH tolerance range. General limits for anaerobic digestion are pH 6.7–7.4 with optimum operation at pH 7.0–7.1.

Biological-treatment systems are adversely affected by toxic substances. Industrial wastes from metal-finishing industries usually contain toxic metal ions, such as copper and chromium. Phenol is an extremely toxic compound found in chemical industry wastes. These inhibiting compounds are commonly removed by pretreatment at the industrial site prior to disposal of the industrial wastes to a municipal sewer.

Environmental conditions that adversely affect the desired microbial growth in an activated-sludge aeration tank can cause production of sludge with poor settling characteristics. This condition, resulting in excessive carryover of activated sludge floc in the clarifier effluent (referred to a sludge bulking), is associated with filamentous growths, either fungi or filamentous bacterial growths.

11-7

POPULATION DYNAMICS

Previous sections described the important characteristics of each group of microorganisms (bacteria, fungi, algae, and protozoans) independently. But in biological-waste-treatment systems, the naturally occurring cultures are mixtures of bacteria growing in mutual association and with other microscopic plants and animals. A general knowledge of the relationships, both cooperative and competitive, between various microbial populations in mixed cultures is essential to understanding biological-treatment processes.

When organic matter is made available to a mixed population of microorganisms, competition arises for this food between the various species. Primary feeders that are most competitive become the dominant microorganisms. Under normal operating conditions, bacteria are the dominant primary feeders in activated-sludge-treating municipal wastewater (Fig. 11-13). Saprobic protozoans, those that feed on dead organic matter (e.g., *Euglena*) are not effective competitors against bacteria.

Species of dominant primary bacteria depend chiefly upon the nature of the organic waste and environmental conditions in aeration tanks. Conditions adverse to bacteria, such as acid pH, low dissolved oxygen, and nutrient shortage, can produce a predominance of filamentous fungi, resulting in sludge bulking. An excessive amount of soluble sugar from industrial wastes can cause a heavy growth of filamentous *Sphaerotilus natans* (Fig. 11-1). These abnormal circumstances are rare in municipal activated-sludge systems treating waste-

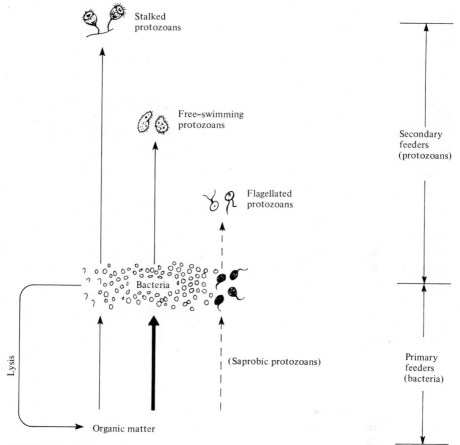

FIGURE 11-13 Schematic diagram of the population dynamics in activated sludge.

water composed chiefly of domestic wastes. However, when bulking in an activated-sludge system does occur, sanitary engineers must be prepared to find the cause and recommend corrective action.

Primary bacteria in an activated-sludge system are maintained in the declining or endogeneous growth phases. Under these conditions, the primary bacteria die and lyse releasing their cell contents to solution. In this process raw organic matter is synthesized and resynthesized by various groups of bacteria.

Holozoic protozoans, which feed on living organic matter, are common in activated sludge. They grow in association with the bacteria in a prey–predator relationship; that is, the bacteria (plants) synthesize the organic matter, and the protozoans (animals) consume the bacteria (Fig. 11-13). For a single reproduction a protozoan consumes thousands of bacteria, with two major beneficial effects of the prey–predator action. Removal of the bacteria stimulates further bacterial growth, resulting in accelerated extraction of organic matter from solution. Second, the flocculation characteristics of activated sludge are improved by reducing the number of free bacteria in solution, and a biological floc with improved settling characteristics results.

There is also competition for food between the secondary feeders. In a solution with high bacterial populations, free-swimming protozoans are dominant,

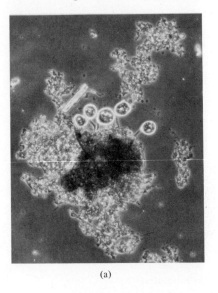

(a)

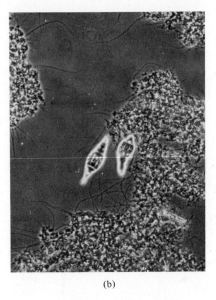

(b)

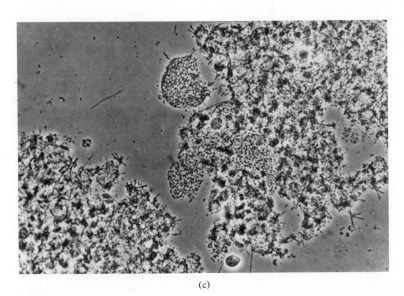

(c)

FIGURE 11-14 Photomicrographs of activated sludge. (a) Activated sludge floc with stalked protozoans (100X). (b) Rotifers in activated sludge (100×). (c) Activated sludge floc showing clusters of bacterial cells. Flagellated protozoan in lower center (400×).

but when food becomes scarce, stalked protozoans increase in numbers. Stalked protozoans do not require as much energy as free-swimming protozoans; therefore, they compete more effectively in a system with low bacterial concentrations. The photomicrographs shown in Fig. 11-14 illustrate the appearance of a healthy activated sludge.

The process of anaerobic digestion is carried out by a wide variety of bacteria, which can be categorized into two main groups, acid-forming bacteria and methane-forming bacteria. (Protozoans do not function in the digester's strict anaerobic environment.) The acid formers are facultative or anaerobic bacteria, which metabolize organic matter, forming organic acids as an end product, along with carbon dioxide and methane (associated with oxidation of fats to organic acids). Acid-splitting methane formers use organic acids as substrate and produce gaseous end products of carbon dioxide and methane. These methane bacteria are strict anaerobes inactivated by the presence of dissolved oxygen and inhibited by the presence of oxidized compounds. The growth medium must contain a reducing agent such as hydrogen sulfide. Acid-splitting methane bacteria are sensitive to pH changes and other environmental conditions.

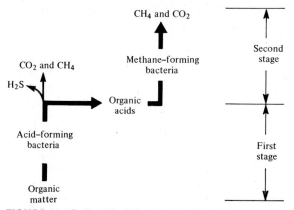

FIGURE 11-15 Simplified diagram of population dynamics in anaerobic digestion.

A simplified diagram (Fig. 11-15) portrays the relationship between the two bacterial stages in digestion of organic matter. Both major groups of bacteria must cooperate to perform the overall gasification of organic matter. The first stage creates food (organic acids) for the second stage where these organic acids are consumed, preventing excess acid accumulation. In addition to producing food for the methane bacteria, acid-formers also reduce the environment to one of strict anaerobiosis by using the oxidized compounds and excreting reducing agents.

Problems in operating anaerobic treatment systems result when an inbalance occurs in the population dynamics. For example, if a sudden excess of organic matter is fed to a digester, acid formers very rapidly process this food developing excess organic acids. The methane formers, whose population had been limited by a previous lower organic acid (food) supply, are unable to

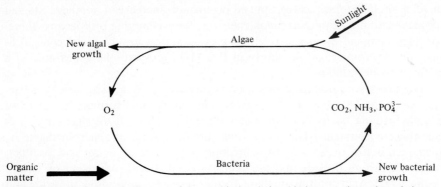

FIGURE 11-16 Schematic diagram of the symbiotic relationship between bacteria and algae.

metabolize the organic acids fast enough to prevent a drop in pH. When the pH drops, the methane bacteria are affected first, further reducing their capacity to breakdown the acids. Under severe or prolonged overloading, the contents of the digester "pickles" in excess acids, and all bacterial activity is inhibited. In addition to organic overloading, the digestion process can be upset by a sudden increase in temperature, a significant shift in the type of substrate, or additions of toxic or inhibiting substances from industrial wastes.

A unique relationship exists between bacteria and algae in small ponds and streams (Fig. 11-16). The bacteria metabolize organic matter, releasing nitrogen and phosphorus nutrients and carbon dioxide. Algae use these compounds, along with energy from sunlight, for synthesis, releasing oxygen into solution. Oxygen released by the algae is taken up by the bacteria, thus closing the cycle. This type of association between organisms is referred to as *symbiosis*, a relationship where two or more species live together for mutual benefits such that the association stimulates more vigorous growth of each species than if growth were separate.

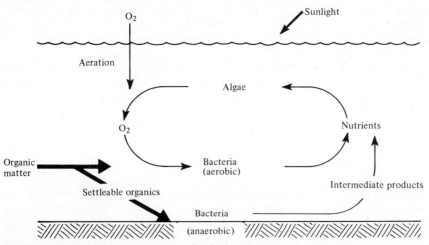

FIGURE 11-17 Operational lagoon as a facultative system.

In a shallow oxidation pond with adequate sunlight and moderate temperatures, the bacteria–algae relationship in Fig. 11-16 depicts the primary biological reactions which take place. In addition, a variety of predators (protozoans, rotifers, and higher animals) feed on the plant growth (algae and bacteria).

At the liquid depth commonly used in stabilization-pond design, bottom waters may become anaerobic while the surface remains aerobic. In terms of general oxygen conditions, these lagoons are commonly referred to as *facultative lagoons* (see Fig. 11-17). During periods when the dissolved oxygen is less than saturation level, the surface water is aerated through wind action. During the winter both bacterial metabolism and algal synthesis are slowed by cold temperatures. The lagoon generally remains aerobic, even under a transparent ice cover. If the sunlight is blocked by a snow cover, the algae cannot produce oxygen, and the lagoon becomes anaerobic. The result is odorous conditions during the spring thaw until the algae become reestablished. This may take from a few days to weeks, depending upon climatic conditions and the amount of organic matter accumulated in the lagoon during the winter.

Wastewater-Treatment Systems

11-8

CHARACTERISTICS OF DOMESTIC WASTEWATER

Wastewater is defined as liquid wastes collected in a sewer system and conveyed to a treatment plant for processing. In most communities storm-runoff water is collected in a separate storm sewer system and conveyed to the nearest watercourse for disposal without treatment. Several large cities have a combined wastewater-collection system where both storm water and sanitary wastes are collected in the same pipe sytem. The dry-weather flow in the combined sewers is collected for treatment, but during storms the wastewater flow in excess of plant capacity is bypassed directly to the receiving watercourse.

Sanitary or domestic wastewater refers to liquid material collected from residences, business buildings, and institutions. The term *industrial* (trade) *wastes* refers to that from manufacturing plants. *Municipal wastewater* is a general term applied to liquid treated in a municipal treatment plant. Municipal wastes from towns frequently contain industrial effluents from dairies, laundries, bakeries, and factories, and in a large city may have wastes from major industries, such as chemical manufacturing, breweries, meat processing, metal processing, or paper mills.

The volume of municipal-wastewater flow varies from 50 to over 200 gal per capita per day (gpcd), depending on sewer uses of the community. A common value for sanitary flow is 100 gpcd (400 liters).

In moderate climates, the summer average daily flow frequently exceeds winter flows by 20–30%. Hourly flow rates range from a minimum to a maximum of 20–250% or more of the average daily rate for small communities, and from 50 to 200% for larger cities. The wastewater flow diagrams in Fig. 11-18 exemplify hourly flow variations for two communities of different sizes.

Per capita contribution of organic matter in domestic wastewater is approximately 0.20 lb (91 g) of suspended solids per day and 0.17 lb (77g) of BOD per day. These values are the population equivalents used to convert total pounds

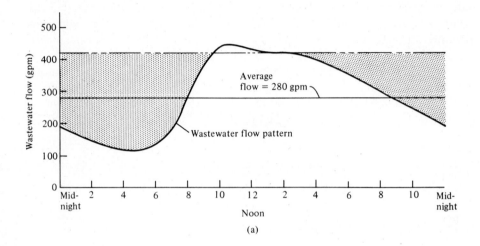

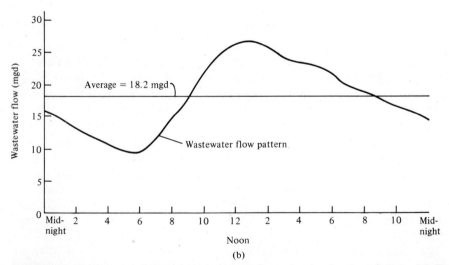

FIGURE 11-18 Diagrams of municipal-wastewater flow, showing hourly variations. (a) Flow diagram from a town with a population of 4500. (b) Flow diagram from a city with a population of 150,000.

of solids or BOD of industrial wastes to equivalent population. Disposal of household kitchen wastes through garbage grinders increases the previous population equivalents approximately 60% for suspended solids, and 30% for BOD.

The data in Table 11-1 represent the approximate composition of domestic wastewater before and after primary sedimentation. BOD and suspended solids (nonfiltrable residue) are the two most important parameters used to define the characteristics of a domestic wastewater. A suspended-solids concentration of 240 mg/l is equivalent to 0.20 lb of suspended solids in 100 gal, and 200 mg/l of BOD is equivalent to 0.167 lb of BOD in 100 gal. Reduction of suspended solids and BOD in primary sedimentation is approximately 50% and 35%, respectively. Approximately 70% of the suspended solids are volatile, defined as those lost upon ignition at 550°C.

TABLE 11-1
APPROXIMATE COMPOSITION OF AN AVERAGE DOMESTIC WASTEWATER (mg/l)

	Before Sedimentation	After Sedimentation	Biologically Treated
Total solids	800	680	530
Total volatile solids	440	340	220
Suspended solids	240	120	30
Volatile suspended solids	180	100	20
BOD	200	130	30
Ammonia nitrogen as N	15	15	20
Total nitrogen as N	35	25	20
Soluble phosphorus as P	7	7	7
Total phosphorus as P	10	8	7

Total solids (residue on evaporation) include organic matter and dissolved salts; concentration of the latter is dependent to a considerable extent on the hardness of the municipal water. Concentration of nitrogen in domestic waste is directly related to the concentration of organic matter (BOD). Approximately 40% of the total nitrogen is in solution as ammonia. If raw wastewater has been retained for a long time in collector sewers, a greater percentage of ammonia nitrogen results from deamination of the proteins and urea in wastewater. Ten mg/l of phosphorus is approximately equivalent to a 3-lb phosphorus contribution per capita per year. More than 2 lb of this is from phosphate builders used in synthetic detergents.

Biodegradable organic matter in wastewater is generally classified in three categories: carbohydrates, proteins, and fats. *Carbohydrates* are hydrates of carbon with the empirical formula $C_nH_{2n}O_n$ or $C_n(H_2O)_n$. The simplest carbohydrate unit is known as a *monosaccharide*, although few monosccharides occur naturally. Glucose is a common monosaccharide in the structure of polysaccharides. *Disaccharides* are composed of two monosaccharide units. Sucrose (table sugar) is glucose plus fructose. Common milk sugar is lactose consisting of glucose plus galactose. *Polysaccharides* are long chains of monosaccharides, such as cellulose, starch, and glycogen. Cellulose is the common polysaccharide in wood, cotton, paper, and plant tissues. Starches are primary nutrient polysaccharides for plant growth and abundant in potatoes, rice, wheat, corn, and other plant forms.

Proteins in simple form are long-chain molecules composed of amino acids connected by peptide bonds, and important in both the structural (e.g., muscle tissue) and the dynamic aspects (e.g., enzymes) of living matter. Twenty-one common amino acids when linked together in long peptide chains form a majority of simple proteins found in nature. A mixture of proteins as a bacterial substrate is an excellent growth medium, since proteins contain all the essential nutrients. On the other hand, pure carbohydrates are unsuitable as a growth medium since they do not contain the nitrogen and phosphorus essential for synthesis.

Lipids, together with carbohydrates and proteins, form the bulk of organic matter of living cells. The term refers to a heterogeneous collection of biochemical substances having the mutual property of being soluble to varying degrees in organic solvents (e.g., ether, ethanol, hexane, and acetone) while being only sparingly soluble in water. Lipids may be grouped according to their shared chemical and physical properties as fats, oils, and waxes. A simple fat when broken down by hydrolytic action yields fatty acids. In sanitary engineering the words *fats* in current usage apparently conveys the meaning of lipids. The term *grease* applies to a wide variety of organic substances in the lipid category that are extracted from aqueous solution or suspension by hexane.

Actually not all biodegradable organic matter can be classed into these three simple groupings. Many natural compounds have structures which are combinations of carbohydrates, proteins and fats, such as lipoproteins and nucleoproteins.

Approximately 20–40% of the organic matter in wastewater appears to be nonbiodegradable. Several organic compounds, although biodegradable in the sense that specific bacteria can break them down, must be considered by sanitary engineers as partially biodegradable because of time limitations in waste-treatment processes. For example, lignin, a polymeric noncarbohydrate material associated with cellulose in wood fiber, is for all practical purposes nonbiodegradable. Cellulose itself is not readily available to the general population of domestic wastewater bacteria. Saturated hydrocarbons are a problem in treatment because of their physical properties and resistance to bacterial action. Alkyl benzene sulfanate (ABS synthetic detergent) is only sparingly biodegradable in wastewater treatment.

●EXAMPLE 11-1

Using values of 0.20 lb of suspended solids and 0.17 lb of BOD/100 gal of domestic wastewater, calculate the suspended solids and BOD concentration in mg/l.

○*Solution*

$$BOD = \frac{0.17\ lb}{100\ gal} \times \frac{1,000,000\ gal}{8.34\ lb} = 204\ mg/l$$

$$suspended\ solids = \frac{0.20\ lb}{100\ gal} \times \frac{1,000,000\ gal}{8.34\ lb} = 240\ mg/l$$

●**EXAMPLE 11-2**

What is the BOD equivalent population for an industry which discharges 0.10 mgd of wastewater with an average BOD of 450 mg/l? What is the hydraulic equivalent population of this watewater?

○*Solution*

$$\frac{BOD\ equivalent}{population} = \frac{0.10\ mg \times 450\ mg/l \times 8.34\ lb/mg}{0.17\ lb/person \qquad mg/l} = 2200$$

$$\frac{hydraulic\ equivalent}{population} = \frac{100,000\ gal/day}{100\ gal/person} = 1000$$

11-9

DEVELOPMENTS IN WASTEWATER TREATMENT

Need for disposal of liquid wastes came with development of the water-carriage waste system (Fig. 11-19). The first disposal system was a cesspool or soak-away, and although satisfactory in porous soils, the walls and bottom plugged with waste solids in fine-grained soils. A septic tank–drain field was developed to overcome this problem. Wastewater is settled in a septic tank and the overflow drains to a tile field, where it percolates into the ground. Septic tanks are currently used for individual households in rural and suburban areas without sewer systems.

A septic tank was used in the early development of municipal treatment systems, but its operation was generally unsatisfactory. The main problem resulted from combining the processes of sedimentation and decomposition of the accumulated sludge in the same tank. Resolution of this problem resulted in development of Imhoff tanks, where the sedimentation zone is separated from the digestion compartment. Solids that settle in the upper portion of the tank pass through a slot into a bottom hopper, are stored and digested in the bottom compartment of the tank, and are periodically withdrawn for disposal. Gases produced from digestion of the solids are vented to the sides and cannot pass up through the slot to interfere with the settling process. Although many Imhoff tanks are operating today, they are generally not considered feasible in new treatment-plant construction.

The final step in development of primary treatment was complete separation of the sedimentation and sludge-disposal processes. Digestion of accumulated waste solids in the bottom of an Imhoff tank was generally poor but significantly improved in a separate tank with a temperature-controlled environment.

For many years after the development of primary treatment for municipal systems, the settled wastewater was disposed of by dilution in the nearest watercourse. Chemical precipitation in the primary-treatment phase was developed and later abandoned, because of the high cost of coagulants and the difficulty of sludge disposal. One of the earliest secondary-treatment processes developed was the intermittent sand filter. In sand filtration the settled wastewater is applied recurrently on sand beds underlain with drain tile to convey the per-

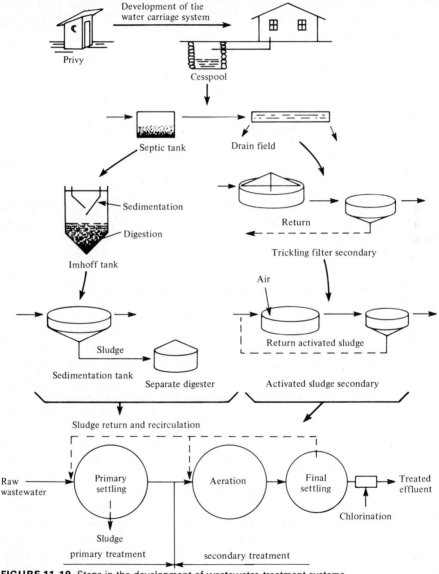

FIGURE 11-19 Steps in the development of wastewater-treatment systems.

colated effluent. The action of the sand filter is a combination of straining and biological action. Because of the large area required, few sand filters were constructed by cities.

The first major breakthrough in the search for a method of secondary treatment occurred in the 1890s, when it was observed that the slow movement of wastewater in films over the surface of gravel media, while in contact with air, resulted in a rapid strength reduction of the wastewater. The descriptive name *trickling filtration* was applied to this process. Perhaps it would be more accurate to say that the process was misnamed trickling filtration, since the treatment process is one of biological oxidation of organic matter by microbial slimes attached to the stone, rather than filtration. (The British use the term *bacteria bed*.) The first municipal installations of trickling filters were constructed in the 1910s.

During laboratory experiments in the 1910s, it was noted that biological solids, which developed in polluted water, had a strong affinity for organic pollutants. These masses of biological solids acted as "activated sludge," significantly increasing the rate of BOD removal when mixed with a polluted water. In the 1920s the initial continuous-flow-treatment plants were constructed using activated sludge to remove the BOD from solution.

A pictorial diagram in Fig. 11-19 summarizes the steps in the development of the wastewater-treatment systems. In succeeding sections of this chapter, the current practice in the design of trickling filters, activated-sludge processes, and anaerobic digestion are presented.

11-10

OBJECTIVES OF DOMESTIC WASTEWATER TREATMENT

Methods of wastewater disposal (ultimate disposition of wastewater) are surface irrigation, seepage and/or evaporation, and dilution in a natural watercourse. Irrigation disposal may be by furrow or spray methods. Control of wastewater in a septic tank drainfield is by seepage. Water loss from stabilization ponds with no outlet (complete retention lagoons) is by evaporation and seepage. No surface-water pollution results from these land distribution systems provided the effluent does not drain to a surface flow. However, groundwater pollution is a potential problem.

Disposal by dilution in rivers, lakes, and the ocean is the most conventional method for eliminating treated municipal wastewater. This is generally the only feasible method and, for some communities on rivers it is the single method which ensures adequate water resources for downstream users during times of drought flows. Wasting of untreated wastewater by dilution is no longer acceptable. Dilution disposal after primary treatment, although still practiced, in most cases does not adequately protect the water resources of a receiving stream. On several major rivers, primary treatment is presently considered adequate by some persons, but the recent water-quality criteria developed by states generally call for implementation of secondary-treatment facilities by all cities and industries. (Section 8-4 discusses effluent standards.)

The objectives of municipal waste treatment are to protect the receiving water body, ensuring adequate quality for beneficial uses such as municipal water supply, recreation, propagation of fish life, irrigation, and aesthetics. To meet these objectives, current practice in wastewater treatment stresses reduction in BOD, suspended solids, and pathogens. Decreased BOD and suspended solids can be achieved by current secondary treatment. Disinfection for further reducton of pathogens is normally accomplished by chlorinating the plant effluent.

If the receiving water is a lake, or a stream that flows into one, nutrient enrichment resulting from the high levels of nitrogen and phosphorus in the wastewater may cause accelerated eutrophication of the lake. Treatment plants designed for standard criteria (BOD reduction, suspended-solids reduction, and disinfection) do not remove large percentages of the inorganic nutrient salts. Secondary-treatment plant effluent normally contains more than one-half the nitrogen and phosphorus found in the raw-wastewater influent. Chapter 13 discusses the elimination of nutrient salts by advanced wastewater-treatment methods.

11-11

WASTEWATER-TREATMENT-PLANT EFFICIENCIES

A report by the Michigan Department of Public Health states that it is "... quite impossible to predict with any high degree of precision what to expect from any single (wastewater treatment) plant on the basis of the performance of another." BOD removal data from the monthly operation reports, submitted by superintendents of municipal plants on the Lower Peninsula, were graphed for primary plants, trickling filter plants, and activated-sludge plants. The results given in Fig. 11-20 show a wide range of BOD reduction.

Activated-sludge plants performed best with half of the plants at or above 91% BOD removal. Similarly, trickling filters were second with half above 83%, and primary plants third with half at or above 40%. Corresponding BOD concentrations in the effluents were 14, 28, and 97 mg/l, respectively.

The report states: "The few extremely low removals indicated are chargeable to unusual circumstances such as the adverse effects of industrial wastes or start-up problems in new plants. The less extreme variations in performance among plants utilizing similar processes are attributable to many factors."

Trickling Filters

Trickling-filter systems are commonly used for secondary treatment of municipal wastewater. Primary effluent is sprayed on a bed of crushed rock, or other

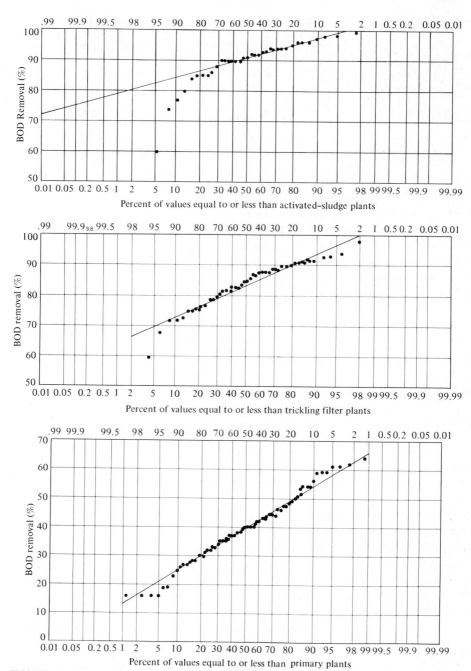

FIGURE 11-20 Graphs showing percentage of BOD removal for municipal wastewater-treatment plants. (Courtesy Michigan Department of Public Health.)

media, coated with biological films. The biological slime layer consists of bacteria, protozoans, and fungi. Sludge worms, filter-fly larvae, rotifers, and other higher animals frequently find the environment suitable for growth. The surface of the bed may support algal growth when temperature and sunlight conditions are optimum. The lower portion of a deep filter frequently supports populations of nitrifying bacteria.

Although classified as an aerobic-treatment device, the microbial film on the filter medium is aerobic to a depth of only 0.1–0.2 mm (Fig. 11-21). The zone next to the medium is anaerobic. As the wastewater flows over the microbial film, the soluble organics are rapidly metabolized and the colloidal organics adsorbed onto the surface.

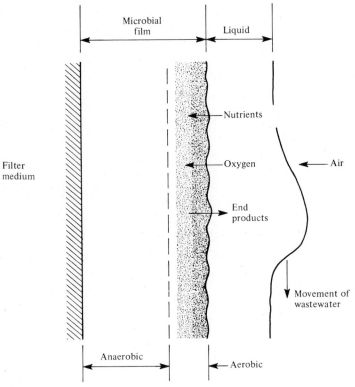

FIGURE 11-21 Schematic diagram showing the form of the biological process in a trickling filter.

Microorganisms near the surface of the bed, where food concentration is high, are in a rapid growth phase, while the lower zone of a bed is in a state of starvation. Overall operation of a trickling filter may be considered in the endogenous growth phase.

Dissolved oxygen extracted from the liquid layer is replenished by reoxygenation from the surrounding air. Undesirable anaerobic conditions can be created in a trickling filter by inhibiting aeration of the bed. Plugging of the air passages with excess microbial growth, as a result of organic overload, can create anaerobiosis and foul odors. Generally trickling filters treating domestic waste operate relatively odor-free; however, certain types (e.g., food-processing waste-waters) have caused foul odors, a condition for which there is frequently no apparent solution.

11-12

DESCRIPTION OF A TRICKLING FILTER

A cutaway view of a rock-filled trickling filter is shown in Fig. 11-22. Major components are the filter media, underdrain system, and rotary distributor. The filter media provide a surface for biological growth and voids for passage of liquid and air. The most common media in existing filters are crushed rock, slag, or field stone, because these materials are durable, insoluble, and resistant to spalling. The preferred range of size for the stone media is 3 to 5 in. in diameter. In recent years several forms of manufactured plastic media have been marketed, their main advantages being light weight, chemical resistance, and high specific surface (ft^2/ft^3 or m^2/m^3) with a large percentage of free space.

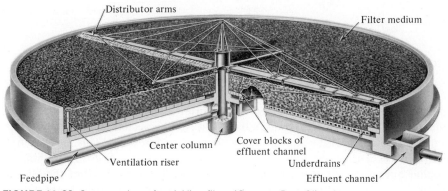

FIGURE 11-22 Cutaway view of a trickling filter. (Courtesy Dorr-Oliver.)

The underdrain system carries away the effluent and permits circulation of air through the bed. The underdrainage system, with provision made for flushing, the effluent channels, and the effluent pipe are designed to permit free passage of air.

A rotary distributor provides a uniform hydraulic load on the filter surface. The most prevalent kind is driven by the reaction of the wastewater flowing out of the distributor nozzles. This generally requires a minimum pressure head of 24 in. measured from the center of the arms.

Normally, trickling filters are preceded by primary treatment. Those receiving unsettled wastes, or others used for pretreatment of settled strong wastes prior to subsequent aeration, are referred to as *roughing filters*. Because of the heavy organic loading on roughing beds, their design must consider the problem of bed plugging. The standard rock-filled filter must be preceded by a sedimentation tank equipped with a scum-collecting device. The high percentage of void volume in manufactured media permits their application and use in roughing filters.

11-13

TRICKLING-FILTER SECONDARY SYSTEMS

A trickling-filter secondary-treatment system includes a final settling tank to remove biological growths that are washed off the filter media. These sloughed solids are commonly disposed of through a drain line from the bottom of the final clarifier to the head end of the plant. This return sludge flow is mixed with the raw wastewater and settled in the primary clarifier.

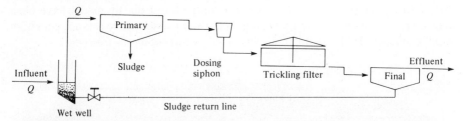

FIGURE 11-23 Profile of a typical low-rate trickling-filter plant.

In a low-rate trickling filter (Fig. 11-23), the wastewater passes through only once and the effluent is then settled prior to disposal. The sludge line is operated, generally once or twice a day, to waste the accumulated settled solids. Low-rate filters operate intermittently, dosing and resting. This operation is required because of the low hydraulic load on the filter. A dosing siphon, or similar alternating flow-control system, must be installed ahead of the filter to provide an adequate flow rate to turn the reaction-type rotary distributor. Otherwise, the distributor arm would stop turning during periods of low flow such as at night, and wastewater would trickle down only under the stalled arm.

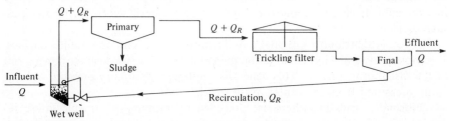

FIGURE 11-24 Profile of a typical high-rate trickling-filter plant.

In a high-rate trickling filter, raw wastewater is diluted with recirculation flow so that it is passed through the filter more than once. A typical high-rate filter design is illustrated in Fig. 11-24. A return line from the final clarifier serves a dual function as a sludge return and a recirculation line. The combined flow ($Q + Q_R$) through the trickling filter is always sufficiently great to turn the distributor, and a dosing siphon is not needed. A high-rate filter is dosed continuously.

Several recirculation patterns used in high-rate filter systems are shown in Fig. 11-25. The recirculation ratio is the ratio of recirculated flow to the quantity

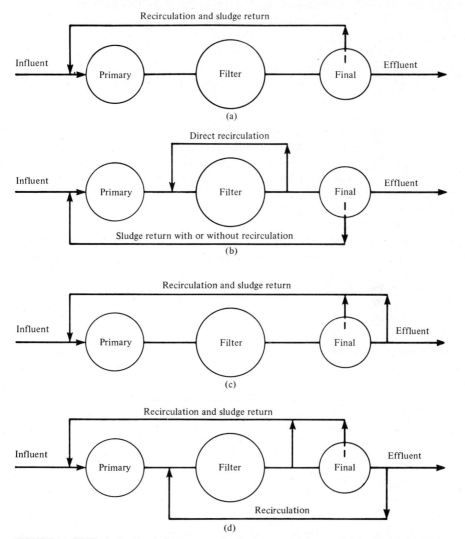

FIGURE 11-25 Typical recirculation patterns for single-stage high-rate trickling filters. (a) Recirculation with sludge return. (b) Direct recirculation around filter. (c) Recirculation of plant effluent. (d) Dual recirculation.

of the raw wastewater. A common range for recirculation ratio values is 0.5 to 3.0. Recirculation is done (1) only during periods of low wastewater flow; (2) at a rate proportional to raw-wastewater flow, (3) at a constant rate at all times, or (4) at two or more constant rates predetermined automatically or by manual control.

For treatment of average-strength wastewater, the gravity-flow recirculation–sludge return system illustrated in Figs. 11-24 and 11-25a is commonly used. The rate of recirculation flow is generally regulated by a valve in the line which is controlled by the liquid level in the wet well. In this manner the rate of return is increased during periods of low raw-wastewater flow and reduced, or stopped, during high-flow periods. This type of recirculation operation is shown graphically on Fig. 11-18a. The shaded area represents recirculated flow. The average raw-wastewater flow is 280 gpm. Flow through the plant $(Q + Q_R)$ is 420 gpm, except from 10 A.M. to 3 P.M., when the raw-wastewater flow exceeds 420 gpm. The recirculation ratio for this illustrated flow-recirculation pattern is 0.5.

Direct recirculation, depicted in Fig. 11-25b, is frequently used in the treatment of stronger wastewaters, where recirculation ratios of 2–3 are desirable. If high rates are used in pattern (a), pattern (c), or pattern (d), the clarifiers must be sized for the increased flow rates created by the greater volume. In other words, if recirculation flow is routed through a clarifier during the peak hourly flows of the raw wastewater, the clarifier must be increased in size to prevent disturbance of the settled solids and resultant loss of removal efficiency in the sedimentation tank. Consider the flow diagram in Fig. 11-18a, and assume that the shaded area is the flow in the sludge-return line shown in Fig. 11-25b.

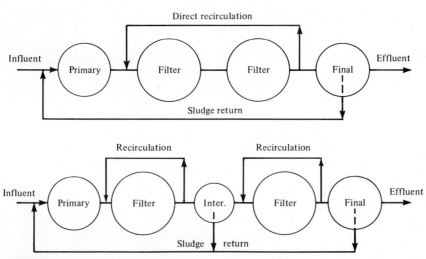

FIGURE 11-26 Typical flow diagrams for two-stage tricking filters.

Then apply a direct recirculation of 420 gpm around the filter using constant-speed pumps. The resultant ratio for the trickling filter is 2.0, 0.5 from indirect recirculation (from the final to the head of the primary) and 1.5 from direct recirculation.

Two-stage trickling-filter secondary systems have two filters in series, with or without an intermediate settling tank. Two typical flow diagrams for two-stage filter installations are sketched in Fig. 11-26. Two-stage filters are used where a high-quality effluent is required, or for treatment of strong wastewater. Although many two-stage filter installations currently exist, the economics of new design have limited their application. Generally, a single-stage activated-sludge secondary can provide at a lower cost the higher degree of treatment obtainable from a two-stage filter secondary.

11-14

BOD AND HYDRAULIC LOADINGS

Standards that have been used in the design and operation of trickling-filter secondaries with stone or slag media are given in Table 11-2. The BOD load on a trickling filter is calculated using the raw BOD in the primary effluent applied to the filter, without regard to the BOD in the recirculated flow.

BOD loadings are expressed in terms of lb of BOD applied per day per unit of volume as 1000 ft^3 or as acre-ft. Current values used in design are 15 lb/1000 ft^3 (240 g/m^3) for low-rate filters, 35 lb/1000 ft^3 (560 g/m^3) for high-rate filters and 55 lb/1000 ft^3 (880 g/m^3) for two-stage filters based on the total volume in both filters.

TABLE 11-2

STANDARDS FOR TRICKLING FILTERS WITH ROCK OR SLAG MEDIA

	Low Rate	High Rate	Two Stage
BOD loading			
lb/1000 ft^3/day[a]	5–25	25–45	45–65
lb/acre-ft/day	200–1100	1100–2000	2000–2800
Hydraulic loading			
mg/acre/day[b]	2–5	10–30	10–30
gpm/ft^2	0.03–0.06	0.16–0.48	0.16–0.48
Operation	Intermittent	Continuous	Continuous
Recirculation ratio	0	0.5–3.0	0.5–3.0
Depth of bed (ft)	5–7	5–7	5–7

[a] 1.0 lb/1000 ft^3/day = 16 g/m^3-day.
[b] 1.0 mg/acre/day = 0.935 m^3/m^2-day.

The hydraulic load is computed using the raw-wastewater flow plus recirculated flow. Hydraulic loadings are expressed in terms of mg applied per acre of surface area per day, or average flow in gpm applied per square foot of surface per day.

In a low-rate trickling filter, the relationship between BOD load and hydraulic load depends upon the strength of the applied wastewater. This relationship exists because there is no recirculation flow and depth of the bed is limited to 5–7 ft. The primary effluent for an average domestic wastewater has a BOD of 130 mg/l (Table 11-1). If this is applied to a 6-ft-deep filter at a BOD load of 15 lb/1000 ft^3/day, the hydraulic load is 3.6 mg/acre/day (see Example 11-3). This hydraulic load is within the range 2–5 mg/acre/day (1.9 to 4.7 m^3/m^2-day) given in Table 11-2.

In the BOD loading range applied to low-rate filters, the biological slimes are reduced significantly by endogenous respiration. Only small quantities of the microbial film are washed off the stone media and collected in the final clarifier. The solids removed by the final clarifier in a low-rate trickling filter secondary account for only about 10% of the overall BOD removal.

As the BOD load increases on a trickling filter, the amount of biological growth developed in the bed increases. At a BOD load of 20–25 lb/1000 ft^3, the voids in the bed tend to fill with biological growth, impeding the passage of liquid and air. This is the upper limit for BOD loadings on low-rate filters. However, the BOD load on a filter can be increased beyond this point if the liquid flow through the bed is increased to continuously flush out the excess growth by hydraulic action. Maintaining an open, well-aerated bed by increasing the hydraulic load is the concept applied in high-rate filtration. The minimum hydraulic loading recommended in high-rate filter operation is 10 mg/acre/day (9.4 m^3/m^2-day). Above 30 mg/acre/day flushing action is excessive, and the contact time of the liquid with filter media becomes too short.

Two-stage filters are designed and operated as high-rate filters in series. The BOD loadings given in Table 11-2 refer to the calculated BOD load based on total volume of both filters. It is customary in the design of two-stage filters to use two filters of equal size.

In the early development of rock-filled trickling filters, low-rate filters had depths greater than 8 ft and high-rate filters were frequently 3–4 ft deep. Experience indicated that two 3-ft-deep filters in series were not significantly different from one 6-ft filter, and that the filter media below 6 ft in a bed did not result in significant BOD removal. Therefore, trickling filters are currently designed within a depth range of 5–7 ft (1.5–2.1 m) for maximum economy in filter construction.

●EXAMPLE 11-3

A settled wastewater with a BOD concentration of 130 mg/l is applied to a 6-ft-deep low-rate trickling filter. Calculate the hydraulic loading on the filter if the BOD loading is 15 lb of BOD/1000 ft^3/day.

○*Solution*

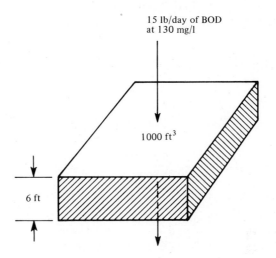

15 lb/day of BOD
at 130 mg/l

1000 ft³

6 ft

$$\text{surface area} = \frac{1000 \text{ ft}^3}{6 \text{ ft}} \frac{\text{acre}}{43,560 \text{ ft}^2} = 0.00383 \text{ acre}$$

$$\text{flow required for 15 lb of BOD/day} = \frac{15 \text{ lb/day}}{130 \text{ mg/l} \times 8.34 \text{ (lb/mg)/(mg/l)}} = 0.0138 \text{ mgd}$$

$$\text{hydraulic load} = \frac{0.0138 \text{ mgd}}{0.00383 \text{ acre}} = 3.6 \text{ mg/acre/day}$$

●**EXAMPLE 11-4**

If a 6-ft-deep high-rate filter is loaded at a rate of 45 lb of BOD per 1000 ft³, calculate the required recirculation ratio to provide a hydraulic load of 20 mg/acre/day. The applied wastewater has a BOD of 130 mg/l.

○*Solution*

From Example 11-3, a 130 mg/l wastewater has a hydraulic load of 3.6 mg/acre/day for a BOD loading of 15 lb of BOD per 1000 ft³. Therefore, at 45 lb of BOD per 1000 ft³, the hydraulic load from the wastewater flow is 10.8 mg/acre/day.

Recirculation ratio required for a hydraulic load of 20 mg/acre/day = $(20 - 10.8)/10.8 = 0.85$.

●**EXAMPLE 11-5**

The raw-wastewater flow from a municipality is 1.5 mgd with an average BOD strength of 180 mg/l. Determine the diameter and depth dimensions for a low-rate trickling filter secondary. Use a design BOD loading of 15 lb per 1000 ft³, a hydraulic loading of 2–4 mg/acre/day and a depth between 5 and 7 ft. Assume overflow rates of 600 gpd/ft² for the primary clarifier and 800 gpd/ft² for the final clarifier.

○*Solution*
   From Fig. 9-20, the BOD removal in the primary tank is 35%. The BOD load on the trickling filter $= 0.65 \times 180$ mg/l $\times 1.5$ mgd $\times 8.34 = 1460$ lb/day. The hydraulic load is 1.5 mgd.

$$\text{volume of filter media required} = 1460 \text{ lb} \times \frac{1000 \text{ ft}^3}{15 \text{ lb}} = 97,400 \text{ ft}^3$$

At 5-ft depth,

$$\text{surface area} = \frac{97,400 \text{ ft}^3}{5 \text{ ft}} = 19,500 \text{ ft}^2 = 0.45 \text{ acre}$$

$$\text{hydraulic load} = \frac{1.5 \text{ mg}}{0.45 \text{ acre}} = 3.4 \text{ mg/acre/day}$$

At 6-ft depth,

   surface area $= 16,200$ sq ft $= 0.37$ acre

hydraulic load $= 4.0$ mg/acre/day

Use two 104-ft-diameter filters with a 5.75-ft depth of rock media.

●**EXAMPLE 11-6**
Calculate the BOD and hydraulic loadings on a single-stage high-rate trickling filter based on the following data:

$$\text{wastewater flow pattern} = \text{as shown in Fig. 11-18a}$$
$$\text{recirculation rate} = \text{as shown in Fig. 11-18a}$$
$$\text{settled wastewater BOD (primary effluent)} = 130 \text{ mg/l}$$
$$\text{diameter of filter} = 12.5 \text{ m}$$
$$\text{depth of media} = 2.1 \text{ m}$$

○*Solution*
raw-wastewater flow $= 280$ gpm $= 1530$ m^3/day
   recirculation flow $= 0.50 \times 1530 = 765$ m^3/day

$$\text{BOD load} = 1530 \frac{\text{m}^3}{\text{day}} \times 130 \frac{\text{mg}}{\text{l}} \times \frac{\text{kg/m}^3}{1000 \text{ mg/l}} = 200 \text{ kg/day}$$

surface area of filter $= 122$ m^2

   volume of media $= 122 \times 2.1 = 256$ m^3

$$\text{BOD loading} = \frac{200,000 \text{ g}}{256 \text{ m}^3} = 781 \text{ g/m}^3$$

$$\text{hydraulic loading} = \frac{1530 \text{ m}^3 + 765 \text{ m}^3}{122 \text{ m}^2} = 18.8 \text{ m}^3/\text{m}^2$$

11-15

TRICKLING-FILTER FORMULAS

The following trickling-filter formulas are based on those given by Eckenfelder.[5]

The mean contact time between the wastewater trickling through a filter and the surface of a filter media is related to filter depth, hydraulic loading, and nature of the filter packing.

$$\frac{t}{D} = \frac{CD}{Q^n} \tag{11-16}$$

where t = mean residence time, minutes
D = filter depth, ft
Q = hydraulic loading, mg/acre/day
C and n = constants related to specific surface and configuration of the media packing

The soluble-BOD removal relationship for trickling filters without recirculation can be represented as follows:

$$\frac{L_e}{L_0} = e^{-KD/Q^n} \tag{11-17}$$

where L_e = effluent BOD, mg/l
L_0 = influent BOD, mg/l
e = napherian base $e = 2.718$
K = reaction-rate constant related to specific surface
D = depth, ft
Q = hydraulic loading, mg/acre/day
n = constant related to specific surface and configuration of packing

When recirculation is used, the influent BOD is diluted by recirculation flow. The BOD applied to the filter becomes

$$L_0 = \frac{L_a + NL_e}{N + 1} \tag{11-18}$$

where L_0 = BOD of the wastewater after mixing with the recirculated flow, mg/l
L_a = influent BOD, mg/l
L_e = effluent BOD, mg/l
N = recirculation ratio, Q_R/Q

The soluble-BOD removal relationship for filters with recirculation is

$$\frac{L_e}{L_a} = \frac{e^{-KD/Q^n}}{(1 + N) - Ne^{-KD/Q^n}} \tag{11-19}$$

TABLE 11-3

SUMMARY OF BOD REMOVAL CHARACTERISTICS OF VARIOUS MEDIA TREATING SETTLED WASTEWATER

Description	Specific Surface (ft²/ft³)	Temperature Range (°C)	Influent BOD Range (mg/l)	Depth (ft)	Hydraulic Loading Range (mg/acre/day)	n^a	K^a at 20°C
1½-in. flexirings	40.0	2–26	65–90	8	12.5–26.9	0.39	0.46
1-in. clinker	61.5	7–17	220–320	6	0.96–1.2	2.56	0.865
2½-in. clinker	37.4	7–17	220–320	6	0.96–1.2	0.84	0.685
1-in. slag	60.0	7–17	220–320	6	0.96–1.2	0.30	0.865
2½-in. slag	33.0	7–17	220–320	6	0.96–1.2	0.75	0.640
1-in. rock	43.3	7–17	220–320	6	0.96–1.2	2.36	0.74
2½-in. rock	27.6	7–17	220–320	6	0.96–1.2	3.80	0.645
1-in. rounded gravel	44.5	7–17	200–320	6	0.96–1.2	3.00	0.625
2½-in. rounded gravel	19.7	7–17	220–320	6	0.96–1.2	5.40	0.57
Surfpac	28.0	24	200	21.6	31–250	0.50	0.395
Surfpac	28.0	24	200	12	62–250	0.45	0.33
2½- and 4-in. rock filter	15.0	24	200	12	31–94	0.49	0.275
1½- and 2½-in. slag	42.0	7–17	112–196	6	5–12.5	1.0	0.87
1- to 3-in. granite	29.0	16–18	186–226	6	2–16	0.4	0.312
¾-in. Raschig rings	75.8	16–18	186–226	6	2–16	0.7	0.55
1-in. Raschig rings	52.2	16–18	186–226	6	2–16	0.63	0.42
1½-in. Raschig rings	35.0	16–18	186–226	6	2–16	0.306	0.28
2¼-in. Raschig rings	22.7	16–18	186–226	6	2–16	0.276	0.25
Straight block	28.2	16–18	186–226	6	2–16	0.345	0.2

^{a} n and K are constants for Eq. 11-17.

Source: S. Balakrishnan, W. W. Eckenfelder, and C. Brown, "Organics Removal by a Selected Trickling Filter Media," *Water and Wastes Eng. 6*, no. 1 (1969); copyright and published by Reuben H. Donnelley Corp.

The previous equations can be applied in trickling-filter design provided constants related to the specific surface and configuration of the packing have been determined for the particular filter medium. The media in the filter must have a uniform, thin biological layer, and the hydraulic load has to be uniformly distributed on the bed. These conditions do not always occur, particularly in rock filters, which may develop unequal biological growths resulting in short-circuiting of liquid flow through the bed.

A summary of BOD removal characteristics of various media treating settled wastewater is given in Table 11-3.

●EXAMPLE 11-7

The purpose of this example is to illustrate how a pilot-plant study was used to determine the constants in Eq. 11-17 for a specific filter medium. Investigations of this type are necessary to provide information useful in the development of new design criteria. Data presented were taken from a pilot plant study reported by Balakrishnan, Eckenfelder and Brown[6] [*Water and Wastes Eng.* 6, no. 1 (1969): A-22 to A-25; copyrighted and published by Reuben H. Donnelley Corp.].

The trickling-filter model was a 20-in.-diameter, 9-ft-deep filter with an air sparger to provide a uniform distribution of air from the bottom and distribution plates on top for uniform hydraulic loading. The filter was packed to a depth of 8 ft using 1.5-in. polypropylene Flexiring (Fig. 11-27) media with a specific surface of 40 ft²/ft³ and 96% free space.

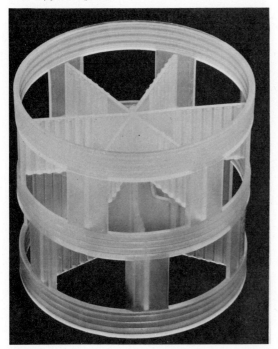

FIGURE 11-27 Flexiring filter medium. (Courtesy Koch Engineering Co., Inc.)

Liquid residence in the filter was found by measuring the time required for concentrated doses of salt solution to flush through the filter. Residence times, with and without the presence of biological slime growth, were related to hydraulic loading as graphed in Fig. 11-28.

After the filter was acclimated to a feed of settled domestic wastewater, samples were collected at various depths in the filter for laboratory analysis. The wastewater samples were settled for 30 minutes and filtered through Whatman No. 42 filter paper prior to BOD testing. Data on BOD removal with respect to depth at various hydraulic loads are shown in Fig. 11-29a.

Slopes of the BOD remaining versus depth curves (Fig. 11-29a) were plotted against their respective hydraulic flow rates to get the constant n (Fig. 11-29b). The constant K was then determined, as illustrated in Fig. 11-29b, by plotting D/Q^n versus BOD remaining. Note that several of the scales on these plots are logarithmic.

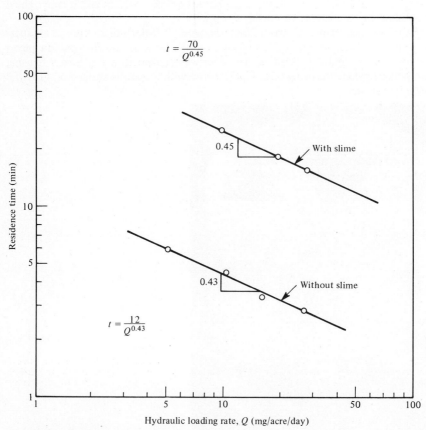

$$t = \frac{70}{Q^{0.45}}$$

With slime

$$t = \frac{12}{Q^{0.43}}$$

Without slime

Residence time (min)

Hydraulic loading rate, Q (mg/acre/day)

FIGURE 11-28 Relationship between hydraulic loading and residence time for Example 11-7. [Source: S. Balakrishnan, W. W. Eckenfelder, and C. Brown, "Organics Removal by a Selected Trickling Filter Media," *Water and Wastes Eng.* 6, no. 1 (1969): A-22–A-25.]

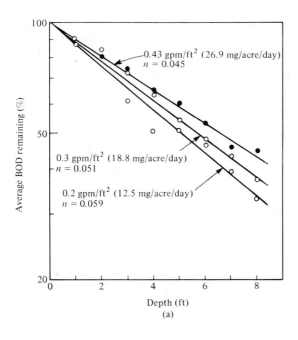

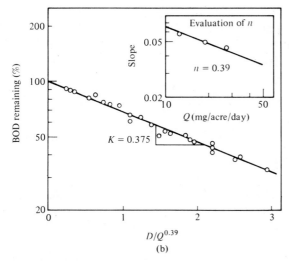

FIGURE 11-29 Plotted data for Example 11-7. (a) Relation between filter depth and percent BOD remaining at various hydraulic loads. (b) Diagrams for the determination of constants *n* and *K*. [Source: S. Balakrishnan, W. W. Eckenfelder, and C. Brown, "Organics Removal by a Selected Trickling Filter Media," *Water and Wastes Eng.* 6, no. 1 (1969): A-22–A-25.]

The average operating temperature of the pilot filter was 14°C. The K rate of 0.375, from Fig. 11-29b, was corrected to a K rate of 0.46 at 20°C using the relationship

$$K_T = K_{20}(1.035)^{T-20} \tag{11-20}$$

Substituting values of $K = 0.46$ and $n = 0.39$ in Eq. 11-17, the BOD removal relationship for the specific filter media tested, under the described study conditions, is

$$\frac{L_e}{L_0} = e^{-0.46D/Q^{0.39}}$$

11-16

TRICKLING-FILTER DESIGN

General practice in trickling-filter design has used empirical relationships to find the required filter volume for a desired degree of wastewater treatment. Several of these associations have been developed from operational data collected at existing treatment plants. One of the first evolved was the National Research Council (NRC) formula, based on data collected from filter plants at military installations in the United States in the early 1940s.

The NRC formula for a single-stage trickling filter is

$$E = \frac{100}{1 + 0.0561(w/VF)^{0.5}} \tag{11-21}$$

where E = percentage BOD removal at 20°C
$\quad w$ = BOD load applied, lb
$\quad V$ = volume of filter media, ft$^3 \times 10^{-3}$
$\quad F$ = recirculation factor

The recirculation factor is calculated from the formula

$$F = \frac{1 + R}{(1 + 0.1R)^2} \tag{11-22}$$

where R is the recirculation ratio (ratio of recirculation flow to raw-wastewater flow).

The NRC formula for the second stage of a two-stage filter is

$$E_2 = \frac{100}{1 + [0.0561/(1 - E_1)](w_2/VF)^{0.5}} \tag{11-23}$$

where E_2 = percentage BOD removal of the second stage at 20°C
$\quad E_1$ = fraction of BOD removed in the first stage
$\quad w_2$ = BOD load applied to the second stage, lb

The solid-line curves in Fig. 11-30 are plots of the NRC formula for a single-stage filter (Eq. 11-21) at various BOD loadings and recirculation ratios. The short dashed line extending from 75% efficiency at 10 lb/1000 ft^3 and 67% efficiency at 50 lb/1000 ft^3 is taken from the 1971 *Recommended Standards for*

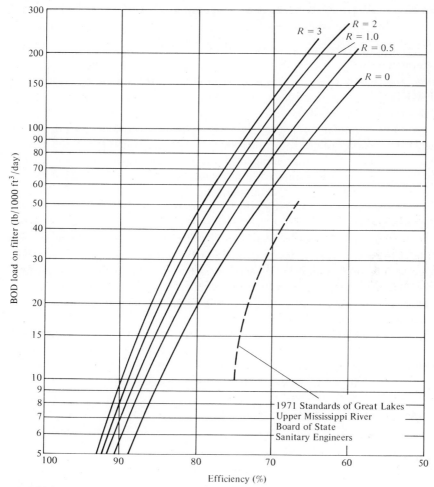

FIGURE 11-30 Curves for single-stage trickling-filter efficiencies based on the National Research Council formula (Eq. 11-21). (1.0 kg/m³-day = 62.4 lb/1000 ft³/day.)

Sewage Works, Great Lakes—Upper Mississippi River Board of State Sanitary Engineers (GLUMRB).[8] The *Standards* state, "Expected reduction of BOD of the primary settling tank effluent by a single-stage filter packed with crushed rock, slag or similar material and subsequent settling tank shall be determined from" the dashed line in Fig. 11-30. "In developing this curve, loading due to recirculated flow has not been considered. Expected performance of filters packed with manufactured media shall be determined from pilot plant and full scale experience."

The effect of waste-water temperature on rock-filled trickling filter efficiency is expressed by Howland[9] as follows:

$$E = E_{20} 1.035^{(T-20)} \tag{11-24}$$

where E = BOD removal efficiency at temperature T, °C

E_{20} = BOD removal efficiency at 20°C

The efficiencies recommended by the GLUMRB *Standards* are 5–10% less than the NRC values without recirculation. These lower figures reflect, in general, the actual operating conditions in the 10-state area represented by GLUMRB. The average efficiency of trickling filter plants in Michigan, based on the data in Fig. 11-20 is 83%. Assuming a primary removal of 35%, the average filter secondary efficiency is 74%, which is the average GLUMRB value in the loading range 15–35 lb/1000 ft^3. Although the GLUMRB recommended efficiencies are not referenced to any particular operating temperature, comparison of the NRC values without recirculation to the GLUMRB results, using Eq. 11-24, reflects a 2–3°C temperature drop. Filters in the 10-state area frequently operate at efficiencies about 5% below the yearly average during cold winter months.

BOD removal efficiencies computed by the NRC formulas include final settling of the filter effluent. In the empirical development of these formulas, the field procedure used in collecting the data sampled the filter influent and final clarifier effluent. Therefore, in evaluating the efficiency of a trickling-filter secondary, the overflow rate and detention time of the final clarifier should be examined for adequacy of design.

For a two-stage filter secondary without an intermediate settling tank (Fig. 11-26), the NRC formulas cannot be used to determine the efficiency of the first stage. In this case it is common to assume that the first-stage efficiency is 50% and find that of the second stage from Eq. 11-23.

●**EXAMPLE 11-8**

Calculate the 17°C BOD removal efficiency for the low-rate filter plant described in Example 11-5. Pertinent data are: primary efficiency = 35%, filter BOD loading = 15 lb/1000 ft^3/day, and $R = 0$.

○*Solution*

Using Eq. 11-21, we obtain

$$E_{20} = \frac{100}{1 + 0.0561(15/1)^{0.5}} = 82.2\%$$

Or using the $R = 0$ curve in Fig. 11-30 at 15 lb/1000 ft^3 yields

$E = 82\%$

From Eq. 11-24.

$E_{17} = 82.2 \times 1.035^{(17-20)} = 74\%$

(*Note:* This is the recommended efficiency of GLUMRB shown on Fig. 11-30.) Plant efficiency

$$E = 100 - 100[(1 - 0.35)(1 - 0.74)] = 83\%$$

●**EXAMPLE 11-9**
Calculate the BOD removal efficiency for the single-stage high-rate trickling filter described in Example 11-6. Pertinent data are: settled wastewater BOD loading = 49.7 lb/1000 ft^3/day of BOD, and recirculation ratio = 0.5.

○**Solution**
By Eqs. 11-22 and 11-21,

$$F = \frac{1 + 0.5}{(1 + 0.1 \times 0.5)^2} = 1.36$$

$$E = \frac{100}{1 + 0.0561(49.7/1.36)^{0.5}} = 74.7\%$$

Or from the $R = 0.5$ curve in Fig. 11-30 at 49.7 lb/1000 ft^3/day of BOD,

$$E = 74\%$$

●**EXAMPLE 11-10**
A single-stage high-rate trickling-filter plant treating wastewater with 220 mg/l of BOD cannot meet the effluent standard of less than 30 mg/l of BOD. Primary settling removes 35% of the raw-waste BOD and the 7-ft-deep secondary filters have an efficiency of 75% under loadings of 45 lb/1000 ft^3/day of BOD and 11.5 mg/acre/day without recirculation. This provides an average effluent BOD during the summer of 35 mg/l, while the winter effluent quality is much poorer. Furthermore, the stone medium plugs periodically, causing ponding and foul odors. One proposal is to remove the rock, replace it with Flexiring medium, and cover the filters with fiberglass domes to confine odors and prevent the wastewater from cooling below 16°C in winter. Calculate the effluent BOD at 16°C for the filters packed with the Flexiring medium described in Example 11-7.

○**Solution**
Settled wastewater BOD = 0.65 × 220 = 143 mg/l. The K value for Flexiring medium from Table 11-3 is 0.46 at 20°C. Correcting to 16°C by Eq. 11-20, the K rate is

$$K_{16} = 0.46^{(1.035)^{16-20}} = 0.40$$

Substituting into Eq. 11-17 with $K = 0.40$, $D = 7$ ft, $n = 0.39$, and $Q = 11.5$ mg/acre/day,

$$\frac{L_e}{L_0} = e^{-0.40 \times 7/11.5^{0.39}} = 0.34$$

Therefore, the effluent BOD = 0.340 × 143 = 49 mg/l, which is greater than the allowable 30 mg/l. Consider modifying the plant to provide a recirculation ratio of 1.0 by direct return flow, as illustrated in Fig. 11-25b. Substituting into Eq. 11-19 for an $N = 1.0$, the fraction of BOD remaining is

$$\frac{L_e}{L_0} \simeq \frac{0.34}{(1 + 1.0) - 1.0 \times 0.34} = 0.20$$

Now the effluent BOD $= 0.20 \times 143 = 29$ mg/l, which meets the effluent maximum of 30 mg/l during cold-weather operation.

●EXAMPLE 11-11
Determine the volume of filter media required for a single-stage high-rate secondary filter based on the following:

$$
\begin{aligned}
\text{plant efficiency} &= 85\% \\
\text{recirculation ratio} &= 1.5 \\
\text{raw-wastewater flow} &= 0.5 \text{ mgd} \\
\text{raw-wastewater BOD} &= 280 \text{ mg/l} \\
\text{primary overflow rate} &= 500 \text{ gpd/ft}^2 \\
\text{secondary overflow rate} &= 700 \text{ gpd/ft}^2
\end{aligned}
$$

○**Solution**
From Fig. 9-20, the BOD removal in the primary tank is 36% for an overflow rate of 500 gpd/ft². The required secondary efficiency is

$$85 = 100 - 100[(1 - 0.36)(1 - E/100)]$$
$$E = 76.6\%$$
BOD load $= 0.64 \times 0.5 \times 280 \times 8.34 = 747$ lb/day

Using Eq. 11-22,

$$F = \frac{1 + 1.5}{(1 + 0.1 \times 1.5)^2} = 1.89$$

Substituting the above values in Eq. 11-21,

$$76.6 = \frac{100}{1 + 0.0561\left(\dfrac{747}{V \times 1.89}\right)^{0.5}}$$

$$V = 13,300 \text{ ft}^3$$

Check solution:

$$\text{BOD loading} = \frac{747}{13.3} = 56 \text{ lb/1000 ft}^3$$

From Fig. 11-30, for 56 lb/1000 ft³ and $R = 1.5$,

$$E = 76\% \quad \text{(OK)}$$

$$\text{hydraulic loading assuming 6-ft depth} = \frac{0.5 + 1.5 \times 0.5}{13,300/(6 \times 43,560)} = 24.6 \text{ mg/acre/day}$$

$$\text{(OK)}$$

The final clarifier overflow rate at 700 gpd/ft² is less than maximum 800 gpd/ft² allowed by GLUMRB. (OK)

●EXAMPLE 11-12

The design flow for a new two-stage trickling-filter plant is 1.2 mgd with an average BOD concentration of 450 mg/l. Determine the dimensions of the sedimentation tanks and trickling filters (surface areas and depths) for the flow scheme shown in Fig. 11-31. Calculate the volume of filter media based on a loading of 50 lb of BOD/1000 ft³, and divide the resulting volume equally between the primary and secondary filters. Estimate the BOD concentration in the plant effluent.

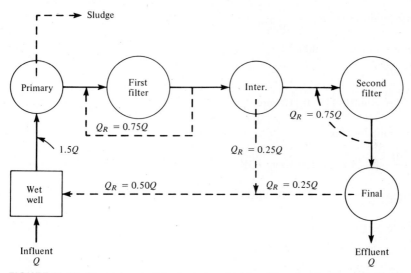

FIGURE 11-31 Flow scheme of the two-stage trickling-filter plant for Example 11-12.

○*Solution*

Primary Tank

Criteria: (1) 500-gpd/ft² overflow rate based on raw Q, or 750-gpd/ft² based on Q plus recirculation flow; (2) minimum depth of 7 ft. However, if accumulated sludge is to be retained in the bottom of the tank, increase the depth to accommodate the necessary sludge storage volume.

$$\text{area required} = \frac{1,200,000}{500} = 2400 \text{ ft}^2$$

or

$$= \frac{1.5 \times 1,200,000}{750} = 2400 \text{ ft}^2 \quad \text{(Use)}$$

Estimate daily sludge accumulation at 4% solids, assuming a sludge-solids accumulation equal to 90% of the BOD load:

$$\text{volume} = \frac{0.9 \times 450 \times 1.2 \times 8.34}{0.04 \times 62.4} = 1620 \text{ ft}^3$$

$$\text{depth of sludge} = \frac{1620}{2400} = 0.7 \text{ ft}$$

Provide a side-wall depth of 8 ft plus freeboard.

primary BOD removal $= 35\%$

Trickling Filters
Criteria: (1) BOD loading $= 50$ lb/1000 ft^3/day; (2) hydraulic loading $= 10$–30 mg/acre/day.

$$\text{volume required} = \frac{0.65 \times 450 \times 1.2 \times 8.34}{0.050} = 58{,}500 \text{ ft}^3$$

volume of each filter $= 29{,}300$ ft^3

Try 6-ft depth,

$$\text{area} = \frac{29{,}300}{6.0} = 4880 \text{ ft}^2 = 0.112 \text{ acre}$$

Check hydraulic loading,

$$\frac{(1.5 + 0.75)1.2}{0.112} = 24.1 \text{ mg/acre/day} \quad \text{(OK)}$$

Use 6-ft-deep filters with a 4880 ft^2 area.

Intermediate Settling Tank
Criteria: (1) 1000-gpd/ft^2 overflow rate; (2) minimum depth of 7 ft.

$$\text{area required} = \frac{1.25 \times 1{,}200{,}000}{1000} = 1500 \text{ ft}^2$$

Use side wall depth of 7 ft.

Final Settling Tank
Criteria: (1) 800-gpd/ft^2 overflow rate; (2) minimum depth of 7 ft.

$$\text{area required} = \frac{1{,}200{,}000}{800} = 1500 \text{ ft}^2$$

Use a side-wall depth of 7 ft.
Calculation of BOD removal efficiency:

primary tank $= 35\%$

First-stage filter:

$$\text{BOD loading} = \frac{0.65 \times 450 \times 1.2 \times 8.34}{29.3} = 100 \text{ lb/1000 ft}^3/\text{day}$$

$$R = \frac{0.50Q + 0.75Q}{Q} = 1.25$$

$$E = 70\%$$

Second-stage filter:

BOD loading $= 0.30 \times 100 = 30$ lb/1000 ft^3

$$R = \frac{0.25Q + 0.75Q}{Q} = 1.0$$

$$F = \frac{1 + 1.0}{(1 + 0.1 \times 1.0)^2} = 1.65$$

By Eq. 11-23,

$$E_2 = \frac{100}{\{1 + [0.0561/(1 - 0.70)]\}(30/1.65)^{0.5}} = 57\%$$

Plant efficiency

$E = 100 - 100[(1 - 0.35)(1 - 0.70)(1 - 0.57)] = 92\%$

Estimated effluent BOD $= 0.08 \times 450 = 38$ mg/l.

11-17

OPERATIONAL PROBLEMS OF TRICKLING FILTERS

Major problems in trickling-filter operation are associated with organic loading, cold-weather operation, emission of odors, and control of filter flies.

Severe organic overloading of a trickling filter can plug the bed and pond wastewater on the surface. The safe organic load for a filter is directly related to size and condition of the stone media. Many early filters were constructed with 1.5- to 2.5-in. crushed stone, providing small void spaces. Current practice specifies 3- to 4.5-in. crushed stone. Plugging frequently occurs in filters that were originally designed as low-rate filters but now receive BOD loads comparable to high-rate loadings without comparable hydraulic loadings to wash out the excess biological growth. One temporary measure to remedy plugging is to periodically dose the filter with chlorine. For an overloaded low-rate trickling filter, a permanent solution may be to convert the operation to high rate by the addition of a recirculation system. In some cases the only answer is to replace the filter media.

During severe cold weather, open trickling filters may form ice on the surface of the bed. Reduction of the recirculation flow, adjustment of the nozzles, or construction of windbreaks are methods used to reduce icing problems.

Emission of foul odors is normally associated with organic overloading, particularly in the spring and fall of the year when air temperatures reduce natural air circulation through the bed.

Filter flies, *Psychoda*, are a nuisance problem around filters during warm weather. They breed under the filter stones on the surface and along the inside of the retaining walls. Periodic spraying of the bed surface and walls with an insecticide is a common method used for controlling *Psychoda*.

Biological Towers

Primary reasons for popularity of rock-filled trickling filters are their simplicity, low operating cost, and production of a waste sludge that is easy to handle. However, when treating high-strength municipal wastewaters, single-stage filters cannot provide sufficient organic removal to meet the effluent standard of 30 mg/l of BOD and, in the case of two-stage systems, the first unit under high loading may have problems of bed plugging and emission of foul odors. Several forms of manufactured media have been developed to overcome these disadvantages of crushed rock by providing a high specific surface (ft^2/ft^3) with a corresponding large percentage of void volume. This permits substantial biological growth without inhibiting passage of air through the bed. A uniform medium also allows even loading distribution, and lightweight packing permits construction of deeper beds with the ability to handle high-strength wastes. Biological towers, generally 14–22 ft in depth, allow a longer wastewater residence time, permitting the biological reaction to proceed further into the endogenous phase.

Biological towers for treating municipal wastewater are generally single-stage units following primary sedimentation, although in special cases they may be used as roughing filters preceding primary settling, or installed for two-stage operation when handling strong industrial wastes discharged to the municipal system. Towers can be circular, using a rotary distributor, or rectangular, with fixed nozzles to spread the wastewater. Direct recirculation is employed to maintain the desired flow through the tower with a recirculation ratio in the range 1:3. Sometimes a portion of the recirculation flow is drawn from the bottom of the final clarifier to develop a microbial floc in the wastewater circulating through the tower. Waste sludge accumulated in the final clarifier is returned to the plant influent for settling in the primary. Tower loadings are usually in the range 50–150 lb/1000 ft^3/day of BOD with hydraulic applications of 1–5 gpm/ft^2. The exact design loading for a biological tower depends on strength, biodegradability, and temperature of the wastewater; type and depth of synthetic medium employed; and wastewater recirculation pattern and ratio.

11-18

BIOLOGICAL TOWERS WITH PLASTIC MEDIUM

Polyvinyl chloride packing is generally constructed of alternate flat and corrugated sheets bonded together in rectangular modules of 16–18 ft^3. The medium illustrated in Fig. 11-32 can be shipped with the plastic sheets nested together and then expanded to about 5 times the shipping volume for installation. The specific surface is approximately 28 ft^2/ft^3 with a void volume of 95%. The weight of packing in a tower, including both the medium and biological growth,

FIGURE 11-32 Circular biological tower packed with plastic Surfpac medium and equipped with a rotary distributor. (Courtesy Eimco BSP Division, Envirotech Corporation.)

is estimated at 15 lb/ft³. Because of this light weight, most plastic media marketed have sufficient structural strength to be self-supporting in packed depths to at least 20 ft; greater depths require intermediate support decks.

BOD removal in biological towers with plastic packing, treating domestic wastewater, can be estimated from the relationship

$$\frac{L_e}{L_0} = e^{-0.088D/Q^{0.5}} \tag{11-25}$$

where L_e = effluent BOD, mg/l

$\quad L_0$ = influent BOD, mg/l

$\quad L_e/L_0$ = fraction of BOD remaining

$\quad D$ = depth of tower, ft

$\quad Q$ = hydraulic loading (without recirculation flow), gpm/ft²

$\quad$ 0.088 = waste treatability factor k for domestic wastewater on plastic Surfpac medium after Germain[10]

$\quad$ 0.5 = coefficient n for plastic medium

This equation has the same form as Eq. 11-17; however, the hydraulic loading Q is expressed in gallons per minute per square foot rather than million gallons per day per acre. The value of 0.5 for coefficient n has been established for plastic-medium filters by several studies. The wastewater-treatability factor k, also referred to as the reaction-rate constant, was established by Germain[10] at 0.088 for open-type plastic packing treating settled domestic wastewater. If the hydraulic loading is in mg/acre/day, as in Eq. 11-17, the reaction rate K equals 0.70 rather than 0.088; the value of n is 0.5 in both formulas. According to Germain, treatability factors may be as low as 0.01 or as high as 0.10 for industrial wastes. Pilot-plant or laboratory treatability studies are recommended to determine the k value for wastes where previous experience is not documented.

The depth of a biological tower has significant affect on the volume of medium required. Filters in one or more stages with a total depth of 20 ft or more can produce efficient BOD removal at high organic and hydraulic loadings. Based on Eq. 11-25, the volume of packing required for a 20-ft-deep tower is one-half the amount needed for a 10-ft-deep bed at a BOD removal efficiency of 80%. To facilitate design, graphs of BOD loading versus efficiency can be

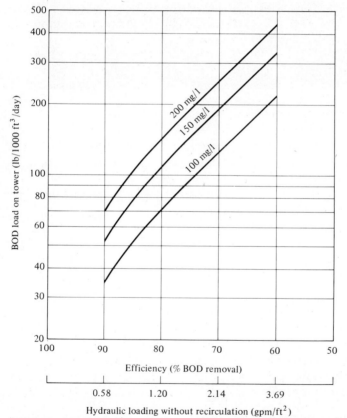

FIGURE 11-33 Efficiency curves for 20-ft-deep, plastic-medium, biological towers treating domestic wastewater based on Eq. 11-25. BOD removal efficiency includes final clarification of the wastewater, and the hydraulic loadings given on the abscissa do not include any recirculation flow. (1.0 kg/m³-day = 62.4 lb/1000 ft³/day and 1.0 m³/m²-day = 0.017 gpm/ft².)

developed for various depths of medium. Figure 11-33 is a chart for a filter depth of 20 ft based on a treatability factor for settled domestic wastewater. BOD removal efficiency includes final clarification, and hydraulic loadings along the abscissa were calculated from wastewater flows without considering recirculation.

Hydraulic loading on a biological tower appears to have no major influence on BOD removal between the normal operating limits of 1 and 5 gpm/ft². Commonly, a recirculation ratio of 1 is used in treating municipal wastewater to maintain a relatively constant liquid dosage. Direct recirculation of the tower effluent during periods of low raw-wastewater flow is necessary to maintain uniform hydraulic loading with both rotary and fixed distributors. From Figure 11-33 at an efficiency of 80%, the average daily settled wastewater flow would be 1.2 gpm/ft², which increases to 2.4 gpm/ft² with a direct 1:1 recirculation.

●**EXAMPLE 11-13**
A circular biological tower with plastic medium is being sized for secondary treatment of 5.0 mgd of domestic wastewater. The average BOD concentration in the raw waste is 230 mg/l, and the effluent BOD must not exceed 30 mg/l. (a) Calculate the design BOD loading, hydraulic application assuming a recirculation ratio of 1.0, and volume of packing required. (b) Perform the same computations for a depth of 10 ft.

○**Solution**
Assuming 35% removal in the primary clarifier, the wastewater applied to the biological tower $= 0.65 \times 230$ mg/l $= 150$ mg/l.

$$\text{BOD removal required in tower} = \left(\frac{150 - 30}{150}\right)100 = 80\%$$

a Substituting in Eq. 11-25 with $D = 20$ ft and solving for Q,

$$\frac{L_e}{L_0} = 0.20 = e^{-0.088 \times 20/Q^{0.5}}$$

$$Q = 1.20 \text{ gpm/ft}^2$$

$$\text{BOD loading} = \frac{1.20 \times 1440 \times 150 \times 8.34}{1,000,000 \times 0.020} = 108 \text{ lb/1000 ft}^3/\text{day}$$

hydraulic loading at a recirculation rate of $1.0 = 2 \times 1.2 = 2.4$ gpm/ft²

These same loading values can be obtained from Fig. 11-33 by entering at 80% efficiency. The hydraulic loading without recirculation is given on the abscissa as 1.20 gpm/ft², and the BOD loading of 108 lb/1000 ft³/day is read on the ordinate from the 150-mg/l BOD concentration line.

$$\text{volume of packing required} = \frac{5.0 \times 8.34 \times 150}{0.108} = 57,900 \text{ ft}^3$$

b   For 80% efficiency and $D = 10$ ft.

$$\frac{L_e}{L_0} = 0.20 = e^{-0.088 \times 10/Q^{0.5}}$$

$$Q = 0.30 \text{ gpm/ft}^2$$

$$\text{BOD loading} = \frac{0.30 \times 1440 \times 150 \times 8.34}{1,000,000 \times 0.010} = 54 \text{ lb/1000 ft}^3/\text{day}$$

hydraulic loading at R of $2.0 = 2 \times 0.30 = 0.60 \text{ gpm/ft}^2$

Therefore, based on Eq. 11-25 for a BOD removal of 80%, twice as much tower packing is required for a 10-ft-deep bed as is needed for a 20-ft depth of medium.

11-19

BIOLOGICAL TOWERS WITH REDWOOD MEDIUM

Redwood packing consists of individual racks made of rough-sawn wooden lath fixed to supporting rails. For installation, the racks are stacked with slats and rails aligned vertically to permit free passage of liquid and air (Fig. 11-34). Alternate layers are interlocked to provide structural stability, and the top of the bed is covered with heavier decking to allow a walking surface. The common depth of medium in a redwood tower is 14 ft. Packing is supported on 4-in. by 6-in. redwood joists laid on piers cast in a concrete floor. Floor spaces between the piers are sloped to serve as effluent channels. Since the medium is self-standing, the outer walls of a tower may be constructed of corrugated metal, plastic, wood, or concrete block, to provide the required degree of thermal insulation and permit architectural design to enhance the appearance of a treatment plant.

Piping and fixed distributors are suspended from steel beams placed approximately 3.5 ft above the top of the packing. The fixed distributors may be either nonclogging spray nozzles or of the type illustrated in Fig. 11-34b. The latter has a movable splash plate between the riser pipe and a fixed upper plate. The rising splash disc has an orifice in the center so that at low flow the plate rests on the riser pipe and wastewater passing through the orifice is spread outward by the fixed upper plate. During high flow, water discharging from the pipe holds the rising splash plate against the top. This design allows uniform wastewater distribution over a wide range of flow rates.

Figure 11-35 is a schematic flow diagram illustrating the process application of redwood-medium towers. One operational mode utilizes the tower as a high-rate trickling filter. Direct recirculation of tower effluent is used to maintain adequate hydraulic flow and enhance BOD removal. Gravity flow from the final clarifier returns settled solids to the wet well of the plant for removal in the primary; this return flow can be used for indirect recirculation during periods of low raw-wastewater influent. For strong municipal waste, two towers may be set in series with or without an intermediate clarifier.

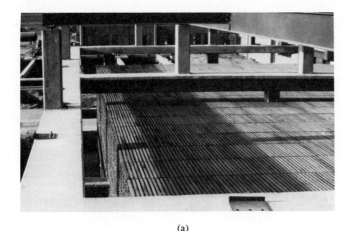

(a)

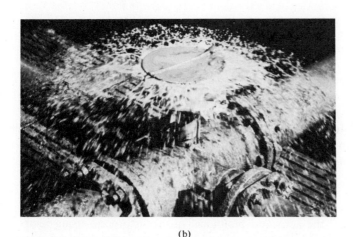

(b)

FIGURE 11-34 Rectangular biological tower packed with redwood Bio-Media and equipped with fixed distributors. (Courtesy Neptune Microfloc Inc., subsidiary of Neptune International Corp.) (a) Redwood packing in a rectangular biological tower under construction. Racks (4 ft by 4 ft) of horizontal wooden slats ($\frac{3}{8}$ in. by 1$\frac{1}{2}$ in.) stapled to rails (1$\frac{3}{4}$ in. deep) are stacked to form a tower medium 14 ft in depth. (b) Fixed wastewater distributor consisting of a riser pipe with self-adjusting splash plate so that the waste is spread uniformly during both high and low flows.

A biological tower can also be operated as an activated biological filtration system by returning settled solids from the final clarifier to the tower influent. This recirculation of settled microbial solids results in a buildup of activated sludge in the flow passing through the tower. Thus the system responds as both a filter, with fixed biological growth, and a mechanical aeration system with suspended microbial floc. The recommended recirculations for this process are

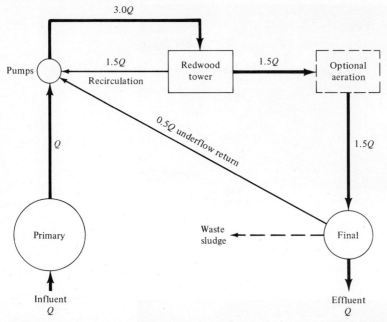

FIGURE 11-35 Flow diagram for a biological-tower installation that can be operated as either a high-rate trickling filter or as an activated biological filtration system with optional supplemental aeration. The later process mode includes return of underflow from the final clarifier to develop mixed-liquor suspended solids in the wastewater circulating through the tower.

$1.5Q$ direct return plus $0.5Q$ indirect flow of settled sludge from the final clarifier, for a total recirculation ratio of 2 (Fig. 11-35). Supplemental aeration between biological filtration and clarification may be used to provide a higher-quality effluent or maintain operating stability in plants that frequently encounter shock loads. The latter may result from industrial waste discharges to the municipal system, or infiltration and inflow during wet weather.

The following general design parameters apply to secondary treatment of domestic wastewater by the activated biological filtration process using recirculation as shown in Fig. 11-35, and a mixed-liquor suspended solids greater than 1000 mg/l. For effluent BOD and suspended-solids concentrations of 20–30 mg/l, the recommended range of BOD loadings is between 75 and 125 lb BOD/1000 ft^3/day (1.2–2.0 kg/m^3-day) with hydraulic loadings of not less than 1 gpm/ ft^2 or greater than 5 gpm/ft^2 (0.17–0.085 m^3/m^2-day), and no supplemental aeration.

●EXAMPLE 11-14
Design a redwood-medium tower for secondary treatment of a municipal waste assuming activated biological filtration operation with an overall recirculation ratio of 3.0. Average daily flow is 2.0 mgd with 240 mg/l, and wet-weather flow is 3.0 mgd with 150 mg/l.

○*Solution*

$$\text{average BOD load} = 2.0 \times 240 \times 8.34 = 4000 \text{ lb/day}$$

BOD load after primary settling $= 4000 \times 0.65 = 2600$ lb/day

Using a design loading of 120 lb/1000 ft³/day, the volume of packing required is

$$V = \frac{2600}{120/1000} = 21{,}700 \text{ ft}^3$$

$$\text{surface area of tower} = \frac{21{,}700}{14} = 1550 \text{ ft}^2$$

Hydraulic loadings with a recirculation ratio of 3.0 are:

At 2.0 mgd: $Q = \dfrac{2{,}000{,}000 \times 3.0}{1440 \times 1550} = 2.7$ gpm/ft²

At 3.0 mgd: $Q = \dfrac{3{,}000{,}000 \times 3.0}{1440 \times 1550} = 4.0$ gpm/ft²

Both of these loadings are within the allowable range of 1–4 gpm/ft².

Biological-Disc Process

A biological-disc unit consists of a shaft of rotating circular plastic plates immersed approximately 40% in a contour-bottomed tank. The discs are spaced so that during submergence, wastewater can enter between the surfaces. When rotated out of the tank, air enters the voids while the liquid trickles out over the fixed films of biological growth attached to the media. Alternating exposure to organics in the wastewater and oxygen in the air is similar to dosing a trickling filter with a revolving distributor. Excess microbial solids slough from the media and are carried out in the process effluent for gravity separation in a final clarifier. Installations in Europe and the United States have revealed several advantages when compared with other biological processes. Rotating-disc treatment provides lower power consumption, greater process stability, and a smaller quantity of waste sludge than activated sludge. Efficient aeration and increased contact time between the biomass and wastewater yields better treatment than trickling filtration.

11-20

DESCRIPTION OF BIOLOGICAL-DISC TREATMENT

Field-scale plastic discs are manufactured either plain or corrugated in 10- to 12-ft diameters. Figure 11-36 pictures a pilot plant with Bio-surf medium and a Biomodule package plant. Settled wastewater is applied to the first in a series of disc chambers separated by baffles, with each containing several stages

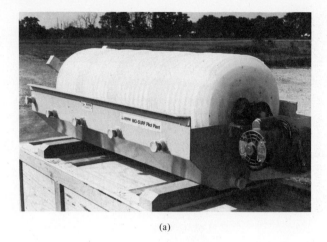

(a)

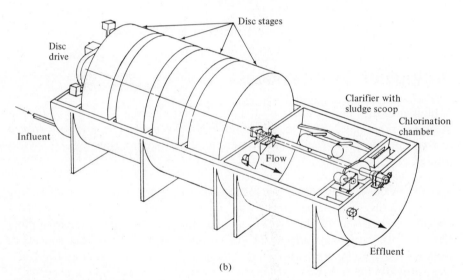

(b)

FIGURE 11-36 Biological-disc pilot plant for field-treatability studies, and diagram of a package plant for use with conventional primary treatment in small communities and commercial institutions such as motels and mobile-home parks. (a) Pilot plant. (b) Package plant. (Courtesy Bio-systems Division, Autotrol Corp.)

of media. Slow movement of wastewater through the steps simulates plug flow. Suspended solids sheared from the disc surfaces are collected during clarification of the effluent. The package plant uses a sludge scoop attached to a revolving shaft for removal of solids from a cylindrical clarifier chamber. In larger installations, the four stages are on separate shafts, with each placed in its own contour-bottomed tank of reinforced concrete. Here the wastewater flows in series through the four units perpendicular to the rotating shafts. Effluent clarification is performed in a conventional circular settling basin. The bio-disc process is a

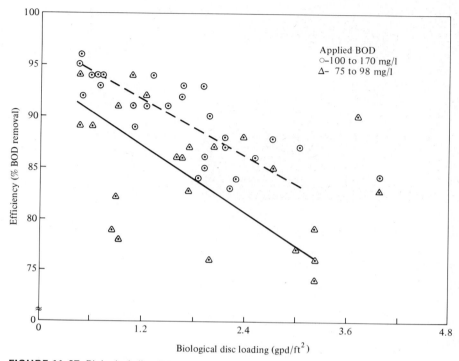

FIGURE 11-37 Biological-disc BOD removal efficiencies from treating settled municipal waste-water at temperatures greater than 55°F. [Source: R. L. Antonie, D. L. Kluge, and J. H. Mielke, "Evaluation of a Rotating Disk Wastewater Treatment Plant," *J. Water Poll. Control Fed.* 46, no. 3 (1974) : 498–511.]

once-through treatment, so in-plant recycling is not practiced. In northern climates, the media must be covered for protection from freezing temperatures. Insulated plastic covers can be used to enclose disc stages, or the units may be placed in a suitable building with adequate ventilation.

A full-scale biological-disc plant at Pewaukee, Wisconsin,[11] was used to demonstrate secondary treatment of a domestic wastewater. Following primary sedimentation, the wastewater was applied to four stages of 10-ft-diameter discs mounted in semicircular concrete tanks. The discs, fabricated from expanded polystyrene beads, were $\frac{1}{2}$-in. thick with $\frac{5}{6}$-in. spaces between plates. Flows and BOD concentrations were highly variable, allowing collection of data over a wide range of loadings. Waste temperatures exceeded 55°F during the summer and dropped to the mid-40s during cold weather. BOD removal proved to be essentially a function of hydraulic loading, as shown in Fig. 11-37. For the BOD concentration range typical of settled municipal wastewater, efficiency was approximately 85% at a surface area loading of 2.8 gpd/ft² of disc area. The effect of temperature on BOD removal is graphed in Fig. 11-38. Data at various temperatures were separated into four hydraulic loading ranges having average values of 1.8, 2.4, 3.0, and 3.8 gpd/ft². Efficiencies were in the range 80–90%

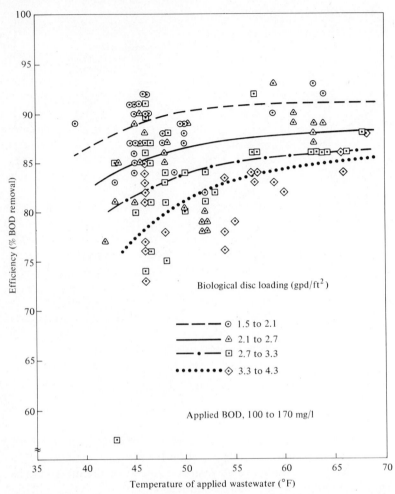

FIGURE 11-38 Effect of wastewater temperature on BOD removal in bio-disc secondary processing of settled municipal waste. [Source: R. L. Antonie, D. L. Kluge, and J. H. Mielke, "Evaluation of a Rotating Disk Wastewater Treatment Plant," *J. Water Poll. Control Fed.* 46, no. 3 (1974): 498–511.]

above 50°F, but below 45°F only the lower hydraulic loadings provided sufficient retention time for efficient BOD removal.

This study demonstrated that the bio-disc process provides stable operation under conditions of fluctuating hydraulic and organic loadings. BOD and susded solids removals proved that it is possible to produce a high-quality effluent from secondary treatment of municipal wastewater; however, decreasing temperatures must be compensated for by reducing the hydraulic loading. High density of the waste-sludge solids indicated that disposal may be accomplished economically by conventional means.

11-21

SUGGESTED DESIGN PROCEDURE

The primary design criterion for the bio-disc process is hydraulic loading expressed as gallons of wastewater flow per unit of disc area (gpd/ft²). This expression is significant since it relates the amount of organics applied to the biomass adhering to the media surface, rather than liquid retention time in the basins, which has little meaning. Figure 11-39 is a suggested chart for selecting a design hydraulic loading based on a desired percentage removal for BOD concentrations of 80–350 mg/l. This graph is based on four stages of media in separate tanks operated in series. Optimum peripheral velocity for all stages is 60 ft/min in treating domestic wastewater. For operating temperatures less than 55°F, the hydraulic loading must be reduced by a correction factor from Table 11-4. Example 11-15 illustrates the use of these suggested design criteria.

Primary and secondary clarifiers employed with the biological-disc process can be sized by the same standards applicable to rock-filled trickling-filter plants. Neither direct nor indirect recirculation is employed in the process flow scheme; therefore, biological growths that wash off the media are disposed of through a drain line from the bottom of the final clarifier to the head of the plant. This small quantity of sludge flow mixes with the raw wastewater and is settled in the

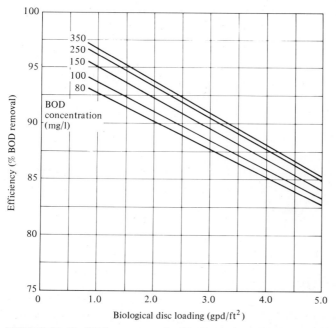

FIGURE 11-39 Efficiency curves for BIO-SURF rotating-disc process treating domestic wastewater at temperatures greater than 55°F using four stages of media. For lower operating temperatures, reduce the hydraulic loading by the appropriate correction factor from Table 11-4. (Courtesy Bio-systems Division, Autotrol Corp.)

TABLE 11-4

CORRECTION FACTORS FOR BIO-DISC
LOADINGS FROM FIG. 11-39 FOR LOW
WASTEWATER TEMPERATURES[a]

BOD Removal Efficiency (%)	Wastewater Temperatures (°F)				
	55	50	45	40	35
95	1.0	0.9	0.8	0.7	0.6
90	1.0	0.7	0.6	0.5	0.4
85	1.0	0.6	0.4	0.3	0.2

[a] Multiply hydraulic loading by appropriate factor based on temperature and the desired efficiency.

primary clarifier. The combined waste solids thicken in primary settling to a sludge of 4–6% concentration. Waste-solids production in a bio-disc secondary treating municipal waste is approximately 0.5 lb of dry solids/lb of BOD applied.

●**EXAMPLE 11-15**
Calculate the rotating-disc surface area required for secondary treatment of 1.0 mgd of raw domestic wastewater having 230 mg/l. The effluent quality specified is 20 mg/l when the wastewater temperature is 45°F.

○*Solution*

$$\text{primary effluent BOD} = 0.65 \times 230 = 150 \text{ mg/l}$$

$$\text{bio-disc efficiency required} = \frac{150 - 20}{150} = 87\%$$

From Fig. 11-39, biological-disc loading = 3.9 gpd/ft²
Applying a correction factor from Table 11-4 for wastewater temperature, the design biological-disc loading = 0.5 × 3.9 = 2.0 gpd/ft².

$$\text{disc surface area required} = \frac{1,000,000}{2.0} = 500,000 \text{ ft}^2$$

Activated Sludge

Activated-sludge processes are used for both secondary treatment and complete aerobic treatment without primary sedimentation. Wastewater is fed continuously into an aerated tank, where the microorganisms metabolize and biologically flocculate the organics (Fig. 11-10). Microorganisms (activated sludge) are settled from the aerated mixed liquor under quiescent conditions in the final clarifier and returned to the aeration tank. Clear supernatant from the final settling tank is the plant effluent.

The primary feeders in activated sludge are bacteria; secondary feeders are holozoic protozoans (Fig. 11-13). Microbial growth in the mixed liquor is maintained in the declining or endogenous growth phase to ensure good settling characterstics (Fig. 11-11). Synthesis of the waste organics results in a buildup of the microbial mass in the system. Excess activated sludge is wasted from the system to maintain the proper food/microorganism ratio (F/M) to ensure optimum operation.

Activated sludge is truly an aerobic treatment process since the biological floc are suspended in a liquid medium containing dissolved oxygen. Aerobic conditions must be maintained in the aeration tank; however, in the final clarifier the dissolved-oxygen concentration can become extremely low. Dissolved oxygen extracted from the mixed liquor is replenished by air supplied to the aeration tank.

11-22

BOD LOADINGS AND AERATION PERIODS

General loading and operational parameters for the activated-sludge processes used in treatment of municipal wastewater are listed in Table 11-5.

The BOD load on an aeration tank is calculated using the BOD in the influent wastewater without regard to that in the return sludge flow. BOD loadings are expressed in terms of lb of BOD applied per day per 1000 ft^3 of liquid volume in the aeration tank and in terms of lb of BOD applied/day/lb of mixed liquor suspended solids (MLSS) in the aeration tank. The latter, the F/M ratio, is expressed by some authors in terms of lb of BOD applied/day/lb of volatile mixed liquor suspended solids (VSS).

The aeration period is the detention time of the raw-wastewater flow in the aeration tank expressed in hours. It is calculated by dividing the tank volume by the daily average flow without regard to return sludge. The activated sludge returned is expressed as a percentage of the raw-wastewater influent. For example, if the return sludge rate is 20% and the raw-wastewater flow into the plant is 10 mgd, the return sludge is 2.0 mgd.

BOD loadings per unit volume of aeration tank vary from greater than 100 to less than 10 lb of BOD per 1000 ft^3, while the aeration periods correspondingly vary from 2.5 to 24 hours. The relationship between volumetric BOD loading and aeration period is directly related to BOD concentration in the wastewater. For example, converting the average BOD concentration of 200 mg/l into units of lb/1000 ft^3 yields a concentration of

$$200 \text{ mg/l} \times \frac{62.4 \text{ lb/1000 ft}^3}{1000 \text{ mg/l}} = 12.5 \text{ lb/1000 ft}^3$$

TABLE 11-5

GENERAL LOADING AND OPERATIONAL PARAMETERS FOR
ACTIVATED-SLUDGE PROCESSES

Process	BOD Loading		Aeration Period (hours)	Average Return Sludge Rates (%)	BOD Efficiency (%)
	lb of BOD/day[a] / 1000 ft^3	lb of BOD/day / lb of MLSS			
High rate (complete mixing)	100 up	0.5–1.0	2.5–3.5	100	85–90
Step aeration	30–50	0.2–0.5	5.0–7.0	50	90–95
Conventional (tapered aeration)	30–40	0.2–0.5	6.0–7.5	30	95
Contact stabilization	30–50	0.2–0.5	6.0–9.0	100	85–90
Extended aeration	10–30	0.05–0.2	20–30	100	85–95
High-purity oxygen	120+	0.6–1.5	1.0–3.0	30	90–95

[a] 1.0 lb/1000 ft^3/day = 16 g/m^3-day.

Therefore, a 200 mg/l wastewater applied to an extended aeration system with a 24-hour (1-day) aeration period results in a BOD loading of 12.5 lb/1000 ft^3/day. If a high-rate aeration period of 3.0 hours is considered, the BOD loading becomes

$$12.5 \text{ lb/1000 ft}^3 \times \frac{24 \text{ hours}}{3.0 \text{ hours}} = 100 \text{ lb/1000 ft}^3/\text{day}$$

The wide range of aeration periods and BOD loadings used in activated-sludge processes tend to contrast one process with another. Also, the variety of physical features, such as aeration-tank size and shape, used in the various processes tends to accent the differences. Actually all activated-sludge processes are biologically similar, as seen in the generalized activated-sludge process, Fig. 11-40.

Comparison of various activated sludge processes can best be explained by examining relative BOD loadings in terms of lb of BOD applied/day/lb of MLSS in the aeration tank. The values of lb of BOD/lb of MLSS listed in Table 11-5 range from 1.0 for high rate to less than 0.1 for extended aeration.

The relative positions of various activated-sludge processes are shown along with the abscissa in Fig. 11-41. The abscissa scale is in terms of lb of MLSS in

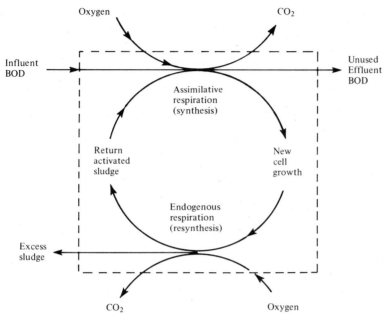

FIGURE 11-40 Generalized biological process reactions in the activated-sludge process.

the aeration tank divided by the lb BOD fed per day (the inverse of the BOD loading). This value, sometimes referred to as the BOD sludge age, is expressed in days and is related to the retention time of activated-sludge solids in the system.

The sum of the biological-process reactions shown in Fig. 11-40 is represented on any vertical line through the diagram, Fig. 11-41. BOD is removed in an activated-sludge system by process reactions of assimilative respiration, endogeneous respiration, and withdrawal of the excess sludge. Influent BOD is equal to unused BOD plus respiration plus excess VSS produced. For example, assume a conventional process operating at a lb of MLSS per lb of BOD/day ratio of 3.0. Imagine a vertical line through the diagram at 3.0. The BOD removal efficiency is approximately 93%. BOD satisfied by respiration is about 65% of the influent and excess sludge produced is approximately 0.28 lb of VSS/lb of influent BOD.

Sludge age is a parameter that relates the quantity of microbial solids in an aeration basin to the amount of influent organic matter. Equations 11-26 and 11-27 establish sludge age on the basis of MLSS in the system relative to the applied wastewater suspended solids. Sludge age can also be related to applied BOD by Eq. 11-28; in this case the term is the inverse of BOD loading expressed as M/F. Some authors apply MLVSS (volatile fraction of MLSS) rather than the

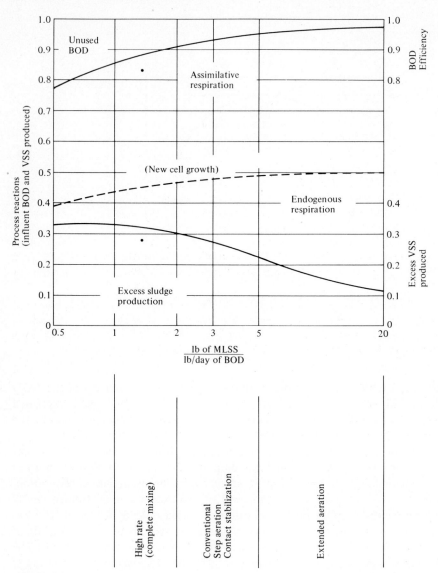

FIGURE 11-41 Schematic diagram of process reactions for various activated-sludge processes.

MLSS in sludge-age formulas. Because of these variations, data in the literature must be carefully evaluated to determine the method of calculation used by a particular author.

$$\text{SS sludge age} = \frac{MLSS \times V}{SS \times Q} \qquad (11\text{-}26)$$

or

$$= \frac{MLSS \times AP}{SS} \tag{11-27}$$

$$\text{BOD sludge age} = \frac{MLSS \times V}{BOD \times Q} \tag{11-28}$$

where sludge age = days
> $MLSS$ = mixed-liquor suspended solids, mg/l
> V = volume of the aeration tank, mg
> SS = suspended solids in the wastewater, mg/l
> Q = influent wastewater, mgd
> AP = aeration period, days
> BOD = BOD in the wastewater, mg/l

The suspended-solids concentration maintained in the mixed liquor of a conventional activated-sludge process ranges from 1500 to 3000 mg/l. The MLSS in high-rate activated sludge is generally 4000–5000 mg/l.

Solids retention in an activated sludge system is measured in days, whereas the liquid aeration period is in hours. For example, a conventional activated-sludge system with a MLSS of 2500 mg/l operating at a 6-hour aeration period, has a sludge age of 5.2 days when the applied wastewater contains 120 mg/l of suspended solids. The suspended solids are cycled in the system from final clarifier back to aeration tank, while liquid flows through the system.

● **EXAMPLE 11-16**

Data from a field study on a high-rate activated sludge secondary are as follows:

> aeration tank volume = 0.64 mg (85,500 ft³)
> settled wastewater flow = 6.53 mgd
> return sludge flow = 2.92 mgd
> waste sludge flow = 20,000 gpd
> MLSS in aeration tank = 3170 mg/l
> VSS in aeration tank = 2560 mg/l
> SS in waste sludge = 25,000 mg/l
> VSS in waste sludge = 20,300 mg/l
> influent wastewater BOD = 227 mg/l
> effluent wastewater BOD = 39 mg/l
> influent suspended solids = 111 mg/l

Using these data, calculate the following: (a) the loading and operational parameters listed in Table 11-5; (b) the sludge age; (c) the excess VSS production. (d) Plot the BOD efficiency and excess sludge production on Fig. 11-41.

○*Solution*

BOD load $= 6.53 \times 227 \times 8.34 = 12,400$ lb/day

MLSS $= 0.64 \times 3,170 \times 8.34 = 16,900$ lb

a   BOD loading $= \dfrac{12,400}{85.5} = 145$ lb/1000 ft^3

BOD loading $= \dfrac{12,400}{16,900} = 0.73 \dfrac{\text{lb of BOD}}{\text{lb of MLSS}}$

aeration period $= \dfrac{0.64 \times 24}{6.53} = 2.35$ hours

return sludge rate $= \dfrac{2.92 \times 100}{6.53} = 45\%$

BOD efficiency $= \dfrac{227 - 39}{227} \times 100 = 83\%$

b   Using Eq. 11-26,

sludge age $= \dfrac{3170 \times 0.64}{111 \times 6.53} = 2.8$ days

c   Excess VSS production $= 0.020 \times 20,300 \times 8.34 = 3390$ lb of VSS

d   BOD sludge age $= \dfrac{16,900}{12,400} = 1.36$ days

excess VSS production $= \dfrac{3390}{12,400} = 0.27 \dfrac{\text{lb of VSS produced}}{\text{lb of BOD applied/day}}$

Efficiency $= 83\%$ and excess VSS sludge production $= 0.27$ are plotted as dots along the vertical line through 1.36 in Fig. 11-41.

11-23

ACTIVATED-SLUDGE-TREATMENT SYSTEMS

Flow diagrams for the most common activated-sludge processes are shown in Fig. 11-42.

The *conventional activated-sludge process* (Fig. 11-42a), an outgrowth of the earliest activated sludge systems constructed, is used for secondary treatment of domestic wastewater. The aeration basin is a long rectangular tank with air diffusers on one side of the tank bottom to provide aeration and mixing. Settled raw wastewater and return activated sludge enter the head of the tank and flow down its length in a spiral flow pattern. An air supply is tapered along the length of the tank to provide a greater amount of diffused air near the head where the rate of biological metabolism and resultant oxygen demand is the greatest. A

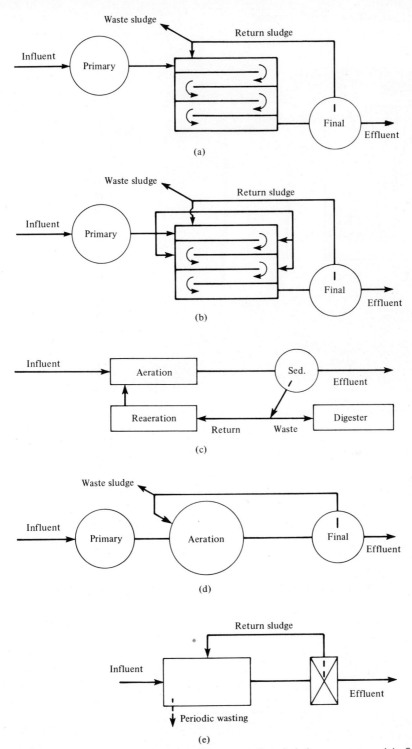

FIGURE 11-42 Flow diagrams for common activated-sludge processes. (a) Conventional activated-sludge process. (b) Step-aeration activated-sludge process. (c) Contact stabilization, without primary sedimentation. (d) High-rate (complete mixing) activated-sludge process. (e) Extended aeration, without primary sedimentation.

569

FIGURE 11-43 Conventional activated-sludge aeration tank with swing air diffusers. (CHICAGO PUMP Products, courtesy Environmental Equipment Div., FMC Corp.)

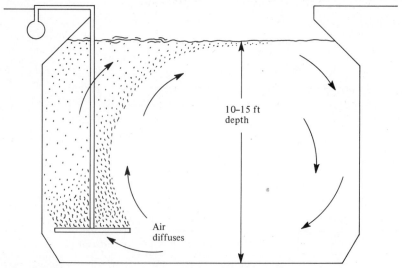

10–15 ft
depth

Air
diffuses

FIGURE 11-44 Cross section of a typical diffused air, spiral flow, conventional activated-sludge aeration tank.

conventional activated-sludge aeration tank is shown in Fig. 11-43; the cross section of a typical tank is illustrated in Fig. 11-44.

The standard activated-sludge process uses fine-bubble air diffusers set at a depth of 8 ft or more to provide adequate oxygen transfer and deep mixing. Two types of fine bubble diffusers are shown in Fig. 11-45. Flexofuser diffusers are attached to an air header pipe of a swing diffuser assembly (Fig. 11-43). The swing diffuser arm is jointed so the diffusers can be swung out of the tank for cleaning and maintenance. Duosparj nozzles attached to removable submerged air header pipes are a type generally less subject to clogging.

The Inka aeration system,[12] which is popular in Sweden, uses a coarse bubble aeration grid with shallow submergence of about 2.5 ft (Fig. 11-46). A fiber glass baffle is extended from the edge of aeration grid to within 3 ft of the

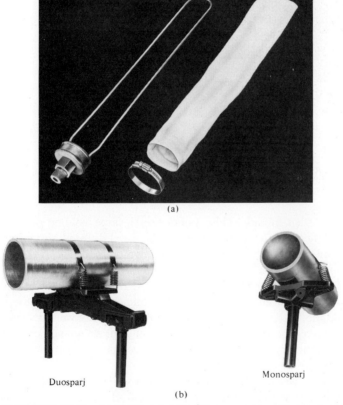

(a)

Duosparj   Monosparj

(b)

FIGURE 11-45 Two types of fine-bubble diffusers used in activated-sludge aeration tanks. (a) The Flexofuser diffuser consists of a synthetic cloth tube that slips over a U-shaped support pin which can be attached to header air pipe. The cloth portion of a diffuser can be easily removed for washing. (CHICAGO PUMP Products, courtesy Environmental Equipment Div., FMC Corp.) (b) Duosparj or Monosparj diffuser nozzles are connected directly to the submerged air header pipe and produce a stream of fine air bubbles. (Courtesy Walker Process Division of Chicago Bridge & Iron Company.)

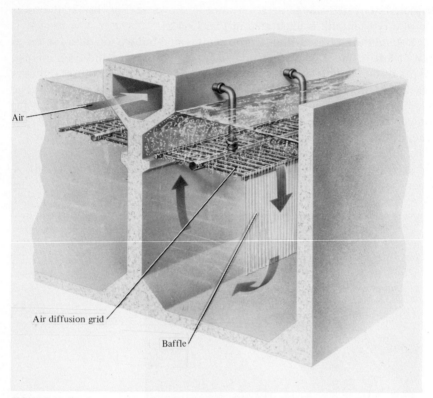

Air

Air diffusion grid

Baffle

FIGURE 11-46 Cross section of an Inka aeration tank. Air is admitted through an aeration grid, and a vertical baffle promotes deep mixing. (Courtesy Dorr-Oliver.)

tank bottom to ensure deep mixing. This system uses a large volume of low-pressure air generated by centrifugal fans, whereas the usual diffused air system resorts to a smaller volume of high-pressure air from compressors.

The *step-aeration activated-sludge process* (Fig. 11-42b) is a modification of the conventional process. Instead of introducing all raw wastewater at the tank head, raw flow is introduced at several points along the tank length. Stepping the influent load along the tank produces a more uniform oxygen demand throughout. While tapered aeration attempts to supply air to match oxygen demand along the length of the tank, step loading provides a more uniform oxygen demand for an evenly distributed air supply.

The *contact stabilization activated-sludge process* (Fig. 11-42c) provides for reaeration of the return activated sludge from the final clarifier allowing this method to use a smaller aeration tank. The sequence of aeration–sedimentation–reaeration has been used as a secondary treatment process. However, current use is primarily in complete aerobic treatment without primary sedimentation (Fig.

11-50). Raw wastewater is aerated in a contact tank and then settled. The supernatant from the clarifier is the plant effluent, and the subnatant is reaerated prior to mixing with raw influent in the contact aeration tank.

The *high-rate* (*complete mixing*) *activated-sludge process.* (Fig. 11-42d) operates with the highest BOD load per unit volume of aeration tank of any activated-sludge system. The BOD loading is approximately three times that of the conventional process, and the aeration period is proportionately shorter. High-rate mixed liquor is in the declining growth phase rather than the endogenous stage. This makes the activated sludge more difficult to settle, and a special clarifier arm with hydraulic pickup pipes (Fig. 9-22) is necessary to ensure adequate suspended-solids separation. The D-O Aerator illustrated in Fig. 11-47 is a high-rate secondary aeration basin which uses a combination of compressed-air aeration and mechanical mixing. Air is introduced through holes in a sparge ring located at the base of the mixer shaft. Turbines churn the mixed liquor, shearing the bubbles and mixing air with the tank contents.

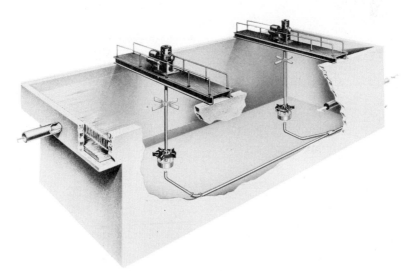

FIGURE 11-47 High-rate activated-sludge aeration tank with a large-bubble air sparge ring mounted under mixing impellers. (Courtesy Dorr-Oliver.)

The *extended aeration activated-sludge process* (Fig. 11-42e) is commonly used to treat small wastewater flows from schools, housing developments, trailer parks, institutions, and small communities. The aeration period is 24 hours or greater, longest of any activated-sludge process. Because of low BOD loading, the extended aeration system operates in the endogenous growth phase.

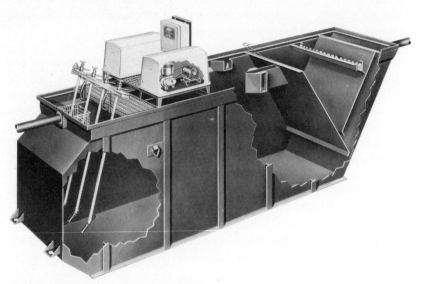

FIGURE 11-48 Factory-built extended aeration activated-sludge system with diffused aeration and a slot for returning settled sludge from the final settling basin. (Courtesy Smith & Loveless Div., Ecodyne Corp.)

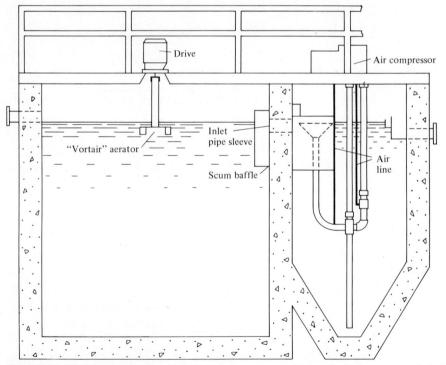

FIGURE 11-49 Extended aeration activated-sludge system with mechanical aerator and air-lift pump for returning settled sludge from the final clarifier. (Courtesy Infilco Degremont Inc.)

The extended aeration process can accept periodic (intermittent) loadings without becoming upset. Stability of the process results from the large aeration volume and complete mixing of the tank contents. Final settling tanks are conservatively designed using low overflow rates and long detention times. Overflow rates generally range from 200 to 600 gpd/ft^2 for aeration tank volumes ranging from 5000 to 150,000 gal. Excess sludge is not generally wasted continuously from an extended aeration system. Instead, the mixed liquor is allowed to increase in suspended-solids concentration and a large volume of the aeration tank contents are periodically pumped to disposal. The MLSS concentration in the aeration chamber of operating plants varies from 1000 to 10,000 mg/l. Under normal loading conditions, the MLSS concentration increases at a rate of 40–60 mg/l/day when treating domestic wastewater.

The high demand for small extended aeration treatment plants in recent years has stimulated the development of a wide variety of manufactured units. The Oxigest unit (Fig. 11-48) is fabricated and assembled in the factory and shipped to the field site for installation. The Accelo-Biox plant shown in Fig. 11-49 can be factory-built, or the mechanical aeration equipment may be installed in cast-in-place reinforced-concrete tanks.

For larger installations, circular steel wastewater treatment plants are commonly used. The design has a center chamber for sedimentation and a segmented cylindrical-shaped outer chamber for aeration. A field-erected Stepaire unit drawn in Fig. 11-50 can be operated using the extended aeration process, step aeration process, or contact stabilization.

The most recent activated-sludge system, introduced to the United States from the Netherlands, is the oxidation ditch with rotor aeration (Fig. 11-51). Unsettled wastewater is fed to mixed liquor in the ditch, where it is mechanically aerated using a caged rotor. The ditch effluent is clarified and the settled sludge returned to maintain a desired MLSS. Oxidation ditch is basically an extended aeration process; however, a shorter aeration period can be used when adequate rotor aeration and efficient suspended-solids separation in the final clarifier are provided. The manufacturers suggested volume for the oxidation ditch shown in Fig. 11-51 is 74 ft^3/lb of BOD applied/day with an aeration period of not less than 12 hours.

●EXAMPLE 11-17

A contact-stabilization plant, similar to the one diagrammed in Fig. 11-50, has compartments with the following liquid volumes:

aeration chamber = 85 m^3

reaeration chamber = 173 m^3

aerobic digester = 153 m^3

sedimentation tank = 122 m^3 (30.7-m^2 surface area)

If the plant is designed for an equivalent population of 2000 persons, calculate the BOD loading, aeration periods, and detention times.

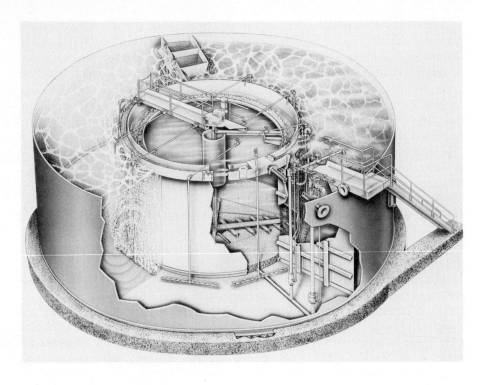

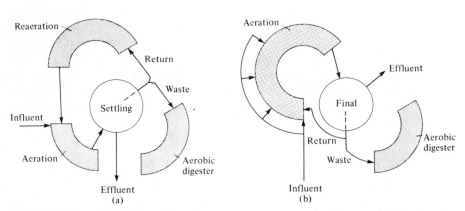

FIGURE 11-50 Field-erected circular steel wastewater treatment plant for extended aeration, step aeration, or contact stabilization processes. Photo shows cutaway view of aeration tank and clarifier. (a) Contact stabilization. (b) Extended or step aeration. (CHICAGO PUMP Products, courtesy Environmental Equipment Div., FMC Corp.)

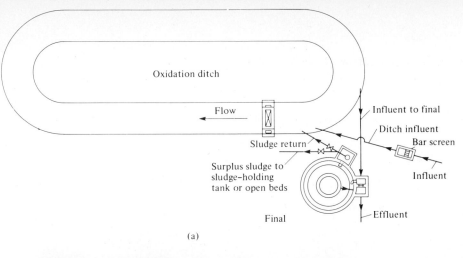

Oxidation ditch

Flow

Influent to final

Ditch influent

Bar screen

Sludge return

Surplus sludge to
sludge-holding
tank or open beds

Influent

Final

Effluent

(a)

(b)

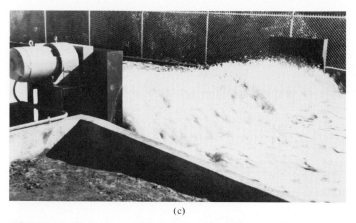

(c)

FIGURE 11-51 Oxidation ditch with rotor aeration. (a) Flow diagram of an oxidation ditch. (b) Picture of rotor and ditch construction. (c) Picture of rotor in operation. (Courtesy Lakeside Equipment Corp.)

○*Solution*

$$\text{hydraulic load} = 2000 \times 379 \text{ liters/person} = 758,000 \text{ liters/day} = 758 \text{ m}^3/\text{day}$$

$$\text{BOD load} = 2000 \times 77 \text{ g/person} = 154,000 \text{ g/day}$$

$$\text{BOD loading on aeration tanks} = \frac{154,000}{85 + 173} = 600 \text{ g/m}^3\text{-day}$$

$$\text{aeration period (based on raw-wastewater flow)} = \frac{85 \times 24}{758} = 2.7 \text{ hours}$$

$$\text{reaeration period (based on raw-wastewater flow)} = \frac{173 \times 24}{758} = 5.5 \text{ hours}$$

detention time for sedimentation

$$\text{(assuming 100\% recirculation flow)} = \frac{122 \times 24}{2 \times 758} = 1.9 \text{ hours}$$

$$\text{overflow rate on final clarifier} = \frac{758}{30.7} = 24.7 \text{ m}^3/\text{m}^2\text{-day}$$

(based on effluent flow)

● **EXAMPLE 11-18**

Calculate dimensions for equal-size aeration basins for secondary treatment of a waste-water flow of 18.2 mgd with 19,700 lb of BOD. (a) For a conventional activated-sludge process with a maximum loading of 35 lb of BOD per 1000 ft³ and minimum aeration period of 6 hours. (b) For a high-rate activated-sludge process with a maximum loading of 100 lb of BOD per 1000 ft³ and minimum aeration period of 2.5 hours.

○*Solution*

   a   Conventional process

$$V \text{ (based on BOD loading)} = \frac{19,700 \times 1000}{35} = 563,000 \text{ ft}^3$$

$$V \text{ (based on aeration period)} = \frac{18,200,000 \times 6}{24 \times 7.48} = 608,000 \text{ ft}^3$$

Use 608,000 ft³ with an aeration period of 6.0 hours at 32 lb of BOD/1000 ft³ loading. Use 10 aeration tanks, with 13 ft liquid depth and 24 ft width.

$$\text{length of each tank} = \frac{608,000}{10 \times 13 \times 24} = 196 \text{ ft}$$

   b   High-rate process

$$V \text{(based on BOD loading)} = \frac{19,700 \times 1000}{100} = 197,000 \text{ ft}^3$$

$$V \text{ (based on aeration period)} = \frac{18,200,000 \times 2.5}{24 \times 7.48} = 253,000 \text{ ft}^3$$

Use 253,000 ft³ with an aeration period of 2.5 hours at 78 lb of BOD/1000 ft³ loading. Use three circular aeration tanks with 16 ft liquid depth.

$$\text{diameter of each tank} = \left(\frac{253,000 \times 4}{3 \times 16 \times \pi}\right)^{0.5} = 82 \text{ ft}$$

11-24

ACTIVATED-SLUDGE-PROCESS STABILITY

Activated-sludge systems which use long rectangular tanks, such as the conventional process, have an apparent lack of stability because of the oscillating microbial growth pattern that develops in a plug-flow system. The relatively high food/microorganism ratio at the head of the aeration tank permits a rapid growth rate, which decreases dramatically by the time the mixed liquor is aerated during the 6- to 8-hour flowing-through period. The population of active microorganisms near the influent end of the tank is high, after the raw wastewater and return activated sludge have been aerated for a short period of time. This population of active microorganisms diminishes during the aeration period. Upon initiation of the next feed cycle, the population is suddenly increased again. The dominant microorganism populations change with time as the mixed liquor is aerated. Primary bacteria and protozoans dominate near the influent part of the tank, secondary bacteria and protozoans control at the effluent end. Therefore, the weakest rather than the strongest microbial populations in the return activated sludge are adapted to the metabolism of raw-waste organics.

Problems of instability due to oscillating microbial populations in a plug-flow activated-sludge system are amplified by the normal cycle flow of domestic wastewater. Usually, the BOD concentration in raw wastewater follows the same pattern—lower than average at night and higher than the median during the daytime. Therefore, the load variation, in pounds of BOD per hour, is more pronounced than the flow diversity. In treating the wastewater flow illustrated in Fig. 11-18b, a conventional aeration tank designed for an average period of 6 hours at 35 lb/1000 ft³/day of BOD would have a night aeration period of approximately 9 hours at 15 lb BOD /1000 ft³/day and a day period of 4.5 hours at 60 lb/1000 ft³/day of BOD. The same type of aeration tank for the flow pattern of a small city would result in about the same deviation in aeration times but a much more pronounced variation in BOD loading. For the wastewater flow graphed in Fig. 11-18a, the BOD concentration for several hours in the early morning dropped below 50 mg/l while during the afternoon it averaged 300 mg/l. The BOD load for a nominal 6-hour aeration period would vary from 5 to 70 lb/1000 ft³/day of BOD. Because of wide hourly variations in waste loads from small cities and towns, the conventional activated-sludge process has been generally restricted to treating wastewater flows greater than 0.5 mgd. Large cities, with long trunk sewers and commercial activities that extend throughout the 24-hour day, generally have few problems of instability in operating a conventional activated-sludge system.

Problems of instability in activated-sludge treatment have been significantly reduced by using completely mixed aeration basins, where the influent raw wastewater is quickly dispersed throughout the mixed liquor. Under uniform-flow conditions of wastewater flow, sludge recirculation, and sludge wasting, the biological process would operate at a constant F/M ratio. Normal wastewater flow patterns tend to shift the F/M ratio somewhat during the daily loading cycle, but the range of F/M ratio variation is relatively narrow for a completely mixed aeration basin. The large mass of suspended solids in the mixed liquor and high sludge recirculation flow dampens the BOD load fluctuations of the influent waste flow.

High-rate activated-sludge and extended aeration processes are complete mixing systems. Complete mixing in high-rate systems permits increased BOD loadings and shortened aeration periods. Thorough mixing in combination with long retention periods used in extended aeration plants provides the assimilative capacity necessary to accept periodic loading without loss of efficiency. For example, extended aeration plants have been used successfully for schools, where the load enters during a 10- to 12-hour period each day, for only 5 days per week.

11-25

OXYGEN TRANSFER AND AIR REQUIREMENTS

In activated sludge, oxygen is supplied to the microorganisms by diffusing air into the mixed liquor. A diagram of the commonly accepted oxygen-transfer scheme is given in Fig. 11-52. Oxygen is dissolved in solution and then extracted from solution by the biological cells.

Direct oxygen transfer from bubble to cell is possible if the microorganisms are adsorbed on the bubble surface. Bennett and Kempe[13] demonstrated direct oxygen transfer in a laboratory fermentor using a culture of *Pseudomonas ovalis* converting glucose to gluconic acid. The extent of direct oxygen transfer in activated-sludge systems is not known; however, it is generally felt to be secondary to oxygen transfer through the intermediate dissolved-oxygen phase.

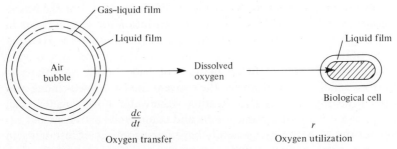

FIGURE 11-52 Schematic diagram of oxygen transfer in activated sludge.

The rate of oxygen transfer from air bubbles to dissolved oxygen in an aeration tank may be expressed as follows:

$$\frac{dc}{dt} = \alpha K(\beta Cs - Ct) \tag{11-29}$$

where dc/dt = rate of oxygen transfer, mg/l/hour
  α = oxygen-transfer coefficient of the wastewater
  β = oxygen-saturation coefficient of the wastewater
  K = oxygen-transfer coefficient, per hour
  Cs = oxygen concentration at saturation, mg/l
  Ct = oxygen concentration in the liquid, mg/l
  $\beta Cs - Ct$ = dissolved-oxygen deficit, mg/l

Equation 11-29 without the α and β coefficients applies to pure water. The α and β factors depend upon the characteristics of the wastewater being aerated, primarily the concentration of dissolved solids. The K depends upon temperature and the aeration system features such as type of diffuser, depth of aerator, type of mixer, and tank geometry. In general, the rate of oxygen transfer increases with decreasing bubble size, longer contact time, and added turbulence.

Because K depends upon the installed aeration equipment, the best method to evaluate aerators is to conduct full-scale field tests. Conway and Kumke[14] present the results of oxygen-transfer studies carried out on three full-scale aeration basins.

The rate of dissolved oxygen utilization by microorganisms in an activated-sludge system can be determined by placing a sample of mixed liquor in a closed container and measuring the dissolved oxygen depletion with respect to time. The slope of the resultant curve, r, is the oxygen-utilization rate. Figure 11-53 is a dissolved oxygen-depletion curve for a mixed liquor from a high-rate activated-sludge aeration tank. The r value depends upon the microorganisms' ability to metabolize the waste organics based upon such factors as the food/microorganism ratio, mixing conditions and temperature. A general range for r in the mixed liquor of conventional and high-rate activated-sludge systems is 30–100 mg/l/hour.

Under steady-state conditions of oxygen transfer in an activated-sludge system, the rate of oxygen transfer to dissolved oxygen (dc/dt) is equal to the rate of oxygen utilization (r). Substituting r in Eq. 11-29 for dc/dt and rearranging, we obtain

$$\alpha K = \frac{r}{\beta Cs - Ct} \tag{11-30}$$

where r is the oxygen-utilization rate by microorganisms in activated sludge, mg/l/hour.

The rate of aerobic microbial metabolism is independent of the dissolved-oxygen concentration above a critical (minimum) value. Below critical value,

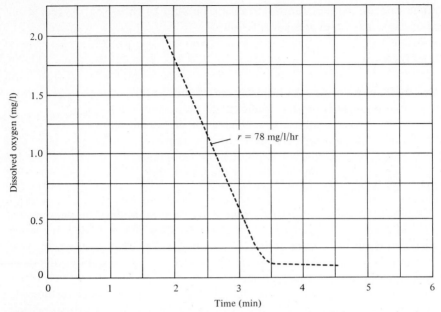

FIGURE 11-53 Oxygen-utilization curve for a sample of mixed liquor from a high-rate activated-sludge aeration basin.

the rate is reduced by the limitation of oxygen required for respiration. Critical dissolved-oxygen concentrations reported in the literature for activated-sludge systems range from 0.2 to 2.0 mg/l, depending upon the type of activated-sludge process and characteristics of the wastewater. The most frequently referenced critical dissolved oxygen value for conventional and high-rate aeration basins is 0.5 mg/l.

The quantity of air required to satisfy microbial oxygen demand and to provide adequate mixing in an aeration tank depends upon the type of activated-sludge process, BOD loading, and oxygen-transfer efficiency of the aeration equipment. In the design of any activated-sludge system, the air requirements should be based on proved performance of the aeration equipment. Capacity of the aeration equipment must furnish sufficient air to meet peak BOD loads without the dissolved-oxygen concentration dropping below the critical level for aerobic metabolism.

Oxygen-transfer efficiency is normally stated in terms of pounds of oxygen transferred per pound of oxygen supplied in the diffused air, and is generally in the range 1–10%. For example, if a supply of 1000 ft^3 of air per pound of BOD load is required, the oxygen-transfer efficiency is approximately 6%. High-rate aeration basins with mechanical mixing, in addition to compressed air supply, may have transfer efficiencies greater than 10%.

The 1971 *Recommended Standards for Sewage Works, Great Lakes–Upper Mississippi River Board of State Sanitary Engineers*[8] considers the following as minimum normal air requirements for diffused air systems: conventional, step

aeration, and contact stabilization 1500 ft³ of air applied per lb of BOD aeration tank load; modified or high-rate 400–1500 ft³/lb of BOD load; and extended
aeration 2000 ft³/lb of BOD load. These demands assume that the aeration
equipment is capable of transferring at least 1.0 lb of oxygen to the mixed liquor
per pound of BOD aeration-tank loading. In any case, aeration equipment shall
be capable of maintaining a minimum of 2.0 mg/l of dissolved oxygen in the
mixed liquor at all times and ensuring thorough mixing of the mixed liquor.

●EXAMPLE 11-19

The following data were collected during field evaluation of a complete mixing activated-sludge secondary treating municipal wastewater. The aeration basin, with a diameter
of 80 ft and a liquid depth of 17 ft, was mixed with four turbine mixers mounted above
air sparge rings. Twenty-four-hour composite BOD analyses were run on the aeration
basin influent, final clarifier effluent, and waste-activated sludge. The oxygen-utilization
rate in the aeration basin was measured each hour throughout the 24-hour sampling
period and individual values averaged for oxygen utilization rate of the mixed liquor.

$$\text{influent wastewater flow} = 6.52 \text{ mgd}$$
$$\text{waste-activated sludge} = 15,000 \text{ gpd}$$
$$\text{influent BOD} = 125 \text{ mg/l}$$
$$\text{effluent BOD} = 18 \text{ mg/l}$$
$$\text{waste-sludge BOD} = 5300 \text{ mg/l}$$
$$\text{air supplied (20°C and 760 mm)} = 1650 \text{ cfm}$$
$$\text{minimum DO in mixed liquor} = 0.8 \text{ mg/l}$$
$$\text{average DO in mixed liquor} = 1.1 \text{ mg/l}$$
$$\text{temperature of mixed liquor} = 24°C$$
$$\text{oxygen utilization rate of mixed liquor} = 74 \text{ mg/l/hr}$$
$$\text{beta factor of mixed liquor} = 0.9$$

Use these data to calculate the following:

a Pounds of BOD load.
b Cubic feet of air applied per lb of BOD load.
c Pounds of oxygen utilized per lb of BOD.
d Oxygen-transfer efficiency.
e αK.

○Solution

a   lb of BOD load $= 6.52 \text{ mg} \times 125 \text{ mg/l} \times 8.34 = 6800 \text{ lb}$

volume aeration tank $= \pi(40)^2 17 = 85,500 \text{ ft}^3$

BOD loading $= 79.5 \text{ lb of BOD}/1000 \text{ ft}^3/\text{day}$

b   air applied $= 1650 \dfrac{\text{ft}^3}{\text{minute}} \times 1440 \dfrac{\text{minutes}}{\text{day}} = 2,380,000 \text{ ft}^3$

$$\frac{\text{ft}^3 \text{ of air applied}}{\text{lb of BOD load}} = \frac{2,380,000}{6800} = 350 \frac{\text{ft}^3}{\text{lb of BOD}}$$

c   lb of oxygen utilized $= r \times$ volume of aeration tank $\times$ time

$$= 74 \frac{mg/l}{hour} \times 28.3 \frac{1}{ft^3} \times 85{,}500 \ ft^3$$

$$\times \frac{lb}{453{,}600 \ mg} \times 24 \frac{hours}{day}$$

$$= 9420 \ lb$$

lb of BOD satisfied $=$ lb of BOD removed $-$ lb of BOD wasted

$$= 6.52(125 - 18)8.34 - 0.015 \times 5300 \times 8.34$$

$$= 5190 \ lb$$

$$\frac{lb \ of \ oxygen \ utilized}{lb \ of \ BOD \ satisfied} = \frac{9420}{5190} = 1.82$$

$$\frac{lb \ of \ oxygen \ utilized}{lb \ of \ BOD \ applied} = \frac{9420}{6800} = 1.39$$

d   lb of oxygen applied $= 2{,}380{,}000 \frac{ft^3}{day} 0.0174 \frac{lb \ of \ oxygen}{ft^3}$

$$= 41{,}400 \ lb$$

$$\text{oxygen-transfer efficiency} = \frac{9420}{41{,}400} \times 100 = 22.8\%$$

e   Using Eq. 11-30 and temperature correction formula from Ref. 14

$$\alpha K \text{ at } 24°C = \frac{74}{0.9 \times 8.5 - 1.1} = 11.1 \text{ per hour}$$

(*Note:* 8.5 mg/l is the saturation dissolved-oxygen concentration of pure water at 24°C.) Correcting to 20°C (assuming that $\alpha = 1.0$),

$$11.1 = K_{20} 1.02^{(24 - 20)}$$

$$K = 10.3 \text{ per hour}$$

●EXAMPLE 11-20

A complete-mixing activated-sludge aeration tank, with a volume of 6080 ft³, was used to treat raw wastewater from a town with a population of 4500. Air was supplied, through fine-bubble diffusers submerged 7.5 ft, at a continuous rate of 1130 cfm. Based on the following field observations, calculate the air applied per lb of BOD load. BOD loadings as lb/1000 ft³/day

24-hour average	134
9 A.M. to 3 P.M. average	280
1 A.M. to 7 A.M.	26

Dissolved oxygen in mixed liquor, mg/l

24-hour average	2.4
10 A.M. to 9 P.M.	1.2–1.6
4 A.M. to 9 A.M.	4.1–4.5

○*Solution*

$$\frac{\text{air applied}}{\text{lb of BOD load}} = \frac{1130 \text{ ft}^3/\text{min} \times 1440 \text{ min/day}}{\dfrac{134 \text{ lb of BOD} \times 6080 \text{ ft}^3}{1000 \text{ ft}^3/\text{day}}} = 2000 \frac{\text{ft}^3}{\text{lb of BOD}}$$

Note: The air supply of 1130 cfm (2000 ft³/lb of BOD applied) was necessary to meet the peak oxygen demand during daytime when the BOD load was considerably above the average BOD load. Between 10 A.M. and 7 P.M., the BOD load was approximately 200% of the 24-hour average. During early morning hours when it was about 20% of average BOD load, the dissolved oxygen concentration in the mixed liquor was well above the critical level for microbial respiration. For routine operation, two air compressors (1130 cfm) were used during daylight hours and only one compressor (565 cfm) at night and on Sundays.

11-26

COMPLETE-MIXING ACTIVATED-SLUDGE EQUATIONS

The following mathematical equations (after McKinney[15,16]) apply to complete-mixing activated-sludge systems operating in the declining growth phase. Fundamental microbiological relationships indicate that growth is controlled by the rate of addition of food in the declining growth phase. In effect, the concentration of microorganisms is sufficiently high, so the rate of growth is controlled by the concentration of food remaining unmetabolized.

The unmetabolized substrate in the effluent is calculated as follows:

$$F = \frac{Fi}{Kmt + 1} \tag{11-31}$$

where F = unmetabolized BOD in the effluent, mg/l
Fi = influent BOD, mg/l
Km = metabolism factor, 7.2/hr at 20°C
t = raw-waste aeration time, hour

The value F represents only the unmetabolized substrate in the effluent and does not include excess microbial solids carryover. Generally, the mixed-liquor suspended solids carried out in the clarifier effluent exert a greater BOD than the unmetabolized substrate.

Effluent BOD is computed from the following equation:

$$\text{BOD}_{\text{eff}} = F + KbMa_{\text{eff}} \tag{11-32}$$

where BOD_{eff} = BOD in effluent, mg/l
F = unmetabolized BOD from Eq. 11-31
Kb = 0.8 (BOD factor)
Ma_{eff} = active microbial mass in the effluent, mg/l of VSS

The Ma_{eff} for Eq. 11-32 can be calculated as follows:

$$Ma_{\text{eff}} = M_{T\,\text{eff}} \times \frac{Ma}{M_T} \tag{11-33}$$

where $M_{T\,eff}$ = total suspended solids in effluent, mg/l
$\quad\quad Ma$ = active microbial mass, mg/l of VSS. (Eq. 11-35)
$\quad\quad M_T$ = mixed-liquor suspended solids, mg/l (Eq. 11-34)

Composition of mixed-liquor suspended solids in the aeration basin is determined by the following equations:

$$M_T = Ma + Me + Mi + Mii \tag{11-34}$$

where M_T = mixed-liquor suspended solids, mg/l
$\quad\quad Ma$ = active microbial mass, mg/l of VSS
$\quad\quad Me$ = endogenous respiration mass, mg/l of VSS
$\quad\quad Mi$ = inert, nonbiodegradable organic suspended solids, mg/l of VSS
$\quad\quad Mii$ = inert, inorganic suspended solids, mg/l of nonvolatile SS

Ma, Me, Mi, and Mii are calculated as follows:

$$Ma = \frac{KsF}{Ke + (1/t_s)} \tag{11-35}$$

$$Me = 0.2\,KeMat_s \tag{11-36}$$

$$Mi = Mi_{inf}\,t_s/t \tag{11-37}$$

$$Mii = Mii_{inf}\,t_s/t + 0.1(Ma + Me) \tag{11-38}$$

where Ks = synthesis factor, 5.0/hr at 20°C

$\quad\quad Ke$ = endogenous respiration factor, 0.02/hr at 20°C
$\quad\quad F$ = unmetabolized BOD from Eq. 11-31
$\quad\quad t_s$ = sludge turnover time, hours

$$t_s = \frac{\text{lb of MLSS in aeration tank}}{\text{lb of SS in effluent and waste sludge/day} \pm \text{lb of SS change in}}$$
$\quad\quad\quad$ mixed liquor/day

$\quad\quad Mi_{inf}$ = nonbiodegradable organic suspended solids in influent, approximately 40% of the VSS in normal domestic wastewater, mg/l of VSS
$\quad\quad Mii_{inf}$ = inert suspended solids in influent, mg/l of nonvolatile SS
$\quad\quad t$ = raw-waste aeration time, hour

The metabolism factor, Km, synthesis factor, Ks, and endogenous respiration factor, Ke, are all temperature-dependent. Values for these factors at temperatures other than 20°C may be determined using the following relationship:

$$K_T = K_{20}(1.072)^{T-20} \tag{11-39}$$

where K_T = Km, Ks, or Ke at temperature T (°C)
$\quad\quad K_{20}$ = Km, Ks, or Ke at 20°C

The oxygen-utilization rate of mixed liquor in the aeration basin is calculated by the following formula:

$$\frac{d0}{dt} = \frac{1.5(Fi - F)}{t} - \frac{1.42(Ma + Me)}{t_s} \qquad (11\text{-}40)$$

where $d0/dt$ = oxygen-utilization rate, mg/l/hr
Fi = influent BOD, mg/l

F, t, Ma, Me, and t_s as above.

● EXAMPLE 11-21
This example shows calculations using the complete-mixing activated-sludge equations. Compute the concentration of mixed-liquor suspended solids, effluent BOD, and oxygen-utilization rate using the following data:

aeration tank volume = 0.64 mg
influent wastewater flow = 6.52 mgd
aeration period = 2.36 hours
waste activated sludge = 2500 lb/day
influent BOD = 125 mg/l
influent suspended solids = 100 mg/l
influent volatile suspended solids = 93 mg/l
effluent suspended solids = 30 mg/l
temperature of mixed liquor = 24°C
nonbiodegradable fraction of organic suspended solids in influent = 35%

○ *Solution*
Correcting Km, Ks, and Ke to 24°C by Eq. 11-39,

Km (24°C) = 7.2 $(1.072)^{24-20}$ = 9.5
Ks (24°C) = 5.0 (1.32) = 6.6
Ke (24°C) = 0.02 (1.32) = 0.026

Unmetabolized BOD in effluent from Eq. 11-31,

$$F = \frac{125}{9.5 \times 2.36 + 1} = 5.3 \text{ mg/l}$$

Sludge turnover time, assuming a MLSS concentration (M_T) = 3900 mg/l,

$$t_s = \frac{\text{lb of MLSS in aeration tank}}{\text{lb of SS in effluent and waste sludge/day}}$$

$$= \frac{0.64 \times 3900 \times 8.34}{6.52 \times 30 \times 8.34 + 2500} \times 24 = 121 \text{ hours}$$

Mixed-liquor suspended solids using Eqs. 11-34 through 11-38,

$$Ma = \frac{6.6 \times 5.3}{0.026 + (1/121)} = 1020 \text{ mg/l}$$

$$Me = 0.2 \times 0.026 \times 1020 \times 121 = 640 \text{ mg/l}$$

$$Mi = 0.35 \times 93 \times \frac{121}{2.36} = 1670 \text{ mg/l}$$

$$Mii = (100 - 93)\frac{121}{2.36} + 0.1(1020 + 640) = 530 \text{ mg/l}$$

$$M_T = 1020 + 640 + 1670 + 530 = 3860 \text{ mg/l}$$

Effluent BOD using Eqs. 11-33 and 11-32,

$$Ma_{\text{eff}} = 30 \times \frac{1020}{3860} = 8 \text{ mg/l}$$

$$BOD_{\text{eff}} = 5.3 + 0.8 \times 8 = 12 \text{ mg/l}$$

Oxygen-utilization rate employing Eq. 11-40,

$$\frac{d0}{dt} = 1.5\frac{125 - 5}{2.36} - 1.42\frac{1020 + 640}{121} = 57 \text{ mg/l/hr}$$

Note: The complete-mixing activated-sludge equations are based on steady-state conditions of uniform BOD loading, uniform sludge wasting, and constant oxygen transfer. Application of these equations in the design of activated-sludge systems assumes that the metabolism, synthesis, and endogenous respiration factors apply to the mixed liquor developed on the wastewater substrate. Since most municipal wastewater includes industrial wastes, characteristics of the composite wastewater must be related to the mathematical equations used in designing joint-treatment activated-sludge systems.

11-27

OPERATION AND CONTROL

Operation of an activated-sludge treatment plant is controlled by regulating the (1) quantity of air supplied to the aeration tank, (2) rate of activated-sludge recirculation, and (3) amount of sludge wasted from the system.

The following tests are used to monitor operation of an activated-sludge system: dissolved-oxygen concentration in the mixed liquor, influent and effluent BOD, concentration of MLSS in the aeration tank, and settleability of the mixed liquor. From these data the BOD removal efficiency, BOD loadings, and sludge volume index can be determined. Sludge volume index (SVI) is the volume in milliliters occupied by 1 gram of suspended solids after 30 minutes settling. It is computed as follows:

$$SVI = \frac{\text{sludge volume after settling (ml/l)} \times 1000}{\text{MLSS (mg/l)}} \qquad (11\text{-}41)$$

Sludge volume is determined by placing a sample of mixed liquor in a 1-liter graduated cylinder and reading the volume occupied by the settled solids after a 30-minute period. The SVI is a measure of activated-sludge settleability and can indicate a bulking sludge. In general, a sludge-volume index in the range 50–150 indicates a good settling sludge. The MLSS in conventional and step aeration basins is normally held between 1500 and 2500 mg/l, while high-rate processes carry approximately 4000 mg/l.

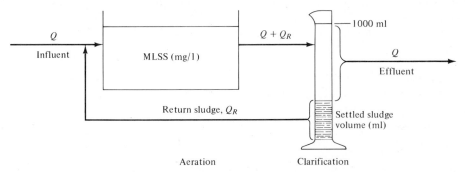

FIGURE 11-54 Hypothetical relationship between settled sludge volume in the SVI test and quantity of return sludge in an activated-sludge process.

Settleability of activated sludge can be hypothetically related to the quantity and solids concentration in the return sludge of an aeration system as depicted in Fig. 11-54. This correlation assumes that the final clarifier responds identically to the graduated cylinder used in the SVI test. This premise appears reasonable with respect to extended aeration and conventional processes, although in high-rate systems there may be considerable deviation in sludge settleability between an actual clarifier and that measured in a laboratory container. Equation 11-42 equates the ratio of returned sludge flow divided by the quantity entering the final clarifier to the settled sludge volume over the capacity of the graduated cylinder (Fig. 11-54). This formula can be used to calculate recirculation ratio (Q_R/Q), or the rate of sludge return for a given influent flow and sludge volume after settling.

$$\frac{Q_R}{Q + Q_R} = \frac{\text{sludge volume after settling (ml/l)}}{1000 \text{ (ml)}} \tag{11-42}$$

where Q = wastewater influent, mgd or gpm
Q_R = return activated sludge, mgd or gpm

The concentration of suspended solids in the return sludge can be calculated from the SVI value as follows:

$$\frac{\text{suspended solids}}{\text{in return sludge}} = \frac{1000 \text{ (mg/g)} \times 1000 \text{ (ml/l)}}{\text{SVI (ml/g)}} = \frac{10^6}{\text{SVI}} \text{ (mg/l)} \tag{11-43}$$

This relationship assumes that sludge recirculation is exactly at the required rate. A greater quantity returns clarified wastewater unnecessarily, while a lesser flow leaves solids in the settling basin and results in their eventual loss in the process effluent. Sometimes the term "sludge density index" (SDI) is used to quantify the solids concentration in return sludge. The SDI is numerically equal to the reciprocal of the SVI expressed in units of g/ml.

●**EXAMPLE 11-22**

The MLSS concentration in an aeration basin is 2200 mg/l and the sludge volume after 30 minutes of settling in a 1-liter graduated cylinder is 180 ml. Calculate the SVI, required return sludge ratio, and suspended-solids concentration in the recirculated sludge.

○*Solution*

Using Eqs. 11-41, 11-42, and 11-43:

$$\text{SVI} = \frac{180 \times 1000}{2200} = 82 \text{ ml/g}$$

$$\frac{Q_R}{Q + Q_R} = \frac{180}{1000}; \text{ thus } \frac{Q_R}{Q} = 0.22$$

$$\text{SS concentration in return sludge} = \frac{1{,}000{,}000}{82} = 12{,}000 \text{ mg/l } (1.2\%)$$

High-Purity-Oxygen Activated Sludge

The primary distinguishing feature of this process is the application of high-purity oxygen instead of air for aerobic treatment by activated sludge. The major deterrent to oxygen use until recently was the production cost. Currently, high-purity oxygen gas can be efficiently generated by cryogenic air separation or pressure-swing adsorption processes. The former is used in large installations, whereas the latter is a reliable method for oxygen production in smaller quantities.

11-28

DESCRIPTION OF HIGH-PURITY-OXYGEN PROCESS

Major components of a system are an oxygen gas generator, covered compartmented aeration tank, final clarifier, pumps for recirculating activated sludge, and sludge disposal facilities. Preliminary treatment must include screening and degritting of the wastewater, with primary sedimentation optional primarily depending on the size of the treatment plant. The aeration tank is divided into three or four stages by means of baffles to simulate plug flow, and covered with

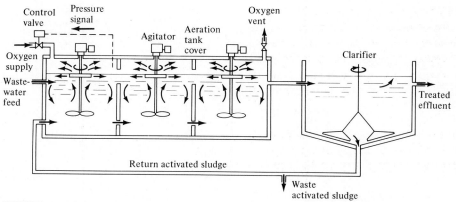

FIGURE 11-55 Schematic diagram of a high-purity-oxygen activated-sludge process with surface aerators in three stages. (Courtesy UNOX System, Union Carbide Corp.)

a gas-tight enclosure (Fig. 11-55). Raw wastewater, return activated sludge, and oxygen gas under a slight pressure are introduced to the first stage and flow concurrently through succeeding sections. Oxygen can be mixed with the tank contents by injection through a hollow shaft to a rotating sparger device, or a surface aerator installed on top of the mixer turbine shaft to contact oxygen gas with the mixed liquor. Successive aeration chambers are connected to each other so that the liquid flows through submerged ports, and head gases pass freely from one stage to the next with only a slight pressure drop. Exhausted waste gas is a mixture of carbon dioxide, nitrogen, and about 10–20% of the applied oxygen. Effluent mixed liquor is settled in either a scraper type or rapid-sludge-return clarifier, and the activated sludge returned to the aeration tank.

Oxygen is supplied from on-site gas generators and liquid oxygen shipped to the treatment plant. Transported liquid oxygen is too costly for regular application; therefore, it is stored for use only as a backup supply or to meet unexpected peak demands. For large installations, standard cryogenic air separation, involving liquefaction of air followed by fractional distillation, is employed to separate the major components of nitrogen and oxygen. The pressure-swing adsorption system (PSA) is more suitable for the majority of treatment-plant sizes because of its efficiency and reliability. The main items of equipment in a PSA unit are a feed air compressor, adsorber vessels with manifold pipes and sequencing valves, and a cycle control system (Fig. 11-56). Air is compressed in vessels filled with granular adsorbent that collects carbon dioxide, water, and nitrogen gas under high pressure, leaving a gas relatively high in oxygen. The adsorber is regenerated by depressurizing and purging. Three vessels are installed for continuous and constant flow of oxygen gas; two units alternate between operation and regeneration while the third is a standby unit. Figure 11-57 shows covered aeration tanks with mixer motors mounted on the deck, three pressure-swing adsorption vessels, and a liquid-oxygen storage tank.

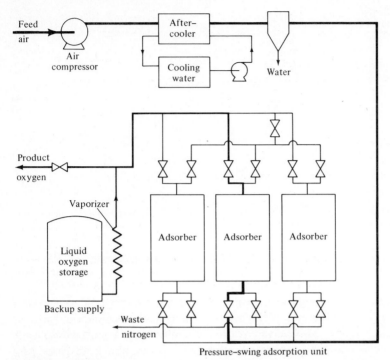

Pressure–swing adsorption unit

FIGURE 11-56 Schematic diagram of a pressure-swing adsorption (PSA) oxygen-generating system for on-site production of high-purity oxygen gas. (Courtesy UNOX System, Union Carbide Corp.)

FIGURE 11-57 High-purity-oxygen aeration system with the mixer drive motors mounted on the tank covers. The three vessels on the right are pressure-swing adsorption units, and the single tank in the background on the left stores liquid oxygen for a backup supply. (Courtesy UNOX System, Union Carbide Corp.)

Several advantages are attributed to high-purity-oxygen activated sludge compared with diffused air systems (Table 11-5). High efficiency is possible at increased BOD loads and reduced aeration periods by maintaining the food/ microorganism ratio with MLSS concentrations of 4000–8000 mg/l. Waste-sludge production in pilot-plant studies was reportedly less than that generated in air activated-sludge sytems. This is ascribed to the greater oxidation of organic matter at high-DO levels maintained in the mixed liquor, and improved settleability, resulting in a return sludge containing 2–3% solids. Even though the process simulates plug flow, shock organic loads do not produce instability, since extra oxygen is supplied to the first stage automatically on demand. Emission of foul odors is virtually eliminated because of the highly aerobic environment and reduced volume of exhaust gases. Covered tanks also help to reduce cooling of the wastewater during cold-weather operation.

11-29

DESIGN CRITERIA FOR HIGH-PURITY-OXYGEN AERATION

Results from pilot-plant studies of oxygen-activated sludge at five different locations are listed in Table 11-6. These installations were operated on a diurnal flow cycle similar to that of the actual wastewater flow pattern for the municipal or combined municipal–industrial waste at the site. At BOD loadings of 160 lb/1000 ft^3/day (2.56 kg/m^3-day) and mean aeration periods of 1.8 hours, BOD and suspended solids effluent concentrations were consistently less than the maximum allowable criterion of 30 mg/l. Resulting BOD removal efficiency ranged from 87 to 97%, depending on the influent concentration. Loadings in terms of food/microorganism ratio were 0.5–0.6 lb of BOD/day/lb of volatile MLSS. A F/M ratio of 0.5 is equivalent to treating a 150-mg/l of BOD waste-water with an aeration period of 1.8 hours, holding 4000 MLVSS in the basin. Dissolved oxygen of 3–9 mg/l was held in the mixed liquor, resulting in oxygen utilization of approximately 90%. Oxygen consumption per unit of BOD removed varies over a considerable range, depending on the food/microorganism ratio as shown in Fig. 11-58. At a F/M of 0.5, oxygen uptake is from 1.0 to 1.5 lb of oxygen/lb of BOD removed.

Settleability of the suspended solids in oxygen-activated sludge, with typical SVI values near 50, is better than aerated mixed liquor. Consequently, thickness of return sludge is generally about 2%, and a normal return sludge rate is 30%. Design parameters for final settling tanks do not differ significantly from those recommended for conventional activated-sludge clarifiers. The maximum allowable recommended overflow rate is 1200 gpd/ft^2 during peak hourly flow, generally resulting in an overflow rate of 600–800 gpd/ft^2 based on average daily flow. Solids loading on a clarifier ranges from 30 to 60 lb/day/ft^2 as a result of the high suspended-solids concentration in the mixed liquor flowing into the basin. It is suggested, therefore, that a rapid-sludge-return clarifier with a hydraulic pickup arm be used to aid in thickening the settled activated sludge. The extent of excess solids production is a function of sludge age, BOD loading,

TABLE 11-6

PILOT-PLANT DATA FROM HIGH-PURITY-OXYGEN ACTIVATED-SLUDGE
EVALUATIONS AT FIVE LOCATIONS

Parameter	Municipal			Combined Municipal–Industrial	
	Primary Range	Primary Effluent	Raw Degritted	Primary Effluent	Raw Degritted
BOD loading (lb/day/1000 ft³)	239	155	162	171	193
Aeration period (hours)	0.8	1.8	1.7	2.0	1.8
Influent concentrations					
BOD (mg/l)	128	184	184	227	229
COD (mg/l)	222	375	380	467	377
SS (mg/l)	93	168	183	76	236
Effluent concentrations					
BOD (mg/l)	17	18	6	8	12
COD (mg/l)	76	87	67	81	66
SS (mg/l)	19	16	17	15	28
Removal efficiencies					
BOD (%)	87	90	97	96	95
COD (%)	66	77	82	82	82
SS (%)	90	90	91	80	88
Food/microorganism ratio lb of BOD/day/lb of MLVSS	1.6	0.6	0.6	0.5	0.6
Mixed liquor					
DO (mg/l)	8	7	6	8	6
MLSS (mg/l)	3630	6080	6200	7010	7350
MLVSS (mg/l)	2500	3860	4650	5400	5450
SVI (ml/g)	33	40	55	42	48
Return sludge rate (%)	11	30	27	30	26
Return sludge density (%)	3.6	2.6	2.7	3.1	3.2
Clarifier overflow rate (gpd/ft²)	1100	650	750	600	650
Oxygen utilization (%)	93	90	91	90	93

Source: *Oxygen Activated Sludge Wastewater Treatment Systems*, Environmental Protection Agency, Technology Transfer, August 1973.

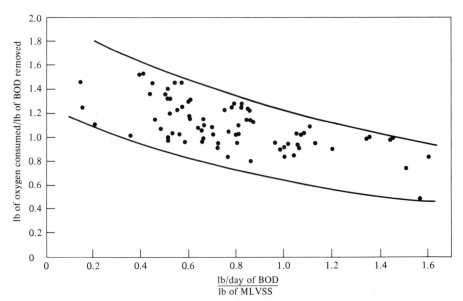

FIGURE 11-58 Oxygen consumption in high-purity-oxygen activated sludge, per pound of BOD removed, as a function of the food/microorganism ratio expressed as lb of BOD/day/lb of volatile MLSS. [Source: *Oxygen Activated Sludge Wastewater Treatment Systems*, Environmental Protection Agency, Technology Transfer (August 1973).]

level of dissolved oxygen held in the mixed liquor, and characteristics of the wastewater. For domestic wastes being treated under normal operating conditions, the excess volatile solids production is 30–50% of the applied BOD.

●EXAMPLE 11-23
A municipality has an average daily wastewater flow of 280 gpm with a peak hourly rate of 450 gpm. The average BOD concentration is 200 mg/l except during several weeks when a seasonal industry increases the mean BOD to 250 mg/l. A high-purity-oxygen system as in Fig. 11-55 without primary clarification of the raw wastewater is being considered. Calculate (a) the volume of aeration tank capacity required, and (b) surface area and depth for a final settling tank. Recommended design criteria are: maximum BOD loading of 160 lb/1000 ft³/day, largest food/microorganism ratio of 0.5 lb of BOD/day/lb of MLVSS, operating MLSS concentration of 5500 mg/l (MLVSS of 4200 mg/l); and highest overflow rate of 1200 gpd/ft² during peak flow.

○Solution

    a Aeration tank volume
      required at 250 mg/l of BOD $= \dfrac{280 \times 1440 \times 250 \times 8.34}{1{,}000{,}000 \times 0.160} = 5250 \text{ ft}^3$

           aeration period
           at average flow $= \dfrac{5250 \times 7.48 \times 24}{280 \times 1440} = 2.34 \text{ hours}$ (OK)

           BOD load at
           200 mg/l of BOD $= \dfrac{280 \times 1440 \times 200 \times 8.34}{1{,}000{,}000 \times 5.25} = 128 \text{ lb/1000}$

                                                  ft³/day   (OK)

Check the F/M ratio for both waste loadings (128 and 160 lb/1000 ft³/day) at a MLVSS concentration of 4200 mg/l.

$$\text{At } \frac{128 \text{ lb of BOD/day}}{1000 \text{ ft}^3} : \frac{F}{M} = \frac{128/1000}{\dfrac{4200 \times 62.4}{1,000,000}} = 0.49 \frac{\text{lb of BOD/day}}{\text{lb of MLVSS}} \quad \text{(OK)}$$

$$\text{At } \frac{160 \text{ lb of BOD/day}}{1000 \text{ ft}^3} : \frac{F}{M} = \frac{160/1000}{\dfrac{4200 \times 62.4}{1,000,000}} = 0.61 \text{ (slightly greater than 0.5)}$$

Therefore, use a three-stage aeration basin with a total volume of 5250 ft³.

b Final clarifier surface area
required based on peak flow $= \dfrac{450 \times 1440}{1200} = 540 \text{ ft}^2$

overflow rate at average flow $= \dfrac{280 \times 1400}{540} = 747 \text{ gpd/ft}^2$

Assume additional final clarifier design parameters of 8.0 ft minimum depth and 2.5 hours as the minimum detention time. Then, the clarifier depth required for an overflow rate of 747 gpd/ft² is

$$\text{depth} = \frac{747 \text{ gal}}{\text{day} \times \text{ft}^2} \times \frac{\text{ft}^3}{7.48 \text{ gal}} \times 2.5 \text{ hours} \times \frac{\text{day}}{24 \text{ hours}} = 10.4 \text{ ft}$$

For a better proportion between surface area and depth, use an overflow rate of 700 gpd/ft² based on average daily flow. For this value the recalculated surface area, tank depth, and maximum overflow rate are 576 ft², 9.7 ft, and 1130 gpd/ft². Use a circular clarifier equipped with a rapid-sludge-removal apparatus.

Stabilization Ponds

Domestic wastewater may be effectively stabilized by the natural biological processes which occur in shallow pools. Those suitable for treating raw or partially treated wastewater are referred to as stabilization ponds, lagoons, or oxidation ponds.

Stabilization ponds have light BOD loadings, in the range 0.2–0.5 lb/1000 ft³/day, and correspondingly long liquid-retention times, 20–120 days. A wide variety of microscopic plants and animals find the environment a suitable habitat. Waste organics are metabolized by bacteria and saprobic protozoans as primary feeders. Secondary feeders include protozoans and higher animal forms, such as rotifers and crustaceans. When the pond bottom is anaerobic, biological activity results in digestion of the settled solids. Nutients released by bacteria are used by algae in photosynthesis. The overall process in a facultative stabilization pond (Fig. 11–17) is the sum of individual reactions of the bacteria, protozoans, and algae.

The degree of stabilization produced in an oxidation pond is significantly influenced by climatic conditions. During warm, sunny weather, decomposition and photosynthetic processes flourish, resulting in rapid and complete stabilization of the waste organics. The pond water becomes supersaturated with dissolved oxygen during the afternoon. Suspended solids and BOD in the pond effluent are primarily from the algae. BOD reductions in the summer usually exceed 95%. During cold weather under ice cover, biological activity is extremely slow and the lagoon-treatment process is, for practical purposes, reduced to sedimentation. Anaerobic conditions can occur from a lack of reaeration by wind action and photosynthesis. Suspended solids and BOD in the pond effluent include organics from raw wastes and intermediate organics issuing from incomplete anaerobic metabolism. Under winter ice cover, BOD reductions are generally about 50%.

11-30

DESCRIPTION OF A STABILIZATION POND

A stabilization pond is a flat-bottomed pond enclosed by an earth dike (Fig. 11-59). It can be round, square, or rectangular with a length not greater than 3 times the width. Operating liquid depth has a range of 2–5 ft with 3 ft of dike

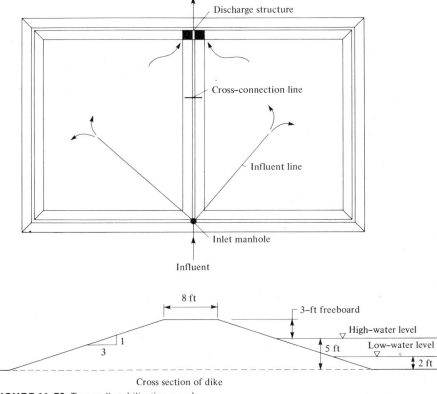

FIGURE 11-59 Two-cell stabilization pond.

freeboard. A minimum of about 2 ft is required to prevent growth of rooted aquatic plants. Operating depths greater than 5 ft can create odorous conditions because of anaerobiosis on the bottom.

Influent lines discharge near the center of the pond and the effluent usually overflows in a corner on the windward side to minimize short-circuiting. The overflow is generally a manhole or box structure with multiple-valved draw-off lines to offer flexible operation. Where the lagoon area required exceeds 6 acres, it is good practice to have multiple cells which can be operated individually, in series or parallel. If the soil is pervious, bottom and dikes should be sealed to prevent groundwater pollution. A commonly used sealing agent is bentonite (clay). Dikes and area surrounding the ponds are seeded with grass, graded to prevent runoff from entering the ponds, and fenced to preclude livestock and discourage trespassing.

11-31

BOD LOADINGS

BOD loadings on stabilization ponds are expressed in terms of lb of BOD applied/acre of surface area/day (kg/ha-day or g/m^2-day) or as design population equivalent per acre of surface area. For a domestic wastewater of 100 gpcd at a BOD of 200 mg/l, a loading of 120 persons/acre is equivalent to 20 lb of BOD /acre/day, or 0.46 lb of BOD/1000 ft^2 (2.2 g/m^2-day).

The area and loadings recommended by the 1971 *Standards of the Great Lakes–Upper Mississippi River Board of State Sanitary Engineers*[8] are as follows:

> One acre of water surface should be provided for each 100 design population or population equivalent. In terms of BOD, a loading of 0.5 pound per day per 1000 square feet should not be exceeded. Higher design loadings will be judged after review of material contained in the engineer's report and after a field investigation of the proposed site by the reviewing authority.
>
> Substantially higher loadings may be approved for the more southerly states and for installations requiring summer operation only. In cold climates, where substantial ice cover may be expected for an extended period, it may be desirable to operate the facility to completely retain wintertime flows.

In the southwestern United States, loadings in the range 40–50 lb of BOD/acre/ day (4.5 to 5.6 g/m^2-day) are used for design of stabilization ponds.

11-32

OPERATION OF STABILIZATION PONDS

In the case of a multiple-pond installation, the sequence of pond operation and liquid operating depth are regulated to provide control of the treatment system. Operating ponds in series generally cause increased BOD reduction by preventing short-circuiting. Conversely, parallel operation may be desirable to distribute the raw BOD load and avoid potential odor problems.

Where discharges of pond effluent in the winter result in pollution of the receiving stream, the operating level can be lowered before ice formation and gradually increased to 5 ft by retention of winter flows. The elevation can then be slowly lowered in the spring when the dilution flow in the receiving stream is high. Shallow operation can be maintained during the spring with gradually increasing depths to prevent emergent vegetation. In the fall, the level can again be lowered to hold winter flows.

Most stabilization ponds emit odors occasionally. This is the primary reason for locating them as far as practicable from present or future developed areas and on the leeward side so that prevailing winds are in the direction of uninhabited areas. Lagoons treating only domestic wastewater normally operate odor-free. Only for a short period of time in the early spring, when the ice melts and the algae are not flourishing, are offensive odors discharged. Lagoons treating certain industrial wastes, in combination with domestic wastewater, are often noted for their persistent obnoxious odors. The cause of these odors is most likely due to continuous or periodic overloading from industrial waste discharges and/or a result of the odorous nature of the industrial waste itself. The author has observed unpleasant odors emitting from lagoons of small municipalities where poultry processing wastes, slaughterhouse wastes, or creamery wastes were discharged to the municipal sewer without pretreatment.

11-33

ADVANTAGES AND DISADVANTAGES OF STABILIZATION PONDS

A general list of items to be considered before selecting the stabilization pond process for treatment of a municipal wastewater is offered. In general, stabilization ponds are suitable for small towns which do not anticipate extensive industrial expansion, and where land with suitable topography and soil conditions is available for siting.

Advantages
1 Probably lower initial cost than a mechanical plant.
2 Lower operating costs.
3 Regulation of effluent discharge possible, thus providing control of pollution during critical times of the year.
4 Treatment system is not significantly influenced by a leaky sewer system which collects storm water.

Disadvantages
1 Extensive land area required for siting.
2 Assimilative capacity for certain industrial wastes is poor.
3 Potential odor problems.
4 Expansion of town and new developments may encroach on the lagoon site.
5 Effluent quality generally cannot meet the standard for suspended solids of 30 mg/l.

Inability to meet an effluent standard of 30 mg/l of suspended solids is a serious problem in lagoon treatment. Algae suspended in the water during summer months generally contribute 50–70 mg/l. In some cases the effluent standard can be met in a lightly loaded pond system using series operation and careful control of effluent discharge. Alternatives are land disposal, or installation of an additional treatment unit to remove a portion of the suspended solids prior to disposal. Spray irrigation on pastureland appears to be one of the best solutions, but equipment installation is generally expensive. Furthermore, the irrigated land area would in most cases be greater than that required for the stabilization ponds. Where evaporation rate exceeds rainfall by a wide margin, lagoons could be large enough to provide complete retention, so water losses would equal influent flow and result in zero discharge. The only other alternative is effluent filtration. Gravity filters similar to those used in water treatment are not feasible, but biflow filters appear to hold some promise. Water passing up through the filter media can be controlled to capture a portion of the suspended solids without being preceded by chemical flocculation.

●**EXAMPLE 11-24**
Design population for a town is 1200 persons, and the anticipated industrial load is 20,000 gpd at 1000 mg/l of BOD from a milk-processing plant. Calculate the surface area required for a stabilization pond system, and estimate the number of days of winter storage available. Assume the following:
(a) Wastewater flow of 100 gpcd with 0.17 lb of BOD per capita.
(b) Design BOD loading of 25 lb of BOD/acre/day.
(c) Water loss from evaporation and seepage of 60 in./yr.
(d) Annual rainfall of 20 in./yr.

○*Solution*
BOD load (domestic + industrial)

$$= 1200 \times 0.17 + 0.020 \times 1000 \times 8.34 = 371 \text{ lb of BOD/day}$$

$$\text{stabilization pond area required} = \frac{371}{25} = 14.8 \text{ acres (use two ponds)}$$

volume available for winter storage between 2-ft and 5-ft depths

$$= (5 - 2)14.8 \times 43,560 = 1,930,000 \text{ ft}^3$$

water loss per day (evaporation + seepage − rainfall)

$$= \frac{(60 - 20)14.8 \times 43,560}{12 \times 365} = 5890 \text{ ft}^3/\text{day}$$

$$\text{wastewater influent per day} = \frac{1200 \times 100 + 20,000}{7.48} = 18,700 \text{ ft}^3/\text{day}$$

$$\text{winter storage available} = \frac{1,930,000}{18,700 - 5900} = 150 \text{ days}$$

●**EXAMPLE 11-25**

An existing 60-acre municipal lagoon system receives an average daily load of 0.87 mg with 2200 lb of BOD. For 8 months of the year, mean effluent discharge is 1.13 mgd at 50 mg/l of BOD. During the remaining 4 months no effluent is released. Compute the following: (a) BOD loading, (b) BOD concentration in the raw wastewater, (c) average annual water loss, and (d) average annual BOD removal efficiency.

○*Solution*

a   BOD loading $= \dfrac{2200}{60} = 36.7$ lb of BOD/acre/day (4.1 g/m²-day)

b   BOD concentration in raw wastewater $= \dfrac{2200}{0.87 \times 8.34} = 303$ mg/l

c   Average annual water loss

$$= \left(0.87 - \frac{8}{12} \times 1.13\right) \times \frac{1{,}000{,}000 \times 365 \times 12}{7.48 \times 60 \times 43{,}560} = 26 \text{ in./yr}$$

d   Mean annual BOD removal efficiency

$$= \frac{\text{lb of BOD}_{\text{inf}} - \text{lb of BOD}_{\text{eff}}}{\text{lb of BOD}_{\text{inf}}}$$

$$= \frac{2200 - (8/12) \times 1.13 \times 50 \times 8.34}{2200} \times 100 = 86\%$$

11-34

COMPLETE-MIXING AERATED LAGOONS

Aerated ponds for pretreatment of industrial wastes, or first-stage treatment of municipal wastewaters, are commonly complete-mixing lagoons 8- to 12-ft deep with floating or platform-mounted mechanical aeration units. A floating aerator consists of a motor-driven impeller mounted on a doughnut-shaped float with a submerged intake cone (Fig. 11-60). Inspection and maintenance is performed using a boat, or by disconnecting the restraining cables and pulling the unit to the edge of the lagoon. Platform-mounted aerators are placed on piles or piers extending into the pond bottom. The impeller is held beneath the liquid surface by a short shaft connected to the motor mounted on the platform. A bridge may be constructed from the lagoon dike to the aerator for ease of inspection and maintenance.

Complete mixing and adequate aeration are essential environmental conditions for lagoon biota. Selection and design of mixing equipment depends on manufacturers' laboratory test data and field experience. Aerators are spaced to provide uniform blending for dispersion of dissolved oxygen and suspension of microbial solids. Their oxygen-transfer capability must be able to satisfy the BOD demand of the waste while retaining a residual dissolved-oxygen

FIGURE 11-60 Floating aerators in a complete-mixing lagoon treating combined domestic and industrial wastewater. (Courtesy Envirex Inc., a Rexnord Company.)

centration. Figure 11-61 illustrates the general relationships between power required for mixing and that required for aeration. There is only one detention time for a given wastewater strength where both stirring and aeration functions are at optimum. Thus deviations in loadings should be considered in the design selection and operational control of mechanical aeration units.

Organic stabilization depends on suspended microbial floc developed within the basin, since there is no provision for settling and returning activated sludge. BOD removal is a function of detention time, temperature, and nature of the waste, primarily biodegradability and nutrient content. The common relationships are:

$$\frac{L_e}{L_0} = \frac{1}{1 + kt} \tag{11-44}$$

$$k_T = k_{20°C}\theta^{T-20} \tag{11-45}$$

where L_e = effluent BOD, mg/l
L_0 = influent BOD, mg/l
k = BOD-removal-rate constant, per day
t = detention time, days
T = temperature, °C
θ = temperature coefficient

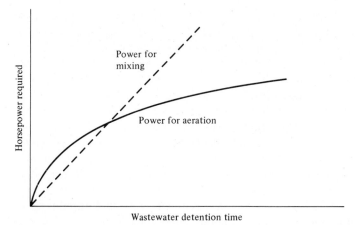

FIGURE 11-61 General relationships between power required for mixing and that needed for aeration relative to wastewater detention time for mechanical aerators in complete-mixing lagoons.

The value of k relates to degradability of the waste organics, temperature, and completeness of aeration mixing. At 20°C, k values have been found to range from 0.3 to over 1.0; the precise value for a particular waste must be determined experimentally. Coefficient θ is a function of biodegradability and generally falls between 1.035 to 1.075, with 1.035 the most common value.

Biological oxygen utilization is equal to assimilative plus endogenous respiration as in the activated-sludge process. However, with the low concentration of microbial suspended solids in the aerating wastewater, oxygen uptake can be simply related to BOD removal by the relationship

$$\text{lb of oxygen/day} = a \times \text{lb of BOD removed/day} \qquad (11\text{-}46)$$

The magnitude of a is determined by laboratory- or field-testing the particular wastewater to be treated. Values are from 0.5 to 2.0, 1.0 being typical.

Oxygen transferred by surface-aeration units can be computed by

$$R = R_0 \, \frac{\beta Cs - Ct}{9.2} \, 1.02^{T-20}(\alpha) \qquad (11\text{-}47)$$

where R = actual rate of oxygen transfer, lb of oxygen/hp-hr
  R_0 = rate of oxygen transfer of manufacturer's unit under standard conditions (water at 20°C, 1 atmosphere pressure, and zero dissolved oxygen), lb of oxygen/hp-hr
  β = oxygen-saturation coefficient of the wastewater
  Cs = oxygen concentration at saturation, mg/l
  Ct = oxygen concentration existing in liquid, mg/l
  T = temperature of lagoon liquid, °C
  α = oxygen-transfer coefficient
  9.2 = saturation oxygen concentration of pure water at 20°C, mg/l

The manufacturer's oxygen-transfer rate R_0 is guaranteed performance based on aerator test data stated in terms of standard conditions. The α and β factors depend on the characteristics of the wastewater being aerated, principally the concentration of dissolved solids. They are unity for pure water and range from 0.8 to 0.9 for most wastewaters.

A facultative, aerated lagoon results if insufficient mixing permits deposition of suspended solids. BOD removal cannot be predicted with certainty in a non-homogeneous system. Anaerobic decomposition of the settled sludge may cause emission of foul odors, particularly in treating certain industrial wastes. Facultative conditions often result from overloaded complete-mixing lagoons or may derive from poorly designed systems with inadequate mixing. To ensure odor-free operation, pond contents must be thoroughly stirred, with dissolved oxygen available throughout the liquid. Aeration equipment installed should be of proved performance purchased from a reputable manufacturer.

Significant increases in effluent BOD can occur from reducing detention time since the biological process is time-dependent. Cooling-water discharges and shock loads of relatively uncontaminated water, for example storm runoff, should be diverted around the lagoon to the secondary ponds. Sudden, large inputs of biodegradable or toxic wastes resulting from industrial spills can also upset the process. Pretreatment and control systems at industrial sites should be furnished to prevent taxing the lagoon's equalizing capacity. Biodegradability studies are essential for municipal wastewater containing measurable amounts of industrial discharges to determine such design parameters as BOD removal-rate constant, influence of temperature, nutrient requirements, oxygen utilization, and sludge production.

●EXAMPLE 11-26

Size an aerated lagoon to treat a domestic plus industrial waste flow of 0.30 mgd with an average BOD of 600 mg/l (1500 lb of BOD/day). Temperature extremes anticipated for the lagoon contents range from 10°C in winter to 35°C in summer. Minimum BOD reduction through the lagoon should be 75%. The surface aerators to be installed carry a manufacturer's guarantee to transfer 2.5 lb of oxygen/hp-hr under standard conditions. During laboratory treatability studies, the wastewater exhibited the following characteristics: $k_{20°C} = 0.68$ per day, $\theta = 1.047$, $\alpha = 0.9$, and $\beta = 0.8$.

○Solution

Required detention time at critical temperature of 10°C using Eqs. 11-44 and 11-45:

$$k_{10°C} = 0.68 \times 1.047^{10-20} = 0.43 \text{ per day}$$

$$\frac{L_e}{L_0} = 1 - 0.75 = \frac{1}{1 + 0.43t}$$

$$t = 7.0 \text{ days}$$

lagoon volume $= 0.30$ mgd $\times 7.0$ days $= 2.1$ mg $= 280,000$ ft³

Use a 10-ft depth with earth side slopes appropriate for soil conditions.
Oxygen utilization using $a = 0.8$,

At 10°C: BOD removal $= 0.75 \times 1500 = 1120$ lb of BOD/day

$$\text{oxygen required} = \frac{1120 \times 0.8}{24} = 37 \text{ lb of oxygen/hr}$$

At 35°: $k = 0.68 \times 1.047^{35-20} = 1.35$ per day

$$\text{BOD removal} = 1500 - \frac{1500}{1 + 1.35 \times 7.0} = 1360 \text{ lb of BOD/day}$$

$$\text{oxygen required} = \frac{1360 \times 0.8}{24} = 45 \text{ lb of oxygen/hour}$$

Aerator power requirements using Eq. 11-47 at a minimum of 2.0 mg/l of dissolved oxygen ($Cs = 11.3$ mg/l at 10°C and 7.1 mg/l at 35°C),

$$R_{10°C} = \frac{2.5(0.8 \times 11.3 - 2.0)}{9.2} 1.02^{10-20} \times 0.9 = 1.4 \text{ lb of oxygen/hp-hr}$$

$$\text{power required} = \frac{37 \text{ lb of oxygen/hr}}{1.4 \text{ lb of oxygen/hp-hr}} = 26 \text{ hp}$$

$$R_{35°C} = \frac{2.5(0.8 \times 7.1 - 2.0)}{9.2} 1.02^{35-20} \times 0.9 = 1.2 \text{ lb of oxygen/hp-hr}$$

$$\text{power required} = \frac{45}{1.2} = 38 \text{ hp}$$

Power requirements at 35°C control design. Use four 10-hp surface aerators.

11-35

AERATED STABILIZATION PONDS

Stabilization ponds overloaded due to industrial wastes or reduced temperature under ice cover often produce odors and insufficient BOD removals. Aeration and mixing provide distribution of dissolved oxygen for decomposition of organic matter when oxygenation by algae and wind mixing are not sufficient. A common system is air diffusion through a plastic hose dispensing network fixed to the pond bottom, although floating surface aerators and air diffusers may be used if ice formation does not impair their operation. As illustrated in Fig. 11-62, compressed air is introduced through rows of plastic tubing strung across the pond bottom. Streams of bubbles rising to the surface induce vertical mixing and oxygenation of the water. Cells normally have 10 ft of liquid depth and operate in series with a total detention time of 25–35 days. Tube aeration systems have been most successful in locations where the pond surfaces are frozen for several months in the winter and an external air supply is needed to maintain aerobic conditions. In this climate the effluent of unaerated facultative

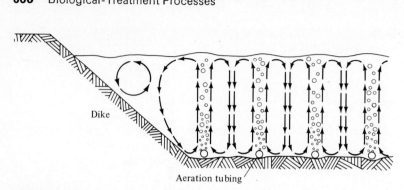

FIGURE 11-62 Stabilization pond aerated and mixed by compressed air emitting from plastic tubing laid on the lagoon bottom.

ponds is generally unsatisfactory and surface aerators are impaired by ice formation. Although aerated lagoons can provide excellent year-round BOD removal, the suspended solids concentration may exceed 30 mg/l in summer months due to algal growth.

●EXAMPLE 11-27

Compute the capacity and power requirements for an aerated stabilization pond to serve a mountain community based on the following data:

design population	600
wastewater flow, gpcd	100
average BOD, mg/l	200
maximum effluent BOD, mg/l	20
elevation above MSL, ft	8200
mean pond temperature, °C	
summer	15
winter	2
minimum dissolved oxygen, mg/l	2.0
BOD removal rate k, per day	0.5
temperature coefficient, θ	1.035
oxygen utilization, a	1.4
oxygen-transfer coefficient, α	0.9
oxygen-saturation coefficient, β	0.9

○*Solution*

Detention times using Eqs. 11-44 and 11-45 at 15°C and 2°C, respectively:

At 15°C: $k = 0.5(1.035)^{15-20} = 0.42$

$$t = \frac{(200/20) - 1}{0.42} = 21 \text{ days}$$

At 2°C: $k = 0.5(1.035)^{2-20} = 0.27$

$$t = \frac{(200/20) - 1}{0.27} = 33 \text{ days}$$

Stabilization pond volume,

At 15°C: $V = 21 \times 100 \times 800 = 1{,}680{,}000$ gal

At 20°C: $V = 33 \times 100 \times 600 = 1{,}980{,}000$ gal

The aeration power requirement is critical during winter under ice cover:

oxygen utilization based on applied BOD $= 1.4 \times 200$ mg/l $\times 0.060$ mgd $\times 8.34$

$$= 140 \text{ lb of oxygen/day}$$

Saturation dissolved oxygen at 2°C is 13.8 mg/l at mean sea level (760 mm of mercury). This value decreases with increasing elevation approximately in proportion to barometric pressure. Under standard conditions the latter drops about 25 mm/1000 ft of increased elevation. Therefore,

$$Cs \text{ at } 8200 \text{ MSL} = 13.8 \, \frac{760 - 25 \times 8.2}{760} = 10.1 \text{ mg/l}$$

Substituting in Eq. 11-47 using a manufacturer's recommended oxygen-transfer rate of 3.2 lb of oxygen/hp-hr,

$$R = 3.2 \frac{0.9 \times 10.1 - 2.0}{9.2} \, 1.02^{2-20} \times 0.9 = 1.6 \text{ lb of oxygen/hp-hr}$$

$$\text{Horsepower required} = \frac{140 \text{ lb of oxygen/day}}{24 \times 1.6 \text{ lb of oxygen/hp-hr}} = 3.6 \text{ hp}$$

Repeating the aeration-power calculations for summer conditions results in a requirement of 4.4 hp, but it is anticipated that a portion of the summer oxygen demand will be met by algal photosynthesis. Use two ponds, each with a capacity of 1.0 mg, 8–10 ft in depth piped for both parallel and series operation. Provide 4.0 hp of aeration capacity, and install plastic air tubing in accordance with manufacturer's instructions.

11-36

ANAEROBIC TREATMENT OF WASTEWATER[18]

The important characteristics of a wastewater if it is to be amenable to anaerobic treatment are: high organic strength, particularly in proteins and fats rather than sugars; sufficient inorganic biological nutrients; adequate alkalinity; absence of toxic substances; and relatively high temperature. Some industrial wastes, such as meat-processing wastewaters, have these characteristics. A typical slaughterhouse or packinghouse wastewater has a BOD of 1400 mg/l, a grease content of 500 mg/l, a pH near 7, and a temperature of 82°F.

Anaerobic processes currently being used to treat meat-processing wastewaters are the anaerobic-contact process and anaerobic lagoons. Lagoons are commonly used at slaughterhouses in rural locations. The anaerobic filter has been demonstrated as an effective treatment process in laboratory studies; however, neither pilot-scale nor full-scale anaerobic-filter systems have been tested.

The process scheme for an anaerobic-contact treatment consists of equalization, complete mixing digestion, vacuum degasifying, and final settling. This flow plan is similar to the activated-sludge process with the addition of equalization and degasifying. Raw-wastewater flow is equalized over a 24-hour period by storage in an equalizing tank. After preheating to 95°F, the wastewater is applied at a uniform rate to covered digesters. Detention time in the digesters is approximately 12 hours with a MLSS concentration of 7000–12,000 mg/l. The mixed-liquor effluent flows through vacuum degasifiers to final settling tanks. Degasifying of the mixed liquor is necessary for good settling. Settled sludge is recycled to the digester at a rate of approximately 3 : 1 by volume of incoming raw wastewater. Excess sludge is wasted to a disposal lagoon and the digestion gases collected and burned to heat the raw wastewater. BOD removal in an anaerobic contact process is approximately 96–98 %.

Anaerobic lagoons are deep (15 ft), earth-diked ponds with steep sidewalls. Raw wastewater enters near the bottom at one end of the lagoon and mixes with the active microbial mass of suspended solids in the sludge blanket which is about 6 ft deep. A discharge pipe is located on the opposite end of the lagoon submerged below the liquid surface. Excess undigested grease floats on the liquid surface of the lagoon, forming a natural cover for the retention of heat and strict anaerobic conditions. In an anaerobic-lagoon system, the wastewater is neither equalized nor heated. Excess sludge is washed out in the wastewater effluent and recirculation is not necessary. Major advantages of anaerobic lagoons are low first and maintenance costs, the ability to accept shock and intermittent loading, and simplicity of operation. Anaerobic lagoons operating at loadings of 15–20 lb of BOD/1000 ft^3/day, at a detention time of 4 or more days, and at a temperature above 75°F, remove 75–85 % of the influent BOD.

The latest anaerobic-treatment process being evaluated is the anaerobic filter, consisting of a covered tank filled with media. Wastewater is applied at the bottom and withdrawn from the top, with filter media submerged. The results of laboratory tests indicate the following advantages: high solids-retention times can be achieved without recirculation; low-strength wastewaters can be treated at hydraulic detention times less than 1 day; satisfactory operation is possible at relatively low temperatures; and the filter can be operated intermittently without significant loss of efficiency.

Odor Control

Increased urbanization has resulted in wastewater-treatment plants being situated in close proximity to housing areas and commercial developments. This has caused complaints about odors and, in serious situations, led to lawsuits against municipalities operating the disposal systems. Although the problem of foul odors emitting from treatment plants is not new, only in recent years have political and legal pressures forced processing facilities to consider abatement.

11-37

SOURCES OF ODORS IN WASTEWATER TREATMENT

Principal odors are hydrogen sulfide and organic compounds generated by anaerobic decomposition. The latter include mercaptans, indole, skatole, amines, fatty acids, and many other volatile organics. Often, industrial wastes in a municipal sewer create odors inherent in the raw materials being processed or the manufactured products (poultry processing, slaughtering and rendering, tanning, and manufacture of volatile chemicals). With the exception of hydrogen sulfide, a specific odor-producing substance is very difficult to identify. Weather conditions, such as temperature and wind velocity, influence the intensity and prevalence of emissions.

Frequently, the initial evolution of malodors is from septic wastewater in the sewer collection system. Flat sewer grades, warm temperatures, and high-strength wastes lead to anaerobiosis. The first sources at the treatment plant are the wet well and grit chamber. Turbulent flow and preaeration of raw waste strip dissolved gases and volatile organics, discharging them into the atmosphere. Odors also may arise from the liquid held in primary clarifiers, particularly if excess activated sludge is returned to the head of the plant, resulting in an active microbial seed being mixed with the settleable raw organic matter. Sludge taken from these tanks has an obnoxious smell. Pumping it into uncovered holding tanks releases the scent previously confined under the water cover. Polymers do not neutralize the olfactory compounds prior to vacuum filtration; therefore, the air drawn through the sludge cake picks up volatile compounds and carries them to the atmosphere. Use of ferric chloride and lime for conditioning chemicals significantly reduces odors, but for most municipal wastes, polymers provide more economical operation. The process of anaerobic digestion takes place in enclosed tanks while digested sludge is dewatered either mechanically or on drying beds. The smell of well-digested sludge is earthy, but if the digestion process is not complete, intermediate aromatic compounds may be released during drying.

Secondary biological processes also yield odors, particularly rock-filled trickling filters. Although referred to as aerobic devices, filters are actually facultative, since the microbial films are aerobic on the surface and anaerobic adjacent to the medium (Fig. 11-21). Because of this potential for anaerobic decomposition, filters under heavy organic loading often reek. Odors are not as likely to be created in biological towers because of thinner biological films and improved aeration. Activated-sludge processes yield a relatively inoffensive musty odor carried by the air passing through the mixed liquor. Foul smells are rare since microbial flocs in the aeration basin are completely surrounded by liquid containing dissolved oxygen.

11-38

METHODS OF ODOR CONTROL

Modern treatment plant design should incorporate the concept of odor prevention. This involves a comprehensive understanding of potential problems possible in the application of certain processes for handling a wastewater. For example, the exhaust from vacuum filtration of a raw sludge coagulated with a polymer is likely to be offensive, particularly if the municipal wastewater contains industrial wastes. Individual unit processes can be designed to minimize the possibility of anaerobiosis, such as the vent and underdrain design of a trickling filter to allow free circulation of air. In siting plants it is wise to provide a reasonable buffer zone to prevent encroachment of activities that will be offended by the essence of a treatment plant.

The first step in analyzing an existing problem is to determine the cause of odorousness and attempt to isolate the sources. Special attention must be paid to industrial wastes entering the sewer system. Overloading often increases malodors, however, expansion of facilities is no guarantee that the situation will change dramatically. Foul emissions can be given off by properly loaded units if the design is poor, if they are not maintained properly, or when the waste includes organics with an inherent smell. Adding secondary treatment to an existing primary plant can unwittingly create obnoxious conditions that did not previously exist. In one case history, waste sludge from a new activated-sludge secondary was disposed of by return to the head of the plant. This resulted in anaerobic biological decomposition of the settled sludge, causing upset of the primary tanks and operational problems in vacuum filtration, with both leading to foul odors.

Chemicals can sometimes be used to oxidize odorous compounds, particularly hydrogen sulfide. Chlorination of the wastewater in main sewers, or prior to primary settling, may prove beneficial. Using lime and ferric chloride as chemical sludge conditioners reduces bacterial activity and oxidizes many products of anaerobic decomposition. In some cases the only feasible solution has been to provide sealed enclosures to prevent odors from reaching the atmosphere (Fig. 11-63). Air confined under clarifier and grit basin covers must be continuously purified to remove corrosive gases, such as hydrogen sulfide. Air must be circulated through a trickling-filter bed for ventilation. This can be accomplished by forcing fresh air under the dome and collecting it from the underdrain system. Recirculated air can be cleansed using wet scrubbers with a chemical solution, or activated-carbon absorption beds. Solutions of either permanganate or hypochlorite have been effective oxidizing agents for some odors, but carbon filters generally perform better. Sometimes carbon beds are used in series with wet scrubbers. At one major installation, emissions from trickling filters are handled by hypochlorite scrubbers followed by activated carbon filters. Air extracted from under the clarifier covers is routed only through the carbon filters.

FIGURE 11-63 Fiberglass cover enclosing a trickling filter to contain odors and reduce cooling of the wastewater in winter. (Courtesy Fiberglass Specialty Company.)

Septic Tanks

Nearly one-fourth of the homes in the United States are located in unsewered areas and must rely on individual household disposal systems. The most popular process uses a septic tank and absorption field because of the low cost and the desirability of underground effluent disposal.

11-39

INDIVIDUAL HOUSEHOLD SEPTIC-TANK SYSTEMS

Where plumbing systems are installed, wastewaters from rural dwellings and other buildings are usually disposed of in the ground. If soil and site conditions are favorable, the septic-tank system can be expected to give satisfactory service.

The first step in the design of a septic-tank system is to determine whether the soil is suitable for taking up the effluent water and, if so, at what rate. The soil must have a satisfactory absorption rate without interference from groundwater or impervious strata. The minimum depth to groundwater should be greater than 4 ft, and impervious strata should be at a depth greater than 4 ft below the bottom of the tile trench or seepage pit. If these conditions cannot be met, the site is generally unsuitable for a septic-tank installation.

Percolation Tests

In the absence of groundwater or subsoil information, subsurface explorations are necessary. This investigation may be carried out with a posthole auger or soil auger with an extension handle. In some cases the examination of road cuts or foundation excavations will give useful information. If subsurface investigation appears suitable, percolation tests should be made at typical points where the disposal field is to be located.

Percolation tests determine the acceptability of the site and serve as the basis of design for the liquid absorption. The following procedure for percolation tests was developed at the Robert A. Taft Sanitary Engineering Center.[20]

1 Six or more tests should be made in separate test holes uniformly spaced over the proposed absorption field site.

2 Dig or bore a hole with horizontal dimensions of from 4 to 12 in. and vertical sides to the depth of the proposed trench (Fig. 11-64).

3 Carefully scratch the bottom and sides of the excavation with a knife blade or sharp-pointed instrument, to remove any smeared soil surfaces and to provide a natural soil interface into which water may percolate. Add 2 in. of coarse sand or fine gravel to the bottom of the hole.

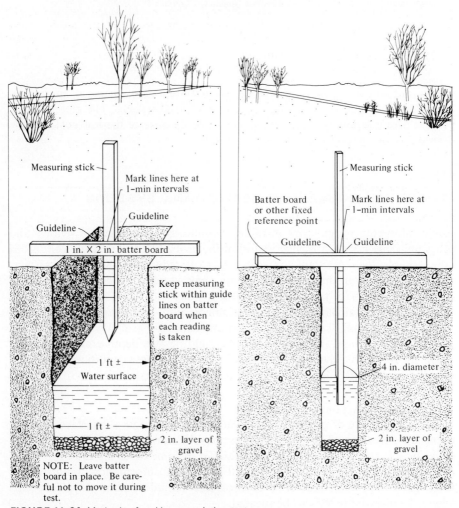

FIGURE 11-64 Methods of making percolation tests.

4 Carefully fill the hole with clear water to a minimum depth of 12 in. above the gravel or sand. Keep water in the hole at least 4 hours, preferably overnight. In most soils it will be necessary to augment the water as time progresses. Determine the percolation rate 24 hours after water was first added to the hole. In sandy soils containing little clay, this prefilling procedure is not essential, and the test may be made as described under item 5c, after water from one filling of the hole has completely seeped away.

5 The percolation-rate measurement is determined by one of the following methods:

a If water remains in the test hole overnight, adjust the water depth to approximately 6 in. above the gravel. From a reference batter board as shown in Fig. 11-64, measure the drop in water level over a 30-minute period. This drop is used to calculate the percolation rate.

b If no water remains in the hole the next day, add clean water to bring the depth to approximately 6 in. over the gravel. From the batter board, measure the drop in water level at approximately 30-minute intervals for 4 hours, refilling to 6 in. over the gravel as necessary. The drop in water level that occurs during the final 30-minute period is used to calculate the percolation rate.

c In sandy soils (or other soils in which the first 6 in. of water seeps away in less than 30 minutes, after the overnight period), the time interval between measurements shall be taken as 10 minutes and the test run for 1 hour. The drop in water level that occurs during the final 10 minutes is used to calculate the percolation rate.

Leaching System

If the percolation rate is slower than 1 in. in 30 minutes, the soil is unsuitable for seepage pits, and if the rate is less than 1 in. in 60 minutes, the area is unsuitable for any type of soil-absorption system. Selection of the leaching system will be dependent to some extent on the site under consideration. The two common absorption systems are trenches and seepage pits. A seepage pit is a covered excavation with an open-joint lining through which water from the septic tank may seep or leach into the surrounding soil. Trench systems utilize tiled porous drains embedded in gravel which distribute the septic-tank waste liquid over a large area of soil.

Where percolation rates are satisfactory, the required absorption area is determined by Table 11-7 and Fig. 11-65. In general, all soil absorption systems should be kept 100 ft away from any water-supply well, 50 ft away from any stream or watercourse, and 10 ft away from dwellings or property lines.

A trench absorption field consists of 12-in. lengths of 4-in. agricultural drain tile, 2- or 3-ft lengths of vitrified clay sewer pipe, or perforated, nonmetallic pipe, laid in such a manner that water from the septic tank will be distributed uniformly over the soil. Individual laterals (straight sections of pipeline) should not be over 100 ft in length and should be laid on a grade of 2–4 in./100 ft. Use of several shorter laterals is preferred because if one should become clogged, most of the field will still be serviceable.

TABLE 11-7

ABSORPTION-AREA REQUIREMENTS FOR PRIVATE RESIDENCES[a]

Percolation Rate (Time Required for Water to Fall 1 in.) (minutes)	Required Absorption Area, in Square Feet per Bedroom[b] Standard Trench[c] and Seepage Pits[d]	Percolation Rate (Time Required for Water to Fall 1 in.) (minutes)	Required Absorption Area, in Square Feet per Bedroom[b] Standard Trench[c] and Seepage Pits[d]
1 or less	70	10	165
2	85	15	190
3	100	30[e]	250
4	115	45[e]	300
5	125	60[e, f]	330

[a] Provides for garbage grinders and automatic-sequence washing machines.
[b] In every case, sufficient area should be provided for at least two bedrooms.
[c] Absorption area for standard trenches is figured as trench-bottom area.
[d] Absorption area for seepage pits is figured as effective side-wall area beneath the inlet.
[e] Unsuitable for seepage pits if over 30.
[f] Unsuitable for leaching systems if over 60.

Source: "Manual of Septic-Tank Practice," *U.S. Public Health Service Publication No. 526* (Washington, D.C.: Government Printing Office, 1957).

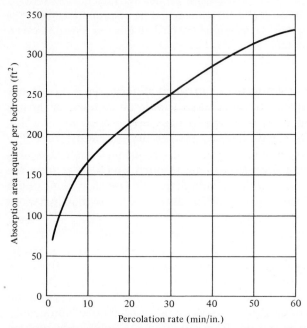

FIGURE 11-65 Absorption-area requirements for private residences.[20]

Several designs have been used in laying out the trench pattern. The choice may depend upon the shape of the available disposal area and the topography. Typical layouts are shown in Figs. 11-66 and 11-67. Details of sections of a trench are shown in Fig. 11-68. The minimum depth of trench should be 18 in.

Additional depth may be necessary because of topography. The trench should be 18–24 in. in width and the tile laid on 6 in. of gravel. The absorption area is proportional to the width of trench up to 36 in. wide.

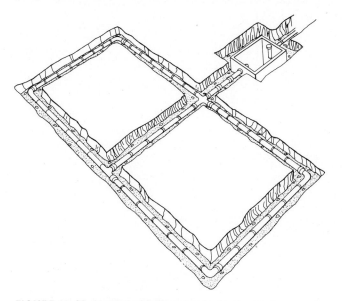

FIGURE 11-66 Absorption-field system for level ground.

A distribution box is usually considered necessary for a trench absorption-field system. The purpose of the box is to ensure equal distribution of effluent to all the laterals. The box is particularly important on sloping ground.

A seepage pit is a covered pit with open-jointed lining through which septic-tank effluent may be absorbed into the surrounding soil. It is normally considered a less desirable method of liquid disposal than the trench method.

Sizing a Septic Tank

A septic tank allows solids to settle out of a waste and permits a clarified effluent to be discharged. The solids are broken down by anaerobic digestion. Remaining solids accumulate in the tank and must be pumped out periodically.

One of the most important considerations in septic-tank design is the liquid capacity. Table 11-8 lists the recommended capacities for private dwellings based upon the number of bedrooms, and allows for the use of all normal household appliances, including garbage grinders.

Septic tanks should be of watertight construction and made of materials not subject to excessive corrosion or decay. Concrete, coated metal, vitrified clay, concrete block, or brick have been successfully used. The tank should be so constructed as to provide adequate access to each compartment of the tank, as well as the inlet and outlet. A 20-in. manhole is considered the minimum size for access. A typical domestic septic tank is shown in Fig. 11-69.

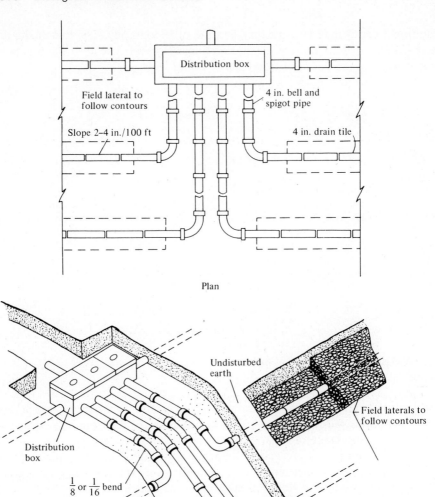

Plan

Detail

FIGURE 11-67 Absorption field for sloping ground.[20]

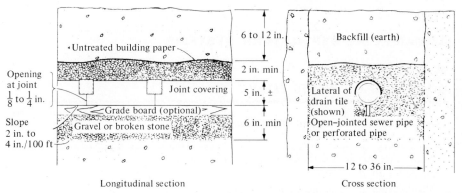

FIGURE 11-68 Absorption trench and lateral.[20] Note: Unevenness of ends of drain tile will usually provide the necessary openings between joints. Special collars may be used if desired.

TABLE 11-8

LIQUID CAPACITY OF TANK[a]

Number of Bedrooms	Recommended Minimum Tank Capacity (gal)	Equivalent Capacity per Bedroom (gal)
Two or less	750	375
Three	900	300
Four[b]	1000	250

[a] Provides for use of garbage grinders, automatic washers, and other household appliances.
[b] For each additional bedroom, add 250 gal.

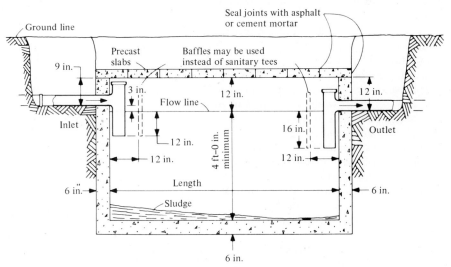

FIGURE 11-69 Longitudinal section of a concrete septic tank.

11-40

COMMERCIAL SEPTIC-TANK SYSTEMS

TABLE 11-9
QUANTITIES OF WASTE FLOW[20]

Type of Establishment	Gallons per Person per Day
Small dwellings and cottages with seasonal occupancy	50
Single-family dwellings	75
Multiple-family dwellings (apartments)	60
Rooming houses	40
Boarding houses	50
Additional kitchen wastes for nonresident boarders	10
Hotels without private baths	50
Hotels with private baths (two persons per room)	60
Restaurants (toilet and kitchen wastes per patron)	7–10
Restaurants (kitchen wastes per meal served)	2.5–3
Additional for bars and cocktail lounges	2
Tourist camps or trailer parks with central bathhouse	35
Tourist courts or mobile home parks with individual bath units	50
Resort camps (night and day) with limited plumbing	50
Luxury camps	100–150
Work or construction camps (semipermanent)	50
Day camps (no meals served)	15
Day schools without cafeterias, gymnasiums, or showers	15
Day schools with cafeterias, but no gymnasiums or showers	20
Day schools with cafeterias, gyms, and showers	25
Boarding schools	75–100
Day workers at schools and offices (per shift)	15
Hospitals	150–250+
Institutions other than hospitals	75–125
Factories (gallons per person per shift, exclusive of industrial wastes)	15–35
Picnic parks with bathhouses, showers, and flush toilets	10
Picnic parks (toilet wastes only) (gallons per picnicker)	5
Swimming pools and bathhouses	10
Luxury residences and estates	100–150
Country clubs (per resident member)	100
Country clubs (per nonresident member present)	25
Motels (per bed space)	40
Motels with bath, toilet, and kitchen wastes	50
Drive-in theaters (per car space)	5
Movie theaters (per auditorium seat)	5
Airports (per passenger)	3–5
Self-service laundries (gallons per wash, i.e., per customer)	50
Stores (per toilet room)	400
Service stations (per vehicle served)	10

Septic tanks are used in providing waste treatment and disposal for many types of establishments, such as schools, motels, rural hotels, trailer parks, housing projects, camps, and others. While many effluents from commercial septic tanks are disposed of by soil absorption, others are discharged to available water-courses after suitable treatment. Where soil absorption is contemplated, it is essential to determine the characteristics of the soil as the first step in design.

Next, it is necessary to obtain information on the volume of waste to be treated. In the absence of definite information, the volume of wastewater can be estimated from Table 11-9. Values listed in this table are averages and should be modified where additional information is available.

With information from percolation tests, the rate at which wastewater may be applied to a trench field may be taken from Table 11-10 or Fig. 11-70. These data do not allow for wastes from garbage grinders and automatic washing machines. If wastes from these appliances are to be handled, the absorption field should be increased by 20% for garbage grinders, 40% for automatic washers, and 60% if both are to be used.

TABLE 11-10

ALLOWABLE RATE OF WASTEWATER APPLICATION TO A SOIL-ABSORPTION SYSTEM

Percolation Rate (time in minutes for water to fall 1 in.)	Maximum Rate of Waste Application for Standard Trenches[b] and Seepage Pits[c] (gal/ft^2/day)[a]	Percolation Rate (time in minutes for water to fall 1 in.)	Maximum Rate of Waste Application for Standard Trenches[b] and Seepage Pits[c] (gal/ft^2/day)[a]
1 or less	5.0	10	1.6
2	3.5	15	1.3
3	2.9	30[d]	0.9
4	2.5	45[d]	0.8
5	2.2	60[d, e]	0.6

[a] Not including effluents from septic tanks that receive wastes from garbage grinders and automatic washing machines.
[b] Absorption area for standard trench is figured as trench-bottom area.
[c] Absorption area for seepage pits is effective side-wall area.
[d] Over 30 unsuitable for seepage pits.
[e] Over 60 unsuitable for leaching systems.

Single-compartment septic tanks are acceptable for private household installations; tanks with two compartments are normally provided for large institutional systems. The first compartment serves as the primary settling basin and should be two to three times the capacity of the second compartment. The relative absence of solid material in the second tank indicates that the system is functioning properly.

For flows between 500 and 1500 gpd, the capacity of the tank below the flowline should be equal to at least 1.5 times the day's waste flow. With flows

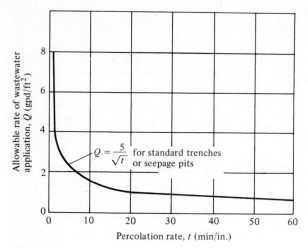

FIGURE 11-70 Graph showing the relation between percolation rate and allowable rate of waste-water applicable.[20]

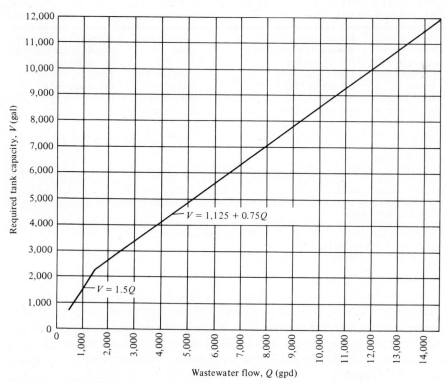

FIGURE 11-71 Effective septic-tank capacities for waste flows up to 14,500 gpd.

greater than 1500 gpd, the tank capacity below the waterline should equal 1125 gal plus 75% of the daily waste flow.

$$V = 1125 + 0.75Q \qquad\qquad (11\text{-}48)$$

where V = volume of tank below waterline, gal
   Q = daily waste flow, gal

Figure 11-71 gives the required tank capacity below the flowline for flow from 500 to 14,500 gpd. When the volume of wastewater exceeds the amount that can be absorbed in 500 lineal feet of trench, a dosing tank should be used. The dosing tank should be equipped with an automatic siphon which discharges the dosing chamber every 3–4 hours. The tank should have a capacity equal to about 70% of the interior volume of the tile to be closed at one time. Additional information on the design of septic tanks and absorption fields will be found in "Manual of Septic-Tank Practice," *Public Health Service Publication No. 526*, and recent literature.

●EXAMPLE 11-28
Determine the volume of a septic tank and the length of absorption trench for a hotel without private baths to handle 100 persons. Results of a percolation test gave a rate of 1 in. in 5 minutes.

○*Solution*
The waste flow for a hotel without private baths is taken from Table 11-9 and is found to be 50 gal/day/person.

total waste flow $= 50 \times 100 = 5000$ gpd

Volume of the septic tank below the waterline is obtained from Fig. 11-71 and is found to be 4850 gal.

Allowable rate of wastewater application to the absorption field is obtained from Fig. 11-60. The dosage rate is found to be 2.2 gpd/ft² of trench.

The trench is assumed to be 2 ft wide.

absorption area $= 5000 \text{ gpd}/2.2 \text{ gpd/ft}^2$

$\qquad\qquad = 2270 \text{ ft}^2$

length of trench $= 2270/2$

$\qquad\qquad = 1135 \text{ ft}$

Volume of the dosing chamber is calculated to be 70% of the volume of 1135 ft of 4-in. tile.

$$\text{volume} = 1135 \text{ ft} \times \frac{\pi}{36} \times 7.48 \times 0.70$$

volume of dosing chamber $= 518$ gal

11-41

INDIVIDUAL HOUSEHOLD AERATION UNITS

Several types of household aeration systems that can be placed underground are manufactured in different shapes and sizes, constructed of concrete, steel, or fiberglass. A typical unit consists of a compartmented tank with clarification following aeration. Some designs have a presedimentation chamber to collect heavy solids and floating grease. The aeration portion is sized for a 24- to 48-hour detention period, and is mixed with either compressed air or some type of mechanical aerator. The final settling chamber has steep walls for gravity separation and return of settled solids through a slot in the hopper bottom. Although these systems have the appearance of an extended aeration process, in reality they seem to function more like a complete-mixing aerated lagoon. The main reason is that gravity return of solids is not effective in maintaining an adequate mixed-liquor solids concentration, particularly with surging influent caused by high rates of discharge from household appliances. Consequently, the process is not efficient in solids capture or retention, which leads to insufficient treatment and a poor-quality effluent.

The main motivation for developing miniature aeration units was an attempt to permit surface effluent discharge and thus avoid the use of absorption fields. However, this raises the question of disinfection to meet the required bacteriological effluent standard. The simplest system would be a chlorine-contact chamber equipped with a tablet-type chlorinator. One such unit consists of a slotted tube containing solid tablets of calcium hypochlorite suspended in the wastewater flow such that a suitable amount of chlorine is dissolved for disinfection. Problems include control of chlorine dosage over a wide range of flows, and the possibility that homeowners may not be conscientious in replacing hypochlorite tablets because of inconvenience and cost. Also, a poorly operating aeration tank may discharge excessive suspended solids that prevent effective disinfection. There is little evidence to substantiate that effluent standards of 30 mg/l of BOD, 30 mg/l of suspended solids, and a fecal coliform concentration of 200/100 ml can be met by small-compartmented aeration tanks. An underground absorption field is acceptable, but then there is no purpose for installing a more expensive aeration tank when an ordinary septic tank produces an equally suitable effluent for leaching. Where drain fields are impossible due to impermeable soils, rather extensive treatment is required to meet the established quality standards for surface discharge. The labor and cost of operating, maintaining, and monitoring to ensure satisfactory performance of individual household aeration–chlorination systems is too costly for adoption by most homeowners.

PROBLEMS

11-1 The empirical formula for bacterial protoplasm is $C_5H_7O_2N$. What is the percentage by weight of nitrogen in this formula?

11-2 Why do *Thiobacillus* convert hydrogen sulfide to sulfuric acid?

11-3 List the major nutritional and environmental conditions necessary to culture algae in a laboratory container.

11-4 What are the differences and similarities between algae and nitrifying bacteria?

11-5 Discuss the relationships among metabolism, energy, and synthesis and the effect of these upon growth under aerobic and anaerobic conditions. Include comments on growth rates, extent of metabolism, and limiting factors under the two environments.

11-6 What nutrients are required for growth of iron bacteria inside water mains?

11-7 Why do some bacteria convert ammonia nitrogen to nitrate (Eqs. 11-1 and 11-2) while others reduce nitrate to nitrogen gas (Eq. 11-12)?

11-8 In a mixed culture fermentation, why is dissolved oxygen the preferred hydrogen acceptor?

11-9 Why is bacterial synthesis for the same quantity of substrate greater in an aerobic environment than under anaerobiosis?

11-10 How does temperature affect biological processes?

11-11 List the characteristics of the endogenous growth phase.

11-12 State the two main reasons why an activated-sludge system is operated at a relatively low food/microorganism ratio.

11-13 Why are bacteria rather than protozoans the primary feeders in activated sludge?

11-14 Based on Eqs. 11-9 and 11-10 and Fig. 11-15, why is impending failure of an anaerobic digestion process forecast by an increase in the percentage of carbon dioxide in the gas produced?

11-15 Describe the role of algae in biological stabilization of wastewater in a stabilization pond.

11-16 What is wrong with the following statement? " The key to waste stabilization in a pond is the algal growth; overloading kills the algae, while they thrive on a reasonable supply of organic matter."

11-17 The wastewater from a candy manufacturer is 97,700 gpd with a BOD concentration of 1560 mg/l. Calculate the BOD equivalent population of this waste. What is the hydraulic equivalent population?

11-18 The municipal wastewater flow from a city with a population of 150,000 is 76,000 m^3/day with a BOD equal to 320 mg/l. Compute the BOD equivalent population for the wastewater flow.

11-19 The domestic and industrial waste from a community consists of 100 gpcd from 7500 persons; 65,000 gpd from a milk-processing plant with a BOD of 1400 mg/l; and a 90,000 gpd containing 450 lb of BOD from potato-chip manufacturing. Calculate the combined wastewater flow, BOD concentration in the composite waste, and BOD equivalent population.

11-20 The combined wastewater flow from a community includes domestic waste from a sewered population of 2000 and industrial wastes from a dairy and a poultry-dressing plant. The poultry plant discharges 125 m³/day and 136 kg of BOD/day. The dairy produces a flow of 190 m³/day with a BOD concentration of 900 mg/l. Estimate the total combined wastewater flow from the community and the BOD concentration in the composite discharge.

11-21 The wastewater flow from a slaughterhouse that processes 500,000 lb of LWK (live-weight kill) per day is 0.53 mgd containing 7840 lb of BOD. Compute the BOD concentration in the wastewater, equivalent populations of the waste flow, and equivalent populations per 1000 lb of LWK.

11-22 The following are design data for the town of Nancy, with a sewered population of 7600. Design flows are: average daily = 0.84 mgd, peak hourly = 1.25 mgd, and minimum hourly 0.12 mgd. Design average BOD equals 1740 lb/day and average suspended solids equal 1530 lb/day. Calculate the following: equivalent population based on 0.17 lb of BOD per capita; design flows in units of gpm, ft³/sec, m³/min, and m³/day; mean BOD and SS concentrations in mg/l.

11-23 A low-rate, single-stage, trickling-filter secondary consists of two 65-ft-diameter, 6-ft-deep filters operating in parallel. Calculate the BOD and hydraulic loads for a settled wastewater flow of 0.50 mgd with a BOD concentration of 140 mg/l. Determine the BOD removal efficiency for this system, using: (a) the NRC formula; (b) the GLUMRB standards; (c) the formula derived in Example 11-7.

11-24 A trickling filter, with a diameter of 26.0 m and depth of 2.0 m, receives a wastewater flow of 6500 m³/day containing 610 kg of BOD. Calculate the BOD loading in g/m³ and the hydraulic loading in m³/m². Determine the NRC efficiency, assuming that the flow includes a $R = 0.5$.

11-25 Determine the dimensions for two equal-size, single-stage, circular, low-rate trickling filters for the town of Nancy (Prob. 11-22). Base your calculations on the following: (a) average daily design flow and average BOD load; (b) a primary clarifier with an overflow rate of 600 gpd/ft²; (c) a design loading rate of 15 lb/1000 ft³/day of BOD; (d) a 6.0-ft depth of rock media. Compute the effluent BOD concentration based on the NRC formula.

11-26 Determine the diameter for a single-stage, circular, high-rate trickling filter for the town of Nancy (Prob. 11-22) based on: (a) design average BOD load; (b) a primary BOD removal efficiency of 35%; (c) a design loading of 45 lb/1000 ft³/day of BOD; (d) a rock-filled filter depth of 7.0 ft. For this filter, calculate the recirculation flow required to attain a hydraulic loading of 15 mg/acre/day. Compute the effluent BOD concentration based on the NRC formula.

11-27 The municipal wastewater flow from a town is 1890 m³/day with an average BOD of 280 mg/l. Assuming 35% BOD removal in the primary, calculate the size required for one single-stage, high-rate trickling filter. Use a depth of 2.1 m and a BOD loading of 480 g/m³-day. Compute the BOD concentration in the effluent at 22°C using the NRC formula and temperature correction relationship in Eq. 11-24.

11-28 A single-stage trickling-filter plant consists of a primary clarifier, trickling filter 70 ft in diameter with a 7-ft depth of Flexiring medium, and final clarification. The raw-wastewater flow is 0.80 mgd with 200 mg/l of BOD, indirect recirculation to the wet well is 0.40 mgd, and direct recirculation around the filter is 400 gpm. Assuming 35 % BOD removal in the primary, calculate the effluent BOD concentration for a wastewater temperature of 20°C.

11-29 Estimate the effluent BOD for the two-stage trickling-filter plant designed in Example 11-12 for a wastewater flow of 1.0 mgd with a BOD concentration of 350 mg/l.

11-30 Determine the NRC BOD removal efficiency for a two-stage trickling-filter plant based on the following: primary clarification with 35 % BOD reduction; first-stage filters loaded at 90 lb/1000 ft³/day; intermediate settling; second-stage filters sized identical to the first-stage units; an operating recirculation ratio of 1.0 for all filters; final clarification of the effluent; and a wastewater temperature of 18°C.

11-31 Size a circular biological tower with plastic medium for secondary treatment for the wastewater from the town of Nancy (Prob. 11-22). Assume the following: 35 % BOD removal in the primary, a required effluent BOD of 30 mg/l, and a tower depth of 20 ft. Calculate the design BOD loading, hydraulic loading using a recirculation ratio of 1.0, and volume of packing required.

11-32 What is the BOD removal efficiency in a biological tower-final clarifier system with plastic medium based on the following: tower depth = 6.1 m (20 ft), wastewater BOD = 150 mg/l, BOD loading of 1.60 g/m³-day, and hydraulic loading without recirculation = 72 m³/m²-day.

11-33 Size a redwood-medium tower for the town of Nancy (Prob. 11-22) based on the criteria used in Example 11-14.

11-34 Calculate the rotating-disk surface area required for secondary treatment of the wastewater from the town of Nancy (Prob. 11-22). Assume 35 % BOD removal in the primary, a required effluent quality of 20 mg/l of BOD, and wastewater temperature of 50°F.

11-35 A step-aeration activated-sludge basin, with a total volume of 100,000 ft³, receives 3.67 mgd of wastewater with a BOD of 130 mg/l. The MLSS concentration in the aerating liquid is 2500 mg/l. Calculate the aeration period and BOD loadings in both lb/1000 ft³/day and lb of BOD/day/lb of MLSS.

11-36 A conventional activated-sludge secondary processes 29,000 m³/day of waste-water with an influent BOD of 170 mg/l. The total volume of the aeration basins is 8480 m³ and the MLSS concentration is 2700 mg/l. Calculate the aeration period and BOD loadings in both g/m³-day and g of BOD/day/g of MLSS.

11-37 The following activated-sludge systems are being considered for the town of Nancy (Prob. 11-22); (a) extended aeration without primary sedimentation (oxidation ditch); (b) step-aeration secondary following primary settling that removes 35 % of the BOD; (c) high-purity oxygen without primary sedimentation. Calculate the aeration-tank capacity in cubic feet required for each system based on the following criteria:

System	Maximum BOD Loading (lb/1000 ft³/day)	Minimum Aeration Period (hours)
Extended aeration	30	24
Step aeration	40	7.5
High-purity oxygen	100	2.5

11-38 A step-aeration activated-sludge system at a loading of 40 lb of BOD/1000 ft³/day requires an air supply of 1200 ft³/lb of BOD applied to maintain an adequate dissolved-oxygen level. The measured average oxygen utilization of the mixed liquor is 36 mg/l/hr. Calculate the oxygen-transfer efficiency.

11-39 An air supply of 1000 ft³ of air is required per pound of BOD applied to a diffused aeration basin to maintain a minimum DO of 2.0 mg/l. Assuming that the installed aeration equipment is capable of transferring 1.0 lb of oxygen to dissolved oxygen per pound of BOD applied, calculate the oxygen-transfer efficiency of the system. (One cubic foot of air at standard temperature and pressure contains 0.0174 lb of oxygen.)

11-40 Using the CMAS equations after McKinney, compute the oxygen-utilization rate of the mixed liquor in a high-rate activated-sludge system based on the following values: $Fi = 280$ mg/l, $t = 3.0$ hours, $Km = 15$, $Ma = 2800$ mg/l, $Ke = 0.02$, and $t_s = 72$.

11-41 The mixed liquor near the discharge end of a spiral-flow aeration tank was tested for suspended-solids concentration and settleability. The MLSS concentration was 2200 mg/l and the volume occupied by the MLSS after 30-minute settling in a 1-liter graduated cylinder was 200 ml. Compute the SVI, theoretical recirculation ratio for system operation, and the estimated solids concentration in the return activated sludge.

11-42 A dairy wastewater of 0.25 mgd with 1000 mg/l of BOD is treated in an aeration tank and clarifier system, without primary settling. The volume of the aeration basin is 69,500 ft³. The MLSS in the aerating liquor is 2000 mg/l, and settles to a volume of 200 ml/l in 30 minutes. Compute the following: aeration period, BOD loadings, BOD sludge age, SVI, and suggested sludge-recirculation rate.

11-43 A complete-mixing activated-sludge secondary with an aeration tank volume of 170 m³ has an applied load of 830 m³/day with an average BOD of 200 mg/l and suspended solids of 240 mg/l. The mixed liquor in the aeration tank is held at 3600 mg/l of suspended solids. Based on these data, calculate the following: aeration period in hours; BOD loading in g/m³-day and g of BOD/g of MLSS; suspended-solids sludge age; BOD sludge age; and excess sludge production in g of VSS produced/g of BOD applied.

11-44 The high-purity-oxygen process illustrated in Fig. 11-55 is being considered to treat an unsettled domestic wastewater flow of 0.30 mgd with an average BOD of 200 mg/l. Compute the aeration volume required based on a maximum allowable BOD loading of 130 lb of BOD/1000 ft³/day and minimum aeration period of 1.8 hours. For this size basin, what is the F/M ratio in terms of lb of BOD/day/lb of MLVSS, assuming a MLSS of 5500 mg/l, which is 75% volatile. Estimate the oxygen consumed based on the data given in Fig. 11-58.

11-45 Calculate the surface area required for a stabilization pond to serve a domestic population of 1000. Assume 80 gpcd at 210 mg/l of BOD. Use a design loading of 20 lb of BOD/acre/day. If the average liquid depth is 4 ft, calculate the retention time of the wastewater based on influent flow. The effluent is spread on grassland by spray irrigation at a rate of 2.0 in./week (54,300 gal/acre/wk). Compute the land area required for land disposal. In these computations assume no evaporation or seepage losses from the ponds.

11-46 Faculative stabilization ponds with a total surface area of 6.0 hectare (1 ha = 10,000 m²) serves a community with a waste discharge of 530 m³/day at a BOD of 280 mg/l. Calculate the BOD loading and days of winter storage available between the 0.6-m and 1.5-m depths assuming a daily water loss of 0.30 cm by evaporation and seepage.

11-47 Stabilization pond computations for the town of Nancy (Prob. 11-22). (a) Calculate the lagoon area required for a design loading of 40 lb of BOD/acre/day.
(b) The percentage of the average design flow that appears as effluent from the lagoons assuming a water loss from the ponds of 60 in./yr (seepage plus evaporation minus precipitation).
(c) Using the water-balance data from part (b), calculate the BOD removal efficiency if the average BOD concentrations in the influent and effluent are 250 mg/l and 25 mg/l, respectively.
(d) How many acres of cropland are needed to dispose of the effluent by spray irrigation if the application rate is 2.0 in./week year-round?

11-48 A complete-mixing aerated lagoon is being considered for pretreatment of a strong industrial wastewater with a $k = 0.70$ at 20°C and $\theta = 1.035$, using a detention time of 4 days. What is the BOD reduction at 20°C based on Eqs. 11-44 and 11-45? If the wastewater temperature is 10°C, compute the detention time required to achieve the same degree of treatment.

11-49 A manufacturer's specified oxygen-transfer capacity of a surface-aeration unit is 3.0 lb of oxygen/hp-hr. Using Eq. 11-47, calculate the oxygen-transfer capability of this unit for an $\alpha = 0.9$, $\beta = 0.8$, temperature = 20°C, and a dissolved oxygen level of 2.0 mg/l in the lagoon water.

11-50 A complete-mixing aerated lagoon with a volume of 1.0 million gallons is to treat a daily wastewater flow of 250,000 gal with a BOD of 400 mg/l. The liquid temperature ranges from 4°C in the winter to 30°C in the summer. There are four 5.0-hp surface aerators rated at 2.0 lb of oxygen/hp-hr. The wastewater characteristics are: $\theta = 1.035$, $\alpha = 0.9$, $\beta = 0.9$, k at 20°C = 0.80 per day, and $a = 1.0$ lb of oxygen utilized/lb of BOD removed. The designer states that this system will remove at least 400 lb of BOD/day and maintain a dissolved-oxygen concentration greater than 2.0 mg/l. Verify these claims by appropriate calculations.

REFERENCES

1. R. E. McKinney, *Microbiology for Sanitary Engineers* (New York: McGraw-Hill Book Company, 1962).
2. H. A. Hawkes, *The Ecology of Waste Water Treatment* (Elmsford, N.Y.: Pergamon Press, 1963).

3. C. M. Palmer, "Algae in Water Supplies," *U.S. Public Health Service Publication No. 657* (Washington, D.C.: Department of Health, Education, and Welfare, 1959).

4. I. C. Gunsalus and R. Y. Stanier, *The Bacteria, Volume II: Metabolism* (New York: Academic Press, Inc., 1961).

5. W. W. Eckenfelder, Jr., *Industrial Water Pollution Control* (New York: McGraw-Hill Book Company, 1966), pp. 188–198.

6. S. Balakrishnan, W. W. Eckenfelder, and C. Brown, "Organics Removal by a Selected Trickling Filter Media," *Water and Wastes Eng.* 6, no. 1 (1969): A-22–A-25.

7. "Sewage Treatment at Military Installations," Report of the Subcommittee on Sewage Treatment in Military Installations, National Research Council, *Sewage Works J.* 18, no. 5 (1946): 787–1028.

8. *Recommended Standards for Sewage Works, Great Lakes–Upper Mississippi River Board of State Sanitary Engineers,* 1971 Edition (Albany, N.Y.: Health Education Service).

9. W. E. Howland, "Effect of Temperature on Sewage Treatment Processes," *Sewage and Industrial Wastes.* 23 (1953): 161–169.

10. J. E. Germain, "Economic Treatment of Domestic Waste by Plastic-Medium Trickling Filters," *J. Water Poll. Control Fed.* 38, no. 2 (1966): 192–203.

11. R. L. Antonie, D. L. Kluge, and J. H. Mielke, "Evaluation of a Rotating Disk Wastewater Treatment Plant," *J. Water Poll. Control Fed.* 46, no. 3 (1974): 498–511.

12. N. C. H. Fischerström, "Low Pressure Aeration of Water and Sewage," *Proc. Am. Soc. Civil Engrs., J. San. Eng. Div.* 86, no. SA5 (1960): 21–56.

13. G. F. Bennett and L. L. Kempe, "Oxygen Transfer in Biological Systems," *Proc. 20th Industrial Waste Conf., Purdue Univ. Ext. Service,* XLIX, no. 4 (1965): 435–447.

14. R. A. Conway and G. W. Kumke, "Field Techniques for Evaluating Aerators," *Proc. Am. Soc. Civil Engrs., J. San. Eng. Div.* 92, no. SA2 (1966): 21–42.

15. R. E. McKinney, "Mathematics of Complete-Mixing Activated Sludge," *Proc. Am. Soc. Civil Engrs., J. San. Eng. Div.* 88, no. SA3 (1962): 87–113.

16. B. L. Goodman and A. J. Englande, Jr., "A Unified Model of the Activated Sludge Process," *J. Water Poll. Control Fed.* 46, no. 2 (1974): 312–332.

17. "Oxygen Activated Sludge Wastewater Treatment Systems," Environmental Protection Agency, Technology Transfer, August 1973.

18. M. J. Hammer and C. D. Jacobson, "Anaerobic Lagoon Treatment of Packinghouse Wastewater," *Proceedings, Second International Symposium for Waste Treatment Lagoons* (Lawrence, Kans.: University of Kansas, 1970).

19. R. Stone, "Sewage Treatment System Odors and Air Pollutants," *Proc. Am. Soc. Civil Engrs., J. San. Eng. Div.* 96, no. SA4 (1970): 905–909.

20. "Manual of Septic-Tank Practice," *U.S. Public Health Service Publication No. 526* (Washington, D.C.: Government Printing Office, 1957).

21. J. A. Cotteral, Jr. and D. P. Norris, "Septic Tank Systems," *Proc. Am. Soc. Civil Engrs., J. San. Eng. Div.* 95, no. SA4 (1969): 715–746.

12
Processing
of Sludges

Combinations of physical, chemical, and biological processes are employed in handling sludges. These operations, being uniquely adapted to eliminating precipitates, are presented as a separate topic rather than partitioning them among the previous chapters. While the purposes in treating water and wastewater are to remove impurities from dilute solution and consolidate them into a smaller volume of liquid, the objectives of processing sludges are to extract water from the solids and dispose of the dewatered residue. Furthermore, the relationships among the processes must be understood since disposal schemes involve sequencing of operations. For example, gravity thickening, anaerobic digestion, chemical conditioning, and mechanical dewatering comprise a physical–biological–chemical–physical sequence. Discussions on the characteristics of wastewater sludges and water-treatment-plant residues are isolated in different sections. Generalized process flow diagrams are also separated, although descriptions of individual unit operations are directed toward both water and wastewater sludges. Comments relating to the applicability of each process are included.

Sources, Characteristics, and Quantities of Waste Sludges

Mathematical relationships for estimating specific gravity and computing sludge volume appear first since they are fundamental calculations applicable to all sludges. Then the characteristics and methods for estimating sludge quantities are presented separately for wastewater and water-treatment-plant residues.

12-1

WEIGHT AND VOLUME RELATIONSHIPS

The majority of sludge solids from biological wastewater processing are organic with a 60–80% volatile fraction. Concentration of suspended solids in a liquid sludge is determined by straining a measured sample through a glass-fiber filter. Nonfilterable residue, expressed in milligrams per liter, is the solids content. Since the filterable portion of a sludge is very small, sludge solids are often determined by total residue on evaporation (i.e., the total deposit remaining in a dish after evaporation of water from the sample and subsequent drying in an oven at 103°C). Volatile solids are found by igniting the dried residue at 550°C in a muffle furnace. Loss of weight upon ignition is reported as milligrams per liter of volatile solids, and the inerts remaining after burning as fixed solids. Waste from chemical coagulation of a surface water contains both organic matter removed from the raw water, and mineral content derived from the chemical coagulants. Most solids are nonfilterable and have a volatile fraction of 20–40%. Precipitate from treated well water is essentially mineral.

The specific gravity of solid matter in a sludge can be computed from the relationship

$$\frac{Ws}{Ss\,\gamma} = \frac{Wf}{Sf\,\gamma} + \frac{Wv}{Sv\,\gamma} \tag{12-1}$$

where Ws = weight of dry solids, lb
Ss = specific gravity of solids
γ = unit weight of water, lb/ft^3 (lb/gal)
Wf = weight of fixed solids (nonvolatile), lb
Sf = specific gravity of fixed solids
Wv = weight of volatile solids, lb
Sv = specific gravity of volatile solids

The specific gravity of organic matter is 1.2–1.4, while the solids in chemically coagulated water vary from 1.5 to 2.5. The value for a solids slurry is calculated from

$$S = \frac{Ww + Ws}{(Ww/1.00) + (Ws/Ss)} \tag{12-2}$$

where S = specific gravity of wet sludge
$\quad\quad Ww$ = weight of water, lb
$\quad\quad Ws$ = weight of dry solids, lb
$\quad\quad Ss$ = specific gravity of dry solids

Consider a waste biological sludge of 10% solids with a volatile fraction of 70%. Their specific gravity can be estimated using Eq. 12-1 by assuming values for the fixed matter of 2.5 and for the volatile residue 1.0.

$$\frac{1.00}{Ss} = \frac{0.30}{2.5} + \frac{0.70}{1.0} = 0.82$$

$$Ss = \frac{1}{0.82} = 1.22$$

Then, the specific gravity of the wet sludge by Eq. 12-2 is 1.02.

$$S = \frac{90 + 10}{(90/1.00) + (10/1.22)} = 1.02$$

These calculations demonstrate that for organic sludges of less than 10% solids the specific gravity may be assumed to be 1.00 without introducing significant error. Example 12-1 illustrates that, even for mineral residue, a high concentration of precipitate is required to increase the specific gravity of a slurry above 1.0.

The volume of waste sludge for a given amount of dry matter and concentration of solids is given by

$$V = \frac{Ws}{(s/100)\gamma S} = \frac{Ws}{[(100 - p)/100]\gamma S} \tag{12-3}$$

where V = volume of sludge, ft^3 (gal) [m^3]
$\quad\quad Ws$ = weight of dry solids, lb [kg]
$\quad\quad s$ = solids content, %
$\quad\quad \gamma$ = unit weight of water, 62.4 lb/ft^3 (8.34 lb/gal) [1000 kg/m^3]
$\quad\quad S$ = specific gravity of wet sludge
$\quad\quad p$ = water content, %

In this formula the volume of a sludge is indirectly proportional to the solids content. Thus, if a waste is thickened from 2% to 4% solids, the volume is reduced by one half, and if consolidation is continued to a concentration of 8% the quantity of wet sludge is only one-fourth of the original amount. During this concentration process, water content is reduced from 98% to 92%. In applying Eq. 12-3, specific gravity of the sludge S is normally taken as 1.0 and therefore not included in computations, as demonstrated in Example 12-2.

●**EXAMPLE 12-1**
Coagulation of a surface water using alum produces 10,000 lb (4540 kg) of dry solids/ day, of which 20% are volatile. Both the settled sludge following coagulation and filter backwash water are concentrated in clarifier–thickeners to a solids density of 2.5%.

Centrifugation can be used to increase the concentration to 20%, a consistency similar to soft wet clay, or the clarifier–thickener underflow can be dewatered to a 40% cake by pressure filtration. (a) Estimate the specific gravities of the thickened sludge, concentrate from centrifugation, and filter cake. (b) Calculate the daily sludge volumes from each process.

○*Solution*

a   Applying Eq. 12-1,

$$\frac{1.00}{Ss} = \frac{0.80}{2.50} + \frac{0.20}{1.00} = 0.52$$

$$Ss = \frac{1}{0.52} = 1.9$$

Using Eq. 12-2,

$$S \text{ (thickened sludge)} = \frac{97.5 + 2.5}{(97.5/1.0) + (2.5/1.9)} = 1.0$$

$$S \text{ (centrifuge discharge)} = \frac{80 + 20}{(80/1.0) + (20/1.9)} = 1.1$$

$$S \text{ (filter cake)} = \frac{60 + 40}{(60/1.0) + (40/1.9)} = 1.2$$

b   Substituting these values into Eq. 12-3 with $Ws = 10{,}000$ lb/day,

$$V \text{ (thickened sludge)} = \frac{10{,}000}{(2.5/100)8.34 \times 1.0} = 48{,}000 \text{ gpd}$$

$$\left[V \text{ (thickened sludge)} = \frac{4540 \text{ kg}}{(2.5/100)1000 \text{ kg/m}^3 \times 1.0} = 182 \text{ m}^3/\text{day} \right]$$

$$V \text{ (centrifuge discharge)} = \frac{10{,}000}{(20/100)8.34 \times 1.1} = 5400 \text{ gpd} \ (20.6 \text{ m}^3/\text{day})$$

$$V \text{ (filter cake)} = \frac{10{,}000}{(40/100)8.34 \times 1.2} = 2500 \text{ gpd} \ (9.5 \text{ m}^3/\text{day})$$

●**EXAMPLE 12-2**

Estimate the quantity of sludge produced by a trickling-filter plant treating 1.0 mgd of domestic wastewater. Assume the following: a suspended-solids concentration of 220 mg/l in the raw wastewater, solids content in the sludge equivalent to 90% removal, and a sludge of 5.0% concentration withdrawn from the settling tanks.

○*Solution*

$$\text{solids in the sludge} = 1.0 \times 220 \times 8.34 \times 0.90 = 1650 \text{ lb/day}$$

$$\text{volume of sludge (using Eq. 12-3)} = \frac{1650}{0.05 \times 62.4} = 530 \text{ ft}^3/\text{day}$$

12-2

CHARACTERISTICS AND QUANTITIES OF WASTEWATER SLUDGES

The purpose of primary sedimentation and secondary aeration is to remove waste organics from solution and concentrate them in a much smaller volume to facilitate dewatering and disposal. Concentration of organic matter in wastewater is approximately 200 mg/l (0.02%), while that in a typical raw-waste sludge is about 40,000 mg/l (4%). Based on these approximate values, treatment of 1.0 mg of wastewater produces about 5000 gal of sludge. This raw odorous and putrescible residue must be further processed and reduced in volume for land disposal, incineration, or barging to sea. Common methods include mechanical thickening, biological digestion, and dewatering after chemical conditioning.

The quantity and nature of sludge generated relates to character of the raw wastewater and processing units employed. Daily sludge production may fluctuate over a wide range, depending on size of municipality, contribution of industrial wastes, and other factors. Both maximum and average daily sludge volumes are considered in designing facilities. A limited quantity of solids can be stored temporarily in clarifiers and aeration tanks to provide short-term equalization of peak loads. Mechanical thickening and dewatering units may be sized to handle sludge quantities as high as double the daily average. Other processes, such as conventional anaerobic digestion, have substantial equalizing capacity and are designed on the basis of maximum average monthly loading. Often, selection of conservative design parameters and liberal estimates of sludge yield take into account anticipated quantity variations. In this manner, designers disguise the fact that maximum sludge yield is being considered in sizing unit processes. For example, in designing a trickling-filter plant, the dry solids may be calculated assuming 0.20 lb/capita/day, excluding household garbage grinders, when the actual amount realized in the treatment process is closer to 0.12 lb/capita/day. Furthermore, the required digester volume may be computed using a conservative figure of 5 ft^3/population equivalent.

Primary sludge is a gray-colored, greasy, odorous slurry of settleable solids, accounting for 50–60% of the suspended solids applied, and tank skimmings. Scum is usually less than 1% of the settled sludge volume. Primary precipitates can be dewatered readily after chemical conditioning because of their fibrous and coarse nature. Typical solids concentrations in raw primary sludge from settling municipal wastewater are 6–8%. The portion of volatile solids varies from 60 to 80%.

Trickling-filter humus from secondary filtration is dark brown in color, flocculent, and relatively inoffensive when fresh. The suspended particles are fragments of biological growth washed from the filter media. Although they exhibit good settleability, the precipitate does not compact to a high density. For this reason and the fact that sloughing is irregular, underflow from the final clarifier containing filter humus is returned to the wet well for mixing with the inflowing raw wastewater. Thus humus is settled with raw organics in the primary clarifier. The combined sludge has a solids content of 4–6% which is slightly thinner than primary residue with raw organics only.

Waste-activated sludge is a dark-brown, flocculent suspension of active microbial masses inoffensive when fresh, but it turns septic rapidly because of biological activity. Mixed-liquor solids settle slowly, forming a rather bulky sludge of high water content. Thickness of return activated sludge is 0.5–2.0% suspended solids with a volatile fraction of 0.7–0.8. Excess activated sludge in most processes is wasted from the return sludge line. High water content, resistance to gravity thickening, and the presence of active microbial floc make this residue difficult to handle. Routing of waste activated to the wet well for settling with raw wastewater is not recommended. Carbon dioxide, hydrogen sulfide, and odorous organic compounds are liberated from the settlings in the primary basin as a result of anaerobic decomposition, and the solids concentration is rarely greater than 4%. Waste-activated sludge can be thickened effectively by flotation or centrifugation, however, chemical additions may be needed to ensure high solids capture in the concentrating process.

Anaerobically digested sludge is a thick slurry of dark-colored particles and entrained gases, principally carbon dioxide and methane. When well digested, it dewaters rapidly on sand drying beds, releasing an inoffensive odor resembling that of garden loam. Substantial additions of chemicals are needed to coagulate a digested sludge prior to mechanical dewatering, owing to the finely divided nature of the solids. Dry residue is 30–60% volatile and compactness of digested sludge ranges from 6 to 12%, depending on the mode of digester operation.

Aerobically digested sludge is a dark-brown, flocculent, relatively inert waste produced by long-term aeration of sludge. The suspension is bulky and difficult to thicken, thus creating problems of ultimate disposal. Since a clear supernatant cannot be decanted, the primary functions of an aerobic digester are stabilization of organics and temporary storage of waste sludge. Thickness of aerobically digested sludge is less than that of the influent, since approximately 50% of the volatile solids are converted to gaseous end products. Stabilized sludge, expensive to dewater, is often disposed of by spreading on land for its fertilizer value. For these reasons, aerobic digestion is generally limited to treatment of waste activated from aeration plants without primary clarifiers.

Mechanically dewatered sludges vary in characteristics based on the type of sludge, chemical conditioning, and unit process employed. Density of dewatered cakes ranges from 15 to 40%. The thinner waste is similar to a wet manure while the latter is a chunky solid. Method of ultimate disposal and economics dictate the degree of moisture reduction that is necessary.

Waste-solids production in primary and secondary processing can be estimated using the following formulas:

$$Ws = Ws_p + Ws_s \qquad\qquad (12\text{-}4)$$

where Ws = total dry solids, lb/day
Ws_p = raw primary solids, lb/day
Ws_s = secondary biological solids, lb/day

$$Ws_p = f \times SS \times Q \times 8.34 \qquad (12\text{-}5)$$

where Ws_p = primary solids, lb of dry weight/day
f = fraction of suspended solids removed in primary settling
SS = suspended solids in unsettled wastewater, mg/l
Q = daily wastewater flow, mgd
8.34 = conversion factor, lb/mg per mg/l

$$Ws_s = k \times BOD \times Q \times 8.34 \qquad (12\text{-}6)$$

where Ws_s = biological sludge solids, lb of dry weight/day
k = fraction of applied BOD that appears as excess biological growth in waste activated or filter humus, assuming about 30 mg/l of BOD remaining in the secondary effluent
BOD = concentration in applied wastewater, mg/l
Q = daily wastewater flow, mgd

The first expression simply states that the total weight of dry solids produced equals the sum of the primary plus secondary residues. Settleable matter removed in primary clarification can be considered to be a function of the suspended-solids concentration (Eq. 12-5). For typical municipal wastes, the value for f is between 0.4 and 0.6. The settleable fraction of suspended solids in a fresh domestic wastewater is about 0.6, but septic conditions and industrial waste contributions are likely to decrease the portion of settlings in a wastewater. For example, many food-processing discharges are high in colloidal matter and exhibit BOD/suspended solids ratios of 2:1, or greater. Thus a combined wastewater may exhibit an f value that is considerably less than the average 0.5 for domestic waste.

Organic matter entering secondary biological treatment is colloidal in nature and represented best by its BOD value. Most is synthesized into flocculent biological growths that entrain nonbiodegradable material. Therefore, excess activated-sludge solids from aeration and humus from biological filtration can be estimated by Eq. 12-6, which relates residue production to BOD load. Coefficient k is a function of process food/microorganism ratio and biodegradable (volatile) fraction of the matter in suspension. For trickling-filter humus, k is assumed to be in the range 0.3–0.5, with the lower value for light BOD loadings and the bigger number applicable to high-rate filters. The k for secondary activated-sludge processes can be estimated using Fig. 11-41. A daily waste volatile-suspended-solids yield can be read from the excess sludge production curve by entering the diagram with a known BOD sludge age. (BOD sludge age is the inverse of the F/M ratio, expressed in lb of MLSS per lb of BOD/day.) The chart value in lb of VSS produced/lb of BOD applied is converted to k for

Eq. 12-6 if divided by the volatile fraction of solids in the waste sludge, thus converting the units to lb of total solids produced/lb of BOD applied. Consider conventional activated sludge at a F/M = 0.33 and a MLSS of 70% volatile solids. Enter Fig. 11-41 at 3.0 along the abscissa and read 0.28 lb of VSS/lb of BOD on the right ordinate. Then $k = 0.28/0.70 = 0.40$.

For activated-sludge processes handling unsettled wastewater, the excess solids determined using this method are increased by approximately 70%. Thus for aeration systems without primary clarifiers, the k factor is the value determined from Fig. 11-41 multiplied by 1.7.

Design of a sludge-handling system is based on the volume of wet sludge as well as dry solids content. Once the dry weight of residue has been estimated, the volume of sludge is calculated by applying Eq. 12-3.

The foregoing formulations are reasonable for sludge quantities from processing domestic wastewater at average daily design flow. Real sludge yields may differ considerably from anticipated values when treating a municipal discharge containing substantial contributions of industrial wastes, and when loading or operational conditions create unanticipated peak sludge volumes.

Peak loads must be assessed for each treatment-plant design based on local conditions such as seasonal industrial discharges, anticipated trends in per capita solids contribution, and the type of unit operations employed in wastewater processing. A rule-of-thumb approach assumes that the maximum weekly dry solids yield will be approximately 25% greater than the yearly mean. Variations in daily sludge volumes may be considerably greater, owing to changes in moisture content. For example, if the concentration of settlings drawn from primary clarifiers shifts from 6% to 4%, the total quantity of wet sludge increases 50% for the same amount of dry solids. These deviations are normally taken into account by selecting conservative design criteria in sizing biological digesters. Where vacuum filtration is practiced, increased quantities of sludge can be handled by higher chemical dosages to improve filter yield, and by extending the time of operation each day. Perhaps the most difficult parameter to predict is the compactness of waste-activated sludge. Bulking can easily reduce the solids content by one-half, say from 15,000 mg/l to 7500 mg/l, thus doubling the volume. Processing of this larger quantity must be evaluated by the designer. One guideline is to size mechanical thickeners at 200% of the estimated volume during normal operation in order to ensure consolidation of the diluted slurry. Also, provisions should be made for the addition of coagulants to aid in concentrating waste-activated sludge.

●EXAMPLE 12-3

One million gallons of municipal wastewater with a BOD of 260 mg/l and suspended solids of 220 mg/l are processed in a primary plus secondary activated-sludge plant. Estimate the quantities and solids contents of the primary, waste-activated, and mixed sludges. Assume the following: 60% SS removal and 40% BOD reduction in clarification, water content of 94.0% in the raw sludge, an operating F/M ratio of 1:3 in the aeration basin, volatile fraction in the MLSS of 0.75, and solids concentration of 15,000 mg/l in the waste-activated sludge. If the sludges are blended and gravity-thickened to a solids content of 6.0%, calculate the consolidated sludge volume assuming 95% solids capture.

○*Solution*

Primary sludge solids and volume based on Eqs. 12-5 and 12-3 are, respectively:

$$Ws_p = 0.60 \times 220 \times 1.0 \times 8.34 = 1100 \text{ lb}$$

$$V = 1100/[(100 - 94)/100]8.34 = 2200 \text{ gal}$$

To determine k for Eq. 12-6, enter Fig. 11-41 with M/F = 3.0, and read 0.27 lb of VSS produced/lb of BOD applied. Then, dividing this by the volatile fraction, $k = 0.27/0.75 = 0.36$.

Waste-activated sludge based on Eq. 12-6 is

$$Ws_s = 0.36 \times 260 \times 1.0 \times 0.60 \times 8.34 = 470 \text{ lb}$$

$$V = \frac{470}{0.015 \times 8.34} = 3700 \text{ gal}$$

The blended sludge volume, solids content, and solids concentration are

$$V = 2200 + 3700 = 5900 \text{ gal}$$

$$Ws = 1100 + 470 = 1570 \text{ lb}$$

$$s = \frac{1570 \times 100}{5900 \times 8.34} = 3.2\%$$

The thickened sludge volume is

$$V = \frac{0.95 \times 1570}{0.06 \times 8.34} = 3000 \text{ gal}$$

12-3

CHARACTERISTICS AND QUANTITIES OF WATER-PROCESSING SLUDGES

Water-treatment residues are derived from sedimentation and filtration of chemically conditioned water. Surface supplies yield wastes containing colloidal matter removed from the raw water and chemical flocs, while groundwater-processing precipitates are mineral with little or no organic material. Sludges vary widely in composition depending on character of the water source and chemicals added during treatment. A typical method for handling a turbid river supply includes presedimentation for reduction of settleable solids, lime softening, alum coagulation, and filtration for removal of colloids, plus addition of activated carbon for taste and odor control. The presedimentation deposit is silt plus detritus; settling-basin sludge is a mixture of inerts, organics, and chemical precipitates, including metal hydroxides; and filter backwash water contains floc from agglomerated colloids and unspent coagulant hydroxides. Lake and reservoir waters are often dosed with alum and flocculation aids, plus activated carbon. Settlings during the summer may include significant quantities of algae. Precipitates from lime-soda ash softening are predominantly calcium carbonate and magnesium hydroxide with traces of other minerals, such as oxides of iron and manganese.

Sludge storage capacity and time intervals between withdrawals are governed by installation design, type of water processing, and operations management. Settled sludge is allowed to accumulate and consolidate in plain rectangular or hopper-bottomed basins. These tanks are cleaned at time intervals varying from a few weeks to several months by draining and removing the compacted sludge. Clarifiers equipped with mechanical scrapers discharge sludge either continuously at a low rate or intermittently, often daily. Backwashing of filters produces a high flow of dilute wastewater for a few minutes usually once a day for each filter. Obviously, any system for handling water-treatment wastes must consider temporary storage and thickening of wash water.

Alum-coagulation sludge is dramatically influenced by the gelatinous nature of the aluminum hydroxides formed in the reaction with raw-water alkalinity. Particles entrained in the floc and other coagulation precipitates do not suppress the jellylike consistency that makes an alum slurry difficult to dewater. Coagulation settlings and backwash water can normally be gravity thickened to about 2%, although polymers may be needed to achieve this consolidation. Studies have shown that centrifugal dewatering can concentrate this waste to a truckable 20% with a consistency similar to a soft wet clay. Pressure filtration will produce a 40% cake that breaks easily. Complete dehydration by drying or freezing results in a granular material that does not revert to its original gelatinous form if again mixed with water. Enmeshed water of hydration, not water of suspension, causes the original jelly consistency. Iron coagulants yield slightly denser sludges that are somewhat easier to handle.

Surface-water wastes are highly variable, owing to changes in raw-water quality. High turbidities during spring runoff and periods of high rainfall result in a decreased percentage of aluminum hydroxide solids. The result is a precipitate that settles better and is easier to dewater. Water-temperature changes affect algal growth in surface supplies, the rate of chemical reactions in treatment, and filterability of the sludge. In designing a waste-handling system, changes in raw-water quality and accompanying variations in sludge characteristics must be investigated by long-term studies of daily and seasonal records.

Accurate estimates of sludge production from surface-water treatment plants are difficult, primarily due to the lack of plant operational data. Equations 12-7 and 12-8 are empirical relationships that were developed from field and laboratory data for a specific plant.[2] The raw water, taken from a reservoir, was normally of high quality, having 2–10 Jackson turbidity units (Jtu). During periods of high rainfall in the winter, this figure was considerably higher, up to

150 Jtu. For this particular water source, suspended solids concentration related to turbidity measurements so that the latter could be used in formulas. Water treatment consisted of alum coagulation, sedimentation, rapid sand filtration, and chlorination. Sedimentation basin underflow averaged 1 % solids, while the filter wash water was less than 0.1 %. Gravity thickening of these wastes in clarifier–thickeners produced consolidated slurries of 3–6 %.

$$\text{Total sludge solids (lb/mg)} = 2.75 \times \text{alum dosage} + 8.34 \times \text{turbidity} \qquad (12\text{-}7)$$

$$\text{sedimentation basin sludge (lb/mg)} = \text{total sludge solids} - \text{solids to filters}$$

$$(12\text{-}8)$$

where alum dosage is in units of mg/l, turbidity is that of raw water, Jackson turbidity units, and solids to filters averaged 26 lb/mg in summer and 60 lb/mg in winter.

Coagulation-softening sludges result from processing hard, turbid surface waters, such as those found in Midwestern rivers. A typical treatment-plant flow arrangement is presedimentation followed by two-stage or split treatment; lime softening and coagulation with alum or iron salts; and rapid sand filtration. Solids concentration in settled sludges varies with turbidity in the raw water, ratio of calcium to magnesium in the softening precipitate, type and dosage of metal coagulant, and filter aids used. In general, filter wash water gravity thickens to about 4 %, alum-lime sludges to densities up to 10 %, and lime-iron precipitates range between 10 and 25 %. The quantity of sludge produced is difficult to predict because the chemical treatment varies with hardness and turbidity of the river water. In-plant modifications to help even out fluctuations in sludge yield may include applying polymers in coagulation, closer pH control to reduce the amount of magnesium hydroxide produced in softening, and providing flexibility to thicken sludges separately or combined.

Lime-soda ash softening sludges produced in treating groundwaters contain calcium carbonate and to a lesser extent magnesium hydroxide. Aluminum hydroxides and other coagulant aids may be present if added in water processing. The quantity of $Mg(OH)_2$ depends on magnesium hardness in the raw water and the softening process employed. In general, dry solids are 85–95 % $CaCO_3$. The residue is stable, dense, inert, and relatively pure, since groundwater does not contain colloidal inorganic or organic matter. Settled solids concentrations range from 2 to greater than 15 %, based on the ratio of calcium to magnesium precipitates, unit operations of the softening process, and method of sludge withdrawal.

Calcium carbonate compacts readily while magnesium hydroxide, like aluminum hydroxide, is gelatinous and does not consolidate as well nor dewater as easily. Slurry wasted from flocculator–clarifiers (upflow units) has a solids content in the range 2–5 %. Stored sludges in manually cleaned sedimentation tanks of straight-line systems compact to 10 % solids concentration or greater. Stoichiometrically, from Eq. 12-9, 3.6 lb of calcium carbonate is precipitated for each pound of lime applied. However, in actual practice the dry solids yield from softening is closer to 2.6 lb/lb of lime applied due to incomplete chemical reaction, impurities in commercial-grade lime, and precipitation of variable amounts of magnesium.

$$CaO + Ca(HCO_3)_2 = 2CaCO_3 + H_2O \qquad (12-9)$$

Iron and manganese oxides removed from groundwater by aeration and chemical oxidation are flocculent particles with poor settleability. The amount of sludge produced in the removal of these metals, without simultaneous precipitation softening, is relatively small. The majority of hydrated ferric and manganic oxides pass through sedimentation tanks, are trapped in the filters, and appear in the dilute backwash water.

Filter wash water is a relatively large volume of wastewater with a low solids concentration of 100–1000 mg/l. The exact amount of water used in backwashing is a function of the type of filter system, cleansing technique, and quality and source of the raw water being treated. Generally, 2–3% of the water processed in a plant is used for filter washing. The fraction of total waste solids removed by filtration depends on efficiency of the coagulation and sedimentation stages, type of treatment system, and characteristics of the raw water. The amount may be a substantial portion, say 30%, of the dry solids resulting from treatment.

●EXAMPLE 12-4

A reservoir water supply with a turbidity of 10 Jtu in the summer is treated by applying an alum dosage of 30 mg/l. For each million gallons of water processed, estimate the total solids production, volume of settled sludge, and quantity of filter wash water. Assume a settled sludge concentration of 1.5% solids, a backwash water with 400 mg/l of suspended solids, and that Eqs. 12-7 and 12-8 are applicable. Compute the composite sludge volume after the two wastes are gravity thickened to 3.0% solids.

○*Solution*

Using Eqs. 12-7 and 12-8,

$$\text{total sludge solids} = 2.75 \times 30 + 8.34 \times 10 = 166 \text{ lb/mg}$$
$$\text{solids to filters} = 26 \text{ lb/mg}$$

$$\text{sedimentation basin sludge} = 166 - 26 = 140 \text{ lb/mg}$$

Applying Eq. 12-3,

$$V \text{ of settled sludge} = \frac{140}{0.015 \times 8.34} = 1100 \text{ gal}$$

$$V \text{ of wash water} = \frac{26}{(400/1,000,000)8.34} = 7800 \text{ gal}$$

$$V \text{ of thickened sludge} = \frac{166}{0.030 \times 8.34} = 660 \text{ gal}$$

●EXAMPLE 12-5

The lime-soda ash softening process described in Example 10-5 requires a lime dosage of 630 lb/mg of CaO (2.7 meq/l) and a soda ash addition of 530 lb/mg of Na_2CO_3 (1.2 meq/l). Based on the appropriate chemical reactions, calculate the calcium carbonate residue produced in the softening of 1 million gallons of water, assuming the practical limit of $CaCO_3$ precipitation is 0.60 meq/l (30 mg/l).

○**Solution**

Based on Eq. 10-36, the 0.4 meq/l of CO_2 is precipitated by addition of 0.4 meq/l of lime to form 0.4 meq/l of $CaCO_3$.

$$Ca(OH)_2 + CO_2 = CaCO_3\downarrow + H_2O \tag{10-36}$$

From Eq. 10-31, the 2.3 meq/l of $Ca(HCO_3)_2$ reacts with 2.3 meq/l of lime to form 4.6 meq/l of $CaCO_3$.

$$Ca(HCO_3)_2 + Ca(OH)_2 = 2CaCO_3\downarrow + 2H_2O \tag{10-31}$$

And, finally 1.2 meq/l of soda ash precipitates 1.2 meq/l of $CaSO_4$ by the reaction

$$CaSO_4 + Na_2CO_3 = CaCO_3\downarrow + Na_2SO_4 \tag{10-35}$$

The residue is theoretically equal to the stoichiometric quantities of $CaCO_3$ formed minus the practical limit of treatment (solubility).

$$0.4 + 4.6 + 1.2 - 0.6 = 5.6 \text{ meq/l}$$

or

$$5.6 \times 50 \times 8.34 = 2340 \text{ lb/mg of } CaCO_3$$

Arrangement of Unit Processes in Sludge Disposal

Many processes for sludge handling are applied to both wastewater sludges and water-treatment-plant residues. Individual unit operations are discussed in the latter sections of this chapter with comments relative to their method of operation and application. Under this heading, selection and arrangement of units are outlined to illustrate how they integrate with each other. Referring back to these discussions while studying the individual operations will help relate them to complete sludge-disposal schemes.

12-4

SELECTION OF PROCESSES FOR WASTEWATER SLUDGES

Techniques selected for processing waste sludges are a function of the type, size, and location of the wastewater plant, unit operations employed in treatment, and method of ultimate solids disposal. The system adopted must be able to accept the primary and secondary sludges produced and economically convert them to a residue that is environmentally acceptable for disposal.

TABLE 12-1

PROCESSES FOR STORAGE, TREATMENT, AND DISPOSING OF WASTEWATER SLUDGES

Storage prior to processing
  In the primary clarifiers
  Separate holding tanks
Thickening prior to dewatering or digestion
  Gravity settling
  Dissolved air flotation
  Centrifugation
Conditioning prior to dewatering
  Stabilization by anaerobic digestion
  Stabilization by long-term aeration
  Chemical coagulation
  Elutriation of anaerobically digested sludge
  Heat treatment or wet oxidation
Mechanical dewatering
  Vacuum filtration
  Centrifugation
  Pressure filtration
  Heat drying
Air drying of digested sludge
  Sand drying beds
  Shallow lagoons
Disposal of liquid digested sludge
  Spreading on agricultural land
  Piping into the ocean
Disposal of dewatered solids
  Burial in sanitary landfill
  Incineration
  Production of soil conditioner
  Barging to sea

Methods for storage, treatment, and disposition are listed in Table 12-1. Settled raw solids may be stored in the bottom of primary clarifiers during the day, or perhaps over a weekend, and then pumped directly to a processing unit. In small plants, sludge is transferred to anaerobic digesters either once or twice a day. Vacuum filtration requires a steady flow of sludge during the operating period, which may extend from 4 to 16 hours/day. Separate holding tanks can be used to receive and blend primary and secondary sludges. The former may be pumped intermittently using an automatic time-clock control, while the latter flow may be either continuous or periodic. Aerated holding tanks for accumulating and biologically stabilizing waste from complete aeration plants are called aerobic digesters.

The economics of chemical conditioning and biological treatment are directly related to sludge density. As a general rule, the solids content of settled waste must be at least 4% for feasible handling. Primary precipitates and mixtures of primary and secondary settlings are amenable to compaction by sedimentation; therefore, gravity thickeners are used to increase the concentration of sludges withdrawn from either clarifiers or holding tanks. Because of the flocculent nature of waste-activated sludge, separate thickening is performed by either dissolved air flotation or centrifugation. The float or concentrate is then blended with primary waste and directed to holding tanks, or the next dewatering or treatment step.

The intention of anaerobic digestion is to convert bulky, odorous, and putrescible raw sludge to a well-digested material that can be rapidly dewatered without emission of noxious odors. In addition to stabilizing and gasifying the organic matter, the volume of residue is significantly reduced by withdrawal of supernatant from digesters to thicken the sludge. Aerobic digestion is almost exclusively used to treat excess sludge from plants without primary clarifiers. While it stabilizes the organic matter, solids thickening and dewatering are troublesome, owing to the bulky nature of overaerated sludge. Chemical coagulation with polymers, or metal coagulants and lime, is required for mechanical dewatering of both raw and digested wastes. Elutriation of digested sludge, still practiced at some treatment plants, improves filterability and reduces the amount of chemical conditioner required. The process consists of adding water to the sludge, mixing, and then allowing separation by gravity. The elutriate is decanted and returned to the head of the plant, and underflow solids are conditioned for filtration. Elutriation is rarely considered when designing new facilities.

Heat treatment of raw sludge for a short period of time under pressure results in coagulation of the solids. This conditioning sterilizes, deodorizes, and prepares the waste for mechanical dewatering without addition of chemicals. Current processes using heat treatment include the Porteus and the low-pressure Zimpro system. These techniques apply steam to heat the reactor vessel to about 300°F under a pressure of 150 psi or greater. Sludge from the reactor is discharged to a decant tank, from which the underflow is withdrawn for dewatering. The supernatant contains high concentrations of water-soluble organic compounds and must therefore be returned to the treatment plant for processing. This may be a serious drawback in handling some sludges, since it can result in cycling of solids through the heating process back to the treatment plant, which extracts them and returns them again to the heat treatment. Wet oxidation, sometimes referred to as the Zimmerman process, can be achieved under high pressure at elevated temperatures. Liquid sludge with compressed air is fed into a pressure vessel, where the organic matter is stabilized. Inert solids are separated from the effluent by dewatering in lagoons or by mechanical means. Pyrolysis (heating without oxygen) and freezing have also been investigated as sludge-conditioning methods. Significant development is still required before either process can be applied effectively.

Vacuum filtration and centrifugation are both popular for dewatering chemically conditioned wastes. The former has been widely adopted for handling raw primary sludge, and mixtures of settled and thickened waste activated. Vacuum filters are not installed in small plants for economic reasons. The minimum size available from one leading manufacturer is a package unit with a nominal filter area of 60 ft^2. Applying typical design criteria, this unit could dewater the domestic sludge from a community with a population of 10,000. The minimum size of centrifuges also dictates their application to larger installations. Although continuous improvement in machine design is broadening the field of centrifugation, the most prevalent current practice is dewatering of anaerobically digested sludge in large cities. Pressure filtration has been recently introduced as a dewatering technique. The main advantages appear to be high cake density, clarity of filtrate, and automated operation making them feasible in smaller treatment plants. Mechanical heat drying is associated with production of soil conditioner, and incineration is used in locations where less costly disposal of dewatered solids is not possible.

The oldest technique for drying digested sludge is on open sand beds contained by short concrete walls. Well-digested slurry is drained or pumped onto the surface of the bed to a depth of 8–12 in. Moisture leaves by evaporation and seepage; the latter is collected in underdrain piping for return to the plant influent. Humus residue is removed manually from the sand surface. Mechanical cleaning equipment is not feasible since it results in disturbance of the bed and loss of sand. As a result, many operators use shallow lagoons for air drying of digested sludge. This permits the use of front-end loaders or buckets on draglines, but operations can be disrupted by inclement weather. Towns in agricultural regions may dispose of liquid digested sludge by spreading on farmland. Often, land disposal is originated by the plant operator to avoid the problem of handling the dried humus cake. Lead by this precedent, engineers are now designing land application schemes rather than open drying beds for new installations.

Dewatered solids may be disposed of in sanitary landfill. Dried anaerobic cake and incinerator ash are relatively inoffensive, but cake composed of raw solids is putrescible and potentially pathogenic. The latter must be covered every day to prevent nuisances and health hazards. Incineration is much more costly than land disposal, yet in highly urbanized areas it is often the only feasible alternative. Rather than burning, digested sludge may be dried and used as a soil conditioner. Unfortunately, the disadvantages as a soil additive outweigh the fertilizer merits, and consequently there is very little demand for the product. Plants that have the option of incineration or heat drying usually burn the majority of the waste solids. Many cities near the ocean find that the best means for sludge disposal are digestion, thickening, and barging to sea. Coastal cities may discharge stabilized sludge to the ocean through an underwater pipeline.

The following are recent trends taking place in sludge processing:

1   Sludge characteristics are changing due to increased contributions of manu-facturing discharges that make residues more difficult to dewater.
2   Waste-activated sludge is being thickened separately and then blended with primary solids for dewatering, rather than returning excess sludge to the plant influent.
3   A variety of new polymeric flocculents have been developed that are easier and more economical to use than inorganic coagulants in sludge conditioning.
4   Mechanical dewatering is being adopted by an increasing number of cities, owing to the rising cost of land and operating problems of biological digesters.
5   The application and use of centrifuges is increasing.
6   Incineration at both large and small treatment plants is growing in popularity.
7   Anaerobic digestion as a means of conditioning prior to incineration is no longer considered necessary; thus raw solids are being burned after dewatering.
8   More coastal cities are disposing of thickened digested sludge by barging to the ocean.
9   Land disposal of liquid digested sludge is becoming more popular at small plants sited in rural areas.
10 Elutriation and heat drying of sludge are rarely considered viable processes in design of new plants.

The following process flow diagrams graphically illustrate the selection and arrangement of common sludge-processing layouts. Figure 12-1 is typical for communities of less than 10,000 population. Raw solids and filter humus are settled and stored in the primary clarifiers. Once or twice a day, sludge and scum are pumped to anaerobic digesters, and supernatant is withdrawn and returned to the treatment-plant influent. Stabilized and thickened sludge accumulates in

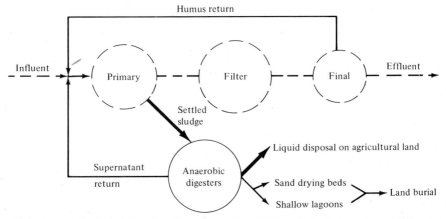

FIGURE 12-1 Typical sludge-handling scheme for a trickling-filter plant serving a community of fewer than 10,000 people.

the digesters for withdrawal when weather conditions permit disposal. The residue may be eliminated by spreading the liquid on agricultural land, or by drying on sand beds or in lagoons followed by hauling to land burial.

Many activated-sludge plants that use a flow scheme returning secondary biological floc to the plant influent experience difficulties resulting from anaerobiosis and thinning of settled sludge in the primary clarifiers. As a result, the two waste streams are often separated and thickening units employed to increase solids concentrations. In Fig. 12-2a the waste activated is thickened independently by dissolved air flotation, a process that gives reliable and effective results. Water of separation is returned for reprocessing, while the float is pumped to mixed holding tanks along with the primary sludge. The tanks are sized so that sludge can accumulate when filters are not being operated, and mixing impellers are installed to ensure a homogeneous feed for chemical conditioning. Dewatered solids are eliminated by sanitary landfill or incineration. An alternative arrangement, Fig. 12-2b, blends waste-activated and raw sludges in a gravity thickener.

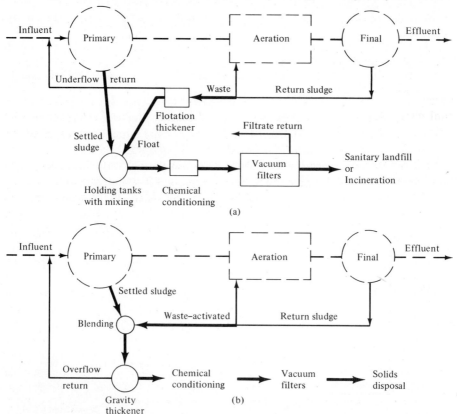

FIGURE 12-2 Alternative schemes for processing activated-sludge plant wastes by thickening in advance of chemical conditioning and mechanical dewatering. (a) Separate flotation thickening of waste-activated sludge before mixing with primary settlings. (b) Gravity thickening of combined raw primary and waste-activated sludges prior to vacuum filtration.

Consolidation of this waste mixture may yield only marginal results, because of carryover of flocculent solids, thus providing poor solids capture. Supernatant from the thickener is returned to the plant inlet, and underflow drawn continuously during the period of vacuum-filter operation.

Aerobic processing of wastewater without primary sedimentation often utilizes aerobic digestion for stabilization (Fig. 12-3). Decanting chambers for return of supernatant, installed in the digesters of package treatment plants,

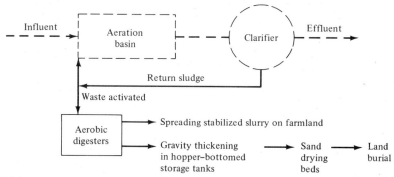

FIGURE 12-3 Common disposal methods for waste-activated sludge from small treatment plants without settling prior to aeration.

rarely provide efficient solids separation. However, separate hopper-bottomed tanks may be effective in concentrating aerobically digested solids. The most efficient method for eliminating the liquid stabilized sludge is by spreading on farmland.

Two possible schemes with anaerobic treatment preceding mechanical dewatering are given in Fig. 12-4. The first, used by large inland cities, burns the stabilized solids after vacuum filtration. Ash is commonly transported from the furnace in a water slurry and stored in lagoons prior to land burial. In the second scheme, the digested sludge slurry is concentrated sufficiently by centrifugation to allow disposal in the ocean through a pipeline or by transporting in barges. Centrifuges may be utilized for thickening the raw waste as well as stabilized sludge.

Design of a sludge-processing system requires a thorough understanding of the characteristics of the waste being produced and the most feasible method for solids disposal. Selection of the latter is dictated by local conditions and practices. Intermediate steps of thickening, treatment, and dewatering must be integrated so that each relates to the prior operation and prepares the residue for subsequent handling. A scheme should have flexibility to allow alternative modes of operation, since actual conditions may differ from those assumed at the time of design. Too often built-in rigidity, by limiting piping and pumping facilities, does not permit plant personnel to vary operations to meet changing conditions. Practice of sanitary design requires both knowledge and foresight to consider all available options.

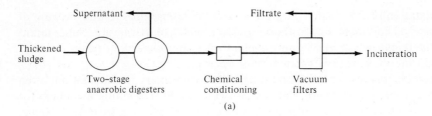

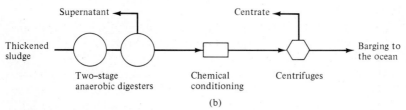

FIGURE 12-4 Sludge-processing diagrams applicable for large cities using anaerobic digestion prior to dewatering. (a) Biological and chemical conditioning prior to dewatering and combustion. (b) Conditioning and thickening for disposal in the ocean.

The following illustrate typical problems that have confronted plant operators:

1 Providing only sand drying beds for residue dewatering of anaerobic sludge at small plants with no provision for spreading on adjacent farmland. Plant operators usually modify the piping and buy a tank wagon for liquid disposal because of the high cost of labor to scoop up the dried cake by hand.
2 Converting anaerobic digesters to aerobic stabilization without modifying the disposal method. Since aerobically digested sludge does not thicken readily, sand drying beds designed for anaerobic sludge are not appropriate, and there is no suitable means for dewatering the voluminous stabilized slurry.
3 Returning excess activated sludge to the head of the treatment plant with no alternative for separate thickening or disposal. The result is upset primary clarifiers, reduced efficiency, thin sludge, and higher chemical costs for conditioning.
4 Incinerating a sludge when sanitary landfill is much less expensive.
5 Producing soil conditioner by drying sludge solids when there is no market for the fertilizer.

12-5

SELECTION OF PROCESSES FOR WATER-TREATMENT SLUDGES

Historically, settled coagulation wastes and backwash waters have been disposed of without treatment. The practice was justified on the basis that the returned material had originally been present in the river or lake. However, this argument is not considered valid, since the discharge also contains polluting chemicals used in processing the water, and the source of supply may be groundwater. More stringent effluent standards now apply to waste discharges from water purification and softening facilities. New water plant designs provide for zero discharge by recycling of wastewaters, reclaiming chemicals, and ultimate disposal of the useless or unmarketable residue remaining. Existing plants lack reclamation facilities and generate appreciable amounts of waste that must be hauled to landfill or barged to sea. Special consideration may be given to old plants with unique problems; otherwise, little or no return discharge is permitted unless it meets effluent standards.

Water-treatment processes can be modified to change the characteristics and reduce the quantities of wastes. Polyelectrolytes as coagulant aids lower the required dosages of alum and auxiliary chemicals. The result is less sludge that is easier to dewater because of the reduced content of hydroxide precipitate. Polymers also enhance presedimentation of turbid river waters, thus controlling carryover of solids to subsequent chemical coagulation. Addition of specially manufactured clays can be used to aid flocculation of relatively clear surface supplies by producing a denser floc that settles more rapidly. Alum substitution is being considered as a cost-effective technique for modifying sludge-handling processes at many surface-water plants. Groundwater softening plants can also change their mode of operation to lessen the volume of waste sludge. Emphasis is placed on preventing magnesium hydroxide precipitation, since it inhibits dewaterability and processing for recovery of lime. Also, hardness reduction can be limited to produce less solid material while still supplying a moderately soft water acceptable to the general public.

The common processes for storage, treatment, and disposal of water-treatment sludges are listed in Table 12-2. Each waterworks is unique in that local conditions and existing facilities tend to dictate techniques applied in waste-sludge disposal. Settled solids from coagulation may be stored in plain sedimentation basins for periods up to several months, then removed by taking the tanks out of service and emptying the contents.

Modern clarifiers equipped with mechanical scrapers have limited storage capacity, however, and must discharge sludge at regular intervals, usually daily. Separate holding tanks can be installed to accumulate this slurry prior to dewatering. Filter backwash can be stored in clarifier–flocculators that serve as both temporary holding tanks and wash-water settling basins. Equalization and settling are generally the only prerequisites if sludges are discharged to a sewer for processing at the municipal wastewater treatment plant.

A typical system for thickening coagulation waste and filter wash water is shown schematically in Fig. 12-5. The two primary sources of waste are sludge from the clarifier, following chemical coagulation, and wash water from back-washing filters. The latter is discharged to a clarifier–holding tank for gravity separation of the suspended solids and flow equalization. Settled solids consolidate to a sludge volume less than one-tenth of the wash-water volume. Supernatant is withdrawn slowly and recycled to the plant inlet. After a sufficient portion

TABLE 12-2

PROCESSES FOR STORAGE, TREATMENT, AND DISPOSING OF WATER-TREATMENT SLUDGES

Storage prior to processing
   Sedimentation basins
   Separate holding tanks
   Flocculator-clarifier basins
Thickening prior to dewatering
   Gravity settling
Chemical conditioning prior to dewatering
   Polymer application
   Lime addition to alum sludges
Mechanical dewatering
   Centrifugation
   Pressure filtration
   Vacuum filtration
Air drying
   Shallow lagoons
   Sand drying beds
Disposal of dewatered solids
   Sanitary landfill
   Barging to sea
Chemical recovery
   Recalcination of lime precipitates
   Alum recovery

has been drained, the holding tank is able to receive the next backwash surge. Settled sludges from both the wash-water tanks and in-line clarifiers are given second-stage consolidation in a clarifier thickener. Polymer is normally added to enhance solids capture, overflow is recycled, and thickened sludge withdrawn for further dewatering and disposal.

Mechanical dewatering of chemical sludges can be accomplished by centrifugation, pressure filtration, and in some instances by vacuum filtration (Fig. 12-6). A major advantage of centrifugation is operational flexibility. Machine variables, such as speed of rotation, allow a range of moisture content in the discharged solids varying from a dry cake to a thickened slurry. Feed rates, original solids content, and chemical conditioning are also variables that influence performance. Polyelectrolytes or other coagulants are normally applied with the slurry feed to enhance solids capture. Lime sludges compact readily producing a cake with 60–70% solids from a feed of 10–20%. In recalcination plants, centrifuges can be operated to selectively thicken calcium carbonate precipitate with the bulk of undesirable magnesium hydroxide appearing in the liquid discharge. Alum sludges do not dewater as readily and discharge with a toothpaste consistency suitable for either further processing or transporting by truck or barge to a disposal site. Gravity-thickened sludge containing about one-

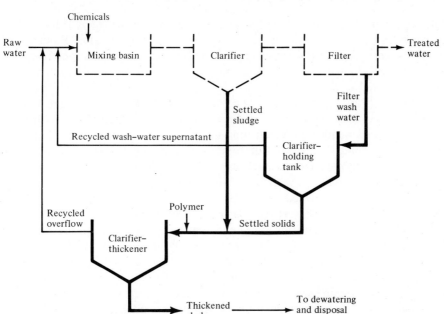

FIGURE 12-5 Thickening coagulation waste and filter wash water in preliminary handling of water-treatment-plant sludges.

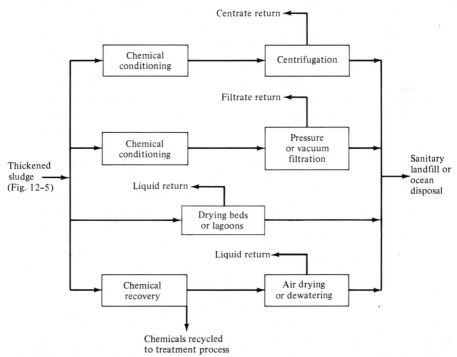

FIGURE 12-6 Alternative methods for disposal of water-treatment sludges by dewatering, drying, or chemical recovery of thickened chemical coagulation wastes.

half aluminum hydroxide slurry can be concentrated to 10–15% solids, while one-quarter hydrate slurry can be dewatered to 20% or greater. Removal of solids by centrifugation varies over a broad range, depending on operating situations and chemical conditioning—in general, solids recovery and density of cake are related to polymer dosage.

Pressure filtration is particularly advantageous for dewatering alum sludges if a high solids concentration in the filter cake is desired. Aluminum hydroxide wastes are often conditioned with lime to improve their filterability. The filter medium is precoated with either diatomaceous earth or fly ash before applying sludge solids. Precoat protects against blinding of the filter cloth by fines, and ensures easy cake discharge without sticking. With proper chemical conditioning, alum sludges can be pressed to a solids content of about 40%, which can be handled as a chunky solid rather than the paste consistency associated with a 15% density. Vacuum filtration is not as extensively used in dewatering chemical sludges. Alum precipitates do not dewater readily under vacuum, although belt filters with synthetic cloth have been adopted rather successfully for pure lime sludges. Consequently, the greatest applicability of vacuum filtration appears to be dewatering softening wastes from groundwater-treatment plants.

Lagooning is an accepted method for dewatering, thickening, and temporary storage of waste sludge where suitable land area is available. The diked pond area needed relates to character of the sludge, climate, design features such as underdrains and decanters, and method of operation. Clarified overflow may be returned to the treatment plant, particularly if filter wash water is directed to the lagoons without prior thickening. Sludges from lime softening consolidate to about 50% solids, which can be removed by a scraper or dragline and hauled to land burial. Alum sludge dewaters more slowly to a density of only 10–15%. Although the surface may dry to a hard crust, the underlying sludge turns to a viscous liquid upon agitation. This slurry must be removed, usually by dragline, and spread on the banks to air dry prior to hauling. Freezing enhances the dewatering of alum sludge by breaking down its gelatinous character. Neither lime nor alum sludges make a good, stable landfill. Air drying at small water plants can be done on sand beds with tile underdrains. Repeated sludge applications over a period of several months can be made in depths up to several feet. Dewatering action is by drainage and air drying, although operation may include decanting supernatant. Dried cake is removed either by hand shoveling or mechanical means.

Recovery of chemicals is considered where dewatering and disposal by landfill or barging to the ocean are not feasible. Recovery of lime or alum must be justified both economically and technologically by careful studies based on local conditions. Major steps in recalcination are gravity thickening, mechanical dewatering with classification of solids to remove magnesium hydroxide and other impurities, and heating the dried solids to produce lime and carbon dioxide. The process appears to be most adaptable at large plants where the precipitate is largely calcium carbonate with little or no magnesium or extraneous material.

Softening–coagulation sludges from treatment of turbid surface waters are difficult, if not impossible, to recalcine due to insolubles, such as clay, enmeshed in the sludge. Recovery of alum is not widely practiced because of low process efficiency and the poor quality of regenerated chemical. The major problem is separation of undesirable impurities from the aluminum hydroxide. From all chemical recovery processes, unwanted liquid and solid residues remain that require handling and disposal.

Gravity Thickening

Gravity thickening is the simplest and least expensive process for consolidating waste sludges. Thickeners in wastewater treatment are employed most successfully in consolidating primary sludge separately or in combination with trickling filter humus. Occasionally, raw primary and waste-activated sludges are blended and concentrated, but results are often marginal because of poor solids capture. Water-treatment wastes from both sedimentation and filter backwashing can be compacted effectively by gravity separation.

12-6

DESCRIPTION OF GRAVITY THICKENERS

A typical waste sludge thickener is illustrated in Fig. 12-7. The tank resembles a circular clarifier except that the depth/diameter ratio is greater and the hoppered bottom has a steeper slope. A bridge fastened to the tank walls supports a truss-type scraper arm mounted on a pipe shaft equipped with a power-lift device for overloads. Increased torque demand on the main drive automatically hoists the scraper mechanism, which can be returned to the bottom

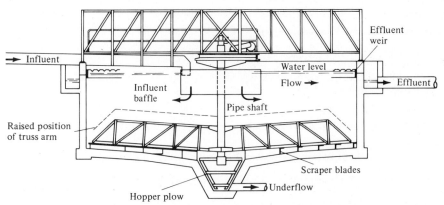

FIGURE 12-7 Cross-sectional view of a gravity sludge thickener. (LINK-BELT Product, courtesy Environmental Equipment Div., FMC Corp.)

by operator control. Sludge enters at the center behind a circular baffle that directs it downward, and supernatant overflows a peripheral weir. Settled solids are gently agitated by slow rotation of the scraper to dislodge gas bubbles, prevent bridging of the solids, and to move the slurry toward a central well for withdrawal. Feed is provided continuously while the underflow may be extracted intermittently for further processing.

Three settling zones in a thickener are the clear supernatant on top, feed zone characterized by hindered settling, and compression near the bottom where consolidation occurs. The theory of zone settling is discussed in Sections 9-12 and 9-13. Settling data may be collected from batch-type laboratory tests conducted in small cylinders. These are influenced by such factors as cylinder diameter, initial height, temperature, effect of stirring, and so on. Vesilind[1] suggests that batch thickening tests use an 8-in.-diameter cylinder, initial height of at least 3 ft, filling the cylinder from the bottom, and slow stirring of the sample throughout the test. Continuous-flow bench-scale experiments can also be conducted, but they are difficult and often yield questionable results. Because of the problems in scale-up from laboratory units to real systems, designers rely extensively upon experience acquired from studies at full-size installations.

Evaluating the performance of a thickener often involves mass-balance calculations. Overflow plus underflow solids equals influent solids. Also, the sum of overflow and underflow volumes is equal to the quantity of applied sludge. These values can be calculated using Eq. 12-3, as illustrated in Example 12-6.

●EXAMPLE 12-6

An alum-lime slurry with 4.0% solids content is gravity thickened to 20% with a removal efficiency of 95%. Calculate the quantity of underflow per 1.0 m³ of slurry applied, and concentration of solids in the overflow. Assume a specific gravity of 2.5 for the dry solids.

○*Solution*

solids applied $= 1.0 \text{ m}^3 \times 1000 \text{ kg/m}^3 \times 0.04 = 40 \text{ kg}$

underflow solids $= 0.95 \times 40 = 38 \text{ kg}$

Specific gravity of underflow using Eq. 12-2 is

$$S = \frac{80 + 20}{(80/1.0) + (20/2.5)} = 1.14$$

volume of underflow (Eq. 12-3) $= \dfrac{38}{0.20 \times 1000 \times 1.14} = 0.17 \text{ m}^3$

volume of overflow $= 1.0 - 0.17 = 0.83 \text{ m}^3$

From Eq. 12-3, concentration of solids in the overflow is

$$s = \frac{0.05 \times 40 \times 100}{0.83 \times 1000 \times 1.0} = 0.24\% = 2400 \text{ mg/l}$$

12-7

DESIGN OF WASTEWATER SLUDGE THICKENERS

The principal design criterion is solids loading expressed in units of pounds of solids applied per square foot of bottom area per day ($lb/ft^2/day$). Typical loading values and thickened sludge concentrations based on operational experience are listed in Table 12-3. These data assume good operation and chemical addi-

TABLE 12-3

GRAVITY-THICKENER DESIGN LOADINGS AND UNDERFLOW CONCENTRATIONS FOR WASTEWATER SLUDGES

Type of Sludge	Average Solids Loading ($lb/ft^2/day$)[a]	Underflow Concentration (% solids)
Primary	20	8–10
Primary plus filter humus	10	6–8
Primary plus activated sludge	8	4–6

[a] 1.0 $lb/ft^2/day$ = 4.88 kg/m^2-day.

tions, such as chlorine, if necessary to inhibit biological activity. Solids recovery in a properly functioning unit is about 95%, with perhaps the exception of a unit handling primary plus waste activated where it is difficult to achieve this degree of solids capture. Most continuous-flow thickeners are designed with a side water depth of approximately 10 ft to provide an adequate clear-water zone, sludge blanket depth, and space for temporary storage of consolidated waste. Sludge blanket depths (feed plus compaction zones) should be 3 ft or greater to ensure maximum compaction, using a suggested solids retention time of 24 hours. This is estimated by dividing the volume of the sludge blanket by the daily sludge withdrawal; values vary from 0.5 to 2 days, depending on operation. Overflow rates should be 400–900 gpd/ft^2 (16 to 37 m^3/m^2-day) and are defined by the quantity of sludge plus supplementary dilution water applied.

Gravity thickeners are normally sized to handle the maximum seasonal or monthly sludge yield anticipated. Peak daily sludge production often requires storage in the thickener or other sludge-processing units. Low liquid overflow rates result in malodors from septicity of the thickener contents. A common remedy is to feed dilution water to the thickener along with the sludge to increase hydraulic loading. An alternative is to apply chlorine to reduce bacterial activity. Design of pumps and piping should be sufficiently flexible to allow regulation of the quantity of dilution water and have the capacity to transport viscous, thickened sludges.

●**EXAMPLE 12-7**

The daily quantity of primary sludge from a trickling-filter plant contains 1130 lb of solids at a concentration of 4.5%. Size a gravity thickener based on a solids loading of 10 lb/ft²/day. Calculate the daily volumes of applied and thickened sludges, assuming an underflow of 8.0% and 95% solids capture. What is the flow of dilution water required to attain an overflow rate of 400 gpd/ft²? If the blanket of consolidated sludge in the tank has a depth of 3.0 ft, estimate the solids retention time.

○**Solution**

$$\text{tank area required} = \frac{1130}{10} = 113 \text{ ft}^2$$

$$\text{diameter} = \left(\frac{113 \times 4}{\pi}\right)^{1/2} = 12.0 \text{ ft; use a depth} = 10.0 \text{ ft}$$

$$\text{volume of applied sludge} = \frac{1130}{0.045 \times 62.4} = 402 \text{ ft}^3/\text{day} = 3010 \text{ gpd}$$

$$\text{overflow rate of applied sludge} = \frac{3010}{113} = 27 \text{ gpd/ft}^2$$

$$\text{supplemental dilution flow to attain 400 gpd/ft}^2$$
$$= (400 - 27)113 = 42{,}000 \text{ gpd}$$

$$\text{volume of thickened sludge} = \frac{1130 \times 0.95}{(8.0/100)62.4} = 215 \text{ ft}^3/\text{day}$$

$$\text{solids retention time} = \frac{3 \times 113 \times 24}{215} = 38 \text{ hours}$$

12-8

DESIGN OF WATER-TREATMENT SLUDGE THICKENERS

Currently few data are available on the performance of thickeners handling water-treatment-plant wastes. Alum sludges from surface-water coagulation settle to a density in the range 2–6% solids. Coagulation–softening mixtures from the treatment of turbid river waters gravity thicken approximately as follows: alum-lime sludge, 4–10%; iron-lime settlings 10–20%; alum-lime filter wash water, about 4%; and iron-lime backwash up to 8%. Density achieved in gravity thickening relates to the calcium-magnesium ratio in the solids, quantity of alum, nature of impurities removed from the raw water, and other factors. Calcium carbonate residue from groundwater softening consolidates to 15–25% solids. In most cases, special studies have to be conducted at a particular waterworks to determine settleability of solids in waste sludges and wash water. Flocculation aids are used to improve clarification in most cases.

Relatively dense chemical slurrys are thickened in tanks similar to the one shown in Fig. 12-7. Thin sludges and backwash waters may be concentrated in clarifier–thickeners that have an inlet well equipped with mixing paddles, where the feed can be flocculated with polymers or other coagulants. Holding tanks are used to dampen hydraulic surges of filter wash water. These units can be plain tanks with mixers or clarifiers equipped for removing settled solids and decanting clear supernatant.

Flotation Thickening

Air flotation is most applicable in concentrating waste-activated sludges, and pretreatment of industrial wastes to separate grease or fine particulate matter. Fine bubbles to buoy up particles may be generated by air dispersed through a porous medium, by air drawn from the liquid under vacuum, or by forcing air into solution under elevated pressure followed by pressure release. The latter, called *dissolved-air flotation*, is the process employed most frequently in thickening sludges because of its reliable performance.

12-9

DESCRIPTION OF DISSOLVED-AIR FLOTATION

The major components of a typical flotation system are sludge pumps, chemical feed equipment to apply polymers, an air compressor, a control panel, and a

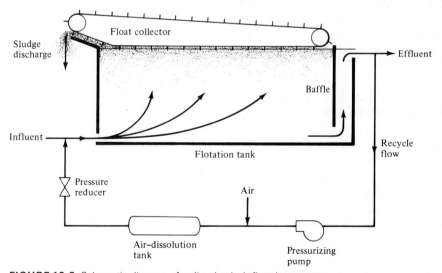

FIGURE 12-8 Schematic diagram of a dissolved-air flotation system.

flotation unit. Figure 12-8 is a schematic diagram of a dissolved-air system. Influent enters near the tank bottom and exits from the base at the opposite end. Float is continuously swept from the liquid surface and discharged over the end wall of the tank. Effluent is recycled at a rate of 30–150% of the influent flow through an air-dissolution tank to the feed inlet. In this manner, compressed air at 60–80 psi is dissolved in the return flow. After pressure release, minute bubbles with a diameter about 80 μm pop out of solution. They attach to solid particles, and become enmeshed in sludge flocs, floating them to the surface. The sludge blanket, varying from 8 to 24 in. thick, is skimmed from the surface. Flotation aids are introduced in a mixing chamber at the tank inlet.

Operating variables for flotation thickening are air pressure, recycle ratio, detention time, air/solids ratio, solids and hydraulic loading rates, and application of chemical aids. Operating air pressure in the dissolution tank influences the size of bubbles released. If too large, they do not attach readily to sludge particles, while too fine a dispersion breaks up fragile floc. Generally, a bubble size less than 100 μm is best; however, the only practical way to establish the proper rise rate is by conducting experiments at various air pressures.

Recycle ratio is interrelated with feed solids concentration, detention time, and air/solids ratio. Detention time in the flotation zone is not critical, providing that particles rise rapidly enough, and the horizontal velocity does not scour the bottom of the sludge blanket. An air/solids ratio of 0.01 to 0.03 lb of air/lb of solids is sufficient to achieve acceptable thickening of waste-activated sludge. Optimum recycle ratio must be determined by on-site studies.

Operating data from plant-scale units indicate solids loadings of 2–4 lb/ft^2/hr, with hydraulic flows of about 1 gpm/ft^2, can produce floats of 4–8% solids. Without polyelectrolyte addition, solids capture is 70–90%. However, removal efficiency increases to a mean of 97%, with a polymer dosage of approximately 10 lb/ton of dry suspended solids. This is the reason most wastewater installations use flotation aids.

12-10

DESIGN OF DISSOLVED-AIR FLOTATION UNITS

Wherever possible laboratory and pilot-scale tests are recommended to help determine specific design criteria for a given waste. Notwithstanding, the suggested design criteria for flotation thickening of typical waste-activated sludges are listed in Table 12-4. A conservative solids design loading is 2 lb/ft^2/hr (10 kg/m^2-hr) with the use of flotation aids. From actual operating data, at least 3 lb/ft^2/hr can be expected, and most thickeners have a built-in capacity for 4–5 lb/ft^2/hr loadings. While a 4% minimum float concentration is specified for design purposes, 5–6% solids can be normally expected. Flotation without polymers generally results in a concentration that is about 1 percentage point less than with chemical aids. Removal efficiency varies from 90 to 98% with poly-electrolyte addition. The maximum hydraulic loading for design is set at 0.8 gpm/ft^2; this is equivalent to applying a waste with a solids concentration of 5000 mg/l at a loading of 2 lb/ft^2/hr. Lesser solids levels or higher hydraulic loadings result in lower removal efficiencies and/or float densities.

The typical design values recommended in Table 12-4 apply to anticipated average sludge production. This procedure provides a significant safety factor and permits flexibility in operations. Peak solids loads at municipal treatment plants can usually be accommodated, since these conservative design criteria allow a maximum loading of nearly 100% greater than the average without a

TABLE 12-4

DESIGN PARAMETERS FOR DISSOLVED-AIR
FLOTATION OF WASTE-ACTIVATED SLUDGE WITH
ADDITION OF POLYELECTROLYTE FLOTATION AIDS

Parameter	Typical Design Value	Anticipated Results
Solids loading (lb/ft^2/hr)^a	2	3–5
Float concentration (%)	4	5–6
Removal efficiency	90–95	97
Polyelectrolyte addition (lb/ton of dry solids)	10	5–10
Air/solids ratio (lb of air/lb of solids)	0.02	
Effluent recycle ratio (% of influent)	40–70	
Hydraulic loading (gpm/ft^2)	0.8 maximum	

^{a} 1.0 lb/ft^2/hr = 4.88 kg/m^2-hr.

serious drop in performance. Perhaps the most critical condition is during a period of sludge bulking when the waste mixed liquor is more difficult to thicken and maximum hydraulic loading is applied to the flotation unit.

Sizing of flotation units for an existing plant can be calculated from available data on sludge quantities, characteristics, and solids concentrations. For new plant design, raw wastewater is often assumed to contain 0.20 lb of dry solids/capita/day. A portion of these solids are removed in primary settling, and a conservative estimate for secondary activated-sludge production is 0.10 lb/capita/day. The actual amount is likely to be closer to one-half of this value, because of biological decomposition. Solids yield in an activated-sludge process without primary settling may be safely assumed to be 0.17 lb/capita/day for domestic wastewater. If the waste sludge from such a system is aerobically digested, the concentration of solids is reduced by about 35%.

Operating hours of a flotation unit depend on size of plant and the working schedule. Although a unit does not require continuous operator attention, periodic checks of the system are scheduled. Generally, 48 hours/week is adequate for plants with capacities less than 2 mgd. For systems of 2–5 mgd, two shifts 5 days per week establishes an operating period of 80 hours/week. Treatment plants handling more than 20 mgd have operators on duty continuously, and thickening units are run on a schedule appropriate for sludge dewatering and disposal.

●**EXAMPLE 12-8**

A dissolved-air flotation thickener is being sized to process waste-activated sludge based on the design criteria given in Table 12-4. The average waste flow is 33,600 gpd at 15,000 mg/l (1.5%) suspended solids, and the maximum daily quantity contains 50% more solids at a reduced concentration of 10,000 mg/l. What is the peak daily hydraulic loading that can be processed? Base all computations on a 14 hr/day operating schedule.

○**Solution**

Flotation tank surface area required for the average daily flow at a design loading of 2.0 lb/ft²/hr for a 14-hour/day schedule is:

$$\text{area} = \frac{33,600 \times 0.015 \times 8.34}{2.0 \times 14} = \frac{4200}{28} = 150 \text{ ft}^2$$

Check the solids loading and overflow rate at maximum daily sludge production

$$\text{maximum solids loading} = \frac{1.5 \times 4200}{150 \times 14} = 3.0 \text{ lb/ft}^2/\text{hr} \quad (\text{OK})$$

$$\text{maximum sludge volume} = \frac{1.5 \times 4200}{0.01 \times 8.34} = 75,500 \text{ gpd}$$

$$\text{maximum overflow rate} = \frac{75,500}{150 \times 14 \times 60} = 0.60 \text{ gpm/ft}^2 \quad (\text{OK})$$

peak hydraulic loading based on 0.80 gpm/ft² = 0.80 × 150 × 14 × 60 = 100,000 gpd

Biological Sludge Digestion

Biological stabilization of settlings from wastewater treatment is still widely practiced, even though mechanical dewatering has gained prominence in recent years. Anaerobic digestion is used extensively in small cities employing primary clarification followed by either trickling filter or activated-sludge secondary treatment. Aerobic digestion stabilizes waste-activated sludge from aeration plants without primary settling tanks. Fundamental differences between aerobic and anaerobic digestion are illustrated in Figs. 11-7 and 11-8. The end product of aerobic digestion is cellular protoplasm, and growth is limited by depletion of the available carbon source. The end products of anaerobic metabolism are methane, unused organics, and a relatively small amount of cellular protoplasm. Growth is limited by a lack of hydrogen acceptors. Anaerobic digestion is basically a destructive process, although complete degradation of the organic matter under anaerobic conditions is not possible.

12-11

ANAEROBIC SLUDGE DIGESTION

Anaerobic digestion consists of two distinct stages which occur simultaneously in digesting sludge (Fig. 11-15). The first consists of hydrolysis of the high-molecular-weight organic compounds and conversion to organic acids by acid-forming bacteria (Eq. 11-9). The second stage is gasification of the organic acids to methane and carbon dioxide by the acid-splitting methane-forming bacteria (Eq. 11-10).

Methane bacteria are strict anaerobes and very sensitive to conditions of their environment. The optimum temperature and pH range for maximum growth rate are limited. Methane bacteria can be adversely affected by excess concentrations of oxidized compounds, volatile acids, soluble salts, and metal cations, and also show a rather extreme substrate specificity. Each species is restricted to the use of only a few compounds, mainly alcohols and organic acids, whereas the normal energy sources, such as carbohydrates and amino acids, are not attacked. An enrichment culture developed on a feed of acetic or butyric acid cannot decompose propionic acid.

The sensitivity exhibited by methane bacteria in the second stage of anaerobic digestion, coupled with the rugged nature of the acid-forming bacteria in the first stage, creates a biological system where the population dynamics are easily upset. Any shift in environment adverse to the population of methane bacteria causes a buildup of organic acids, which in turn further reduces the metabolism of acid-splitting methane formers. Pending failure of the anaerobic digestion process is evidenced by a decrease in gas production, a lowering in the percentage of methane gas produced, an increase in the volatile acids concentration, and eventually a drop in pH when the accumulated volatile acids exceed the buffering capacity created by the ammonium bicarbonate in solution. Digester failure may be caused by any of the following: significant increase in organic loading, sharp decrease in digesting sludge volume (i.e., when digested sludge is withdrawn), sudden increase in operating temperature, or accumulation of a toxic or inhibiting substance.

General conditions for mesophilic sludge digestion are given in Table 12-5.

TABLE 12-5

GENERAL CONDITIONS FOR SLUDGE DIGESTION

Temperature	
Optimum	98°F (35°C)
General range of operation	85°–95°F
pH	
Optimum	7.0–7.1
General limits	6.7–7.4
Gas production	
Per pound of volatile solids added	6–8 ft³
Per pound of volatile solids destroyed	16–18 ft³
Gas composition	
Methane	65–69%
Carbon dioxide	31–35%
Hydrogen sulfide	Trace
Volatile acids concentration as acetic acid	
Normal operation	200–800 mg/l
Maximum	approx. 2000 mg/l
Alkalinity concentrations as Ca CO₃	
Normal operation	2000–3500 mg/l

12-12

SINGLE-STAGE FLOATING-COVER DIGESTERS

The cross section of a typical floating-cover digestion tank is shown in Fig. 12-9. Raw sludge is pumped into the digester through pipes terminating either near the center of the tank or in the gas dome. Pumping sludge into the dome helps to break up the scum layer which forms on its surface.

Digested sludge is withdrawn from the tank bottom. The contents are heated in the zone of digesting sludge by pumping them through an external heater and returning the heated slurry through the inlet lines. The tank contents stratify with a scum layer on top and digested, thickened sludge on the bottom. The middle zones consist of a layer of supernatant (water of separation) underlain by the zone of actively digesting sludge. Supernatant is drawn from the digester through any one of a series of pipes extending out of the tank wall. Digestion gas from the gas dome is burned as fuel in the external heater, or wasted to a gas burner.

The weight of the cover is supported by sludge, and the liquid forced up between the tank wall and the side of the cover provides a gas seal. Gas rises out of the digesting sludge, moves along the ceiling of the cover, and collects in the gas dome. The cover can float on the surface of the sludge between the landing brackets and the height of the overflow pipe. Rollers around the circumference of the cover keep it from binding against the tank wall.

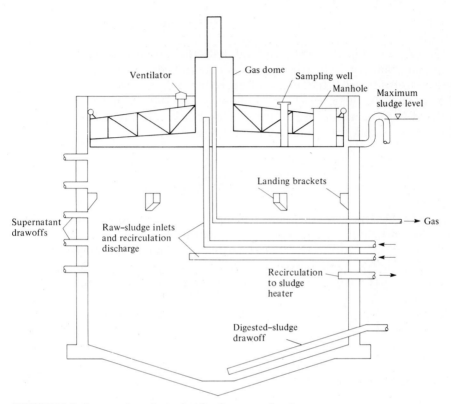

FIGURE 12-9 Cross section of a typical floating-cover digester.

Three functions of a single-stage floating-cover digester are (1) anaerobic digestion of the volatile solids; (2) gravity thickening; and (3) storage of the digested sludge. A floating-cover feature of the tank provides for a storage volume equal to approximately one-third that of the tank. The unmixed operation of the tank permits gravity thickening of sludge solids and withdrawal of the separated supernatant. Anaerobic digestion of the sludge solids is promoted by maintaining near optimum temperature and stirring the digesting sludge through the recirculation of heated sludge. However, the rate of biological activity is inhibited by the lack of mixing; on the other hand, good mixing would prevent supernatant formation. Therefore, in single-tank operation, the biological process is compromised to allow both digestion and thickening to occur in the same tank.

In the operation of an unmixed digester, raw sludge is pumped to the digester from the bottom of the settling tanks once or twice a day. Supernatant is withdrawn daily and returned to the influent end of the treatment plant. It is normally returned by gravity flow to the wet well during periods of low raw-wastewater flow, or, in the case of an activated-sludge plant, it may be pumped to the head end of the aeration basin. Because of the floating cover, supernatant does not have to be drawn off simultaneously with the pumping of raw sludge into the digester.

Digested sludge is stored in the tank and withdrawn periodically for disposal. In small plants, it is frequently dried on sand beds or in lagoons and then hauled to land burial. In the case of open sludge-drying beds, weather dictates the schedule for digested sludge disposal. In northern climates, the cover is lowered as close as possible to the landing brackets in the fall of the year to provide maximum volume for winter sludge storage.

12-13

HIGH-RATE (COMPLETE MIXING) DIGESTERS

The biological process of anaerobic digestion is significantly improved by complete mixing of the digesting sludge either mechanically or by use of compressed digestion gases. Mechanical mixing is normally accomplished by an impeller suspended from the cover of the digester (Fig. 12-10a). Three common methods of gas mixing are by injection of compressed gas through a series of

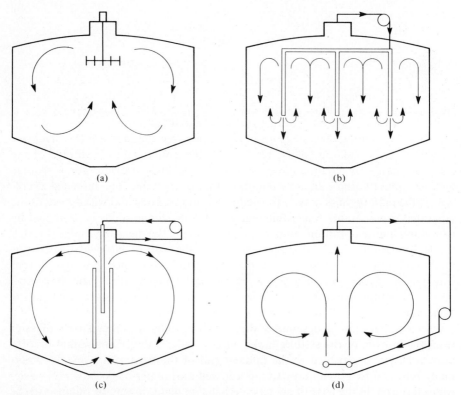

(a)   (b)

(c)   (d)

FIGURE 12-10 High-rate digester—mixing systems (a) Mechanical mixing; (b) gas mixing using a series of gas discharge pipes; (c) gas mixing using a central draft tube; (d) gas mixing using diffusers mounted on the tank bottom.

small-diameter pipes hanging from the cover into the digesting sludge (Fig. 12-10b); use of a draft tube in the center of the tank, with compressed gas injected into the tube to lift recirculating sludge from the bottom and spill it out on top (Fig. 12-10c); or supplying compressed gas to a number of diffusers mounted in the center at the bottom of the tank (Fig. 12-10d).

A complete mixing digester may have either a fixed- or a floating-cover tank. Digesting sludge is displaced when raw sludge is pumped into the digester. By use of a floating cover, tank volume for storage of digesting sludge is available, and withdrawals do not have to coincide with the introduction of raw sludge.

The homogeneous nature of the digesting sludge in a high-rate digester does not permit formation of supernatant. Therefore, thickening cannot be performed in a complete mixing digester. High-rate-digestion systems normally consist of two tanks operated in series (Fig. 12-11). The first stage is a complete mixing, heated, floating, or fixed-cover digester fed as continuously as possible, whose function is anaerobic digestion of the volatile solids. The second stage may be heated or unheated, and it accomplishes gravity thickening and storage of the digested sludge. Two-stage systems often consist of two similar floating-cover tanks with provisions for mixing in one tank.

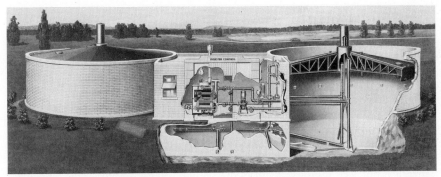

FIGURE 12-11 Two-stage digester system, showing piping in control room and external sludge heater. (Courtesy Water Quality Control Div., Envirex, a Rexnord Co.)

12-14

VOLATILE SOLIDS LOADINGS AND DETENTION TIMES

Typical ranges of loadings and detention times employed in the design and operation of heated anaerobic–digestion tanks treating domestic waste sludge are listed in Table 12-6. Values given for volatile solids loading and digester capacity for conventional, single-stage digesters are based on the total sludge volume available in the tank (i.e., the volume with the floating cover fully raised). Figures given for high-rate digestion apply only to the volume needed for the first-stage tank. There are no established design standards for the tank capacity required in second-stage thickening and supernatant separation.

The loading applied to a digester is expressed in terms of pounds of volatile solids applied per day per cubic foot of digester capacity. Detention time is the volume of the tank divided by the daily raw-sludge pumpage. Digester capacity in Table 12-6 is given in terms of cubic feet of tank volume provided per design population equivalent of the treatment plant.

TABLE 12-6

LOADINGS AND DETENTION TIMES FOR HEATED ANAEROBIC DIGESTERS

	Conventional Single-Stage (Unmixed)	First Stage High-Rate (Complete Mixing)
Loading (lb/ft^3/day of VS)^a	0.02–0.05	0.1–0.2
Detention time (days)	30–90	10–15
Capacity of digester (ft^3/population equivalent)		
Primary only	2–1	0.4–0.6
Primary and secondary	4–6	0.7–1.5
Volatile solids reduction (%)	50–70	50

^{a} 1.0 lb/ft^3/day = 16.0 kg/m^3-day.

12-15

ANAEROBIC DIGESTER CAPACITY

The 1971 *Standards of the Great Lakes–Upper Mississippi River Board of State Sanitary Engineers*[5] states:

> In recent years a number of modifications to the conventional anaerobic-sludge-digestion process have been developed, especially in the area commonly known as "high-rate digestion." Design standards, operating data, and experience are not well established for some of these modifications. This should be considered in the selection and design of the process modification. The total digestion-tank capacity should be determined by rational calculations based upon such factors as volume of sludge added, its percent solids, and character, the temperature to be maintained in the digesters, the degree of extent of mixing to be obtained, the degree of volatile solids reduction required, and the size of the installation with appropriate allowances for sludge and supernatant storage. Calculations should be submitted to justify the basis of design.

If rational computations are not submitted GLUMRB recommends a maximum of 0.08 lb/ft^3/day of VS (1.3 kg/m^3-day) for high-rate digestion and a maximum of 0.04-lb (0.6-kg) VS loading for single-stage operation. These loadings assume that the raw sludge is derived from domestic wastewater, the digestion temperature is in the range of 85–95°F (29–35°C), volatile solids reduction is 40–50%, and the digested sludge will be removed frequently.

The capacity required for a single-stage floating-cover digester can be determined by the rational formula

$$V = \frac{V_1 + V_2}{2} T_1 + V_2 \times T_2 \qquad (12\text{-}10)$$

where V = total digester capacity, ft^3
V_1 = volume of average daily raw sludge feed, ft^3/day
V_2 = volume of daily digested sludge accumulation in tank, ft^3/day
T_1 = period required for digestion, days (approximately 25 days at a temperature of 85–95°F)
T_2 = period of digested sludge storage, days (normally 30–120 days)

The diagram in Fig. 12-12 is a pictorial representation of Eq. 12-10.

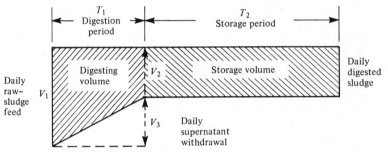

FIGURE 12-12 Pictorial presentation of Eq. 12-10.

It is often difficult to predict the daily volumes of raw sludge produced and the digested sludge accumulated as required in Eq. 12-10. Therefore, the capacity of conventional digesters frequently is based on empirical values relating digester capacity to the equivalent population design of the plant (Table 12-6). Values of 4, 5, and 6 $ft^3/capita$ are frequently used for low-rate trickling-filter plants, high-rate trickling-filter plants, and activated-sludge plants, respectively.

The minimum detention time for satisfactory high-rate digestion at 95°F is in the range 7–10 days. In general, this limiting period is dependent upon the minimum time required to digest the grease component of raw sludge. Also, too short a detention time results in depletion of the methane-bacteria populations since they are washed out in the effluent. The maximum volatile solids loadings, at a 10-day detention time, vary from 0.2 to 0.3 $lb/ft^3/day$ for adequate volatile solids destruction and gas production. For larger treatment plants with uniform loading conditions, design values of a 10-day minimum detention time and maximum 0.2 $lb/ft^3/day$ of VS loading appear to be satisfactory. Digesters for small treatment plants with wider variations in daily sludge production should be planned using more conservative loading rates.

Capacities required for a high-rate digestion system can be determined by the following equations

$$V_1 = V_1 \times T \qquad (12\text{-}11)$$

where V_I = digester capacity required for first-stage high-rate, ft³
V_1 = volume of average daily raw sludge feed, ft³/day
T = period required for digestion, days

and

$$V_{II} = \frac{V_1 + V_2}{2} T_1 + V_2 \times T_2 \qquad (12\text{-}12)$$

where V_{II} = digester capacity required for second-stage digested sludge thickening and storage, ft³
V_1 = volume of digested sludge feed = volume of average daily raw sludge, ft³/day
V_2 = volume of daily digested sludge accumulation in tank, ft³/day
T_1 = period required for thickening, days
T_2 = period of digested sludge storage, days

The diagrams in Fig. 12-13 explain Eqs. 12-11 and 12-12.

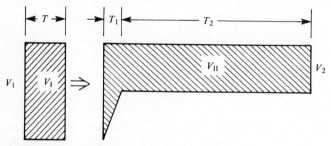

FIGURE 12-13 Diagrams for Eqs. 12-11 and 12-12.

● **EXAMPLE 12-9**
A high-rate trickling-filter plant treats a domestic wastewater flow of 0.40 mgd. Characteristics of the wastewater are identical to those in Table 11-1. Determine the digester capacities required for (a) a single-stage floating-cover digestion system; (b) a two-stage high-rate digestion system, operating at near optimum temperature. Digested sludge is to be dried on sand beds, and the longest anticipated storage period required is 90 days.

○ **Solution**
 a Equivalent population = 4000 at 0.20 lb/capita
Assume the following:
　　　water content in raw sludge = 96%
 volatile solids in raw sludge solids = 70%
　water content of digested sludge = 94%
　　　　volatile solids reduction = 50%
Volume of raw sludge, using Eq. 12-3,

$$V = \frac{4000 \times 0.20}{[(100 - 96)/100]62.4} + 321 \ \text{ft}^3/\text{day}$$

Volume of digested sludge,

$$V = \frac{0.30(4000 \times 0.20) + 0.70 \times 0.50(4000 \times 0.20)}{[(100 - 94)100]62.4}$$

$$= 139 \text{ ft}^3/\text{day}$$

Substituting into Eq. 12-10,

$$V = [(321 + 139)/2]25 + 139 \times 90 = 18,250 \text{ ft}^3$$

Check the volatile solids loading,

$$= \frac{0.70 \times 4,000 \times 0.20}{18,250} = 0.031 \text{ of lb/ft}^3/\text{day of VS}$$

Verify the digester capacity per capita

$$= \frac{18,250}{4000} = 4.56 \text{ ft}^3/\text{EP}$$

(This value is only slightly less than the empirical design figure of 5 ft^3/day/ equivalent population.)

b   Assume the following:

maximum allowable loading $= 0.15$ of lb/ft^3/day of VS
minimum detention time $= 15$ days
period required for thickening $= 10$ days
quantity of raw sludge $=$ same as a
water content of digested sludge $= 92\%$
volatile solids reduction $=$ same as a

First-stage digester capacity using Eq. 12-11,

$$V_1 = 321 \times 15 = 4810 \text{ ft}^3$$

Check volatile solids loading,

$$= \frac{0.70 \times 4000 \times 0.20}{4810} = 0.116 \text{ lb/ft}^3/\text{day of VS}$$

Volume of digested sludge,

$$V_2 = \frac{0.30(4000 \times 0.20) + 0.70 \times 0.50(4000 \times 0.20)}{[(100 - 92)/100]62.4}$$

$$= 104 \text{ ft}^3/\text{day}$$

Second-stage digester capacity, using Eq. 12-12,

$$V_{II} = [(321 + 104)/2]10 + 104 \times 90 = 11,500 \text{ ft}^3$$

12-16

STARTUP AND MONITORING OF DIGESTERS

Anaerobic digesters are much more difficult to start operating than are aerobic systems. This is due to the slow growth rate and sensitivity of methane-forming bacteria. If a substantial amount of digesting sludge from an operating digester is used as seed, a new digester can be in operation within a few weeks. However, if only raw sludge is available, startup may take months. Normal procedure for startup of a digester is to fill the tank with wastewater and apply a one-tenth sludge feed rate. Lime may be added with the raw sludge to maintain the pH near 7.0. After the digestion process has been established, the feed rate is gradually increased by small increments to full loading.

Operation of a digester can be monitored by any of the following methods: plotting the daily gas production per unit raw sludge fed; percentage of carbon dioxide in the digestion gases; or concentration of volatile acids in the digesting sludge. A reduction in gas production, increase in carbon dioxide percentage, and rise in volatile acids concentration all indicate reduced activity of the acid-splitting methane-forming bacteria.

12-17

AEROBIC SLUDGE DIGESTION

The function of aerobic digestion is to stabilize waste sludge solids by long-term aeration, thereby reducing the BOD and destroying volatile solids. The most common application of aerobic digestion is in handling waste activated sludge from wastewater treatment systems without primary settling tanks (Fig. 11-50).

Aerobic digestion is accomplished in one or more tanks mixed by diffused aeration. Since aerating solids have a low rate of oxygen demand, the need for effective mixing rather than microbial metabolism usually governs the air supply required. The volume of air supplied for aerobic digestion is normally in the range of 15–30 cfm/1000 ft^3 of digester. Generally, a supernatant separation chamber is provided in the digestion tank to decant the supernatant of digesting sludge and return it to the aeration tank. Customary methods for disposal of aerobically digested sludge are spreading on farmland, dumping in landfill, lagooning, or drying on sand beds.

There are no precise established standards for designing aerobic digesters. Design criteria applied vary with the type of activated-sludge system, BOD loading, and the means provided for ultimate disposal of the digested sludge. Small contact stabilization and complete-mixing activated-sludge plants without primary sedimentation are generally provided with 2–3 ft^3 of aerobic digester volume per design population equivalent of the plant. The volatile solids loading of an aerobic digester should be less than 0.07 lb/day/ft^3 of digester volume and the aeration period greater than 5 days. Generally, 10 days at 20°C or 15 days at 10°C is adequate. Volatile solids and BOD reduction at these loadings are in the range 35–50%, and the digested sludge can be dried without causing odorous or nuisance conditions. Drainability of the sludge is good and it dries readily on sand beds.

Long-term aeration of waste-activated sludge creates a bulking material difficult to thicken. Typically, an aerobically digesting sludge with a suspended solids concentration in excess of 5000 mg/l cannot be settled by gravity to decant a clear supernatant. This poor settleability frequently creates problems in disposing of the large volume of aerobically digested sludge. Thickening by flotation, centrifuging, or other mechanical methods is too expensive for incorporation in small treatment plants. Therefore, plant design should take care of storage and elimination of a relatively large volume of aerobically digested sludge produced.

●EXAMPLE 12-10

Refer to the data noted in Example 11-16. Calculate the cubic feet of aerobic digester volume provided per design population equivalent, and estimate the volatile solids loading on the aerobic digester.

○*Solution*

Data from Example 11-17,

$$\text{volume of aerobic digester} = 153 \text{ m}^3 = 5400 \text{ ft}^3$$
$$\text{design population of plant} = 2000$$

Therefore, the volume provided is $= \dfrac{5400}{2000} = 2.7 \dfrac{\text{ft}^3 \text{ of digester volume}}{\text{design population equivalent}}$

Data from Example 11-16,

$$\text{BOD load on plant} = 154 \text{ kg/day} = 340 \text{ lb/day}$$

Assuming a BOD loading of 0.33 lb of BOD/lb of MLSS applied to the aeration tank, the estimated excess sludge produced per day from Fig. 11-41 is 0.27 lb of VSS/lb of BOD load. Then the estimated volatile solids loading applied to the aerobic digester is $= 1.7(340 \times 0.27)/5400 = 0.029$ lb/ft^3/day of VS.

12-18

OPEN-AIR DRYING BEDS

Historically, small communities have dewatered digested sludge on open beds because of their simplicity, rather than operating more complex mechanical systems. Their disadvantages include poor drying during inclement weather, manual labor required for cleaning, potential odor problems, and the relatively large land area required. A typical sand bed consists of 6–9 in. of coarse sand supported on a graded gravel bed that incorporates tile or perforated pipe underdrains. These are spaced 20 ft apart and return seepage to the treatment-plant influent. Individual sections, nominally 25 ft by 100 ft, are contained by watertight walls extending 18 in. above the surface; concrete tracks are constructed in the bed to support a vehicle used to haul away the dried cake. A pipe header with a gated opening to each cell is used to apply liquid slurry in depths of 8–10 in.

Rational design for sludge beds is difficult, owing to the multitude of variables that affect drying rate. These include climate and atmospheric conditions, such as temperature, rainfall, humidity, and wind velocity; sludge characteristics, including degree of stabilization, grease content, and solids concentration; depth and frequency of sludge application; and condition of the sand stratum and drainage piping. The bed area furnished for desiccating anaerobically digested sludge is from 1 to 2 ft^2/BOD design population equivalent of the treatment plant. Solids loadings average about 20 lb/ft^2/yr (100 kg/m^2-yr) in northern states, while unit loading may be as high as 40 lb/ft^2/yr in southern climates. Drying time ranges from several days to weeks, depending on drainability of the sludge and suitable weather conditions for evaporation. Dewatering may be improved and exposure time shortened by chemical conditioning, such as addition of a polyelectrolyte. Traditionally, dried cake has been removed manually using a shovel-like fork. Attempts to employ mechanical equipment often lead to disturbance of the bed and excessive loss of sand.

Paved drying areas with limited drainage systems can be constructed to permit mechanical cleaning. But climatic conditions must be favorable, since the major water loss is through evaporation. Wedge-wire beds have been used in England to increase the rate of sludge dewatering and for ease of cleaning. During sludge application, the underdrain is filled with water and the wedge-wire bottom is submerged 1 in. to serve as a cushion, permitting the sludge to float into position without contacting the surface of the wedge wire. Later, the water is drained and the sludge dries by seepage and evaporation.

Air drying of digested sludge may be practiced in shallow lagoons where permitted by soil and weather conditions. Water removal is by evaporation, and the groundwater table must remain below the bottom of the lagoon to prevent contamination by seepage. Sludge is normally applied to a depth of about 2 ft and residue removed by a front-end loader after an extended period of consolidation. Because of long holding times, odor problems are more likely to occur. Design data and operational techniques are defined by local experience.

12-19

LAND APPLICATION OF DIGESTED SLUDGE

Direct land disposal of liquid digested sludges alleviates the problems associated with drying beds and lagoons, and is usually less costly in agricultural areas. In the past, plant operators have initiated land application for waste discharge to save labor and treatment costs, while avoiding the use of more expensive, sophisticated dewatering methods installed at treatment plants. Manson and Merritt[6] suggest that the critical factors in design are site isolation, slope of the ground surface, soil conditions, application rates, selection of ground cover, spreading equipment, and monitoring.

The site for distributing liquid sludge should be isolated from residence and water wells by at least 300 ft, and from surface streams by plowing or constructing a berm area on the downslope end of the field to prevent runoff. Minimum depth of soil above the high water table, underlying rock, or field tiles is 4 ft. Hedge rows of shrubs or trees are desirable for esthetics. A maximum slope of 5% is recommended but depends on rate of application, crop cover, and other factors. Soil tests are a prerequisite in choosing a site since soils differ greatly in their ability to assimilate sludges.

Application rates are determined by soil type, the nitrogen and heavy metal contents of the waste, and the nutrient-uptake characteristics of the cover crop. For relatively impermeable soils the maximum liquid rate suggested is 500 gpd/acre. A safe long-term solids loading rate is in the order of 10 dry tons/acre/yr, which converts to 1.5–3.0 acres/1000 population equivalent load on the treatment plant. The maximum application rate at a particular site may vary considerably from this suggested value, based on soil characteristics, heavy metals concentration, nutrient content, and crop requirements.

A harvestable crop as ground cover is most desirable; nevertheless, the main considerations should be nitrogen uptake, salability of the product, and flexibility of spreading operations. Alfalfa ensures a maximum uptake of nitrogen, allows 3 or 4 cuttings per growing season, and is a highly palatable livestock feed. But it is susceptible to damage from saturated soils and infection from direct sludge application to a mature crop. Corn is the most widely used cover crop, with the advantages of high nitrogen uptake and good salability. Flexibility in sludge application is limited since only spray, or ridge-and-furrow methods, are applicable during the growing season. Another vegetation is forage grasses, such as timothy, reed canary, rye, red top, fesque, and sudan grass. Their greatest advantage is site accessibility, permitting tank-truck operation during both inclement weather and the active growing season.

Tank trucks with capacities of 1000–2000 gal are commonly used to haul and spread liquid sludge. In small plants, tractor-drawn tank wagons may be adequate. For wet weather, a paved haul road may be used for side discharge. However, the first choice for operational flexibility is a fixed or movable irrigation system. Storage capacity at the plant in digesters or holding tanks must be sufficient to hold sludge during equipment breakdown and bad weather.

Surface water, groundwater, site soil, and sludge should be monitored on a regular basis. Testing is an expensive regulatory function and should be minimized as much as possible. This may be done by preventing direct runoff, testing existing wells nearby, and analyzing seepage that collects in the tile underdrains. Surface soils and applied sludge may be tested for common parameters, including fecal coliforms, nutrients, heavy metals, and pH.

Vacuum Filtration

Rotary vacuum filters are widely used for dewatering both raw and digested wastewater sludges from plants with design flows greater than 5 mgd. Adoption in handling water-treatment-plant wastes is limited to thickening lime-softening precipitates. Alum coagulation sludge, being more gelatinous, does not dewater readily by suction; thus pressure filtration is more successful.

12-20

DESCRIPTION OF ROTARY VACUUM FILTRATION

The principal components of a vacuum filter system are illustrated in Fig. 12-14. Positive-displacement pumps draw sludge from clarifiers or holding tanks and discharge it into a conditioning tank. Here the waste is mixed with chemical coagulants metered by solution feeders, and then applied through a feed chute to a vat under the filter. The cylindrical drum, covered with a porous medium, is partially submerged in the liquid sludge. As it slowly rotates, vacuum applied immediately under the filter medium draws solids to form a cake on the surface. Suction continues to dewater the solids adhering to the belt as it rotates out of the liquid, then vacuum is stopped while the belt rides over a small-diameter roller for removal of the cake, and the medium is washed by water sprays before reentering the vat. Collecting channels behind the belt in the drum surface are

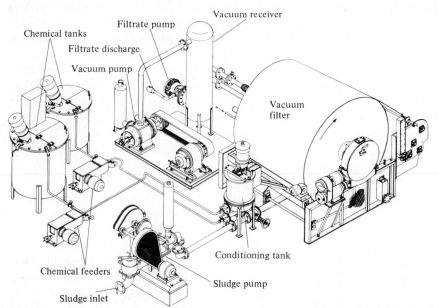

FIGURE 12-14 Rotary vacuum-filter system. (Courtesy Envirotech Corporation, Eimco BSP Division, Salt Lake City, Utah.)

connected by pipes to a combination vacuum receiver and filtrate pump. The principal purpose of the receiver is air–liquid separation. Air taken from the top is discharged through a wet-type vacuum pump, while water from the bottom is removed by a filtrate pump.

Three categories of rotary vacuum filters are defined by the type of medium used and mechanism for cake discharge. A belt-type unit (Fig. 12-14) can be fitted with a variety of synthetic and natural-fiber filter cloths of differing porosities, such as wool, nylon, Orlon, or Dacron. Each material has advantages and disadvantages related to durability, cake production, solids pickup, and response to washing. A tightly woven medium removes a high percentage of fines but can also lead to premature blinding of the pores and prevent further dewatering. Conversely, an open weave may produce a filtrate high in suspended solids. Careful consideration must be given to proper selection of a fabric for each application. For cake discharge, the medium leaves the drum surface at the end of the drying zone and passes over a small-diameter roller. Normally, a curved bar is placed under the belt between the drum and discharge roller to support the cloth weighted down with solids. This support bar also aids discharge by breaking the cake free from the belt. The cloth then passes under a wash roller, rises vertically, traveling over the top of a takeup roller, and returns to the filter drum. The wash roller is immersed in a trough of water, and spray jets rinse the belt before it is drawn over the takeup roller.

A coil filter has a medium consisting of two layers of stainless-steel helically coiled springs about 0.4 in. in diameter. They are placed around the filter drum in corduroy fashion with the upper layer resting on the bottom springs which are held in place by grooved division strips attached to the drum surface. As shown in Fig. 12-15, the top layer of coil springs separates from the drum and travels over a small-diameter roller for removal of dewatered cake. Release is rarely a problem, since fork tines positioned between the springs ensure complete removal of the solids mat. The lower coils pass over a separate roller, and both layers are washed by spray nozzles and reapplied to the drum by grooved aligning rollers. The coil-spring medium is very effective in drying fibrous wastewater sludge, but slurries with fine particles that resist flocculation dewater poorly.

A drum filter differs from the previous two types described in that the cloth covering does not leave the drum for solids discharge or washing. Cake is scraped from the fabric on the cylinder surface after being loosened by compressed air blown through the medium from the inside. In handling waste sludges, dewatering may have to be stopped periodically to wash the drum cloth to prevent blinding. For this reason, belt and coil filters are preferred.

The objectives of vacuum filtration are to obtain an acceptable filter yield, relatively clear filtrate, high solids concentration in the cake, and to minimize operational costs. Filter yield, expressed in pounds of dry sludge solids discharged per square foot of filter area per hour, varies from 2 to 15 $lb/ft^2/hr$. Yield generally does not include the weight of conditioning chemicals added to the

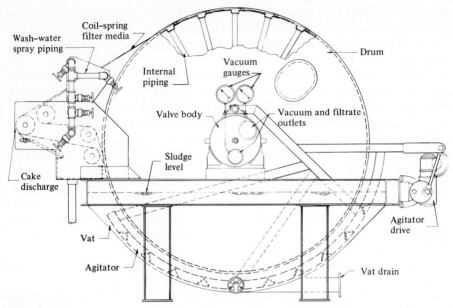

FIGURE 12-15 Cross-sectional view of a coil-type vacuum filter. (Courtesy Komline-Sanderson, Peapack, N.J.)

sludge. Output increases with rising dosage of coagulants, and for the same chemical conditioning yield increases with solids concentration in the feed sludge. Solids capture varies from 85 to 99%, depending on the type of filter covering, character and density of the applied sludge, and chemical conditioning. Cake solids content is affected by the same factors as well as the machine variables of vacuum pressure, drum submergence, and speed of rotation. Optimum suction relates to cake compressibility—a relatively incompressible solids layer dewaters better under high vacuum, while compressing an organic cake may decrease its porosity and filterability. Drum submergence and speed can be varied to adjust the form and drying cycles for optimum performance. All these parameters must be viewed as economical factors to determine cost-effective operation. Higher chemical consumption can increase yield and reduce operating time, thus raising chemical costs while decreasing labor and power. Disposal costs may be directly related to cake density. For example, the expense of hauling a relatively wet sludge to a distant landfill may be more costly than operating the filter to produce a drier cake.

12-21

THEORY OF VACUUM FILTRATION

The theoretical filtration equation is

$$\frac{dV}{dt} = \frac{PA^2}{\mu(rwV + R_f A)} \tag{12-13}$$

where V = volume of filtrate, m³ (ml)
   t = time, sec (sec)
   P = pressure difference, newtons/m² (dynes/cm²)
   A = filter area, m² (cm²)
   μ = viscosity of filtrate, Newton-sec/m² (poise)
   r = specific resistance of sludge cake, m/kg (sec²/g)
   w = mass of dry solids per unit volume of filtrate, kg/m³ (g/cm³)
   R_f = resistance of filter medium (sec²/cm²)

This basic formula assumes laminar flow, uniform solids deposition during filtration, and a constant increase in filtrate flow resistance as the cake increases in thickness. For constant pressure throughout filtration, integration of Eq. 12-13 yields

$$\frac{t}{V} = \frac{\mu r w V}{2PA^2} + \frac{\mu R_f}{PA} \tag{12-14}$$

Equation 12-14 is that of a straight line, and may be written as

$$\frac{t}{V} = bV + a \tag{12-15}$$

where $b = \mu rw/2PA^2$.

Finally, Eq. 12-16 can be written for specific resistance r by equating the coefficients of V from Eqs. 12-14 and 12-15.

$$r = \frac{2PA^2 b}{\mu w} \tag{12-16}$$

where r = calculated specific resistance, m/kg (sec²/g)
   b = slope of the line passing through a plot of t/V versus V data from a laboratory filterability test, sec/m⁶ (sec/ml²)
   w = cake solids deposited per volume of filtrate, kg/m³ (g/cm³)
   P, A, μ = same as above

Specific resistance for sludge is determined using the laboratory apparatus illustrated in Fig. 12-16. The test is conducted by pouring a chemically treated sludge sample in the Büchner funnel, a vacuum imposed, and the amount of filtrate drawn into the graduated cylinder recorded with respect to time. From these data, t/V is plotted versus V on arithmetic paper. The value of b is the slope of a straight line drawn through the plotted points. Funnel size defines A, a vacuum gage measures P, and the value for viscosity is normally taken as that of water at filtrate temperature. Cake solids w can be determined by drying the material accumulated on the filter and relating it to filtrate volume, or w may be calculated from known solids concentrations in the cake and filtrate. Various authors use different metric units to express specific resistance; this practice has

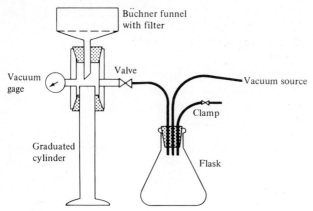

FIGURE 12-16 Büchner-funnel apparatus for laboratory determination of the specific resistance of a sludge.

led to considerable confusion in relating data found in literature.

Specific resistance varies with filter area, solids concentration, liquid viscosity, and pressure. For compressible sludge solids, r has been found to vary empirically as

$$r = r_0 \left(\frac{P}{P_0} \right)^s \tag{12-17}$$

where r = specific resistance at pressure P
r_0 = specific resistance at pressure P_0
s = coefficient of cake compressibility

The coefficient of compressibility is determined by conducting a series of filtration tests at various pressures. The value of s equals the slope of the line through r versus P data plotted on log-log paper.

Laboratory Büchner-funnel filtration analyses and resulting specific resistance data are used to measure differences in sludge filterability. The best chemical conditioning for specified conditions can be determined by a series of tests conducted on sludge portions flocculated with increasing chemical additions. A plot of specific resistance versus chemical dosage yields a dipping curve. Optimum dosage is at the point of least resistance, while smaller or greater conditioning decreases filterability. Although filter-yield data can be derived from these tests, they are not recommended for design purposes.

●**EXAMPLE 12-11**
A waste sludge was tested for filterability using a Büchner funnel apparatus (Fig. 12-16) with a filter diameter of 7.5 cm and a 10-psi vacuum drawn on the sample. For 50 ml of filtrate, the dried cake solids collected on the filter were 3.15 g. The slope of the t/V versus V data plotted on graph paper was 0.72 sec/ml² and filtrate temperature was 50°F. Calculate the specific resistance.

○*Solution*

$P = 10$ psi $= 703$ dynes/cm$^2 = 6.90 \times 10^4$ N/m^2
$A = 3.14(3.75)^2 = 44.2$ cm$^2 = 0.00442$ m^2
$b = 0.72$ sec/ml^2 (100 cm/m)$^6 = 72 \times 10^{10}$ sec/m^6

From Table 9-1, μ at 50°F equals 1.308 centipoise,

$\mu = 0.013$ poise $\times 0.1 = 0.0013$ N-sec/m^2

$$w = \frac{3.15\ g}{50\ \text{ml}} = 0.063 \text{ g/ml} = 63 \text{ kg/m}^3$$

Substituting into Eq. 12-16,

$$r = \frac{2 \times 6.90 \times 10^4 \times 0.00442^2 \times 72 \times 10^{10}}{0.0013 \times 63} = 2.4 \times 10^{13} \text{ m/kg } (2.4 \times 10^{17} \text{ sec}^2/\text{g})$$

12-22

SIZING VACUUM FILTERS

Vacuum-filter installations may be designed based on experience, pilot-plant tests, filter-leaf testing, or a combination of these. In sizing equipment for dewatering wastewater sludges, experience indicates that filter yield will be about 1 lb/ft^2/hr (5 kg/m^2-hr) for each percentage of solids in the feed sludge. Consequently, common design yields are 6–8 lb/ft^2/hr for raw sludge drawn from primary clarifiers and 5–7 lb/ft^2/hr for anerobically digested waste. Thin sludges containing less than 4% solids should be concentrated for economical filter operation.

Equipment manufacturers lease portable pilot-plant vacuum filters. These scaled-down units, with a drum area of about 10 ft^2, can be operated on site for sludge-dewatering evaluations. Pilot-plant studies are superior to laboratory analyses since both machine and operational variables are considered.

A leaf-filter apparatus for laboratory evaluation of sludge filterability is shown in Fig. 12-17. The filter holder is a circular device with a drainage grid to firmly support a fitted cloth medium with an effective area of 0.10 ft^2 having characteristics similar to a full-scale filter medium. After applying suction, the unit is immersed upside down in the sludge sample to simulate cake formation for 0.5 to 1.5 minutes. It is then carefully withdrawn and held upright for dewatering for the same length of time. Form and drying times relate to the speed of drum rotation and submergence of a full-scale unit. Immediately after dewatering, the cake is removed from the face of the filter, and both cake and filtrate can be tested for solids content. Filter yield is computed using the following relationship:

$$\text{yield (lb/ft}^2/\text{hr)} = \frac{\text{dry sludge solids (lb)} \times \text{cycles per hour}}{\text{filter-leaf area (ft}^2)} \tag{12-18}$$

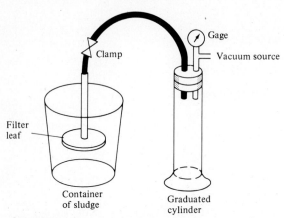

FIGURE 12-17 Filter-leaf apparatus for laboratory evaluation of sludge filterability.

Filter-leaf analyses are conducted at varying chemical dosages and selected operating conditions, sludge solids concentration, vacuum pressure, and cycle time. For a given set of conditions, values of filter yield, sludge solids content, and suspended solids in the filtrate are plotted versus chemical dosage for graphical presentation. Air flow used in dewatering the cake per cycle can be calculated if the air flow through the filter leaf is measured.

● EXAMPLE 12-12

A wastewater sludge was tested for filterability at various polymer dosages using a leaf-filter apparatus (Fig. 12-17). Laboratory analyses included cake solids concentration, filtrate suspended solids, and total solids accumulation on the filter leaf. Equation 12-18 was used to calculate the yield for each test. The results were:

Polymer Dosage (%)	Filter Yield (lb/ft²/hr)	Cake Solids (%)	Filtrate SS (mg/l)
0	2.8	20	2000
0.2	4.0	26	1000
0.4	5.8	28	600
0.6	7.0	29	500
0.8	7.8	28	400

Plot the test data. Select the polymer dosage and cake density for a yield of 6.0 lb/ft²/hr. Calculate the chemical consumption per ton of dry solids. And for this yield, calculate the weight of cake produced and volume of filtrate per 1000 gal of 6.0% sludge applied.

○*Solution*

From the graphs plotted in Fig. 12-18, the polymer dosage required for a 6.0 lb/ft²/hr yield is 0.46%, and the cake density is 28% solids.

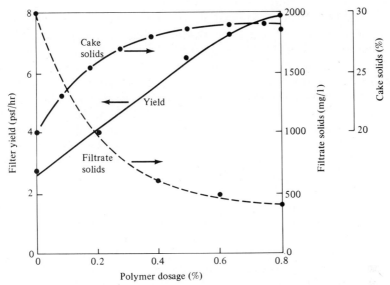

FIGURE 12-18 Plot of filter-leaf data for Example 12-12.

$$\text{polymer dosage} = 0.0046 \times 2000 = 9.2 \text{ lb/ton of dry solids}$$

$$\text{dry solids/1000 gal of sludge} = 0.06 \times 1000 \times 8.34 = 500 \text{ lb}$$

$$\frac{\text{weight of cake produced}}{\text{1000 gal of sludge filtered}} = \frac{500}{0.28} = 1790 \text{ lb}$$

$$\text{volume of cake} = \frac{1790}{1.05 \times 8.34} = 204 \text{ gal}$$

$$\text{volume of filtrate} = 1000 - 204 = 796 \text{ gal}$$

12-23

VACUUM FILTRATION OF WASTEWATER SLUDGES

Efficient dewatering requires a minimum sludge concentration of about 4% solids for acceptable yield and reasonable chemical conditioning. Many trickling-filter plants pump settled sludge directly from primary clarifiers to vacuum filtration. Gravity thickening, although not common in small plants, may be used in large installations to increase performance of vacuum filters. Unthickened waste-activated sludge is not dewatered separately because of its low solids concentration and high percentage of fine particles. It may be vacuum-filtered after mixing with raw primary, but the blended sludges should preferably be thickened for economical operation. One successful scheme for pretreatment of wastes from an aeration plant is separate flotation thickening of the waste-activated sludge followed by blending with primary wastes in a mixed holding tank.

Coil filters and belt-type units are manufactured with surface areas ranging from 50 to more than 300 ft². Based on typical design criteria, the smallest unit can dewater sludge from 1 mgd of domestic wastewater in an operating period of 6–8 hours/day. Sizing of filters is based on an anticipated yield, which is related to sludge characteristics and number of hours of operation. Thirty hours per week are commonly assumed for small plants, while at larger ones operations may extend for two shifts plus cleanup, for a total of 20 hours/day. For new installations, it is popular to assume a conservative design loading such as 0.2 lb/day of dry solids per population equivalent design load on the treatment plant.

Chemical conditioning of wet sludge is necessary to achieve satisfactory yield and a clear supernatant. Fine particles in untreated sludge tend to blind the medium by plugging the pores. Coagulation agglomerates the very fine particles, thus reducing filter resistance and clarifying the filtrate. Minimum chemical conditioning requires greater than 90% solids capture in dewatering. Recycling of matter in the filtrate can lead to excessive solids circulating within the treatment plant, so their continuous and adequate removal is essential to efficient operation.

Common chemicals in conditioning are ferric chloride and lime, or poly-electrolytes. All are mixed with water and applied by solution feeders, with typical results for various types of sludge listed in Table 12-7. Chemical dosages are expressed as percentages of the dry solids filtered, for example, 2% conditioner means 2 lb of coagulant per 100 lb of dry solids in the filter cake. The tabulated values illustrate the relative chemical conditioning for various residues and are not intended to be used for design or operation. Effectiveness of proprietary polyelectrolytes differs considerably, and actual polymer additions range from approximately 0.2 to 2 times the dosages listed in Table 12-7.

TABLE 12-7

TYPICAL RESULTS FROM VACUUM FILTRATION OF CHEMICALLY CONDITIONED WASTEWATER SLUDGES

Type of Sludge	Filter Yield (lb/ft²/hr[a])	Cake Solids (%)	Chemical Dosage		
			Ferric Chloride (%)	Lime (%)	Polyelectrolytes[b] (%)
Raw primary	6–10	25–40	1–4	6–8	1–4
Primary plus filter humus	4–8	20–35	2–4	6–8	3–6
Primary plus waste activated	3–5	15–25	3–6	5–10	5–10
Anaerobically digested	5–10	20–35	3–6	5–10	4–10

[a] 1.0 lb/ft²/hr = 4.88 kg/m²-hr.
[b] Polyelectrolyte dosages are given as relative numbers; actual dosages range from 0.2 to 2 times these values.

Choice of chemicals is based on economics, desired operating conditions, and method of ultimate cake disposal. New design should have the flexibility to apply either inorganic metal coagulants and lime, or polyelectrolytes. The advantages of ferric chloride and lime are disinfection and stabilization, thus reducing health hazards and odors. Conversely, they are more difficult to handle and feed, and may be more costly. Polymers are easy to apply and frequently more economical but do not provide disinfection. A minimum cake density of 20–25% solids is generally satisfactory for hauling to landfill. Nevertheless, if long-distance trucking is involved, a lower moisture content reduces the total weight of cake transported. Incineration requires a low moisture content for burning with a minimum of auxiliary fuel. Laboratory Büchner-funnel and filter-leaf analyses are advantageous in determining suitable chemical dosages. Special full-scale studies may be conducted periodically to ensure economical performance and selection of the best coagulant, particularly if various brands of polyelectrolytes are being compared for possible adoption.

●EXAMPLE 12-13

A wastewater treatment plant produces 50,000 gpd of sludge containing 5.5% solids. What size vacuum filters are required, assuming a yield of 5.0 lb/ft^2/hr and 12 hours of operation per day? Chemical conditioning is 8.0% lime and 2.0% ferric chloride. The cake is 23% solids, including organics plus chemical additions. Calculate the chemical dosages per ton of dry solids, and weight of filter cake produced per day.

○*Solution*

dry sludge solids $= 50,000 \times 8.34 \times 0.055 = 22,900$ lb/day

$$\text{filter area required} = \frac{22,900}{5.0 \times 12} = 382 \text{ ft}^2$$

Use two 200-ft^2 filters, each 8 ft in diameter by 8 ft wide.

lime dosage $= 0.08 \times 2000 = 160$ lb/ton of dry solids

ferric chloride dosage $= 0.02 \times 2000 = 40$ lb/ton of dry solids

total cake solids $=$ sludge solids $+$ chemical additions

$= 22,900 + (0.08 + 0.02)22,900 = 25,200$ lb/day

$$\text{weight of wet filter cake} = \frac{25,200}{0.23 \times 2000} = 54.8 \text{ tons/day}$$

Pressure Filtration

Small, manually operated filter presses have been used extensively in various industries to dewater slurries. Recent improvements in machine design, including automated operation, improved filter media, and larger capacity presses, have increased their application in handling waste sludges. Pressure filtration is par-

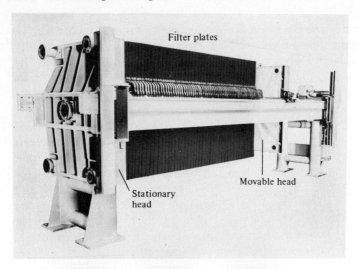

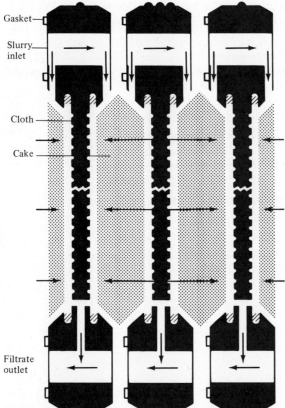

FIGURE 12-19 Photograph of a pressure filter and diagram of the filter chamber assembly showing slurry channel and filtrate discharge. (Shriver-Envirotech Filter Press, courtesy Shriver Division, Envirotech Corporation.)

ticularly suited to dewatering wastes with a high specific resistance such as coagulation precipitates from surface-water treatment. Major advantages are a very dry cake, clear filtrate, and high solids capture.

12-24

DESCRIPTION OF PRESSURE FILTRATION

A pressure filter consists of depressed plates held vertically in a frame for proper alignment and pressed together by a hydraulic cylinder (Fig. 12-19). Each plate is constructed with a drainage surface on the depressed portion of the face. Filter clothes are caulked onto the plate and peripheral gaskets seal the frames when the press is closed. Influent and filtrate ports are formed by openings that extend through the press. Sludge is pumped under pressure into the chambers between the plates of the assembly, and water passing through the media drains to the filtrate outlets. Solids retained form cakes between the cloth surfaces and ultimately fill the chambers. High pressure consolidates the cakes by applying air to the sludge inlet at about 200 psi. After this filter cycle, which requires from 1 to 2 hours, compressed air blows the feed sludge remaining in the influent ports back to a holding tank. The filter plates are separated and dewatered cakes drop out of the chambers into a hopper equipped with a conveyor mechanism. Cake release is assisted by introducing compressed air behind the filter cloths.

Chemical conditioning improves sludge filterability by agglomerating fine particles so that the cake remains reasonably porous, allowing passage of water under high pressure. Precoating the media with diatomaceous earth or fly ash helps to protect against blinding and ensures easy separation of the cake for discharge. Filter aid for the precoat is placed by feeding a water suspension through the filter before applying sludge. In some cases the aid may be added to the conditioned sludge mixture to improve porosity of the solids as they collect. Dewatered cake densities are normally in the range of 40–50% dry solids.

12-25

APPLICATION OF PRESSURE FILTRATION

Water-treatment-plant wastes are suited to pressure filtration since they are often difficult to dewater, particularly alum sludges and softening precipitates containing magnesium hydroxide. Figure 12-20 shows a schematic flow diagram for a pressure filtration process. Gravity-thickened alum wastes are conditioned by the addition of lime slurry. A precoat of diatomaceous earth or fly ash is applied prior to each cycle, and conditioned sludge then fed continuously to the pressure filter until filtrate ceases and the cake is consolidated under high pressure. A power pack holds the chambers closed during filtration, and transports the movable head for opening and closing. An equalization tank provides uniform pressure across the filter chambers as the cycle begins. Prior to cake

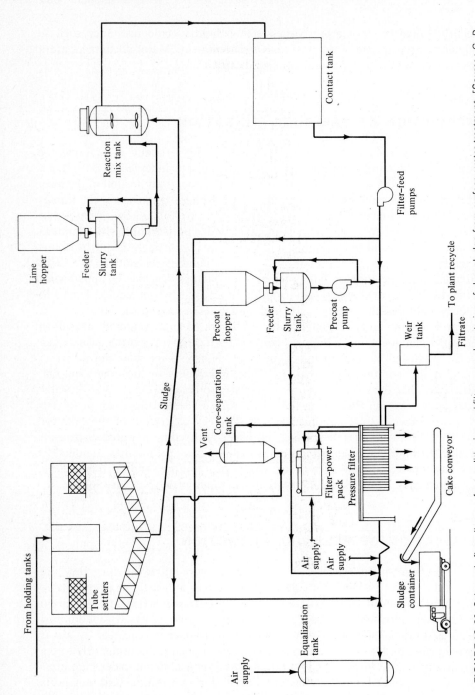

FIGURE 12-20 Schematic flow diagram for conditioning and filter-press dewatering of alum sludge from surface-water treatment. [Source: G. P. Westerhoff and M. P. Daly, "Water-Treatment-Plant Waste Disposal," *J. Am. Water Works Assoc.* 66, no. 7 (1974): 443. Copyrighted 1974 by he American Water Works Association, Inc.)

discharge, excess sludge in the inlet ports of the filter is removed by air pressure to a core-separation tank. Filtrate is measured through a weir tank and recycled to the inlet of the water-treatment plant. Cake is transported by truck to a disposal site.

Alum sludges are conditioned using lime and/or fly ash. Lime dosage is in the range 10–15% of the sludge solids. Ash from an incinerator, or fly ash from a power plant, is applied at a much higher dosage, approximately 100% of dry sludge solids. Polyelectrolytes may also be added to aid coagulation. Fly ash and diatomaceous earth are used for precoating; the latter requires about 5 lb/100 ft^2 of filter area. Under normal operation, cake density is 40–50% solids and has a dense, dry, textured appearance.

Wastewater sludges are also amenable to dewatering by pressure filtration after conditioning with ferric chloride and lime, or fly ash. For small treatment plants the primary advantage, relative to vacuum filtration, is less operator supervision. The filtration cycle can be automated so that batching of feed chemicals and cake discharge are the only operations requiring immediate attention. Where sludge incineration is practiced, a major benefit is a drier cake for burning. Ferric chloride and lime dosages are 5 and 10%, respectively, for a waste concentration of 5% solids or greater. Conditioning with fly ash requires 100–150% additions. Either diatomaceous earth or fly ash is used for precoating the filter media. Cake solids are generally 45–50%, and cycle time between 1.5 and 2 hours. Raw primary and digested sludges yield similar results except that the filtrate from pressing raw organic sludges has a higher BOD than that from dewatering anaerobically digested wastes.

Centrifugation

Centrifuges are employed for both dewatering sludges and thickening waste slurries for further processing. Historically, they have been used to concentrate wastewater settlings. Recently, centrifugation has been applied to sludges difficult to dewater by gravity separation, such as alum coagulation residues and waste-activated sludge. Centrifugal classification separates calcium carbonate precipitate from magnesium hydroxide floc in pretreatment for recalcination of lime-softening sludges.

12-26

DESCRIPTION OF CENTRIFUGATION

All centrifuges have the same basic operating principle. Solids are removed from the waste stream flowing through the machine under the influence of a centrifugal field of 100–600 times the force of gravity. Particles are deposited against the spinning solid bowl while the overflow is a clear liquid supernatant. The fundamental difference in centrifuges is the manner of solids collection and discharge—the method of discharge determines the size and nature of the particles removed by a particular unit. Material encountered in wastes includes a broad range of granular, fibrous, flocculent, and gelatinous solids that differ in settling and compaction characteristics. Therefore, the type of centrifuge adopted is determined by the particular waste, as well as discharge requirements of supernatant clarity and cake dryness. The two most popular types for handling sanitary wastes are the scroll or conveyor centrifuge and the imperforate basket.

   Scroll centrifuges can handle large quantities of fairly coarse solids (Fig. 12-21). Two principal elements are a rotating solid bowl in the shape of a cylinder with a cone section on one end, and an interior rotating screw conveyor. Feed slurry enters at the center and is spun against the bowl wall. Settled solids are moved by the conveyor to one end of the bowl and out of the liquid for drainage

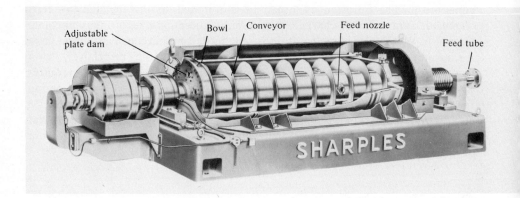

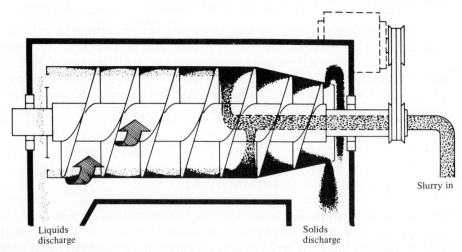

FIGURE 12-21 Solid bowl, scroll centrifuge. (Courtesy Sharples-Stokes Division, Pennwalt Corporation.)

before discharge while clarified effluent discharges at the other end over a dam plate. This system is best suited for separating solids that compact to a firm cake and can be conveyed easily out of the water pool. If solids compact poorly, moving a soft cake causes redispersion, resulting in poor clarification and a wet concentrate. Flocculent solids can generally be made scrollable by chemical conditioning of the sludge, and the redispersing action that occurs in a scroll machine is advantageous in classification of particles by centrifugation.

A major advantage of scroll dewatering is operational flexibility. Machine variables include pool volume, bowl speed, and conveyor speed. Depth of liquid in the bowl and pool volume can be controlled by an adjustable plate dam. Pool volume adjustment varies the liquid retention time and changes the drainage deck surface area in the solids-discharge section. Bowl speed affects gravimetric forces on the settling particles, and conveyor rotation controls the solids retention time. The driest cake results when bowl speed is increased, pool depth is the minimum allowed, and differential speed between the bowl and conveyor is the maximum possible. Flexibility of operation allows a range of densities in the solids discharge varying from a dry cake to a thickened liquid slurry. Feed rates, solids content, and prior chemical conditioning can also be varied to influence performance. Solids removal and cake consolidation can both be enhanced by adding polyelectrolytes or other coagulants with the feed sludge.

The imperforate basket centrifuge illustrated in Fig. 12-22 does not have facilities for continuous discharge of solids. Sludge feed is interrupted to release accumulated cake, a process that takes 1–2 minutes of the overall 10- to 30-minute cycle time. Clarification is more efficient than in a scroll centrifuge, since there is no conveyor disturbing the liquid pool. Thickening even difficult sludges, such as alum precipitates or activated sludge, can achieve over 90% solids capture without chemical conditioning. Slurry is applied at the bottom of the basket behind an accelerator wheel that brings the feed up to maximum bowl speed. Solids accumulate on the bowl wall while clarified effluent overflows a lip ring at the top of the basket. A cake-thickness detector indicates when it is necessary to discharge solids. Soft cakes may be removed by submerging a skimmer into the solids layer while the bowl is spinning at full speed—under these conditions, the sludge flows up the bowl wall to the skimmer inlet. Dense deposits can be scraped from the inside of the basket with a knife unloader at reduced bowl speed, the knife being used with each discharge cycle, or less frequently, to remove accumulations of heavy solids that do not skim readily.

Performance of centrifugal dewatering, for given feed and machine-operating conditions, depends on dosage of chemical coagulants. Suspended solids removal and usually cake dryness increase with greater chemical additions, while carryover of solids in the centrate decreases. There is, however, a saturation point at which flocculent dosage does not significantly improve centrate clarity. Optimum chemical conditioning without overdosing can be determined most reliably by full-scale or pilot-plant tests. For some wastes centrate recycling can improve overall suspended solids removal, but for others it may cause upset,

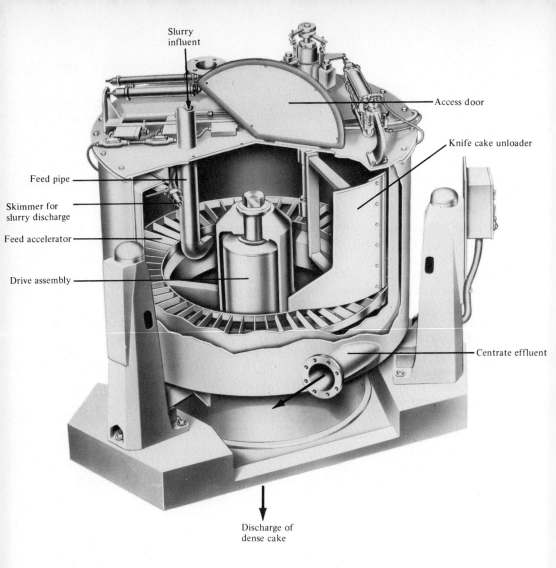

Slurry influent

Access door

Knife cake unloader

Feed pipe

Skimmer for slurry discharge

Feed accelerator

Drive assembly

Centrate effluent

Discharge of dense cake

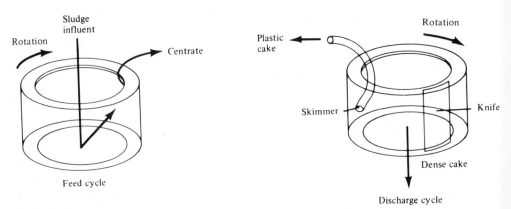

Sludge influent

Rotation

Centrate

Feed cycle

Rotation

Plastic cake

Skimmer

Knife

Dense cake

Discharge cycle

FIGURE 12-22 Imperforate basket centrifuge. (Courtesy Sharples-Stokes Division, Pennwalt Corporation.)

owing to an accumulation of fine particles. The following formulas are helpful in calculating efficiency and feed rates from study data:

$$\text{suspended solids removal} = \frac{SSc(SSi - SSe)}{SSi(SSc - SSe)} \tag{12-19}$$

$$\frac{\text{feed rate}}{\text{centrate rate}} = \frac{Qi}{Qe} = \frac{SSc - SSe}{SSc - SSi} \tag{12-20}$$

where SSc = cake suspended-solids concentration
SSi = feed (influent) suspended-solids concentration
SSe = centrate (effluent) suspended-solids concentration
Qi = feed (influent) flow rate
Qe = centrate (effluent) flow rate

● **EXAMPLE 12-14**

A scroll centrifuge dewaters an alum-lime sludge containing 8.0% at a feed rate of 20 gpm. The cake produced has a solids content of 55% and the centrate contains 9000 mg/l of suspended matter. Calculate the suspended-solids removal efficiency and the centrate flow.

○*Solution*

Using Eq. 12-19,

$$\text{solids removal} = \frac{55(8.0 - 0.9)}{8.0(55 - 0.9)} = 0.90 = 90\%$$

Substitution into Eq. 12-20 yields

$$\frac{20 \text{ gpm}}{Qe} = \frac{55 - 0.9}{55 - 8.0} \qquad Qe = 17.4 \text{ gpm}$$

12-27

APPLICATIONS OF CENTRIFUGATION

Alum sludge from surface-water treatment is amenable to centrifuge dewatering. Performance must be verified by testing at each location, since sludge characteristics vary considerably. In general, aluminum hydroxide slurries from coagulation settlings and gravity-thickened backwash waters can be concentrated to a truckable pasty sludge of about 20% solids. Removal efficiency in a scroll centrifuge ranges from 50 to 95% based on operating conditions and polymer dosage, and the centrate is correspondingly turbid or clear. Pretreatment of the sludge with a cyclonic degritter is usually required. A basket centrifuge thickening the same waste can provide higher solids capture and a clearer overflow even without polymer addition, but the cake is often less dense and cycle time longer than that of a scroll machine.

Lime-softening precipitates compact more readily than alum floc in a scroll centrifuge. A settled sludge input with 15–25% solids can be dewatered to a solidified cake of 65%. Suspended-solids recovery is often 85–90% with polyelectrolyte flocculation.

Scroll centrifuges are used to classify lime sludges for selective thickening of calcium carbonate precipitate. The process provides pretreatment for calcining to remove magnesium, which is detrimental in the reuse of recovered lime. Granular calcium carbonate particles settle and compact readily for efficient scrolling and compact readily for efficient scrolling and drainage in a conveyor centrifuge, whereas magnesium hydroxide floc is gelatinous and about one-half as dense. Thus a machine can be operated so that most of the undesirable magnesium hydroxide is in the liquid discharge while a purer sludge cake is provided for recalcining. Centrifugal classification can also be used to isolate organics and phosphate precipitates from sludges produced in lime treatment of wastewaters.

Scroll centrifuges are employed for dewatering wastewater sludges, although they are not as popular as vacuum filters. Anaerobically digested solids separate effectively because of their particulate nature. A moist cake of 20–30% solids can be produced with 80–95% recovery and a reasonable level of chemical conditioning. Raw primary sludge is more difficult to dewater efficiently, since it contains organics that are difficult to clarify and scroll. For an average cake density of 30%, solids recovery is 50–70% without chemical addition but increases to the 70–90% range with proper chemical coagulation. Return of dirty centrate to the head of a treatment plant can cause operational problems, the most common result being a load of fine solids recirculating around through the centrifuge and wastewater-processing units of the treatment plant. The choice between centrifugation and vacuum filtration is based on both performance and economics.

Basket centrifuges are becoming competitive with dissolved-air flotation for thickening waste biological sludge since they can be operated with high efficiency. Solids capture in excess of 90% is possible without chemical addition while producing a concentrated slurry up to 10% solids. However, waste-activated-sludge thickening by scroll centrifugation is generally considered unfeasible, owing to complications in producing a clear centrate.

Recovery of Chemicals

12-28

LIME RECOVERY

Calcination for recovery of lime involves burning calcium carbonate precipitate at temperatures of 600–2000°F to yield calcium oxide and carbon dioxide,

$$CaCO_3 \longrightarrow CaO \text{ (lime)} + CO_2 \uparrow \qquad (12\text{-}21)$$

Water-softening sludges are pretreated to remove magnesium by carbonation and/or centrifugation Carbon dioxide preferentially solubilizes magnesium hydroxide precipitate without affecting the calcium carbonate solids. Carbonation also improves centrifugal classification by providing better calcium recovery and a denser cake. Water-softening precipitates include both calcium hardness removed from the raw water and calcium from the added lime; therefore, recalcination should theoretically produce more recycled lime than needed. Impurities in the sludge and calcium losses, during dewatering and dust recovery in the recalciner, however, reduce the actual yield such that practical recovery is usually only about 25% greater than the original lime applied.

A lime-recovery process generally consists of the following steps: gravity thickening to 25% solids or greater; sludge carbonation using stack gases from the recalcining furnace to dissolve magnesium hydroxide; centrifugation to classify precipitates and increase solids content to greater than 60%; flash drying of dewatered sludge using hot off-gases from the recalciner; and conversion of calcium carbonate to calcium oxide in a rotary kiln or fluidized bed calciner. In the rotary-kiln process, dewatered paste is conveyed to a sloping, cylindrical enclosure for drying and firing. The kiln shell has a refractory lining, rotates at approximately 1 rpm, and is sloped so that dried sludge travels toward the firing end. Solids retention time is approximately 1.5 hours, and the unit is heated by burning oil or natural gas. Burned lime is cooled and discharged to a conveyor for transfer to a storage bin, exhaust gases are scrubbed to remove dust; and a portion is diverted back to the sludge processing step for use in carbonation.

The fluidized-bed calciner requires pretreatment of the wet cake with soda ash and previously dried calcium carbonate to form lumps that move and roll freely. Particles, now 80% solids, are applied to a cage mill, where they are dried and beaten into a fine powdery dust that becomes suspended in the exhaust gases. Calcium carbonate dust, removed from the gases by a cyclone separator, is fed to a tall cylindrical furnace, where it is rapidly burned, forming hard spherical pellets of calcium oxide that are held suspended in the vertical air stream passing through the furnace. After growing to sufficient size, they drop out of suspension and are removed from the reactor. Recalcining requires a fuel rate of about 10 million Btu per ton of lime produced.

Lime recovery from wastewater coagulation requires greater pretreatment since the precipitate contains organics, phosphorus, and other undesirable constituents in addition to magnesium. Calcium recovery is less efficient and makeup lime is required. A multiple hearth furnace is employed for calcining the dewatered cake.

Adoption of recalcination depends on the relative expense and feasibility of alternative sludge-disposal methods. Lime recovery may not be technologically feasible for many treatment plants, owing to the presence of undesirable impurities. Appreciable quantities of magnesium, aluminum, iron, or clay may hinder dewatering operations and limit the purity and value of the product lime. Economical operation is generally limited to large treatment plants that can process more than 20 tons/day, although in certain locations smaller plants may be feasible. The best method for disposal of lime sludges is related to plant size, nature of the sludge, and available disposal options.

12-29

ALUM RECOVERY

Aluminum sulfate applied in coagulation of water precipitates as aluminum hydroxide. The alum can be recovered by reacting sulfuric acid with the settled sludge as follows:

$$2Al(OH)_3 + 3H_2SO_4 \longrightarrow Al_2(SO_4)_3(alum) + 6H_2O \qquad (12\text{-}22)$$

A complete recovery process includes concentrating the alum sludge, reacting with sulfuric acid, and removal of impurities. Pretreatment by gravity thickening is generally sufficient to produce the desired minimum solids concentration of 2%, thereby assuring adequate strength of the recycled aluminum sulfate solution (a solution with less than 1.0% alum is not considered suitable for reuse). Recovery reaction with sulfuric acid is carried out at a pH of about 2.0. Acid in excess of the chemical reaction is needed to establish this lower pH and allow for chemical decomposition of organic matter present in the sludge. The final step of removing undesirable suspended matter is accomplished by either gravity clarification or filtration. Recovered alum solution is held in tanks and metered to points of application in the treatment plant. Waste residue from processing is treated with lime, dewatered, and disposed of by land burial.

The most critical problem in alum regeneration is elimination of impurities that cannot be separated by gravity. Recovered alum solution may carry over resolubilized iron and manganese, inert solids such as clay, carbon added for odor control, and colloidal organics charred by the sulfuric acid. One proposed scheme is pressure filtration of the reclaimed alum solution using fly ash as a filter aid. Near the end of each filter cycle lime is applied to neutralize the cake. Gravity clarification recovers only about 70% of the original alum fed, while pressure filtration allows nearly 100% recovery. Most treatment plants should be equipped to supply new alum in case of failure in the alum recovery process For example, high turbidity in the raw water during spring runoff may result in contaminated sludge that cannot be processed to recover the alum.

Ultimate Disposal

Final sludge disposal possibilities are discharge to the ocean, spreading or burial on land, or burning. Coastal cities may barge or pipe digested sludge into the sea. Land application of stabilized sludge is popular in agricultural areas (Sec. 12-19), whereas sanitary landfill of dewatered cake, either raw or digested, and incinerator ash is common where a site is available within trucking distance. Sludge combustion is really only a means of volume reduction and stabilization, since it leaves an ash residue. Although practiced in only a few cities, wet sludge may be dried to yield a product with some value as a soil conditioner.

12-30

COMBUSTION OF ORGANIC SLUDGES

Incineration involves drying sludge cake to evaporate the water, followed by burning for complete oxidation of the volatile matter. Drying occurs at a temperature of approximately 700°F, and burning is sustained in the range 1200–1400°F. A minimum temperature of 1350°F is needed to deodorize exhaust gases. Excess air is required to ensure complete combustion of organics and minimize the escape of odor-producing compounds in stack gases. The amount needed is 25–100% over the stoichiometric air requirement, and varies with the nature of the sludge and type of incineration equipment. Supplying excess air has the adverse effects of reducing the burning temperature and increasing heat losses from the furnace. Heat emitted from burning volatile solids, minus losses, is available for drying the incoming sludge and heating the air supply. Self-sustained combustion is often possible with dewatered raw sludges after the incinerator temperature has been raised to the ignition point by burning an auxiliary fuel.

Heat yield from sludge combustion is related directly to moisture content and volatile solids concentration. Several theoretical equations have been proposed for calculating calorific values based on elemental composition, principally the carbon, hydrogen, oxygen, and sulfur contents. However, experience has shown that calculated results often are inaccurate. A laboratory calorimeter test is the only reliable method for determining the heat value of a waste. Equation 12-23[9] is an empirical formula for fuel values of different types of vacuum-filtered sludges, taking into account the amount of coagulant added before filtration:

$$Q = a\left(\frac{100Pv}{100 - Pc} - b\right)\left(\frac{100 - Pc}{100}\right) \tag{12-23}$$

where Q = fuel value, Btu/lb of dry solids
   a = coefficient (131 for raw primary sludge, 107 for waste-activated sludge)
   b = coefficient (10 for raw primary sludge, 5 for activated sludge)
   Pv = volatile solids in sludge, %
   Pc = inorganic conditioning chemicals applied for dewatering the sludge, % ($Pc = 0$ for organic polymers)

Heat-balance calculations for an incinerator include the heat evolved from burning of volatile solids and auxiliary fuel, the quantity absorbed for evaporation of moisture in the wet cake, losses in the stack gases and ash, and radiation through the furnace walls. The fuel value of raw sludge ranges from 6500 to 9500 Btu/lb of dry solids, while the calorific value for digested dry solids is 2500–5500 Btu, depending on the fraction of inerts. Evaporation of moisture from wet sludge varies from 1800 to 2500 Btu/lb of water. Information on heat losses from a furnace are available from the equipment manufacturer. The amount of excess air used during operation defines the heat loss in exhaust gases. Natural gas, or fuel oil, is frequently required for complete combustion, at least for warming up the furnace to operating temperature. Moisture content and heat value of the volatile solid determine whether the burning process is self-sustaining. A minimum volatile solid concentration of 16–18 % is needed for complete combustion in an efficient system that includes heat recovery in the combustion air.

●EXAMPLE 12-15

Raw primary sludge is dewatered by vacuum filtration prior to incineration. The cake contains a total of 24% dry solids, of which 65% are volatile. The combined ferric chloride and lime dosage in sludge conditioning was 8.0% of the dry solids. Calculate fuel values for the dry solids and the wet sludge. Assuming 2200 Btu to evaporate each pound of water, compute the heat required to dry the sludge.

○*Solution*

Fuel value of dry solids using Eq. 12-19,

$$Q = 131\left(\frac{100 \times 65}{100 - 8} - 10\right)\left(\frac{100 - 8}{100}\right) = 7310 \text{ Btu/lb of dry solids}$$

Fuel value of wet sludge,

$$Q = 0.24 \times 7300 = 1750 \text{ Btu/lb of wet sludge}$$

Heat required to dry 1 lb of wet sludge,

$$Q = 0.76 \times 2200 = 1670 \text{ Btu/lb}$$

These calculations indicate that the heat from combustion of the dry solids is slightly greater than that required to evaporate moisture from the wet cake. Auxiliary fuel will be needed to heat the incinerator to ignition temperature, and may be required to maintain an operating temperature high enough to deodorize the stack gases.

12-31

SLUDGE INCINERATORS

Two major incineration systems employed in the United States are the multiple hearth furnace and the fluidized bed reactor. The multiple hearth unit has received widest adoption because of its simplicity and operational flexibility. Its furnace consists of a lined circular shell containing several hearths arranged in a vertical stack, and a central rotating shaft with rabble arms (Fig. 12-23). Operating capacity is related to the total hearth area (diameter and number of stages), and varies from 200 to 8000 lb/hr of dry sludge. Sludge cake is fed onto the top hearth and is raked slowly in a spiral path to the center. Here, it falls to the second level, is pushed to the periphery and drops, in turn, to the third hearth, where it is again raked to the center. The upper levels allow for evaporation of water, the middle hearths burn the solids, and the bottom zone cools the ash prior to discharge.

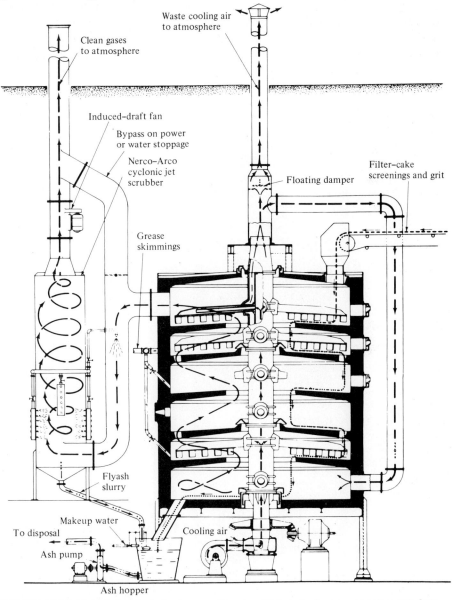

FIGURE 12-23 Multiple hearth sludge furnace. (Courtesy Nichols Engineering and Research Corp., subsidiary of Neptune International Corp.)

The hollow central shaft is cooled by forced air vented out the top. A portion of this preheated air from the shaft is piped to the lowest hearth and is further heated by the hot ash and combustion as it passes up through the furnace. The gases are then cooled as heat is absorbed by the incoming sludge. The counter-current flow pattern of air and sludge solids reduces heat losses and increases incineration efficiency. Stack gases are discharged through a wet scrubber to remove fly ash and other air pollutants. Emission-control requirements specify that the exhaust cannot contain particulate matter in excess of 70 mg/m^3 or exhibit 10% or greater opacity (shadiness), except for 2 minutes in any 1-hour period. The presence of uncombined water is the only reason for failure to meet the latter requirement.

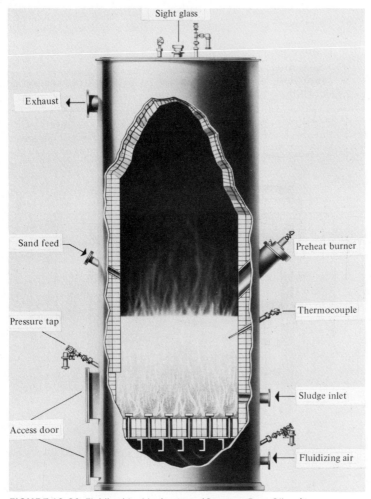

FIGURE 12-24 Fluidized bed incinerator. (Courtesy Dorr-Oliver.)

Multiple hearth furnaces can also be designed to dry organic sludge or recalcine calcium carbonate precipitate. The solids and air-flow system for calcination is similar to the pattern used for sludge incineration, with heat being supplied by burning natural gas or oil. For sludge drying, hot gases from an external furnace flow concurrent with the sludge down through the multiple hearths in order to dry the organics without scorching.

Sludge can also be dried in a flash dryer system. First, wet sludge cake is mixed with recycled dried solids and pulverized in a cage mill. Hot gases from a separate incinerating furnace suspend the dispersed sludge particles up into a pipe duct, where they are dried, and a cyclone separator removes the dried solids from the moisture-laden hot gas. Part of the dried sludge returns to the mixer for blending with incoming wet cake, while the remainder is either withdrawn for a soil conditioner or burned in the furnace for fuel.

A fluidized bed incinerator uses a sand bed as a heat reservoir to promote uniform combustion of sludge solids (Fig. 12-24). The bed is expanded by upflow of air through the sand. Dewatered sludge is injected into the fluidized sand above the grid. Violent mixing of the solids and gases in the hot sand promotes rapid drying and burning at a temperature of 1400–1500°F. The quantity of excess air needed is about 25%. The sand bed acts as a heat reservoir, enabling reduced startup time when the unit is operated only 4 to 8 hours/day. When necessary the bed is preheated by using an auxiliary fuel. Water vapor and ash are carried out of the bed by combustion gases. A cyclonic wet scrubber removes ash from the exhaust and, finally, separates it from the scrubber water in a cyclone separator.

12-32

SANITARY LANDFILL

Raw or digested wastewater sludges and chemical residues from water treatment may be buried if a suitable site is available. Except in highly urbanized areas, land and transportation costs are less expensive than incineration or chemical recovery. Sludges are often buried at municipal sanitary landfills along with other solid wastes, requiring systematically depositing, compacting, and covering the wastes. Usually 6–12 in. of earth are placed over each 2 ft of compacted fill. The top earth cover should have a minimum depth of 2 ft and be grassed to prevent erosion.

Site selection considers soil conditions, groundwater levels, location relative to populated areas, and future land-use planning. Conditions must be such that gas leachate, water seepage, and runoff do not cause pollution, nuisance, or health hazards. A monitoring program should be established to ensure that an adequate environment is maintained at the site. Projected land use may be a park with recreational facilities that are not affected by gradual subsidence of the ground surface.

12-33

OCEAN DISPOSAL

Coastal cities have discharged digested sludge into the sea for many years. Dewatered waste may be transported to offshore sites in barges and dumped, as in New York City, or sludge slurry may be pumped to deep water through a submarine outfall, as practiced by Los Angeles. A marine environment can be adversely affected if decomposing solids accumulate on the bottom. Disposal sites should have adequate current velocities for initial dilution and waste dispersion to prevent formation of sludge banks. Some waste slurries have a lower specific gravity than seawater and rise to the surface under certain conditions. Currents may then transport these buoyant solids onto beaches, interferring with recreational use. Ocean disposal may be banned in the future because of these potential health and nuisance problems.

PROBLEMS

12-1 The settled sludge from coagulation of a surface water is 1.5% solids, of which 30% are volatile. Compute the specific gravity of the dry solids and specific gravity of the wet sludge. Assume specific gravity values of 2.5 for the fixed matter and 1.0 for the volatile. What is the volume of waste in cubic meters per 1000 kg of dry solids?

12-2 A primary wastewater sludge contains 6.0% dry solids that are 65% volatile. Calculate the specific gravities of the solid matter and the wet sludge. If this residue is thickened to a cake of 22.0% solids, what is the specific gravity of the moist cake?

12-3 Compute the volume of a waste sludge with 96% water content containing 1000 lb of dry solids. If the moisture content is reduced to 92%, what is the sludge volume?

12-4 A municipal wastewater with 220 mg/l of BOD and 260 mg/l of suspended solids is processed in a two-stage trickling-filter plant. Calculate the following per 1.0 mg of wastewater treated: (a) the dry solids production in primary and secondary treatment assuming 50% SS removal and 35% BOD reduction in primary settling and a k value of 0.34 applicable for the trickling-filter secondary; (b) the daily sludge volume, both primary and secondary solids, assuming 5.0% solids content. (A specific gravity of 1.0 can be assumed for the wet sludge.)

12-5 An activated-sludge wastewater plant with primary clarification treats 10 mgd of wastewater with a BOD of 240 mg/l and suspended solids of 200 mg/l. (a) Calculate the daily primary and waste-activated sludge yields in pounds of dry solids and gallons, assuming the following: 60% suspended solids removal and 35% BOD reduction in primary settling; a primary sludge concentration 6.0% solids; an operating food/ microorganism ratio of 1:3 in the aeration basin; and a solids concentration of 15,000 mg/l (70% volatile) in the waste-activated sludge. (b) What would be the solids content of the sludge mixture if the two waste volumes were blended together? (c) If the waste activated is thickened separately to 4.5% solids before blending, what would be the combined sludge volume and solids content?

12-6 What is the estimated waste-activated sludge from a conventional aeration process treating 7.7 mgd with 173 mg/l of BOD operating at a F/M of 0.24 lb of BOD/day/lb of MLSS. Assume a suspended solids of 9800 mg/l in the waste sludge. Express answers in units of lb of dry solids/day and gpd.

12-7 A river-water-treatment plant coagulates raw water having a turbidity of 12 Jtu with an alum dosage of 40 mg/l. (a) Estimate the total sludge solids production in lb/mg of water processed. (b) Compute the volumes of waste from the settling basins and filter backwash water assuming that 70% of the total solids are removed in sedimentation and 30% in filtration, a settled sludge-solids concentration of 1.2%, and a solids content of 600 mg/l in the filter wash water. (c) Calculate the composite sludge volume after the two wastes are gravity-thickened to 3.5% solids.

12-8 Example 10-4 describes softening of a water by excess lime treatment in a two-stage system. Based on the appropriate chemical reactions, both precipitation softening and recarbonation removal of excess lime, calculate the total residue produced per million gallons of water processed.

12-9 The proposed sludge-processing scheme for a conventional step-aeration activated-sludge plant is as follows: return of waste-activated sludge to the head of the plant, withdrawal of the combined sludge (raw and waste activated) from the primary clarifier, concentration in a gravity thickener, plant effluent applied to the thickener for dilution water, return of the thickener overflow to the plant inlet, and vacuum filtration of the thickened underflow. Briefly comment on the operating problems you would anticipate with this system. What type of sludge handling and thickening would you recommend to replace the proposed scheme?

12-10 Outline the alternative methods for disposal of wastes from a precipitation softening plant processing groundwater.

12-11 A gravity thickener handles 33,000 gpd of wastewater sludge, increasing the solids content from 3.0% to 7.0% with 90% solids recovery. Calculate the quantity of thickened sludge.

12-12 A waste sludge flow of 40 m³/day is gravity-thickened in a circular tank with a diameter of 6.8 m. The solids concentration is increased from 4.5% to 7.5% with 95% suspended-solids capture. Calculate the solids loading in kg/m²/day and the quantity of thickened sludge in m³/day.

12-13 Size a gravity thickener based on 10 lb/ft²/day for a waste-sludge flow of 25,000 gpd with 5.0% solids. Assume a side water depth of 10.0 ft. After installation, operation at design flow yields an underflow of 8.0% with 90% solids removal. What is the flow of dilution water needed to maintain an overflow rate of 400 gpd/ft²? If the consolidated sludge blanket in the tank is 4.0 ft thick, compute the estimated solids retention time in the thickener.

12-14 Settled sludge and filter wash water from water treatment are thickened in the clarifier–thickeners prior to mechanical dewatering. The settled sludge volume is 1150 gpd with 1.0% solids, and the backwash is 9800 gpd containing 500 mg/l of suspended solids. Calculate the daily quantity of thickened sludge assuming a concentration of 3.0% solids.

12-15 A flotation thickener processes 250 m³ of waste-activated sludge in a 16-hour operating period. The solids content is increased from 10,000 mg/l to 40,000 mg/l with 92% solids capture. Calculate the quantity of float in cubic meters per 16-hour period. If the solids loading is 12 kg/m²-hr, what is the hydraulic loading in m³/m²-day?

12-16 Waste-activated sludge processed by dissolved-air flotation is concentrated from 9800 mg/l to 4.7% with 95% suspended-solids capture. During a 24-hour operating period, 50,000 gal of sludge was applied with a polyelectrolyte dosage of 32 mg/l. The thickener surface area is 75 ft². Calculate the solids loading, volume of float produced in 24 hours, and polyelectrolyte addition in lb/ton of dry solids.

12-17 A single-stage anaerobic digester has a capacity of 13,800 ft³, of which 10,600 ft³ is below the landing brackets. The average raw-waste sludge solids fed to the digester are 580 lb of solids/day. (a) Calculate the digester loading in lb of volatile solids fed/ft³ of capacity below the landing brackets/day. Assume that 70% of the solids are volatile. (b) Determine the digester capacity required based on the rational formula using the following data:

average daily raw-sludge solids = 580 lb
raw-sludge moisture content = 96%
digestion period = 30 days
solids reduction during digestion = 45%
digested-sludge moisture content = 94%
digested-sludge storage required = 90 days

12-18 Calculate the digester capacity in cubic meters required for conventional single-stage anaerobic digestion based on the following parameters:

daily raw-sludge solids production = 630 kg
volatile solids in raw sludge = 70%
moisture content of raw sludge = 95%
digestion period = 30 days
volatile solids reduction during digestion = 50%
moisture content of digested sludge = 93%
storage volume required for digested sludge = 90 days

12-19 The average daily quantity of thickened raw waste sludge produced in a municipal treatment plant is 15,000 gal containing 10,000 lb of solids. The solids are 70% volatile. (a) Calculate the percentage of moisture in the thickened sludge. (b) Determine the volume required for a first-stage high-rate digester based on the following criteria: a maximum loading of 100 lb of volatile solids per 1000 ft³/day and a minimum detention time of 15 days. (c) If the volatile solids reduction in the complete mixing digester is 60%, what is the percentage of moisture in the digested sludge?

12-20 The waste-sludge production from a trickling-filter plant is 12.5 m³/day containing 620 kg of solids (70% volatile). The proposed anaerobic digester design to stabilize this waste consists of two floating-cover heated tanks, each with a volume of 480 m³ when the covers are fully raised and a volume of 310 m³ with the covers resting on the landing brackets. Calculate the following: (a) the digester volume per equivalent population with the covers fully raised. Assume 90 g of solids production per capita. (b) Digester loading in terms of kg volatile solids fed/m³ of tank volume below the landing brackets/day. (c) Digested-sludge storage time available between lowered and raised cover positions. Assume a volatile solids reduction of 60% during digestion, and a digested-sludge moisture content of 93%.

12-21 A wastewater sludge is stabilized and thickened in anaerobic digestion. The daily raw-sludge feed is 100,000 lb containing 5500 lb of dry solids. Forty percent of the matter applied is converted to gases during digestion, and the digested residue is increased to 10% solids by supernatant withdrawal. Calculate the volumetric reduction of wet sludge achieved in the digestion process.

12-22 Aerobic digesters with a total capacity of 50,000 ft³ stabilize waste-activated sludge from an extended aeration treatment plant without primary clarification. The average daily sludge flow pumped to the digestion tanks is 32,000 gal with 1.5% solids, of which 65% are volatile. The estimated solids production per capita is 0.17 lb of VS/ person/day. Calculate the sludge detention time in the digesters, VS loading lb/ft³/day and volume provided per capita.

12-23 Anaerobic digestion provides for stabilization and reduction of volatile solids, storage of digested sludge, and thickening of the waste by withdrawal of supernatant. Which of these functions are accomplished in aerobic digestion?

12-24 A waste activated with a total solids concentration of 12,500 mg/l and volatile solids content of 8800 mg/l is applied to an aerobic digester with a detention time of 25 days. The volatile solids destruction is 50% during aeration. Calculate the VS loading in units of kg/m³-day and the solids content of the digested sludge.

12-25 A primary sludge feed of 10,000 gpd, containing 6.0% dry solids that are 65% volatile, is dewatered by vacuum filtration, producing a cake with 22.0% solids. Compute the specific gravities of the wet sludge and filter cake. Assuming 100% solids capture in filtration and neglecting the addition of conditioning chemicals, calculate the volume and weight reductions achieved in the dewatering process.

12-26 The smallest vacuum filter produced commercially by a leading manufacturer has a 7-ft-diameter drum that is 2 ft 10 in. in width; the nominal surface area is 60 ft². Estimate the equivalent population that can be served by this unit assuming 0.20 lb of dry solids/capita/day of sludge production, a filter yield of 5.0 lb/ft²/hr, and a 7-hour operating period each day.

12-27 A vacuum filter with an area of 23 m² processes 60 m³ of sludge containing 65,000 mg/l of solids in 6 hours of operation. Calculate filter yield and estimate the volume of cake with 28% solids, assuming a specific gravity of 1.3 for the dry solids. The weight of polymer added to the raw sludge for conditioning may be neglected.

12-28 The raw sludge produced by a trickling filter plant is 6600 gpd containing 2700 lb of dry solids. What area of vacuum-filter surface would you recommend assuming a 7-hour operating period each day? Calculate the weight of the cake produced assuming 95% capture of the waste solids, chemical additions of 10%, and a moisture content of 75%.

12-29 An alum sludge is dewatered by pressure filtration. A daily volume of 40 m³ of slurry is pressed from 2.0% solids concentration to a cake of 40% solids, after conditioning with 10% lime. Calculate the volume of filtrate and weight of the cake produced per day. (The 40% concentration of cake solids includes the alum precipitate and the lime addition.)

12-30 A lime precipitate from a water-softening plant is concentrated in a scroll centrifuge from 10% to 60%. The suspended solids capture is 90%. Calculate the suspended-solids concentration in the centrate and the volumetric reduction of the waste slurry as a percentage of the feed.

12-31 A basket centrifuge is employed to thicken a waste activated sludge from 1.0% solids to 5.0%. The flow rate is 70 gpm and the suspended solids in the centrate is 1200 mg/l. Compute the suspended solids removal and centrate flow rate.

REFERENCES

1. P. A. Vesilind, *Treatment and Disposal of Wastewater Sludges* (Ann Arbor Mich.: Ann Arbor Science Publishing Inc., 1974).

2. H. L. Nielsen, K. E. Carns, and J. N. DeBoice, "Alum Sludge Thickening and Disposal," *J. Am. Water Works Assoc.* 65, no. 6 (1973): 385–394.

3. R. J. Calkins and J. T. Novak, "Characterization of Chemical Sludges," *J. Am. Water Works Assoc.* 65, no. 6 (1973): 423–428.

4. Metcalf & Eddy, Inc., *Wastewater Engineering* (New York: McGraw-Hill Book Company, 1972).

5. *Recommended Standards for Sewage Works, Great Lakes–Upper Mississippi River Board of State Sanitary Engineers*, 1971 Edition (Albany, N.Y.: Health Education Service).

6. R. J. Manson and C. A. Merritt, "Land Application of Liquid Municipal Wastewater Sludges," *J. Water Poll. Control Fed.* 47, no. 1 (1975); 20–29.

7. C. O'Donnell and F. Keith, Jr., "Centrifugal Dewatering of Aerobic Waste Sludges," *J. Water Poll. Control Fed.*, 44, no. 11 (1972): 2162–2171.

8. G. P. Fulton, "Recover Alum to Reduce Waste-Disposal Costs," *J. Am. Water Works Assoc.* 66, no. 5 (1974): 312–318.

9. G. M. Fair, J. C. Geyer, and D. A. Okun, *Water and Wastewater Engineering* (New York: John Wiley & Sons, Inc., 1968).

13
Advanced Wastewater Treatment Processes

Advanced wastewater treatment (AWT) refers to methods and processes that remove more contaminants from wastewater than are usually taken out by present conventional techniques. The term *tertiary treatment* is often used as a synonym for advanced waste treatment, but the two are not precisely the same. Tertiary suggests an additional step applied only after conventional primary and secondary waste processing, for example, chemical coagulation of a secondary effluent for removal of phosphorus. Advanced treatment actually means any process or system which is used after conventional treatment, or to modify or replace one or more steps, to remove refractory contaminants. Several of these pollutants not taken out by secondary biological methods can adversely affect aquatic life in streams, accelerate eutrophication of lakes, and hinder reuse of surface waters for domestic needs.

The term *water reclamation* implies that the combination of conventional and advanced treatment processes employed returns the wastewater to nearly original quality (i.e., reclaims the water). Reclamation is an introduction to renovating wastewater by improving the quality to such an extent that it may be directly reused. Renovation and reuse of wastewater are becoming more important as water needs cannot be met by the relatively fixed natural supply available. The envisioned reuse of wastewater ranges from agricultural and industrial process water to potable supplies, with the degree of purification required varying according to the specific use.

Limitations of Secondary Treatment

Contamination of municipal water results from human excreta, food-preparation wastes, and a wide variety of organic and inorganic industrial wastes. Conventional treatment employs physical–biological processes, followed by chlorination to reduce biochemical oxygen demand, suspended solids, and pathogens. Water-quality deterioration from municipal use results in the following approximate increases of pollutants after ordinary secondary treatment: 300 mg/l of total solids, 200 mg/l of volatile solids, 30 mg/l of suspended solids, 30 mg/l of BOD, 20 mg/l of nitrogen, and 7 mg/l of phosphorus. From industrialized cities, treated wastewater may also contain traces of organic chemicals, heavy metals, and other contaminants.

13-1

EFFLUENT STANDARDS

The Environmental Protection Agency defines the minimum acceptable level of effluent quality in terms of BOD, suspended solids, fecal coliform bacteria, and pH. Average values for BOD and suspended solids cannot exceed 30 mg/l, the geometric mean of fecal coliform bacteria should not be greater than 200 per 100 ml, and the limit for pH values is 6.0–9.0. In some cases, these standards do not adequately protect the receiving water body to ensure adequate quality for beneficial uses, such as municipal water supply, recreation, propagation of wildlife, irrigation, and esthetics. Advanced waste-treatment techniques can be used to reduce BOD and suspended-solids levels, remove plant nutrients, and diminish the concentration of persistent, nonbiodegradable organics.

Phosphorus is a key nutrient for growth of algae and aquatic plants associated with eutrophication. Where effluents are discharged to lakes and estuaries, either directly or via rivers, standards have been adopted by most states to limit phosphorus input. A commonly established maximum effluent concentration is 1.0 mg/l. This criterion requires a reduction of about 90% in domestic wastewater treatment.

Ammonia nitrogen, the most common form in the effluent from biological treatment, is toxic to fish at relatively low concentrations and can exert a significant oxygen demand. Theoretically, 4.6 lb of oxygen is used to bacterially nitrify 1.0 lb of ammonia nitrogen. For disposal in a receiving stream with limited dilutional flow, oxidation of a major portion of the ammonia is sometimes needed, along with more stringent BOD and suspended-solids removals.

For example, effluent limits may be established at 10 mg/l of BOD, 10 mg/l of SS, and 5 mg/l of NH_3–N. Denitrification is occasionally specified where the receiving watercourse is withdrawn for a public water supply, or where nitrogen removal is deemed necessary to retard eutrophication. The limit for nitrate concentration in drinking water is 10 mg/l of nitrogen. At present, emphasis is being placed on phosphorus extraction rather than nitrogen removal for retarding fertilization of lake waters since phosphorus is generally considered the limiting nutrient for controlling eutrophication. Perhaps equally important is the fact that nitrogen compounds are more difficult and costly to eliminate in wastewater processing.

Many organic and inorganic chemicals are refractory to conventional treatment, and local conditions often dictate discharge limits for specific compounds. Nonbiodegradable organics can cause taste and odor problems in downstream water supplies, while foam and color are unesthetic. Toxic metal ions are injurious to aquatic life and, in sufficiently high concentrations, prevent reuse of the water as a public supply source. Removal of these contaminants may warrant advanced waste-treatment operations, such as chemical coagulation, filtration, and carbon adsorption.

13-2

FLOW EQUALIZATION

Wastewater flows have a diurnal variation ranging from less than one-half to more than 200% of the average flow rate. In addition, daily volumes are increased by inflow and infiltration into the sewer collection system during wet weather. Strength of a municipal waste also has a pronounced diurnal variation resulting from nonuniform discharge of domestic and industrial wastes. Treatment plants traditionally have been sized to handle these deviations without significant loss of efficiency. For conventional primary and secondary treatment, designing units to accommodate peak hourly flow is more economical than installing flow-equalization basins to eliminate the diurnal influent pattern. Industrial wastes entering a municipal system can cause excessively large flows and peak organic loads, so it is better to install facilities at the industrial site for flow smoothing prior to discharge.

Many advanced waste-treatment operations, such as filtration and chemical clarification, are adversely affected by flow variation and sudden changes in solids loading. Maintaining a relatively uniform influent allows improved chemical feed control and process reliability. Costs saved by installing smaller units for chemical precipitation and filtration, together with reduced operating expenses, may be less than the added costs for flow-equalization facilities.

The AWT scheme for the Walled Lake–Novi Waste-Water Treatment Plant[1] employing side-line flow equalization is illustrated in Fig. 13-1. This facility, in operation since 1971, uses biological–chemical processing followed

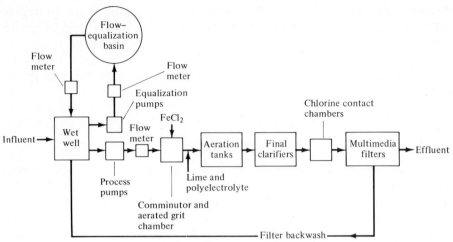

FIGURE 13-1 Process diagram for biological–chemical treatment followed by filtration using side-line flow equalization, Walled Lake–Novi Waste-Water Treatment Plant. [Source : *Flow Equalization*, Environmental Protection Agency, Technology Transfer (May 1974), p. 19.]

by multimedia filters. Ferric chloride and lime are added ahead of aeration for phosphate precipitation with the biological floc. Waste sludge is dewatered on drying beds after aerobic digestion. The flow-equalization basin is a circular concrete tank with a volume of 315,000 gal, which is equivalent to 15% of the 2.1-mgd design flow. Diffused aeration with a capacity of 2 cfm/1000 gal of storage provides mixing. Chlorination is available for odor control, and a sludge scraper prevents consolidation of settled solids. The process pumps transfer a constant preset flow from the wet well for treatment, and variable-speed pumps deliver excess flow to the equalization basin. During periods of low influent flow, wastewater is released from the basin to the wet well to maintain the established flow rate through the plant.

Equalization chambers can also be designed as in-line units to pass all the wastewater through the basins. Mechanical or diffused aeration is required to keep solids in suspension and prevent septicity. Although normal placement is between grit removal and primary settling, holding tanks for flow smoothing can be placed at other points in the treatment scheme. For example, a basin serving as a pump suction pit can be located just ahead of filters to dampen hydraulic surges without providing complete flow equalization.

Basin volume required for flow equalization is determined from mass diagrams based on average diurnal flow patterns. The capacity needed is usually equivalent to 10–20% of the average daily dry-weather flow. Extra volume should be provided below the low-water level, since both mechanical and diffused aeration systems must have a minimum depth to maintain mixing. Example 13-1 includes sample calculations for determining equalization basin volume.

●EXAMPLE 13-1

Determine the basin volume needed to equalize the diurnal wastewater flow pattern diagrammed in Fig. 11-18b.

○*Solution*

Hourly flow rates, measured from Fig. 11-18b, and calculated cumulative volumes are listed in Table 13-1; the mass diagram of wastewater flow from Table 13-1 is drawn in Fig. 13-2. The slope of the line connecting the origin and final point on the mass curve equals the average 24-hour rate of 18.2 mgd. To find the required volume for equalization, construct lines parallel to the average flow rate and tangent to the mass curve at the high and low points. The vertical distance between these two parallels is the required basin capacity; in this case, the value is 2.5 mg or 13.7% of the 18.2-mg daily influent.

TABLE 13-1

WASTEWATER FLOWS FROM FIGURE 11-18b

Time	Flow Rate (mgd)	Cumulative Volume (million gallons)
Midnight	—	0
2 A.M.	13.1	1.1
4	10.7	2.0
6	9.7	2.8
8	14.8	4.0
10	22.5	5.9
Noon	26.6	8.1
2 P.M.	25.6	10.3
4	23.5	12.2
6	21.8	14.0
8	18.9	15.6
10	16.7	17.0
Midnight	14.4	18.2
	Average = 18.2	

Selection of Advanced Wastewater Treatment Processes

Table 13-2 lists popular AWT methods. Upgrading treatment to increase BOD and suspended-solids removal is usually accomplished by tertiary operations, such as filtration and carbon adsorption. Emphasis is also being placed on unit operations for nutrient removal, particularly precipitation of phosphates by either separate chemical coagulation or biological–chemical aeration, and nitrogen extraction by air stripping or biological nitrification–denitrification.

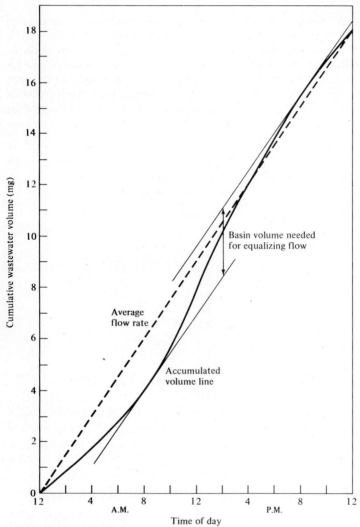

FIGURE 13-2 Mass diagram of wastewater flow, from Table 13-1 for Example 13-1, to determine the basin volume needed to equalize flow.

TABLE 13-2

SELECTED ADVANCED WASTEWATER TREATMENT PROCESSES

Suspended-solids removal
    Filtration through granular beds
    Microscreening
    Chemical coagulation and clarification
Organic removal
    Adsorption on granular activated carbon
    Extended biological oxidation
Phosphorus removal
    Biological–chemical precipitation and clarification
    Chemical coagulation and clarification
    Irrigation of cropland
Nitrogen removal
    Biological nitrification–denitrification
    Ammonia reduction by air stripping
    Breakpoint chlorination
    Ion-exchange extraction
    Irrigation of cropland

13-3

SELECTING AND COMBINING UNIT PROCESSES

Culp and Culp[3] present the following major factors affecting unit process selection: influent wastewater characteristics, effluent quality required, reliability, sludge handling, process compatibility, and costs. Degrees of treatment needed at present and in the future are the primary considerations, because neither raw-wastewater characteristics nor effluent quality specifications can be safely considered as permanent. When considering alternative AWT schemes, the best choice for the original plant allows for expansion and change to meet future needs. Maximum flexibility for modifying operations to improve performance is of major importance.

The reliability of a process is directly related to the experience gained from operation of plant-scale systems. Only limited data are available on actual operating results and costs of new treatment techniques. Pilot-plant studies can provide satisfactory information on liquid processing, but they do not often adequately evaluate sludge problems or compatibility with previous and subsequent unit operations. Problems involved in going from pilot to full scale are not easily recognized; thus, for maximum reliability, future plant designers should take advantage of prior art and practice developed from existing full-scale installations.

Sludge handling and disposal dictate to a considerable extent the selection of processes that are most feasible for separating contaminants from wastewater. A unit operation, even though successful in extracting pollutants from water, can be unacceptable if the waste sludge produced is difficult and costly to dewater; hence sludge disposal must always be considered an integral part of any treatment technique. Elimination of residue by spreading on land, burial, incineration, or ocean dumping necessitates concentration of waste slurries by biological or chemical stabilization and mechanical dewatering methods. Advanced waste-treatment processes require greater study because they often include chemical precipitates as well as organic sludges. A critical question is whether organic and inorganic wastes should be kept separate or combined for thickening and disposal. The answer often determines if chemical treatment is applied as a tertiary step or combined with conventional operations of primary sedimentation and biological aeration.

A fourth consideration involves compatibility of unit processes applied in the overall treatment scheme. Optimum pH is often important and can influence the sequence of operations. For instance, ammonia stripping is most efficient from alkaline water, while disinfection by chlorine is more effective at low pH values. Possible effects of recycled waste streams from individual unit operations in the overall treatment process must be considered. Segregation or blending of centrate, filtrate, backwash water, and other return flows, and their point of return, demand evaluation in design.

Both capital and operating costs influence process selection and often dictate design decisions. Common factors include electrical-power consumption; recovery and reuse of chemicals; carbon regeneration; choice of methods for sludge disposal; and separation or combination of biological and chemical treatment. Regarding the latter, phosphorus-removal schemes incorporated in present-day conventional processes may appear to be cheaper than tertiary treatment until the cost of handling the biological–chemical sludge mixture is included in the evaluation, which makes separate disposal of the two sludges more economical, except in locations where the blended sludge can be disposed of by land burial or barging to sea.

Advanced waste treatment costs vary from a few cents per 1000 gal for increased BOD and suspended solids removal to more than 10 cents/1000 gal for nutrient reduction. Cost of tertiary treatment by chemical coagulation appears to be in the range of conventional secondary processing. Estimates for water reclamation increase to 300% or more of secondary-treatment costs. The precise percentage increase for any given treatment plant depends on size and location, type of AWT processes instituted, and expense of existing conventional treatment. Current figures for various systems are available in literature from the Environmental Protection Agency.

Suspended-Solids Removal

13-4

GRANULAR-MEDIA FILTRATION

Design criteria for wastewater filters cannot be derived directly from experience in potable water systems. Waterworks filters are generally operated at constant rates under relatively steady suspended-solids loading. Unless equalization is provided, a wastewater plant must handle a varying rate of flow with peak hydraulic and solids loadings occurring simultaneously. Particulate matter found in typical wastewaters are less predictable and much more "sticky" than water-plant solids, thus making filter backwashing more difficult. Also, the nature of suspended matter is not consistent and varies with the preceeding treatment processes. Microbial flocs are the dominant suspended solids following secondary biological treatment, while carryover from biological–chemical and physical–chemical methods contain a significant quantity of coagulant residue.

Several filter configurations are shown schematically in Fig. 13-3. Upflow filtration through a relatively deep, coarse filter medium has been adopted in the United Kingdom but is limited in American practice. The major advantage is that straining takes place from the coarse to fine sand in upward flow with backwashing performed in the same direction but at a higher rate. The chief disadvantage of upflow filtration occurs at high head losses when uplift forces exceed the weight of the sand, allowing solids carryover in the effluent. Hence the filter may not be compatible with the higher upflow rates during peak hydraulic loadings. Bed expansion is minimized by a restraining grid placed within the sand, near the surface, to reduce uplifting during filtration. A biflow filter is an upflow unit modified to prevent bed expansion (Fig. 13-3b). Water applied to both top and bottom is drawn from a strainer located in the upper sand layer. About 80% of filtration is upward through the major portion of the media, while downward flow through the surface layer of finer sand prevents bed expansion. Backwashing is accomplished by applying the entire flow to the underdrain and flushing the bed from bottom to top.

Dual- and tri-media filters with downward flow dominate in American practice. This design permits production of a high-quality effluent with reasonable filter runs between bed cleanings. Usually a coarse anthracite coal overlies a layer of finer silica sand, and occasionally a still finer garnet sand is used at the bottom. These media have respective specific gravities of 1.35–1.75, 2.65, and 4–4.2. After backwashing, the media are arranged with the coarse, lighter coal on top; the finer, heavier garnet sand on the bottom; and middle-size silica sand between the two layers. The actual distribution and degree of intermixing depend on both relative particle-size gradations and specific gravities of the

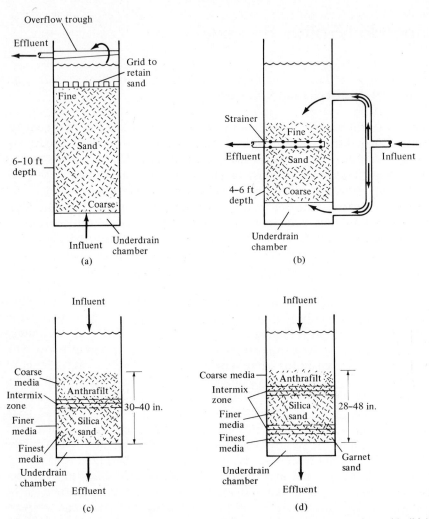

FIGURE 13-3 Granular-media filter configurations: (a) upflow filter with surface grid; (b) biflow filter; (c) dual-media, downward-flow filter; (d) mixed-media (triple-media) filter. [Source: *Wastewater Filtration*, Environmental Protection Agency, Technology Transfer (July 1974), p. 12.]

media. A filter bed with decreasing particle size from top to bottom allows both surface straining and "in-depth" filtration, without causing premature surface plugging or solids breakthrough.

Gravity-flow and pressure filters are usually designed to operate at a constant rate of filtration. A standard-gravity unit in water treatment has both constant-rate and constant-water-level controls (refer to Sec. 9-24). The maximum head available for forcing water through the bed is equal to the difference between the water surface above the filter and the level in the clear well underneath, normally 9–12 ft. A rate controller throttles flow in the discharge pipe, restraining the quantity of filtration when the bed is clean. As the medium collects impurities, resistance to flow increases and the controller valve opens wider to maintain a preset rate (Fig. 9-31). Disadvantges of effluent control for constant-rate operation are the high costs associated with a complex mechanical system, and the fact that suction in the bed releases dissolved gases that tend to collect in the filter pores, inhibiting passage of water.

Gravity filters can be designed to operate without rate controllers as diagrammed in Fig. 13-4. The driving force for percolation is only the pressure head, since the effluent elevation in the clear well is above the surface of the bed. Rate of filtration is established by the quantity of water applied. This is achieved by splitting the plant influent flow equally to all operating units by means of weir boxes at the filters. The water level above the medium is shallow when the bed is clean, but rises to the influent-weir-box elevation as the filter becomes dirty. Filter-box walls must be high enough above the effluent weir to provide the desired terminal head loss. The additional box depth is the major disadvantage of a system without effluent flow control.

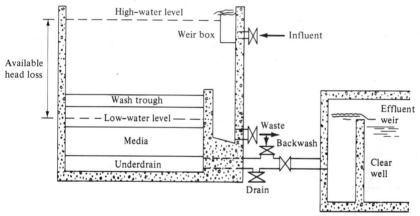

FIGURE 13-4 Schematic diagram of a gravity filter and clear well that operates by influent-flow-splitting filtration. [Source: *Wastewater Filtration*, Environmental Protection Agency, Technology Transfer (July 1974), p. 15.]

Variable declining-rate filtration is another method of operation for gravity filters.[5] The principal difference between this technique and influent flow-splitting is the location and type of inlet arrangements. Water enters the filter box below the wash-trough level rather than at the top of the wall. The portion flowing to each unit is determined by the water level in the filter box. As discharge from the dirtiest beds decreases, inflows are automatically redistributed so that the cleaner filters pick up the capacity lost by the dirtier ones. This method of operation causes a gradual declining rate of percolation near the end of a filter run.

Pressure filters (Fig. 9-26) are normally operated at a constant rate of flow, although some units use a constant applied pressure without effluent control, resulting in declining-rate filtration. Two advantages of enclosed filters are that high pressure can be used to overcome a greater head loss, and negative head conditions are not created in the filter.

Cleaning solids from granular media requires air or mechanical scouring in addition to fluidizing the bed by upward flow of wash water. One of the best methods for filter cleaning is air scour for 3 to 5 minutes followed by back-washing with 75–100 gal/ft^2 (3–4 m^3/m^2).[4] Auxiliary surface-wash devices operated before and during water fluidization can also provide the vigorous cleansing action needed. The best source of backwash water is effluent from tertiary filters. If the filtrate is disinfected, the chlorine contact tank can serve as a supply. Otherwise, a storage tank is needed with sufficient capacity to clean the filters during their peak usage. Dirty wash water should be collected in an equalizing tank and returned to the plant influent at a nearly constant rate for treatment.

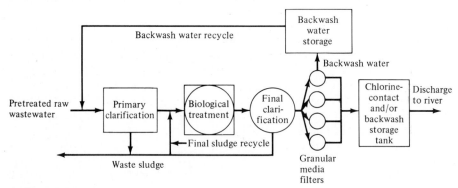

FIGURE 13-5 Typical layout of a biological treatment plant with tertiary granular-media filters. [Source: *Wastewater Filtration*, Environmental Protection Agency, Technology Transfer (July 1974), p. 6.]

Design considerations for granular-media filters[4, 6] include: filter configuration; types, size, gradation, and depths of filter media; method of flow control; backwash requirements; filtration rate; terminal head loss; quantity and characteristics of the water applied; and desired effluent quality. Figure 13-5 is a schematic diagram of a biological treatment plant with tertiary filtration. Normally, at least two and usually four filter cells are provided for operational flexibility. The bed area should be sufficient to allow peak design flows with one unit out of service for backwashing or repair.

Filtration rate, terminal head loss, length of filter run, and solids loading are interrelated parameters. Production from gravity-flow beds is normally 2–6 gpm/ft^2 (120 to 350 m^3/m^2-day), and terminal head loss 8–10 ft (2.5–3.0 m) for gravity filters and 20–30 ft when using pressure filters. The length of filter runs should be a minimum of 6 hours to avoid excessive use of wash water, and shorter than 40 hours to reduce bacterial decomposition of organics trapped in the media. These design limits provide an average run length of about 24 hours. The highest solids loading on tertiary filters coincides with maximum hydraulic flow, since clarification following secondary treatment is least efficient at peak overflow. Even in well-operated plants, suspended-solids content during high flows can range from 30–50 mg/l (15–25 turbidity units). Therefore, the critical design condition is often based on the maximum 4-hour flow rate at the worst expected suspended-solids concentration.

Experience indicates that filtration following secondary biological treatment can reduce suspended solids to a level of 4–9 mg/l so the expected performance of a well-designed and properly operated system is an effluent with SS and BOD concentrations of less than 10 mg/l. However, chemical treatment prior to filtration is required to consistently produce an effluent of 5 mg/l of SS and 5 mg/l of BOD. (In the State of Florida, advanced wastewater treatment is defined as the effluent standards of 5 mg/l of BOD and suspended solids, 3 mg/l of total nitrogen, and 1 mg/l of phosphorus.)

●EXAMPLE 13-2
Compute the area of granular-media filters required for suspended solids removal from a trickling-filter plant effluent. The average daily flow is 18.2 mgd (68,900 m^3/day), and the maximum wet-weather flow for a 4-hour period is 31.0 mgd (117,000 m^3/day). Data from a pilot-plant study are plotted in Fig. 13-6. Filters are dual-media, gravity-flow beds, downtime for backwashing is 30 minutes, and water usage equals 150 gal/ft^2 (6.1 m^3/m^2). Assume a nominal filtration rate of 3 gpm/ft^2 (54 m^3/m^2-day), and check the peak rate with one filter cell out of service.

○*Solution*
For the average daily wastewater flow and nominal filtration rate of 3.0 gpm/ft^2

$$\text{filter area} = \frac{18.2 \text{ mgd} \times 694 \text{ gpm/mgd}}{3.0 \text{ gpm/ft}^2} = 4200 \text{ ft}^2$$

Use four filters, each 1050 ft^2 (97.5 m^2).
From Fig. 13-6, the length of filter run is 21 hours for 3.0 gpm/ft^2 and 30 mg/l of SS.

$$\text{filter cycle time} = \text{run length} + \text{backwash time}$$
$$= 21.0 + 0.5 = 21.5 \text{ hours}$$
$$\text{volume of filtrate/cycle} = 21 \times 3.0 \times 60 = 3800 \text{ gal/ft}^2$$
$$\text{backwash volume/cycle} = 150 \text{ gal/ft}^2$$

$$\text{backwash as a percentage of filtrate} = \frac{150}{3800} = 4.0\%.$$

needed filtration rate to account for downtime and return of backwash water

$$= 3.0 \times 1.04 \times \frac{21.5}{21} = 3.2 \text{ gpm/ft}^2$$

$$(57 \text{ m}^3/\text{m}^2\text{-day})$$

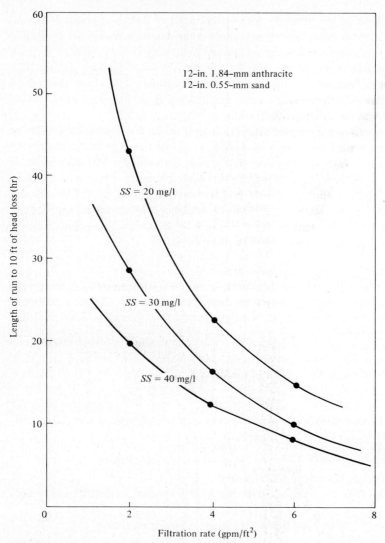

FIGURE 13-6 Length of filter run at various filtration rates and suspended-solids concentrations for a bed of anthracite coal and sand media, and an applied wastewater of trickling-filter-plant effluent. [Source: E. R. Baumann and J. Y. C. Huang, "Granular Filters for Tertiary Wastewater Treatment," *J. Water Poll. Control Fed.* 46, no. 8 (August 1974).]

The quantity of suspended solids removed assuming 30 mg/l applied and 5 mg/l remaining in the effluent

$$= (30 - 5) \times 18.2 \times 8.34 = 3800 \text{ lb of dry SS/day (1700 kg)}$$

For the maximum 4-hour flow of 31.0 mgd, estimate the required filtration rate with three filter cells operating and the fourth out of service, if the return backwash water is 10% of the wastewater flow.

$$\text{filtration rate} = \frac{31.0 \times 1.1 \times 694}{3 \times 1050} = 7.5 \text{ gpm/ft}^2 \text{ (130 m}^3\text{/m}^2\text{-day)}$$

The length of filter run from Fig. 13-6 at 40 mg/l is 6 hours, which should be sufficient to allow operation when one cell is out of operation for cleaning or emergency repair.

13-5

MICROSCREENING

Microscreens, referred to in Section 9-19 as microstrainers, are rotating drum filters that operate under gravity conditions (Fig. 13-7). Peripheral drum speed is controllable up to a maximum of 150 ft/min with normal operation in the lower range. Filtering fabrics of finely woven stainless steel fitted to the drum have mesh openings of 20–25 μm for tertiary wastewater screening. Influent enters the open end of the drum and flows outward through the rotating fabric. Solids collected on the screen form a mat that strains out suspended particles having dimensions smaller than the screen openings. High-pressure water jets, located outside at the top of the drum, continuously backwash the sludge mat into an

FIGURE 13-7 Microscreen (microstrainer) operating as a tertiary filter to remove suspended solids from the effluent of a wastewater-treatment plant. (Courtesy Crane Co., Cochrane Environmental Systems Division.)

effluent trough within the drum. Hydraulic loadings vary from 5 to 10 gpm/ft^2 of submerged drum surface area, providing about 65% of the total area. Wash water is 2–5% of the filtered effluent, varying with the jet pressure and suspended solids removal. A typical drum is 10 ft in diameter.

Performance of a microscreen depends primarily on the nature and concentration of applied suspended solids, hydraulic loading, mesh size, and drum speed. In general, drum rotation should be at the slowest rate possible consistent with the feed flow, and provide an acceptable head differential across the fabric, usually 3–6 in. Effluent suspended solids content varies with operating conditions in the range of 5–15 mg/l. Major problems associated with microstraining are difficulty in maintaining a consistent filtrate quality with fluctuating influent loading, and buildup of biological slime on the microfabric.

Carbon Adsorption

Activated carbon effectively removes dissolved organics at low concentrations in water and wastewater. Powdered carbon, commonly used in water treatment, has not received widespread application in wastewater processing, owing to the difficulty of regeneration. The current trend in AWT is installation of granular-carbon columns either as tertiary conditioning following biological treatment, or as the second phase in physical–chemical systems. When the adsorptive capacity is exhausted, the spent carbon can be regenerated for reuse.

13-6

GRANULAR CARBON COLUMNS

Dissolved organics not extracted by conventional biological treatment can be removed to a considerable degree by adsorption and biodegradation on activated carbon. The large surface area of activated carbon assimilates organics while microbial degradation reopens pores in the granules. Because of this biological contribution, toxic substances in the applied wastewater can reduce removal capacity. Some readily biodegradable substances are difficult to adsorb on carbon, thereby making it difficult to predict the quality of effluent achievable by the physical–chemical process for a given wastewater. The ability to remove soluble organics should be demonstrated by pilot-plant tests as well as experience from existing full-scale plants.

The configuration and operation of a carbon contactor, in addition to the nature of waste organics and character of the granular carbon, influence the effectiveness of adsorption. Contact times are generally less than 30 minutes since longer periods do not substantially enhance removals but do produce a favorable environment for generating hydrogen sulfide. Anaerobiosis can be controlled by decreased contact time, more frequent backwashing, oxygen addition, or prechlorination.

The two alternative carbon-contacting systems are downward passage through the bed, either under pressure or gravity flow and upflow through a packed or expanded column. Granular carbon in a downflow unit adsorbs organics and filters out suspended materials. However, this dual-purpose approach generally results in an unsatisfactory effluent owing to losses of efficiency in both filtration and adsorption. Downflow beds must be periodically backwashed to remove accumulated solids, and therefore have an underdrain system and water piping similar to granular-media filters.

Figure 13-8 diagrams the upflow, countercurrent, packed-bed carbon columns installed in the Orange County Water District AWT plant. After chemical clarification and granular-media filtration, water enters the bottom of the column through a screen manifold and overflows via outlet screens at the top. Fresh carbon slurry is added by gravity, and the spent packing is withdrawn from the bottom. With the column in service, replacement can be performed continuously at a slow rate, or intermittently by replacing 5–10% of the column contents at a time. About 10% of the contactor volume is void space at the top of the tank. This permits upward flow to be increased for bed expansion and flushing particulate matter to waste to reduce head loss through the bed. The upflow-to-waste cycle can also be used to clean the medium of excess carbon fines, if necessary. Countercurrent operation allows carbon near the inlet to become fully saturated with impurities prior to withdrawal for regeneration.

Expanded-bed upflow contactors can be either pressurized or constructed with open tops. Tank volume includes a space of about 50% for bed enlargement during operation. Expanded-bed units have the advantage of being able to treat wastewaters relatively high in suspended solids without the excessive head losses that occur in a downflow unit.

The general ranges of design criteria for both tertiary and physical–chemical carbon columns are listed in Table 13-3. Granular-media filters are generally

TABLE 13-3

GENERAL DESIGN PARAMETERS FOR GRANULAR CARBON COLUMNS

Carbon dosage (regeneration requirement)	
Tertiary treatment	200–400 lb/mg of wastewater
Physical–chemical process	500–1800 lb/mg of wastewater
Contact time (empty bed basis)	10–50 minutes
Hydraulic loading	2–10 gpm/ft^2
Backwash rate	15–20 gpm/ft^2
Flow configuration	Upflow or downflow; one-stage or multistage
Contactor configuration	Gravity or pressure vessels; steel or concrete construction

Source: *Process Design Manual for Carbon Adsorption*, Environmental Protection Agency, Technology Transfer (October 1973), p. 6–1.

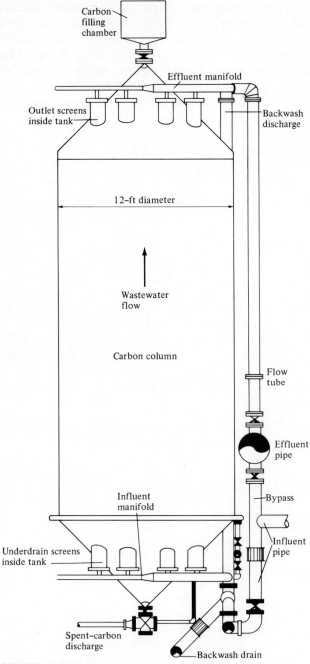

FIGURE 13-8 Upflow countercurrent carbon column, Orange County Water District, California.

placed ahead of carbon units to reduce the suspended-solids loading on the columns. Upsets in pretreatment, which produce a rapid rise in turbidity of the influent, can suddenly increase the head loss through downflow and packed-upflow beds. Installation of upflow expanded-bed contactors employs down-stream filtration to remove bacterial floc and other suspended matter flushed from the carbon columns.

13-7

CARBON REGENERATION

Exhausted activated carbon must be regenerated and reused to make adsorption economically feasible. Restoration is accomplished by heating in a multiple hearth furnace with a low-oxygen steam atmosphere (Fig. 12-23). Adsorbed organics are volatilized and released in gaseous form at a temperature of about 1700°F. With proper control, granular carbon can be restored to near virgin adsorptive capacity with only 5–10% weight loss. Thermal regeneration involves drying, baking (pyrolysis of adsorbates), and activating by oxidation of the remaining residue. The time required is about 30 minutes—15 minutes for drying, 5 minutes for gasifying the volatiles, and 10 minutes for reactivation. Furnace controls include temperature, rate of steam feed, and rotational speed of the rabble arms. The latter two determine carbon depth on the hearths and residence time in the furnace. Temperature is the most critical factor since insufficient heat will not volatilize the organic matter, while too high a temperature burns the carbon.

Figure 13-9 shows a process flow diagram for carbon contacting and re-generation. Spent carbon, withdrawn from the bottom of the columns, is trans-ported in water slurry to a dewatering tank. After draining, the wet carbon feeds by a screw conveyor into the top of a multihearth furnace. Regenerated carbon, discharging from the bottom of the furnace, is quenched in water. The slurry is then washed to remove carbon fines and hydraulically transported to filling chambers of the carbon columns, or to storage. Makeup carbon is also trans-ported in water suspension to the contactors. Regeneration furnaces should be sized to allow for a substantial maintenance downtime. The fuel requirement for regeneration, including both furnace heating and steam generation, approximates 4000–4500 Btu/lb of carbon.

Phosphorus Removal

Most phosphorus entering surface waters is from man-generated wastes and land runoff. Contributions from nonpoint sources in surface drainage vary from 0 to 15 lb of phosphorus/acre/year, depending on land use, agricultural practice, fertilizer additions, topography, soil conservation practices, and others.

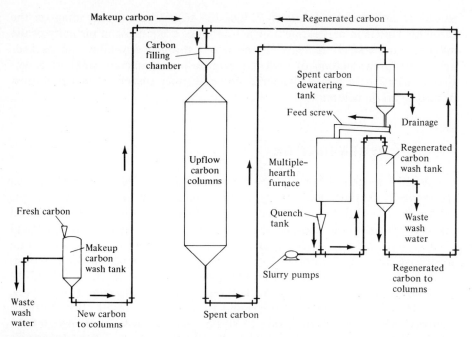

FIGURE 13-9 Carbon regeneration diagram for upflow columns. [Source: *Process Design Manual for Carbon Adsorption*, Environmental Protection Agency, Technology Transfer (October 1973), pp. 3–66.]

Domestic waste contains approximately 3.5 lb of phosphorus/capita/year, of which about 60% is from phosphate builders used in synthetic detergents.

Several years ago, a national ban on phosphate builders in laundry products was considered the best method of phosphorus control. Subsequent studies revealed that only about 15% of domestic wastewater is discharged to inland lakes in danger of eutrophication. Approximately one-half is disposed of in the ocean, either directly or via major rivers, plus another major portion applied to seepage beds of septic-tank systems. Phosphate additives were also found to be the safest, most effective detergent builders. Sodium nitrilotriacetate, which appeared to be a suitable substitute, is now viewed as a hazard to human health. Carbonate additives do not clean as well and can result in calcium carbonate deposits on fabrics and washing-machine surfaces. A few brands are sufficiently caustic to cause damage to eyes and mucus membranes if inhaled or eaten. The negative effects of carbonate laundry products can be offset to some extent by homewater softening and use of chemical additives to reduce carbonate buildup. Several major cities in complying with a stringent phosphorus effluent standard have restricted sale of phosphate detergents to reduce the cost of municipal-wastewater treatment. One case history is the city of Lackawanna, located on Lake Erie, where analyses conducted before and after banning the sale of phosphate detergents in Erie County showed a phosphorus reduction of 50–60%

in the raw municipal wastewater.[9] Although this reduction is not sufficient to meet the established effluent standard of 0.5 mg/l of phosphorus, significant savings in chemical costs for treatment were realized.

The most common forms of phosphorus are organic phosphorus, ortho-phosphates ($H_2PO_4^-$, HPO_4^{2-}, PO_4^{3-}), and polyphosphates. Typical polyphosphates are sodium hexametaphosphate, $Na_3(PO_3)_6$; sodium tripolyphosphate, $Na_5P_3O_{10}$; and tetrasodium pyrophosphate, $Na_4P_2O_7$. All polyphosphates gradually hydrolyze in aqueous solution and revert to the ortho form. Domestic wastewater contains approximately 10 mg/l of total phosphorus, of which about 70% is soluble.

Reactions involving phosphorus are:

$$PO_4 + NH_3 + CO_2 \xrightarrow{\text{sunlight}} \text{green plants} \tag{13-1}$$

$$\text{organic P} \xrightarrow[\text{decomposition}]{\text{bacterial}} PO_4 \tag{13-2}$$

$$\text{polyphosphates} \xrightarrow[\text{in water}]{\text{hydrolysis}} PO_4 \tag{13-3}$$

$$PO_4 + \text{multivalent metal ions} \xrightarrow[\text{coagulant}]{\text{excess}} \text{insoluble precipitates} \tag{13-4}$$

Reaction 13-1 is photosynthesis. Equations 13-2 and 13-3 are decomposition and hydrolysis reactions in which complex phosphates are converted to the stable orthophosphate forms. Equation 13-4 is the precipitation of orthophosphate by chemical coagulation. Substantial excess concentrations of hydrolyzing coagulants (aluminum or iron) or lime are required for effective phosphate precipitation.

13-8

BIOLOGICAL PHOSPHORUS REMOVAL

Taking phosphates out of solution by photosynthesis led to the concept of removing nutrients from wastewaters by growing algae in stabilization ponds and then separating the cells from suspension by physical or chemical means. Shortcomings of culturing algae, however, have prevented practical application of this technique. Biological problems include the imbalance of carbon to nitrogen to phosphorus ratio in wastewater, adequate sunlight intensity, proper pH; and temperature control. The large land area required for adequate liquid detention time and costly mechanical harvesting techniques are major physical limitations.

Primary sedimentation in conventional treatment settles only a small percentage of the phosphorus in wastewater, since the majority is in solution. Secondary biological processing involves removal of soluble phosphate taken up by the microbial floc. The amount synthesized into growth is related to the concentration of phosphates in the wastewater relative to the BOD content.

Treating waste with a high BOD/P ratio eliminates a large percentage of the phosphorus, whereas processing with phosphorus in excess of biological needs results in lower removal efficiency. Domestic wastewater has a surplus of phosphorus relative to the quantities of nitrogen and carbon necessary for synthesis. In general, the amount of P embodied in the biological floc of a conventional activated-sludge process is equal to about 1% of the BOD applied. Anticipated removal in treatment of a typical wastewater with 200 mg/l of BOD is 2 mg/l of P, or a 20% phosphorus reduction.

A study by Menar and Jenkins[10] concluded that conventionally designed primary and activated-sludge secondary treatment can remove a maximum of 20–30% of an influent 10 mg/l of phosphorus by biological means. These authors attributed greater phosphate reduction, reported in some activated-sludge plants, to hard-water wastes where phosphates are complexed by calcium followed by enmeshing the precipitate into the biological floc. Another conclusion was that operational parameters such as organic loading, mixed-liquor suspended solids, and dissolved oxygen have no effect on enhanced removal of phosphate by activated sludge. Others contend that environmental control of the biological process can result in improved phosphorus removal in wastewater aeration. Carberry and Tenney[11] concluded, based on laboratory investigations, that luxury uptake of phosphate by activated sludge occurs by a biological mechanism.

The method of processing and disposal of sludge withdrawn from primary and secondary settling tanks is an important consideration in nutrient removal. The only phosphorus considered extracted is that portion which does not end up in surface waters, namely, the amount in solids disposed of or hauled away from the treatment-plant site. An extended aeration system operating without sludge wasting extracts no phosphorus. Vacuum filtration of raw waste sludge followed by land burial of solids results in maximum phosphorus removal. Conventional sludge stabilization by anaerobic or aerobic digestion returns to the influent of the treatment plant a supernatant liquid containing nutrients.

13-9

BIOLOGICAL–CHEMICAL PHOSPHORUS REMOVAL

Chemical precipitation, using aluminum and iron coagulants or lime, is effective in phosphate removal. Three popular alternatives employed with conventional biological treatment are illustrated in Fig. 13-10.

Chemical Precipitation and Biological Treatment

Coagulation with biological aeration is used both to upgrade existing plants and in new design. For this process (Fig. 13-10a) chemicals are added to the activated-sludge tank or to the effluent of the aeration basin before final settling. Proper mixing at the point of addition and a few minutes of flocculation prior

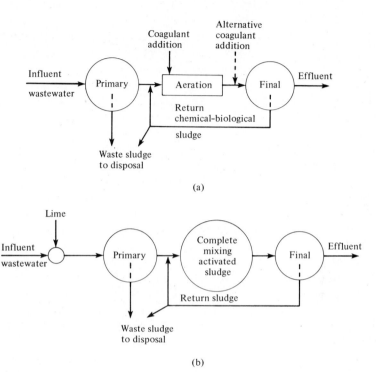

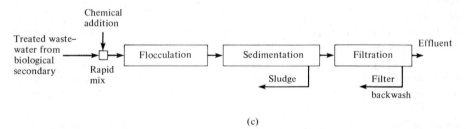

FIGURE 13-10 Phosphorus-removal schemes. (a) Chemical precipitation with activated-sludge treatment; (b) lime precipitation in primary sedimentation followed by secondary complete mixing activated sludge; (c) tertiary treatment by chemical precipitation.

to clarification are essential for maximum effectiveness. The location for best floc formation and subsequent settling is determined experimentally in the field by varying the position of chemical application and monitoring settleability of the solids.

Both alum and ferric chloride are used in combined chemical–biological flocculation, with lime and polyelectrolytes occasionally applied as coagulation aids. The theoretical chemical reaction between alum and phosphate is

$$Al_2(SO_4)_3 \cdot 14.3H_2O + 2PO_4^{3-} = 2AlPO_4\!\downarrow + 3SO_4^{2-} + 14.3H_2O \qquad (13\text{-}5)$$

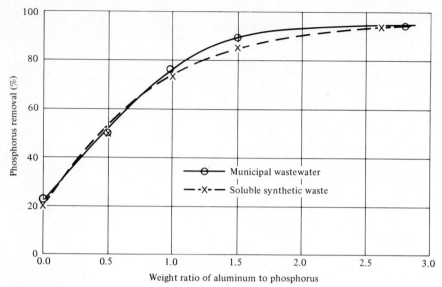

FIGURE 13-11 Biological–chemical phosphorus removal in a bench-scale, activated-sludge unit with alum added to the aeration chamber. The substrates applied were settled municipal wastewater and a synthetic medium of glucose and glutamic acid, with phosphorus contents of approximately 10 mg/l of P. [Source: D. T. Anderson and M. J. Hammer, "Effects of Alum Addition on Activated Sludge Biota," *Water and Sewage Works* 120, no. 1 (January 1973).]

The molar ratio of aluminum to phosphorus is 1 : 1, equivalent to a weight ratio of 0.87 : 1.00. Since alum contains 9.0 % Al, 9.7 lb of coagulant is theoretically required to precipitate 1.0 lb of P. The actual coagulation reaction in wastewater is only partially understood and more complex than Eq. 13-5 because of secondary reactions with colloidal solids and alkalinity.

Alum demand is also a function of the degree of phosphorus removal as diagrammed in Fig. 13-11.[13] The data are from operation of a bench-scale activated-sludge unit consisting of an aeration chamber and separate settling tank with a sludge return line. The wastewaters applied were settled domestic and a glucose–glutamic acid medium, both with a BOD of 150–190 mg/l and average phosphorus level of 10 mg/l. Without alum added to the aeration basin, phosphorus removal averaged 23 % (Fig. 13-11). Chemical–biological processing improved phosphate extraction with increased Al to P dosages, indicating a drop in coagulation efficiency as the amount of phosphorus remaining decreases. The aluminum to phosphorus dosages for 90 % P removal from the municipal and synthetic wastes were 1.5 to 2.0 Al to P, respectively, equivalent to 170 and 220 mg/l of alum. In full-scale activated-sludge and trickling-filter plants, alum applications vary from 50 to 200 mg/l for 80–95 % phosphorus removal. Diffused aeration provides more efficient flocculation and, for the same degree of phosphorus removal, uses less alum than trickling-filter installations.

A large addition of aluminum coagulant has a marked influence on the biota of activated sludge. Anderson and Hammer[13] reported that free-swimming and stalked protozoans are adversely affected to the extent that higher life forms are virtually absent with alum dosages in excess of 150 mg/l. Under these conditions it appears that chemical flocculation replaces the role of protozoans in clarifying settled effluent. BOD and suspended-solids removals are enhanced by coagulant addition to aeration basins and trickling filters, and generally result in effluent concentrations in the range of 10–20 mg/l.

Ferric chloride can also be applied with biological aeration to complex the phosphate ion. The hypothetical chemical reaction is

$$FeCl_3 + PO_4^{3-} = FePO_4 \downarrow + 3Cl^- \qquad (13\text{-}6)$$

The molar and weight ratios of Fe to P are $1:1$ and $1.8:1$, respectively. Theoretically, 5.2 lb of $FeCl_3$ is needed to precipitate 1 lb of P since ferric chloride is 34% iron. The actual dosage is usually greater than this equation predicts for 85–95% phosphate removal, and lime is commonly applied to maintain optimum pH and aid coagulation.

Characteristics of waste chemical–biological sludge are influenced by co-agulant dosage, nature of the wastewater solids, system design, and operating conditions. The quantity of chemical precipitate produced can be estimated using the theoretical equations for a given dosage, and the organic residue cal-culated as described in Section 12-2. Addition of metal salts improves settleability of microbial floc resulting in a denser waste slurry. Still, it is difficult to predict a specific value for the sludge-volume index because of many other influencing factors. Chemical residues do not hamper sludge thickening and stabilization by any conventional methods; in fact, mechanical dewatering is improved. Studies indicate that coagulation precipitates do not release phosphate back into solution during biological digestion.[12,14] Of course, the greater inert content of inorganic–organic sludge will influence ultimate disposal by incineration.

Coagulation of Raw Wastewater

Aluminum and iron coagulants mixed with raw wastewater precipitate phosphates in primary clarification. This point of chemical addition, however, is not as popular in upgrading existing treatment plants as the previously des-cribed secondary chemical–biological processing. The advantage of first-stage settling is increased suspended-solids removal, thus reducing the organic load to a biological secondary and permitting use of impure chemicals derived from industrial wastes. Ferrous sulfate produced in pickling steel with sulfuric acid and ferrous chloride from hydrochloric acid brining are the two most common waste liquors from metal manufacturing. Their iron content varies from 5 to 10% with free acid in the range 0.5–15%; the latter necessitates addition of lime or sodium hydroxide for satisfactory coagulation. Amounts of coagulant and alkali required vary with the pickle liquor and wastewater characteristics and must be determined experimentally—typical dosages are 40 mg/l of iron and 70 mg/l of lime for precipitation of approximately 80% P and 60% of the BOD in primary sedimentation.

Lime Precipitation of Raw Wastewater

Lime applied prior to primary clarification (Fig. 13-10b) precipitates phosphates and hardness cations along with organic matter. Reaction with alkalinity (Eq. 13-7) consumes most of the lime and produces calcium carbonate residue that aids in settling suspended solids. Calcium ion also combines with ortho-phosphate in an alkaline solution to form gelatinous calcium hydroxyapatite (Eq. 13-8). Treating domestic wastewater requires a dosage of 100–200 mg/l as calcium hydroxide to remove 80% of the phosphate. The actual amount applied depends primarily on phosphorus concentration and hardness of the wastewater.

$$Ca(HCO_3)_2 + Ca(OH)_2 = 2CaCO_3 \downarrow + 2H_2O \tag{13-7}$$

$$5Ca^{2+} + 4OH^- + 3HPO_4^{2-} = Ca_5(OH)(PO_4)_3 \downarrow + 3H_2O \tag{13-8}$$

The system sequence of lime precipitation followed by activated-sludge treatment, rather than reversing their order, uses less lime when an effluent of low phosphorus concentration is desired. The principal reason is that a biological system can readily extract low concentrations of phosphorus which would require excessive lime addition to precipitate. Secondary treatment by complete-mixing activated sludge is not adversely affected by chemical pretreatment. Conclusions from a laboratory study by Schmid and McKinney[15] included: addition of 150 mg/l of calcium hydroxide resulted generally in a pH of 9.5; removals of total phosphorus 80%, BOD 60%, and suspended solids 90%; operation of a complete-mixing activated-sludge system following lime treatment is not hindered, with proper control exercised; microbial production of carbon dioxide in the activated-sludge unit is sufficient to maintain a pH near neutral in the aeration compartment; and, this chemical–biological system of phosphate extraction can remove 90–95% of the total phosphorus from a domestic wastewater containing 40–50 mg/l as total phosphate (13–16 mg/l of P), with lime dosages generally less than 150 mg/l as calcium hydroxide.

Use of excess lime in chemical treatment has two potential problems: scale formation on tanks, pipes, and other equipment, and disposal of the large quantity of lime sludge produced. Only operation of a full-scale installation will reveal the significance of these possible troubles. The quantity of sludge produced is about 1.5–2 times that obtained by conventional treatment.[15]

Tertiary Treatment by Chemical Precipitation

This process, delineated in Fig. 13-10c, is biological secondary treatment followed by chemical treatment with a flow diagram similar to that used in processing surface-water supplies. The mixing and sedimentation system can consist of either separate rapid mix, flocculation, and sedimentation units in series, or a flocculator–clarifier with these three operations in a single-compartmented tank. Filters are usually multimedia beds operated by pressure or gravity flow. Possible chemical additives are lime, alum, ferric chloride, and ferric sulfate, with polyelectrolytes as flocculation aids. A major design consideration of tertiary treatment is processing and disposal of settled sludge and filter back-wash water.

The water reclamation plant at Lake Tahoe is a tertiary processing installa-tion constructed to prevent fertilization of this deep, oligotrophic, alpine lake. In 1961 consulting engineers were retained to investigate alternatives and recom-mend a plan for permanent disposal of the South Tahoe wastewater-treatment-plant effluent. Effluent disposal at this time was accomplished by spray irrigation on land. After a detailed investigation, engineers recommended the following: continue with land disposal for the next few years; develop an advanced method of waste treatment that would permit removal of the effluent from the drainage basin, or allow disposal within the basin; and, study available routes for piping the effluent from the basin.

An advanced waste-treatment plant was put in service in 1965 and has since been modified and expanded. The present plant has a design capacity of 7.5 mgd to serve a city of 100,000 population. An effluent export system has also been constructed with the reclaimed water being pumped to a reservoir located in a mountain pass 26 mi from the treatment plant. A schematic flow and process diagram for the South Tahoe Public Utility District water reclamation plant is given in Fig. 13-12.[3,16] Primary and secondary treatment processes comprise a conventional activated-sludge system. The primary and secondary sludges are incinerated in a multiple hearth furnace. Prior to incineration, the waste activated sludge is dewatered by a concurrent-flow solid-bowl centrifuge.

The phosphorus and nitrogen removal processes consist of lime precipita-tion and air-stripping of the secondary effluent. A lime dosage of about 400 mg/l as CaO is necessary for maximum phosphate precipitation (Fig. 13-13). The resulting high pH of about 11 requires recarbonation to a pH of 7.5 to prevent deposition of calcium carbonate on the piping and filter beds and is accomplished by a two-stage system using CO_2 obtained from the furnace stack gas. Prior to lowering the pH, the lime-treated effluent is pumped through a cooling tower, where ammonia nitrogen is stripped from the water and released to the atmo-sphere as a gas.

The lime sludge is thickened in a gravity-flow thickener and dewatered in a centrifuge. Lime is then reclaimed by recalcining in a multiple-hearth furnace fired by natural gas. Reuse of lime in the treatment does not result in significant economy in chemical costs but produces substantial savings by reducing the volume of sludge requiring final disposal.

A detailed diagram of the tertiary treatment process for removal of phos-phorus, nitrogen and organics is shown in Fig. 13-14. The most recent method of operation involves classification of the precipitated phosphorus by the lime centrifuge. The centrate, which contains the phosphorus, is returned to the head of the plant for sedimentation in the primary tank. Primary sludge is incinerated, therefore, this phosphorus is permanently removed from the system and does not return as a recycling load to the lime recalcining system. The remaining reclaimed lime plus makeup lime is applied to the tertiary chemical treatment

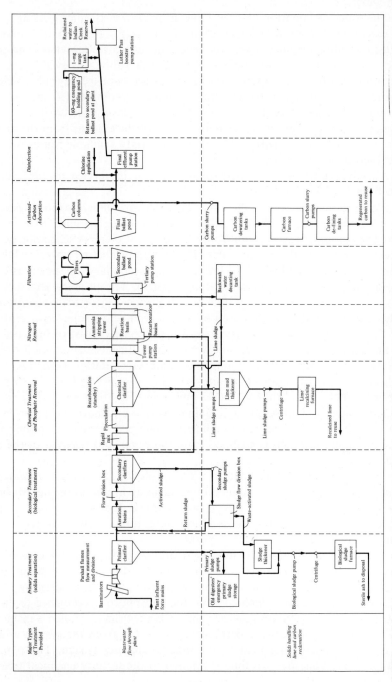

FIGURE 13-12 Schematic flow and process diagram, Lake Tahoe Water Reclamation Plant. [Source: R. L. Culp, "Water Reclamation at South Tahoe, "*Water and Wastes Eng.* 6, no. 4 (1969): 36; copyrighted and published by Reuben H. Donnelly Corp.; courtesy South Tahoe Public Utilities District.]

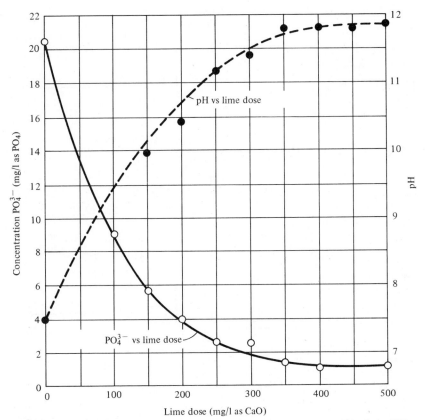

FIGURE 13-13 Phosphate concentration, pH, and lime dose. [Source : R. L. Culp, "Waste-Water Reclamation at South Tahoe Public Utilities District," *J. Am. Water Works Assoc.* 60, no. 1 (1968) : 91. Copyright 1968 by the American Water Works Association, Inc.]

process. Lime sludge from the tertiary chemical clarifier after classification in the lime centrifuge goes to the recalcining process for recovery and reuse.

The separation beds are mixed-media pressure filters with a total depth of 33 in. composed of 21 in. of anthracite coal (sp gr 1.5), 9 in. of graphite (sp gr 2.4) and 3 in. of fine garnet (sp gr 4.2) underlain by 3 in. of coarse garnet supported by a conventional silica gravel bed.

Activated carbon columns take out refractory-soluble organics not removed by lime coagulation. These refractory substances include nonbiodegradable organics, color, COD, taste, and odor-producing compounds and residual BOD. The exhausted granulated carbon is reactivated and reused. Thermal reactivation consists of heating to about 1700°F in a steam-air atmosphere to burn off the absorbed organics and restore the activated carbon.

Table 13-4 is a summary of the water quality expected from the new treatment system using lime precipitation. (The system used alum coagulation prior to expansion in 1968.) Values shown in Table 13-4 are based on operational data from 1969 and prior pilot-plant studies.

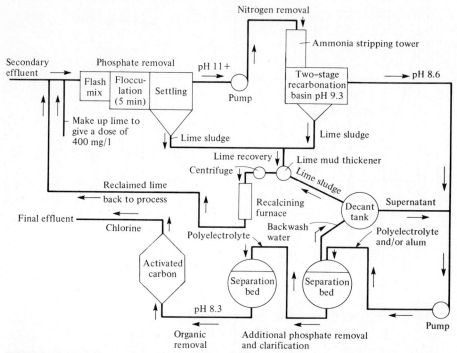

FIGURE 13-14 Tertiary treatment process for removal of phosphorus, nitrogen, and organics, Lake Tahoe. [Source: R. L. Culp, "Waste-Water Reclamation at South Tahoe Public Utilities District," *J. Am. Water Works Assoc.* 60, no. 1 (1968): 93. Copyright by the American Water Works Association, Inc.]

TABLE 13-4

AVERAGE-WATER-QUALITY DATA, SOUTH TAHOE PUD WATER RECLAMATION PLANT

Parameter	Raw Wastewater	Secondary Effluent	Separation Bed Effluent	Carbon Column Effluent
BOD (mg/l)	144	18	3	<1
COD (mg/l)	156	53	20	10
Suspended solids (mg/l)	156			<1
MBAS[a] (mg/l)	5	2	0.6	0.1
Color (units)			9	5
Turbidity (Jtu)				0.5
Phosphate (mg/l of PO₄)	20	17	0.3	0.05
Nitrogen				
NH₃ (mg/l of N)		15		2.0
NO₂ and NO₃ (mg/l of N)				0.2
Coliforms (MPN)				<2.2
Hardness (mg/l as CaCO₃)	69		144	124

[a] Methylene blue active surfactant [*Standard Methods for the Examination of Water and Wastewater*, 13th ed. (1971), p. 339].

Although the advanced wastewater-treatment system at South Lake Tahoe reclaims the wastewater to nearly drinking quality, a small quantity of nutrients and trace elements still exist in the effluent. Therefore, the decision was to transport the effluent entirely out of the natural drainage area of Lake Tahoe. The export system for reclaimed water is shown in Fig. 13-15. A pipeline conveys the water over Luther Pass (elevation: 7735 ft) to a reservoir 26 mi away, where it is used for irrigation and recreation. A fishery has been established in the receiving reservoir, which contains 1 billion gal of reclaimed water.

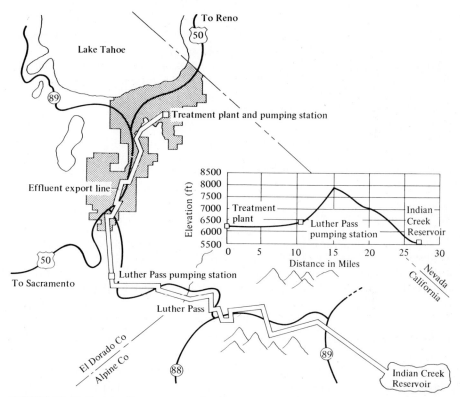

FIGURE 13-15 Reclaimed water export pipeline, Lake Tahoe Water Reclamation Plant. (Courtesy South Tahoe Public Utilities District.)

●**EXAMPLE 13-3**

The effluent standard for a conventional, primary plus activated-sludge secondary, treatment plant is 1.0 mg/l of phosphorus. Biological–chemical aeration applying alum is being considered for phosphorus removal following primary settling which reduces the concentration to 10 mg/l of P. The waste-activated sludge, containing 400 lb of solids, without chemical addition is 4800 gal/mg of wastewater treated; this is equivalent to a solids content of 1.0% and SVI equal to 100 ml/g. A laboratory study using chemical–biological treatment resulted in the alum dosage–phosphorus removal curve for municipal wastewater shown in Fig. 13-11. Alum coagulation improved the activated-sludge settleability, with a resulting SVI of 60 ml/g. Calculate the dosage of alum required based on the laboratory data, and the quantity of waste chemical–biological sludge produced per million gallons of wastewater treated.

○**Solution**

Required phosphorus removal is 9.0 mg/l, or 90%. From Fig. 13-11, the aluminum/phosphorus weight ratio for 90% P reduction is 1.5.

$$\text{alum dosage} = 10 \frac{\text{mg of P}}{1} \times 1.5 \frac{\text{mg of Al}}{\text{mg of P}} \times \frac{600 \text{ mg of alum}}{54 \text{ mg of Al}} = \frac{15}{0.09}$$

$$= 167 \text{ mg/l}.$$

Based on Eq. 13-5, 9 mg of phosphorus reacts with 87 mg of alum to precipitate 36 mg. The remaining 80 mg of alum combines with alkalinity to form 21 mg of $Al(OH)_3$ (Eq. 10-20). Therefore, assuming complete removal of coagulation solids, the chemical residue per million gallons of wastewater processed = $(36 \text{ mg/l} + 21 \text{ mg/l}) \times 1.0 \times 8.34 = 470$ lb of $AlPO_4$ and $Al(OH)_3$.

Waste chemical–biological sludge concentration

$$= \frac{1,000,000}{\text{SVI}} = \frac{1,000,000}{60} = 17,000 \text{ mg/l} = 1.7\%$$

Volume of waste sludge, including both biological floc and alum residue, per mg of wastewater treated

$$= \frac{400 + 470}{(1.7/100)8.34} = 6100 \text{ gal}$$

Thus, the volumetric increase in waste sludge is anticipated to be 27% (4800 gal/mg to 6100 gal/mg) even though the alum coagulant adds about 120% to the quantity of dry solids (400 lb/mg to 870 lb/mg). This is due to improved settleability of the chemical–biological floc observed by the SVI reduction from 100 ml/g to 60 ml/g.

13-10

PHYSICAL–CHEMICAL PROCESSING

This treatment is chemical coagulation, filtration, and activated-carbon adsorption of raw wastewater, or primary effluent, to eliminate the need for biological processing. Phosphates are precipitated with the flocculated suspended solids in primary sedimentation. The second-stage granular-carbon adsorption extracts the remaining soluble organics, and filtration either by the carbon columns, or separate multimedia beds, to clarify the plant effluent. Physical–chemical treatment appears to be most applicable where the presence of toxic wastes, or space limitations for plant construction, prevent use of biological processes. Also, in some situations, AWT by physical–chemical methods can provide a higher-quality effluent than biological techniques.

Chemicals applied, either singly or in combination, for raw wastewater coagulation are lime, iron salts, aluminum, and polymers. Lime, often preferred because it is less expensive, does not add soluble ions to the water, provides an alkaline pH for ammonia stripping, and produces a sludge that is easier to handle. Even with the greater quantity of precipitate than for either iron or alum residues, disposal cost is less, owing to the superior thickening and dewatering characteristics of lime slurries. Recalcination and reuse of the lime, however, is usually not practical because of the large amount of inert matter present in the sludge from raw-wastewater solids.

One of the largest physical–chemical plants treats combined industrial and sanitary wastes at Niagra Falls, New York. Manufacturing discharges containing toxic and refractory compounds cannot be biologically processed. Pilot-plant studies[18] demonstrated, however, that lime coagulation followed by carbon adsorption could provide the required effluent quality of 35 mg/l of SS, 112 mg/l of COD, 1.0 mg/l of P, and 0.23 mg/l of phenols.

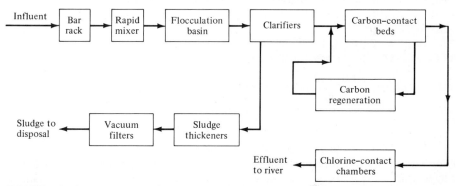

FIGURE 13-16 Flow diagram for physical–chemical treatment, Niagara Falls, N.Y. [Source: *Physical–Chemical Wastewater Treatment Plant Design*, Environmental Protection Agency, Technology Transfer (August 1973), p. 36.]

After screening, raw wastewater is chemically precipitated by rapid mixing, flocculation, and sedimentation (Fig. 13-16). Lime, a metal salt (ferric chloride, alum, or chlorinated ferrous sulfate), a coagulant aid, return sludge from the primary sedimentation basins, acid (if depressing the pH is necessary), and spent backwash water from the activated-carbon beds are blended in the rapid-mix basins. The choice between lime and a metal coagulant depends on pH of the raw waste, cost and availability of chemicals, and ease of sludge dewatering. Primary sludge is gravity-thickened and dewatered on vacuum filters, with a selection of ferric chloride, polymers, and lime for conditioning chemicals, and disposed of by land burial since heat drying (incineration) and lime recovery are more expensive. The granular activated-carbon beds are downflow gravity units providing both adsorption and filtration, with the backwash water being returned to the rapid mixer and spent carbon regenerated in a six-hearth furnace. Prior to discharging in the Niagara River, the effluent passes through a chlorine contact chamber.

●**EXAMPLE 13-4**

Estimate the quantity of sludge generated per cubic meter of wastewater processed by physical–chemical treatment employing lime precipitation at a dosage of 400 mg/l as $Ca(OH)_2$, followed by filtration and carbon adsorption. Raw wastewater and effluent characteristics are:

Component	Influent	Effluent
Suspended solids (mg/l)	250	20
Calcium (mg/l of Ca^{2+})	70	40
Magnesium (mg/l of Mg^{2+})	Negligible	
Phosphorus (mg/l of P)	11	1

Assume that the settled sludge has a solids content of 10% and is concentrated to 30% by mechanical dewatering prior to disposal by land burial.

○**Solution**

Based on Eq. 13-8, 1.0 mg of Ca^{2+} reacts with 0.46 mg of P to precipitate 2.5 mg of $Ca_5(OH)(PO_4)_3$. Hence, for the removal of 10 mg/l of P,

$$\text{calcium reacted} = \frac{1.0 \times 10}{0.46} = 22 \text{ mg/l of } Ca^{2+}$$

$$\text{precipitate} = 2.5 \times 22 \text{ mg/l} \times 1.0 \frac{g/m^3}{mg/l} = 55 \text{ g/m}^3$$

The calcium reacting to form calcium carbonate equals the amounts in the applied lime plus raw wastewater minus that complexed by phosphorus and discharged in the effluent.

$$\text{calcium in applied lime} = 400 \frac{40.1}{74.1} = 216 \text{ mg/l of } Ca^{2+}$$

$$\text{reactive calcium} = 216 + 70 - 22 - 40 = 224 \text{ mg/l of } Ca^{2+}$$

From Eq. 13-7, 1.0 mg of Ca^{2+} precipitates as 2.5 mg of $CaCO_3$. Therefore,

$$CaCO_3 \text{ precipitate} = 224 \times 2.5 = 560 \text{ g/m}^3$$

Waste residue calculated from suspended solids reduction is

$$(250 - 20) \text{ mg/l} \times 1.0 \frac{g/m^3}{mg/l} = 230 \text{ g/m}^3$$

$$\text{total solids} = 55 + 560 + 230 = 845 \text{ g/m}^3$$

$$\text{weight of wet sludge} = \frac{845}{1000 \times 0.10} = 8.45 \text{ kg/m}^3$$

$$\text{volume of wet sludge} = \frac{8.45 \text{ kg}}{1000 \text{ kg/m}^3} = 0.00845 \text{ m}^3/\text{m}^3$$

$$\text{weight of dewatered cake} = \frac{845}{1000 \times 0.30} = 2.82 \text{ kg/m}^3$$

Nitrogen Removal

Most nitrogen found in surface waters is derived from land drainage (3 to 24 lb of N/acre/year) and dilution of wastewater effluents. Feces, urine, and food-processing discharges are the primary sources of nitrogen in domestic waste with a per capita contribution in the range of 8–12 lb of N/yr. About 40% is in the form of ammonia and 60% bound in organic matter. Conventional primary and secondary processing extracts approximately 40% of the total nitrogen, leaving most of the remainder as ammonia in the effluent.

The nitrogen forms of interest in AWT are organic, inorganic, and gaseous nitrogen. Bacterial decomposition releases ammonia by deamination of nitrogenous organic compounds (Eq. 13-9), and continued aerobic oxidation results in nitrification (13-10). Equation 13-11 is biochemical denitrification that occurs with heterotrophic metabolism in an anaerobic environment. These three reactions in sequence define the biological nitrification–denitrification process. Water-soluble inorganic nitrogens (NH_3, NO_2^-, NO_3^-) serve as plant nutrients in photosynthesis (Eq. 13-12). Finally, ammonia can be air-stripped from solution at high pH.

$$\text{organic N} \xrightarrow[\text{decomposition}]{\text{bacterial}} NH_3 \tag{13-9}$$

$$NH_3 + O_2 \xrightarrow[\text{bacteria}]{\text{nitrifying}} NO_3^- \tag{13-10}$$

$$NO_3^- \xrightarrow[\text{denitrification}]{\text{bacterial}} N_2\uparrow \tag{13-11}$$

$$\text{inorganic N} + CO_2 \xrightarrow{\text{sunlight}} \text{green plants} \tag{13-12}$$

$$NH_4OH \xrightarrow[\text{at basic pH}]{\text{air stripping}} NH_3\uparrow \tag{13-13}$$

13-11

BIOLOGICAL NITRIFICATION

Nitrification does not remove ammonia but converts it to the nitrate form, thereby eliminating problems of toxicity to fish and reducing the nitrogen oxygen demand (NOD) in streams. Ammonia oxidation to nitrate is a diphasic process performed by autotrophic bacteria with nitrite as an intermediate product (Eq. 13-10). These aerobic reactions yield energy for metabolic functions such as synthesis of carbon dioxide into new cell growth. Conversion of ammonia to nitrite is the rate-limiting step that controls the overall reaction; therefore, nitrite concentrations normally do not build up to significant levels. The rate of nitrification in wastewater, being essentially linear, is a function of time and independent of ammonia–nitrogen concentration (zero-order kinetics).

$$NH_4^+ + 1.5O_2 \xrightarrow{\text{Nitrosomonas}} NO_2^- + 2H^+ + H_2O + \text{energy}$$

$$NO_2^- + 0.5O_2 \xrightarrow{\text{Nitrobacter}} NO_3^- + \text{energy}$$

(13-14)

Temperature, pH, and dissolved oxygen concentration are important parameters in nitrification kinetics studied by Wild, Sawyer, and McMahon.[19] The relationship of the nitrification rate at all temperatures studied to the rate at 30°C is shown in Fig. 13-17. Because 30°C is a high wastewater temperature

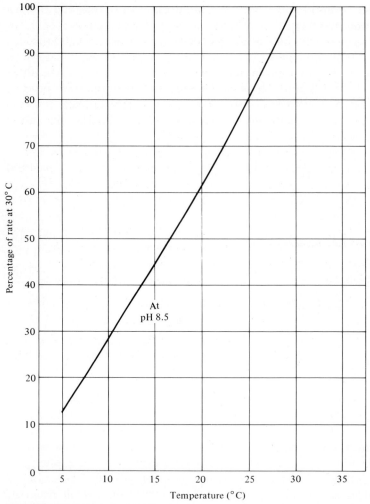

FIGURE 13-17 Rate of nitrification at all temperatures compared to the rate at 30°C. [Source: H. E. Wild, Jr., C. N. Sawyer, and T. C. McMahon, "Factors Affecting Nitrification Kinetics," *J. Water Poll. Control Fed.* 43, no. 9 (1971) : 1852.]

TABLE 13-5

RELATIVE RATES OF NITRIFICATION AT VARIOUS
TEMPERATURES BASED ON FIG. 13-17

30°C	25°C	20°C	15°C	10°C	5°C
100	80	60	48	27	12
	100	75	60	34	16
		100	80	45	21

for all but the most southern states in the United States, a summary of relative rates, from this diagram, in terms of other maximum temperatures is given in Table 13-5. These data indicate that the process decreases by one-half for every 10–12°C temperature drop above 10°C, and then decreases more rapidly in cold wastewater, such that lowering the temperature from 10°C to 5°C halves the rate of ammonia oxidation. Based on temperature alone, a winter aeration period would have to be serveral times longer than in the summer; however, this seasonal effect can be overcome to a considerable degree by increasing mixed-liquor suspended solids (MLSS) and adjusting pH to a more favorable level. Optimum pH for nitrification is 8.2–8.6, with 90% of the maximum occurring at 7.8 and 8.9, and less than 50% of optimum below 7.0 and above 9.8 (Fig. 13-18). The laboratory studies showed further that there was no detectable inhibition of nitrification at dissolved-oxygen levels exceeding 1.0 mg/l, or at ammonia–nitrogen concentrations up to 60 mg/l.

Sludge age and temperature are interrelated factors in establishing and maintaining healthy nitrifier populations essential to efficient ammonia oxidation. In continuous-flow aeration systems, a long sludge age (retention time) is required to prevent excessive loss of viable bacteria (i.e., the growth rate must

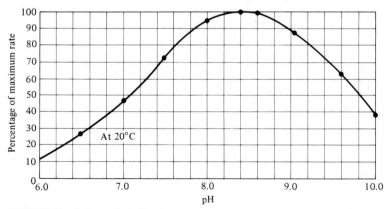

FIGURE 13-18 Rate of nitrification versus pH at constant temperature. [Source: H. E. Wild, Jr., C. N. Sawyer, and T. C. McMahon, " Factors Affecting Nitrification Kinetics," *J. Water Poll. Control Fed.* 43, no. 9 (1971) : 1852.]

be rapid enough to replace microbes lost through sludge wasting and washout in the plant effluent). The supply of organic matter controls growth of heterotrophic organisms, while the quantity of ammonia applied governs synthesis of nitrifiers. Increased sludge wasting, as a result of organic loading, reduces the sludge retention time and removes both heterotrophs and nitrifiers from the system. On a diet of domestic waste, which has a BOD/total nitrogen ratio of approximately 200 mg/l to 35 mg/l, growth rates of nitrifying bacteria are substantially lower than those of decomposers. Therefore, in activated-sludge processes under normal operating conditions, nitrification is limited because of loss of the autotrophic populations, and current design for biological nitrification calls for two-step treatment. The first stage reduces BOD without oxidation of the ammonia nitrogen to produce an effluent having a lower BOD/ammonia ratio—about 40 mg/l to 25 mg/l. Applying this flow to a second-stage nitrification unit provides an adequate growth potential for nitrifiers relative to heterotrophs, since the system can be operated at an increased sludge age.

The relative reproduction rates of heterotrophic and nitrifying bacteria are also influenced to a measurable extent by temperature. In southern climates, nitrification may be possible in a single-stage extended aeration unit treating domestic waste if pH and sludge wasting are carefully controlled to compensate for the relatively high BOD/ammonia-N feed. However, winter wastewater temperature in northern states is often 10–15C, requiring a two-stage system with the best combination of aeration-tank capacity, MLSS concentration, and pH control. Operation is possible at a less favorable pH level and lower mixed-liquor solids at a warm temperature provided the first stage is properly controlled. If nitrification is allowed to occur in carbonaceous aeration, the reduced ammonia supply to the secondary leads to starving the nitrifier populations. Sludge age, rate of recirculation, and air supply can be adjusted to minimize ammonia oxidation in the first stage; nevertheless, provisions for chlorinating the effluent of the aeration tank prior to clarification is recommended in design. Addition of 2–8 mg/l of chlorine is effective in inhibiting nitrifying bacteria and can also help control sludge bulking caused by denitrification in the carbonaceous-phase clarifier.

Nitrification by Suspended-Growth Systems

This discussion is based on tentative design criteria for ammonia oxidation based on available experience.[20] The optimum aeration tank is either a long, narrow, spiral-flow basin with diffused aeration, or a shorter tank divided into a series of at least three compartments for diffused or mechanical aeration equipment, with intervening ports (Fig. 13-19). This tank configuration simulates plug flow compatible with the zero-order kinetics of ammonia oxidation.

Biological nitrification destroys alkalinity, which can result in a drop of pH when processing wastewaters of moderate hardness, or where alum precipitation has been used for phosphate removal in the preceding activated-sludge

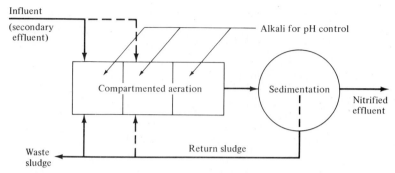

FIGURE 13-19 Flow diagram for nitrification by diffused or mechanical aeration of wastewater following conventional biological treatment.

phase. Theoretically, 7.2 lb of alkalinity is destroyed per pound of ammonia nitrogen oxidized to nitrate, as follows:

$$2NH_4HCO_3 + 4O_2 + Ca(HCO_3)_2 = Ca(NO_3)_2 + 4CO_2 + 6H_2O \qquad (13\text{-}15)$$

Whether pH should be controlled by chemical addition depends on the rate of nitrification desired, as limited by other environmental conditions. For example, when operating at a low temperature, lime can be applied to maintain oxidation efficiency for the aeration-tank capacity available. New plant design should provide for installation of chemical feeders and instrumentation for monitoring pH in the aeration basin.

The recommended design mixed-liquor concentration for a nitrification process receiving normal secondary effluent is in the range 1500–2000 mg/l of volatile suspended solids. Currently, insufficient data are available to accurately select a design sludge age. Preliminary results have shown that solids retention times must be greater than those practiced in carbonaceous activated-sludge processes, and quantitative values available indicate that a sludge age up to 20 days is needed.

Ammonia content in the flow entering second-stage aeration equals the amount of nitrogen in the raw waste, minus removals in the primary and secondary processes. About 10 mg/l of the 35 mg/l of total nitrogen in average domestic wastewater is taken out by sedimentation, and an additional 5 mg/l synthesized in biological treatment, leaving about 20 mg/l of N in solution. Most is in the form of ammonia, unless excessive aeration in the secondary produces nitrates, and a small portion is organic nitrogen bound in suspended solids carried out in the effluent. Methods used in stabilizing and dewatering waste sludge withdrawn from settling tanks effect nutrient extraction since only that nitrogen in the solids hauled away from the treatment-plant site is actually eliminated. Anaerobic and aerobic digestion return to the plant influent supernatant liquid containing a substantial concentration of soluble nitrogen released in bacterial decomposition. Vacuum filtration of raw waste sludge followed by

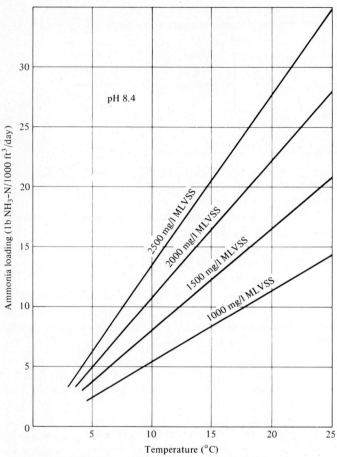

FIGURE 13-20 Permissible nitrification-tank loadings at optimum pH of 8.4. [Source: *Nitrification and Denitrification Facilities*, Environmental Protection Agency, Technology Transfer (August 1973), p. 22.]

land burial, or incineration, of the residue results in maximum nitrogen removal.

Loading on nitrification basins is expressed in units of lb of ammonia-N/ 1000 ft^3/day of aeration-tank volume. Figure 13-20 shows recommended loadings for various temperatures and volatile mixed-liquor concentrations (MLVSS) at pH 8.4, based on studies at Marlboro, Massachusetts. Corrections for permissible loadings at pH values other than 8.4 can be taken from Fig. 13-18. The value selected for design loading should include a factor to allow for reasonable peaking of the influent ammonia content. A commonly adopted peak loading is 1.5 times the average daily nitrogen load under low-temperature conditions.

Stoichiometrically, nitrification of 1.0 lb of ammonia-N in the form of ammonium bicarbonate requires 4.6 lb of oxygen (Eq. 13-15); however, additional oxygen allowance must be made for carbonaceous BOD carried over to the nitrification stage. A dissolved oxygen concentration of 3.0 mg/l in the mixed liquor is suggested under average loading conditions with a lower concentration permitted during peak loads, but not below 1.0 mg/l

Suggested design criteria for final clarifiers following nitrification are an overflow rate of 400–500 gpd/ft^2 based on average daily discharge, with a maximum permissible value of 1000 gpd/ft^2 at peak hourly flow, and a side-water depth of at least 10 ft. Because of the relatively slow settling velocities of nitrifying sludges, more than two clarifiers are desirable to ensure satisfactory operation when one tank is out of service for maintenance. Hydraulic-type collector arms are recommended for rapid sludge return since denitrification can occur in settled sludge, creating problems of floating solids. When present, the float should be collected by skimmers and returned to the aeration basins. A desirable capacity for sludge-recirculation pumps is 100% of the raw-wastewater influent, although normal operation will probably be at a rate of only 50% return. Accumulation of biological floc is limited by the low organic loading and slow growth of nitrifying bacteria. Consequently, the volume of excess sludge produced is small, with a reasonable estimate being less than 1% of the quantity of wastewater processed.

Nitrification by Fixed-Growth Systems

Although suspended-growth aeration is used more extensively, fixed-growth systems have certain advantages, such as ease of operation and greater process stability. Biological discs, described in Section 11-20, are effective for nitrification in multistage operation at hydraulic loadings reduced to compensate for low-temperature conditions.[21] Nitrifying bacteria adhering to the media must be in sufficient numbers to provide ammonia oxidation, since the rotating-disc process does not include sludge recirculation. Therefore, to prevent inhibition by heterotrophic growths covering the disc surfaces, BOD of the applied wastewater should be relatively low, preferably, not exceeding 20–30 mg/l.

Biological towers with plastic packing (Sec. 11-18) or redwood medium (Sec. 11-19) can also be employed for nitrification. Underflow from clarification of the tower effluent is recirculated for uniform hydraulic loading on the packing and return of settled floc to increase the mass of nitrifying bacteria in contact with the wastewater.[22]

●EXAMPLE 13-5

Calculate the aeration basin volume for suspended growth nitrification following conventional secondary treatment. The wastewater characteristics are:

$$\text{average daily design flow} = 10 \text{ mgd}$$
$$\text{ammonia nitrogen} = 20 \text{ mg/l}$$
$$\text{BOD} = 40 \text{ mg/l}$$
$$\text{minimum operating temperature} = 10°C$$
$$\text{operating pH} = 7.8$$
$$\text{design volatile MLSS} = 1500 \text{ mg/l}$$

○*Solution*

average ammonia load $= 10 \times 20 \times 8.34 = 1670$ lb/day of N
maximum ammonia load $= 1.5 \times 1670 = 2500$ lb/day of N

Permissible nitrification-tank loading from Fig. 13-20 for a temperature of 10°C and 1500 mg/l of MLVSS is 8.1 lb of NH_3–N/1000 ft³/day. Correcting this to a pH of 7.8 using Fig. 13-18, the allowable loading $= 8.1 \times 0.88 = 7.1$ lb/1000 ft³/day.

$$\text{aeration basin volume} = 2500 \, \frac{1000}{7.1} = 350,000 \text{ ft}^3$$

$$\text{resulting aeration period} = \frac{350,000 \times 7.48 \times 24}{10,000,000} = 6.3 \text{ hours}$$

BOD load on aeration basin $= 10 \times 40 \times 8.34 = 2300$ lb/day

Oxygen uptake using 4.6 lb of O_2/lb of NH_3–N and 1.0 lb of O_2/lb of BOD
$= 2500 \times 4.6 + 2300 \times 1.0 = 13,800$ lb/day

13-12

BIOLOGICAL DENITRIFICATION

Nitrite and nitrate are bacterially reduced to gaseous nitrogen by a variety of facultative heterotrophs in an anaerobic environment. An organic carbon source, such as acetic acid, acetone, ethanol, methanol, or sugar, is needed to act as a hydrogen donor (oxygen acceptor) and to supply carbon for biological synthesis. Certain autotrophic bacteria are also capable of denitrification by oxidizing an inorganic compound for energy and using carbon dioxide for synthesis. While denitrification is considered an anaerobic process because it occurs in the absence of dissolved oxygen, strict anaerobiosis characterized by hydrogen sulfide and methane production is not necessary.

Methanol is the preferred carbon source because it is the least expensive synthetic compound available that can be applied without leaving a residual BOD in the process effluent—but this does not imply that methanol treatment is cheap. Introduction of methanol first reduces the dissolved oxygen present by Eq. 13-16; then biological reduction of nitrate and nitrite occurs (Eqs. 13-17 and 13-18).

$$3O_2 + 2CH_3OH = 2CO_2\uparrow + 4H_2O \tag{13-16}$$

$$6NO_3^- + 5CH_3OH = 3N_2\uparrow + 5CO_2\uparrow + 7H_2O + 6OH^- \tag{13-17}$$

$$2NO_2^- + CH_3OH = N_2\uparrow + CO_2\uparrow + H_2O + 2OH^- \tag{13-18}$$

From these reactions, the amount of methanol required as a hydrogen donor for complete denitrification is

$$CH_3OH = 0.7DO + 1.1NO_2\text{–}N + 2.0NO_3\text{–}N \tag{13-19}$$

where CH_3OH = methanol, mg/l
DO = dissolved oxygen, mg/l
$NO_2–N$ = nitrite nitrogen, mg/l
$NO_3–N$ = nitrate nitrogen, mg/l

Approximately 30 % excess methanol feed is needed for synthesis; hence chemical consumption to satisfy both energy and synthesis can be estimated from the relationship

$$CH_3OH = 0.9DO + 1.5NO_2–N + 2.5NO_3–N \qquad (13\text{-}20)$$

Limited information is available at present on the kinetics of the denitrification reaction. Optimum pH falls in the same range as most heterotrophic bacteria (between 5.5 and 7.5) with the rate reducing to about 80 % of maximum

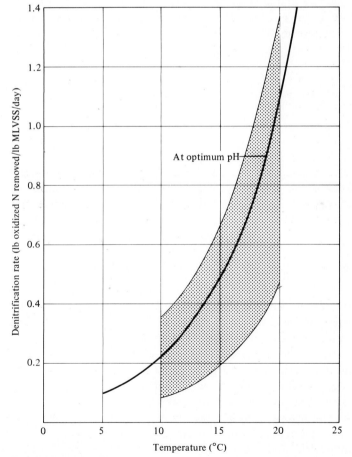

FIGURE 13-21 Effect of temperature on the rate of denitrification. [Source: *Nitrification and Denitrification Facilities*, Environmental Protection Agency, Technology Transfer (August 1973), p. 28.]

when the pH is lowered to 6.1 or raised to 7.9. Nitrified wastewaters, which tend to be basic, are naturally controlled from excessively high pH by carbon dioxide generated in a denitrification unit; thus there appears to be no need for addition of chemicals to control pH in actual systems. Effect of temperature on the rate of denitrification after Mulbarger[20] is sketched in Fig. 13-21.

Denitrification by Suspended-Growth Systems

The process studied most extensively consists of a complete-mixing basin followed by a clarifier for sludge separation and return (Fig. 13-22). Although a single, mixed chamber is common in laboratory studies, plug flow minimizes short circuiting and becomes more suitable for the relatively short detention periods required. Underwater stirrers, comparable to those used in waterworks flocculation tanks, mix the contents sufficiently to keep microbial floc in suspension without producing undue aeration. Power supply in the range of 1 hp for each 2000–4000 ft^3 of tank volume appears to be adequate. Whether basins should be covered to minimize absorption of oxygen is a matter of conjecture, but certainly airtight covers should be avoided.

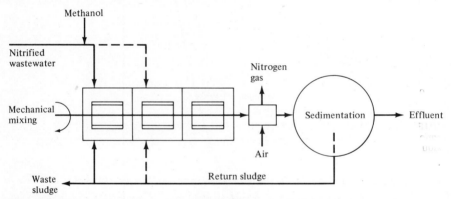

FIGURE 13-22 Flow diagram for a complete-mixing, compartmented, denitrification basin and clarifier.

Denitrification reactions form carbon dioxide and nitrogen gas bubbles that inhibit gravity settling by adhering to the biological floc. Supersaturation of the mixed liquor with gases can be relieved by short-term aeration in an open channel, or tank, between the denitrification basin and final clarifier. Settleability of sludge solids following this air stripping appears to be similar to that of other biological sludges. Recommended clarifier depths and overflow rates are the same as those suggested for final settling tanks following the nitrification process. Basins should be equipped with rapid-sludge-return collector arms and skimming devices. A sludge-recirculation capacity equal to the average wastewater flow is recommended, and scum may be returned to the denitrification tank or routed to disposal. Withdrawal of excess microbial solids, to keep the biological system in balance, ranges from 0.2 to 0.3 lb/lb of methanol applied.

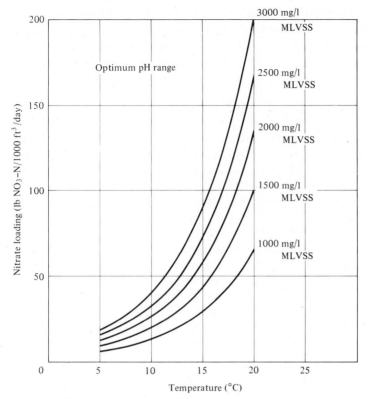

FIGURE 13-23 Permissible denitrification-tank loadings in the pH range 6.5–7.5. [Source: *Nitrification and Denitrification Facilities*, Environmental Protection Agency, Technology Transfer (August 1973), p. 30.]

The volumetric capacity needed for denitrification can be estimated using Fig. 13-23, which is based on pilot-plant studies.[20] Denitrifying sludges, after degasification, have good settling characteristics, allowing a design mixed-liquor solids of 2000–3000 mg/l which are approximately 65% volatile.

Denitrification by Fixed-Growth Systems

Submerged (anaerobic) filters for nitrate reduction have been evaluated in laboratory and pilot-plant studies, but no general design criteria are proposed for full-scale units. Upflow beds appear to have the advantages of efficient operation at cold temperature and greater resistance to shock loadings than suspended-growth systems, but clogging of the media with biological growth is an unresolved problem. An extensive study of anaerobic filters, by Tamblyn and Sword[26] for removal of nitrate from irrigation return water containing about 20 mg/l nitrate-N reported 88–98% reduction by upflow through a gravel bed using a detention time of 1 hour and methanol as a carbon source. Experimentation with a variety of media (sand, gravel, cinders, coal, and plastic) demonstrated that their sorptive characteristics had no effect on nitrogen removal and the larger 1-inch-diameter medium functioned as well as finer material with lower head loss. Operation of the filters showed that backwashing to clean the voids of accumulated solids could have an adverse effect on performance if the populations of nitrifying bacteria are excessively reduced by washout.

Biological Nitrification–Denitrification
A two-stage system composed of the units illustrated in Figs. 13-19 and 13-22, following secondary biological treatment, can achieve about 90% inorganic nitrogen reduction and 80–85% total nitrogen removal under normal operating conditions. The biological cultures performing ammonia oxidation are more sensitive to heavy metals and organic toxins than conventional activated sludge. Therefore, industrial wastes discharged to municipal sewers must be carefully monitored and necessary controls established to ensure that the nitrifying microorganisms are not inhibited.

Advantages of biological treatment for nitrogen removal are: the system is adaptable as an addition to existing secondary plants, and the nitrification portion can be built to meet a current requirement for ammonia removal with the denitrification step added in the future, if needed. Also, production of waste sludge solids is minimal, and nitrogen is converted to an ecologically harmless gas. Reliable cost estimates for nitrification and denitrification processes are possible, since these operations are very similar to those used in conventional treatment.

●**EXAMPLE 13-6**
Based on the following data, calculate the volume needed for suspended-growth denitrification.

$$
\begin{aligned}
\text{average daily design flow} &= 10 \text{ mgd} \\
\text{nitrate nitrogen} &= 20 \text{ mg/l of N} \\
\text{dissolved oxygen} &= 8 \text{ mg/l} \\
\text{minimum operating temperature} &= 8°C \\
\text{operating pH} &= 7.8 \\
\text{design volatile MLSS} &= 2000 \text{ mg/l}
\end{aligned}
$$

○*Solution*

$$
\begin{aligned}
\text{average nitrate load} &= 10 \times 20 \times 8.34 = 1670 \text{ lb/day of N} \\
\text{maximum nitrate load} &= 1.5 \times 1670 = 2500 \text{ lb/day of N}
\end{aligned}
$$

Permissible denitrification-tank loading from Fig. 13-23 for a temperature of 8°C and 2000 mg/l is 20 lb of NO_3–N/1000 ft³/day. Correcting this to a pH of 7.8, the allowable loading $= 20 \times 0.9 = 18$ lb/1000 ft³/day.

$$
\text{denitrification basin volume} = 2500 \frac{1000}{18} = 140,000 \text{ ft}^3
$$

$$
\text{resulting detention time} = \frac{140,000 \times 7.48 \times 24}{10,000,000} = 2.5 \text{ hours}
$$

Average methanol dosage for 8 mg/l of DO and 20 mg/l of NO_3–N, using Eq. 13-20, is

$$CH_3OH = 0.9 \times 8 + 2.5 \times 20 = 57 \text{ mg/l}$$

13-13

AMMONIA STRIPPING

Equilibrium of the ammonium ion and dissolved ammonia gas in water is controlled by both pH and temperature, as shown in Fig. 13-24. Only NH_4^+ ions are present in neutral solution at ambient temperatures, while at a pH of 11, essentially all the ammonia appears as NH_3 gas. The percentage of dissolved ammonia, relative to ammonium ion, decreases with cooling of the wastewater.

The physical–chemical process of ammonia stripping consists of (1) raising the wastewater pH to a value in the range 10.8–11.5 (Fig. 13-24), with lime applied for preceding phosphate precipitation, (2) formation and re-formation of water droplets in a stripping tower, and (3) providing air–water contact and droplet agitation by circulation of large quantities of air through the tower. The rate of gas transfer from liquid to air is influenced by pH, temperature, relative ammonia concentrations, and agitation at the air–water interface. The latter

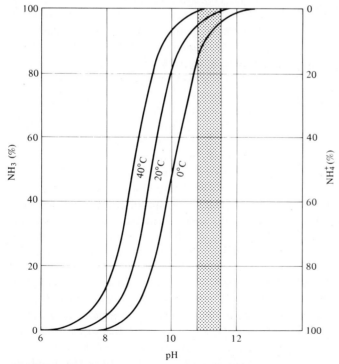

FIGURE 13-24 Distribution of ammonia and ammonium ion in water relative to pH and temperature. The normal pH range of 10.8–11.5 for ammonia stripping is shown shaded.

two factors depend primarily on hydraulic loading, quantity of air flow, and configuration of the column packing. Countercurrent towers, where air enters at the bottom and exhausts from the top while the water drips and splatters down through the medium, have been found to be most efficient. In cross-flow units, air is pulled in through louvered sides, drawn through the packing, and discharged from the top.

Advantages of air stripping for nitrogen removal are simplicity of operation, ease of control, and low cost relative to other extraction processes. Disadvantages are the inability to operate at an ambient air temperature below 32°F, and deposition of calcium carbonate scale on the packing.

Towers cannot be employed in northern climates where cold-weather nitrogen removal is needed without providing an alternative method for winter operation. Experience with a full-scale cross-flow tower at Lake Tahoe demonstrated both the excellent reduction possible during warm weather (Table 13-4) and the serious problems that occur during low-temperature operation. Efficiency decreased from 90% removal at 20°C to a maximum of 75% at 10°C. When air temperature dropped below freezing, the process was shut down, owing to ice formation in the tower.[27]

The most serious operating problem at Tahoe was accumulation of calcium carbonate scale on the tower packing of rough-hewn hemlock slats. Cleaning was extremely difficult because of the construction, and ultimately removal efficiency decreased as a result of scale interference with both droplet formation and air flow. Recently designed countercurrent towers with smooth plastic fill are less susceptible to scaling, and removable panels can be placed along corridors within the tower for access to clean or remove packing modules.

The fate of ammonia discharged to the atmosphere should be considered for each location although air pollution and washout by precipitation are not normally considered problems. Concentration in the stack discharge, before dilution in the atmosphere, is about 6 mg/m^3. This is significantly lower than the odor threshold of 35 mg/m^3 for ammonia. The United States has no emission standard, or air-quality criterion, for ammonia, since it is not considered an air pollutant. The ammonia content in rainfall is directly related to the atmospheric concentration; thus washout can occur where tower discharge increases in natural background concentration in the air. Ammonia in rainfall decreases to about the natural level at a downwind distance of 3 mi from a tower processing 15 mgd of wastewater.[27] Unless the precipitation falls directly on a lake, or is conveyed by runoff over paved surfaces, the nitrogen will most likely be retained by surface soils.

Recarbonation of lime-treated water is necessary to convert hydroxide ions to bicarbonates for control of scaling in subsequent treatment units. Carbon dioxide neutralizes excess lime, precipitating calcium carbonate at a pH of about 10 (Eq. 13-21); additional recarbonation reduces the pH further and stabilizes the CaCO$_3$ to soluble calcium bicarbonate (Eq. 13-22). A two-stage system with intermediate clarification can be used to separate these two reactions and extract a major portion of the calcium carbonate precipitate from solution. Advantages are a finished water lower in hardness and total dissolved solids, and a sludge of almost pure calcium carbonate suitable for recalcining. Single-stage recarbonation, which performs Eqs. 13-21 and 13-22 in a single basin, stabilizes water, leaving all the calcium in solution.

$$Ca(OH)_2 + CO_2 = CaCO_3 \downarrow + H_2O \tag{13-21}$$

$$CaCO_3 + CO_2 + H_2O = Ca(HCO_3)_2 \tag{13-22}$$

South Lake Tahoe, California

A full-scale experimental tower with a design flow of 7.5 mgd has been operated at the water reclamation plant on an intermittent basis since 1969. Scale formation in the packing adversely affected removal efficiency, and ice formation during the winter season made operation impossible. A modified flow scheme to reduce the impact of these limitations, illustrated in Fig. 13-25, consists of three basic steps: (1) holding in high-pH, surface-agitated ponds; (2) stripping in a modified cross-flow, forced-draft tower through air sprays installed in the chamber without packing; and (3) breakpoint chlorination. Effluent from lime clarification will be held in equilization ponds for 7–18 hours and recycled through spray nozzles installed above the basins. An ammonia removal of approximately one-third is anticipated, even in cold weather. Pond contents will then be sprayed into the forced-draft tower constructed from the existing cross-flow unit by removing packing and installing pipe headers with nozzles to direct the water upward into the chamber. A substantial portion of the ammonia is expected to be released from solution by several spraying cycles during both winter and summer operation. Downstream breakpoint chlorination will be used to oxidize the remaining ammonia, which will vary from 5 to 16 mg/l, depending on plant flow and temperatures. First-step chlorination in the primary recarbonation chamber will be used to reduce the pH to about 9.6, thus eliminating the need for carbon dioxide addition at this point. The balance of chlorine needed to reach breakpoint will be added in a chamber immediately downstream from the secondary recarbonation basin.[28]

Orange County, California

A 15-mgd wastewater reclamation plant and a 3-mgd seawater desalting facility have been constructed on the same site at Santa Ana by the Orange County Water District. Reclaimed waters from the two plants are blended and pumped into injection wells for groundwater recharge, forming a barrier against seawater intrusion into the freshwater aquifer. The 15-mgd trickling filter effluent is processed as follows: (1) lime coagulation to a pH of 11 followed by clarification; (2) ammonia stripping in countercurrent towers followed by two-stage recarbonation with intermediate settling; (3) gravity filtration through mixed-media beds, applying alum and/or polymers as needed; (4) adsorption in upflow granulated-carbon columns with facilities for carbon regeneration; and

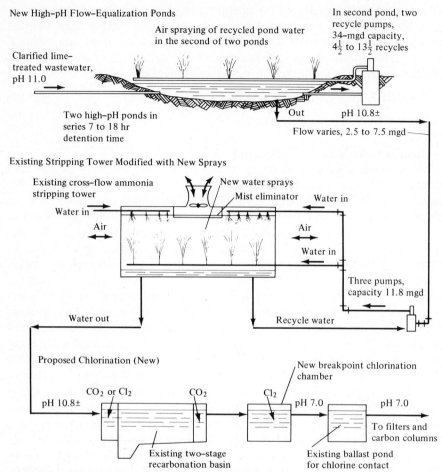

New High–pH Flow–Equalization Ponds

Air spraying of recycled pond water in the second of two ponds

In second pond, two recycle pumps, 34-mgd capacity, $4\frac{1}{2}$ to $13\frac{1}{2}$ recycles

Clarified lime–treated wastewater, pH 11.0

Two high–pH ponds in series 7 to 18 hr detention time

Out   pH 10.8±

Flow varies, 2.5 to 7.5 mgd

Existing Stripping Tower Modified with New Sprays

Existing cross–flow ammonia stripping tower

New water sprays

Mist eliminator   Water in

Water in

Air

Air

Water in

Three pumps, capacity 11.8 mgd

Water out

Recycle water

Proposed Chlorination (New)

New breakpoint chlorination chamber

pH 10.8± CO_2 or Cl_2 CO_2 Cl_2 pH 7.0   pH 7.0

To filters and carbon columns

Existing two–stage recarbonation basin

Existing ballast pond for chlorine contact

FIGURE 13-25 Proposed new and modified ammonia–nitrogen removal processes at the Lake Tahoe Water Reclamation Plant. [Source: *Physical–Chemical Nitrogen Removal,* Environmental Protection Agency, Technology Transfer (July 1974), p. 8.]

(5) breakpoint chlorination for disinfection and removal of any remaining ammonia. Sludge from the clarifier and recarbonation basins is dewatered by thickening and centrifuging prior to incineration in a multiple hearth furnace. The resulting calcium oxide is reused, and the remaining ash hauled to disposal.

The ammonia-stripping towers, shown in Fig. 13-26, also serve as cooling units for two process flows from the seawater desalting plant—concentrated brine disposed of in the ocean, and water from the barometric condensers returned to the desalting plant. The cross-flow cooling section is located around the periphery, near the base of the tower. Inlet air warmed to approximately 32°C increases ammonia-stripping efficiency while reducing the temperature of the desalting process waters to about 28°C. The countercurrent stripping units

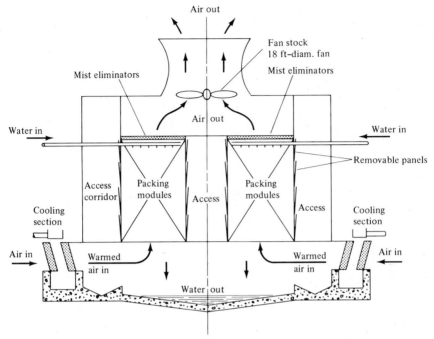

FIGURE 13-26 Countercurrent ammonia-stripping towers at the wastewater reclamation plant, Orange County Water District, Santa Ana, California. [Source: J. G. Gonzales and R. L. Culp, "New Developments in Ammonia Stripping," *Public Works* 104, no. 6 (June 1973): p. 82.]

are designed for a hydraulic loading of 1.0 gpm/ft² and an air flow of 400 ft³/gal.[27]

The tower packing is constructed of 0.5-in.-diameter plastic pipe laid in a criss-cross pattern on 3-in. centers horizontally. Prefabricated modules of this fill, about 6 by 6 by 4 ft high, are positioned behind removable air baffle panels. Corridors around the stripping columns in the tower allow access for replacing modules, if necessary, and removing excessive calcium carbonate scale by hosing with water.

More than 90% of the ammonia is removed by air stripping, with the remainder oxidized by subsequent breakpoint chlorination. The maximum concentration of ammonia–N in injection water is 1.0 mg/l.

Closed-Loop Ammonia Stripping

This process, still in its initial stages of development, appears to overcome the limitations of temperature, scale formation, and discharge to the atmosphere, while having the advantage of recovering ammonia as a byproduct. Stripping and absorption units are connected by appropriate ducting sealed from outside air (Fig. 13-27). Ammonia, transported from the stripping tower in a recycled gas stream (initially air), is removed by an absorbing liquid maintained at a low pH to convert it to the ammonium ion. When using sulfuric acid, ammonium

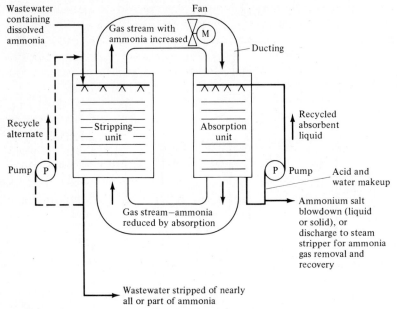

FIGURE 13-27 Closed-loop process for ammonia removal and recovery. [Source: L. G. Kepple, "Ammonia Removal and Recovery Becomes Feasible," *Water and Sewage Works* 121, no. 4 (April 1974): 42.]

sulfate formed can be discharged from the absorption device by liquid blowdown. Steam stripping provides another potential method of ammonia removal from the absorbent without acid, since only the ammonia gas is extracted.

The usual scaling problem associated with air-stripping towers does not occur in a closed system, since the gas stream is devoid of carbon dioxide, thus preventing formation of calcium carbonate precipitate. Eliminating freezing by excluding nearly all outside air, operating temperature is governed by the wastewater. Recovery of nitrogen as ammonium sulfate, or aqua ammonia, can result in a salable fertilizer product. This process, if it becomes feasible, will be of significant benefit to physical–chemical treatment of wastewater in cold climates.

13-14

BREAKPOINT CHLORINATION

Oxidation of ammonia–nitrogen by the addition of excess chlorine can be represented by the following unbalanced chemical reaction:

$$NH_3 + HOCl \longrightarrow N_2\uparrow + N_2O\uparrow + NO_2^- + NO_3^- + Cl^- \qquad (13\text{-}23)$$

Possible products formed, in order of importance, are nitrogen gas, nitrous oxide, and nitrite–nitrate nitrogen. Analyses in the pH range 6.5–7.5, for initial ammonia–nitrogen concentrations of 8–15 mg/l, have shown that breakpoint chlorination can yield 95% removal, with nitrate and nitrogen trichloride residuals never exceeding 0.5 mg/l.[30] The rate and extent of this reaction depend on pH, temperature, contact time, and initial chlorine/ammonia ratio. The weight ratio of chlorine to nitrogen needed for ammonia destruction ranges between 8 : 1 and 10 : 1 of Cl_2 to N, with the lower value applicable to pretreated wastewater.

Breakpoint chlorination is adaptable to physical–chemical treatment and has the advantages of low capital cost, a high degree of efficiency and reliability, insensitivity to cold weather, and release of nitrogen as a gas. The main disadvantage is that essentially all the chlorine added is reduced to chloride ion, thus contributing to dissolved-solids concentration in the treated water. For example, at a 10 : 1 dosage ratio, oxidation of 20 mg/l of ammonia–nitrogen contributes 200 mg/l of chloride ion. Also, required equipment and process controls are relatively complex (Fig. 13-28).[27] Subbreakpoint dosages for partial ammonia oxidation produce chloramines that may present problems if discharged directly to receiving waters. One possible solution is subsequent carbon adsorption to destroy the remaining combined residual.

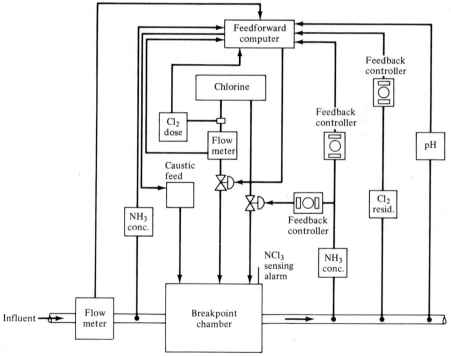

FIGURE 13-28 Breakpoint-chlorination control system. [Source: *Physical–Chemical Nitrogen Removal*, Environmental Protection Agency, Technology Transfer (July 1974), p. 20.]

13-15

ION EXCHANGE FOR AMMONIUM IONS

Ammonium ions are present in low concentrations relative to other cations in wastewater, and are consequently difficult to selectively extract by ion exchange. Clinoptilolite is a natural inorganic zeolite that has unusual selectivity for the ammonium ion—no synthetic resin with this characteristic is currently available. Ammonium ions are sorbed on clinoptilolite in preference to calcium, magnesium, and sodium, but not potassium. Effectiveness of ion exchange depends on the initial NH_4^+ content, competing cation concentrations, hydraulic loading of the bed, pH, and method of regeneration, but not temperature.[27, 31] Extensive pretreatment, including clarification by chemical coagulation and granular-media filtration, is required prior to passing a wastewater through an ion-exchange column.

Selective removal with clinoptilolite has the advantages of high efficiency, insensitivity to temperature, and removal of ammonia with minimal addition of dissolved solids. However, this process is still experimental and applicability in wastewater denitrification depends to a considerable extent on developing a method for handling spent regenerant. Ion-exchange capacity of the zeolite is regained by passing a concentrated salt solution through the bed to remove the sorbed nitrogen. The original regenerant was a lime slurry that converted ammonium ions, extracted from the bed, to gaseous ammonia removable by air stripping. Later, it was observed that a solution of NaOH and NaCl provided better ammonia elution from clinoptilolite and allowed recovery of the regenerant solution by electrolytic treatment. However, the most recent concept uses a closed-loop ammonia stripping unit in regeneration.

Upper Occoguan District, Virginia

Advanced treatment of the secondary wastewater effluent consists of lime precipitation, recarbonation, filtration, carbon adsorption, selective ion exchange for ammonia extraction, and chlorination for disinfection and residual ammonia removal. The ion exchangers, with a 4-ft depth of clinoptilolite medium, are constructed similar to waterworks pressure filters. The design hydraulic loading rate is 5.25 gpm/ft^2 (10.8 bed volumes/hr) for an anticipated ammonia reduction from 20 to 1.0 mg/l of N.

The zeolite will be operated on a sodium cycle with the ammonium ions absorbed being replaced by sodium ions from a regenerant solution of 2% NaCl. The beds will be backwashed prior to rejuvenating with salt solution. Spent regenerant will be restored by removing the eluted ammonium ions and unwanted bivalent cations. Sodium hydroxide added to a pH of about 11 precipitates magnesium and converts NH_4^+ to NH_3. After clarification, the brine is to be air-stripped in a closed-loop system to remove the ammonia (Fig. 13-27). Sodium chloride will be added to increase the salt content of the regenerant prior to reuse.

Land Treatment of Wastewater

Land disposal is a potential technique for treatment of municipal wastes where an available site has suitable soil conditions and groundwater hydrology, and the climate is favorable. Hundreds of efficient systems are currently operating in regions with limited water resources to boost growth of grass, crops, or forests. While the impetus behind these installations is primarily water reuse, future applications will also emphasize land disposal for AWT. In addition to determining potential dangers of contaminating the soil mantle and groundwater, unprejudiced cost evaluations must prove that irrigation is a feasible alternative.[32] Recycling of nutrients to land, instead of polluting surface waters, has a strong ecological appeal; however, the requirements for large tracts of land and wastewater storage are major disadvantages in humid climates, northern states, and metropolitan areas.

13-16

METHODS OF APPLYING WASTEWATER TO LAND

Techniques for spreading of wastewaters are normally classed into the three categories, illustrated in Fig. 13-29—irrigation, overland flow, and infiltration–percolation. Selecting the type of system at a given site is governed primarily by drainability of the soil, which determines the allowable liquid loading rate, and required degree of water reclamation. Other factors, such as wastewater quality, climate, and land availability, in combination with soil conditions and pollution potential, may exclude any form of land treatment at a viable selection. Major features of the three application methods are compared in Table 13-6. The difference in land area required to handle irrigation relative to infiltration and the greater probability of groundwater contamination by applying the latter method are of particular interest. For the purpose of wastewater renovation, irrigation is the process most widely practiced in the United States, while overland flow is still experimental. Rapid infiltration is presently used for groundwater recharge and aquifer injection as a barrier against seawater intrusion. Some times, seepage from septic-tank drainfields is included in the infiltration–percolation category.

Irrigation

In western states, conventionally treated effluent is often applied to supplement rainfall for maximizing crop production. This process has the greatest tertiary treatment potential, owing to the low loading rates and wide dispersal of pollutants. However, because of the high water loss through evapotranspiration, the concentration of dissolved solids in percolate to the groundwater can be undesirable. Irrigation for the primary purpose of effluent disposal is greater

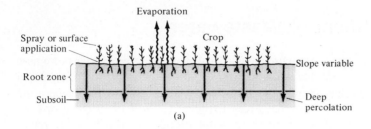

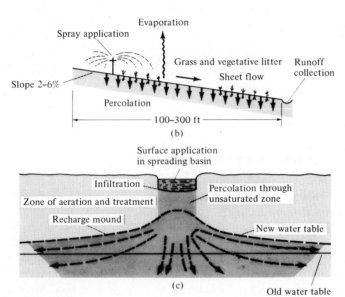

FIGURE 13-29 Three types of land-treatment systems. (a) Irrigation; (b) overland flow; (c) infiltration-percolation. [Source: C. E. Pound and R. W. Crites, *Wastewater Treatment and Reuse by Land Application, Vol. II*, Environmental Protection Agency, Office of Research and Development (August 1973).]

than the minimum hydraulic loading for crop yield; thus water is allowed to penetrate below the root zone. The soil profile and type of cover crop are primary factors controlling nutrient removal and infiltration. Ideally, the surface soil should be a silt loam texture with adequate permeability and a minimum depth of 5 ft to groundwater.

Water distribution is by fixed or moving sprinkling systems, or surface spreading. Fixed nozzles are attached to risers from either surface or buried pipe networks. The most popular moving sprinkling system is a center-pivot spray boom that rotates around a central tower with the distribution piping suspended between wheel supports riding on circumferential tracks. On flat land, having less than 1 % slope, surface irrigation is possible by the ridge-and-furrow method. Water applied to furrows, spaced about 3 ft apart, flows down slope by gravity and seeps into the ground. Border-strip irrigation, the second method for surface

TABLE 13-6

COMPARATIVE CHARACTERISTICS OF IRRIGATION, OVERLAND FLOW, AND INFILTRATION–PERCOLATION SYSTEMS

Parameter	Irrigation	Overland flow	Infiltration-percolation
Liquid loading rate	0.5–4 in./week	2–5.5 in./week	4–120 in./week
Annual application	2–8 ft/yr	8–24 ft/yr	18–500 ft/yr
Land required for 1-mgd flow	140–560 acres plus buffer zones	46–140 acres plus buffer zones	2–62 acres plus buffer zones
Application techniques	Spray or surface	Usually spray	Usually surface
Soils	Moderately permeable soils with good productivity when irrigated	Slowly permeable soils, such as clay loams and clay	Rapidly permeable soils, such as sands, loamy sands, and sandy loams
Probability of influencing groundwater quality	Moderate	Slight	Certain
Needed depth to groundwater	About 10 ft	Undetermined	About 15 ft
Wastewater lost to:	Predominantly evaporation or deep percolation	Surface discharge dominates over evaporation and percolation	Percolation to groundwater

Source: C. E. Pound and R. W. Crites, *Wastewater Treatment and Reuse by Land Application*, Vol. I, Environmental Protection Agency, Office of Research and Development (August 1973).

spreading, uses parallel soil ridges constructed in the direction of slope. Water introduced between the ridges at the upper end flows down the 20- to 100-ft wide strips several hundred feet long.

Overland Flow

This method, an adaptation of spray irrigation to impermeable soils, results in surface drainage that must be collected for reuse or discharged to a surface watercourse. Waste is applied by sprinklers to the upper one-third of terraces that are 200–300 ft in length and on grades of 2–4%. Collection trenches at the bottom of each slope carry away the treated runoff. Renovation is achieved by filtration and bacterial decomposition as the water moves slowly through the grass cover.

Spray-runoff systems have been used in treating food-processing wastes and are currently being evaluated for municipal discharges. An 18-month pilot study at Ada, Oklahoma, used overland flow to treat raw domestic wastewater.[34] After removal of solids that would plug the spray nozzles, water was applied to experimental plots at rates of 3–4 in./week. Effluent BOD and suspended solids were less than 20 mg/l all year round. Phosphorus removal was about 50 % with relatively minor seasonal variation, while the nitrogen efficiency of 90 % during the summer dropped substantially in winter.

Two major restrictions to overland runoff are difficulty in maintaining consistent quality in the renovated water and site-preparation costs. A surface-flow process is very dependent on weather, such as sunshine and temperature, so year-round use is practical only in mild climates. Even in suitable regions it is difficult to predict the degree of renovation possible by a spray-runoff system without conducting pilot-plant studies.

Infiltration-Percolation

In this process, restricted to rapidly permeable soils such as sands and sandy loams, applied wastewater passes through the soil profile to the ground-water table. Recharge basins are the usual practice (Fig. 13-29), although high-rate spray systems can be used for effluent spreading. The arbitrary dividing point between irrigation and infiltration–percolation processes is 4 in./week (Table 13-6). The majority of water renovation occurs by physical, chemical, and biological mechanisms in the soil matrix, with vegetation playing a relatively minor role. While a grass cover helps to remove suspended solids and organic matter, only small portions of nutrient salts are photosynthesized. Multivalent ions appear to be extracted rather easily, while monovalent species tend to be flushed down through the soil mantle. There is also a significant risk of carrying pathogens, particularly viruses, to underground water. Thus the chief limitation of infiltration–percolation appears to be pollution of ground-water with refractory contaminants.

The Flushing Meadows Project, located in the Salt River bed west of Phoenix, Arizona, applied secondary wastewater effluent through infiltration basins at a hydraulic loading rate of about 70 in./week. The underlying soil consisted of 3 ft of fine loamy sand underlain by coarse sand and gravel extending through the groundwater table at a depth of about 10 ft. One of the most interesting observations was the fate of the applied nitrogen, which amounted to about 30 mg/l of ammonia–N. Estimated overall removal, based on average nitrogen concentration in renovated water withdrawn from observation wells, was about 30%. Denitrification was probably the main mechanism for this nitrogen reduction, since only a small quantity could be removed by harvesting the surface grasses.

The typical year-round operating cycle was 2 weeks of flooding followed by a 10- to 20-day rest period. Ammonium ions in the percolate were absorbed by cation exchange in the soil until saturation; then ammonia was carried into the groundwater. During the drying period, complexed ammonium ions were converted to nitrate under aerobic soil conditions. Anaerobiosis, from reflooding with water, resulted in a portion of the nitrate being converted to nitrogen gas while the majority was flushed into the underlying groundwater because nitrate

TABLE 13-7
GENERAL DESIGN CONSIDERATIONS FOR LAND-TREATMENT SYSTEMS

Wastewater Characteristics	Climate	Geology	Soils	Plant Cover	Topography	Application
Flow	Precipitation	Groundwater	Type	Indigenous to region	Slope	Method
Constituent load	Evapotranspiration	Seasonal depth	Gradation	Nutrient-removal capability	Aspect of slope	Type of equipment
	Temperature	Quality	Infiltration/permeability	Toxicity levels	Erosion hazard	Application rate
	Growing season	Points of discharge	Type and quantity of clay	Moisture and shade tolerance	Crop and farm management	Types of drainage
	Occurrence and depth of frozen ground	Bedrock	Cation-exchange capacity	Marketability		
	Storage requirements	Type	Phosphorus adsorption potential			
	Wind velocity and direction	Depth	Heavy metal adsorption potential			
		Permeability	pH			
			Organic matter			

Source: R. D. Johnson, "Land Treatment of Wastewater," *The Military Engineer* 65, no. 428 (1973): 375.

anions are not significantly affected by ion-exchange capacity of a soil. After 3 years of rapid infiltration, the ammonia–N content of the renovated water under the site averaged 10 mg/l, with nitrate–N concentrations varying from near zero to greater than 20 mg/l during inundation. These data dramatize the limited capability of removing inorganic nitrogen, which can lead to serious problems if the percolated water is reused as a domestic supply.

Water Balance and Storage Calculations

The water balance for a land application system can be calculated by the relationship

precipitation + wastewater loading
$$= \text{evapotranspiration} + \text{percolation} + \text{runoff} \qquad (13\text{-}24)$$

Runoff is zero for most irrigation and infiltration–percolation installations, while precipitation and evapotranspiration are of little significance compared to wastewater loading and percolation for rapid infiltration systems. In overland flow, percolation is minimal and runoff represents 40–80% of the applied wastewater.

Storage required can be calculated using monthly, weekly, or daily water-balance determinations. If the sum of evapotranspiration and percolation is less than precipitation plus available effluent, the balance must be stored. Conversely, when losses exceed the quantity available, water can be drawn from storage to supplement irrigation. These relationships are summarized in the following statement:

$$
\begin{bmatrix} \text{precipitation} \\ + \\ \text{effluent available} \end{bmatrix}
\pm
\begin{bmatrix} \text{change} \\ \text{in} \\ \text{storage} \end{bmatrix}
=
\begin{bmatrix} \text{evapotranspiration} \\ + \\ \text{percolation} \end{bmatrix}
\qquad (13\text{-}25)
$$

Table 13-7 summarizes the general considerations in design of land-treatment systems.

13-17

SPRAY IRRIGATION[36,37]

Grassland irrigation is the best land-treatment system in terms of predictability of renovated water quality and ease of operation. Moderately permeable soils able to support fodder grasses, or cash crops such as corn, are available in most regions of the country. Specified pretreatment for sprinkling is usually biological processing plus chlorination to ensure a water suitable for irrigation and to satisfy environmental concerns. Major constraints on spray irrigation are climate, since water cannot be spread on frozen or water-saturated ground, and the large land area required (e.g., for a year-round application rate of 2 in./week, a town of 10,000 population would require about 130 irrigated acres plus buffer zones). In northern states, large ponds must be provided for wastewater storage when spraying is suspended during winter months. Runoff is also collected and returned to storage for reapplication.

Nitrogen and phosphorus can be effectively removed by a cover crop and soil mantle. Pretreating and spraying converts essentially all the nitrogen in wastewater to nitrate form. With proper management, the majority of this nitrate is taken up by the vegetative cover. Perennial grasses are preferred to annual crops, since their root system is fully established throughout the spraying season. Denitrification may release a portion of the nitrate as gas, and, of course, ions can percolate through the root zone to the groundwater. Although the amount of phosphorus in domestic wastewater normally exceeds the need for photosynthetic uptake, excess orthophosphate is readily taken up in the soil by adsorption or ion exchange. In a well-operated system, spray irrigation should remove at least 80% of the applied nitrogen and 90% of the phosphorus.

Heavy metals, sodium, and trace elements, such as boron, can be toxic to plant growth, decrease porosity of soil structure, or appear in reclaimed water. Most heavy metals are removed in conventional wastewater treatment, and the trace quantities remaining in irrigation water are extracted by soil adsorption and plant growth. Based on current information, heavy metal uptake from irrigating with municipal waste does not appear to be a serious problem. Salinity, related to the sodium content of a wastewater, adversely affects both soil and plants. High evaporation rates in arid climates can produce critical salt concentrations in irrigated soils, while in humid regions it is unlikely that salt accumulations will be significant. Trace quantities of inorganics, essential growth factors in very low dosages, can cause toxicity in higher concentrations.

Most bacteria and viruses are trapped in the first few feet of a moderately permeable soil profile, although the latter may reach greater depths in some instances. Survival of pathogens in soil has not been extensively explored, but it appears that the environment is not conducive to their longevity. Micro-organisms can also be transported in air as a result of spraying wastewaters and may be a greater health hazard than water transmission, although it is difficult to demonstrate that breathing aerosols from treatment processes is dangerous. Respiratory illnesses among wastewater-treatment plant workers appear to be no higher than for the public at large.[36]

Muskegon, Michigan

This project is a pioneer in integrating tertiary treatment with crop irrigation and soil infiltration operation for a municipal wastewater containing about 60% papermill discharge. The system consists of stabilization in aerated treatment cells, spray irrigation of cropland, filtration through sandy soil, and collection of the percolated water in a buried pipe network (Fig. 13-30). Raw wastewaters are collected and transported by force main to the reclamation site 11 mi east of the city. After biological processing in aerated lagoons, the wastewater enters storage basins sealed to prevent exfiltration. Storage volume

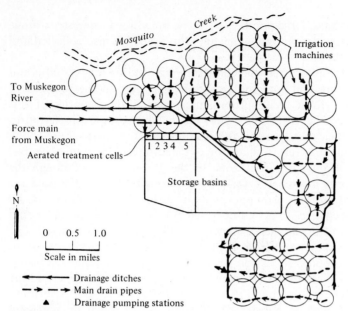

FIGURE 13-30 Plan view of the spray irrigation system at Muskegon, Michigan. [Source: *Civil Engineering* (May 1973).]

is sufficient to accumulate the flow for about 4 months during winter when sprinkling is not possible. After chlorination, the effluent is pumped through a piping system to center-pivot spray irrigation units, and a network of underground pipes collects the filtrate and pumps it to the Muskegon River for disposal. Berms around the edge of the site prevent surface runoff. Anticipated efficiencies in treatment are 99% removal of BOD and suspended solids, 90% reduction of phosphorus, and 76% or more reduction of nitrogen.

●EXAMPLE 13-7

These data are from sample calculations by Pound and Crites.[33] Compute storage requirements using monthly water balances for a 1-mgd irrigation system based on the following: (1) Design evapotranspiration and precipitation data, listed in columns (2) and (5) of Table 13-8, are for the wettest year in 25, with average monthly distribution. (2) A perennial grass is grown and irrigated year-round. (3) Runoff is contained and reapplied. (4) The design year begins in October with the storage reservoir empty. (5) Nitrogen is the limiting factor for a land area of 120 field acres to balance the quantity applied with the amount of nitrogen in harvested grass. (6) Design allowable percolation rate is 10 in./month from March through November and 5 in./month for the remaining months [column (3), Table 13-8].

○*Solution*

The effluent available per month

$$= \frac{1.0 \text{ mg/day} \times 30.4 \text{ days/month} \times 36.8 \text{ acre-in./mg}}{120 \text{ acres}} = 9.3 \text{ in.}$$

TABLE 13-8

WATER BALANCE AND STORAGE CALCULATIONS FOR EXAMPLE 13-7 (all values given in inches)

(1) Month	(2) Evapotranspiration	(3) Percolation	(4) Water Losses: (2) + (3)	(5) Precipitation	(6) Effluent Applied	(7) Effluent Available	(8) Total Water Available: (5) + (7)	(9) Change in Storage: (8) − (4)	(10) Total Storage
Oct.	2.3	10.0	12.3	1.6	10.7	9.3	10.9	−1.4	0
Nov.	1.0	10.0	11.0	2.4	8.6	9.3	11.7	0.7	0.7
Dec.	0.5	5.0	5.5	2.7	2.8	9.3	12.0	6.5	7.2
Jan.	0.2	5.0	5.2	3.0	2.2	9.3	12.3	7.1	14.3
Feb.	0.3	5.0	5.3	2.8	2.5	9.3	12.1	6.8	21.1
March	1.1	10.0	11.1	3.4	7.7	9.3	12.7	1.6	22.7
April	3.0	10.0	13.0	3.0	10.0	9.3	12.3	−.7	22.0
May	3.5	10.0	13.5	2.1	11.4	9.3	11.4	−2.1	19.9
June	4.8	10.0	14.8	1.0	13.8	9.3	10.3	−4.5	15.4
July	6.0	10.0	16.0	0.5	15.5	9.3	9.8	−6.2	9.2
Aug.	5.7	10.0	15.7	1.1	14.6	9.3	10.4	−5.3	3.9
Sept.	3.9	10.0	13.9	2.0	11.9	9.3	11.3	−2.6	1.3
TOTAL ANNUAL	32.3	105.0	137.3	25.6	111.7	111.8	137.2		

Source: C. E. Pound and R. W. Crites, *Land Treatment of Municipal Wastewater Effluents*, Environmental Protection Agency, Technology Transfer (August 1975).

Water losses [column (4), Table 13-8] are found by summing evapotranspiration and percolation. Effluent applied [column (6)] is the difference between water losses and precipitation. The total water available [column (8)] equals the sum of effluent available and precipitation.

The monthly change in storage [column (9)] using Eq. 13-25 is the difference between total water available [column (8)] and water losses [column (4)]. Total accumulated storage [column (10)], computed by summing the monthly changes, reaches a maximum of 22.7 in. in March.

$$\text{The basin capacity equivalent to this value} = \frac{22.7 \text{ in.} \times 120 \text{ acre}}{12 \text{ in./ft}}$$

$$= 227 \text{ acre-ft} = 74 \text{ mg}$$

The wastewater effluent applied is 111.7 in. (9.3 ft) on an annual basis, with about 3.5 in./week being the maximum rate in July.

PROBLEMS

13-1 List the common characteristics of a wastewater discharged from conventional biological processing. Which contaminants are most likely to cause pollution if the effluent is discharged to a stream with limited dilutional flow? Which constituents adversely affect a lake or reservoir receiving the wastewater discharge?

13-2 What are the potential benefits of flow equalization in municipal wastewater processing?

13-3 Compute the basin volume required to equalize the diurnal wastewater flow pattern diagrammed in Fig. 11-18a. Follow the procedure described in Example 13-1.

13-4 Discuss how sludge handling and disposal influence selection of unit processes for removing pollutants from wastewater.

13-5 What are basic differences between the characteristics of water and wastewater that influence design of granular-media filters?

13-6 How can gravity filters be designed to operate without rate of flow controllers commonly used in waterworks?

13-7 Assume that flow equalization is provided prior to filtration in Example 13-2, thus allowing a nominal application rate of 5.0 gpm/ft^2 for design. Compute the following: filter area required, cycle time, backwash as a percentage of filtrate, and rate of filtration with three of four cells operating.

13-8 Calculate the area of granular-media filters needed for suspended solids removal from a wastewater effluent with an average daily flow of 1500 m^3/day and peak 4-hour discharge of 3300 m^3/day. The nominal filtration rate should exceed neither 180 m^3/m^2-day with all beds operating nor 350 m^3/m^2-day at maximum discharge with only three of four beds in service.

13-9 What kinds of pollutants are removed by granular carbon columns?

13-10   Discuss the advantages of upflow, countercurrent, packed-bed carbon columns relative to downflow, and gravity contactors.

13-11   How is granular activated carbon regenerated?

13-12   A domestic wastewater, with 200 mg/l of BOD, 240 mg/l of SS, and 10 mg/l of P, is treated by extended aeration without primary sedimentation at a loading of 12.5 lb of BOD/1000 ft^3/day and aeration period of 24 hours. Alum dosing for phosphate precipitation is graphed in Fig. 13-11. What is the alum addition in mg/l required for 80% phosphorus removal? Estimate the biological–chemical solids produced per million gallons of wastewater processed using the technique presented in Sec. 12-2 for biological waste and Eq. 13-5 to approximate the amount of chemical precipitate. Compute the volume of sludge in gal/mg of wastewater treated assuming a solids content of 1.5%.

13-13   Chemical coagulation and sedimentation of a raw municipal wastewater using 60 mg/l of $FeCl_3$ and 50 mg/l of CaO precipitates 230 mg/l of suspended solids and 8 mg/l of phosphorus. What is the weight ratio of iron applied to phosphorus removed? Calculate the weight of organic-chemical sludge solids per cubic meter of wastewater treated using the method in Sec. 12-2 for organic matter and Eqs. 13-6 and 13-7 for chemical precipitates. If the sludge has an 8.0% solids concentration, compute the volume of waste in liters/m^3 of wastewater processed.

13-14   Estimate the quantity of sludge generated per million gallons of wastewater processed by physical–chemical treatment employing lime precipitation at a dosage of 300 mg/l as CaO. Raw wastewater and effluent characteristics are:

Component	Influent	Effluent
Suspended solids (mg/l)	300	40
Calcium (mg/l of Ca^{2+})	90	40
Magnesium (mg/l of Mg^{2+})	negligible	removal
Phosphorus (mg/l of P)	10	1

Assume that the settled sludge has a solids content of 10% and is concentrated to 30% by mechanical dewatering. Refer to Example 13-4 for a sample solution to this type of problem.

13-15   Why is it impractical to perform biological nitrification concurrently with BOD reduction in an activated-sludge aeration basin?

13-16   What is the reduction in rate of biological nitrification: (a) if the wastewater temperature decreases from 20°C to 10°C; (b) if the operating pH is 7.0 rather than 8.5? How can reduced temperature and pH be compensated in operating a suspended-growth nitrification system?

13-17   Calculate the aeration basin volume for suspended-growth nitrification of an average daily design flow equal to 1500 m^3/day containing 25 mg/l of ammonia nitrogen. Assume a minimum operating temperature of 16°C, operating pH 7.4, and design MLVSS of 1500 mg/l. If flow equalization is provided for the plant influent, what basin volume would you recommend?

13-18 Calculate the methanol addition needed for biological denitrification of 1500 m^3/day containing 25 mg/l of nitrate nitrogen and 8 mg/l of dissolved oxygen.

13-19 Tertiary nitrification–denitrification processing is being considered for a 30-mgd secondary-treatment plant. The wastewater characteristics selected for nitrification process design are: 30 mg/l of NH_3–N, 30 mg/l of BOD, temperature of 18°C, pH 7.4, and 1500 mg/l of MLVSS. Using these parameters, compute the aeration basin volume required, recommended clarifier surface area, and estimated oxygen utilization. Assume the following characteristics for subsequent denitrification design: 27 mg/l of NO_3–N, 5 mg/l of DO, temperature 16°C, pH 7.0, and 1500 mg/l of MLVSS. Calculate the denitrification basin volume, recommended clarifier surface area, and methanol dosage.

13-20 Discuss the major factors that limit the use of air-stripping towers for nitrogen removal.

13-21 Why is two-stage recarbonation preferred to single-stage for neutralization of wastewater following excess lime treatment?

13-22 Describe the proposed new ammonia-removal scheme at the Lake Tahoe Water Reclamation Plant.

13-23 Estimate the chlorine dosage needed to oxidize 15 mg/l of ammonia nitrogen. For this dosage, how many mg/l of chloride ion is added to the wastewater?

13-24 List some potential disadvantages of using ion exchange to remove nitrogen from wastewater.

13-25 What are the major disadvantages of the (a) overland flow and (b) infiltration–percolation methods of land disposal relative to irrigation?

13-26 List the environmental concerns of spray irrigation for disposal of treated municipal wastewater.

13-27 Using Eq. 13-24, estimate the land area in acres required for irrigation of 1.0 mgd based on the following: precipitation of 20 in./yr, evapotranspiration of 40 in./yr, percolation equal to 10 in./month, and zero runoff.

REFERENCES

1. *Flow Equalization*, EPA Technology Transfer Seminar Publication, Environmental Protection Agency (May 1974).

2. M. D. LaGrega and J. D. Keenan, "Effects of Equalizing Wastewater Flows," *J. Water Poll. Control Fed.* 46, no. 1 (January 1974): 123–132.

3. Russell L. Culp and Gordon L. Culp, *Advanced Wastewater Treatment* (New York: Van Nostrand Reinhold Company, 1971).

4. *Wastewater Filtration*, EPA Technology Transfer Seminar Publication, Environmental Protection Agency (July 1974).

5. J. L. Cleasby, "Filter Rate Control without Rate Controllers," *J. Am. Water Works Assoc.* 61, no. 4 (April 1969): 181–185.

6. E. R. Baumann and J. Y. C. Huang, "Granular Filters for Tertiary Wastewater Treatment," *J. Water Poll. Control Fed.* 46, no. 8 (August 1974): 1958–1972.

7. *Process Design Manual for Suspended Solids Removal,* Environmental Protection Agency, Technology Transfer (January 1975).

8. *Process Design Manual for Carbon Adsorption,* Environmental Protection Agency, Technology Transfer (October 1973).

9. P. Pieczonka and N. E. Hopson, "Phosphorus Detergent Ban—How Effective?," *Water and Sewage Works* 121, no. 7 (July 1974): 52–55.

10. A. B. Menar and D. Jenkins, "Fate of Phosphorus in Waste Treatment Processes: Enhanced Removal of Phosphate by Activated Sludge," *Environ. Sci. Technol.* 4, no. 12 (December 1970): 1115–1121, and *Proceedings, 24th Industrial Waste Conference* (Lafayette, Ind.: Purdue University, 1969): 655–673.

11. J. B. Carberry and M. W. Tenney, "Luxury Uptake of Phosphate by Activated Sludge," *J. Water Poll. Control Fed.* 45, no. 12 (December 1973): 2444–2462.

12. *Process Design Manual for Phosphorus Removal,* Environmental Protection Agency, Technology Transfer (April 1976).

13. D. T. Anderson and M. J. Hammer, "Effects of Alum Addition on Activated Sludge Biota," *Water and Sewage Works* 120, no. 1 (January 1973): 63–67.

14. J. C. O'Shaughnessy, J. B. Nesbitt, D. A. Long, and R. R. Kountz, "Digestion and Dewatering of Phosphorus-Enriched Sludges," *J. Water Poll. Control Fed.* 46, no. 8 (August 1974): 1914–1926.

15. L. A. Schmid and R. E. McKinney, "Phosphate Removal by a Lime-Biological Treatment Scheme," *J. Water Poll. Control Fed.* 41, no. 7 (July 1969): 803–815.

16. R. L. Culp, "Water Reclamation at South Tahoe," *Water and Wastes Eng.* 6, no. 4 (June 1969): 36–39.

17. *Physical–Chemical Wastewater Treatment Plant Design,* EPA Technology Transfer Seminar Publication, Environmental Protection Agency (August 1973).

18. T. F. X. Flynn and J. C. Thompson, "Physical–Chemical Treatment for Joint Municipal–Industrial Wastewater at Niagara Falls, New York," *Proceedings, 27th Industrial Waste Conference* (Lafayette, Ind.: Purdue University, 1972): 180–190.

19. H. E. Wild, Jr., C. N. Sawyer, and T. C. McMahon, "Factors Affecting Nitrification Kinetics," *J. Water Poll. Control Fed.* 43, no. 9 (September 1971): 1845–1854.

20. *Nitrification and Denitrification Facilities,* EPA Technology Transfer Seminar Publication, Environmental Protection Agency (August 1973).

21. R. L. Antonie, "Nitrification of Activated Sludge Effluent: BIO-SURF Process," *Water and Sewage Works* 121, no. 11 (November 1974): 44–47.

22. G. A. Duddles, S. E. Richardson, and E. F. Barth, "Plastic-Medium Trickling Filters for Biological Nitrogen Control," *J. Water Poll. Control Fed.* 46, no. 5 (May 1974): 937–946.

23. W. K. Johnson, "Process Kinetics for Denitrification," *Proc. Am. Soc. Civil Engrs., J. San. Eng. Div.* 98, no. SA4 (August 1972): 623–634.

24. P. M. Sutton, K. L. Murphy, and R. N. Dawson, "Low-Temperature Biological Denitrification of Wastewater," *J. Water Poll. Control Fed.* 47, no. 1 (January 1975): 122–134.

25. J. N. English et al., "Denitrification in Granular Carbon and Sand Columns," *J. Water Poll. Control Fed.* 46, no. 1 (January 1974): 28–42.

26. T. A. Tamblyn and B. R. Sword, "The Anaerobic Filter for the Denitrification of Agricultural Subsurface Drainage," *Proceedings, 24th Industrial Waste Conference* (Lafayette, Ind.: Purdue University, 1969): 1135–1150.

27. *Physical–Chemical Nitrogen Removal,* EPA Technology Transfer Seminar Publication, Environmental Protection Agency (July 1974).

28. J. G. Gonzales and R. L. Culp, "New Developments in Ammonia Stripping," *Public Works* 104, no. 6 (June 1973): 82.

29. L. G. Kepple, "Ammonia Removal and Recovery Becomes Feasible," *Water and Sewage Works* 121, no. 4 (April 1974): 42.

30. T. A. Pressley, D. F. Bishop, and S. G. Roan, "Ammonia-Nitrogen Removal by Breakpoint Chlorination," *Environ. Sci. Technol.* 6, no. 7 (July 1972): 622.

31. J. H. Koon and W. J. Kaufman, "Ammonia Removal from Municipal Wastewaters by Ion Exchange," *J. Water Poll. Control Fed.* 47, no. 3 (March 1975): 448–464.

32. D. R. Egeland, "Land Disposal I: A Giant Step Backward," *J. Water Poll. Control. Fed.* 45, no. 7 (July 1973): 1465–1475.

33. C. E. Pound and R. W. Crites, *Wastewater Treatment and Reuse by Land Application*, Vols. I and II, Office of Research and Development, Environmental Protection Agency (August 1973).

34. *Feasibility of Overland Flow for Treatment of Raw Domestic Wastewater*, Office of Research and Development, Environmental Protection Agency (July 1974).

35. H. Bouwer, J. C. Lance, and M. S. Riggs, "High-Rate Land Treatment II: Water Quality and Economic Aspects of the Flushing Meadows Project," *J. Water Poll. Control Fed.* 46, no. 5 (May 1974): 844–860.

36. *Land Treatment of Municipal Wastewater Effluents, Design Factors, Part I*, EPA Technology Transfer Seminar Publication, Environmental Protection Agency (August 1975).

37. *Land Treatment of Municipal Wastewater Effluents, Design Factors, Part II*, EPA Technology Transfer Seminar Publication, Environmental Protection Agency (August 1975).

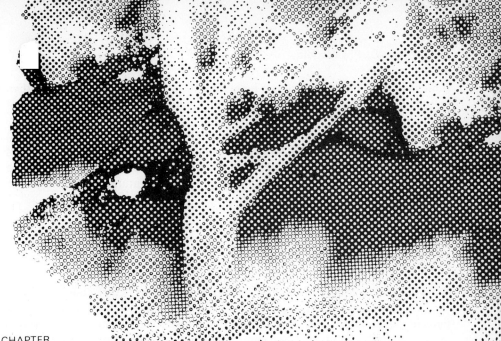

14

Water-Quality Models

Environmental systems are often delicately balanced, and perturbations imposed on them may produce either positive or negative effects. An understanding of the reaction of these systems under stress is fundamental to sound planning. Quantitative models can provide valuable insights to systems performance and, if used in the proper context, can be effective tools in the hands of water resources planners.

The use of models to guide the decision-making process is not new but is of a different, more sophisticated and more encompassing form than in the past. Models are representations of actual or proposed systems and their use permits manipulation over real times of seconds which would take years for the proto-type. This is the feature which makes use of these tools so attractive and which holds such potential for the analysis of even the largest, most complex systems. It is also the feature that makes this approach so well suited to water resources planning.

Models can be used to determine end results of various courses of action and translate questions being asked into predictive answers. Questions such as —What will be the net effect of a given level of treatment on the water quality of a stream? or, How will a particular watershed management practice affect water quality?—can be answered. Quantitative predictions of short- and long-run effects of conceived plans can be made with a higher level of certainty than before, and this enables decision makers to chart more rational courses of action.

The Federal Water Pollution Control Act Amendments of 1972 (PL 92-500) requires in Sections 201, 208, and 303e the development and implementation of waste-treatment management plans and practices. Water-quality models are widely employed to determine these practices, and the future holds even greater promise for their use.

14-1

WATER-QUALITY MODELS

A water-quality model is a mathematical statement or set of statements that equate water quality at a point of interest to causative factors. Mathematical modeling is an effective way to approach quantitative problem solving. Models may be simple or complex, the degree of sophistication depending on the complexity of the problem, the nature of objectives, and the ability to describe the system in mathematical terms.

Water-quality models should provide for (1) the determination of constituent concentration versus time at points of entry to the system, (2) the determination of the mixing and reaction kinetics of the system, and (3) the synthesis of a time-distributed output at the system outlet.

Either stochastic (containing probabilistic elements) or deterministic approaches may be taken in developing methods for predicting pollutional loads. The former technique is based upon determining the likelihood (frequency) of a particular output quality response by statistical means. This is similar to frequency analysis of floods or low flows. Water-quality records should be available for at least 5 years (preferably much longer) for this approach to have great fidelity.

The deterministic approach (output explicitly determined for a given input) requires that a model be developed to relate the water-quality loading to a known or assumed hydrologic input. Such a model may vary from an empirical concentration–discharge relationship to a soundly based physical equation representing the hydrochemical cycle. The ultimate modeling technique is that which best defines the actual mechanism triggering the water-quality response. The cause of a given state of pollution can then be specifically identified.

Water-Quality Constituents

Water-quality constituents may be classified as conservative or nonconservative; somewhat more narrowly as organic, inorganic, radiological, thermal, and biological; and finally further subdivided into specific forms, such as BOD, nitrogen, phosphorous, and so on. Pollutants of interest include fecal organisms, silt, pesticides, fertilizers, solid wastes, nitrates, and phosphates.[1-4]

Hydrologic data necessary to compute pollutional loadings of various constituents must be obtained concurrently with water-quality data to ensure proper calibration and verification of models. A knowledge of the frequency and time distribution of loading may be more important than a knowledge of the total loading. This is particularly true in cases where the objective is to determine the impact of a waste flow on a receiving stream. For example, a short, high-peaked surface runoff hydrograph of suspended matter could be expected to more seriously affect a stream than a hydrograph which released the same volume of suspended matter over an extended period of time. On the other hand, when waste flows enter a lake, the annual volume of contaminants takes on special importance. In general, it is desirable that water-quality data be recorded in a continuous manner so that the time rate of delivery of the constituent loading can be determined. If this is not done, only very gross estimates of the impact of water-quality inputs on receiving waters can be obtained.

Nonconservative Water Quality

Unstable pollutants such as radioactive wastes and heat which have a time-dependent decay are nonconservative in nature. Biochemical oxygen demand is a good example of such a pollutant.

Several nonconservative models have been proposed for rivers in the United States.[5-7] The application of nonconservative BOD water-quality models is clearly illustrated by Smith et al.,[5] Thayer and Krutchkoff,[6] and Loucks et al.[7] The classic dissolved-oxygen depletion model is given by the Streeter–Phelps equation (Chapter 7).

Development of useful nonconservative models is dependent upon (1) reliable input data related to surface runoff and other discharges, (2) representative mixing models, and (3) knowledge of the reaction kinetics involved.

Conservative Water Quality

Some inorganic pollutants can be considered conservative. The essential features of a conservative water-quality model include (1) the determination of the input to the system by relating it to the study-area hydrology and/or other factors, and (2) the determination of a mixing model to be employed. Useful studies can be found in references 5–11. Steele[8] gives a very good summary of previous studies designed to relate water quality to various defining parameters. Models representing the transport of solutes from a watershed may be developed according to known physical laws or may be largely empirical in nature.

The strict physical approach attempts to evaluate the change in the water's chemical character as it moves through the phases of the hydrochemical cycle. For each phase, the quantity of runoff involved, the nature of the chemical reactions governing the changes, and the character of any natural constraints must be known or simulated. The hydrologic simulation or modeling can be carried out according to various schemes.

The problem of the chemical change is difficult to resolve.[8] Few reliable studies of the chemical phenomena of dissolution and ion exchange under natural conditions are available. Physical conditions affecting chemical reactions occurring in the field are not subject to the simple control of the laboratory. In addition, the chemical, physical, and hydraulic character of the total hydrochemical system of a watershed defies an exact determination of the contribution to total solute load by individual phases of the cycle.[8]

Fortunately, it has been demonstrated that the complexities of natural water-quality systems can sometimes be circumvented. Several studies have shown that such systems frequently exhibit consistent chemical behavior which can be explained rationally in terms of adequate field data.[8] An empirical approach can often be used to identify the more important factors controlling water-quality changes on a watershed and to develop useful working relationships for predicting water-quality inputs.

Some studies have indicated that chemical quality data can be satisfactorily related to discharge by an equation or series of equations of the form

$$C = kQ^n \qquad (14\text{-}1)$$

where C is the constituent concentration, Q is the stream flow, and k and n are regression parameters.[5,8,10] Figure 14-1 is typical of such a relationship.[8]

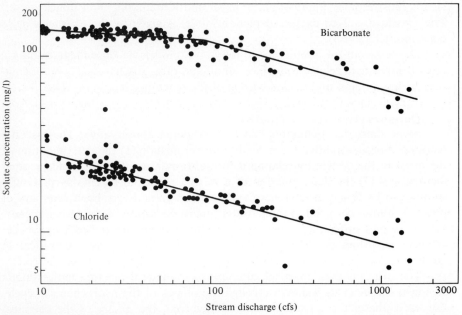

FIGURE 14-1 Concentration–discharge relations for selected solutes—San Lorenzo River at Big Trees, California. [Source: T. D. Steele, "Seasonal Variations in Chemical Quality of Surface Water in the Pescadero Creek Watershed, San Mateo County, California," Ph.D. Dissertation (Stanford, Calif.: Stanford University, May 1968).]

Based on somewhat limited data, Weibel, Anderson, and Woodward have published graphical correlations between rain (in./storm) and BOD (lb/storm), solids (mg/l) and flow ranges (cfs) for urban data on a 27-acre area in Cincinnati.[12] Although these relationships are on a total storm basis, they do indicate apparent trends.

14-2

SIMPLE WATER-QUALITY MODEL FOR DETERMINING LAKE FLUSHING RATES

To illustrate the conceptualization and construction of a water-quality model, the following example relative to estimation of lake flushing rates is given.[14]

Individuality of Lakes

Lakes have certain common general features but often strikingly different performance characteristics. This individuality has its virtues, but to the analyst it presents the problem of having to understand both the general system functioning plus variations of the generalized performance due to important local traits.

Zumberge and Ayers[13] define a lake as "an inland basin filled or partially filled by a water body whose surface dimensions are sufficiently large to sustain waves capable of producing a barren wave-swept shore." This is the generalization. Of significance is the fact that hydrologic characteristics of lake systems vary considerably because of differences in depth, length, width, surface area, basin material, surrounding ground cover, reservoir, prevailing winds, climate, surface inflows and outflows, and other factors. What this means in terms of analysis is that each lake will require its own model and that these models will be characterized by different degrees of variance from a generalized conceptual model.

Water Budget

A *water budget* is an accounting of the disposition of storage, inflow, and outflow of a system during some period of time. The budget may be static or dynamic, depending upon the choice of a single interval for investigation or of a sequence of intervals. The water budget, constituent budget (to be defined later), and mixing process of the lake jointly define the flushing rates for both water and pollutants or other constituents.

Figure 14-2 illustrates conceptually the various components of the water budget. Classifying these as inputs, outputs, and storage elements permits writing the water budget in the form

$$\text{input–output} = \text{change in storage} \qquad (14\text{-}2)$$

$$(I_c + I_o + I_g + P + R) - (E + T + G_s + O_c + W) = \Delta S \qquad (14\text{-}3)$$

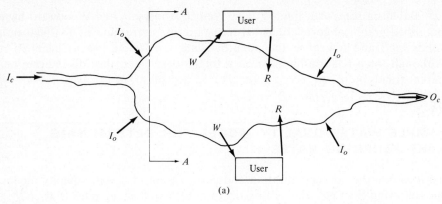

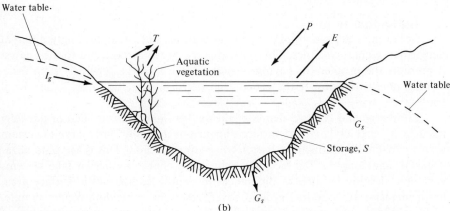

FIGURE 14-2 Conceptual lake illustrating variables in the water budget. (a) Plan view; (b) section A-A.

where I_c = channel inflow
I_o = overland inflow
I_g = groundwater inflow
P = precipitation
R = return flow
E = evaporation
T = transpiration
G_s = seepage
O_c = channel outflow
W = withdrawal
S = lake storage at time t
t = time

Rearranging Eq. 14-3 yields

$$(I_c + I_o + I_g + P + R) - (E + T + W) = f(\Delta S, \ldots) \qquad (14\text{-}4)$$

where

$$f(\Delta S, \ldots) = G_s(S_1 + \Delta S, \ldots) + O_c(S_1 + \Delta S, \ldots) + S_2 - S_1 \qquad (14\text{-}5)$$

and the subscripts on S denote time periods.

Conceptually, Eq. 14-4 may be solved for ΔS if all the terms on the left-hand side can be evaluated and the initial storage S_1 is known or can be estimated. A knowledge of ΔS will permit determination of the water flushing rate $(O_c + G_s)$ if the functional relationships given in Eq. 14-5 are defined.

Terms requiring quantification for solution of Eq. 14-4 may be evaluated as follows:

1 Channel inflow (I_c): by gaging, deterministic relationship to causal factors, generation of stochastic records.
2 Overland inflow (I_o): by rainfall–runoff relationships, physics of flow equations, gaging.
3 Groundwater inflow (I_g): by seepage equations, flow-net analyses, gaging.
4 Precipitation (P): by gaging, generation of stochastic traces.
5 Return flow (R): by gaging, estimation.
6 Evaporation (E): by gaging, evaporation-prediction relationships.
7 Transpiration (T): by consumptive use equations, gaging.
8 Withdrawal (W): by gaging, estimation.

Estimation or measurement of any of these variables may present a challenge in itself, and the reliability of estimated values will vary with availability and accuracy of prediction techniques, quality of gaging, time period involved, and other factors. As a result, calculated flushing rates must be qualified as to expected range of errors. In practice, actual computations are not as straightforward as the previous remarks might indicate. A good deal of modeling capability is usually a requisite.

Solutions to functional relationships between water-budget variables are dependent upon quantification of parameters descriptive of characteristics of the particular lake system to be studied. For example, seepage rates are related to permeability of the surrounding soils and water temperatures. To solve the water-budget equation, the initial storage volume of the lake is needed. Changes in water surface level can only be determined if the geometry of the basin is known. All types of descriptive data are required. It may also be necessary to continuously monitor precipitation, stream flow, evaporation, return flows, and withdrawals. These records may serve as a historic basis for synthetic data generation or be used directly in "real-time" control systems. The system of interest must be understood and described quantitatively if projections or estimates are to be made.

Constituent Budget
Water quantity and quality considerations are directly linked. In order to account for the movement of pollutants in a lake, it is necessary to combine the

water budget with a constituent budget and a model of the mixing mechanics of the system. Unfortunately, the complexities of the water budget pale in comparison to those of the constituent budget. This is because the constituent budget includes conservative and nonconservative components and is highly dependent upon the circulation of the lake. The ultimate disposition of constituents and the reaction kinetics of both chemical and biological components are affected by circulation. An incomplete understanding of the mixing mechanics of large water bodies and of various aspects of the hydrochemical and hydrobiological cycles compounds the problem. Models combining these aspects are—to say the least—difficult to construct.

Conceptually, the constituent budget is similar to the water budget—an accounting of inputs, storages, and outputs. Basically, a separate budget must be struck for each constituent of interest. A simple conceptual model structure for a single constituent (I) would be

$$C_{II} - C_{IO} = \frac{\Delta C_{IC}}{\Delta t} \tag{14-6}$$

where C_{II} = constituent input to the system per unit time
   C_{IO} = constituent output from the system per unit time
   ΔC_{IC} = change in concentration of constituent stored within the system
   Δt = time interval of interest

Note that in general,

$$C_{II} = \sum_{i=1}^{n} C_{IIi} \tag{14-7}$$

$$C_{IO} = \sum_{j=1}^{m} C_{IOj} \tag{14-8}$$

where i and j represent the total number of sources and outlets of constituent I, respectively. The notation C_{II3} would represent the rate of input of constituent I from source 3, for example. It should be understood that the values of the C_{IIi} and C_{IOj} must be determined by monitoring, the use of empirical or theoretical relationships, or some form of projection. These determinations in themselves may be quite complex. Also,

$$C_{IOj} = f(C_{IIi}, \text{mixing}, \ldots) \tag{14-9}$$

Data needs for constituent budgeting are extensive. They range from simple monitoring to the determination of reaction constants for functional relationships. An initial value of the concentration of each constituent in the system must be obtained for use in the budget equation. Data on the thermal properties of the lake are vital. Unless input prediction equations can be developed, source monitoring is essential.

Mixing

Interest in flushing rates centers around the need to determine how rapidly undesirable constituents may be removed from a lake system under varying conditions of constituent loading and hydrology. The rate of removal of a constituent is highly dependent on mixing in the lake. To explain this, consider the following situation. Water-budget calculations indicate an outflow equal to the entire volume of a lake during some time period. Does this mean that an undesirable constituent would be totally flushed during this same interval, assuming that the input had ceased immediately prior to the beginning of the time period involved? No it does not. In fact, if the system inflow were primarily a surface current, little constituent removal would occur. The greater the mixing near the pollutants, the more rapid the flushing. Principal mixing mechanisms for lakes are the actions or interactions of wind, temperature changes, and atmospheric pressure. In shallow lakes with high volumes of boat traffic, propeller stirring may also be significant.

Thermal Considerations

Water has its temperature of maximum density (4°C) above the freezing point and, as a result, most lakes in the temperate zone have two thermal circulation periods, one in the spring and another in the fall. During these periods a condition of temperature homogeneity results and vertical circulation occurs.

In the fall, surface waters become cooled and these dense waters sink, to become replaced by warmer water from lower elevations. As surface temperatures approach 4°C, surface layers become denser than any underlying layers and sink to the bottom. This convective circulation finally establishes a lake of uniform density which can be mixed at all levels by the action of the wind. Continued cooling below 4°C establishes an inverse temperature structure with colder water on the top. After 4°C and prior to freeze-up, the overturn may cease at any time once the surface density is reduced enough to preclude wind mixing to any considerable depth. When ice cover is established, the mixing potential of the wind is shielded against, and wind mixing is ineffective until spring.

When springtime surface temperatures again approach 4°C, convective sinking and wind action again bring the lake to a uniform temperature and a state of complete vertical circulation. As the surface waters warm further, a density gradient develops which ultimately rules out complete wind mixing and again the lake tends to stagnate. This is the period of summer stratification.

Complete mixing of lake waters is in general limited to the periods of spring and fall overturn. During other periods, a more limited mixing due to various currents may occur.

Currents

Most lakes have an inlet and outlet and the resulting flow-through generates what is known as the *gradient* or *slope current*. The relative dimensions of the inlet and outlet channels and those of the lake are usually such as to rule out any significant gradient currents. When mixing is considered, these currents are usually of negligible importance when compared to those generated by the wind.

The principal classifications of lake currents are periodic and non-periodic.[13,15] *Periodic currents* are associated with *seiches* (water-surface oscillations), and although these may have important local effects on mixing, they are not often as important in dispersing constituents as nonperiodic currents. *Nonperiodic* currents are the product of wind-stress energy transfer to the surface water. The exact mechanism of transfer is not fully known, but it is considered that wind-stress energy input is a function of velocity to some power (usually considered to be from about 1.8 to 3).[13,15] A frequently used relationship is

$$T = 3.2 \times 10^{-6} W^2 \tag{14-10}$$

where T is wind stress in g/cm/sec^2 and W is the wind velocity in cm/sec.[13]

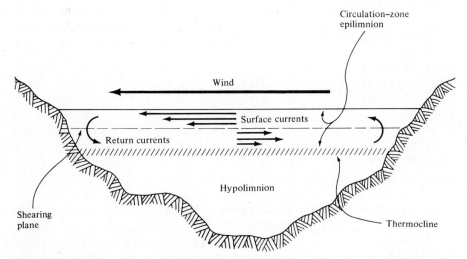

FIGURE 14-3 Relative velocities of wind-induced currents.

The wind-stress energy at the water surface produces an acceleration of water particles. Some of this acceleration generates waves and the remainder produces current. Wind-produced surface currents are usually on the order of 1–3 % of the wind velocity generating them. Along-shore and open-lake currents may be produced by wind action. The distinction is based on the point of departure from the influence of shoreline and bottom topography. Figures 14-3 and 14-4 indicate some possible wind-induced circulation patterns and the role played by multiple density strata.

Circulation Models

Exacting analyses of circulation in stratified lakes are elusive. This is because of the highly complex nature of the circulation process. Such problems can be

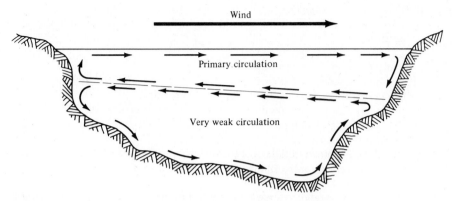

FIGURE 14-4 Hypothetical circulation pattern in a stratified model.

attacked by large computer programs,[23] but even these incorporate many simplifying assumptions. Various models have been proposed and studied for special cases and particular solution orientations.[16-24] Some are research steps and others offer the potential for direct practical application to the determination of circulation in any given lake when combined with field data on velocities.[19] A thorough discussion of circulation mechanics is beyond the scope of this book and the interested reader should consult the references cited.

Crude Flushing Model

Crude estimates often play a useful role in planning. They can usually be arrived at quickly and indicate the order of magnitude or range of values likely to be estimated by more sophisticated methods. Sometimes this information alone will prove adequate; in other cases, useful insight into the efforts required or degree of refinement needed to obtain better results will be the by-product.

A rough estimate of both the maximum and minimum flushing rates of any constituent of interest can be obtained from a minimal knowledge of the lake characteristics and inflow and outflow hydrology.

A   *Maximal flushing (annual)*
   1   Using samples, estimate the average concentration of constituent in the lake at the beginning of the year ($t = T_1$).
   2   Use projections, gaging records, or other means to estimate added inflow concentration during the year ($t_2 - t_1$).
   3   Assume complete mixing of the lake.
   4   From a hydrologic analysis, or based on historic records, estimate average outlet flow over the period ($t_2 - t_1$).
   5   Combine the mean initial constituent concentration at $t = t_1$ with the added mean concentration to obtain the total mixed mean concentration of the constituent.
   6   Multiply the total mixed mean concentration by volume of outflow to arrive at constituent quantity flushed.

B *Minimal flushing (annual)*

1 Using samples, estimate the average concentration of constituent in the lake at the start of the year ($t = t_1$).

2 Use projections, gaging records, or other means to estimate added inflow concentration during the year ($t_2 - t_1$).

3 Estimate average length of time span of spring and fall overturn from observations or records of similar lakes in the area.

4 Assume complete mixing during spring and fall overturns only.

5 Use hydrologic analysis or historic records to estimate mean outlet flow during overturns.

6 Combine the mean initial constituent concentration at $t = t_1$ with the added mean concentration to obtain the total mixed mean concentration of the constituent.

7 Multiply the total mixed mean concentration by volume of water discharged during overturns to arrive at constituent quantity flushed.

These are crude approximations and do not include such considerations as fixation of constituents to bottom deposits, recycling, and others, but they could serve as preliminary planning guidelines and illustrate the basis for the formation of more complex water-quality models, such as those discussed in following sections.

14-3

CHEN–ORLOB MODEL

Chen and Orlob have developed water-quality models for estuarine and lake (reservoir) systems.[25, 26, 28] The lake model is based on a multilayered system, as shown in Fig. 14-5. Horizontal sections are assumed to be completely mixed and thus unstratified. The lake's hydrodynamic behavior is considered density-dependent, with temperatures being the principal determining variable.

Tributary inflows are considered to enter the lake at a level where the surrounding density is their equivalent. As a result of incompressibility, the inflow is assumed to generate an advective flow between the elements above the level of entry. Orlob and Selna have demonstrated that this approach is satisfactory for temperature simulation of lakes and reservoirs.[33] The differential equations describing the quality of the lake ecosystem are solved implicitly.[26, 28]

To use the model, various input data are required. They include hydrologic and water-quality parameters of tributary inflows, the amount and nature of waste discharges, the character of outflows, and meteorologic conditions. Computations can be performed over time periods ranging from hours to 1 day. Table 14-1 gives input requirements of the model and shows the quality variables that can be simulated.

For computational facility, each lake layer is considered as a discrete continuously stirred tank reactor.[26, 28] Hydrodynamic calculations determine the inflow and outflow from each hydraulic element. Changes in concentration of

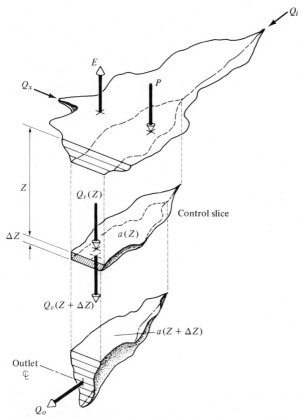

FIGURE 14-5 One-dimensional idealization of a reservoir. [Source: After G. T. Orlob and L. G. Selna, "Temperature Variations in Deep Reservoirs," *Proc. Am.. Soc. Civil Engrs., Hydraulics Div.* 96, HY2, Proc. Paper 7063 (February 1970): 391–410.]

water-quality constituents or biota are related to conditions existing in the assumed reactor. Mass-balance equations describing the physical, chemical, and biological processes affecting the ecosystem variables are used to determine rates of change. Constituent concentrations as a function of time are computed using numerical techniques.

14-4

LOMBARDO–FRANZ MODEL

Lombardo and Franz have developed a model (QUALITY) which simulates the water-quality dynamics of rivers and impoundments.[27, 28] It is used jointly with the Hydrocomp Hydrologic Simulation Program (HSP) for calculating the hydrologic response of a watershed.[34,35] By using both models, the hydrologic and water-quality interactions of a watershed can be simulated.

TABLE 14-1

VARIABLES OF LAKE AND ESTUARY ECOLOGIC MODEL

Parameters	Model Variables[a]			
	Estuarial System		Lake System	
	Hydraulic	Ecologic	Hydraulic	Ecologic
Climate (Zones)				
Latitude	I	—	I	—
Longitude	I	—	I	—
Atmospheric pressure	I	—	I	—
Cloud cover	I	—	I	—
Wind speed	I	—	I	—
Dry-bulb temperature	I	—	I	—
Wet-bulb temperature	I	—	I	—
Wind direction	I	—	I	—
Evaporation	0	C	0	C
Short-wave radiation	0	C	0	C
Net solar radiation	0	C	0	C
Geometry				
Function				
Surface area	I	C	I	C
Side slope	I	I	I	I
Elevation	I	—	I	I
Volume	C	C	C	C
Channel				
Length	I	I	—	—
Width	I	I	—	—
Friction factor	I	—	—	—
Hydrology				
External flow				
Rivers	I	I	I	I
Tide (one point)	I	—	—	—
Waste discharges	I	I	—	—
Outflow	I	I	I	I
Internal flow				
Channel flows	0	C	—	—
Surface overflow	0	C	0	C
Quality constituent[b]				
Temperature	—	0	0	0
Toxicity	—	0	—	—
Total dissolved solid	—	0	—	0
Coliform	—	0	—	0
BOD	—	0	—	0
Oxygen	—	0	—	0
Total carbon (inorganic)	—	C	—	C
PO_4–P	—	0	—	0
Alkalinity	—	0	—	0

Footnotes on p. 788.

TABLE 14-1 (*continued*)

Parameters	Estuarial System		Lake System	
	Hydraulic	Ecologic	Hydraulic	Ecologic
NH$_3$–N	—	0	—	0
NO$_2$–N	—	0	—	0
NO$_3$–N	—	0	—	0
Algae (two groups)	—	0	—	0
Zooplankton	—	0	—	0
Fish (two groups)	—	0	—	0
Benthic animal	—	0	—	0
Detritus	—	0	—	0
CO$_2$	—	0	—	0
PH	—	0	—	0
System coefficients				
Light extinction				
Background	—	I	—	I
Algal suspension	—	I	—	I
Reaeration				
Oxygen	—	I	—	I
CO$_2$–C	—	I	—	I
Decay rate				
BOD	—	I	—	I
Detritus	—	I	—	I
Coliform	—	I	—	I
NH$_3$–N	—	I	—	I
NO$_2$–N	—	I	—	I
Temperature coefficient	—	I	—	I
Algae				
Respiration	—	I	—	I
Settling velocity	—	I	—	I
Oxygenation factor	—	I	—	I
Chemical composition				
(C, N, P)	—	I	—	I
Maximum specific growth	—	I	—	I
Half-saturation constants				
(light, carbon, nitrogen,				
phosphorus)	—	I	—	I
Zooplankton				
Respiration	—	I	—	I
Mortality coefficients	—	I	—	I
Digestive efficiency	—	I	—	I
Chemical composition				
(C, N, P)	—	I	—	I

Footnotes on p. 788.

TABLE 14-1 (*continued*)

| Parameters | Model Variables[a] | | | |
| | Estuarial System | | Lake System | |
	Hydraulic	Ecologic	Hydraulic	Ecologic
Preference for algae	—	I	—	I
Maximum specific growth	—	I	—	I
Half-saturation constants				
(algae, detritus)	—	I	—	I
Fish (two groups)				
Respiration		I		I
Mortality coefficients		I		I
Digestive efficiency		I		I
Chemical composition				
(C, N, P)		I		I
Maximum specific growth		I		I
Half-saturation constants				
(zooplankton, benthic				
animal)		I		I
Benthic animal				
Respiration	—	I	—	I
Mortality coefficients	—	I	—	I
Digestive efficiency	—	I	—	I
Chemical composition (C, N, P)	—	I	—	I
Maximum specific growth	—	I	—	I
Half-saturation constant				
(detritus)	—	I	—	I
Others				
Chemical composition (C, N, P)				
Zooplankton pellet	—	C	—	C
Fish pellet	—	C	—	C
Benthic pellet	—	C	—	C
Detritus	—	I	—	I

[a] I, input; C, calculated; 0, output; —, not applicable.
[b] Requires initial conditions.

Source: C. W. Chen and G. T. Orlob, "Ecologic Simulation for Aquatic Environments, Annual Report," *OWRR Project No. C–2044*, Office of Water Resources Research (Washington, D.C.: Department of the Interior, August 1971).

The Hydrocomp model is based on the subdivision of a watershed into land segments and stream reaches. The HSP system includes three models: LIBRARY, LANDS, and CHANNELS. LIBRARY is the data-handling component, LANDS calculates channel inflow volumes from input rainfall and evapotranspiration, and CHANNELS routes the computed channel inflows for each reach progressively downstream, using the kinematic wave routing method. Point sources and diversions may be specified along the stream system. The model is not designed to account for tidal effects.

The water-quality model (QUALITY) is a separate module used with LIBRARY and LANDS and CHANNELS if desired. QUALITY can simulate water-quality changes in channel flows and, when used in conjunction with LANDS, surface-runoff quality as well. Streams are represented by a series of reaches in the model and lakes by a system of three layers (Figs. 14-6 and 14-7). Each reach and lake layer is modeled as a continuously stirred tank reactor. The partial differential equations that describe the water-quality dynamics are solved using a multiple-step explicit solution. Dispersion is assumed to be negligible. The model can accommodate a watershed with any number of lakes.

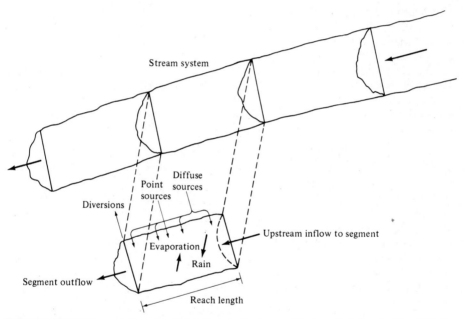

FIGURE 14-6 Representation of a stream hydrologic system. [Source: P. S. Lombardo and D. D. Franz, "Mathematical Model of Water Quality in Rivers and Impoundments" (Palo Alto, Calif.: Hydrocomp, Inc., December 1972).]

To operate the model, stream reach and system parameters are input. Water-quality indices which can be simulated are: temperature, biochemical oxygen demand, coliforms (total, fecal, fecal streptococci), algae–chlorophyll *a*, zooplankton, sediment, organic nitrogen, dissolved oxygen, total dissolved solids, nitrite–N, orthophosphate, potential phosphorus, ammonia–N, conservative constituents. A complete discussion of the model may be found in references 27 and 28.

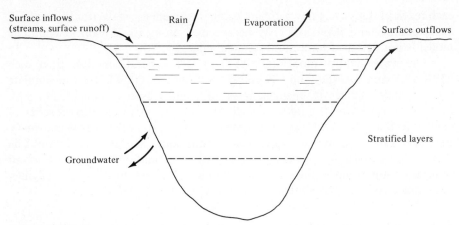

FIGURE 14-7 Representation of an impoundment system. [Source: P. S. Lombardo and D. D. Franz, "Mathematical Model of Water Quality in Rivers and Impoundments," (Palo Alto, Calif.: Hydrocomp, Inc., December 1972).]

14-5

QUAL-I AND DOSAG-I SIMULATION MODELS

QUAL-I and DOSAG-I were developed by the Texas Water Development Board for use in stream quality-simulation studies.[32, 36] QUAL-I is designed to simulate spatial and temporal variations of temperatures, biochemical oxygen demand, dissolved oxygen, and conservative minerals in streams and canals. DOSAG is used to simulate the spatial variation of BOD/DO in a system of streams and canals. It is not as accurate as QUAL-I and cannot be used for routing temperature and conservative minerals. QUAL-I is best suited for detailed studies of a stream under a prescribed condition, whereas DOSAG is better designed for use over a wide range of conditions when less detail is required. QUAL-I will be described in some detail and its use illustrated by an application of it to the Salt Creek Watershed in Nebraska. The program for QUAL-I was written by William A. White.[32] The following discussion is similar to that of the Texas Water Development Board given in reference 32.

QUAL-I is a set of interrelated quality routing models, the operation of which produces a time history and spatial distribution of temperature, biochemical oxygen demand/dissolved oxygen (BOD/DO), and up to three conservative minerals. The framework for its use is a one-dimensional, fully mixed, branching stream or canal system with multiple waste inputs and withdrawals. The models may be operated independently or simultaneously, with one model providing input to another if it is necessary. There are seven options:

1. Route temperature, BOD/DO, and conservative minerals.
2. Route temperature and BOD/DO.
3. Route BOD/DO.

4 Route conservative minerals and temperature.
5 Route temperature.
6 Route BOD/DO and conservative minerals.
7 Route conservative minerals.

Flow augmentation requirements based on preselected minimum allowable dissolved-oxygen concentrations can be set by the user. Restrictions on use of the program are as follows:

1 Maximum number of reaches = 25.
2 Maximum number of waste inputs = 25.
3 Maximum number of headwaters = 5.
4 Maximum number of junctions = 5.
5 Maximum number of computational elements = 500.

In QUAL-I, routing calculations are initiated at the most upstream locations (headwaters) of the stream or canal system. Incremental inflows and waste inputs or withdrawals are entered into the calculations as they occur. A set of simultaneous equations equal in number to the number of computational elements in the system is thus generated at the downstream point. These are solved, advancing the solution forward in time. An iterative process is continued until steady-state conditions are reached. This is the approximate flow time from the uppermost point in the system to its terminus.

To use the program it is first necessary to determine which segments of the stream or canal system are to be simulated. A schematic diagram of the stream system similar to the one shown in Fig. 14-8 facilitates this. The next step is to determine the degree of detail needed. This should be based on the availability and reliability of stream-flow and water-quality data, stream geometry, and the number and location of waste inputs or withdrawals. The stream is then subdivided into reaches having nearly uniform characteristics and reaches, headwaters, waste discharges or withdrawals, and junctions are ordered sequentially from the uppermost point in the system.

The final step is to determine the degree of resolution required. This decision should reflect some knowledge of the prototype behavior. For example, if the dissolved-oxygen concentration changes from saturated to critical and returns to saturated through an interval of about 5 river miles, a degree of resolution of less than 1 mi is appropriate.[32] This decision permits each reach to be divided into computational elements or control volumes, as shown on Fig. 14-8. Complete program details and procedures for use are given in reference 32.

Application of QUAL-I to the Salt Creek Watershed

To illustrate the QUAL-I model in more detail and to provide a better insight into the usefulness of model output in water resources planning processes, an example of the application of QUAL-I to an actual watershed follows. The information presented follows the discussion of use of the model by the Nebraska Natural Resources Commission.[30]

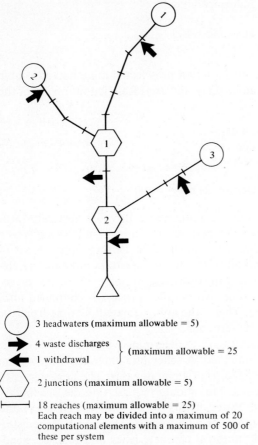

FIGURE 14-8 A schematic diagram of a hypothetical stream system.

The Nebraska Natural Resources Commission, in cooperation with the Environmental Protection Agency, developed the Salt Creek water-quality model for determining the assimilative capacity of Salt Creek.[30] The QUAL-I modeling technique was selected. Water-quality parameters examined in the Salt Creek system were biochemical oxygen demand (BOD) and dissolved oxygen (DO). Sulfate, chloride, and conductivity were also considered but are not discussed herein.

QUAL-I routes these parameters through the stream system on an hourly basis. It assumes that the major transport mechanisms, advection and dispersion, are only along the principal direction of flow of the stream. It allows for multiple waste discharges, withdrawals, tributary flows, and incremental runoff, and has the capability to compute required dilution flows for flow augmentation to meet specified dissolved-oxygen levels. The model was applied to the lower Platte River network (including Salt Creek), as shown on Fig. 14-9.

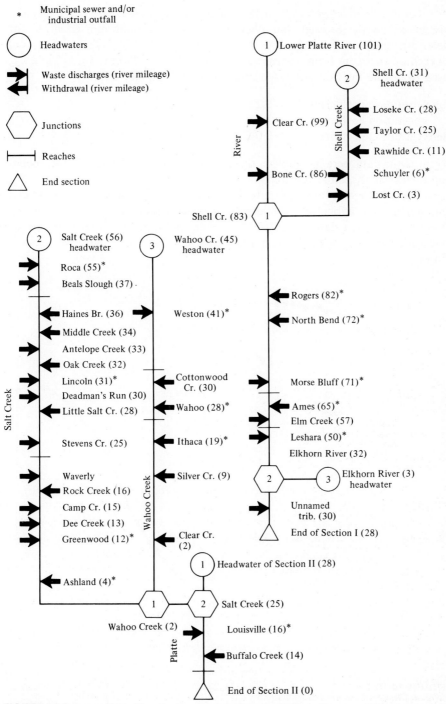

FIGURE 14-9 Lower Platte River basin schematic diagram QUAL-I. (Courtesy Nebraska Natural Resources Commission.)

The dissolved oxygen and biochemical oxygen demand portions of the model are based on the Streeter and Phelps mass-action equilibrium equation for determining the oxygen-sag curve. This curve represents the deficit in dissolved oxygen downstream from a wastewater source. The coefficients k_1, k_2, and k_3, used to denote the rate of oxygen utilization and renewal, were:

deoxygenation coefficient, $k_1 = 0.2$ to base 10/day
    reaeration coefficient, $k_2 = 1.6$–10.8 to base 10/day
deoxygenation coefficient, $k_3 = 0.0$–0.4 to base 10/day

Because the data were incomplete, these values were considered tentative, pending further analysis and research.

The Salt Creek Model was developed principally for use with low flows and did not include agricultural pollutants or urban runoff. It was verified using flows and sampling data from 1972. Constants and flow rates were adjusted for use with the 1970 7-day, 10-year low flows and the 1985 7-day, 10-year low flows.

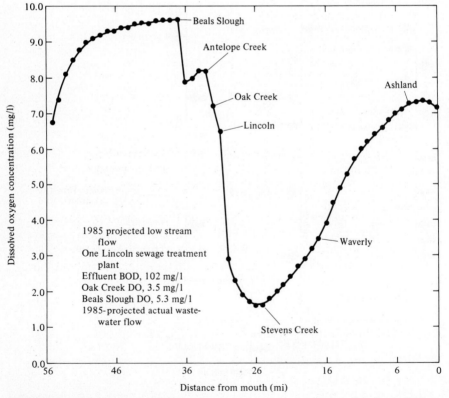

FIGURE 14-10 Results of simulation 1 : dissolved oxygen concentration on Salt Creek. (Courtesy Nebraska Natural Resources Commission.)

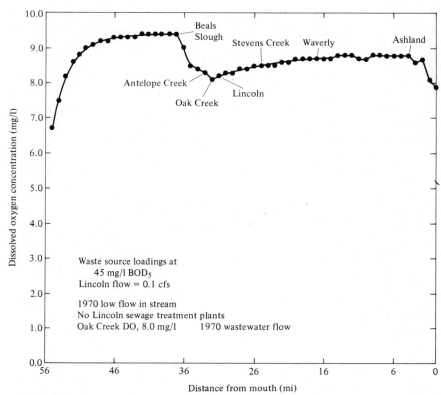

FIGURE 14-11 Results of simulation 2: dissolved-oxygen concentration on Salt Creek. (Courtesy Nebraska Natural Resources Commission.)

The results of two simulation runs are given on Figs. 14-10 and 14-11. The first illustrates the impact of limited wastewater treatment at Lincoln, Nebraska, during a drought. The second shows the effects of eliminating Lincoln's waste flows from Salt Creek.

The point to be made here is that while the model is not an exact replica of Salt Creek, it does give a reliable interpretation of how the stream would act under various conditions of stress. Such information is of considerable value in the planning and management of local and regional wastewater treatment facilities.

14-6

EPA STORM WATER MANAGEMENT MODEL

The EPA Storm Water Management Model was developed by Metcalf and Eddy, Inc.; the University of Florida; and Water Resources Engineers, Inc.[38] The objective was to produce a combined sewer system model which would simulate the quantity and quality of storm-water flows from urbanized areas.

The program consists of a control segment and five computational blocks: (1) Executive; (2) Runoff; (3) Transport; (4) Storage; and (5) Receiving Water.

The Executive block is the control for computational blocks, with all access and transfers between the other blocks routed through its subroutine MAIN. The Runoff block accepts rainfall data and a description of the drainage system as inputs and provides both hydrographs and time-dependent pollutional graphs (pollutographs). The Transport block routes through the distribution system flows which are corrected for both dry-weather flow and infiltration. It is also capable of producing hydrographs and pollutographs at specified points. In the Storage segment, flows from the Transport block are received and the effect of any treatment provided is considered. Effects of the discharge on a receiving stream are determined in the Receiving Water block. Segments may be run separately to facilitate adjustments.

The drainage area is divided into subcatchments, gutters, and pipes. Subcatchments are divided into three parts: pervious, impervious with surface detention, and impervious without surface detention. Subcatchments are defined by area, width, slope, and ground cover, while gutters and pipes are described by slope, length, and Manning's roughness coefficient. This information and rainfall data in the form of hyetographs are read into the Runoff block. A step-by-step computation of runoff volume is then initiated. The mathematical procedure is as follows:

1 The water depth on the subcatchment is found from the hyetograph as follows:

$$D_1 = D_t + R_t \, \Delta t \tag{14-11}$$

where D_1 = water depth after rainfall
D_t = water depth at time t
R_t = intensity of rain, time interval Δt

2 Horton's exponential function is used to account for infiltration.

$$I_t = f_o + (f_i - f_o)e^{-at} \tag{14-12}$$

where I_t = infiltration intensity
f_o, f_i, and a = Horton's coefficients

3 Infiltration is subtracted from water depth according to the following equation:

$$D_2 = D_1 - I_t \, \Delta t \tag{14-13}$$

where D_2 = depth after infiltration
I_t = infiltration rate at time t

4 The water depth D_2 is compared with a specific detention value D_d and, if found greater, the outflow for the catchment is found by using a form of Manning's equation.

$$V = \frac{1.49}{n(D_2 - D_d)^{2/3}s^{1/2}} \tag{14-14}$$

where $V =$ velocity
$n =$ Manning's coefficient
$D_d =$ detention requirement
$s =$ ground slope

The outflow is computed as

$$Q_w = VW(D_2 - D_d)$$

where W is the width of the area.

5 Water depths on the subcatchments are computed by the continuity equation:

$$D_{t+\Delta t} = D_2 - \frac{Q_w}{A\,\Delta t} \tag{14-15}$$

where A is the surface area of the subcatchment.

6 The preceding steps are continued for all subcatchments.

7 Gutter inflow is found by adding the outflow of the subcatchments tributary to it and the flow from all upstream gutters.

$$Q_{in} = \sum Q_{w,i} + \sum Q_{g,i} \tag{14-16}$$

where $\sum Q_{w,i} =$ sum of flow from subcatchments
$\sum Q_{g,i} =$ sum of flow from upstream gutters

8 Depth of flow in gutters is calculated as follows:

$$Y_1 = Y_t + \frac{Q_{in}}{A_s}\Delta t \tag{14-17}$$

where $Y_1,\ Y_t =$ water depths in gutter
$A_s =$ mean water surface area between Y_1 and Y_t

9 Outflow from the gutters is computed from Manning's equation:

$$V = \frac{1.49}{n(R)^{2/3}(S_i)^{1/2}} \tag{14-18}$$

where $R =$ hydraulic radius
$S_i =$ invert slope

$$Q_g = VA_c$$

where A_c is the cross-sectional area at Y_1.

10 Water depth in gutters is found using the continuity equation:

$$Y_{t+\Delta t} = Y_1 + \frac{(Q_{in} - Q_g)\,\Delta t}{A_s} \tag{14-19}$$

where all symbols are as defined previously.

11 Gutter computations are carried out for all gutters in the system and summed to yield runoff.

Input to the runoff-water-quality model consists of flow hydrographs developed in the quantity model. Output takes the form of pollutographs for each pollutant modeled. Pollutographs and hydrographs are introduced into the Transport block, where they are summed and modified by addition of dry-weather flow and infiltration to produce final outfall characteristics. As of 1975, only some water-quality parameters had been modeled.

It is assumed that the amount of pollutant which can be removed during a rainfall is dependent on storm duration and initial quantity of pollutant. This can be modeled by a first-order differential equation of the form

$$-\frac{dP}{dt} = kP \tag{14-20}$$

or

$$P_o - P = P_o(1 - e^{-kt}) \tag{14-21}$$

where P_o = pollutant originally on ground, lb
   P = pollutant after time t, lb
   k = constant

The value of k is assumed to be directly proportional to the rate of runoff. Therefore, $k = br$, where b is a constant and r is the runoff intensity. Based upon available data, a value of 4.6 is assigned b. For each time step, the runoff rate is determined from the hydrograph and a value of P, which becomes the new value of P_o for the next step, is calculated.

14-7

CINCINNATTI URBAN RUNOFF MODEL

The Cincinnati Urban Runoff Model (CURM) was developed by the Division of Water Resources, Department of Civil Engineering, University of Cincinnati.[39] Calculation of pollutant concentrations in runoff is based on the assumption that the rate of pollutant removal depends on the amount of pollutant initially in the drainage area and the rainfall intensity. The equation used is

$$-\frac{dP}{dt} = KqP \tag{14-22}$$

or

$$-\frac{dP}{P} = Kq\,dt \tag{14-23}$$

After integration, this becomes

$$P = P_o e^{-KV}t \tag{14-24}$$

where P = amount of pollutant remaining on the ground at time t, lb.
P_o = initial amount of pollutant
K = constant
V_t = volume of runoff up to time t, ft^3
q = discharge, in./hr/unit area

During a storm, the amount of pollutant washed into the sewer system in a time interval t is

$$P_1 - P_2 = P_o(e^{-KV_1} - e^{-KV_2}) \tag{14-25}$$

The amount of solids washed into a storm sewer during a rainstorm is assumed proportional to the square of the runoff intensity. This equation is as follows:

$$r = \lambda q^2$$

where r = fraction of solids carried off
λ = proportionality factor
q = runoff intensity, in./hr

The amount of solids brought into the system in a time interval may be expressed as

$$P_1 - P_2 = P_o \lambda \bar{q}^2 (e^{-KV_1} - e^{-KV_2}) \tag{14-26}$$

where $\bar{q}$ = mean runoff intensity
V_1, V_2 = runoff volumes at times t_1 and t_2, ft^3

Soluble pollutants are routed downstream at the same velocity as the flow, and are summed in the same manner as flows to determine final values. Provision is made for sediment transport.

14-8

GROUNDWATER-QUALITY MODELING

Modeling of groundwater flow systems is complex, owing to the nature of aquifer structures and the scarcity of data on aquifer properties in many locations. This complexity is even greater for groundwater-quality models and, as a result, the state-of-the-art of these is much less advanced than for surface-water-quality models. Nonetheless, the importance of groundwater quality in water resources management and planning must be taken into account, and reliable models are needed. A discussion of these is beyond the scope of this book, and the interested reader is referred to the work of Gelhar and Wilson, and others.[37]

PROBLEMS

14-1 Select two water-quality models for conservative constituents and compare their characteristics.

14-2 Select two water-quality models for nonconservative constituents and compare their characteristics.

14-3 Obtain the computer program for QUAL-I and make several simulation runs for a watershed of your choice.

14-4 Discuss how water-quality models can complement traditional approaches to water-quality planning.

REFERENCES

1. S. R. Weibel, R. B. Weidner, J. M. Cohen, and A. G. Christianson, "Pesticides and Other Contaminants in Rainfall and Runoff as a Factor in Stream Pollution," *J. Am. Water Works Assoc.* (August 1966).

2. M. G. Wolman, and A. P. Schick, "Effects of Construction on Fluvial Sediment, Urban and Suburban Areas of Maryland," *Water Resources Res.* 3, no. 2 (Second Quarter 1967).

3. C. N. Sawyer, "Fertilization of Lakes by Agricultural and Urban Drainage," *J. New England Water Works Assoc.* 61 (1947).

4. R. O. Sylvester, and G. C. Anderson, "A Lake's Response to Its Environment," *Proceedings ASCE, Journal of Sanitary Engineering Division*, Vol. 90, No. SA1, February 1964.

5. R. L. Smith, W. J. O'Brien, A. R. LeFeuvre, and E. C. Pogge, "Development and Evaluation of a Mathematical Model of the Lower Reaches of the Kansas River Drainage System," Civil Engineering Department (Lawrence, Kans.: University of Kansas, January 1967).

6. R. P. Thayer, and R. G. Krutchkoff, "A Stochastic Model for Pollution and Dissolved Oxygen in Streams," *Water Resources Research Center Publication* (Blacksburg, Va.: Virginia Polytechnic Institute, August 1966).

7. D. P. Loucks, et al., "Linear Programming Models for Water Pollution Control," *Management Sci.* 14, no. 4 (December 1967).

8. T. D. Steele, "Seasonal Variations in Chemical Quality of Surface Water in the Pescadero Creek Watershed, San Mateo County, California," Ph.D. Dissertation (Stanford, Calif.: Stanford University, May 1968).

9. J. O. Ledbetter and E. F. Gloyna, "Predictive Techniques for Water Quality Inorganics," *Proc. Am. Soc. Civil Engrs., J. San. Eng. Div.* 90, no. SA1 (February 1964), part 1.

10. W. H. Wischmeier, "Cropping-Management Factor Evaluations for a Universal Soil-Loss Equation," *Proc. Soil Sci. Soc.* (1960).

11. P. C. Woods, "Management of Hydrologic Systems for Water Quality Control," *Water Resources Center, Contribution No. 121* (Berkeley, Calif.: University of California, June 1967).

12. S. R. Weibel, R. J. Anderson, and R. L. Woodward, "Urban Land Runoff as a Factor in Stream Pollution," *J. Water Poll. Control Fed.* 36, no. 7 (July 1964).

13. J. H. Zumberge, and J. C. Ayers, "Hydrology of Lakes and Swamps," *Handbook of Applied Hydrology.* edited by V. T. Chow (New York: McGraw-Hill Book Company, 1964).

14. Warren Viessman Jr., "Estimation of Lake Flushing Rates for Water Quality Control Planning and Management," *Proceedings of Conference on the Reclamation of Maine's Dying Lakes* (Orono, Me.: University of Maine, March 1971).

15. G. E. Hutchinson, *A Treatise on Limnology*, Vol. I (New York: John Wiley & Sons, Inc., 1957).

16. J. A. Liggett, "Circulation in Shallow Homogeneous Lakes," *Proc. Am. Soc. Civil Engrs., Hydraulics Div.*, no. HY2 (March 1969).

17. J. A. Liggett, "Unsteady Circulation in Shallow Homogeneous Lakes," *J. Hydraulics Div.*, no. HY4 (July 1969).

18. J. A. Liggett, "Cell Method In Computing Lake Circulation," *Proc. Am. Soc. Civil Engrs., Hydraulics Div.*, no. HY3 (March 1970).

19. J. A. Liggett and K. K. Lee, "Properties of Circulation in Stratified Lakes," *Proc. Am. Soc. Civil Engrs., J. Hydraulics Div.*, no. HY1 (January 1971).

20. G. T. Csanady, "Wind Driven Summer Circulation in the Great Lakes," *J. Geophys. Res.* 73, no. 8 (April 1968).

21. D. R. F. Harleman, et al., "The Feasibility of a Dynamic Model of Lake Michigan," *Great Lakes Research Divisions, Publication No. 9* (Ann Arbor, Mich.: University of Michigan, 1962).

22. E. B. Henson, A. S. Bradshaw, and D. C. Chandler, "The Physical Limnology of Cayuga Lake, New York," *Agriculture Experiment Station, Memoir 378*, New York State College of Agriculture (Ithaca, N.Y.: Cornell University, August 1961).

23. K. K. Lee, and J. A. Liggett, "Computation of Wind Driven Circulation in Stratified Lakes," *Proc. Am. Soc. Civil Engrs., J. Hydraulics Div.* 96, no. HY10, Proc. Paper 7634 (October 1970).

24. J. L. Verber, "Current Profiles to Depth in Lake Michigan," *Great Lakes Research Division, Publication No. 13* (Ann Arbor, Mich.: University of Michigan, 1965).

25. C. W. Chen and G. T. Orlob, "Ecologic Simulation for Aquatic Enivronments, Annual Report," *OWRR Project No. C-2044*, Office of Water Resources Research (Washington, D.C.: Department of the Interior, August 1971).

26. C. W. Chen and G. T. Orlob, "Ecologic Simulation for Aquatic Environments, Final Report," *OWRR Project No. C-2044*, Office of Water Resources Research (Washington, D.C.: Department of the Interior, December 1972).

27. P. S. Lombardo and D. D. Franz, "Mathematical Model of Water Quality in Rivers and Impoundments" (Palo Alto, Calif.: Hydrocomp, Inc., December 1972).

28. P. S. Lombardo, "Critical Review of Currently Available Water Quality Models" (Palo Alto, Calif.: Hydrocomp, Inc., July 1973).

29. B. C. Dysart III, "Water Quality Planning in the Presence of Interacting Pollutants," 42nd Annual Conference, Water Pollution Control Federation, Dallas, Texas (October 1969).

30. Nebraska Natural Resources Commission, "Lower Platte River Basin Water Quality Management Plan," Lincoln, Nebraska (June 1974).

31. Warren Viessman Jr., "Assessing the Quality of Urban Drainage," *Public Works* (October 1969).

32. Texas Water Development Board, "QUAL-I—Simulation of Water Quality in Streams and Lands," Program Documentation and Users' Manual, *National Technical Information Service, PB 202973*, Springfield, Va. (1970).

33. G. T. Orlob and L. G. Selna, "Temperature Variations in Deep Reservoirs," *Proc. Am. Soc. Civil Engrs., J. Hydraulics Div.* 96, no. HY2 (February 1970).

34. N. H. Crawford and R. K. Linsley, "The Synthesis of Continuous Streamflow Hydrographs on a Digital Computer," *Department of Civil Engineering Technical Report No. 12* (Palo Alto, Calif.: Stanford University, 1966).

35. Hydrocomp, Inc., "The Hydrocomp Simulation Network Operations Manual," Palo Alto, California, 1969.

36. Texas Water Development Board, "DOSAG-I Simulation of Water Quality in Streams and Canals, Program Documentation and Users' Manual," *National Technical Information Service*, Springfield, Virginia, 1970.

37. L. W. Gelhar and J. L. Wilson, "Ground Water Quality Modeling," *Proc. of 2nd Nat. Ground Water Quality Symposium*, U.S. Environmental Protection Agency, Washington, D.C. 1974.

38. Metcalf and Eddy, Inc., University of Florida, Water Resources Engineers, Inc., "Storm Water Management Model," Environmental Protection Agency, Vol. 1, 1971.

39. Division of Water Resources, Department of Civil Engineering, University of Cincinnati, "Urban Runoff Characteristics," Water Pollution Control Research Series, EPA, 1970.

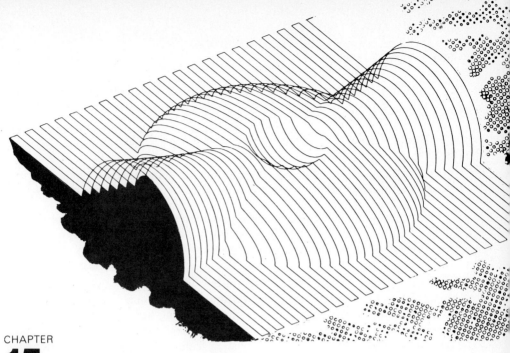

15
The Reuse of
Water

The reuse of water is one of the means of meeting future water demands. It is projected that by 1980, municipalities will withdraw 34 billion gallons per day (bgd) and return 23 bgd, or about 68 %. Industry (not including mining and flow-through cooling) is expected in that year to withdraw 55 bgd and return 50 bdg, or about 91 %.[1] The quantity of withdrawals is expected to increase substantially by the year 2000, but the percentage of returns is expected to remain about the same for both municipalities and industries.

As one-third of the nation's population currently depends on municipal water supplies drawn from streams, reuse is taking place. These withdrawals from streams contain, on the average, 1 gal of previously used water in each 30 gal. In some cases, as much as 20 % of the municipal raw water has been used before.[2]. These supplies receive only conventional purification treatment.

As the goals of P.L. 92-500, the Federal Water Pollution Act of 1972, are met, more reuse will take place.[3] In the Act, Congress has shown clear intent to restore and maintain the chemical, physical, and biological integrity of the nation's waters. Two significant national goals are enunciated:

1 That by July 1, 1983, wherever attainable, water quality will be maintained that will provide for the protection and propagation of fish, shellfish, and wildlife, and provide for recreation in and on water.

803

2 That by December 31, 1985, the discharge of pollutants into all navigable waters will be eliminated.

The act contains a number of policy statements, including:

1 That the discharge of toxic pollutants in toxic amounts be prohibited.
2 That federal financial assistance be provided to construct public owned waste-treatment facilities.
3 That area-wide planning be implemented for waste-treatment management.
4 That public participation be provided for in the development, revision, and enforcement of any regulation, standard, effluent limitation, plan, or program.

The rate at which more reuse of wastewater depends on advances in treatment technology and costs. Present technology is adequate to permit reuse of municipal effluent for many purposes not involving human consumption. The direct reuse of municipal effluent for human consumption depends both upon technology and public acceptance. Existing technology can produce water that meets federal drinking-water standards in terms of physical, chemical, and bacteriological criteria. However, these standards do not take into account all possible toxic ingredients sometimes found in wastewater. Possible viral hazards and new chemicals are of concern with some public health groups. As these public health concerns are mitigated through research and technology, large-scale direct reuse of wastewater for human consumption will become a real possibility.

Natural or Indirect Reuse of Water

It has already been pointed out that surface bodies of water (streams, rivers, lakes) are often used by numerous municipalities, industries, and agricultural organizations for both water supply and wastewater disposal. In this manner water in any surface source is generally reused many times before it finally flows to the sea.

Even though municipal and industrial waste-treatment processes effectively reduce the quantities of pollutants entering the receiving waters, varying quantities of polluting materials still pass through these units and tend to degrade the quality of the water to which they are returned. Fortunately, the forces of natural purification which are continually at work in natural waters result in further purification. The degree of natural purification accomplished in a given river reach is dependent, however, upon numerous factors, which include the degree of dilution, the character of the waste, the time involved, and other elements. Nevertheless, the net result of discharging wastes into natural waters is normally an increase in the wastewater quality.

Before the diluted waste flows from upstream sources can be reused for potable water supplies, still further treatment is required. The type of purification process employed will depend primarily on the degree of pollution of the source waters. Some of the major treatment operations utilized in this respect are storage, coagulation, sedimentation, filtration, and disinfection.

While industry and municipalities are accelerating programs to cope with waste-abatement problems. the nature and dimension of these problems is also rapidly expanding. Nevertheless, higher degrees of waste treatment are going to be required in the future as the number and variety of contaminants discharged into our surface and groundwaters increases and as the need for reusing these waters becomes more pressing.

Direct Reuse of Wastewaters

It has been pointed out that water demands are rapidly increasing. Many areas are finding that they are imposing overdrafts on existing supplies. New supplies must often be developed at considerable distances from their expected point of use, with the result that transportation costs often become prohibitive. Such factors as these are instrumental in focusing greater attention on the modification of past practices of wastewater treatment and disposal. Considerable interest is now being brought to bear on the practicality of direct utilization of treated waste effluents.

Treated wastewater effluents are not presently generally used directly as sources of municipal water supply. At the same time, however, a definite trend in the industrial use of such flows is in evidence. In the field of agriculture, both treated and untreated waste effluents are used. Treated effluent is also finding use in the development of artificial lakes for recreational purposes.

15-1

PUBLIC HEALTH CONSIDERATIONS

The two considerations of primary significance in planning wastewater reuse projects are public health and economics. If the results of a project are expected to prove hazardous to the public's health, changes should be made or the project dropped.

Public health concerns are, for the most part, restricted to those uses in which drinking or bodily contact is planned. There are at present no cities in the United States processing effluent for direct potable reuse. Metzler et al.[4] describe a dramatic case of wastewater reuse for domestic purposes at Chanute, Kansas, in 1956 and 1957. During a severe drought, the city's water-supply impoundment on the Neosho River was forecast to go dry unless additional water was supplied to the reservoir. After considering alternative water-supply sources, it was decided to impound the treated wastewater from the municipal trickling-filter plant and return it to the water-supply reservoir. The mean cycle time through the reservoir, water-treatment plant, distribution system, waste-treatment plant, and back to the reservoir was approximately 20 days.

The recycling system was operated for 5 months during which the water at Chanute was reused approximately 8–15 times. The bacteriological quality of the tap water was maintained within the Drinking Water Standards, primarily by extensive chlorination. However, the water gradually developed a pale yellow color and an unpleasant taste and odor. It contained undesirable concentrations of dissolved minerals and organic substances, and foamed readily when agitated. It should be noted that the reuse of the water practiced in Chanute far exceeds any indirect use that occurs as a result of pollution and reuse of river-water supplies. The only example, however, of planned direct reuse for domestic water supply that can be cited is in Windhoek, South-West Africa. At Windhoek, conventional primary and secondary treatment were utilized, followed by retention in maturation ponds (to reduce nutrients, viruses, and bacteria) and several other processes, including free residual chlorination and activated-carbon filtration. In this instance, the final plant effluent actually contributed an average of 13% of the total community water supply, with a high of 28%. Under normal conditions, no viruses were present in the algae ponds, and even in the case of test injection, none were noticed past the chlorination point.[5]

Perhaps the most obvious problem to be faced during reuse is the control of microbiological water quality. *Disinfection*, as the term is used in the water-supply industry, refers to the destruction of pathogenic bacteria. In conventional water purification, it is assumed that the absence of coliform organisms, together with an adequate free-chlorine residual and a low turbidity, is *prima facie* evidence of the absence of animal viruses. This asumption is probably true insofar as most groundwaters are concerned, and is true in very high quality surface waters. Ammonia–nitrogen is generally present only in trace concentrations; therefore, during chlorination free residual chlorine is retained, increasing the bacterial efficiency of chlorination. It is known, both from experience and from authoritative laboratory tests, that the presence of a free residual in low-turbidity water will provide essentially a disinfected water. When reuse enters the case, however, this is not true. Ammonia–nitrogen may appear, and does appear in rather high concentrations in wastewater effluents. More significantly, infectious viruses are present in even higher concentrations than in poor-quality surface water. Therefore, the problem of quality control where reuse is involved is much more significant than with pristine and protected sources.

Waterborne disease outbreaks are a problem in public water supply and distribution systems. Figure 15-1 shows the outbreaks of waterborne diseases in the United States, 1938–1972. The reasons for the increase in outbreaks since 1950 are probably due in part to reduced raw-water quality.

Another problem associated with direct reuse is the effect of prolonged intake or exposure to low concentrations of toxic materials. There is a con-

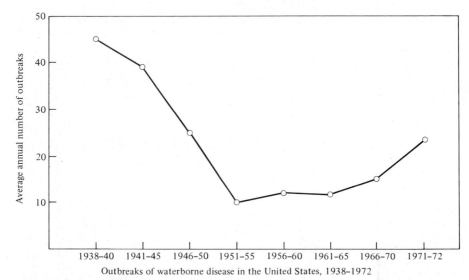

Outbreaks of waterborne disease in the United States, 1938-1972

FIGURE 15-1 Outbreaks of waterborne disease in the United States, 1938–1972. [Source: Committee Report, "Status of Waterborne Diseases in the U.S. and Canada," *J. Am. Water Works Assoc.* 67 (February 1975) : 96.]

tinuing series of research discoveries of unknown quantities of chlorinated hydrocarbons, pesticide and herbicide derivatives, benzene derivatives, new exotic chemicals from industrial processes, and other toxic products which raise doubts as to the acceptability of direct wastewater reuse for public supplies. No practice that increases the possible toxic load on a water supply should be encouraged or permitted until positive assurances of water safety have been obtained.

15-2

INDUSTRIAL REUSE OF MUNICIPAL EFFLUENTS

It has been estimated by the National Water Commission[7] that by 1980, continuous reuse by industry of only 20% of the projected total of 73 bgd of municipal and manufacturing effluents would completely satisfy the projected 1965–1980 increase in water withdrawals for manufacturing (15 bgd). Reuse by industry of an additional 14% of municipal and manufacturing effluents in 1980 would release enough potable water from manufacturing use to meet the 1965–1980 growth in municipal-water-withdrawal needs as well. To the extent that water supplies are limited, the desirability of water reuse will expand. Table 15-1 shows the water use and reuse for various industrial groups.

Present direct reuse of treated municipal wastewaters by industry is relatively small, but the increasing number of examples where this is carried out attests to the enormous potential that exists. Treatment-plant effluent can be

TABLE 15-1

INDUSTRIAL PLANT AND INVESTOR-OWNED THERMAL ELECTRIC PLANT WATER INTAKE, REUSE, AND CONSUMPTION, 1968

SIC	Industrial Group	Water Intake (bgy)				Water Recycled (bgy)	Gross Water Use, Including Recycling (bgy)	Water Consumed (bgy)	Water Discharged (bgy)
		Purpose							
		Cooling and Condensing	Boiler Feed, Sanitary Service, etc.	Process	Total				
20	Food and kindred products	427	93	290	810	535	1,345	57	753
22	Textile mill products	24	22	109	155	174	329	19	136
24	Lumber and wood products	62	20	37	119	87	206	26	93
26	Paper and allied products	652	123	1,478	2,253	4,270	6,523	175	2,078
28	Chemicals and allied products	3,533	210	733	4,476	4,940	9,416	301	4,175
29	Petroleum and coal products	1,230	111	95	1,436	5,855	7,291	219	1,217
31	Leather and leather products	1	1	14	16	4	20	1	15
33	Primary metal industry	3,632	165	1,207	5,004	2,780	7,784	308	4,696
	Subtotal	9,561	745	3,963	14,269	18,645	32,914	1,106	13,163
	Other industries	574	291	332	1,197	1,569	2,786	84	1,113
	Total industry	10,135	1,036[a]	4,295	15,466	20,234	35,700	1,190	14,276
	Thermal electric plants	68,200		—	68,200	8,525	76,725	100	68,100
	TOTAL	78,335	1,036[b]	4,295	83,666	28,759	112,425	1,290	82,376

[a] Boiler feedwater use by thermal electric plants estimated to be equivalent to industrial sanitary service, etc., water use.
[b] Total boiler feed water.

Source: *Water Policies for the Future*, Report by the National Water Commission (June 1973).

used successfully for general plant application, for cooling water, and also for boiler feedwater, if properly treated. Since about 50% of all industrial water use falls within the latter two categories, it can easily be seen that the value of reclaimed wastewater to industry is significant. Treated effluents are used industrially in those areas of the United States where economic conditions make this a favorable choice.

Veatch[9] and Keating and Calise[8] have listed some factors which must be considered in deciding on the practicality of the reuse of wastewater effluent in industry:

1   Local industry must exhibit the need for a process water which does not involve the public health.
2   The treatment plant must be of an adequate size to meet the required flows.
3   Processing costs, which include transportation and pumping, must not exceed those of an alternative supply.
4   The quality of the wastewater effluent should be consistent enough to allow its reuse by the particular industry.

An outstanding example of industrial use of waste effluent is the Bethlehem Steel Company's Sparrows Point plant in Maryland. In 1942 the steel mill first contracted to take 20 mgd of water from the Baltimore treatment plant. This proved to be a better supply than the saline Chesapeake Bay water or the available groundwater supplies, which are increasing in salinity due to overdraft. In 1965 about 22% of the total plant requirement for cooling was reused effluent, at a cost of 3–4 cents/1000 gal.

Another example of industrial reuse is that of the Cosden Oil and Chemical Company at Big Spring, Texas. In 1943 Cosden was experiencing serious problems because of inadequate supplies of water, which were, at best, of poor quality. As a result, full rights for water from the Big Spring contact aeration plant were contracted for. About 15 mg/month is taken from the municipal treatment plant. The cost of this treated effluent is about 25 cents/1000 gal. The water is used as boiler feed and is of better quality than that of the existing wells. In the next 5–10 years it is expected that further needs will arise and that reclaimed waste effluent will be used for cooling purposes as well as for boiler feed. To anticipate this need, Cosden has recently expanded its own facilities for treating wastewater.

Since 1955 a dependable and satisfactory source of boiler makeup and cooling water has been provided the Texas Company's Amarillo Refinery by the water reclamation and treatment plant of the city of Amarillo, Texas. Many factors prompted the use of this reclaimed waste by the refinery. In 1954 the company and municipality signed a contract which provided for the construction of a new treatment plant by the city, the provision of additional water-treatment facilities by the industry, and the delivery of an effluent by the wastewater plant which would meet certain specifications.

15-3

AGRICULTURAL REUSE OF MUNICIPAL EFFLUENTS

Irrigation of lands with wastewater has been practiced for centuries. In some cases the object is simply for disposal, whereas in other cases the primary purpose is production of crops and raising of animals. In the United States, the practice of irrigating with municipal effluents is confined largely to the arid or semiarid regions.

The primary consideration to be made regarding the use of reclaimed domestic wastewater for agricultural purposes is whether or not a public health hazard will result. The risks associated with wastewater farming are dependent to a great extent on the type of crops being irrigated and on the degree of treatment of the effluent used. In general, it is believed that disease can be transmitted by the ingestion of uncooked vegetables that have been irrigated with wastewater.

The suitability of municipal effluents for irrigation depends mainly on the chemical composition of the municipal water supply and how this composition is affected by the municipal use cycle. A one-use domestic cycle may contribute 100–450 mg/l of total dissolved solids. Increases greater than this can usually be attributed to industrial sources. Any wastewater derived from a domestic water-use cycle will usually be acceptable for most irrigation applications (see Sec. 7-3).

The characteristics of municipal wastewater are shown in Table 15-2 and may be classified as physical, chemical, and biological. Column (3) lists a range of actual values obtained from site visits where land application of municipal wastes was being utilized.

The most important physical characteristic of municipal effluent is its total-solids content. This includes floating, colloidal, suspended, and dissolved matter. Suspended solids tend to clog the soil pores and coat the land surface. This limits the ability of the water to infiltrate the surface and percolate through the soil pores.

Generally the most important chemical characteristics are total dissolved solids, nitrogen, heavy metals, and sodium. Soil does not provide a long-term removal mechanism for the removal of salts. Through irrigation, the evapo-transpiration process concentrates the salts in the soil and the subsequent concentrations may be injurious to plants or may be leached to the groundwater. Nitrogen, in the nitrate form, can pass through the soil matrix to the groundwater. Nitrogen can be removed from the soil through plant uptake with harvest and by denitrification. Phosphorus and heavy metals are easily fixed in many soils by precipitation and absorption. Sandy soils provide fewer sites in the soil matrix, and therefore their retention capacities are relatively short. Concentrations of heavy metals, such as copper and zinc, can build up to phytotoxic levels in time, depending upon the soil type. High sodium concentrations relative to calcium and magnesium can reduce the permeability of clay soils.

TABLE 15-2
MUNICIPAL-WASTEWATER CHARACTERISTICS

Constituent	(1) Untreated Sewage (mg/l)	(2) Typical Secondary- Treatment Effluent (mg/l)	(3) Actual Quality Applied to Land (mg/l)
Physical			
Total solids	700	425	760–1200
Total suspended solids	200	25	10–100
Chemical			
Total dissolved solids	500	400	750–1100
pH	7.0 ± 0.5	7.0 ± 0.5	6.8–8.1
BOD	200	25	10–42
COD	500	70	30–80
Total nitrogen	40	20	10–60
Nitrate–N	0	—	0–10
Ammonia–N	25	—	1–40
Total phosphorus	10	10	7.9–25
Chlorides	50	45	40–200
Sulfate	—	—	107–383
Alkalinity (CaCO$_3$)	100	—	200–700
Boron	—	1.0	0–1.0
Sodium	—	50	190–250
Potassium	—	14	10–40
Calcium	—	24	20–120
Magnesium	—	17	10–50
Sodium adsorption ratio	—	2.7	4.5–7.9
Biological			
Coliform organisms (MPN/100ml)	10^6	—	$2.2–10^6$

Source: C. E. Pound and R. W. Crites, "Waste Treatment and Reuse by Land Application," Vol. I, *EPA-660/2-73-006a* (August 1973).

Table 15-3 lists characteristics for a municipal-effluent irrigation system. In order to limit nitrogen buildup in the groundwater, the total nitrogen applied in a year should not greatly exceed the pounds of nitrogen removed by crop harvest. With loamy soils, the liquid loading rate will usually be limiting, but with sandy soils the nitrogen loading rate may be the controlling factor. Site-selection factors are shown in Table 15-4.

Renovation of the effluent water quality usually occurs after passage through the first 2–4 ft of soil. Removals are found to be on the order of 99% for

TABLE 15-3

CHARACTERISTICS OF AN IRRIGATION SYSTEM USING MUNICIPAL EFFLUENT[a]

Factor	Irrigation
Liquid loading rate	0.5–4 in./week
Annual application	2–8 ft/yr
Land required for 1-mgd flow	140–560 acres plus buffer zones
Application techniques	Spray or surface
Soils	Moderately permeable soils with good productivity when irrigated
Probability of influencing groundwater quality	Moderate
Needed depth to groundwater	About 10 ft
Wastewater lost to:	Predominantly evaporation or deep percolation

Source: D. E. Pound and R. W. Crites, "Waste Treatment and Reuse by Land Application," Vol. I, *EPA-660/2-73-006a* (August 1973).

TABLE 15-4

SITE-SELECTION FACTORS AND CRITERIA FOR EFFLUENT IRRIGATION

Factor	Criterion
Soil type	Loamy soils preferable, but most soils from sands to clays are acceptable.
Soil drainability	Well-drained soil is preferable; consult experienced agricultural advisors.
Soil depth	Uniformly 5–6 ft or more throughout sites is preferred.
Depth to groundwater	Minimum of 5 ft is preferred. Drainage to obtain this minimum may be required.
Groundwater control	May be necessary to ensure renovation if the water table is less than 10 ft from surface.
Groundwater movement	Velocity and direction must be determined.

(Continued)

TABLE 15-4 (*continued*)

Factor	Criterion
Slopes	Up to 15% are acceptable with or without terracing.
Underground formations	Should be mapped and analyzed with respect to interference with groundwater or percolating water movement.
Isolation	Moderate isolation from public preferable, degree dependent on wastewater characteristics, method of application, and crop.
Distance from source of wastewater	A matter of economics.

Source: C. E. Pound and R. W. Crites, "Waste Treatment and Reuse by Land Application," Vol. I, *EPA-660/2-73-006a* (August 1973).

BOD, suspended solids, and fecal coliform. It is technically feasible to transfer water now utilized for irrigation to municipal use in exchange for treated sewage effluent. In effect, irrigation water would first be cycled through municipal systems prior to reuse on farms where reasonable.

In semiarid and arid regions water is often difficult to obtain, and in very dry years a sizable supplementary water supply can often be derived from waste flows. One of the difficulties is that the large centers of population and therefore collection points for municipal and industrial wastes are often located at considerable distances from farmlands. Under these circumstances transportation costs might make it uneconomical to use the water for agricultural purposes. On the other hand, where isolated communities are situated near good farmlands, the practice might prove to be highly economical. Table 15-5 lists some examples of the use of municipal effluents for agriculture.

TABLE 15-5

EXAMPLES OF THE USE OF SEWAGE EFFLUENT FOR AGRICULTURAL PURPOSES

Tucson, Ariz.	Cotton, grain, and pastures
Bakersfield, Calif.	Alfalfa, cattle, cotton, and grasses
San Francisco, Calif.	Pastures and park lawns
Abilene, Tex.	Cattle, grain, grasses, and maize
Lubbock, Tex.	Barley, cotton, sorghum, and wheat
Ephrata, Wash.	Corn and hay

15-4

RECREATIONAL USES

A striking demonstration of the reuse of wastewaters for recreational purposes is the project at Santee, California.[13-15]. The Santee County Water District in San Diego County, California, has undertaken a reclamation project which has provided for the development of a series of artificial lakes fed by the effluent of a standard activated-sludge treatment plant. The effluent from this plant is held in a 16-acre lagoon for 30 days (Lake 1). It is then chlorinated and spread onto a percolation area of sand and gravel through which it travels for about 0.5 mi. The percolated water is collected in an interceptor trench and then flows into Lake 4 (11 acres), then into Lake 3 (7.5 acres), and ultimately into Lake 2 (6.8 acres), after which the overflow passes on to Sycamore Canyon Creek. Lakes 3 and 4 are used for fishing and boating. Playgrounds and picnic areas line the shores. In 1965, Lake 5 was approved for swimming by the California Department of Public Health. This approval followed 3 years of study by an Epidemiology Advisory Committee of the California Department of Public Health. The state health director indicated that biological, hydrological, bacteriological, and virological information supported approval by the committee. This community recreational area is provided through the utilization of wastewaters and permits an aquatic environment in an area which affords little attraction of this type. The public acceptance of these facilities has been exceptional. Considering the important emphasis being placed on the recreational use of our water resources, there is little doubt but that expanded development of facilities of this type will be forthcoming in the very near future.

Reclaimed wastewaters are also being used successfully for the irrigation of golf courses in many areas of the Southwest. Some locations where this is being practiced are the El Toro Marine Air Base near Santa Ana, California; Camp Pendleton Marine Base near San Diego; the Naval Ordance Test Station near China Lake in California; Eagle Pass, Texas; Prescott, Arizona; Santa Fe, New Mexico; and Los Alamos, New Mexico.

15-5

REUSE OF EFFLUENT THROUGH RECHARGE TO GROUNDWATER

The replenishment of groundwater supplies through artificial recharge has been given considerable attention in recent years. Impetus to this movement has been derived through a realization that in many areas of this country groundwater levels have been rapidly falling and are not being recovered through natural means. Waters which are suitable for artificial recharge may be classified generally as floodwaters, industrial wastes, and municipal wastewater. Such waters may be introduced into the ground by surface spreading or by pumping underground.

There are many advantages to be gained by groundwater recharge. Some of these are:

1   Supplementation of natural recharge where man's activities have reduced or eliminated this opportunity.
2   Reduction of aquifer overdrafts.
3   Provision for concurrent development of surface and groundwater supplies.
4   Control of saline-water intrusion.
5   Salt-balance control.
6   Control of or aid in combating land-subsidence problems.

A considerable number of municipalities, industries, and agricultural users rely heavily on groundwater supplies to satisfy their needs. Unfortunately, overdrafts on these supplies have been imposed for many years in numerous locations. As a result of this, water tables have dropped to critical levels in some regions, and the continued use of many of these groundwater supplies is becoming uneconomical.

The mounting evidence that many aquifers are being mined has led to the conclusion that control of our groundwater resources is vital. Recharge operations are one means of rectifying many aspects of the groundwater problem. In recognition of this, many communities and industries are undertaking projects to recharge groundwater supplies. In addition, Arkansas, California, Illinois, New Mexico, New York, Texas, Utah, Washington, and Wisconsin have become involved in regional projects. In 1957, a total of 42 recharge projects, designed specifically for municipal, industrial, irrigational, and regional supplies, were under way. As groundwater quantity and quality problems become more pressing, it is certain that large-scale recharge operations will become more common.

Waste flows from treatment plants constitute a potentially valuable and significant source of supply for groundwater recharge. Millions of gallons of municipal effluent are being wasted daily and could easily be used to replenish overdrawn aquifers. Indirect recharge with effluent is actually being accomplished where septic tanks are used or wastewater farming is the practice.

Early studies (1930) in Los Angeles indicated that a well-treated wastewater could be used for groundwater recharge by spreading. The operation was considered to be completely safe. In 1949 studies indicated that groundwater recharge with treated domestic or industrial wastes was practical and economical.[16] Spreading basins of an experimental nature were set up by the Los Angeles County Flood Control District at Azusa and Whittier. In both cases the wastes received secondary treatment before spreading. No significant bacterial contamination below a depth of 7–10 ft was reported. There was no significant contamination of the groundwater. In addition, it was proved that a substantial organic mat was formed at the soil surface. This mat exhibited an increasing ability to remove bacteria from the effluent being filtered, thereby significantly reducing the number of bacteria passing into the soils.

In 1955 the University of California completed a 3-year project whose objective was the determination of the practicality and public-health aspects of injecting treated effluent directly into aquifers.[17] The wastewater was injected into a 5-ft-thick confined aquifer through a recharge well. Findings were that bacterial pollutants moved a maximum of 100 ft in the direction of normal groundwater movement. Additional findings of the study were that the injection was one-half that of the safe yield. Periodic chlorine injection and redevelopment of the well were found to be a requirement to assure continued operation. The important conclusion of the study was that groundwater recharge by direct injection was safe and hydraulically feasible.

Experience at the Hyperion Sewage Treatment Plant in Los Angeles and in the other areas indicate that an economical solution to the problem of groundwater replenishment can be obtained by the surface spreading of effluents. One advantage of the procedure is that less treatment is required than if the effluent is injected or directly discharged to a receiving body of water. Actually, the spreading basin functions as an additional treatment operation in which oxidation is provided for waste stabilization.

An outstanding example of the potential of water reclamation and groundwater recharge using wastewater effluent is the Whittier Narrows water-reclamation project in Los Angeles County, California. The objective of this project is to conserve and assist in restoring the water resources of the agricultural area of Southern California. Since the initiation of plant operation in 1962 the project has reclaimed about 12,000 acre-ft annually. The treated wastewater is distributed over spreading grounds in the San Gabriel and Rio Hondo river basins. Once spread, the reclaimed waters percolate through the ground to the groundwater table and are then pumped for use in the irrigation of agricultural lands.

The plant itself has an initial design capacity of 10 mgd. It is basically an activated-sludge plant and is superimposed on the existing system. The trunk sewer providing the inflows to the reclamation plant carries about 50 mgd of wastewater, which is primarily derived from domestic sources. This large supply source makes possible uniform operation of the plant and also permits variation to meet mechanical and biological treatment requirements. There are no solids-processing units in the plant. All solids are returned to the main trunk sewer after separation from the feed. This precludes the considerable expense of processing and disposing of the solid wastes at the reclamation plant site. Operation of the Whittier Narrows plant is being closely watched by various agencies interested in public health, education, pollution control, and industry.

Wastewater effluent contains higher concentrations of suspended matter and bacteria than fresh water, and thus its spreading rates are lower. For raw-wastewater farms, typical rates are about 0.01–0.09 ft/day. For treated effluents, results have indicated a range in rates of from about 0.2 to 1.2 ft/day for various field and lysimeter tests.[17] In general, it is recommended that alternate wetting and drying periods of from 7 to 14 days be used.

Groundwater recharge by spreading is highly practical where sufficient low-cost land is available. Where land is expensive, injection is the preferable practice. Unless highly treated effluents are used, however, injection wells will frequently clog and operation and maintenance costs will be high.

As in the case of reusing waste flows directly, the buildup of chlorides and other salts through continual recycling must be considered in recharge operations. Very little information is currently available on the effects of recycling waste flows to the ground. To provide the needed information, water-quality studies of aquifers will have to be conducted over long periods of time. A great deal of additional information on chemical and bacterial pollution of groundwaters is also needed. Some important work in this area has been done by Butler, Orlob, and McGauhey.[17] Considerable future research and evaluation of field results is a prerequisite to the maximum development of groundwater recharge as a method of wastewater reuse.

15-6

ECONOMICS OF REUSE

If the "no-discharge" goal of the 1972 Federal Water Pollution Control Act is achieved, the quality of effluent discharge would exceed that of most "natural" water supplies. Thus the reuse of all effluent waters would be achieved without additional cost.

The true direct net cost of treatment for reuse can be expressed as follows[7]:

1 The cost of advanced treatment to make wastewater suitable for reuse.
2 Minus the cost of pollution-control treatment measures otherwise necessary to achieve water-quality standards.
3 Minus the cost of water treatment of the supply being considered as an alternative to reuse.
4 Plus or minus the difference in conveyance costs between the reusable supply and its alternative, including allowance for the cost of separate supply lines if reuse is contemplated for the industrial water supply only.

The approximate costs of secondary and advanced waste treatment of wastewater for reuse are given in Table 15-6. The cost of reuse will vary widely from one locality and situation to another, and reuse may not be the best water management alternative in all cases.

TABLE 15-6

APPROXIMATE COSTS OF SECONDARY AND ADVANCED TREATMENT (June 1967 Cost Levels)

Capacity of Plant (mgd)	Secondary Treatment		Costs of Advanced Treatment Processes in Addition to Costs of Secondary Treatment					
			Nutrient Removal (Including Suspended Solids)[a]		Removal of Nutrients Plus Nonbiodegradable Organics		Removal of Nutrients and Nonbiodegradable Organic Plus Demineralization[b]	
	Capital Costs ($ million)	Total Unit Treatment Costs[c] (¢/1000 gal)	Capital Costs ($ million)	Total Unit Treatment Costs[c] (¢/1000 gal)	Capital Costs ($ million)	Total Unit Treatment Costs[c] (¢/1000 gal)	Capital Costs ($ million)	Total Unit Treatment Costs[c] (¢/1000 gal)
1	0.54	19	0.43	26.8	0.81	58	—	—
10	3.2	11	1.8	14.0	3.4	24	6.8	36
100	20	6.5	10.9	8.6	26	15.6	—	—

[a] Costs based on air stripping. If biological nitrification–dentrification is required, as is presently indicated, the costs would undoubtedly be greater. Costs of this process are not currently available, but some researchers have expressed the view that its use could raise the total cost of nutrient removal by as much as 40%.

[b] Based on assumed mineral concentration of 850 ppm in effluent, reduced to 500 ppm (drinking-water standard), thus providing for one cycle of reuse. Costs of brine disposal, which may be substantial, are not included in above demineralization costs because of variability between sites.

[c] Includes operation and maintenance and interest and amortization on capital investment (at 4.5% interest over 25 years for comparative purposes only; not intended as a recommendation for financing assumption).

Source: U.S. Water Resources Council, *The Nation's Water Resources*, (Washington, D.C.: Government Printing Office, 1968) pp. 4-1–4-4.

REFERENCES

1. U.S. Water Resources Council, *The Nation's Water Resources*, (Washington, D.C.: Government Printing Office, 1968), pp. 4–1–4–4.

2. Jerome Gavis, "Wastewater Reuse," *PB 201 535* (Springfield, Va.: National Technical Information Service, 1971), p. 1.

3. Public Law 92–500, 92 Congress, S-2770 (October 18, 1972).

4. D. F. Metzler et al., "Emergency Use of Reclaimed Water for Potable Supply at Chanute, Kansas," *J. Am. Water Works Assoc.* 51 (August 1958): 1021.

5. G. C. Cillie et al., "The Reclamation of Sewage Effluents to Domestic Use," *Third International Conference on Water Pollution Research*, Washington, D.C. (1966).

6. Committee Report, "Status of Waterborne Diseases in the U.S. and Canada," *J. Am. Water Works Assoc.* 67 (February 1975): 96.

7. *Water Policies for the Future*, Report by the National Water Commission (June 1973).

8. R. J. Keating and V. J. Calise, "Treatment of Sewage Plant Effluent for Industrial Re-Use," *Sewage Ind. Wastes* (July 1955).

9. N. T. Veatch, "Industrial Uses of Reclaimed Sewage Effluents," *Sewage Works J.* 20 (January 1948).

10. C. E. Pound and R. W. Crites, "Waste Treatment and Reuse by Land Application," Vol. I, *EPA–660/2–72–006a* (August 1973).

11. Metcalf and Eddy, Inc., *Wastewater Engineering* (New York: McGraw-Hill Book Company, 1972).

12. "Assessment of the Effectiveness and Effects of Land Disposal Methodologies of Wastewater Management," *Wastewater Management Report 72–1*, Department of the Army, Corps of Engineers (January 1972).

13. J. Merrell and R. Stoyer, "Reclaimed Sewage Becomes a Community," *The American City* (April 1964): 97.

14. F. M. Middleton, "Advanced Treatment of Waste Waters for Re-Use," *Water Sewage Works* (September 1964).

15. J. C. Merrell, Jr., W. F. Jopling, R. F. Bott, A. Katko, and H. E. Pintler, "Santee Recreational Project, Santee California, Final Report, *FWPCA Report WP–20–7* (Cincinnati, Ohio: Publication Office, Ohio Basin Region, Federal Water Pollution Control Administration, 1967).

16. "Studies in Water Reclamation," *Sanitary Engineering Research Laboratory Technical Bulletin No. 13* (Berkeley, Calif.: University of California, July 1955).

17. R. G. Butler, T. T. Orlob, and P. H. McGauhey, "Underground Movement of Bacterial and Chemical Pollutants," *J. Am. Water Works Assoc.* 46 (1954).

16
Legal Considerations

The interrelated and competing uses of water often give rise to varied and complex problems in regions where water supply is inadequate to meet the needs of potential users. The combined development of the water resources of an area is necessary for maximum beneficial use in most drainage basins. Some states in their water laws have recognized an integrated action of all interests, while other states have framed their laws with reference to individual action. Large-scale multiple-purpose projects have created novel problems of reshuffling rights between users and uses, between watersheds, between states, and between countries.

In many areas the virgin opportunities for water development have been largely exhausted. It is becoming increasingly necessary for cities to acquire rights to additional waters needed for their growth through legal action. These additional waters may be available in the local areas, or the water may be conveyed for considerable distance to the places of use. The Feather River project in California takes water from the northern Sierra Nevada and conveys it almost to the Mexican Border. This project covers a distance of several hundred miles down the Sacramento Valley, up the San Joaquin Valley, over a mountain range, and then into Southern California. This is an extreme example, but the expansion of water supply and waste-treatment facilities today requires that the engineer have some knowledge of the legal problems involved.

Water rights play an important role in determining the availability and use of water in many parts of the country. Because water law is a complex subject, the purpose of this chapter is to acquaint the reader with some of the legal problems which may be encountered in water resources engineering.

16-1

LEGAL CLASSIFICATION OF WATER[1]

Legal differences between waters on the surface of the earth and between various classes of groundwaters have been drawn since early times. Some western states have abolished the distinctions between these waters, but in many states they still exist. It is necessary, therefore, to investigate these legal distinctions before discussing water-rights doctrines.

1 *Surface watercourses:* consist essentially of a definite stream, formed by nature, flowing in a definite natural channel; it includes the underflow.

2 *Diffused surface waters:* waters originating from rain and melting snow and flowing vagrantly over the surface before becoming concentrated in watercourses or percolating into the ground.

3 *Underground streams:* have the essential characteristics of a surface watercourse, with the exception that they are buried in the ground. Necessarily, the proof of physical aspects with reasonable certainty is difficult.

4 *Underflow of surface streams:* underflow through the sands, gravel, and other subsoil over which the stream flows. This water is moving in the same direction as and in intimate contact with the surface stream. The lateral boundaries may extend for considerable distances beyond the banks of the surface channel. From a legal point of view, the surface stream and the underflow are a single watercourse.

5 *Percolating groundwater:* water below the surface of the ground that is not part of an underground stream or the underflow of a surface stream. This water is free to move by gravity, and hence to enter wells.

16-2

BASIC WATER DOCTRINES

Ownership is the right of one or more persons to possess and use property to the exclusion of others. Before rules can be prescribed governing the use or the rights to the use of water, its ownership must be determined.

Riparian Rights

This doctrine came to America from the Roman law by several lines of descent via Spain, France, and England. The riparian philosophy was introduced into Texas by the Spanish and Mexican governments and was upheld by the succeeding governments of the Republic and State of Texas.

The riparian principles of the French civil law, which were also based on the Roman law, were brought to the Atlantic seaboard by two eminent American jurists, Joseph Story and James Kent, in the early part of the nineteenth century. After Story and Kent had taken the riparian doctrine from French sources it was adopted in England as a part of its common law and chief reliance was placed on these American authorities.[2] The doctrine became part of the law of numerous states when they adopted the English common law.

Under the philosophy of riparian rights the owner of land containing a natural stream or abutting a stream is entitled to receive the full natural flow of the stream without change in quality or quantity. Physical contact of land and water is an essential factor. The riparian owner is protected against the diversion of water except for domestic purposes upstream from his property, and from the diversion of excess floodwaters toward his property. The riparian concept has been modified to provide for actual use of the water for beneficial purposes, such as irrigation, industry, and for the dilution of sewage effluent.

The riparian doctrine is generally followed in all states east of the Mississippi River. It is recognized to some degree in the six western states that extend from North Dakota to Texas on the 100th meridian, and in the three states that border on the Pacific Ocean. In all other western states it has been completely repudiated or has never been recognized. The supreme courts of New Mexico and Arizona have declared that in those states the riparian doctrine has never existed.

Where the riparian doctrine is in effect, the owner of a tract of land contiguous to the channel of a surface stream, called riparian land, has certain rights to the water flowing in the stream. The owner of the land may divert what water he needs for domestic use. Water for irrigation and other commercial purposes must be reasonable with respect to the requirements of all others. This same principle applies to the ownership of land that overlies an underground stream or the underflow of a surface stream. In each case the water right arises out of the physical contact of land and water source. A limited use of the water on nonriparian land may be permitted in some states. This nonriparian use is considered as a use without formal right rather than an exercise of the riparian right.

The Appropriation Doctrine

The idea of the appropriation doctrine has various antecedents, both ancient and modern. One view as to the origin of this principle in the Southwest is that it had its roots in Roman civil law; another is that it developed from local customs in that area. The idea that it was brought to the New World by the Spaniards, who adopted it from Roman civil law, is supported by the fact that this doctrine has materially influenced the early development of water law in several states in which the Spaniards and Mexicans settled.

The Mormons, settling on public lands in Utah during the middle of the nineteenth century, developed their own system of appropriation. They diverted water from streams for irrigation and applied the principle of priority based on first in time.

During this same period, gold was discovered in California on what were then Mexican public lands. The miners improvised local rules and regulations for taking and holding mining claims and claims for the use of water. Water was essential to hydraulic and placer mining, widely practiced in the area.

The principal feature of these doctrines of appropriation is the concept of "first in time, first in line." The first to appropriate the water for beneficial use has the first right to the water and his right must be completely supplied prior to any other water use. Each appropriator is entitled to use all the water he needs up to the limit of his appropriation prior to any one lower in line receiving any water. This is true regardless of his geographic position on the waterway. In periods of short water supply, water might be flowing in the stream crossing a person's property, but because of his appropriation in point of time he might not be allowed to divert any of the water for any purpose.

The appropriation doctrine gives no preference to the use of the water on land solely because of the contact of the land and the water supply. This right may be acquired and utilized in connection with land whether or not the land is contiguous to the water supply. In most western states water may be appropriated for use outside the watershed in which it naturally flows. In some instances the watersheds of origin have preferential rights even though they have not been exercised. Generally the appropriator does not need to be the owner of the land in connection with which he proposes to appropriate water. He must be in lawful possession such as an entryman on public land or as a lessee on private land to establish water rights.

The appropriative right is to a specific amount of water. The element of priority negates any obligation on the part of the appropriator to share the water with others. However, the appropriator is not entitled to divert and use more water than he can put to beneficial use, even though the quantity he needs may be much less than that to which his right relates.

The appropriation doctrine in the United States was first based on custom, which, in the absence of legislation, the courts recognized as controlling. Generally throughout the West the early statutes codified pervailing local rules and regulations. Later, as development progressed and complications ensued, the appropriation doctrine became both statutory and judicial, with many high-court decisions and constitutional provisions.

16-3

DIFFUSED SURFACE WATER

Basically, custom has established that the capture, by the occupant of the land, of surface water resulting from precipitation prior to this water's reaching a definite natural stream consitutes ownership of the water. Generally, the owner of land on which such water occurs is entitled to the capture and use of such water while it is on his land, without reservation. The right to this water has not been very seriously contested. Most of the law referring to diffused surface water has to do with the avoidance or riddance of such waters. The changing of natural drainage to prevent such waters from entering upon one's property, or changing the manner in which the natural drainage leaves one's property to the detriment of others, is normally prevented.

This is an exception to most water law. The common-law doctrine of percolating groundwater and diffused surface water gives the exclusive ownership of the water while it is on one's land. This doctrine affords no right of protection against depletion by one's neighbor.

16-4

GROUNDWATER

Groundwater poses an extremely difficult legal problem. It is more complex than that for surface water in that it is in many cases difficult to determine the source and rates of recharge, the extent and variation of quality in storage, and the water movements. Three basic rules cover the use of groundwater.

The first, or English, rule is of absolute ownership and allows the overlying landowner to take groundwater from his land at any time and in any quantity, regardless of the effect on the water table of his neighbor's land. Under the rule it would be possible for a landowner to exhaust the total groundwater supply of an area by heavy pumping. This rule has been qualified in some areas to limit the malicious and wasteful use of the water.

The American rule, or rule of reasonable use, recognizes that the landowner has rights to the water under his land but that these rights may be limited. His rights to the water are limited to its reasonable use in relationship to the overlying land.

The third rule covering groundwater is the appropriative principle, whereby the water is specifically allocated.

The English doctrine is followed in some states, both eastern and western. The American principle of reasonable use is followed by many states. In several of the western states the appropriative principle has been applied by statute, court decision, or both, to the use of groundwater.

16-5

COMPARISON OF DOCTRINES

Each of the doctrines has certain advantages and disadvantages, dependent upon the point of view. The problem is further complicated by a marked trend in point of view away from the individual water user to water use by large groups encompassing whole river basins. In many instances the present doctrines and projected uses are in direct conflict and are irreconcilable.

The appropriation doctrine provides for acquiring rights to water by putting it to beneficial use in accordance with procedures set forth in state statutes and judicial decisions. This assures the holder of the water right of his full appropriative supply, in order of priority, whenever the water is available. Early priority holders thus have reasonable assurance of a water supply each year. This enables development to be predicted on a reasonably assured annual supply and allows major investments in water-dependent operations.

Water can be appropriated and stored for beneficial use on either a temporary or a seasonal basis. The operation of the doctrine allows a considerable measure of flexibility for changes in the exercise of water rights. The appropriative right is gained by use and lost by disuse. Therefore, theoretically, all water is available for beneficial use.

It is probable that the appropriation doctrine has allowed the multiple appropriation of the same water in some instances, and it is possible for a late priority to gain in time because of this condition. The problem arises where groundwaters and surface waters are appropriated and the excess has come about from groundwater storage. Late appropriators could use the percolating groundwater and materially diminish stream flow. An earlier (in point of time) water user is displaced therefore by a later appropriation. This problem is further complicated by state boundaries and the later developments of groundwater across a state line in areas where the groundwater is not subject to appropriation. Holders of later priorities have little assurance of water supply in most seasons and the very late ones have no assurance except in very wet years.

Under the riparian doctrine, all landowners are assured of some water when it is available. This doctrine tends to freeze a large proportion of the water to lands whether it is being used or not. This unused right can be held indefinitely without forfeiture and thus can materially affect water-related development of an area. The riparian doctrine does not allow provision for storage of water that is essential for the full development of many water uses.

The unrestricted use of groundwater in many states is in direct conflict with both doctrines, as it is impossible to separate these waters under some conditions. There appears to be a need for modification of all existing doctrines in the future to meet the changing economic conditions in both the eastern and western portions of the United States.

●EXAMPLE 16-1

Under Oklahoma law, priorities are determined under three separate divisions: First, those claiming beneficial use of water prior to statehood, November 15, 1907; second, those claiming right to beneficial use of water because they have made application to the Water Resources Board; and third, those making beneficial use commencing after November 15, 1907, but without application to the Board.

Assume that six users, A, B, C, D, E, and F, are beneficial users of water from "Green River." A can prove that he began to use water for irrigation on March 15, 1904, and has used water from Green River for irrigation continuously since that date. B can prove that he began to use water for cooling purposes on January 1, 1900. C produces a copy of an application approved by the Board, dated October 1, 1910. D shows an affidavit and proves that he began to make beneficial use of water on July 1, 1910. E has an application approved by the Board, dated December 1, 1926. F brings forth an affidavit and demonstrates that he was making beneficial use of water commencing April 15, 1930.

○*Solution*

In this example, the first priority would go to B and the second priority to A. The first in time showing beneficial use will have the better right. Both of these claims were instituted prior to statehood, November 15, 1907, and since B began use prior in time to A, his is a prior right.

C and E both are making beneficial use of water and submitted applications to the Board. C began using water on December 1, 1926, and therefore has a higher priority right than E.

D and F are both making beneficial use of water, but have not made an application to the Board. D began using water on July 1, 1910, F on April 15, 1930. D therefore has a higher priority right than F. D began using water before both C and E; however, he did not make application to the Board. C and E, by virtue of their applications, then, have a higher priority than D. The priorities would accordingly be as follows: B first, A second, C third, E fourth, D fifth, and F sixth.

If B and A are taking more water now than they did on November 15, 1907, their right would be to the amount of water they were using on that date. If either, presently consuming more water, had made application for the additional water to the Board, the priority to the additional water would be dated from the date of application.

Had any of the six parties failed to beneficially utilize all or any part of the water claimed by him for which a right of use was vested for a period of two consecutive years, such unused water would revert to the public and become unappropriated water.

16-6

THE FEDERAL GOVERNMENT

The authority of the federal government over waters has been based on several sections of the U.S. Constitution. Article I, section 8, number 3, gives Congress authority over navigable waters which cross state boundaries or connect with the ocean. The term "navigation" has been rather loosely defined and could include most all waters in the broadest sense. Article IV, section 3, number 2, authorizes Congress to "dispose of and make all needful rules and regulations respecting the territory or other property belonging to the United States...." The treaty powers of the President and Senate under Article VI, number 2, have been used to designate and regulate water rights. Waters originating in the Rio Grande River in the United States are appropriated to Mexico according to a formula based on stream flow, as set up in a treaty between the United States and Mexico. Article I, section 8, number 1, "The Congress shall have power to lay and collect taxes, duties, imports and excises, to pay the debts and provide for the common defense and general welfare of the United States...." Basin development has contributed to the general welfare in the Tennessee Valley

Authority and resulted in construction of the Wilson Dam on the Tennessee River during World War I to produce nitrate for ammunition.

A liberal interpretation of the Constitution allows for considerable federal regulation over the nation's waters. The Supreme Court of the United States has ruled that every state has the power, within its dominion, to change the rule of the common law referring to the rights of the riparian owner to the continual natural flow of the stream and to permit appropriation of the water for such purposes as it deems wise.[2]

Water pollution legislation originated in Congress with a bill passed in 1886. This bill forbid the dumping of impediments to navigation in New York harbor. In 1889, Congress passed the Rivers and Harbors Act, which prohibited deposit of solid wastes into navigable waters. These early concerns with water pollution were strictly in the interests of navigation. The Public Health Service Act of 1912 had a section on waterborne diseases, and the Oil Pollution Act of 1924 was designed to prevent oil discharges from vessels into coastal waters that would damage aquatic life. In 1948 Congress took a national approach when it passed the first Water Pollution Control Act (PL 80-845). This act was extended in 1952 and again in 1955. It gave pollution enforcement authority to the federal government if local efforts failed and included provision for matching grants for waste-treatment facilities. Policy was strengthened with the Water Quality Act of 1965, which set water-quality standards for interstate waters. This act, and the Clean Waters Restoration Act of 1966, provide a vehicle for a broad national approach to water pollution.

On October 18, 1972, the Congress enacted Public Law 92-500, the Federal Water Pollution Control Act Amendments of 1972. Responding to public demand for cleaner water, the law it enacted culminated 2 years of intense debate, negotiation, and compromise, and resulted in the most assertive step in the history of national water-pollution-control activities.

The act, P.L. 92-500, departed in several ways from previous water-pollution-control legislation. It expanded the federal role in water pollution control, increased the level of federal funding for construction of publicly owned waste-treatment works, elevated planning to a new level of significance, opened new avenues for public participation, and created a regulatory mechanism requiring uniform technology-based effluent standards, together with a national permit system for all point-source dischargers as the means of enforcement.

In the strategy for implementation, Congress stated requirements for achievement of specific goals and objectives within specified time frames. The objective of the act is to restore and maintain the chemical, physical, and biological integrity of the nation's waters. In addition, two goals and eight policies are articulated. The goals are:

1   To reach, wherever attainable, a water quality that provides for the protection and propagation of fish, shellfish, and wildlife and for recreation in and on the water by July 1, 1983.
2   To eliminate the discharge of pollutants into navigable waters by 1985.

The policies are:

1 To prohibit the discharge of toxic pollutants in toxic amounts.
2 To provide federal financial assistance for construction of publicly owned treatment works.
3 To develop and implement area-wide waste-treatment management planning.
4 To mount a major research and demonstration effort in wastewater-treatment technology.
5 To recognize, preserve, and protect the primary responsibilities and roles of the states to prevent, reduce, and eliminate pollution.
6 To ensure, where possible, that foreign nations act to prevent, reduce, and eliminate pollution in international waters.
7 To provide for, encourage, and assist public participation in executing the act.
8 To pursue procedures that drastically diminish paperwork and interagency decision on procedures and prevent needless duplication and unnecessary delays at all levels of government.

The act provides for achieving its goals and objectives in phases, with accompanying requirements and deadlines. Phase I, an extension of the program embodied in many state laws and federal regulations, requires industry to install the best practicable control technology currently available; and publicly owned treatment works to achieve secondary treatment by July 1, 1977, as well as any more stringent limitations, including those to meet (state or federal) water-quality standards.

Phase II requirements are intended to be more rigorous and more innovative. Industries are to install the best available technology economically achievable which will result in reasonable further progress toward the national goal of eliminating the discharge of all pollutants; and publicly owned treatment works are to achieve best practicable waste-treatment technology, including reclaiming and recycling of water, and confined disposal of pollutants, by July 1, 1983, as well as any water-quality-related effluent limitation. Ultimately, all point-source controls are directed toward achieving the national goal of the elimination of the discharge of pollutants by 1985.

The act was intended to be more than a mandate for point-source discharge control. It embodied an entirely new approach to the traditional way Americans have used and abused their water resources. Some of these mechanisms are found in Title I, the broad policy title; others are woven throughout the act in grants and planning in standards and enforcement and in permits.

The second section of the act requires the development of comprehensive programs for preventing, reducing, and eliminating pollution, and further asks for research and development aimed at eliminating unnecessary water use. In Section 208, the statute directs the destination of area-wide institutions to plan, control, and maintain water quality and reduce pollution from all sources through land use or other methods.

Construction grants for publicly owned treatment works are made available to encourage full waste-treatment management, providing for:

1   The recycling of potential sewage pollutants through the production of agriculture, silviculture, and aquaculture products, or any combination thereof.
2   The confined and contained disposal of pollutants not recycled.
3   The reclamation of wastewater.
4   The ultimate disposal of sludge in a manner that will not result in environmental hazards.

The grantees are encouraged to combine with other facilities and utilize each other's processes and wastes. Facilities are to be designed and operated to produce revenues.

These statutory provisions outline a long-term program to reduce water use, reduce the generation of wastes, and establish financially self-sustaining public owned pollution control facilities.

The National Safe Drinking Water Act was signed on December 16, 1974. The purpose of the legislation is to assure that water-supply systems serving the public meet minimum national standards for the protection of public health. The act is designed to achieve uniform safety and quality of drinking water in the United States by identifying contaminants and establishing maximum acceptable levels. Prior to this act, the Environmental Protection Agency (EPA) was authorized to prescribe federal drinking-water standards only for water supplies used by interstate carriers. In contrast, this bill permits EPA to establish federal standards to control the levels of all harmful contaminants in the drinking water supplied by all public water systems. It also establishes a joint federal–state system for assuring compliance with these standards. The major provisions of the act are:

1   The establishment of primary regulations for the protection of the public health.
2   The establishment of secondary regulations that are related to taste, odor, and appearance of drinking water.
3   The establishment of regulations to protect underground drinking-water sources by the control of surface injection.
4   The initiation of research on health, economic, and technological problems related to drinking-water supplies.
5   The initiation of a survey of rural water supplies.
6   The allocation of funds to states in improving their drinking-water programs through technical assistance, training of personnel, and grant support.

16-7

ACQUIRING WATER RIGHTS

The riparian right to water accrues when the land title passes. There is no formality in acquiring the water right other than acquiring the land. Rights to the use of percolating groundwater and to vagrant surface water are secured in the same way.

Each of the 17 western states has a statutory procedure under which surface water may be appropriated, and most of the western states have some formal procedure for groundwater appropriation. Nearly all the procedures contemplate applications to state officials for water, and permits or licenses are issued when the request is approved.

In some eastern states, legislation requires a permit from a state agency to take water for irrigation or other purposes from a watercourse or groundwater supply. It is doubtful whether, in most states, these permits establish any substantial water right. They have little effect on riparian rights, but they do provide some administrative restraint upon the exercise of water rights. They do provide a record of use that should be extremely useful in considering water-rights legislation in the future.

Several of the complications of water use and development do not result from engineering considerations. Water can be transported to any place if one has the ability to pay for it. Many of the vital considerations arise from priorities of allocation. New regulations to meet new problems are the result of a process that takes place over a period of time. The one best or ideal water law cannot be passed once and for all. Water legislation should not be discussed only as an academic ideal but also as a practical solution to a given problem at a specific time.

It seems inevitable that the interest of individual water users will come to be represented less and less by individual water rights acquired from the state and more and more by contracts with basin-management districts which will hold mass water rights in trust for the users.

16-8

WASTE TREATMENT

The law has always made a distinction in the liability of a government when it is acting as a government and its liability when it is acting in a business enterprise. A city or political subdivision is liable for its torts committed in its proprietary capacity. It is generally recognized that if the function is for the benefit of the public at large it is governmental, and if solely for the people of the community, it is proprietary. It seems to be generally recognized that a municipality is liable to the same extent as a private corporation for injuries resulting from the creation or maintenance of a nuisance.

The authorities are in conflict as to whether the operation of a waste-treatment plant is a governmental or proprietary function. The prevailing view seems to be that a municipality may be held liable for death or injury resulting from negligence or other tort in the operation of such a plant.

The owner of land taken for the construction of a waste-disposal plant is entitled to an award of the value of the land taken and damages to his business. The municipality has no more right to create a nuisance to the injury of another than has an individual. Authority to install a sewer system or treatment works carries no implication of authority to create or maintain a nuisance. It does not matter whether the nuisance results from negligence or from the design adopted.

Engineers and utility operating personnel frequently act in the capacity of an agent for the city or political subdivision. An agent is subject to liability if, by his acts, he creates an unreasonable risk of harm to the interests of others. This includes situations in which an agent causes harm to a third person by his activities, and also those in which the agent's performance of his duties to his principal results in harm to third persons.

Lest an impression be left that waste-treament plants are per se a nuisance, the courts seem to uniformly hold they are not. Most courts have taken the view that injunctive relief against proposed construction and operation of a waste-treatment plant does not appear warranted on the grounds of anticipatory nuisance.

PROBLEMS

16-1 Prepare a summary of the procedure for appropriating surface water in your state. How are these water rights to be administered?

16-2 Prepare a summary of the procedure for appropriating groundwater in some state. Outline the most important rules for governing the use of this groundwater.

16-3 Investigate a particular state's laws concerning stream pollution and outline those provisions which should be considered in the design of a municipal waste-treatment plant.

16-4 Prepare an outline of an interstate pollution-control compact. Is your state a party to any interstate pollution-control compact?

16-5 Investigate the local ordinances concerning the disposal of industrial waste into the municipal system for a large city in your state. Prepare a summary of those provisions which should be considered in the design of an industrial waste-treatment plant to be located in that city.

REFERENCES

1. W. A. Hutchins and H. A. Steele, "Basic Water Rights Doctrines and Their Implications for River Basin Development," *Law and Contemp. Problems* XXII, no. 2 (Spring 1957), Duke University School of Law.
2. *United States v. Rio Grande Dam and Irrigation Co.*, 174 U.S. 690, 43 L., ed. 1136, 19 Supp. 770.
3. See hearings before a special subcommittee on air and water pollution of the Senate Committee on Public Works, 87 (88th Congress, 1st Session, 1963), "Testimony of Murray Stein."
4. State of Oklahoma, "Summary Water Laws of Oklahoma," Oklahoma Planning and Resources Board, Division of Water Resources (August 1960).
5. *United States v. Fallbrook Utility District*, 1965F. Supp. 806.

Appendix

TABLE OF RELATIVE ATOMIC WEIGHTS

Name	Symbol	Atomic Number	Atomic Weight	Name	Symbol	Atomic Number	Atomic Weight
Actinium	Ac	89	—	Mercury	Hg	80	200.59
Aluminum	Al	13	26.9815	Molybdenum	Mo	42	95.94
Americium	Am	95	—	Neodymium	Nd	60	144.24
Antimony	Sb	51	121.75	Neon	Ne	10	20.183
Argon	Ar	18	39.948	Neptunium	Np	93	—
Arsenic	As	33	74.9216	Nickel	Ni	28	58.71
Astatine	At	85	—	Niobium	Nb	41	92.906
Barium	Ba	56	137.34	Nitrogen	N	7	14.0067
Berkelium	Bk	97	—	Nobelium	No	102	—
Beryllium	Be	4	9.0122	Osmium	Os	76	190.2
Bismuth	Bi	83	208.980	Oxygen	O	8	15.9994
Boron	B	5	10.811	Palladium	Pd	46	106.4
Bromine	Br	35	79.904	Phosphorus	P	15	30.9738
Cadmium	Cd	48	112.40	Platinum	Pt	78	195.09
Calcium	Ca	20	40.08	Plutonium	Pu	94	—
Californium	Cf	98	—	Polonium	Po	84	—
Carbon	C	6	12.01115	Potassium	K	19	39.102
Cerium	Ce	58	140.12	Praseodymium	Pr	59	140.907
Cesium	Cs	55	132.905	Promethium	Pm	61	—
Chlorine	Cl	17	35.453	Protactinium	Pa	91	—
Chromium	Cr	24	51.996	Radium	Ra	88	—
Cobalt	Co	27	58.9332	Radon	Rn	86	—
Copper	Cu	29	63.546	Rhenium	Re	75	186.2
Curium	Cm	96	—	Rhodium	Rh	45	102.905
Dysprosium	Dy	66	162.50	Rubidium	Rb	37	85.47
Einsteinium	Es	99	—	Ruthenium	Ru	44	101.07
Erbium	Er	68	167.26	Samarium	Sm	62	150.35
Europium	Eu	63	151.96	Scandium	Sc	21	44.956
Fermium	Fm	100	—	Selenium	Se	34	78.96
Fluorine	F	9	18.9984	Silicon	Si	14	28.086
Francium	Fr	87	—	Silver	Ag	47	107.868
Gadolinium	Gd	64	157.25	Sodium	Na	11	22.9898
Gallium	Ga	31	69.72	Strontium	Sr	38	87.62
Germanium	Ge	32	72.59	Sulfur	S	16	32.064
Gold	Au	79	196.967	Tantalum	Ta	73	189.948
Hafnium	Hf	72	178.49	Technetium	Tc	43	—
Helium	He	2	4.0026	Tellurium	Te	52	127.60
Holmium	Ho	67	164.930	Terbium	Tb	65	158.924
Hydrogen	H	1	1.00797	Thallium	Tl	81	204.37
Indium	In	49	114.82	Thorium	Th	90	232.038
Iodine	I	53	126.9044	Thulium	Tm	69	168.934
Iridium	Ir	77	192.2	Tin	Sn	50	118.69
Iron	Fe	26	55.847	Titanium	Ti	22	47.90
Krypton	Kr	36	83.80	Tungsten	W	74	183.85
Lanthanum	La	57	138.91	Uranium	U	92	238.03
Lead	Pb	82	207.19	Vanadium	V	23	50.942
Lithium	Li	3	6.939	Xenon	Xe	54	131.30
Lutetium	Lu	71	174.97	Ytterbium	Yb	70	173.04
Magnesium	Mg	12	24.312	Yttrium	Y	39	88.905
Manganese	Mn	25	54.9380	Zinc	Zn	30	65.37
Mendelevium	Md	101	—	Zirconium	Zr	40	91.22

Source: "Report of the International Commission on Atomic Weights—1961" *J. Am. Chem. Soc.* 84 (1962): 4192.

TABLE A-2
WEIGHTS AND MEASURES

Length

Units		cm	m	in.	ft	yd	mi
1 cm	=	1	0.01	0.3937008	0.03280840	0.01093613	6.213712×10^{-6}
1 m	=	100.	1	39.37008	3.280840	1.093613	6.213712×10^{-4}
1 in.	=	2.54	0.0254	1	0.08333333….	0.02777777….	1.578283×10^{-5}
1 ft	=	30.48	0.3048	12.	1	0.3333333….	$1.893939 \times 10^{-4}…$
1 yd	=	91.44	0.9144	36.	3.	1	$5.681818 \times 10^{-4}…$
1 mi	=	1.609344×10^{5}	1.609344×10^{3}	6.336×10^{4}	5280.	1760.	1

Area

Units		cm^2	m^2	in.^{2}	ft^2	acre	mi^2
1 cm^2	=	1	10^{-4}	0.1550003	1.076391×10^{-3}	2.5×10^{-8}	3.861022×10^{-11}
1 m^2	=	10^{4}	1	1550.003	10.76391	2.5×10^{-4}	3.861022×10^{-7}
1 in.^{2}	=	6.4516	6.4516×10^{-4}	1	6.944444×10^{-3}	1.59×10^{-7}	2.490977×10^{-10}
1 ft^2	=	929.0304	0.09290304	144.	1	2.3×10^{-5}	3.587007×10^{-8}
1 acre	=	40.47×10^{6}	4047	6.27×10^{6}	43560	1	1.56×10^{-3}
1 mi^2	=	2.589988×10^{10}	2.589988×10^{6}	4.014490×10^{9}	2.78784×10	640	1

Volume

Units	cm³	liter	in.³	ft³	gal	acre-ft
1 cm³ =	1	10^{-3}	0.06102374	3.531467×10^{-5}	2.641721×10^{-4}	8.1×10^{-10}
1 liter =	1000.	1	61.02374	0.03531467	0.2641721	8.1×10^{-7}
1 in.³ =	16.38706	0.0163706	1	5.787037×10^{-4}	4.329004×10^{-3}	1.33×10^{-8}
1 ft³ =	28,316.85	28.31685	1728.	1	7.480520	2.3×10^{-5}
1 gal (U.S.) =	3875.412	3.785412	231.	0.1336806	1	3.07×10^{-6}
1 acre-ft =	1.23×10^9	1.23×10^6	75.3×10^6	43,560	325,851	1

Mass

Units	g	kg	oz	lb	ton
1 g =	1	10^{-3}	0.03527396	2.204623×10^{-3}	1.102311×10^{-6}
1 kg =	1000.	1	35.27396	2.204623	1.102311×10^{-3}
1 oz (avdp) =	28.34952	0.02834952	1	0.0625	$5. \times 10^{-4}$
1 lb (avdp) =	453.5924	0.4535924	16.	1	0.0005
1 ton =	907,184.7	907.1847	32,000.	2000.	1

TABLE A-3
PROPERTIES OF WATER

Temper- ature (°F)	Specific Weight, γ (lb/ft³)	Mass Density, ρ (lb-sec²/ft⁴)	Dynamic Viscosity, $\mu \times 10^5$ (lb-sec/ft²)	Kinematic Viscosity, $\nu \times 10^5$ (ft²/sec)	Surface Energy,[a] $\sigma \times 10^3$ (lb/ft)	Vapor Pressure, p_v (lb/in.²)	Bulk Modulus, $E \times 10^{-3}$ (lb/in.²)
32	62.42	1.940	3.746	1.931	5.18	0.09	290
40	62.43	1.938	3.229	1.664	5.14	0.12	295
50	62.41	1.936	2.735	1.410	5.09	0.18	300
60	62.37	1.934	2.359	1.217	5.04	0.26	312
70	62.30	1.931	2.050	1.059	5.00	0.36	320
80	62.22	1.927	1.799	0.930	4.92	0.51	323
90	62.11	1.923	1.595	0.826	4.86	0.70	326
100	62.00	1.918	1.424	0.739	4.80	0.95	329
110	61.86	1.913	1.284	0.667	4.73	1.24	331
120	61.71	1.908	1.168	0.609	4.65	1.69	333
130	61.55	1.902	1.069	0.558	4.60	2.22	332
140	61.38	1.896	0.981	0.514	4.54	2.89	330
150	61.20	1.890	0.905	0.476	4.47	3.72	328
160	61.00	1.896	0.838	0.442	4.41	4.74	326
170	60.80	1.890	0.780	0.413	4.33	5.99	322
180	60.58	1.883	0.726	0.385	4.26	7.51	318
190	60.36	1.876	0.678	0.362	4.19	9.34	313
200	60.12	1.868	0.637	0.341	4.12	11.52	308
212	59.83	1.860	0.593	0.319	4.04	14.7	300

[a] In contact with air.

Source: Adapted from "Hydraulic Models," *Manual of Engineering Practice No. 25*, American Society of Civil Engineers (1942).

SYSTÈME INTERNATIONAL D'UNITÉS (SI)

The SI metric system is based on meter-kilogram-second units. The principal units applicable to water and wastewater engineering are listed in Tables A-4 and A-5. Derived SI units are consistent, since the conversion factor among various units is unity (e.g., 1 joule = 1 newton × 1 meter). Prefixes given in Table A-6 may be added to write large or small numbers and thus avoid the use of exponential values of 10. Groups of three digits, on either side of the decimal point, are separated by spaces. Table A-7 lists common conversion factors from customary units to SI metric. [Reference: *Units of Expression for Wastewater Treatment*, Manual of Practice No. 6, 1976 (Washington, D.C.: Water Pollution Control Federation).]

TABLE A-4
BASIC SI UNITS

Quantity	Unit	Symbol
Length	meter	m
Mass	kilogram	kg
Time	second	s
Thermodynamic temperature	Kelvin	K
Molecular weight	mole	mol
Plane angle	radian	rad

TABLE A-5
DERIVED SI UNITS

Quantity	Unit	Symbol	Formula
Energy	joule	J	$N \cdot m$
Force	newton	N	$kg \cdot m/s^2$
Power	watt	W	J/s
Pressure	pascal	Pa	N/m^2

TABLE A-6
MULTIPLES AND SUBMULTIPLES OF SI UNITS

Multiplier		Prefix	Symbol
1 000 000.	$= 10^6$	mega	M
1 000.	$= 10^3$	kilo	k
0.001	$= 10^{-3}$	milli	m
0.000 001	$= 10^{-6}$	micro	μ

TABLE A-7
CONVERSION FACTORS FROM CUSTOMARY UNITS TO SI METRIC

Customary Units		Multiplier	Metric Units	
Description	Symbol Multiply	By	Symbol To Obtain	Reciprocal
Acre	ac	0.404 7	ha	2.471
British thermal unit	Btu	1.055	kJ	0.947 0
British thermal units per cubic foot	Btu/ft³	37.30	J/l	0.026 81
British thermal units per pound	Btu/lb	2.328	kJ/kg	0.429 5
British thermal units per square foot per hour	Btu/ft²/hr	3.158	J/m² sec	0.316 7
Cubic foot	ft³	0.028 32	m³	35.31
Cubic foot	ft³	28.32	l	0.035 31
Cubic feet per minute	cfm	0.471 9	l/sec	2.119
Cubic feet per minute per thousand cubic feet	cfm/1000 ft³	0.016 67	l/m³ sec	60.00
Cubic feet per second	cfs	0.028 32	m³/sec	35.31
Cubic feet per second per acre	cfs/ac	0.069 98	m³/sec ha	14.29
Cubic inch	in.³	0.016 39	l	61.01
Cubic yard	yd³	0.764 6	m³	1.308
Fathom	f	1.839	m	0.546 7
Foot	ft	0.304 8	m	3.281
Feet per hour	ft/hr	0.084 67	mm/sec	11.81
Feet per minute	fpm	0.005 08	m/sec	196.8
Foot-pound	ft-lb	1.356	J	0.737 5
Gallon, U.S.	gal	3.785	l	0.264 2
Gallons per acre	gal/ac	0.009 35	m³/ha	106.9
Gallons per day per linear foot	gpd/lin ft	0.012 42	m³/m day	80.53
Gallons per day per square foot	gpd/ft²	0.040 74	m³/m² day	24.54
Gallons per minute	gpm	0.063 08	l/sec	15.85
Grain	gr	0.064 80	g	15.43

Grains per gallon	gr/gal	17.12	mg/l	0.058 41
Horsepower	hp	0.745 7	kW	1.341
Horsepower-hour	hp-hr	2.684	MJ	0.372 5
Inch	in.	25.4	mm	0.039 37
Knot	knot	1.852	km/h	0.540 0
Knot	knot	0.514 4	m/sec	1.944
Mile	mi	1.609	km	0.621 5
Miles per hour	mph	1.609	km/h	0.621 5
Million gallons	mil gal	3 785.0	m^3	0.000 264 2
Million gallons per day	mgd	43.81	l/sec	0.022 82
Million gallons per day	mgd	0.043 81	m^3/sec	22.82
Ounce	oz	28.35	g	0.035 27
Pound (force)	lbf	4.448	N	0.224 8
Pound (mass)	lb	0.453 6	kg	2.205
Pounds per acre	lb/ac	1.121	kg/ha	0.892 1
Pounds per cubic foot	lb/ft³	16.02	kg/m^3	0.062 42
Pounds per foot	lb/ft	1.488	kg/m	0.672 0
Pounds per horsepower-hour	lb/hp-hr	0.169 0	mg/J	5.918
Pounds per square foot	lb/ft²	4.882	kgf/m^2	0.204 8
Pounds per square inch	psi	703.1	kgf/m^2	0.001 422
Pounds per square inch	psi	6.895	kN/m^2	0.145 0
Pounds per thousand cubic feet per day	lb/1000 ft³/day	0.016 02	kg/m^3 day	62.43
Square foot	ft²	0.092 90	m^2	10.76
Square inch	in.²	645.2	mm^2	0.001 550
Square mile	mi²	2.590	km^2	0.386 1
Square yard	yd²	0.836 1	m^2	1.196
Ton, short	ton	0.907 2	t	1.102
Yard	yd	0.914 4	m	1.094

Acceleration of gravity, $g = 32.174$ ft/s^2 $= 9.806\ 65$ m/s^2.

mgd $\times$ 1.547 $\longrightarrow$ cfs

TABLE A-8

SELECTED ENGLISH–METRIC CONVERSION FACTORS

	English Unit	Multiplier	Metric Unit
Mass	lb	0.453 6	kg
Length	ft	0.304 8	m
Area	ft^2	0.092 90	m^2
	acre	4 047.	m^2
	acre	0.404 7	ha
Volume	gal	0.003 785	m^3
	gal	3.785	liter
	ft^3	0.028 32	m^3
	ft^3	28.32	liter
Velocity	fpm	0.005 08	m/sec
Flow	mgd	3 785.	m^3/day
	gpm	5.450	m^3/day
	gpm	0.063 09	l/sec
	cfs	0.028 32	m^3/sec
BOD loading	lb/1000 ft^3/day	16.02	g/m^3 day
	lb/acre/day	1.121	kg/ha day
Solids loading	lb/ ft^2/day	4.883	kg/m^2 day
	lb/ ft^3/day	16.02	kg/m^3 day
Hydraulic loading	gpd/ft^2	0.040 75	m^3/m^2 day
	mg/acre/day	0.935 3	m^3/m^2 day
	mg/acre/day	9 353.	m^3/ha day
Concentration	lb/mg	0.119 8	mg/l

	Metric Unit	Multiplier	English Unit
Mass	kg	2.205	lb
Length	m	3.281	ft
Area	m^2	10.76	ft^2
	m^2	0.000 247	acre
	ha	2.471	acre
Volume	m^3	264.2	gal
	m^3	35.31	ft^3
	liter	0.264 2	gal
	liter	0.035 31	ft^3
Velocity	m/sec	196.8	fpm
Flow	m^3/day	0.000 264 2	mgd
	m^3/day	0.183 5	gpm
	m^3/sec	35.31	cfs
	l/sec	15.85	gpm
BOD loading	g/m^3 day	0.062 43	lb/1000 ft^3/day
	kg/ha day	0.892 1	lb/acre/day
Solids loading	kg/m^2 day	0.204 8	lb/ft^2/day
	kg/m^3 day	0.062 42	lb/ft^3/day
Hydraulic loading	m^3/m^2 day	24.54	gal/ft^2/day
	m^3/m^2 day	1.069	mg/acre/day
Concentration	mg/l	8.345	lb/mg

SATURATION VALUES OF DISSOLVED OXYGEN IN WATER EXPOSED TO WATER-SATURATED AIR CONTAINING 20.90% OXYGEN UNDER A PRESSURE OF 760 mm OF MERCURY[a]

Temperature (°C)	Chloride Concentration in Water (mg/l) Dissolved Oxygen (mg/l)			Difference per 100 mg Chloride	Temperature (C°)	Vapor Pressure (mm)
	0	5000	10,000			
0	14.6	13.8	13.0	0.017	0	5
1	14.2	13.4	12.6	0.016	1	5
2	13.8	13.1	12.3	0.015	2	5
3	13.5	12.7	12.0	0.015	3	6
4	13.1	12.4	11.7	0.014	4	6
5	12.8	12.1	11.4	0.014	5	7
6	12.5	11.8	11.1	0.014	6	7
7	12.2	11.5	10.9	0.013	7	8
8	11.9	11.2	10.6	0.013	8	8
9	11.6	11.0	10.4	0.012	9	9
10	11.3	10.7	10.1	0.012	10	9
11	11.1	10.5	9.9	0.011	11	10
12	10.8	10.3	9.7	0.011	12	11
13	10.6	10.1	9.5	0.011	13	11
14	10.4	9.9	9.3	0.010	14	12
15	10.2	9.7	9.1	0.010	15	13
16	10.0	9.5	9.0	0.010	16	14
17	9.7	9.3	8.8	0.010	17	15
18	9.5	9.1	8.6	0.009	18	16
19	9.4	8.9	8.5	0.009	19	17
20	9.2	8.7	8.3	0.009	20	18
21	9.0	8.6	8.1	0.009	21	19
22	8.8	8.4	8.0	0.008	22	20
23	8.7	8.3	7.9	0.008	23	21
24	8.5	8.1	7.7	0.008	24	22
25	8.4	8.0	7.6	0.008	25	24
26	8.2	7.8	7.4	0.008	26	25
27	8.1	7.7	7.3	0.008	27	27
28	7.9	7.5	7.1	0.008	28	28
29	7.8	7.4	7.0	0.008	29	30
30	7.6	7.3	6.9	0.008	30	32

[a] Saturation at barometric pressures other than 760 mm (29.92 in.), C_s' is related to the corresponding tabulated values, C_s, by the equation

$$C_s' = C_s \frac{P-p}{760-p}$$

where C_s' = solubility at barometric pressure P and given temperature, mg/liter
C_s = saturation at given temperature from table, mg/liter
P = barometric pressure, mm
p = pressure of saturated water vapor at temperature of the water selected from table, mm

Printer and Binder: Halliday Lithograph

82 83 84 85 86   13 12 11 10 9